Methods in Cell Biology

VOLUME 65
Mitochondria

Series Editors

Leslie Wilson

Department of Biological Sciences
University of California, Santa Barbara
Santa Barbara, California

Paul Matsudaira

Whitehead Institute for Biomedical Research and
Department of Biology
Massachusetts Institute of Technology
Cambridge, Massachusetts

Methods in Cell Biology

Prepared under the Auspices of the American Society for Cell Biology

VOLUME 65

Mitochondria

Edited by

Liza A. Pon

Department of Anatomy and Cell Biology
Columbia University
New York, New York

Eric. A. Schon

Neurology Department and Genetics and Development
Columbia University
New York, New York

San Diego San Francisco New York Boston London Sydney Tokyo

This book is printed on acid-free paper. ∞

Academic Press
A Harcourt Science and Technology Company
525 B Street, Suite 1900, San Diego, California 92101-4495, USA
http://www.academicpress.com

Academic Press
Harcourt Place, 32 Jamestown Road, London NW1 7BY, UK
http://www.academicpress.com

International Standard Book Number: 0-12-544169-X (casebound)
International Standard Book Number: 0-12-561285-0 (paperback)

PRINTED IN THE UNITED STATES OF AMERICA
01 02 03 04 05 06 EB 9 8 7 6 5 4 3 2 1

CONTENTS

CONTRIBUTORS

Numbers in parentheses indicate the pages on which the authors' contributions begin.

Edward K. Ainscow (353), Department of Biochemistry, University of Bristol, Bristol BS8 1TD, United Kingdom

Giuseppe Attardi (114), Division of Biology, California Institute of Technology, Pasadena, California 91125

Antoni Barrientos (397), Department of Biological Sciences, Columbia University, New York 10032

Mark A. Birch-Machin (97), Department of Dermatology, Medical School, University of Newcastle upon Tyne, Newcastle upon Tyne NE2 4HH, United Kingdom

Istvan R. Boldogh (159), Columbia University College of Physicians and Surgeons, New York, New Yark 10032

Eduardo Bonilla (311), Department of Neurology, Columbia University, New York, New Yark 10032

Nathalie Bonnefoy (381), Centre de Gé nétique Moléculaire, Laboratoire propre du CNRS associé à l'Université Pierre et Marie Curie, 91198 Gif-sur-Yvette Cedex, France

Ronald A. Butow (277, 439), Department of Molecular Biology, University of Texas Southwestern Medical Center, Dallas, Texas 75235

Nicoletta Checcarelli (133), H. Houston Merritt Clinical Research Center for Muscular Dystrophy and Related Disorders, Department of Neurology, Columbia University, College of Physicians and Surgeons, New York, New York 10032

Anna Chiesa (353), Department of Experimental and Diagnostic Medicine, Section of General Pathology, 44100 Ferrara, Italy

Mauro Degli Esposti (75), Cancer Research Campaign Molecular Pharmacology Group, School of Biological Sciences, University of Manchester, Manchester M13 9PT, United Kingdom

Anton I. P. M. de Kroon (37), Center for Biomembranes and Lipid Enzymology, Department of Biochemistry of Membranes, University of Utrecht, 3584 CH Utrecht, The Netherlands

Runu Dey (397), Department of Neurology, University of Miami School of Medicine, Miami, Florida 33136

Kerstin Diekert (37), Institut für Zytobiologie und Zytopathologie der Philipps-Universität Marburg, 35033 Marburg, Germany

James A. Dykens (285), MitoKor, San Diego, California 92121

Olof Emanuelsson (175), Stockholm Bioinformatics Center, Stockholm University, S-10691 Stockholm, Sweden

Charles B. Epstein (439), Department of Molecular Biology, University of Texas Southwestern Medical Center, Dallas, Texas 75390

Kammy Fehrenbacher (453), Columbia University, New Yark 10032

Thomas D. Fox (381), Department of Molecular Biology and Genetics, Cornell University, Ithaca, New York 14853

Walker Hale IV (439), Department of Molecular Biology, University of Texas Southwestern Medical Center, Dallas, Texas 75390

Johannes M. Herrmann (217, 429), Institut für Physiologische Chemie, Universität München, 80336 München, Germany

Gyula Kispal (37), Institut fur Zytobiologie und Zytopathologie der Philipps-Universitat Marburg, 35033 Marburg, Germany; and Institute of Biochemistry, University Medical School of Pecs, Hungary

Guido Kroemer (147), Centre National de la Recherche Scientifique, F-94801 Villejuif, France

C. J. Leaver (53), Department of Plant Sciences, University of Oxford, Oxford OX1 3RB, United Kingdom

Giorgio Lenaz (1), Dipartimento di Biochimica "G. Moruzzi," Universitá di Bologna, 40126 Bologna, Italy

A. Liddell (53), Department of Plant Sciences, University of Oxford, Oxford OX1 3RB, United Kingdom

Roland Lill (37), Institut für Zytobiologie und Zytopathologie der Philipps-Universität Marburg, 35033 Marburg, Germany

Carine Maisse (147), Centre National de la Recherche Scientifique, F-94801 Villejuif, France

Giovanni Manfredi (133), Department of Neurology and Neuroscience, Weill Medical College of Cornell University, New York, Ney York 10021

Carmen A. Mannella (245), Resource for the Visualization of Biological Complexity, Wadsworth Center, Albany, New York 12201

Didier Métivier (147), Centre National de la Recherche Scientifique, F-94801 Villejuif, France

A. H. Millar (53), Department of Plant Sciences, University of Oxford, Oxford OX1 3RB, United Kingdom

Carlos T. Moraes (397), Department of Neurology, University of Miami School of Medicine, Miami, Florida 33136

Ali Naini (133), H. Houston Merritt Clinical Research Center for Muscular Dystrophy and Related Disorders, Department of Neurology, Columbia University, College of Physicians and Surgeons, New York, New York 10032

Walter Neupert (217, 429), Institut für Physiologische Chemie, Universität München, 80336 München, Germany

Dan W. Nowakowski (257), Department of Anatomy and Cell Biology, Columbia University, New York, New York 10032

Koji Okamoto (277), Department of Molecular Biology, University of Texas Southwestern Medical Center, Dallas, Texas 75235

Nannette Orme-Johnson (483), Tufts University, Boston, Massachusetts 02111

Francesco Pallotti (1), Department of Neurology, Columbia University, New York, New York 10032

Philip S. Perlman (277), Department of Molecular Biology, University of Texas Southwestern Medical Center, Dallas, Texas 75235

Nikolaus Pfanner (189), Institut für Biochemie und Molekularbiologie, Universität Freiburg, D-79104 Freiburg, Germany

Paolo Pinton (353), Department of Biomedical Sciences and C.N.R. Center for the Study of Biomembranes, 35121 Padova, Italy

Liza A. Pon (159, 257, 333), Department of Anatomy and Cell Biology, Columbia University, New York, New York 10032

Anna M. Porcelli (353), Department of Biology, University of Bologna, 40126 Bologna, Italy

Rosario Rizzuto (353), Department of Experimental and Diagnostic Medicine, Section of General Pathology, 44100 Ferrara, Italy

Michela Rugolo (353), Department of Biology, University of Bologna, 40126 Bologna, Italy

Guy A. Rutter (353), Department of Biochemistry, University of Bristol, Bristol BS8 1TD, United Kingdom

Michael T. Ryan (189), Institut für Biochemie und Molekularbiologie, Universität Freiburg, D-79104 Freiburg, Germany

Hermann Schägger (231), Institut für Biochemie I, Universitätsklinikum Frankfurt, 60590 Frankfurt am Main, Germany

Gisbert Schneider (175), F. Hoffmann–La Roche Ltd., Pharmaceuticals Division, CH-4070 Basel, Switzerland

Eric A. Schon (457, 461, 463), Columbia University, New York 10032

Bonnie L. Seidel-Rogol (413), Department of Biochemistry, Emory University School of Medicine, Atlanta, Georgia 30322

Gerald S. Shadel (413), Department of Biochemistry, Emory University School of Medicine, Atlanta, Georgia 30322

Viviana Simon (333), Banting and Best Department of Medical Research, University of Toronto, Toronto, Ontario, Canada M5G 1L6

Antonella Spinazzola (133), H. Houston Merritt Clinical Research Center for Muscular Dystrophy and Related Disorders, Department of Neurology, Columbia University, College of Physicians and Surgeons, New York, New York 10032

Amy K. Stout (285), MitoKor, San Diego, California 92121

Theresa C. Swayne (257, 333), Department of Anatomy and Cell Biology, Columbia University, New York, New York 10032

Kurenai Tanji (311), Department of Neurology, Columbia University, New York, New York 10032

Douglass M. Turnbull (97), Department of Neurology, Medical School, University of Newcastle upon Tyne, Newcastle upon Tyne NE2 4HH, United Kingdom

Gaetano Villani (119), Division of Biology, California Institute of Technology, Pasadena, California 91125

Gunnar von Heijne (175), Stockholm Bioinformatics Center, Stockholm University, S-10691 Stockholm, Sweden

Wolfgang Voos (189), Institut für Biochemie und Molekularbiologie, Universität Freiburg, D-79104 Freiburg, Germany

Benedikt Westermann (217,429), Institüt für Physiologische Chemie, Universität München, 80336 München, Germany

Hyeong-Cheol Yang (333), Department of Anatomy and Cell Biology, Columbia University, New York, New York 10032

Naoufal Zamzami (147), Centre National de la Recherche Scientifique, F-94801 Villejuif, France

PREFACE

The field of "mitochondriology" has undergone a renaissance in the last decade. Compared to state of the field as recently as 1985, entirely new areas of investigation have emerged, such as the role of mitochondrial function in apoptosis and mitochondrial dysfunction in human disease. Furthermore, our understanding of basic phenomena—mitochondrial import, movement, fusion, fission, inheritance, and interactions with the nucleus and with other organelles—is passing rapidly from the realm of speculation to that of fact.

Concomitant with these conceptual advances has been the remarkable increase in the types of tools available to study mitochondrial structure and function. In addition to the traditional methods for isolating mitochondria and assaying their biochemical properties, we now have new and powerful ways to visualize, monitor, and perturb mitochondrial function and to assess the genetic consequences of those perturbations.

This volume is designed to bring together those methods—both "classic" and modern—that will enable anyone to study this organelle, be it from a biochemical, morphological, or genetic point of view. A noteworthy, and we believe useful, adjunct to the methods are the Appendices, which bring together fundamental information regarding mitochondria not found in any single source. These include lists of every known or suspected mitochondrial protein, maps of mitochondrial genomes from several commonly used organisms, and information on agents that perturb respiratory function.

We dedicate this volume to Gottfried Schatz, a pioneer in the field of mitochondria. His achievements include the first purification of mitochondrial DNA and the discovery of mitochondrial signal sequences. His work on elucidating the components and mode of action of the mitochondrial import machinery has been of exceptional beauty and inordinate impact. Jeff is an inspiration to us all.

Liza A. Pon
Eric A. Schon

CHAPTER 1

Isolation and Subfractionation of Mitochondria from Animal Cells and Tissue Culture Lines

Francesco Pallotti* **and Giorgio Lenaz†**

* Department of Neurology
Columbia University
New York, New York 10032

† Dipartimento di Biochimica "G. Moruzzi"
Universita' di Bologna
40126 Bologna, Italy

METHODS IN CELL BIOLOGY, VOL. 65

I. Introduction

Past and current mitochondrial research has been performed on mitochondria prepared from rat liver, rat heart, or beef heart because these tissues can be obtained readily and in quantity. Toward the end of the 1980s, a new branch of human pathology began with the discovery of human disorders linked to mitochondrial dysfunction (mitochondrial encephalomyopathies). Therefore, it became necessary to investigate and to study the status of mitochondria in human tissue. As it is not always possible to obtain large amounts of human tissue for extracting mitochondria, interest arose in developing smaller scale mitochondrial isolation methods to determine mitochondrial enzyme profiles.

This chapter describes the methods for mitochondrial isolation used in experiments from both human and animal tissues. Further manipulations of isolated mitochondria give the possibility for better investigations of some enzymatic pathways related to the mitochondrial function.

II. General Properties of Mitochondrial Preparations

Almost all preparation methods present major similarities, but unfortunately the details are often tissue or laboratory specific (Nedergaard and Cannon, 1979). All the steps for mitochondrial isolation should be performed on ice or, if possible, in a cold room.

A. Isolation Medium

Principal osmotic support:

a. Nonionic, e.g., sucrose, mannitol, or sorbitol; usually about 0.25 *M* for mammalian tissues.

b. Ionic, e.g., KCl (100–150 m*M*) for those tissues that assume a gelatinous consistency upon homogenization. For special preparations (e.g., mitoplasts), hypotonic KCl buffers can also be used for swelling mitochondria as a valid alternative method to incubation in digitonin.

Possible additions:

a. EDTA (1 m*M*), in order to chelate Ca^{2+} ions, which can function as uncouplers and which are also cofactors for certain phospholipases.
b. Bovine serum albumin (BSA, 0.1–1%). It binds free fatty acids, acyl-CoA esters, lysophospholipids, or other detergents.
c. Buffer: Usually Tris–HCl or Tris–acetate (5–20 m*M*).

B. Homogenization

The initial destruction of intercellular connections, cell walls, and plasma membranes is necessary for mitochondrial isolation. The choice of a destruction technique should be dictated by the type of cells and intercellular connections, and is either mechanical or enzymatic. For soft tissue homogenization of a tissue mince, the use of a Dounce hand homogenizer or a power-driven Potter–Elvehjem glass–Teflon homogenizer is suitable. For more fibrous tissues, the use of a tissue blender (Waring, UltraTurrax) is necessary. The ratio of tissue to isolation medium is usually 1:5 to 1:10 (w/v); the use of proteases softens the tissue, especially muscle, prior to homogenization. The homogenization step is always carried out on ice in order to inactivate cellular ezymes that could damage mitochondria.

C. Differential Centrifugation

Differential centrifugation consists of two-step centrifugation carried out at low and at high speed, consecutively. The low-speed centrifugation is necessary to remove intact cells, cellular debris, and nuclei.

It is likely that mitochondria in the lower part of the tube could remain entrapped in the pellet, therefore resuspension of the latter and recentrifugation at low speed increases the mitochondrial yield. The two supernatants obtained from the low-speed centrifugation undergo a subsequent high-speed centrifugation to sediment mitochondria.

The low-speed centrifugation is usually carried out at 1000*g* for 10 min and the subsequent high-speed centrifugation at 10,000*g* for 10 min. In many tissues, a light-colored pellet or "fluffy" layer may sediment over the darker brown mitochondrial pellet after high-speed centrifugation. This layer consists of broken mitochondrial membranes and of mitochondria with structural alterations. The fluffy layer can be removed by gentle shaking in the presence of a few drops of medium and discarded. In some tissues, two distincts mitochondrial fractions can be isolated. Usually, the heavier pellet contains the metabolically most intact mitochondria.

D. Gradient Centrifugations

The pellet obtained from Section II,C is usually considered "crude," and in some cases a purification based on size and density is necessary. This can be achieved by using discontinuous gradients (sucrose, Ficoll, or metrizamide).

E. Storage

Mitochondrial pellets should be suspended in a minimal volume of isolation buffer. It is preferable to use nonionic media as they prevent the loss of peripheral proteins. Prior to storage or to use, the protein concentration of the mitochondrial preparation should be determined using either the biuret method (Gornall *et al.,* 1949) or the Lowry method (Lowry *et al.,* 1951). The first method is recommended if high yields of mitochondria are obtained, whereas the second method is more sensitive for determining small protein concentrations. It is recommended to store mitochondria at concentrations no lower than 40 mg protein per milliliter of resuspension buffer and kept frozen at −70°C.

F. Contaminants of Mitochondrial Fraction

The nature of the contaminants that may be present in a mitochondrial preparation depends on the biological material from which they are isolated (de Duve, 1967). Except for erythrocytes, few cells are small enough to escape sedimentation during the low-speed centrifugation. However, it is possible to find pinched-off cell fragments; these can be eliminated by gradient centrifugation. Because nuclei are often damaged by homogenization, they also represent a possible contaminant. Nuclei are observed as a grayish sediment, often associated with red blood cells, at the bottom of the mitochondrial pellet. Care should be taken in analyzing mitochondrial DNA from mitochondrial fractions because phagocytized nuclear DNA could also contaminate these preparations.

Secretion granules and Golgi vesicles can contaminate mitochondrial preparations derived from exocrine and endocrine glands (Palade *et al.,* 1962) and can be eliminated by gradient centrifugation.

Lysosomes are almost ubiquitous, but they are particularly abundant in liver, kidney, spleen, leukocytes, and macrophages. They contain a variety of substances at various stages of proteolytic and acidic digestion, as well as indigestible residues. In some cases, it is preferable to separate these contaminants from mitochondrial preparations by gradient centrifugation on a discontinuous metrizamide gradient (Genova *et al.,* 1994).

Peroxisomes (de Duve and Baudhuin, 1966) contain large amount of catalase, hydrogen-producing oxidases, and L-lactate. They are always present in mitochondrial fractions from liver, but having a higher median equilibrium density than mitochondria, they can be separated easily from the latter fraction by gradient centrifugation.

Melanosomes can also be separated from mitochondrial fractions by gradient centrifugation. Particulate glycogen may be present in mitochondrial fractions from liver of fed animals, but an overnight starving prior to sacrifice is sufficient to eliminate this contaminant from mitochondria.

Microsomes (typical endoplasmic reticular membranes with attached dense granules, cell membrane fragments, and Golgi membranes) are present in the "fluffy" layer that covers the mitochondrial pellet after the high-speed centrifugation. "Fluffy" layers are easily dislodged and washed away, but we usually retain this layer as it could contain fragments of mitochondrial membrane as well. The activity of the enzymatic complexes of the respiratory chain is usually assayed both in the absence and in the presence of a specific inhibitor for the complex examined, revealing the specificity for the mitochondrial assay.

G. Criteria of Purity and Intactness

Mitochondria free from contaminant membranes should have negligible activities of the marker enzymes for other subcellular fractions, such as glucose-6-phosphatase for endoplasmic reticulum, acid hydrolases for lysosomes, and catalase and D-amino acid oxidase for peroxisomes. However, mitochondria should be highly enriched for cytochrome *c* oxidase and succinate dehydrogenase. Citrate synthase seems to be a rather stable mitochondrial enzyme whose activity is not subjected to fluctuations and pathological changes; for this reason, when homogenates or impure mitochondrial fractions have to be used for enzymatic determinations, activities are best compared by normalization to citrate synthase in order to prevent artifacts due to differences in the content of pure mitochondria.

Intact mitochondria, when investigated in an oxygen electrode, should have high respiratory control ratios (RCR) with both NAD-linked substrates and succinate; ratios of state-3 to state-4 respiration higher than 4–5 are considered to be diagnostic of tightly coupled mitochondria. Coupled mitochondria usually exhibit ADP/O ratios higher than 2.5, and approaching 3 with NAD-linked substrates, and ratios higher than 1.8, approaching 2, with succinate, glycerol-1-phosphate, or acyl-carnitines as substrates. The ADP/O ratios with 2-ketoglutarate are usually higher than 3 because of substrate-level phosphorylation. Intact mitochondria also exhibit respiratory control at site 3 using ascorbate and tetramethyl-phenylenediamine (TMPD) to reduce cytochrome *c* and cytochrome oxidase in the presence of a complex III inhibitor, such as antimycin A; the RCR, however, is usually lower. The ADP/O ratio at site 3 is usually higher than 0.5, theoretically approaching 1.

III. Mitochondria from Beef Heart

Mitochondria prepared from heart muscle present some peculiar advantages over those from other mammalian tissues. They are stable for up to a week with respect to oxidation and phosphorylation when stored at 4°C and for up to a year at −20°C. Generally, mitochondria prepared from slaughterhouse material are functionally intact. While beef heart is the tissue of choice for this preparation, pig heart may also be used.

A. Small-Scale Preparation

The method used is essentially the one described by Smith (1967), with modifications to allow large-scale isolation (i.e., from one or two beef hearts).

Hearts are obtained from a slaughterhouse within 1–2 h after the animal is slaughtered and placed in ice. All subsequent procedures are carried out at 4°C.

The tissue is trimmed from fat and connective tissue and cut into small cubes. Two hundred grams of tissue is passed through a meat grinder and placed in 400 ml of sucrose buffer (0.25 *M* sucrose, 0.01 *M* Tris–HCl, pH 7.8). The suspension is homogenized in a Waring blender for 5 sec at low speed, followed by 25 sec at high speed. At this stage, the pH of the suspension must be adjusted to 7.5 with 1 *M* Tris.

The homogenate is centrifuged for 20 min at 1200*g* to remove unruptured muscle tissue and nuclei. The supernatant is filtered through two layers of cheesecloth to remove lipid granules and is then centrifuged for 15 min at 26,000*g*.

The mitochondrial pellet obtained is resuspended in sucrose buffer and is homogenized in a tight-fitting, Teflon–glass Potter–Elvejhem homogenizer (clearance of 0.006 inch) and is then centrifuged at 12,000*g* for 30 min. The pellet is resuspended in the sucrose buffer and stored at −80°C, at a protein concentration of 40 mg/ml. The protein concentration is determined using the biuret method (Gornall *et al.,* 1949).

The heavy fraction of beef heart mitochondria obtained by the procedure just described was reported to have a high content of mitochondrial components with an excess of cytochrome oxidase over the other complexes (Capaldi, 1982) (Table I) and with high

Table I
Main Components of Oxidative Phosphorylation System in Bovine Heart Mitochondrial Membrane[a]

	Concentration range		Molecular	Number of	
	nmol/mg protein	μ*M* in lipids[b]	mass (kDa)	polypeptides	Prosthetic groups[c]
Complex I	0.06–0.13	0.12–0.26	700	42	FMN, 7Fe-S
Complex II	0.19	0.38	200	4–5	FAD, 3Fe-S
Complex III	0.25–0.53	0.50–1.06	250	12	2*b*, c_1, Fe-S
Complex IV	0.60–1.0	1.20–2.0	160	12	*a*, a_3, 2Cu
Cytochrome *c*	0.80–1.02	1.60–2.04	12	1	*c*
Ubiquinone-10	3.0–8.0	6.0–16.0	0.75	—	—
NADH-NADP transhydrogenase	0.05	0.1	120	1	—
ATP synthase	0.52–0.54	1.04–1.08	500	23[d]	—
ADP/ATP translocator	3.40–3.60	6.8–9.2	30	1	—
Phospholipids	440–587	—	0.7–1.0	—	—

[a] Modified from Capaldi (1982).
[b] Assuming phospholipids to be 0.5 mg/mg protein.
[c] Fe-S, iron–sulfur clusters; *b*, c_1, *c*, *a*, and a_3 are the corresponding cytochromes.
[d] Fourteen types of subunits.

Table II
Oxidation of Various Substrates by Beef Heart Mitochondria[a]

Substrate	Rate of oxidation[b]	P:O
Pyruvate + malate	0.234	2.9
Glutamate	0.181	3.1
2-Oxoglutarate	0.190	3.8
3-Hydroxybutyrate	0.123	3.0
Succinate	0.050	2.0

[a] Modified from Hatefi *et al.* (1961).
[b] Expressed as micrograms of atoms oxygen taken up per minute per milligram of mitochondrial protein.

respiratory activities accompanied by high RCR and by P/O ratios approaching the theoretical values (Table II).

In our experience, mitochondria obtained using the just-described isolation method do not exhibit good respiratory control. However, they have rates of substrate oxidation comparable to those described by Smith (1967). This is likely to be due to the quality of distilled water used for the preparation as well as to the extent of mitochondrial rupture by the harsh treatment. Therefore, we only use beef heart mitochondria (BHM) for manipulations that do not require coupled mitochondria.

When the activity of individual enzymes, and not overall respiration, has to be determined, impermeable substrates, such as NADH, should be allowed to cross the inner membrane, whereas cytochrome *c* should enter the intermembrane space. BHM preparations should undergo two to three freeze–thaw cycles in order to disrupt the outer and the inner mitochondrial membranes, thus allowing NADH to enter the matrix and cytochrome *c* to become available at the outer side of the inner membrane. Alternatively, small amounts of detergents, such as deoxycholate, may be added to break the permeability barrier. Many individual and combined activities can be assayed, such as NADH CoQ reductase (complex I), succinate CoQ reductase (complex II), ubiquinol cytochrome *c* reductase (complex III), cytochrome *c* oxidase (complex IV), NADH-cytochrome *c* reductase (complexes I + III), and succinate cytochrome *c* reductase (complexes II + III).

B. Preparation of Coupled Submitochondrial Particles

BHM preparations obtained with the method just described are not maximally efficient for the coupling of oxidation to phosphorylation. However, further manipulations of BHM can yield a preparation of submitochondrial particles fully capable of undergoing coupling of oxidative phosphorylation. These are known as phosphorylating electron transfer particles derived from beef heart mitochondria (ETPH).

ETPH are inside-out vesicles formed by pinching and resealing of the cristae during sonication (Fig. 1) and are obtained using the methods of Beyer (1967) and Hansen and

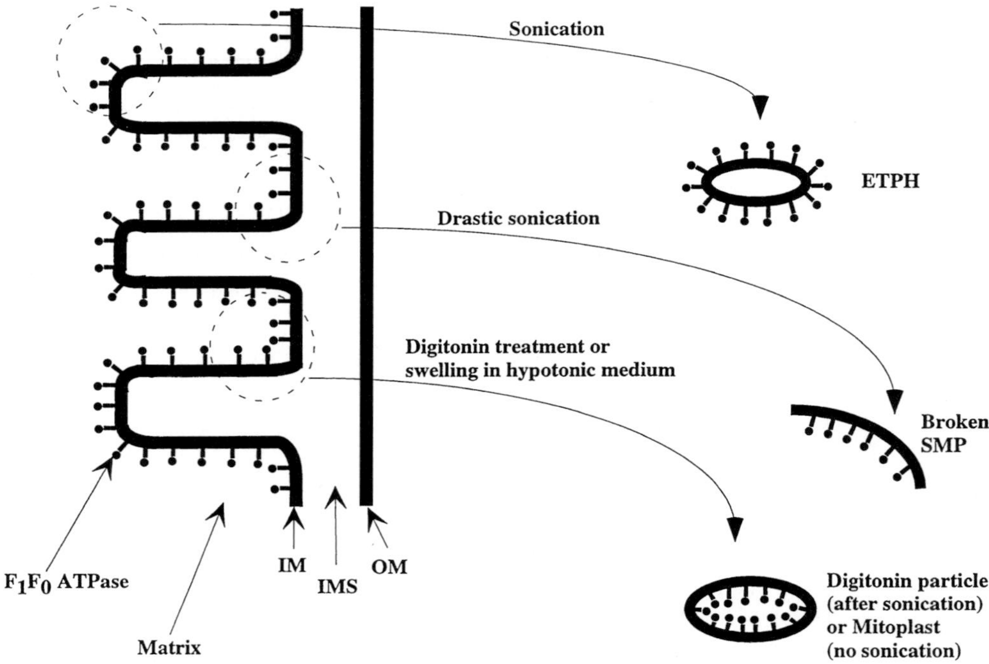

Fig. 1 Derivation of ETPH, digitonin particles, mitoplasts, and SMP from mitochondrial cristae (IM, inner membrane; IMS, intermembrane space; OM, outer membrane).

Smith (1964), with minor modifications. These particles are often better sealed, with a more intact permeability barrier, and therefore they have higher coupling capacity than their parent mitochondria.

BHM are prepared as described in Section II,A. However, the mitochondrial pellet, either freshly prepared or thawed from −80°C storage, is resuspended in STAMS buffer (0.25 *M* sucrose, 0.01 *M* Tris–HCl, pH 7.8, 1 m*M* ATP, 5 m*M* $MgCl_2$, 10 m*M* $MnCl_2$, and 1 m*M* potassium succinate). The final protein concentration is adjusted to 40 mg/ml with STAMS. Aliquots of 20–25 ml of this mitochondrial suspension are subjected to sonic irradiation for 30 s using a probe sonicator (we use either the Branson sonicator at 20kc or the Braun sonifier set at 150).

The suspension is then centrifuged at 20,000*g* for 7 min in order to remove big particles. The supernatant is decanted and centrifuged at 152,000*g* for 25 min, and the pellet is rinsed with STAMS buffer and resuspended to a final protein concentration of 20 mg/ml in a preserving mix containing 0.25 *M* sucrose, 10 m*M* Tris–HCl, pH 7.5, 5 m*M* $MgCl_2$, 2 m*M* GSH, 2 m*M* ATP, and 1 m*M* succinate (Linnane and Titchener, 1960).

ETPH exhibit high rates of NADH and succinate oxidation, but no respiratory control: they are therefore loosely coupled. However, respiratory control can be shown in these particles by the addition of oligomycin; this ATPase inhibitor slows down respiration rates, significantly, indicating that the lack of respiratory control is probably due to the backflow of protons through the ATPase membrane sector.

Coupled ETPH exhibit fluorescence quenching of 9-aminoacridine and derivatives, such as atebrine and ACMA (9-amino-6-chloro-2-methoxy acridine) in the presence of substrates and in the presence of valinomycin plus potassium (added in order to collapse the membrane potential), indicating the formation of ΔpH in closed inverted vesicles (the ΔpH usually obtained with succinate is 3.15) (Lenaz *et al.,* 1982). The precise determination of the pH difference across the external and internal compartment of ETPH is given by the distribution of 9-aminoacridine following the development of transmembrane pH differences, as only the uncharged species of the amine is freely permeable across the membrane (Casadio *et al.,* 1974). The fluorescence intensity of 9-aminoacridine is dependent on the concentration of the amine (Fig. 2).

C. Broken Submitochondrial Particles

Submitochondrial particles (SMP) are broken membrane fragments and therefore can react with exogenous by-added cytochrome *c*. They are not coupled, but they have good rates of individual enzymatic activities, such as NADH-coenzyme Q oxidoreductase and ubiquinol-cytochrome *c* oxidoreductase. SMP have been used in our laboratory for the kinetic characterization of complex I (Estornell *et al.,* 1993) and complex III (Fato *et al.,* 1993).

SMP are prepared by sonication of BHM, as obtained in Section III,A, in an MSE sonifier for 5 min, at 30-s intervals, thus allowing the preparation to cool down. Preparations are kept on ice and under nitrogen in order to avoid lipid peroxidation.

The mitochondrial suspension is then centrifuged at 20,000*g* for 10 min; the supernatant is collected and ultracentrifuged at 152,000*g* for 40 min. SMP are resuspended in sucrose buffer and kept frozen at -80°C at a protein concentration of 40–50 mg/ml until needed.

In these particles, succinate evokes a slight quenching of 9-aminoacridine fluorescence that was roughly calculated to correspond to <5–8% sealed inverted vesicles (Casadio *et al.,* 1974). SMP can be used to assay NADH-, succinate-, and ubiquinol-cytochrome *c* reductase (or cytochrome *c* oxidase) without adding detergents (Table III).

The use of detergents (i.e., deoxycholate) is important in evaluating the status of the particles; if detergents stimulate cytochrome *c* reduction by upstream substrates, the preparation contains high levels of closed particles (Degli Esposti and Lenaz, 1982). However, ETPH must have very low cytochrome *c* reductase activities that are strongly stimulated by adding detergents.

D. Cytochrome *c*-Depleted and Cytochrome *c*-Reconstituted Mitochondria and Electron Transfer Particles Derived from Beef Heart Mitochondria

Cytochrome *c* is a mobile component of the respiratory chain and in order to estimate the amount of this component in mitochondria it is necessary to release it from the organelle. The quantitative extraction of cytochrome *c* can be accomplished with

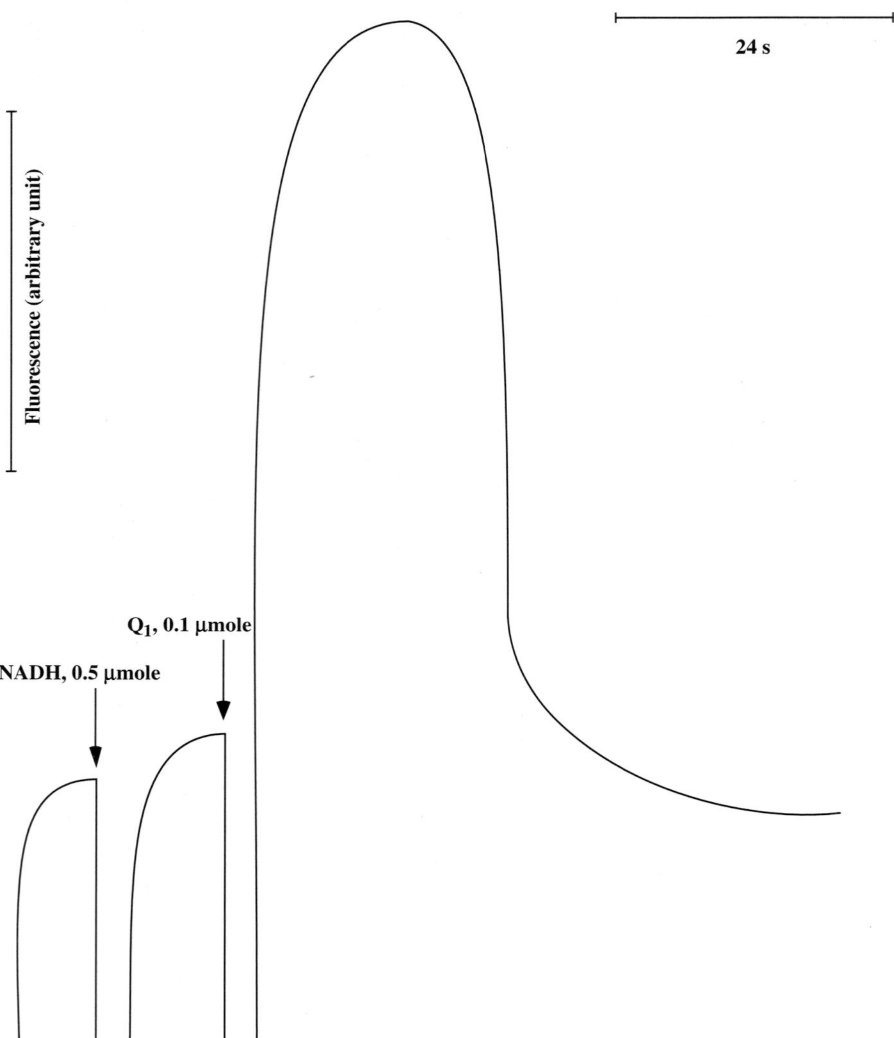

Fig. 2 Quenching of fluorescence of atebrine in an ETPH preparation induced by the oxidation of NADH by coenzyme Q_1(modified from Melandri *et al.,* 1974). The assay is performed in a final volume of 2.5 ml containing 100 µmol glycylglycine, pH 8.0; 1.25 nmol sucrose; 12.5 µmol $MgCl_2$; 250 µmol KCl; 2.5 µmol EDTA; 4 µg valinomycin; 10 nmoles atebrine; and ETPH corresponding to 480 µg of protein.

Table III
Some Individual Respiratory Chain Activities in Broken SMP

Activity	Substrates	V_{max} (μmol/min.mg)	k_{cat} (sec^{-1})
NADH-CoQ[a]	NADH, CoQ_1	0.90	380
	NADH, DB[d]	0.58	225
Ubiquinol-cytochrome *c*[b]	Ubiquinol-1, cytochrome *c*	2.77	220
	Ubiquinol-2, cytochrome *c*	4.65	370
Succinate-CoQ[c]	Succinate, CoQ_1	1.08	63

[a] From Fato *et al.* (1996).
[b] From Fato *et al.* (1993).
[c] Unpublished data from R. Fato and G. Lenaz.
[d] 6-Decylubiquinone.

salts, yielding a good relative amount of nondenatured mitochondria. The method for cytochrome *c* extraction from mitochondrial preparations has been described previously by MacLennan and colleagues (1966). It is summarized briefly in this section.

Suspensions of freshly isolated BHM at a concentration of 40 mg/ml are frozen at −20°C in a preserving medium containing 0.25 *M* sucrose, 0.01 *M* Tris–acetate, pH 7.5, 1 m*M* ATP, 1 m*M* $MgCl_2$, and 1 m*M* succinate. When thawed, the suspension is adjusted to pH 7.8 and centrifuged at 26,000*g* for 10 min. The supernatant and the residual light layer, composed of light mitochondria, are removed and the dark pellet is suspended in a solution of 0.015 *M* KCl (Jacobs and Sanadi, 1960), a hypotonic medium for intact mitochondria, to a protein concentration of 20 mg/ml. Mitochondria are allowed to swell for 10 min on ice before centrifugation at 105,000*g* for 15 min.

The colorless supernatant is discarded and the pellet is resuspended in 0.15 *M* KCl, an isotonic medium for intact mitochondria, and left on ice for 10 min. The last centrifugation step is carried out at 105,000*g* for 15 min. The resulting supernatant is red in color, representing cytochrome *c*. The mitochondrial pellet undergoes two further extractions in the isotonic medium.

The final centrifugation in the isotonic medium yields a fraction of light mitochondria with low phosphorylative capacity, which should be removed. The pellet of cytochrome *c*-depleted mitochondria is resuspended in 0.25 *M* sucrose and 0.01 *M* Tris–acetate, pH 7.5. This procedure usually removes about 85% of cytochrome *c* content from mitochondria, and the rates of substrate oxidation of these mitochondria are 15% of the rates of the mitochondrial preparation before the extraction procedure.

Preparation of cytochrome *c*-depleted ETPH is achieved by the sonication of cytochrome *c*-depleted mitochondria (Lenaz and MacLennan, 1967), using the conditions for sonication described in Section III,A. Usually the ETPH preparation undergoes two sonication cycles: after the first low-speed centrifugation (26,000*g*) the supernatant is collected and the pellet is resonicated and recentrifuged at low speed. The resulting supernatant is combined with the first one and centrifuged at high speed (105,000*g*). Cytochrome *c*-depleted ETPH preparations show reductions in substrate oxidation and in P/O ratios in comparison with normal ETPH.

Table IV
Oxidative and Phosphorylating Properties of Cytochrome c-Depleted and -Reconstituted Mitochondria and ETPH

A. BHM (data obtained from MacLennan *et al.*, 1966)

Substrate	Cytochrome *c* removed oxygen uptake[a]	P/O ratio	Cytochrome *c* added oxygen uptake[a]	P/O ratio
Pyruvate + malate	0.022	1.14	0.189	2.40
Succinate	0.035	0.71	0.136	1.06
Ascorbate + TMPD	0.037	0.24	0.169	0.63

B. ETPH (data obtained from Lenaz and MacLennan, 1966)

Treatment	Oxygen uptake with succinate[a]	P/O ratio
None	0.163	1.02
ETPH (− cytochrome *c*)[b]	0.064	0.77
ETPH (+ cytochrome *c*)[c]	0.155	1.15
ETPH (− cytochrome *c*) + cytochrome *c* added after sonication	0.093	0.55

[a] Expressed as μg atoms/min·mg protein.
[b] By sonication of cytochrome *c*-depleted mitochondria.
[c] Cytochrome *c* added after sonication.

Sonication of cytochrome *c*-depleted mitochondria can also be performed in isotonic KCl instead of in the classic medium for ETPH preparation (see earlier discussion) (Lenaz and MacLennan, 1966), with further reduction of both oxidation rates and P/O ratios.

Cytochrome *c* can be reincorporated in cytochrome *c*-depleted mitochondria through incubation in the presence of high levels of exogenous cytochrome *c*.

Reincorporation of cytochrome *c* into cytochrome *c*-depleted ETPH is complicated by the fact that ETPH are inside-out vesicles that cannot react with exogenous cytochrome *c*. For this reason, KCl-extracted BHM are resupended in the preserving mix for ETPH, with the exception that cytochrome *c* is added at a concentration of 10 μg per milligram of protein, before BHM are disrupted by sonication. The same procedure as for preparation of the cytochrome *c*-depleted ETPH is then followed.

Table IV summarizes the properties of cytochrome *c* depleted and reconstituted preparations.

E. Coenzyme Q-Depleted and Coenzyme Q-Reconstituted Mitochondria

Coenzyme Q (CoQ) can be extracted from BHM preparations using organic solvents. Coenzyme Q-depleted mitochondria are usually reconstituted with coenzyme Q homologues and analogs. Polar solvents, such as acetone, were first used for CoQ extraction, but they irreversibly damage complex I; thus nonpolar solvents are preferred, but they can extract neutral lipids, such as CoQ, only from dry material. The extraction is performed on lyophilized BHM preparations following the method of Szarkowska (1966) with modifications.

The BHM preparation (Section III,A) is thawed on ice and diluted to a final protein concentration of 20 mg/ml in sucrose buffer (0.25 *M* sucrose, 0.01 *M* Tris–HCl, pH 7.5). The suspension is centrifuged at 35,000*g* for 10 min (17,000 rpm in a Sorvall SS34 rotor). The pellet is resuspended in 0.15 *M* KCl, frozen at −80°C, and lyophilized.

The extraction of coenzyme Q is performed using pentane, homogenizing mitochondria in a Potter–Elvejhem homogenizer with a Teflon pestle, and centrifuging the suspension at 1100*g* for 10 min (3000 rpm in a Sorvall SS34 rotor). The supernatant, containing pentane, is removed and extraction–homogenization is repeated four times. Finally, pentane is removed from the extracted mitochondria, first in a rotary evaporator under reduced pressure at 30°C and then under high vacuum at room temperature for 2 h.

Lyophilized mitochondria are homogenized in sucrose buffer, centrifuged at 35,000*g* for 10 min, and resuspended in the same buffer. This suspension can either be used immediately or stored at −80°C for later use.

Reconstitution of coenzyme Q-depleted mitochondria is attained by treating the dry extracted mitochondrial powder with either the pentane extract or coenzyme Q homologues and analogs in pentane, following some modification in the method described by Norling *et al.* (1974).

Depleted particles are homogenized in a small volume of pentane (usually 2 ml) in the presence of a known amount of coenzyme Q. The amount of CoQ is usually adjusted to a final concentration between 0.4 and 25 nmol CoQ/mg mitochondrial protein. The protein content of CoQ-depleted mitochondria is estimated from the dry weight, considering that 3 mg dry weight corresponds to 1 mg mitochondrial protein, as measured by the biuret method.

The particle suspension is transferred in a rotary evaporator at 4°C under a slightly reduced pressure. Gradual removal of pentane is essential in order to get a good rate of incorporation (pentane should evaporate over a period of 30 min). After complete removal of the pentane, particles are dried at 4°C for an additional 30 min under vacuum and resuspended in sucrose buffer.

Coenzyme Q-depleted mitochondria have negligible contents of residual CoQ (less than 20 pmol/mg protein) and exhibit very low rates of both NADH and succinate oxidation in comparison with controls (either intact or lyophilized mitochondria); reconstitution with long-chain ubiquinones (CoQ_5 to CoQ_{10}) reconstitutes maximal rates of both enzymes, whereas short-chain homologs are not able to recover NADH oxidation and behave as complex I inhibitors. Some properties of these mitochondria are shown in Table V.

IV. Mitochondria from Rat Liver

Liver mitochondria prepared from rats are usually suitable for biochemical assays in studies on the pharmacological effects of different drugs or on effects of specific diets on mitochondrial membrane composition. We used rat liver mitochondria in studies after

Table V
Kinetic Constants of NADH- and Succinate-Cytochrome *c* Reductase in Lyophilized and Pentane-Extracted Beef Heart Mitochondria Reconstituted with Some Representative CoQ Homologues Having Different Lengths of the Isoprenoid Side Chain

	NADH-cytochrome *c* reductase		Succinate-cytochrome *c* reductase	
Quinone	V_{max}[a]	K_m[b]	V_{max}[a]	K_m[b]
CoQ_{10}	0.64 ± 0.46	1.53 ± 1.16	0.26 ± 0.17	0.42 ± 0.57[c]
CoQ_3	0.12 ± 0.04	1.42 ± 0.92	0.20 ± 0.01	0.53 ± 0.33[c]
CoQ_5	0.15 ± 0.02	1.46 ± 0.38	0.13 ± 0.05	0.37 ± 0.24[d]
6-Decylubiquinone	0.55	25	0.52	3.8

[a] Expressed as μmol/min/mg protein.
[b] Expressed as nmol quinone/mg protein, except for DB, whose K_m was calculated as m*M* quinone in the phospholipids.
[c] Statistically significant with respect to the corresponding K_m of NADH-cytochrome *c* reductase, $p < 0.05$.
[d] As above, $p < 0.001$.

perfusion and in studies where quinones were incorporated in liver fractions. Rat liver contains a considerable amount of mitochondria and is easier to manipulate than skeletal muscle. Generally, a high mitochondrial yield is obtained from rat liver.

A. Standard Preparation

Our method used for the extraction of mitochondria from rat liver is a modification of the method described by Kun and colleagues (1979). This method is based on the typical differential centrifugation procedure used for other mitochondrial preparations.

Fresh tissue is chilled on ice and washed in 0.22 *M* mannitol, 0.07 *M* sucrose, 0.02 *M* HEPES, 2 m*M* Tris–HCl, pH 7.2, and 1 m*M* EDTA (solution A). Subsequently, it is minced with scissors and washed three times in solution A with 0.4% BSA to remove blood and connective tissue and weighed in a prechilled glass petri dish.

The suspension is then homogenized in a prechilled Potter–Elvehjem glass homogenizer using a Teflon pestle and filtered. The homogenate is centrifuged at 3000*g* for 1.5 min (5000 rpm in a Sorvall SS34 rotor); the supernatant is decanted and the pellet is resuspended in solution A and is subjected to a second centrifugation step.

The two supernatants are combined and centrifuged at 17,500*g* for 2.5 min (12,000 rpm in a Sorvall SS34 rotor). The resulting pellet undergoes a wash in solution A and then it is centrifuged at 17,500*g* for 4.5 min. The pellet is resuspended in 0.22 *M* mannitol, 0.07 *M* sucrose, 0.01 *M* Tris– HCl, pH 7.2, and 1 m*M* EDTA (solution B) and centrifuged at 17,500*g* for 4.5 min. The pellet is finally resuspended in solution B at a ratio of 10 ml solution B per 7 g of starting material in order to standardize the protein content of the mitochondrial fraction.

Mitochondria obtained with this method also contain a lysosomal fraction, but if used immediately they show a good respiratory control with glutamate–malate and with

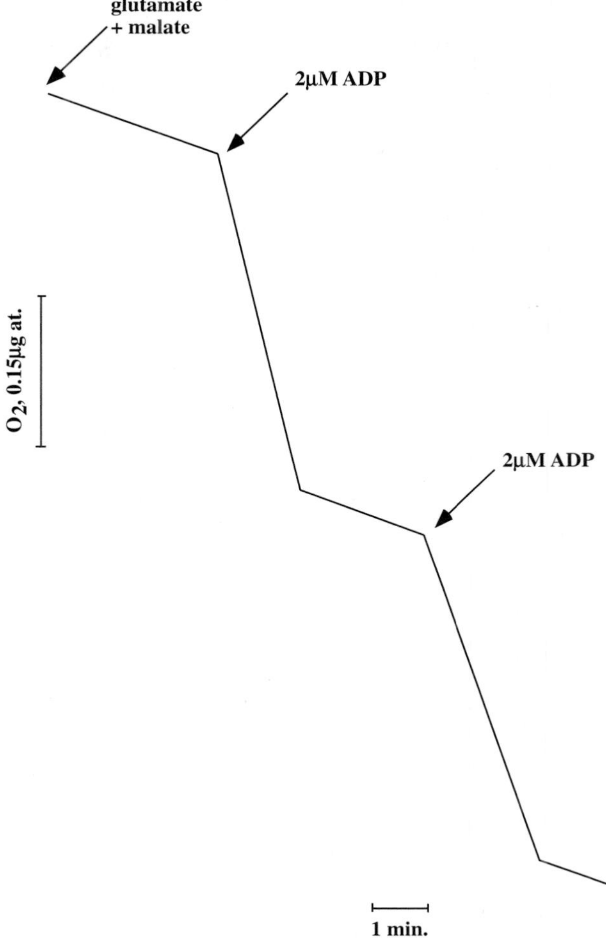

Fig. 3 Respiratory control in the presence of glutamate + malate in a mitochondrial preparation from rat liver. This preparation presented a respiratory control ratio (RCR) equal to 11.

succinate (Fig. 3) and good enzymatic activities of the four mitochondrial complexes (after membrane permeabilization by freeze–thaw cycles or by detergents).

B. Gradient-Purified Rat Liver Mitochondria

The preparation of rat liver mitochondria free from lysosomal contamination involves the purification of mitochondria in metrizamide gradient. The method described here is a modification of the one described by Kalen *et al.* (1990). This procedure for purification of mitochondria has been used for monitoring the incorporation of exogenous CoQ by rat liver fractions (Genova *et al.,* 1994).

The liver is removed, chopped, weighed, and washed at least 10 times in ice-cold solution A (0.33 *M* sucrose, 0.01 *M* Tris–HCl, pH 7.4, 1 m*M* EDTA, 0.4% BSA). The tissue is then homogenized in a prechilled Potter–Elvehjem glass homogenizer with a Teflon pestle to a ratio 1:4 (w/v) tissue to solution A.

The homogenate is centrifuged at 310*g* for 10 min; the supernatant is decanted and the pellet is resuspended in solution B (0.33 *M* sucrose, 0.01 *M* Tris–HCl, pH 7.4, 1 *M* EDTA) and recentrifuged in order to obtain a clean nuclear fraction. The supernatant, collected previously, is centrifuged at 2800*g* for 20 min, and the pellet (washed twice) is recovered. This pellet contains a fraction of mitochondria defined as "heavy mitochondria" (HM). The supernatant is then centrifuged at 11,400*g* for 20 min, thus giving a pellet of mitochondria defined as "light mitochondria" (LM) containing lysosomes. The resulting supernatant, containing microsomes and cytosol, is discarded. The LM and HM fractions are resuspended in solution B at a volume three times the weight of the starting rat tissue.

Aliquots of HM and LM are then added to 2 volumes of metrizamide solution as described (Wattiaux *et al.,* 1978), usually 0.4 ml of mitochondria in 0.8 ml metrizamide 85.6% (w/v).

One milliliter of HM/LM-metrizamide solution is then layered at the bottom of a 5-ml ultracentrifuge tube, and a discontinuous metrizamide gradient is stratified on it by adding 0.6 ml 32.82% metrizamide, 0.6 ml 26.34% metrizamide, 0.6 ml 24.53% metrizamide, and 0.6 ml 19.78% metrizamide. Centrifugation is then performed in a swinging bucket rotor at 95,000*g* for 2 h at 4°C.

Mitochondrial subfractions can be collected using a Pasteur pipette. Usually, three subfractions from HM preparations and four subfractions from LM preparations (Genova *et al.,* 1994) are observed.

Mitochondria prepared with this procedure preserve good activities of the respiratory chain complexes, but little respiratory control.

C. Liver Submitochondrial Particles (SMP)

It is also possible to obtain submitochondrial particles from rat liver. Usually, SMP from rat liver are prepared after intensive sonication of mitochondria; thus they are not coupled. However, they can be used for studying NADH oxidation because intact mitochondria are not permeable to NADH.

The method used in our experiments is essentially the one described by Gregg (1967). Mitochondria are prepared from 20–30 g of rat liver and the buffer used is the same as that used for the preparation of broken SMP particles from BHM (Section III,C), with the addition of 0.4% BSA from the beginning of the isolation procedure.

D. Mitochondria from Rat Hepatocytes

Preparation of isolated rat liver cells is usually performed following the two-step collagenase liver perfusion technique of Seglen (1976). The purpose of this technique is to obtain intact hepatocytes separated from nonparenchymal cells (up to 40% of total liver tissue); we have used this method to treat isolated cells with adriamycin in order to induce oxidative stress by reduction of this potent anticancer agent to semiquinone, thus releasing superoxide anion and hydrogen peroxide. In order to evaluate cellular integrity (or viability), the trypan blue exclusion test is performed. Mitochondria are prepared from hepatocytes (30×10^6 cells) incubated, at a starting concentration of 1×10^6 cells/ml, for 2 h (Wells *et al.,* 1987; Barogi *et al.,* 2000). Cells are suspended in a buffer containing 0.25 *M* sucrose, 0.01 *M* Tricine, 1 m*M* EDTA, 10 m*M* NaH_2PO_4, 2 m*M* $MgCl_2$, pH 8 (solution A) with 0.4% BSA, frozen at −80°C for 10 min to break the plasma membrane, and centrifuged at 760*g* for 5 min.

The supernatant is kept while the pellet undergoes a second homogenization step, using a UltraTurrax homogenizer for 10 min, followed by centrifugation at 760*g* for 5 min. The supernatants from the previous two steps are combined and centrifuged for a further 20 min at 8000*g*. The mitochondrial pellet is washed once with the same buffer in the absence of BSA and finally resuspended in the same buffer.

Mitochondria obtained with this method show good respiratory activities, but the NAD-dependent activities are rather low (Barogi *et al.,* 2000).

E. Isolation of Mitochondria from Frozen Tissues

Generally, mitochondrial isolations are performed on fresh tissue. Under these conditions, isolated mitochondria are pure and suitable for most of the biochemical assays and for further modifications of mitochondria used in particular assays. However, it is often necessary to prepare mitochondria from frozen tissues in cases where fresh tissues are not available.

The most important step in the procedure is the freeze–thawing of the tissue. Mitochondrial membranes are very sensitive to temperature variations, and the rupture of the membranes (cellular and mitochondrial) prior to homogenization not only makes it impossible to achieve separation of the subcellular fractions, but also modifies biochemical parameters.

In our experience, the most suitable method to preserve organs prior to extraction of mitochondria is that described by Fleischer and Kervina (1974) for liver tissue. We have also applied the procedure for extractions from other organs, such as heart, muscle, and kidney (Castelluccio *et al.,* 1994; Barogi *et al.,* 1995). This method allows for tissue storage for prolonged periods prior to the extraction of mitochondria.

After sacrificing the animal, organs are weighed and 1 volume of storage medium [0.21 *M* mannitol, 0.07 *M* sucrose, 20% dimethyl sulfoxide (DMSO), pH 7.5] is added. DMSO, used as an antifreeze agent, has the advantage over glycerol in that it is less viscous and penetrates the tissue rapidly. It is important to prevent ice crystal formation; in order to minimize this phenomenon, the organs should be frozen rapidly, and the thawing of them should be performed as quick as possible. Once in storage medium, the organ can be either diced into small pieces and homogenized or it can be maintained intact before quick freezing in liquid nitrogen. In our experience, intact organs can be stored in liquid nitrogen with no apparent alterations to mitochondrial function.

In soft tissues, such as liver, the storage medium diffuses through the organ rapidly. However, it is recommended to cut more fibrous tissues, such as kidney, heart, and muscle, into two to three pieces (1 cm^3) to facilitate medium penetration into the tissue. After rapid freezing in liquid nitrogen, the organs are stored in liquid nitrogen until mitochondrial isolation is required.

The thawing procedure should be performed quickly. It is necessary to preheat the thawing medium (0.25 *M* sucrose, 0.01 *M* Tris–HCl, pH 7.5) at 45°C before adding it to the frozen tissue at a 4:1 ratio of medium to tissue. The thawing medium for liver should also contain 0.4% BSA. The extraction of mitochondria is then performed using the method described in Section IV,A, which is also suitable for heart, kidney, and muscle.

F. Phospholipid-Enriched Mitoplasts and SMP with/without Excess Coenzyme Q

The addition of phospholipid (PL) to mitoplasts or SMP increases the inner mitochondrial membrane surface area and dilutes intramembrane proteins. In the case of SMP, PL enrichment facilitates the study of diffusion control of electron transfer by ubiquinone.

Mitochondria are prepared from rat liver as described in Section IV,A. The mitochondrial pellet, however, is resuspended in 0.22 *M* mannitol, 0.07 *M* sucrose, 2 m*M* HEPES, pH 7.4 (solution A). After centrifugation at 20,500*g* for 20 min, the pellet is resuspended in 15 m*M* KCl and kept on ice for 10 min, thus allowing the external membrane to swell. The suspension is centrifuged at 31,000*g* for 45 min and resuspended in solution A. Alternatively, mitochondria from rat liver are washed in solution A in the presence of 0.5 mg/ml BSA, centrifuged at 4500*g*, and resuspended in solution A.

An alternative method for removing the outer membrane and for purifying the inner membrane–matrix fraction is given by the "controlled digitonin incubation" method (Schnaitman and Greenawalt, 1968). Digitonin is diluted in solution A at a concentration of 2% (w/v) and 0.5 mg/ml BSA is added after the digitonin is dissolved. The rat liver mitochondrial suspension is then treated with 1.1 mg of digitonin solution per 10 mg of mitochondrial protein. The suspension is stirred gently for 15 min and then diluted in 3 volumes of solution A. This suspension is then centrifuged at 31,000*g* for 20 min and then resuspended in solution A.

Unilamellar vesicles are prepared from soybean phospholipids (asolectin) by suspending the phospholipids in solution A at a concentration of 200 mg PL/ml solution A. The suspension is then sonicated using a MSE sonifier for 30 min with 30-s cycles.

Unilamellar PL vesicles enriched with coenzyme Q_{10} are prepared drying, under nitrogen atmosphere, both CoQ_{10} suspended in ethanol solution and phospholipids in chlorophorm/methanol. Dried PL and CoQ_{10} are then resuspended in solution A and sonicated with an MSE sonifier for 30 min with cycles of 30-s intervals.

Reincorporating mitoplasts or SMP with PL- or PL/CoQ_{10}-enriched vesicles is carried out as follows (Schneider *et al.,* 1980): 3 mg of mitochondrial protein is mixed with 1.5 ml PL or PL + CoQ_{10} vesicles (150 mg/ml). The mixture is frozen in liquid nitrogen and is then allowed to thaw at room temperature. This step is repeated three times. The suspension is then layered on a discontinuous sucrose density gradient (0.6, 0.75, 1, and 1.25 *M* of sucrose) and centrifuged in a swinging bucket rotor (SW 28) at 90,000*g* for

16 h. The fractions obtained are diluted in 0.25 *M* sucrose, 0.01 *M* Tris–HCl, pH 7.8 (SMP solution) for PL-enriched SMP and in solution A for PL-enriched mitoplasts. Fractions are then centrifuged twice at 160,000*g* for 45 min and resuspended in the appropriate buffers (SMP solution and solution A).

Protein content is assayed by the biuret method (Gornall *et al.,* 1949), and lipid content is measured by the method of Marinetti (1962).

V. Mitochondria from Muscle

Biochemical analysis of mitochondria isolated from muscle tissue provides valuable insights into the pathological and physiological characteristics of human diseases related to impaired mitochondrial metabolism (mitochondrial encephalomyopathies). Muscle tissue is a postmitotic tissue, and accumulation of exogenous and endogenous damage during the life span of the organism makes this tissue one of the best characterized tissues bioenergetically.

Skeletal muscle mitochondria are usually isolated from several grams of tissue. This may not be a problem if the animal is sufficiently large or if several muscles can be pooled. However, human samples are usually obtained from surgical biopsies and needle biopsies, the latter yielding less than 1 g of muscle tissue.

When dealing with experimental animals, the method described by Kun and colleagues (1979) is generally used for the isolation of skeletal muscle mitochondria. In our experience, this method is suitable for isolation from human muscle biopsies (Zucchini *et al.,* 1995) and from rat gastrocnemius (Barogi *et al.,* 1995).

The method is based on the use of mannitol and sucrose medium, BSA, and EDTA as a complexing agent. Essentially, muscle tissue is freed from collagen and nerves, weighed and homogenized in a Teflon–glass Potter–Elvehjem homogenizer in solution A (0.22 *M* mannitol, 0.07 *M* sucrose, 2 m*M* Tris, 1 m*M* EDTA, and 20 m*M* HEPES, pH 7.2) with 0.4% BSA, and centrifuged at 600 *g* for 80 s. The supernatant is collected and the crude nuclear fraction is reextracted by the same technique. The two supernatants are combined and centrifuged at 17,000*g* for 2.5 min. The mitochondrial pellet is then washed twice in solution A and resuspended in the same solution.

We have exploited this method to investigate some individual respiratory chain activities of muscle mitochondria from young and old individuals; the activities could be transformed into actual turnover numbers related to complex III content, established on the basis of the antimycin A inhibition titer (Zucchini *et al.,* 1995) (Table VI).

Lee and colleagues (1993) have described a valid alternative method for the isolation of mitochondria from 3 to 5 g of skeletal muscle using KCl medium and proteinase K to soften the tissue and facilitate cell disruption during homogenization.

The muscle is trimmed, minced, and placed in medium A (0.1 *M* KCl, 50 m*M* Tris–HCl, pH 7.5, 5 m*M* $MgCl_2$, 1 m*M* EDTA, 1 m*M* ATP) in a proportion of 10 ml medium A/gram of muscle. Three to 5 mg proteinase K per gram of muscle tissue is then added to the muscle suspension. After 5 min of incubation (2 min are sufficient for human muscle) at room temperature, the mixture is diluted 1:2 with medium A and homogenized for 15 s

Table VI
Enzymatic Activities of Skeletal Muscle Mitochondria from Young and Old Individuals Expressed as Specific Activities and Turnover Numbers (TN)[a,b]

	NADH-DB reductase		Succinate-cytochrome *c* reductase		Ubiquinol-cytochrome *c* reductase	
Age range (years)	nmol/ min·mg protein	TN(s^{-1})	nmol/ min·mg protein	TN(s^{-1})	nmol/ min·mg protein	TN(s^{-1})
18–29 ($n = 5$)[c]	52.44 ± 25.35	8.78 ± 1.86	102.82 ± 57.97	16.44 ± 4.92	876.88 ± 437.93	142.16 ± 18.37
69–90 ($n = 9$)	49.31 ± 31.92	8.45 ± 7.04	90.79 ± 48.86	12.70 ± 8.31	939.72 ± 449.47	124.65 ± 37.29

[a] Mann–Whitney–U nonparametric text: $p > 0.05$ for all parameters evaluated.
[b] From Zucchini *et al.* (1995).
[c] Number of samples in the age range.

by UltraTurrax. The homogenate (pH 7.3) is centrifuged at 600*g* for 10 min; the pellet is discarded and the supernatant, filtered through a two-layer cheesecloth, is centrifuged at 14,000*g* for 10 min. The resulting pellet is resuspended in medium B (0.1 *M* KCl, 50 m*M* Tris–HCl, pH 7.5, 1 m*M* $MgCl_2$, 0.2 m*M* EDTA, 0.2 m*M* ATP) with 1% BSA and centrifuged at 7000*g* for 10 min. The pellet is resuspended in medium B, centrifuged at 3500*g* for 10 min, and resuspended in 0.25 sucrose to give an approximate content of 40–50 mg mitochondrial protein/ml.

Another method for isolation of mitochondria from 25 to 100 mg skeletal muscle tissue has been described by Rasmussen and colleagues (1997) using special equipment. Briefly, muscle tissue is weighed and incubated for 2 min in 500 μl proteinase medium (100 m*M* KCl, 50 m*M* Tris, 5 m*M* $MgSO_4$, 1 m*M* ATP, 1 m*M* EDTA, pH 7.4, 0.5% BSA, and 2 mg proteinase K/ml). The proteinase is then diluted with 3 ml ATP medium (100 m*M* KCl, 50 m*M* Tris, 5 m*M* $MgSO_4$, 1 m*M* ATP, 1 m*M* EDTA, pH 7.4, 0.5% BSA) and the liquid is discarded. ATP medium is added to the digested muscle and the tissue homogenized. The homogenate is centrifuged at 300*g* for 5 min (1600 rpm in a Sorvall SS34 rotor), and the supernatant obtained is centrifuged at 4500*g* (6000 rpm in a Sorvall SS34 rotor) for 10 min. The pellet is washed with 100 m*M* KCl, 50 m*M* Tris, 5 m*M* $MgSO_4$, 1 m*M* EDTA, pH 7.4, and centrifuged at 7000*g* for 10 min (6800 rpm in a Sorvall HB-4 swing-out rotor). The pellet is finally resuspended in 0.225 *M* mannitol, 75 m*M* sucrose. The relative yield is 40–50%, and mitochondria obtained with this method are well coupled and exhibit high rates of phosphorylating respiration.

VI. Synaptic and Nonsynaptic Mitochondria from Different Rat Brain Regions

Mitochondrial preparations from brain are often heterogeneous and a number of different methods have been used in the past to separate the populations of brain mitochondria. These methods, however, involve lengthy centrifugation in hypertonic sucrose gradients (Clark and Nicklas, 1970), thus limiting extensive metabolic studies.

The method for the isolation and characterization of functional mitochondria from rat brain was first described by Lai and colleagues (1977). Three distinct populations of mitochondria from rat brain can be isolated from a single homogenate preparation: two from the synaptosomal fraction (HM and LM, for heavy and light mitochondria, respectively) and one from nonsynaptic origin, the so-called "free mitochondria" (FM).

It is possible to separate the different types of mitochondria from the cerebral cortex, hippocampus, and striatum. The protocol followed in our experiments has been described by Battino and colleagues (1995) with further modifications (Genova *et al.,* 1997; Pallotti *et al.,* 1998). The animal is sacrificed by decapitation and the skull is opened rapidly. The brain is placed on an ice-chilled glass plate and dissected according to the procedure described by Glowinski and Iversen (1966). First, the posterior region containing pons and cerebellum is eliminated. The reminder is divided into two hemispheres; the two brain cortexes are freed from the hippocampus and the striatum, and each one is cut into two pieces. Brain areas should be dissected rapidly (<20 s) and placed in buffer A (0.32 *M* sucrose, 1 m*M* EDTA, 10 m*M* Tris–HCl, pH 7.4). The tissue is then homogenized using a Teflon–glass homogenizer at 800 rpm by five up-and-down passes of the pestle (Villa *et al.,* 1989) and subsequently centrifuged. Centrifugation is carried out at 1000*g*, by gradually increasing speed for 4 min, and at the final speed for an additional 11 s. The pellet is resuspended in buffer A and centrifuged again; this step is repeated once again and the three supernatants are pooled and centrifuged at 15,000*g* for 20 min to obtain the "crude" mitochondrial pellet containing synaptosomes. Isolation of free mitochondria from synaptosomes is obtained by layering the pellet, resuspended in buffer A, on a discontinuous Ficoll–sucrose two-step gradient [12 and 7.5% (w/w) Ficoll in 0.32 *M* sucrose, 50 μ*M* EDTA, 10 m*M* Tris–HCl, pH 7.4]. The gradient is then centrifuged at 73,000*g* for 24 min (24,000 rpm in a swinging bucket rotor Sorvall SW 50.1), resulting in two bands (myelin and synaptosomes) and a pellet containing FM. The myelin is removed by aspiration, and the synaptosomal band at the 7.5–12% (w/w) Ficoll interphase is collected by aspiration, diluted in buffer A, and centrifuged at 15,000*g* for 20 min. In the original method (Battino *et al.,* 1991), buffer A at this stage also contained protease inhibitors. We omit protease inhibitor in our buffers because they interfere with complex I activity.

The pellet obtained is lysed by resuspension in 6 m*M* Tris–HCl, pH 8.1, homogenized, and centrifuged at 14,000*g* for 30 min. The pellet is resuspended in 3% Ficoll in 0.12 *M* mannitol, 30 m*M* sucrose, 25 μ*M* EDTA, 5 m*M* Tris–HCl, pH 7.4, and layered on a discontinuous Ficoll gradient consisting of two layers of 6 and 4.5% Ficoll in 0.24 *M* mannitol, 60 m*M* sucrose, 50 μ*M* EDTA, 10 m*M* Tris–HCl, pH 7.4. After centrifugation at 10,000*g* for 30 min, a pellet is obtained containing the HM fraction. The intermediate fraction is diluted in buffer A and centrifuged at 15,000*g* for 30 min to pellet the LM fraction. FM, LM, and HM pellets are resuspended in minimal volumes of 0.22 *M* mannitol, 0.07 *M* sucrose, 50 m*M* Tris–HCl, 1 m*M* EDTA, pH 7.2, and stored at −80°C.

Several respiratory chain enzymatic activities from three different regions of the brain are reported in Table VII. In addition, we have characterized complex I activity in aging

Table VII
Respiratory Chain Activities of Nonsynaptic (FM) and Synaptic Light (LM) and Heavy (HM) Mitochondria from Rat Brain Regions[a]

	Cortex	Hippocampus	Striatum
Succinate-cytochrome *c*			
FM	0.196 ± 0.076	0.159 ± 0.048	0.134 ± 0.039
LM	0.171 ± 0.050	0.154 ± 0.051	0.159 ± 0.045
HM	0.075 ± 0.019	0.062 ± 0.011	0.063 ± 0.019
CoQ_2H_2-cytochrome *c*			
FM	2.397 ± 0.445	1.513 ± 0.175	1.776 ± 0.666
LM	2.647 ± 1.097	2.153 ± 0.995	2.028 ± 0.823
HM	1.137 ± 0.471	0.505 ± 0.129	0.766 ± 0.415
Cytochrome *c* oxidase			
FM	2.217 ± 0.342	1.677 ± 0.259	2.021 ± 0.380
LM	2.328 ± 0.498	1.870 ± 0.427	1.819 ± 0.688
HM	1.125 ± 0.307	0.698 ± 0.203	0.762 ± 0.219

[a] Enzymatic activities are expressed in μmol/min/mg mitochondrial protein. Results are means ±SD for number of rats >10.

in mitochondrial preparations from cerebral cortex (see Table VIII) and studied the flux control coefficient for NADH–CoQ reductase with respect to NADH oxidase (Lenaz *et al.*,1998) (Fig. 4). To assure availability of NADH, the mitochondrial fractions were pulse sonicated five times for 10 s/min at 150 W in an ice bath under nitrogen gas prior to enzymatic assays.

Table VIII
Biochemical Parameters in Nonsynaptic Mitochondria (FM) and in Synaptic Light (LM) and Heavy Mitochondria (HM) from Rat Brain Cortex

	FM		LM		HM	
	4 months	24 months	4 months	24 months	4 months	24 months
Mitochondrial yield (mg protein/g tissue)	4.46 ± 1.20	3.40 ± 1.82	3.06 ± 0.45	3.26 ± 0.82	5.83 ± 2.04	5.42 ± 1.58
NADH oxidase activity ($nmol \cdot min^{-1} mg^{-1}$)	203 ± 72 [b]	130 ± 51 [b]	109 ± 41	100 ± 55	104 ± 18 [b]	83 ± 18 [b]
NADH-ferricyanide reductase ($\mu mol \cdot min^{-1} mg^{-1}$)	4.29 ± 1.26	3.49 ± 0.73	3.91 ± 0.81	4.36 ± 2.36	3.89 ± 0.47	3.78 ± 0.58
I_{50} of rotenone (pmol rotenone/mg protein)	29 ± 15	41 ± 26	40 ± 14	33 ± 11	25 ± 8	26 ± 9
I_{50} of rotenone (corrected)[a]	6.8 ± 1.7 [b]	10.8 ± 4.8 [b]	9.4 ± 2.8	7.9 ± 4.6	6.3 ± 2.4	6.6 ± 2.4

[a] (pmol rotenone/mg protein)/NADH-ferricyanide reductase.
[b] $p < 0.05$.

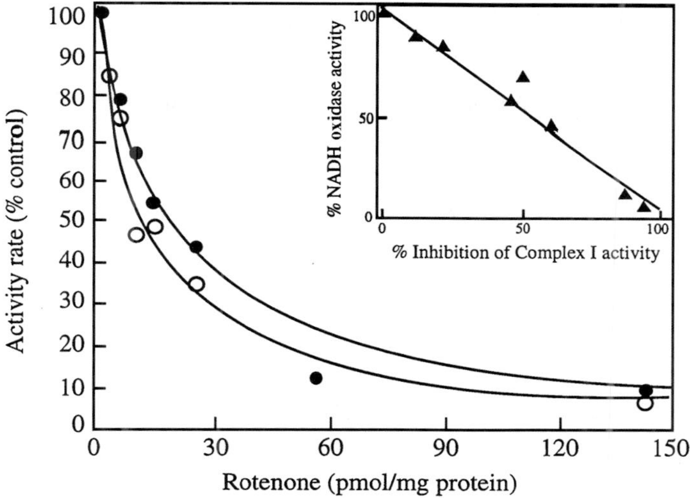

Fig. 4 Flux control of NADH oxidation in rat brain cortex nonsynaptic mitochondria [reprinted from Lenaz *et al.* (1999), with permission from Acta Biochimica Polonica]. The stepwise inhibition by rotenone of NADH-CoQ reductase (○) and of NADH oxidase (●) is shown. (Inset) A plot of NADH oxidase rates against inhibition of complex I activity after rotenone titration. Flux control coefficients were calculated as described elsewhere (Lenaz *et al.*, 1998).

VII. Mitochondria from Hamster Brown Adipose Tissue

The simplest branch of the mitochondrial respiratory chain connected with the CoQ pool is glycerol-3-phosphate dehydrogenase, which is tightly bound to the outer surface of the inner mitochondrial membrane. The amount of this enzyme varies in mitochondrial preparations from different tissues. The highest activity of this enzyme was found in insect flight muscle (Estabrook and Sacktor, 1958), but is also present in brown adipose tissue of either newborn or cold-adapted adult mammals (Chaffee *et al.*, 1964).

We have studied the activities of glycerol-3-phosphate dehydrogenase and glycerol-3-phosphate cytochrome *c* reductase in brown adipose tissue mitochondria from cold-adapted hamsters (*Mesocricetus auratus*) (Rauchova *et al.*, 1992, 1997). Mitochondria are prepared following the method of Hittelman and colleagues (1969). Brown fat is excised rapidly, placed in ice-cold 0.25 *M* sucrose, 10 m*M* Tris–HCl, and 1 m*M* EDTA, pH 7.4 (STE buffer), and carefully cleaned of extraneous tissue. The tissue is homogenized in a Teflon–glass homogenizer followed by a high-speed centrifugation (14,000*g* for 10 min) (Smith *et al.*, 1966). The supernatant is carefully aspirated from beneath the overlying lipid layer, and the latter removed. The pellet is resuspended in STE buffer and centrifuged at 8500*g* for 10 min (Schneider, 1948). The final resuspension of mitochondrial pellet is then carried out in STE buffer.

Table IX
Glycerol-3-phosphate CoQ Reductase Activity in Hamster Brown Adipose Tissue Mitochondria with Different Acceptors at Saturating Concentrations of Both Donor and Acceptor Substrates[a,b]

Acceptor	Specific activity (nmol·min^{-1}mg^{-1}protein)
CoQ_1	225 ± 41
CoQ_2	109 ± 11
Duroquinone	121 ± 31
6-Decylubiquinone	133 ± 27
2,6-Dichlorophenol-indophenol	133 ± 16
Cytochrome *c*	367 ± 44

[a] Results are given as the mean ±SD. Activities are referred to two-electron reduction, except for cytochrome *c*, a one-electron acceptor. Single assays on NADH-cytochrome *c* and succinate-cytochrome *c* reductases gave specific activities of 405 and 265 nmol·min^{-1}mg^{-1}protein, respectively.
[b] From Rauchova *et al.* (1997).

Table IX reports the activities of mitochondrial glycerol-3-phosphate dehydrogenase and glycerol-3-phosphate cytochrome *c* reductase of brown adipose tissue mitochondria of cold-adapted hamster in comparison with the corresponding activities of NADH and succinate oxidation.

VIII. Mitochondria from Insect Flight Muscle

Mitochondria isolated from cockroach flight muscle have been studied for their high content in glycerol-3-phosphate dehydrogenase activity (Rauchova *et al.,* 1997). Cockroach (*Periplaneta americana*) flight muscle mitochondria are prepared according to Novak and colleagues (1979). Red metathoracic muscles are homogenized in a Teflon–glass homogenizer in 0.32 *M* sucrose, 0.01 *M* EDTA, pH 7.4, using 10 ml of medium per 3 g of tissue. Homogenization consists of 12 up-and-down strokes in a 1-min period. Fractions are then separated by low-speed centrifugation (2000*g* for 20 min) or by high-speed centrifugation (10,000*g* for 10 min). Fractions obtained with these centrifugations are composed mainly of mitochondria. Centrifugation of the homogenate at 18,500*g* for 20 min yields a fraction containing mitochondria and membranes of other origin.

IX. Mitochondria from Porcine Adrenal Cortex

Mitochondria from adrenal cortex contain three monooxygenase systems involved in corticosteroidogenesis: 11β-and 18-hydroxylation of deoxycorticosterone (DOC) to corticosterone and side chain cleavage of cholesterol to pregnenolone (Simpson and Boyd,

1971; Simpson and Estabrook, 1969). These involve a second electron transport chain: the primary electron donor is NADPH; electrons are then transferred to a flavoprotein, to adrenodoxin, and to cytochrome P450, where hydroxylation of the steroid precursor occurs. The two electron transport chains are not independent, and steroid hydroxylation can inhibit ATP synthesis.

Mitochondria are extracted from porcine adrenal cortex according to Popinigis and colleagues (1990). The adrenal glands are freed from all connective and fatty tissue. The central medulla is scraped away and the cortex is cut up and resuspended in 0.33 *M* sucrose, 20 m*M* Tris–HCl, pH 7.4, 2 m*M* EDTA, and 0.2% BSA. The suspension is homogenized using first a loose-fitting glass–glass homogenizer and then a tight-fitting, Teflon–glass homogenizer.

The homogenate is centrifuged at 600*g* for 8 min and the resulting supernatant is further centrifuged at 2000*g* for 8 min. The mitochondrial pellet is resuspended in 0.33 *M* sucrose, 20 m*M* Tris–HCl, pH 7.4, and 0.5 m*M* EDTA and washed three times. The final pellet is resuspended in the same washing medium at a final concentration of approximately 50–60 mg/ml.

Mitochondria prepared using the method just described have respiratory control with both succinate and glutamate, but ADP/O ratios are not very high (Popinigis *et al.*, 1990).

X. Mitochondria from Human Platelets

A. Crude Mitochondrial Membranes

Preparation of crude mitochondrial membranes from platelets has an advantage in that a large amount of blood is not required. The platelet-enriched fraction is prepared according to Blass and colleagues (1977). Blood is obtained by venipuncture. Forty to 60 ml of blood is sufficient for few respiratory activities (Merlo Pich *et al.*, 1996); for more complex experiments, involving kinetic detrminations or inhibitor titrations (e.g., rotenone titration), 100 ml of blood is required (Degli Esposti *et al.*, 1994). Erythrocytes (from 100 ml venous blood) are precipitated in 25 ml of 5% Dextran 250,000, 0.12 *M* NaCl, 10 m*M* EDTA, pH 7.4, at 4°C for 45 min (Blass *et al.*, 1977; Degli Esposti *et al.*, 1994). The upper phase is centrifuged at 5000*g* for 10 min. The pellet is resuspended in 3 ml H_2O for 60 s and adding 1 ml of 0.6 *M* NaCl blocks the osmotic shock. After centrifugation at 5000*g* for 10 min, the pellet undergoes a second lysis step. The pellet obtained after the second osmotic shock is resuspended in 3 ml 10 m*M* NaH_2PO_4, 10 m*M* Na_2HPO_4, 0.12 *M* NaCl, pH 7.4 (phosphate buffer), and is centrifuged twice at 15,000*g* for 20 min.

The platelet pellet is resuspended in 0.1 *M* Tris–HCl, 1 m*M* EDTA, pH 7.4, and sonicated five times for 10 s each (150 Hz) with intervals of 50 s. After sonication, the homogenate is centrifuged at 8000*g* for 10 min (10,000 rpm in a table-top Eppendorf centrifuge), and the supernatant, diluted 1:1 with 0.25 *M* sucrose, 50 m*M* Tricine–Cl, 2.5 m*M* $MgCl_2$, 40 m*M* KCl, 0.5% BSA, pH 8, is centrifuged at 100,000*g* for 40 min

Table X
Biochemical Parameters in Platelet Membranes from Young and Aged Female Individuals[a]

	Young ($n = 19$)[b]	Aged ($n = 18$)
NADH-DB reductase (nmol $min^{-1}mg^{-1}$protein)	4.6 ± 1.6	4.5 ± 0.9
Complex I turnover (s^{-1})	9.8 ± 4.6	7.4 ± 2.1
% rotenone sensitivity	76.1 ± 9.9	71.8 ± 10.0
I_{50} of rotenone (pmol mg^{-1} protein)	64.8 ± 44.8	126.1 ± 106.2[c]

[a] Reprinted from Merlo Pich *et al.* (1996), with permission from Elsevier Science.
[b] Number of pools (two individuals each).
[c] $p < 0.03$.

to pellet mitochondrial particles. The pellet is suspended in 0.125 *M* sucrose, 50 m*M* Tricine–Cl, 2.5 m*M* $MgCl_2$, 40 m*M* KCl, pH 8.

Alternatively (Merlo Pich *et al.,* 1996), 40 to 60 ml blood is mixed with 5% Dextran 70,000, 0.9% NaCl at room temperature for 30 min and the upper phase is centrifuged at 3000*g* for 3 min. The upper phase is collected and centrifuged at 4000*g* for 25 min and then washed in phosphate buffer. After sonication, as described earlier, platelet membrane fragments are diluted 1:1 in 0.25 *M* sucrose, 30 m*M* Tris, 1 m*M* EDTA, pH 7.7 (STE buffer), and then separated from the heavier cell debris by centrifugation at 33,000*g* for 10 min. The supernatant is then ultracentrifuged at 100,000*g* for 40 min and the pellet is resuspended in STE buffer.

This crude mitochondrial fraction has been used to study complex I activity in patients affected by Leber's hereditary optic neuropathy (Degli Esposti *et al.,* 1994; Carelli *et al.,* 1999) and in aged individuals (Merlo Pich *et al.,* 1996) (Table X).

B. Coupled Mitochondrial Particles from Platelet Mitochondria

This method involves the preparation of mitochondria followed by sonication in order to obtain coupled inside-out submitochondrial particles of the ETPH type (see Section III,B). Baracca and colleagues (1997) have described a method for preparation of coupled submitochondrial particles from horse platelets. This method is also suitable for human platelets with minor modification.

Horse platelets are isolated and purified from 200 to 500 ml of venous blood according to Blass and colleagues (1977). To isolate mitochondria, platelets are suspended in a hypotonic medium (10 m*M* Tris–HCl, pH 7.6) for 7 min; the osmotic shock is blocked by the addition of 0.25 *M* sucrose, and the suspension is centrifuged at 600*g* for 10 min in an Eppendorf microcentrifuge. The pellet undergoes a second osmotic shock, and the supernatant is collected, pooled with the one obtained from the first osmotic step, and centrifuged at 12,000*g* (12,000 rpm in a table-top Eppendorf centrifuge) for 20 min to sediment mitochondria. The mitochondrial pellet is then suspended in 0.25 *M* sucrose,

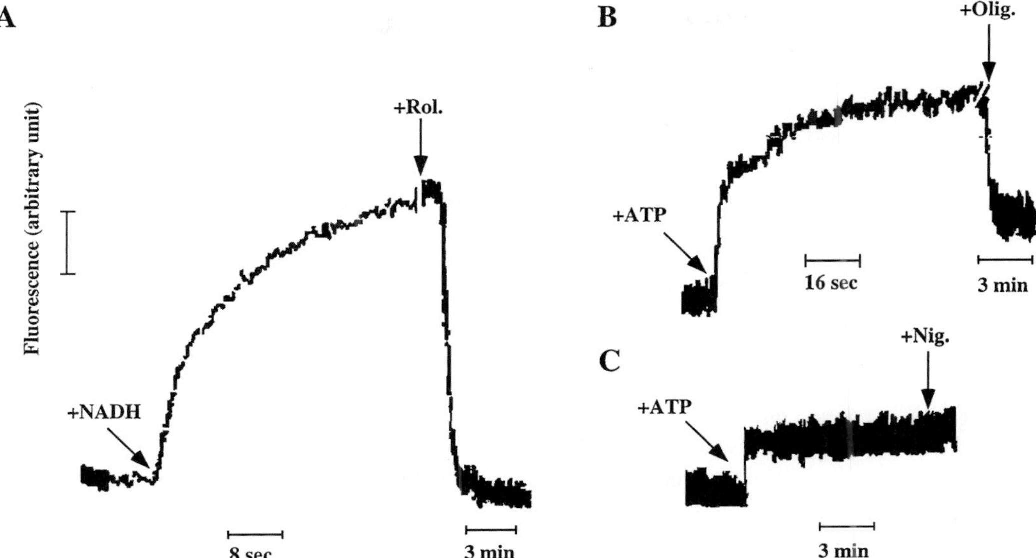

Fig. 5 Quenching of ACMA fluorescence induced upon addition of NADH (A) or ATP (B and C) to platelet-coupled submitochondrial particles (reprinted from Baracca *et al.* 1997). (A) Quenching induced by protophonoric activity of complex I in the presence of 100 μ*M* NADH; the addition of 1 μ*M* rolliniastatin-2 completely restores the initial ACMA fluorescence. (B) Quenching induced by the protonophoric activity of complex V in the presence of 0.8 m*M* ATP; the addition of 1 μ*M* oligomycin partially restores the initial ACMA fluorescence. (C) Addition of ATP to a sample containing 1 μ*M* oligomycin induces a significant quenching of ACMA fluorescence, which is insensitive to nigericin.

2 m*M* EDTA, pH 8, and subjected to sonic oscillation under partial N_2 atmosphere for 1 min at low output (40 W). The suspension is centrifuged at 12,000*g* for 10 min; the resulting supernatant is decanted and centrifuged at 105,000*g* for 40 min to precipitate the particles. The pellet is finally suspended in 0.25 *M* sucrose to give a protein concentration of 10 mg/ml as assayed by the Lowry method (Lowry *et al.,* 1951) in the presence of 1% deoxycholate. These inverted vesicles can be used to oxidize NADH and to measure ΔpH formation by fluorescence quenching of suitable acridine probes (Baracca *et al.,* 1997) (Fig. 5).

This method has been adapted for the preparation of coupled submitochondrial particles from human platelets. Platelets are isolated from 100 ml of venous blood; osmotic shock is kept for 4 min instead of 7 min, and the sonic oscillation step lasts 20 s rather than 1 min. In human platelets, we were not able to measure ΔpH formed by NADH, but the method is suitable for the assay of the protonophoric activity of ATP synthase. We have also applied this method for assaying ATP-dependent activities in patients with mutations in ATP synthase (Baracca *et al.,* 2000).

XI. Mitochondria from Fish Liver

Extraction of mitochondria from fish has been used for purification and characterization of different biochemical parameters. We applied the method for the preparation and characterization of eel (*Anguilla anguilla*) liver mitochondria and submitochondrial particles (Baracca *et al.,* 1992). It has also been applied to isolate mitochondria from sea bass (*Morone labrax*) liver (Ventrella *et al.,* 1982). The method is also suitable for extractions from other types of fish and amphibians (Degli Esposti *et al.,* 1992).

After blotting, the liver is immediately washed in 0.25 *M* sucrose, 0.01 *M* Tris–HCl, 1 m*M* EDTA, pH 7.4, and homogenized in 0.25 *M* sucrose, 24 m*M* Tris–HCl, 1 m*M* EDTA, pH 7.4, and 0.5 mg/ml BSA (homogenizing solution) using an UltraTurrax homogenizer.

The homogenate is then centrifuged at 750*g* for 10 min and the supernatant is filtered and centrifuged at 13,000*g* for 10 min. The pellet is washed once in the homogenizing solution and the suspension is centrifuged at 11,000*g* for 10 min. The mitochondrial pellet is resuspended in the homogenizing medium at a concentration of 40 mg mitochondrial protein/ml of solution.

XII. Mitochondria from Sea Urchin Egg

Mitochondria can be extracted from eggs of the sea urchin *Paracentrotus lividus* (Cantatore *et al.,* 1974). Eggs are collected and suspended in 0.25 *M* sucrose, 0.1 *M* Tris–HCl, pH 7.6, 1 m*M* EDTA, 0.24 M KCl (TEK buffer) and homogenized in Teflon–glass Potter–Elvehjem homogenizer. The homogenate is centrifuged at 600*g* for 10 min and the supernatant is centrifuged at 2600*g* for 10 min. Finally mitochondria are pelleted by centrifugation at 7500*g* for 10 min, suspended in TEK buffer, and centrifuged for 10 min at 15,000*g*. The pellet is suspended in TEK buffer and layered on a double 1.5–1 *M* sucrose discontinuous gradient in 0.1 *M* Tris–HCl, pH 7.6, 1 m*M* EDTA, 0.24 *M* KCl and centrifuged at 50,000*g* for 3 h. The band at the 1.5 *M* sucrose region is collected and centrifuged at 15,000*g* for 10 min. Finally, the pellet obtained is suspended in TEK buffer.

This method has been applied by Degli Esposti and colleagues (1990) to study the natural resistance of the sea urchin mitochondrial complex III to cytochrome *b* inhibitors.

XIII. Mitochondria and Kinetoplasts from Protozoa

Mitochondria from ciliate and trypanosome protozoans have been characterized regarding the sensitivity of ubiquinol:cytochrome *c* reductase to inhibitors (Ghelli *et al.,* 1992). Mitochondria of ciliates, such as *Tetrahymena piriformis,* are resistant to antimycin A and rotenone, whereas mitochondria of trypanosomes are quite resistant to stigmatellin. Both ciliates and trypanosomes are highly resistant to myxothiazol.

Mitochondria from *T. piriformis* are prepared according to Kilpatrick and Erecinska (1977). Cultured cells are harvested by low-speed centrifugation and washed once in cold 0.25 *M* sucrose. The pellet is then resuspended in 6 volumes of 0.25 *M* sucrose, 10 m*M* KCl, 5 m*M* MOPS, and 0.2 m*M* EDTA, pH 7.2 (solution A), and homogenized in a Teflon–glass Potter–Elvehjem homogenizer. The cell homogenate is centrifuged at 600*g* for 6 min and the supernatant is collected. The pellet is resuspended in half the previous volume, homogenized, and centrifuged again as described earlier. The supernatant collected from the previous two steps is centrifuged at 5000*g* for 10 min to pellet the mitochondria. The mitochondrial pellet is then washed twice in solution A containing 0.2% BSA and is finally sedimented by centrifugation at 8000*g* for 10 min. The colorless pellet overlaying the mitochondrial fraction and the black sediment that adheres firmly to the bottom of the centrifuge tube are discarded. Pellets from both extractions are suspended in solution A. According to Kilpatrick and Erecinska (1977), the pellet obtained from the second extraction yields more mitochondria with higher P/O ratios.

Mitochondria from *Paramecium tetraurelia* are prepared essentially as described by Doussiere and colleagues (1979). Cells are harvested by centrifugation at 20°C for 3 min at 100*g* and washed once in 0.5 *M* mannitol and 5 m*M* MOPS. Packed cells are then resuspended in 0.5 *M* mannitol, 5 m*M* MOPS, and 1 m*M* EDTA, pH 7.3, with 0.5% BSA (homogenization medium) and homogenized in a Potter–Elvehjem Teflon–glass homogenizer. The homogenate is centrifuged at 600*g* for 5 min. The supernatant is then collected and centrifuged at 600*g* for 10 min. The mitochondrial pellet is resuspended in the homogenization medium and centrifuged at 600*g* for 4 min to eliminate trichocystis. The supernatant is centrifuged at 5000*g* for 10 min to obtain the crude mitochondrial fraction. The latter is purified by centrifugation through a gradient of sorbitol (40–60%, w/w) containing 5 m*M* MOPS and 1 m*M* EDTA, pH 7.3, at 150,000 *g* for 1 h. After centrifugation, the interphase layer, corresponding to the mitochondrial particles, is removed and diluted in 10 volumes of homogenization medium. The suspension is then centrifuged at 10,000*g* for 15 min, and the mitochondrial pellet is finally resuspended in the homogenization medium.

Mitochondria from *Crithidia lucilae* and *Leishmania infantum* are prepared following the method of Renger and Wolstenholme (1972). Protozoans of the order Kinetoplastida, which includes *Trypanosoma, Leishmania,* and *Crithidia* genera, all contain kinetoplasts. Because this organelle is a modified mitochondrion, mitochondria from trypanosomes are purified from kinetoplast-enriched fractions.

Cells are harvested by centrifugation at 1000*g* for 10 min and washed three times in SSC, pH 7.5. Cells are then resuspended in 0.3 *M* sucrose, 10 m*M* Tris–HCl, pH 7.4, and 1 m*M* EDTA (STE buffer), homogenized in a Waring blender for 15–20 s at high speed, and centrifuged at 700*g* for 10 min. This centrifugation step should be repeated until the supernatant becomes free of all cells. The supernatant is then centrifuged at 8000*g* for 10 min. The pellet is resuspended in STE buffer and incubated with 200 μg/ml of DNase I at 37°C for 30 min in the presence of 7 m*M* $MgCl_2$. The DNase is then removed by washing three times with 40 m*M* EDTA. The kinetoplast-enriched fraction is then resuspended in 0.15 *M* NaCl, 0.1 *M* EDTA, and 0.05 *M* Na_2HPO_4.

XIV. Mitochondria and Mitoplasts from Cultured Cells

Human cultured cells represent a valid experimental model for investigating different biochemical parameters, both in the physiology and in the physiopathology of different disorders. Cells are usually easy to cultivate and to obtain in large amounts; intact mitochondria are usually required for biochemical and genetic analyses.

Mitochondria from cell lines have been isolated as described by Yang and colleagues (1997) with minor modifications. Cells are harvested by centrifugation at 600*g* for 10 min, washed with phosphate-buffered saline, and resuspended with 5 volumes of solution A (0.25 *M* sucrose, 20 m*M* HEPES–KOH, pH 7.5, 10 m*M* KCl, 1.5 m*M* $MgCl_2$, 1 m*M* EDTA, 1 m*M* EGTA, 1 m*M* dithiotreithol, 0.1 m*M* phenylmethylsulfonyl flouride). The cellular suspension is homogenized with a Teflon–glass homogenizer with 20 up-and-down passes of the pestle. In our experience, the mitochondrial yield is increased by the use of a glass–glass pestle. The homogenate is then centrifuged at 750*g* for 10 min, the supernatant is collected, and the pellet is resuspended in solution A and recentrifuged at low speed. The pooled supernatants are then centrifuged at 10,000*g* for 15 min and the crude mitochondrial pellet is resuspended in solution A.

Purified mitochondrial fractions can be obtained by separation on a sucrose gradient. A valid method comprising modifications of the "two-step" procedure described by Tapper and colleagues (1983) has been described by Magalhães and colleagues (1998). Cells are harvested by low-speed centrifugation and resuspended in 10 m*M* NaCl, 1.5 m*M* $MgCl_2$, and 10 m*M* Tris–HCl, pH 7.5. Cells are allowed to swell for 4–5 min on ice and briefly homogenized in a Teflon–glass Potter–Elvehjem homogenizer. Sucrose concentration is then adjusted to 250 m*M* by adding 2 *M* sucrose in 10 m*M* Tris–HCl and 1 m*M* EDTA, pH 7.6 ($T_{10}E_{20}$ buffer); the suspension is then centrifuged at 1300*g* for 3 min and the supernatant is recentrifuged at low speed. Mitochondria are then collected by centrifugation at 15,000*g* for 15 min and the pellet is washed three times with 250 m*M* sucrose in $T_{10}E_{20}$ buffer and resuspended in the same solution. The mitochondrial suspension is layered on a discontinuous sucrose gradient (1–1.7 *M*) in $T_{10}E_{20}$ buffer and centrifuged at 70,000*g* for 40 min. The mitochondrial fraction is recovered from the interface, diluted in an equal volume of 250 m*M* sucrose in $T_{10}E_{20}$ buffer, and washed twice in the same solution. The mitochondrial pellet is finally resuspended in 250 m*M* sucrose in $T_{10}E_{20}$ buffer and protein concentration is determined by the Lowry method.

An alternative method to this procedure consists of the "one-step" procedure described by Bogenhagen and Clayton (1974). Cells are harvested and resuspended as described earlier, but osmotic swelling is avoided by immediately adding 2 *M* sucrose in $T_{10}E_{20}$ buffer. Cell debris and nuclei are pelleted as described in the "two-step" procedure and the supernatant is layered on 3 volumes of 1.7 *M* sucrose in $T_{10}E_{20}$ buffer and centrifuged at 70,000*g* for 40 min. The mitochondrial fraction is then washed and resuspended as described earlier.

A "no gradient procedure" method is based on slight modifications to the method described by Schneider (1948). Instead of isotonic sucrose used in the original method, the buffer used consists of 0.21 *M* mannitol, 0.07 *M* sucrose, 5 m*M* Tris–HCl, 5 m*M* EDTA, pH 7.5 (mannitol–sucrose buffer) (Bogenhagen and Clayton, 1974). Cells are homogenized and nuclei are pelleted as described in the "one-step" procedure. The mitochondrial fraction is collected by centrifugation at 20,000*g* for 20 min and washed twice in mannitol–sucrose buffer. Alternatively, cell debris and nuclei are pelleted by centrifugation at 400*g* for 4 min and the crude mitochondrial pellet is obtained from centrifugation of the supernatant at 27,000*g* for 10 min (Bestwick *et al.,* 1982).

Mitoplasts from gradient purified mitochondria can be obtained by modifications of the original method, as described by Magalhães and colleagues (1998).

In the modification of the "swell–contract" method (Murthy and Pande, 1987), gradient-purified mitochondria are resuspended in 20 m*M* potassium phosphate (pH 7.2) containing 0.02% BSA and are allowed to swell for 20 min on ice. After addition of 1 m*M* ATP and 1 m*M* $MgCl_2$, incubation is prolonged for an additional 5 min and mitoplasts are collected by centrifugation at 15,000*g* for 10 min.

The modification of the "digitonin method" for the production of mitoplasts (Greenawalt, 1974) consists of incubating the resuspended gradient-purified mitochondria with 0.1 mg digitonin/mg of mitochondrial protein. The suspension is then stirred on ice for 15 min, and 3 volumes of 250 m*M* sucrose in $T_{10}E_{20}$ buffer and mitoplasts are pelleted after centrifugation at 15,000*g* for 15 min. Mitoplasts are washed once in 250 m*M* sucrose in $T_{10}E_{20}$ buffer and are finally resuspended in 250 m*M* sucrose, 10 m*M* Tris–HCl, pH 7.6.

Acknowledgments

The authors thank Dr. J. Ojaimi for helpful comments and for critical revising of the manuscript. This work is supported by grants from the Muscular Dystrophy Association (to F.P.) and for "PRIN Bioenergetics and Membrane Transport" from MURST, Rome, Italy (to G.L.).

References

Baracca, A., Barogi, S., Carelli, V., Lenaz, G., and Solaini, G. (2000). Catalytic activities of the mitochondrial ATP synthase in patients with the mtDNA T8993G mutation in the ATPase 6 gene encoding for the a subunit. *J. Biol. Chem.* **275,** 4177–4182.

Baracca, A., Bucchi, L., Ghelli, A., and Lenaz, G. (1997). Protonophoric activity of NADH coenzyme Q reductase and ATP synthase in coupled submitochondrial particles from horse platelets. *Biochem. Biophys. Res. Commun.* **235,** 469–473.

Baracca, A., Degli Esposti, M., Parenti Castelli, G., and Solaini, G. (1992). Purification and characterization of adenosine-triphosphatase from eel liver mitochondria. *Comp. Biochem. Phys. B* **101,** 421–426.

Barogi, S., Baracca, A., Cavazzoni, M., Parenti Castelli, G., and Lenaz, G. (2000). Effect of oxidative stress induced by adriamycin on rat hepatocyte bioenergetics during ageing. *Mech. Aging Dev.* **113,** 1–21.

Barogi, S., Baracca, A., Parenti Castelli, G., Bovina, C., Formiggini, G., Marchetti, M., Solaini, G., and Lenaz, G. (1995). Lack of major changes in mitochondria from liver, heart, and skeletal muscle of rats upon ageing. *Mech. Ageing Dev.* **84,** 139–150.

Battino, M., Bertoli, E., Formiggini, G., Sassi, S., Gorini, A., Villa, R. F., and Lenaz, G. (1991). Structural and functional aspects of the respiratory chain of synaptic and nonsynaptic mitochondria derived from selected brain regions. *J. Bioenerg. Biomembr.* **23,** 345–363.

Battino, M., Gorini, A., Villa, R. F., Genova, M. L., Bovina, C., Sassi, S., Littarru, G. P., and Lenaz, G. (1995). Coenzyme Q content in synaptic and non-synaptic mitochondria from different brain regions in the ageing rat. *Mech. Ageing Dev.* **78,** 173–187.

Bestwick, R. K., Moffett, G. L., and Mathews, C. K. (1982). Selective expansion of mitochondrial nucleoside triphosphate pools in antimetabolite-treated HeLa cells. *J. Biol. Chem.* **257,** 9300–9304.

Beyer, R. E. (1967). Preparation, properties, and conditions for assay of phosphorylating electron transport particles (ETPH) and its variations. *In* "Methods in Enzymology" (R. W. Estabrook and M. E. Pullman, eds.), Vol. 10, pp. 186–194, Academic Press, San Diego.

Blass, J. P., Cederbaum, S. D., and Kark, R. A. P. (1977). Rapid diagnosis of pyruvate and ketoglutarate dehydrogenase deficiencies in platelet-enriched preparations from blood. *Clin. Chim. Acta* **75,** 21–30.

Bogenhagen, D., and Clayton, D. A. (1974). The number of mitochondrial deoxyribonucleic acid genomes in mouse L and human HeLa cells: Quantitative isolation of mitochondrial deoxyribonucleic acid. *J. Biol. Chem.* **249,** 7991–7995.

Cantatore, P., Nicotra, A., Loria, P., and Saccone, C. (1974). RNA synthesis in isolated mitochondria from sea urchin embryos. *Cell Differ.* **3,** 45–53.

Capaldi, R. A. (1982). Arrangement of proteins in the mitochondrial inner membrane. *Biochim. Biophys. Acta* **694,** 291–306.

Carelli, V., Ghelli, A., Bucchi, L., Montagna, P., De Negri, A., Leuzzi, V., Carducci, C., Lenaz, G., Lugaresi, E., and Degli Esposti, M. (1999). Biochemical features of mtDNA 14484 (ND6/M64V) point mutation associated with Leber's hereditary optic neuropathy (LHON). *Ann. Neurol.* **45,** 320–328.

Casadio, R., Baccarini-Melandri, A., and Melandri, B. A. (1974). On the determination of the trans-membrane pH difference in bacterial chromatophores using 9-aminoacridine. *Eur. J. Biochem.* **47,** 121–128.

Castelluccio, C., Baracca, A., Fato, R., Pallotti, F., Maranesi, M., Barzanti, V., Gorini, A., Villa, R. F., Parenti Castelli, G., Marchetti, M., and Lenaz, G. (1994). Mitochondrial activities of rat heart during ageing. *Mech. Ageing Dev.* **76,** 73–88.

Chaffee, R. R. J., Allen, J. R., Cassuto, Y., and Smith, R. E. (1964). Biochemistry of brown fat and liver of cold-acclimated hamsters. *Am. J. Physiol.* **207,** 1211–1214.

Clark, J. B., and Nicklas, W. J. (1970). The metabolism of rat brain mitochondria. Preparation and characterization. *J. Biol. Chem.* **245,** 4724–4731.

de Duve, C. (1967). Criteria for homogeinity and purity of mitochondria. *In* "Methods in Enzymology" (R. W. Estabrook and M. E. Pullman, eds.), Vol. 10, pp. 7–18. Academic Press, San Diego.

de Duve, C., and Baudhuin, P. (1966). Peroxisomes (microbodies and related particles). *Physiol. Rev.* **46,** 323–357.

Degli Esposti, M., Carelli, V., Ghelli, A., Ratta, M., Crimi, M., Sangiorgi, S., Montagna, P., Lenaz, G., Lugaresi, E., and Cortelli, P. (1994). Functional alterations of the mitochondrially encoded ND4 subunit associated with Leber's hereditary optic neuropathy. *FEBS Lett.* **352,** 375–379.

Degli Esposti, M., Ghelli, A., Butler, G., Roberti, M., Mustich, A., and Cantatore, P. (1990). The cytochrome *c* of the sea urchin *Paracentrotus lividus* is naturally resistant to myxothiazol and mucidin. *FEBS Lett.* **263,** 245–247.

Degli Esposti, M., Ghelli, A., Crimi, M., Baracca, A., Solaini, G., Tron, T., and Meyer, A. (1992). Cytochrome *b* of fish mitochondria is strongly resistant to funicolosin, a powerful inhibitor of respiration. *Arch. Biochem. Biophys.* **295,** 198–204.

Degli Esposti, M., and Lenaz, G. (1982). Kinetics of ubiquinol-1-cytochrome *c* reductase in bovine heart mitochondria and submitochondrial particles. *Biochim. Biophys. Acta* **682,** 189–200.

Doussiere, J., Sainsard-Chanet, A., and Vignais, P. V. (1979). The respiratory chain of *Paramecium tetraurelia* in wild type and the mutant. C_1. I. Spectral properties and redox potentials. *Biochim. Biophys. Acta* **548,** 224–235.

Estabrook, R. W., and Sacktor, B. (1958). α-Glycerophosphate oxidase of flight muscle mitochondria. *J. Biol. Chem.* **233,** 1014–1019.

Estornell, E., Fato, R., Pallotti, F., and Lenaz, G. (1993). Assay conditions for the mitochondrial NADH: coenzyme Q oxidoreductase. *FEBS Lett.* **332,** 127–131.

Fato, R., Cavazzoni, M., Castelluccio, C., Parenti Castelli, G., Palmer, G., Degli Esposti, M., and Lenaz, G. (1993). Steady-state kinetics of ubiquinol-cytochrome *c* reductase in bovine heart submitochondrial particles: Diffusional effect. *Biochem. J.* **290,** 225–236.

Fato, R., Estornell, E., Di Bernardo, S., Pallotti, F., Parenti Castelli, G., and Lenaz, G. (1996). Steady-state kinetics of the reduction of Coenzyme Q analogs by Complex I (NADH:ubiquinone oxidoreductase) in bovine heart mitochondria and submitochondrial particles. *Biochemistry (US),* **35,** 2705–2716.

Fleischer, S., and Kervina, M. (1974). Long-term preservation of liver for subcellular fractionation. *In* "Methods in Enzymology" (S. Fleischer and L. Packer, eds.), Vol. 31, pp. 3–6. Academic Press, San Diego.

Genova, M. L., Bovina, C., Formiggini, G., Ottani, V., Sassi, S., and Marchetti, M. (1994). Uptake and distribution of exogenous CoQ in the mitochondrial fraction of perfused rat liver. *Mol. Aspects Med.* **15,** (Suppl.), s47–s55.

Genova, M. L., Bovina, C., Marchetti, M., Pallotti, F., Tietz, C., Biagini, G., Pugnaloni, A., Viticchi, C., Gorini, A., Villa, R. F., and Lenaz, G. (1997). Decrease of rotenone inhibition is a sensitive parameter of complex I damage in brain non-synaptic mitochondria of aged rats. *FEBS Lett.* **410,** 467–469 .

Ghelli, A., Crimi, M., Orsini, S., Gradoni, L., Zannotti, M., Lenaz, G., and Degli Esposti, M. (1992). Cytochrome *b* of protozoan mitochondria: Relationships between function and structure. *Comp. Biochem. Phys. B* **103,** 329–338.

Glowinski, J., and Iversen, L. L. (1966). Regional studies of catecholamines in the rat brain. I. The disposition of [^{3}H]norepinephrine, [^{3}H]dopamine and [^{3}H]dopa in various regions of the brain. *J. Neurochem.* **13,** 655–669.

Gornall, A. G., Bardawill, C. J., and David, M. M. (1949). Determination of serum proteins by means of the biuret reaction. *J. Biol. Chem.* **177,** 751–766.

Greenawalt, J. W. (1974). The isolation of outer and inner mitochondrial membranes. *In* "Methods in Enzymology" (S. Fleisher and L. Packer, eds.), Vol. 31, pp. 310–323. Academic Press, San Diego.

Gregg, C. T. (1967). Preparation and assay of phosphorylating submitochondrial particles: Particles from rat liver prepared by drastic sonication. *In* "Methods in Enzymology" (R. W. Estabrook and M. E. Pullman, eds.), Vol. 10, pp. 181–185. Academic Press, San Diego..

Hansen, M., and Smith, A. L. (1964). Studies on the mechanism of oxidative phosphorylation. VII. Preparation of a submitochondrial particle (ETP_H) which is capable of fully coupled oxidative phosphorylation. *Biochim. Biophys. Acta* **81,** 214–222.

Hatefi, Y., Jurtshuk, P., and Haavik, A. G. (1961). Studies on the electron transport system. XXXII. Respiratory control in beef heart mitochondria. *Arch. Biochem. Biophys.* **94,** 148–155.

Hittelman, K. J., Lindberg, O., and Cannon, B. (1969). Oxidative phosphorylation and compartmentation of fatty acid metabolism in brown fat mitochondria. *Eur. J. Biochem.* **11,** 183–192.

Jacobs, E. E., and Sanadi, D. R. (1960). The reversible removal of cytochrome *c* from mitochondria. *J. Biol. Chem.* **235,** 531–534.

Kalen, A., Soderberg, N., Elmberger, P. G., and Dallner, G. (1990). Uptake and metabolism of dolichol and cholesterol in perfused rat liver. *Lipids* **25,** 93–99.

Kilpatrick, L., and Erecinska, M. (1977). Mitochondrial respiratory chain of *Tetrahymena piriformis:* The termodynamic and spectral properties. *Biochim. Biophys. Acta* **460,** 346–363.

Kun, E., Kirsten, E., and Piper, W. N. (1979). Stabilization of mitochondrial functions with digitonin. *In* "Methods in Enzymology" (S. Fleischer and L. Packer, eds.), Vol. 55, pp. 115–118. Academic Press, San Diego.

Lai, J. C. K., Walsh, J. M., Dennis, S. C., and Clark, J. B. (1977). Synaptic and non-synaptic mitochondria from rat brain: Isolation and characterization. *J. Neurochem.* **28,** 625–631.

Lee, C. P., Martens, M. E, and Tsang, S. H. (1993). Small scale preparation of skeletal muscle mitochondria and its application in the study of human disease. *In* "Methods in Toxicology" (D. Jones and L. Lash, eds.), Vol. 2, pp. 70–83. Academic Press, New York.

Lenaz, G., Bovina, C., Formiggini, G., and Parenti Castelli, G. (1999). Mitochondria, oxidative stress, and antioxidant defences. *Acta Biochim. Polon.* **46,** 1–21.

Lenaz, G., Cavazzoni, M., Genova, M. L., D'Aurelio, M., Merlo Pich, M., Pallotti, F., Formiggini, G., Marchetti, M., Parenti Castelli, G., and Bovina, C. (1998). Oxidative stress, antioxidant defences and aging. *Biofactors* **8,** 195–204.

Lenaz, G., Degli Esposti, M., and Parenti Castelli, G. (1982). DCCD inhibits proton translocation and electron flow at the second site of the mitochondrial respiratory chain. *Biochem. Biophys. Res. Commun.* **105,** 589–595.

Lenaz, G., and MacLennan, D. H. (1966). Studies on the mechanism of oxidative phosphorylation. X. The effect of cytochrome *c* on energy-linked processes in submitochondrial particles. *J. Biol. Chem.* **241,** 5260–5265.

Lenaz, G., and MacLennan, D. H. (1967). Extraction and estimation of cytochrome *c* from mitochondria and submitochondrial particles. *In* "Methods in Enzymology" (R. W. Estabrook and M. E. Pullman, eds.), Vol. 10, pp. 499–504. Academic Press, San Diego.

Linnane, A. W., and Titchener, E. B. (1960). Studies on the mechanism of oxidative phosphorylation. VI. A factor for coupled oxidation in the electron transport particle. *Biochim. Biophys. Acta* **39,** 469–478.

Lowry, O. H., Rosenbrough, N. J., Farr, A. L., and Randall, R. J. (1951). Protein measurement with the Folin phenol reagent. *J. Biol. Chem.* **193,** 265–275.

MacLennan, D. H., Lenaz, G., and Szarkowska, L. (1966). Studies on the mechanism of oxidative phosphorylation. IX. Effect of cytochrome *c* on energy-linked processes. *J. Biol. Chem.* **241,** 5251–5259.

Magalhães, P. J., Andreu, A. L., and Schon, E. A. (1998). Evidence for the presence of 5S rRNA in mammalian mitochondria. *Mol. Biol. Cell* **9,** 2375–2382.

Marinetti, G. V. (1962). Chromatographic separation, identification and analysis of phosphatides. *J. Lipid Res.* **3,** 1–20.

Melandri, B. A., Baccarini Melandri, A., Lenaz, G., Bartoli, E., and Masotti, L. (1974). The site of inhibition of dibromothymoquinone in mitochondrial respiration. *J. Bioenerg.* **6,** 125–133.

Merlo Pich, M., Bovina, C., Formiggini, G., Cometti, G. G., Ghelli, A., Parenti Castelli, G., Genova, M. L., Marchetti, M., Semeraro, S., and Lenaz, G. (1996). Inhibitor sensitivity of respiratory complex I in human platelets: A possible biomarker of ageing. *FEBS Lett.* **380,** 176–178.

Murthy, M. S. R., and Pande, S. V. (1987). Malonyl-CoA binding site and the overt carnitine palmitoyltransferase activity reside on the opposite sides of the outer mitochondrial membrane. *Proc. Natl. Acad. Sci. USA* **84,** 378–382.

Nedergaard, J., and Cannon, B. (1979). Overview: Preparation and properties of mitochondria from different sources. *In* "Methods in Enzymology" (S. Fleischer and L. Packer, eds.), Vol. 55, pp. 3–28. Academic Press, San Diego.

Norling, B., Glazek, E., Nelson, B. D., and Ernster, L. (1974). Studies with ubiquinone-depleted submitochondrial particles: Quantitative incorporation of small amounts of ubiquinone and its effects on the NADH and succinate oxidase activities. *Eur. J. Biochem.* **47,** 475–482.

Novak, F., Novakova, O., and Kubista, V. (1979). Heterogeneity of insect flight muscle mitochondria as demonstrated by different phospholipid turnover *in vivo*. *Insect Biochem.* **9,** 389–396.

Palade, G. E., Siekevitz, P., and Caro, L. G. (1962). Structure, chemistry and function of the pancreatic exocrine cell. *In* "Ciba Foundation Symposium on the Exocrine Pancreas: Normal and Abnormal Function" (A. V. S. de Reuck and M. P. Cameron, eds.), pp. 23–55. Little, Brown, and Co., Boston.

Pallotti, F., Genova, M. L., Merlo Pich, M., Zucchini, C., Carraro, S., Tesei, M., Bovina, C., and Lenaz, G. (1998). Mitochondrial dysfunction and brain disorders. *Arch. Gerontol. Geriatr.* **6** (Suppl.), 385–392.

Popinigis, J., Antosiewicz, J., Mazzanti, L., Bertoli, E., Lenaz, G., and Cambria, A. (1990). Direct oxidation of glutamate by mitochondria from porcine adrenal cortex. *Biochem. Int.* **21,** 441–451.

Rasmussen, H. N., Andersen, A. J., and Rasmussen, U. F. (1997). Optimization of preparation of mitochondria from 25–100 mg skeletal muscle. *Anal. Biochem.* **252,** 153–159.

Rauchova, H., Battino, M., Fato, R., Lenaz, G., and Drahota, Z. (1992). Coenzyme Q-pool function in glycerol-3-phosphate oxidation in hamster brown adipose tissue mitochondria. *J. Bioenerg. Biomembr.* **24,** 235–241.

Rauchova, H., Fato, R., Drahota, Z, and Lenaz, G. (1997). Steady-state kinetics of reduction of coenzyme Q analogs by glycerol-3-phospate dehydrogenase in brown adipose tissue mitochondria. *Arch. Biochem. Biophys.* **344,** 235–241.

Renger, H. C., and Wolstenholme, D. R. (1972). The form and structure of kinetoplast DNA of *Crithidia*. *J. Cell Biol.* **54,** 346–364.

Schnaitman, C., and Greenawalt, J. W. (1968). Enzymatic properties of the inner and outer membranes of rat liver mitochondria. *J.Cell Biol.* **38,** 158–175.

Schneider, H., Lemasters, J. J., Hochli, and Hackenbrock, C. R. (1980). Fusion of liposomes with mitochondrial inner membranes. *Proc. Natl. Acad. Sci. USA* **77,** 442–446.

Schneider, W. C. (1948). Intracellular distribution of enzymes. III. The oxidation of octanoic acid by rat liver fractions. *J. Biol. Chem.* **176,** 259–267.

Seglen, P. O. (1976). Preparation of isolated rat liver cells. *In* "Methods in Cell Biology" (D. M. Prescott, ed.), Vol. 13, pp. 29–83. Academic Press, New York.

Simpson, E. R., and Boyd, G. S. (1971). The metabolism of pyruvate by bovine-adrenal-cortex mitochondria. *Eur. J. Biochem.* **22,** 489–499.

Simpson, E. R., and Estabrook, R. W. (1969). Mitochondrial malic enzyme: The source of reduced nicotinamide adenine dinucleotide phosphate for steroid hydroxylation in bovine adrenal cortex mitochondria. *Arch. Biochem. Biophys.* **129,** 384–395.

Smith, A. L. (1967). Preparation, properties, and conditions for assay of mitochondria: Slaughterhouse material, small-scale. *In* "Methods in Enzymology" (R. W. Estabrook and M. E. Pullman, eds.), Vol. 10, pp. 81–86. Academic Press, San Diego.

Smith, R. E., Roberts, J. C., and Hittelman, K. J. (1966). Nonphosphorylating respiration of mitochondria from brown adipose tissue. *Science* **154,** 653–654.

Szarkowska, L. (1966). The restoration of DPNH oxidase activity by coenzyme Q (ubiquinone). *Arch. Biochem. Biophys.* **113,** 519–525.

Tapper, D. P., Van Etten, R. A., and Clayton, D. A. (1983). Isolation of mammalian mitochondrial DNA and RNA and cloning of the mitochondrial genome. *In* "Methods in Enzymology" (S. Fleischer and B. Fleisher, eds.), Vol. 97, pp. 426–434. Academic Press, San Diego.

Ventrella, V., Pagliarani, A., Trigari, R., and Borgatti, A. R. (1982). Respirazione e fosforilazione ossidativa in mitocondri epatici di branzino (*Morone labrax*) e loro dipendenza dalla temperatura. *Boll. Soc. It. Biol. Sper.* **58,** 1509–1515.

Villa, R. F., Gorini, A., Geroldi, D., Lo Faro, A., and Dell'Orbo, C. (1989). Enzyme activities in perikaryal and synaptic mitochondrial fractions from rat hippocampus during development. *Mech. Ageing Dev.* **49,** 211–225.

Wattiaux, R., Wattiaux-De Coninck, S., Ronveaux-Dupal, M. F., and Dubois, F. (1978). Isolation of rat liver lysosomes by isopycnic centrifugation in a metrizamide gradient. *J. Cell Biol.* **78,** 349–368.

Wells, W. W., Seyfred, M. A., Smith, C. D., and Sakai, M. (1987). Measurement of subcellular sites of polyphosphoinositide metabolism in isolated rat hepatocytes. *In* "Methods in Enzymology" (P. M. Conn and A. R. Means, eds.), Vol. 141, pp. 92–99. Academic Press, San Diego.

Yang, J., Liu, X., Bhalla, K., Kim, C. N., Ibrado, A. M., Cai, J., Peng, T. I., Jones, D. P., and Wang, X. (1997). Prevention of apoptosis by Bcl-2: Release of cytochrome *c* from mitochondria blocked. *Science* **275,** 1129–1132.

Zucchini, C., Pugnaloni, A., Pallotti, F., Solmi, R., Crimi, M., Castaldini, C., Biagini, G., and Lenaz, G. (1995). Human skeletal muscle mitochondria in aging: Lack of detectable morphological and enzymic defects. *Biochem. Mol. Biol. Int.* **37,** 607–616.

CHAPTER 2

Isolation and Subfractionation of Mitochondria from the Yeast *Saccharomyces cerevisiae*

Kerstin Diekert,[*] Anton I. P. M. de Kroon,[†] Gyula Kispal,[*,‡] and Roland Lill[*]

[*] Institut für Zytobiologie und Zytopathologie der Philipps-Universität Marburg
35033 Marburg, Germany

[†] Center for Biomembranes and Lipid Enzymology
Department of Biochemistry of Membranes, University of Utrecht
3584 CH Utrecht, The Netherlands

[‡] Institute of Biochemistry
University Medical School of Pecs
7624 Pecs, Hungary

I. Introduction

The budding yeast *Saccharomyces cerevisiae* has proven to be an excellent model organism for examination of numerous genetic, biochemical, and cell biological problems. In many cases, studies on this yeast have provided the first glimpse into cellular processes and molecular function of proteins. Indeed, despite its relative simplicity, yeast represents an important model organism for the study of the molecular principles underlying some human diseases (see, e.g., Foury, 1997). The advantages of using yeast as a model system are manifold. For instance, the short generation time allows the preparation of biological material in amounts sufficient for subsequent biochemical and biophysical examinations. Further, the genetic amenability of *S. cerevisiae* and the availability of the entire genomic DNA sequence of the organism open a plethora of approaches that are not readily feasible with higher organisms (Attardi and Chomyn, 1995). The yeast system is particularly well suited for investigations on mitochondria. In both lower and higher eukaryotes the protein composition and the cellular tasks performed by these organelles appear to be highly conserved (Attardi and Schatz, 1988; Wallace, 1999; Scharfe *et al.,* 2000).

Numerous investigations on isolated yeast mitochondria have set the grounds for understanding the biological function of this organelle. It is easy to predict that such "in organello" studies will continue to be fruitful in many respects. It is crucial in these *in vitro* experiments to employ mitochondria that maintain their functional integrity. This requires standardized methods for the isolation of mitochondria, and quantitative criteria for the assessment of their purity and integrity. In addition, reliable methods should be used for the unambiguous determination of the submitochondrial localization of proteins, e.g., fractionation of the organelles into the various subcompartments, namely the outer and inner membranes, the intermembrane space, and the matrix. In the past two decades, such standard methods have been developed and optimized further (Daum *et al.,* 1982; Söllner *et al.,* 1989; Glick *et al.,* 1992; Mayer *et al.,* 1993), and several excellent articles have been published that describe detailed protocols that have been used successfully by various groups (Glick, 1991; Herrmann *et al.,* 1994; Mayer *et al.,* 1994, 1995; Glick and Pon, 1995).

This chapter provides a compilation of these methods as used by our group. Section II describes three routine protocols for the isolation of yeast mitochondria. The methods differ by the relative speed of obtaining the mitochondrial fractions and by the purity of the preparations. First, we outline a simple protocol for the rapid purification of a "crude mitochondrial fraction" by glass bead lysis of yeast cells (Lange *et al.,* 1999). This fraction may be most useful for routine testing of the expression of mitochondrial proteins. Second, a standard method is detailed for obtaining "isolated mitochondria" by gentle lysis of spheroplasts and by differential centrifugation (Daum *et al.,* 1982). This material is suitable for numerous biochemical investigations, such as protein and metabolite import reactions, protein folding and degradation studies, protein translation measurements, and the examination of various metabolic processes. Because this preparation of mitochondria contains a significant amount of contaminating membranes, such as those of the endoplasmic reticulum and the plasma membrane (Glick and Pon, 1995), we finally describe a procedure for further purification of isolated mitochondria using Nycodenz density gradient centrifugation (Lewin *et al.,* 1990). The resulting "purified mitochondria" are largely devoid of contaminating membranes and may be useful for studying the subcellular localization of proteins, as this fraction allows for the discrimination between components localized within mitochondria and contaminating membranes. Further, this preparation of mitochondria may be advantageous for *in vitro* studies in which contaminating membranes would contribute a significant background signal in the particular assays performed.

Section III compiles methods for the analysis of quality of isolated and purified mitochondria and for determining the submitochondrial localization of proteins. First, we outline a simple immunostaining procedure to determine the purity of the various preparations. Second, a rapid protocol is described for assessing the quality of the obtained mitochondrial fractions. The assay follows (i) the intactness of the mitochondrial outer membrane and (ii) the ability of the organelles to undergo swelling on hypotonic treatment in a quantitative fashion (Glick *et al.,* 1992). These two criteria have proven to be a reliable measure for the biological activity of the organelles. We routinely take advantage of this test for estimating the suitability of each preparation of mitochondria for subsequent investigations. Additionally, this method can be used for the subfractionation of the organelles and for the determination of the submitochondrial localization of proteins. Third, a method is explained for distinguishing between soluble and membrane-bound proteins, particularly those of the mitochondrial intermembrane space.

Finally, we provide a protocol for the preparation of highly purified outer membrane vesicles (OMV) from yeast mitochondria using both sedimentation and flotation centrifugation steps on sucrose density gradients. This method is based on a purification scheme developed for OMV from *Neurospora crassa* (Mayer *et al.,* 1993). This material may be beneficial for studying processes performed by constituents of the outer membrane, such as protein import through the *t*ranslocase of the *o*uter membrane of *m*itochondria (TOM) complex (Lill and Neupert, 1996) or the interaction of mitochondria with the cytoskeleton. Further, the material is invaluable for determining the exact composition of this biological membrane with respect to the qualitative and quantitative content of lipids (de Kroon *et al.,* 1997) and proteins.

II. Isolation of Yeast Mitochondria of Different Purity

Stock Solutions

1 *M* Tris–SO_4, pH 9.4, autoclave stock solution and store at room temperature

1 *M* HEPES–KOH, pH 7.4, autoclave stock solution and store at room temperature

1 *M* potassium phosphate, pH 7.4, autoclave stock solution and store at room temperature

200 m*M* phenylmethylsulfonyl fluoride (PMSF) in ethanol; prepare fresh before use and keep on ice. [*Note:* PMSF is an effective inhibitor for the preparation of mitochondria exhibiting active protein import or respiratory activity. For other applications, we recommend a protease inhibitor cocktail consisting of benzamidine (10 m*M*), 1,10-phenanthroline (1 mg/ml), antipain (0.5 mg/ml), chymostain (0.5 mg/ml), leupeptin (0.5 mg/ml), pepstatin (0.5 mg/ml), and aprotinin (0.5 mg/ml).]

1 *M* dithiothreitol (DTT); prepare fresh and keep on ice

2.4 *M* sorbitol; autoclave stock solution and store at room temperature

50% (w/v) Nycodenz [5-(*N*-2,3-dihydroxypropylacetamido)-2,4,6,triiodo-*N,N′*-bis(2, 3-dihydroxypropyl)isophthalimide] (Sigma); store frozen at −20°C. Add the Nycodenz powder slowly to the water to avoid the formation of lumps.

A. Preparation of a Crude Mitochondrial Fraction

Working Solution

Buffer 1× SHP: 0.6 *M* sorbitol, 20 m*M* HEPES–KOH, pH 7.4, 1 m*M* PMSF (10 ml)

Procedure

1. Grow the yeast cells in 100 ml of the desired growth media (see later and Sherman, 1991) to an OD_{600} of 1–2. Alternatively, collect cells from an agar plate containing the desired growth media after growth for 2–3 days at 30°C. A cell lawn of ca. 8 cm^2 (corresponding to 0.5 g cells) is sufficient for obtaining 40–50 μg of a crude mitochondrial fraction.
2. Cells are collected by centrifugation (3000*g*, 5 min), washed once in H_2O, and resuspended in 1 ml ice-cold 1× SHP at a density of 0.5 g cells/ml.
3. Glass beads (0.4–0.5 mm diameter) are added to occupy two-thirds of the final volume. Cells are broken by vortexing for 15 s at maximum speed.
4. After 15 s of chilling on ice, vortexing is repeated once.
5. The resulting suspension is subjected to a low-speed centrifugation step to pellet unbroken cells, cell debris, and nuclei. To this end, cells in Eppendorf tubes are centrifuged for 5 min at 600*g* at 2°C. The supernatant harbors mitochondria, other cellular membranes, and the cytosol.
6. The supernatant is removed carefully from the pellet with a pipette and centrifuged for 10 min at 10,000*g* at 2°C. The resulting pellet, containing mitochondria and some contaminating cellular membranes, is resuspended in 10 μl of buffer 1× SHP.
7. The protein concentration is determined by the Coomassie dye-binding assay (Bio-Rad).

Remarks: Due to the crude nature of the preparation, proteolysis may be a problem. Therefore it is crucial to keep the temperature of the preparation close to 0°C and to include freshly prepared PMSF as a protease inhibitor. The crude mitochondrial fraction is largely devoid of cytosolic proteins.

B. Preparation of Isolated Mitochondria by Differenting Centrifugation

Before describing the method to prepare isolated mitochondria from yeast cells, we summarize the growth media and conditions for cultivation of the yeast cells (see Sherman, 1991).

1. Growth of Yeast Cells

Growth medium YP: Dissolve 10 g/liter yeast extract (Difco) and 20 g/liter Bacto-Peptone (Difco) in water and autoclave. The broth is supplemented with various the carbon sources glucose (2% w/v, for YPD), galactose (2% w/v, for YPGal), raffinose (2% w/v, for YPR), glycerol (3% v/v, for YPG). Carbon sources are added from autoclaved 20% stock solutions.

Lactate medium: 3 g/liter yeast extract, 0.5 g/liter glucose, 1 g/liter KH_2PO_4, 1 g/liter NH_4Cl, 0.5 g/liter $CaCl_2 \cdot 2H_2O$, 0.5 g/liter NaCl, 0.6 g/liter $MgSO_4 \cdot H_2O$, 3 mg/liter $FeCl_3$, 2% (v/v) lactate. Adjust pH to 5.5 with NaOH pellets (about 7.5 g/l) and autoclave.

Procedure

1. Yeast cells are grown for 2–3 days on YPD agar plates or on appropriate selective media that will ensure the maintenance of plasmids.

2. Single colonies are used to inoculate 25 ml of YPD medium or selective media in a 100-ml Erlenmeyer flask. The cells are cultivated overnight at 30°C with good aeration (120–150 rpm in a shaker with a 5-cm rotation diameter). In none of the incubations should the yeast cells approach stationary phase, i.e., OD_{600} values should not exceed 2 in lactate media.

3. Cells are diluted into 100–200 ml medium in a 1-liter Erlenmeyer flask and incubated for 6 to 10 h at 30°C. Glucose represses mitochondrial biogenesis by up to 10-fold in some genetic backgrounds. Therefore, for the isolation of mitochondria from wild-type cells it is best to use either YPGal or lactate media. The latter medium is less expensive, but its preparation may be more time-consuming. YPR is effective as a growth medium for yeast bearing respiration-deficient mitochondria. For convenience, the duplication time of the cells may be calculated from the OD_{600} and used to estimate the amount of cells needed for inoculation of the main culture used for the preparation of isolated mitochondria.

4. The main culture consists of 2-liters of growth medium per 5-liter Erlenmeyer flask. Growth of the cells at 30°C is performed overnight and should yield a cell density of OD_{600} of 1 to 2 in lactate media.

2. Isolation of Mitochondria

The following procedure is outlined for 10 g of yeast cells (wet weight), which are usually obtained from 2 liters of main culture. The typical yield of isolated mitochondria is 50 mg protein. The method can be up- and downscaled by at least a factor of 5.

Working Solutions

Buffer TD: 100 m*M* Tris–SO_4, pH 9.4, 10 m*M* DTT; prepare fresh before use (50 ml)

Buffer SP: 1.2 *M* sorbitol, 20 m*M* potassium phosphate, pH 7.4 (500 ml)

Buffer 2× SHP: 1.2 *M* sorbitol, 40 m*M* HEPES–KOH, pH 7.4, 1 m*M* PMSF (100 ml)

Buffer 1× SH: 0.6 *M* sorbitol, 20 m*M* HEPES–KOH, pH 7.4 (500 ml)

Procedure

1. Collect cells by centrifugation for 5 min at 2500*g* at room temperature in tared beakers (in steps 1 through 9 it is best to use a Beckman JA-10 or Sorvall GSA rotor).

2. Resuspend the cells in 200 ml H_2O and spin at 2500*g* for 5 min. Pour off the supernatant and determine the wet weight of the cells.

3. Dissolve zymolyase 100T (1.5 mg/g cells) or 20T (2 mg/g cells) in 10 ml buffer SP. The activity of the zymolyase will be higher when it is dissolved early.

4. Resuspend the cells in 30 ml buffer TD using a rubber policeman attached to a glass rod. Incubate for 5 min at 30°C with gentle shaking.

5. Centrifuge the cells at 2500*g* for 5 min and resuspend the pellet in 40 ml buffer SP.

6. Repeat step 5 and add the zymolyase solution.

7. Incubate the cells at 30°C for 20–40 min with gentle shaking. To test for spheroplast formation, add 20 μl of the cell suspension to 1 ml H_2O or 1 ml buffer SP. Upon transfer into H_2O the spheroplasts break and the suspension should clear up.

8. All subsequent operations are conducted at 0°C or 2°C. Collect the spheroplasts by centrifugation at 2500*g* for 5 min.

9. Resuspend the pellet carefully in 40 ml ice-cold buffer SP and repeat centrifugation. Repeat this step once.

10. Resuspend the spheroplasts in 30 ml buffer 2× SHP. Measure the final volume and dilute with 1 volume of ice-cold H_2O containing 1 m*M* PMSF.

11. Homogenize the suspension in a 50-ml glass homogenizer with a glass piston (Braun-Biotech). For tightly fitting pistons, 15 strokes are sufficient; with loose pistons, one may need up to 25 strokes. This step is critical because too few strokes decrease the yield of mitochondria and too many strokes may damage the mitochondrial membranes (see Section III,C).

12. Spin twice at 2500*g* for 5 min and pour supernatants into fresh 50-ml tubes. In this and all subsequent steps it may be best to use a Beckman JA-20 or Sorvall SS-34 rotor, or any equivalent.

13. Centrifuge the tubes containing the supernatant at 12000*g* for 10 min. The resulting supernatant is removed and may be used as a "postmitochondrial supernatant" for further studies.

14. Carefully resuspend the pellet (containing mitochondria) in 1 ml of buffer 1× SH using pipette tips with a wide opening. For this purpose, cut tips may be used. Alternatively, a small-scale homogenizer may be used to obtain a homogeneous suspension of mitochondria. Dilute the solution with 20–30 ml of buffer 1× SH.

15. Centrifuge at 2500*g* for 5 min. Pour supernatant into a fresh tube and centrifuge at 12000*g* for 10 min.

16. Resuspend the pellet with cut pipette tips in 0.5 ml of buffer 1× SH.

17. Divide the solution into aliquots (0.1–0.5 mg protein) and immediately freeze isolated mitochondria in liquid nitrogen. Store samples at −70°C. Before use, thaw mitochondria rapidly and keep them on ice. Depending on the assay system, mitochondria

may be frozen again. For instance, protein import activity is decreased upon repeated freezing and thawing, whereas the activity of oxidative phosphorylation is not readily compromised.

18. An aliquot is used to measure the protein concentration by the Coomassie dye-binding assay. The typical yield of mitochondria isolated from a 2-liter main culture is 50 mg protein.

C. Purification of Isolated Mitochondria by Nycodenz Density Gradient Centrifugation

Working Solutions

Buffer 1× SH: 0.6 *M* sorbitol, 20 m*M* HEPES–KOH, pH 7.4

Buffer 2× SH: 1.2 *M* sorbitol, 40 m*M* HEPES–KOH, pH 7.4

18.5% Nycodenz: For 10 ml mix 5 ml buffer 2× SH with 3.7 ml 50% Nycodenz solution and 1.3 ml H_2O

14.5% Nycodenz: For 10 ml mix 5 ml buffer 2× SH with 2.9 ml 50% Nycodenz solution and 2.1 ml H_2O

Procedure

1. Conduct all steps on ice. For optimal purification, no more than 50 mg of mitochondrial protein should be loaded on the 10-ml Nycodenz gradient of step 2.

2. Prepare the Nycodenz density step gradients in 14× 89-mm Ultra-Clear centrifuge tubes (Beckman No. 344059) suitable for a Beckman SW-41 or equivalent rotor. Form gradients just before use by overlaying 5 ml of 18.5% Nycodenz solution with 5 ml of 14.5% Nycodenz solution (Fig. 1). To maintain the sharp density interface between layers, the 14.5% Nycodenz solution should be overlaid using cut pipette tips.

3. Overlay the step gradient with a suspension containing freshly isolated mitochondria (in 1 ml of buffer 1× SH) taken from step 14 of the procedure described earlier.

4. Centrifuge the tubes at 270,000*g* for 30 min. Purified mitochondria should appear as a brownish band at the 18.5%/14.5% interface (Fig. 1).

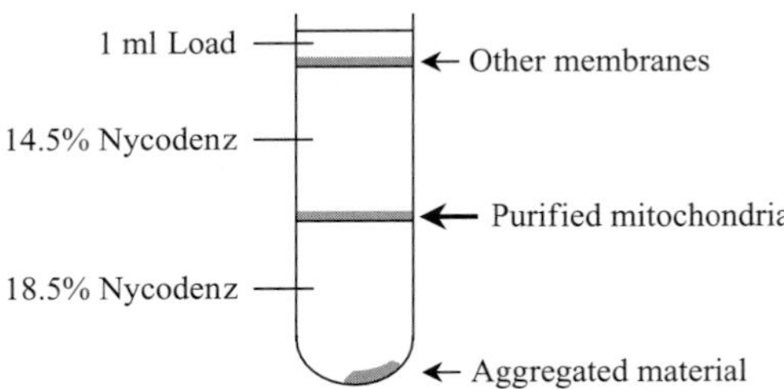

Fig. 1 Purification of mitochondria by Nycodenz density gradient centrifugation. (Left) Nycodenz concentrations before centrifugation. (Right) The position of purified mitochondria and material separated by this purification step.

5. Aspirate off the solution above the mitochondria and harvest purified mitochondria with a pipette. Dilute the sample with 5–10 volumes of buffer 1× SH. Centrifuge the solution for 10 min at 12,000*g*.

6. Resuspend the pellet in 250 μl 1× SH using cut pipette tips.

7. Divide into aliquots (0.1–0.2 mg) and immediately freeze the material in liquid nitrogen.

III. Analysis of Mitochondria Preparations

A. Determination of the Purity of Mitochondrial Fractions

The various preparations of mitochondria contain different amounts of contaminating proteins of other cellular organelles, e.g., of the endoplasmic reticulum, the plasma membrane, and vacuoles. Generally, mitochondrial fractions are devoid of contaminating cytosolic proteins. The purity of the mitochondrial fractions can be tested readily by immunostaining analysis using antibodies directed against proteins of various cellular compartments. For quantitative assessment of the degree of purification achieved by density gradient centrifugation, it is necessary to analyze various amounts of isolated and purified mitochondria (and postmitochondrial supernatants) and to compare the relative signals for a number of marker proteins.

B. Immunostaining Analysis

Isolated or purified mitochondria and postmitochondrial supernatant (PMS, from step 13 in the mitochondrial isolation protocol) are separated by SDS–PAGE followed by blotting of proteins onto nitrocellulose membranes. Immunostaining is performed for marker proteins of various cellular compartments using specific antibodies. A typical example for isolated and purified mitochondria is shown in Fig. 2. The depletion of proteins of the rough endoplasmic reticulum is evident from the at least 10-fold lower amounts of Scj1p in Nycodenz-purified mitochondria as compared to isolated organelles (Schlenstedt *et al.,* 1995). It should be noted, however, that components of the smooth endoplasmic reticulum are hardly depleted by this purification procedure (K. Diekert, unpublished), indicating a tight association of a fraction of this cellular compartment with mitochondria (see Achleitner *et al.,* 1999).

C. Determining the Quality of Mitochondria by Testing Their Integrity and Their Ability to Undergo Swelling under Hypotonic Conditions

The quality of a preparation of mitochondria largely depends on the integrity of the organelles and their ability to undergo swelling in hypotonic media. Both quality criteria are mainly dependent on the intactness of the outer membrane. It is therefore recommended to routinely analyze the integrity and the ability to undergo swelling for each preparation of mitochondria. The assay for integrity of mitochondria involves

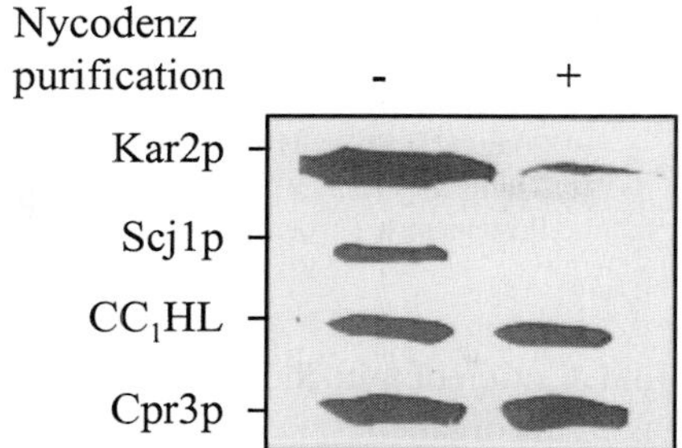

Fig. 2 Purification of mitochondria by Nycodenz density gradient centrifugation. Preparations of isolated mitochondria before (−) and after (+) Nycodenz gradient purification were analyzed by immunostaining using antibodies raised against proteins of the endoplasmic reticulum (Kar2p, Scj1p) and mitochondria [cytochrome c_1 heme lyase (CC_1HL) and the cyclophilin Cpr3p].

protease treatment of the organelles followed by analysis of the degree of protection of protease-sensitive marker proteins of the intermembrane space and the matrix by immunostaining. Proteins of these compartments will not be sensitive to proteolytic degradation when the organelles are intact.

The ability to undergo swelling can be tested by the formation of mitoplasts (i.e., organelles with selectively ruptured outer membranes) upon hypotonic treatment of mitochondria. This procedure should render proteins of the intermembrane space, but not of the matrix, sensitive to proteolytic attack (Diekert *et al.,* 1999). Hence, a preparation of mitochondria suitable for subsequent biochemical studies shows hardly any or almost complete protease degradation of intermembrane space proteins in mitochondria or mitoplasts, respectively.

The assay is also suitable for determining the submitochondrial localization of proteins, provided that they are protease sensitive. The localization of these proteins can be judged from a comparison with the behavior of known components of the mitochondrial subcompartments. For localization of protease-resistant proteins, however, the assay may not be suitable. In the case of soluble proteins, their release into the supernatant of mitoplasts (see later) may be taken as a criterion for intermembrane space localization. Release may require increased ionic strength, i.e., the addition of salt. Soluble proteins of the matrix may be released upon sonication of mitochondria or mitoplasts. In the case of protease-resistant, membrane-bound proteins, localization to the outer or inner membrane may take advantage of immunostaining analysis of purified OMV (see later). The absence of the investigated protein in this fraction may be taken as an indication for localization to the inner membrane.

Procedure

Mitochondria (diluted with buffer 1× SH to 5 mg/ml) are mixed with buffers, detergent, and proteinase K according to the following pipetting scheme (volumes are given in microliters).

Sample	1	2	3	4	5	6
Mitochondria (5 mg/ml)	10	10	10	10	10	10
Buffer 1× SH	90	90	—	—	90	90
20 m*M* HEPES, pH 7.4	—	—	90	90	—	—
Triton X-100 (20%)	—	—	—	—	1	1
Proteinase K (1 mg/ml)	—	5	—	5	—	5

A 10-fold dilution of mitochondria in 20 m*M* HEPES buffer (samples 3 and 4) results in swelling of the organelles, i.e., selective rupture of the outer membrane without opening of the inner membrane. The addition of 0.2% Triton X-100 detergent causes the complete lysis of both mitochondrial membranes. After incubation on ice for 20 min, proteinase K is inactivated by the addition of 1 m*M* PMSF. To avoid further proteolysis, it is best to precipitate the proteins with 12% trichloroacetic acid before separation by SDS–PAGE. The proteins are blotted onto nitrocellulose membranes and immunostained with specific antisera.

Figure 3 shows a typical example of an integrity test using antibodies directed against protease-sensitive proteins of the outer membrane (Tom70p), the intermembrane space (CC_1HL), and the matrix (Tim44p and Mge1p; cf. Diekert *et al.,* 1999). Outer membrane proteins are accessible by proteinase K in both mitochondria (Fig. 3, lane 2) and mitoplasts (lane 4), whereas proteins of the intermembrane space are digested only after opening of the outer membrane upon swelling (lane 4). Usually the opening of the outer membrane upon hypotonic treatment occurs at an efficiency of greater than 90%. The intactness of the inner membrane after the swelling procedure is evident from the protease inaccessibility of matrix proteins in mitoplasts (lane 4). In the presence of 0.2% Triton X-100 detergent, the proteins of all mitochondrial compartments are accessible to proteolytic attack (lane 6).

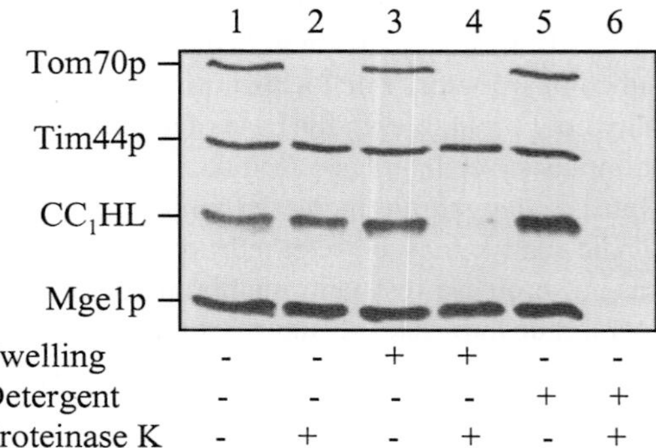

Fig. 3 Analysis of the integrity of mitochondrial membranes and assay for the ability of mitochondria to undergo swelling on hypotonic treatment. See text for details. The assay can be used to determine the submitochondrial localization of proteins.

D. Determining Whether Mitochondrial Proteins Are Soluble or Membrane Bound by Salt or Sodium Carbonate Extraction

In many studies it is important to know whether a mitochondrial protein is soluble or membrane bound. In the latter case, peripheral membrane proteins may be released from the membrane by raising the ionic strength, or may require alkaline conditions for dissociation from the membrane. We describe a method to discriminate between soluble and membrane-bound proteins located in the intermembrane space or the matrix.

Soluble proteins of the intermembrane space are released into the supernatant when the outer membrane of mitochondria has been ruptured by osmotic swelling (see earlier discussion). Examples of soluble proteins of the intermembrane space include cytochromes b_2 and c (Glick *et al.,* 1992). In the case of proteins such as cytochrome c, it may be necessary to treat the mitoplasts with increasing concentrations of salt (e.g., NaCl) to reduce the membrane association of the proteins (Kispal *et al.,* 1996). In other cases, proteins may be tightly bound to the surface of the inner membrane without being integrated into the lipid bilayer. Usually, such proteins can be released by treatment of the mitochondria with sodium carbonate at alkaline conditions (Fujiki *et al.,* 1982). It should be mentioned as a note of caution that not all peripherally bound proteins are released from the membranes by this treatment (e.g., the cytochrome heme lyases; see Steiner *et al.,* 1995).

Working Solutions

Depending on the number of samples to be analyzed, 1–5 ml of the following solutions is needed:

20 m*M* HEPES–KOH, pH 7.4
200 m*M* NaCl
200 m*M* Na_2CO_3 (prepare fresh and do not adjust pH)
Buffer 1× SH: 0.6 *M* sorbitol, 20 m*M* HEPES–KOH, pH 7.4

Procedure

1. Mix 10 μl mitochondria (5 mg/ml) with 90 μl of 20 m*M* HEPES, pH 7.4, and incubate for 20 min on ice.
2. Add 100 μl of 200 m*M* NaCl or 200 m*M* Na_2CO_3 on ice. Vortex the mixture and incubate another 2 min.
3. Centrifuge the solution for 10 min at 35,000*g* at 2°C. Pour supernatant into a new Eppendorf tube.
4. Resuspend the pellet in 100 μl of buffer 1× SH and precipitate the proteins of the supernatant and pellet fractions with 12% (w/v) trichloroacetic acid.
5. Proteins of the supernatant and pellet fractions are separated by SDS–PAGE and analyzed by immunostaining.

IV. Purification of Outer Membrane Vesicles from Yeast Mitochondria

The mitochondrial outer membrane represents about 6% of the total mitochondrial protein content (Mayer *et al.*, 1993). To obtain outer membranes that are largely devoid of inner membranes, isolated mitochondria are purified further by sucrose density gradient centrifugation. The organelles are subjected to hypotonic treatment and homogenized to shear off the outer membrane from the mitoplasts. The outer membranes are purified by sedimentation and flotation centrifugation using sucrose density gradients (Mayer *et al.*, 1993, 1995). The resulting outer membrane vesicles are tightly sealed and exist almost quantitatively in a right-side-out orientation.

Stock Solutions

1 *M* MES–HCl, pH 6.0

2.4 *M* sorbitol

2 *M* sucrose

1 *M* MOPS–KOH, pH 7.2

1 *M* EDTA, pH 7.0

Working Solutions

Buffer M: 1 *M* MES, pH 6.0 (100 ml)

Buffer EM: 2.5 m*M* EDTA, pH 7.2, and 10 m*M* MOPS. It is best to dilute the buffer from a 10× stock solution (100 ml)

Buffer 1× SH: 0.6 *M* sorbitol, 20 m*M* HEPES–KOH, pH 7.4 (100 ml)

Buffer $S^{2.5}$EM: 2.5 *M* sucrose dissolved in buffer 1× EM (50 ml)

Buffer $S^{1.0}$EM: 1.0 *M* sucrose dissolved in buffer 1× EM (25 ml)

Buffer $S^{0.9}$EM: 0.9 *M* sucrose dissolved in buffer 1× EM (45 ml)

Buffer $S^{0.5}$EM: 0.5 *M* sucrose dissolved in buffer 1× EM (20 ml)

30% sucrose: 30% sucrose (w/v) in 0.6 *M* sorbitol and 10 m*M* MES, pH 6.0 (40 ml)

45% sucrose: 45% sucrose (w/v) in 0.6 *M* sorbitol and 10 m*M* MES, pH 6.0 (40 ml)

Procedure

1. As a starting material, use isolated mitochondria (from step 16 or 17 of the procedure described in Section II,B,2). Because the yield of highly purified OMV is low (Mayer *et al.*, 1993), the material obtained from a 10-liter culture should be used as a minimum input. Prepare a sucrose gradient consisting of 15 ml 45% sucrose and 15 ml 30% sucrose in tubes for a Beckman SW28 rotor. Overlay the step gradient with up to 25 mg of mitochondrial protein per tube.
2. Centrifuge the tubes for 1 h at 100,000*g* at 2°C.

3. Collect mitochondria from the 30%/45% sucrose interface and dilute with 5 volumes of buffer 1× SH. Centrifuge the sample at 12,000*g* (using a Beckman JA-20 or equivalent rotor) for 10 min at 2°C.

4. If necessary, the pellet can be frozen at this step in liquid nitrogen and stored overnight at −70°C.

5. Thaw mitochondria and keep on ice. Add water to a final concentration of 1–4 mg mitochondrial protein ml (final volume about 30 ml) and incubate the suspension on ice for 1 h under continuous stirring.

6. Homogenize the suspension using a glass homogenizer (Braun-Biotech) with a tightly fitting glass pestle for 20 strokes.

7. Load 15 ml of the homogenate on top of a sucrose gradient consisting of 12 ml of buffer $S^{1.0}$EM and 9 ml of buffer $S^{0.5}$EM in a tube for a Beckman SW28 rotor.

8. Centrifuge samples for 1 h at 141,000*g* at 2°C.

9. Harvest the outer membrane fraction from the 0.5/1.0 *M* sucrose interface. (*Note*: the pellet can be used for isolation of mitochondrial inner membranes. It can be frozen in liquid nitrogen and stored at −70°C.) Adjust the sucrose concentration to 1.05 *M* by adding about 0.2 volumes of buffer $S^{2.5}$EM.

10. Ten milliliters of the solution is filled into tubes for a SW28 rotor. The solution is overlaid with 20 ml of buffer $S^{0.9}$EM and 5 ml buffer EM.

11. Centrifuge the samples for 16 h at 141,000*g* at 2°C.

12. Collect the purified outer membrane vesicles from the 0/0.9 *M* sucrose interface. Dilute the sample in 5 volumes of buffer EM and centrifuge for 1 h at 200,000*g* in a Beckman Ti60 or equivalent rotor at 2°C.

13. Resuspend the pellet in buffer EM and freeze aliquots (10–50 μg protein) in liquid nitrogen. Store the outer membrane vesicles at −70°C.

Acknowledgements

Our work was supported by grants from the Sonderforschungsbereich 286 of the Deutsche Forschungsgemeinschaft, the Volkswagen-Stiftung, the Fonds der chemischen Industrie, and the Hungarian Funds OKTA.

References

Achleitner, G., Gaigg, B., Krasser, A., Kainersdorfer, E., Kohlwein, S. D., Perktold, A., Zellnig, G., and Daum, G. (1999). Association between the endoplasmic reticulum and mitochondria of yeast facilities interorganelle transport of phospholipids through membrane contact. *Eur. J. Biochem.* **264,** 545–553.

Attardi, G., and Schatz, G. (1988). The biogenesis of mitochondria. *Annu. Rev. Cell Biol.* **4,** 289–333.

Attardi, G. M., and Chomyn, A. (1995). "Mitochondrial Biogenesis and Genetics, Part A." Academic Press, San Diego.

Daum, G., Böhni, P. C., and Schatz, G. (1982). Import of proteins into mitochondria Cytochrome b_2 and cytochrome *c* peroxidase are located in the intermembrane space of yeast mitochondria. *J. Biol. Chem.* **257,** 13028–13033.

de Kroon, A. I. P. M., Dolis, D., Mayer, A., Lill, R., and de Kruijiff, B. (1997). Phospholipid composition of highly purified mitochondrial outer membranes of rat liver and *Neurospora crassa:* Is cardiolipin present in the mitochondrial outer membrane? *Biochim. Biophys. Acta* **1325,** 108–116.

de Kroon, A. I. P. M., Koorengevel, M. C., Geordayal, S. S., Mulders, P. C., Janssen, M. J., and de Kruijiff, B. (1999). Isolation and characterization of highly purified mitochondrial outer memberanes of the yeast. *Saccharomyces cerevisiae. Mol. Membr. Biol.* **16,** 205–211.

Diekert, K., Kispal, G., Guiard, B., and Lill, R. (1999). An internal targeting signal directing proteins into the mitochondrial intermembrane space. *Proc. Natl. Acad. Sci. USA* **96,** 11746–11751.

Foury, F. (1997). Human genetic diseases: A cross-talk between man and yeast. *Gene* **195,** 1–10.

Fujiki, Y., Hubbard, A. L., Fowler, S., and Lazarow, P. B. (1982). Isolation of intracellular membranes by means of sodium carbonate treatment: Application to ER. *J. Cell Biol.* **93,** 97–102.

Gaigg, B., Simbeni, R., Hrastnik, C., Paltauf, F., and Daum, G. (1995). Characterization of a microsomal subfraction associated with mitochondria of the yeast *Saccharomyces cerevisiae:* Involvement in synthesis and import of phospholipids into mitochondria. *Biochim. Biophys. Acta* **1234,** 214–220.

Glick, B. S. (1991). Protein import isolated yeast mitochondria. *Methods Cell Biol.* **34,** 389–397.

Glick, B. S., Brandt, A., Cunningham, K., Müller, S., Hallberg, R. L., and Schatz, G. (1992). Cystochromes c_1 and b_2 are sorted to the intermembrane space of yeast mitochondria by a stop-transfer mechanism. *Cell* **69,** 809–822.

Glick, B. S., and Pon, L. A. (1995). Isolation of highly purified mitochondria from *Saccharomyces cerevisiae. Methods Enzymol.* **260,** 213–223.

Hermann, G. J., and Shaw, J. M. (1998). Mitochondrial dynamics in yeast. *Annu. Rev. Cell. Dev. Biol.* **14,** 265–303.

Hermann, J. M., Fölsch, H., Neupert, W., and Stuart, R. A. (1994). "Isolation of Yeast Mitochondria and Study of Mitochondrial Protein Translation" (J. E. Celis, ed.), Vol. 1, pp. 538–544. Academic Press, San Diego.

Kispal, G., Steiner, H., Court, D. A., Rolinski, B., and Lill, R. (1996). Mitochondrial and cytosolic branched-chain amino acid transaminases from yeast, homologs of the *myc* oncogene-regulated Eca39 protein. *J. Biol. Chem.* **271,** 24458–24464.

Kozlowski, M., and Zagorski, W. (1988). Stable preparation of yeast mitochondria and mitoplasts synthesizing specific polypeptides. *Anal. Biochem.* **172,** 382–391.

Lange, H., Kispal, G., and Lill, R. (1999). Mechanism of iron transport to the site of heme synthesis inside yeast mitochondria. *J. Biol. Chem.* **274,** 18989–18996.

Lewin, A. S., Hines, V., and Small, G. M. (1990). Citrate synthase encoded by the *CIT2* gene of *Saccharomyces cerevisiae* is peroxisomal. *Mol. Cell. Biol.* **10,** 1399–1405.

Lill, R., and Neupert, W. (1996). Mechanisms of protein import across the mitochondrial outer membrane. *Trends Cell Biol.* **6,** 56–61.

Mayer, A., Driessen, A., Neupert, W., and Lill, R. (1994). "Inclusion of Proteins into Isolated Mitochondrial Outer Membrane Vesicles" (J. E. Celis, ed.), Vol. 1, pp. 545–549. Academic Press, San Diego.

Mayer, A., Driessen, A., Neupert, W., and Lill, R. (1995). Purified and protein-loaded mitochondrial outer membrane vesicles for functional analysis of preprotein transport. *Methods Enzymol.* **260,** 252–263.

Mayer, A., Lill, R., and Neupert, W. (1993). Translocation and insertion of precursor proteins into isolated outer membranes of mitochondria. *J. Cell Biol.* **121,** 1233–1243.

Scharfe, C., Zaccaria, P., Hoertnagel, K., Jaksch, M., Klopstock, T, Dembowski, M., Lill, R., Prokisch, H., Gerbitz, K. D., Neupert, W., Mewes, H. W., and Meitinger, T. (2000). MITOP, the mitochondrial proteome database: 2000 update. *Nucleic Acids Res.* **28,** 155–158.

Schlenstedt, G., Harris, S., Risse, B., Lill, R., and Silver, P. A. (1995). A yeast DnaJ homolog, Scj 1 p, can function in the endoplasmic reticulum with BiP/Kar2p via a conserved domain that specifies interactions with Hsp70s. *J. Cell Biol.* **129,** 979–988.

Sherman, F. (1991). Getting started with yeast. *Methods Enzymol.* **194,** 3–21.

Söllner, T., Griffiths, G., Pfaller, R., Pfanner, N., and Neupert, W. (1989). MOM19, an import receptor for mitochondrial precursor proteins. *Cell* **59,** 1061–1070.

Steiner, H., Zollner, A., Haid, A., Neupert, W., and Lill, R. (1995). Biogenesis of mitochondrial heme lyases in yeast: Import and folding in the intermembrane space. *J. Biol. Chem.* **270,** 22842–22849.

Wallace, D. C. (1999). Mitochondrial diseases in man and mouse. *Science* **283,** 1482–1488.

Yaffe, M. P. (1999). The machinery of mitochondrial inheritance and behavior. *Science* **283,** 1493–1497.

CHAPTER 3

Isolation and Subfractionation of Mitochondria from Plants

A. H. Millar, A. Liddell, and C. J. Leaver

Department of Plant Sciences
University of Oxford
Oxford, OX1 3RB, United Kingdom

I. Introduction

Mitochondria from plants share significant similarities with those from other eukaryotic organisms in terms of both their structure and function. Plant mitochondria provide energy in the form of ATP, provide a high flux of metabolic precursors in the form of tricarboxylic acids to the rest of the cell for nitrogen assimilation and biosynthesis of amino acids, and have roles in plant development, productivity, fertility and susceptibility to disease. To engage in novel pathways, plant mitochondria contain a range of enzymatic functions that are not found in the mammalian organelles and which influence assays of mitochondrial function (Douce, 1985). The presence of a rigid cell wall and different intra- and extracellular structures also influence methods for mitochondrial purification from plants.

Techniques for the isolation of plant mitochondria, without changing the morphological structure observed *in vivo* and maintaining their functional characteristics, have required methods that avoid osmotic rupture of membranes and protect organelles from harmful products released from other cellular compartments. Several extensive methodology reviews (Douce, 1985, Neuburger, 1985) and more specific methodology papers (Day *et al.,* 1985; Leaver *et al.,* 1983; Neuburger *et al.,* 1982) are already available on plant mitochondrial purification. This chapter provides a basic appraisal of these classic methods for the reader and some more detailed examples of recent protocols for specialized purposes.

Study of the composition, metabolism, transport processes, and biogenesis of plant mitochondria requires an array of procedures for the subfractionation of mitochondria to provide information on localization and association of proteins or enzymatic activities. This chapter also presents subfractionation procedures to separate mitochondria into four compartments. Further techniques for separation and partial purification of membrane protein complexes are outlined. Finally, two-dimensional profiling of the mitochondrial proteome and identification of proteins by the use of antibodies, tandem mass spectrometry (MS), and *in organello* [^{35}S]Methionine translation assays are discussed with examples.

II. Growth and Preparation of Plant Material

Mitochondria can be isolated from virtually any plant tissue; however, depending on the requirements of amount, purity, and yield, particular tissues have their advantages. (1) Nonphotosynthetic fleshy root and tuber tissues allow for large scale, but low percentage yield, plant mitochondrial isolation (~300 mg mitochondrial protein from 5 to 10 kg FW). Typically, potato, sweet potato, turnip, and sugar beet have been used for this purpose. (2) Etiolated seedling tissues such as hypocotyls, cotyledons, roots, or coleoptiles are used due to the lack of chloroplasts and dense thylakoid membranes. Etiolated tissues also have lower phenolic content than green tissues and produce higher yields of functional mitochondria (~20–50 mg mitochondrial protein from 100 to 200 g FW). (3) Studies of the function and metabolism of mitochondria during photosynthesis, and interest in mitochondria from plant species without significant storage organs, have led to the development of methods for the isolation of chlorophyll-free mitochondria preparations from light-grown leaves and cotyledons (10–25 mg mitochondrial protein from 50 to 100 g FW).

Once the plant material has been chosen, the tissue must be free from obvious fungal or bacterial contamination and should be washed in cold sterile H_2O or, in cases of potential contamination, surface sterilized for ca 2 min in 1 : 15 dilution of a 14% (w/v) sodium hypochlorite stock solution. The major source of contamination is from the seed stocks, and dry seeds should be surface sterilized in sodium hypochlorite before imbibition and grown in sterile soil. These steps ensure that minimal contamination is carried into the further steps of the protocol. These precautions are particularly important in any studies involving isotope labeling. All tissues should then be cooled to 2–4°C before proceeding.

III. Isolation of Mitochondria by Differential Centrifugation

Tissues should be processed as soon as possible following harvesting and cooling in order to maintain cell turgor and thus ensure maximal mitochondrial yields following homogenization. All operations should be carried out in a cold room in which all glassware has been precooled. A basic homogenization medium consists of 0.3–0.4 *M* of an osmoticum (sucrose or mannitol), 1–5 m*M* of a divalent cation chelator (EDTA or EGTA), 25–50 m*M* of a pH buffer (MOPS, TES, or Na-pyrophosphate), and 5–10 m*M* of a reductant (cysteine or ascorbate), which is added just prior to homogenization. The osmoticum maintains the mitochondrial structure and prevents physical swelling and rupture of membranes, the buffer prevents acidification from the contents of ruptured vacuoles, and the EDTA inhibits the function of phospholipases and various proteases requiring Ca^{2+} or Mg^{2+}. The reductant prevents damage from oxidants present in the tissue or produced upon homogenization. This basic medium is sufficient for tuber tissues, but a variety of additions have been suggested to improve yields and protect mitochondria from damage during isolation from some tissues. We have found that the addition of Tricine (5–10 m*M*) to all solutions aids integrity and yield of mitochondria from seedling tissues. However, the presence of Tricine can inhibit *in organello* protein synthesis. Etiolated seedling tissues and green tissues also often require the addition of 0.1–1% (w/v) bovine serum albumin (BSA) to remove free fatty acids and 1–5% (w/v) polyvinyl pyrrolidone (PVP) to remove phenolics that damage organelles in the initial homogenate. As a result, we have routinely used a homogenization medium consisting of 0.4 *M* mannitol, 1 m*M* EGTA, 25 m*M* MOPS–KOH (pH 7.8), 10 m*M* Tricine, 8 m*M* cysteine, 0.1% (w/v) BSA, and 1% (w/v) PVP-40 with success on a variety of plant tissues.

Following preparation of the medium and the plant material, the three most critical steps in maximizing the yield of metabolically active mitochondria are often found to be (1) the method of homogenization, (2) the pH of the medium following homogenization, and (3) the ratio of homogenization medium to plant tissue.

Depending on the tissue, homogenization can be accomplished by one of several methods: (a) grinding in a precooled mortar and pestle; (b) homogenization in a square-form beaker with either a Moulinex mixer for 20–60 s or a Polytron or Ultra-Turrax blender at 50% full speed for 5 × 2 s; (c) homogenization in a Waring blender for 3 × 5 s at low speed; (d) juicing tuber material directly into 5X homogenization medium using a commercial vegetable/fruit juice extractor; or (e) grating of tissue by hand using a vegetable grater submerged under homogenization medium (useful for mitochondrial extraction from soft fruits).

The pH value of the homogenization medium should be adjusted to approximately pH 7.8 before homogenization to allow for the further acidification of the suspension following cell breakage. For some plant tissues, the pH of buffer may need to be increased to ensure that the pH of the final homogenate remains at or above pH 7.0. Alternatively, the pH may need to be adjusted following homogenization with a stock of 5 *M* KOH or NaOH. The homogenizing medium should be added in a ratio of at least 2 ml for each gram FW of nongreen tissue or at least 4 ml per gram FW of green tissue. Decreases in these ratios can result in dramatic losses in yield. Lower ratios of medium to tissue (as low as 0.25 ml per gram FW) do work in the case of tuber tissues extracted by method d.

Once homogenization has been completed, the final suspension must be filtered to remove the majority of the starch and cell debris, and kept at 4°C. Filtration through four to eight layers of muslin, Miracloth (Calbiochem, La Jolla, CA), or disposable clinical sheeting (available from most medical suppliers) is effective. The filtrate is collected in a cooled beaker or conical flask on ice and the remainder of the fluid in the filter is recovered by wringing the filtration cloth from the top into the beaker. The filtered homogenate is then poured into 50-, 250-, or 500-ml centrifuge tubes, depending on the volume of the preparation, and centrifuged for 5 min at $1000g_{max}$ in a fixed angle rotor fitted to a Beckman or Sorvall preparative centrifuge at 4°C. The supernatant following centrifugation is decanted gently into another set of centrifuge tubes, taking care not to transfer the pelleted material, which contains starch, nuclei, and cell debris. The supernatant is then centrifuged for 15–20 min at $12{,}000g_{max}$ at 4°C, and the resulting high-speed supernatant is discarded. The tan or green colored pellet in each tube is resuspended in 5–10 ml of a standard wash medium with the aid of a clean, soft-bristle paint brush and/or a Teflon–glass homogenizer. A standard wash medium consists of 0.4 *M* mannitol, 1 m*M* EGTA, 10 m*M* MOPS–KOH, pH 7.2, and 0.1% (w/v) BSA. If sucrose is the osmoticum used in the preparation, then sucrose can be used to replace mannitol in the standard wash medium. The resuspended organelles are transferred to 50-ml centrifuge tubes, and the volume in each tube is adjusted to 40 ml with wash medium and the samples are centrifuged at $1000g_{max}$ for 5 min. The supernatants are transferred into another set of tubes and the organelles are sedimented by centrifugation at $12{,}000g_{max}$ for 15–20 min. The high-speed supernatant is discarded once again and the washed organelles are resuspended uniformly in approximately 2 ml of wash medium as described earlier. Preparation of the mitochondrial fraction to this point should be undertaken as quickly as possible given the steps involved, without samples warming above 4°C and without storage for extended periods between centrifugation runs.

IV. Density Gradient Purification of Mitochondria

The mitochondrial preparation described earlier is often adequate for a variety of respiratory measurements; however, depending on the plant tissue, it is frequently contaminated by chloroplasts, thylakoid membranes, peroxisomes, and/or glyoxysomes and occasionally by bacteria. Further purification can be carried out using either sucrose or Percoll (Amersham-Pharmacia Biotech) density gradients. Many of the early isolations of plant mitochondria were performed on sucrose gradients. Sucrose offers the advantage in that it is inexpensive and gradients of sucrose are more effective in the separation of bacteria from mitochondria than are Percoll equivalents. However, Percoll gradients offer more rapid purification, allow separation of mitochondria and thylakoid membranes from green tissues, and also ensure isoosmotic conditions in the gradient. The latter eliminates the potential of the extreme osmotic shock encountered in sucrose gradients that can rupture mitochondria from some plant sources.

A. Sucrose Gradients

Washed mitochondria in 1–2 ml of wash medium (from 75 g tissue) are layered on a 13- or 30-ml continuous sucrose gradient (0.6–2.0 *M*) and centrifuged in a swing-out rotor (e.g., Beckman SW28 or SW28.1) at 72,000–77,000g_{max} for 60 min. The gradients can be prepared by using a gradient maker with equal volumes of solution A [0.6 *M* sucrose, 10 m*M* Tricine, 1 m*M* EGTA, pH 7.2, 0.1% (w/v) BSA] and B [2.0 *M* sucrose, 10 m*M* Tricine, 1 m*M* EGTA, pH 7.2, 0.1% (w/v) BSA]. Alternatively, five or more steps of different sucrose concentrations (2.0, 1.45, 1.2, 0.9, 0.6 *M*) can be layered in centrifuge tubes and left overnight at 4°C to equilibrate. After centrifugation, mitochondria form a band at approximately 1.25–1.35 *M* sucrose that can be recovered by aspirating with a Pasteur pipette into a graduated centrifuge tube on ice (Fig. 1A). The molarity of the suspension is then estimated using a refractometer, and the suspension is slowly diluted over 15–20 min by the addition of 0.2 *M* mannitol, 10 m*M* Tricine, and 1 m*M* EGTA (pH 7.2) to give a final osmotic concentration of 0.6 *M*. The diluted suspension is centrifuged at 12,000g_{max} for 15 min, the supernatant is discarded, and the purified mitochondrial pellet is resuspended gently in resuspension medium (0.4 *M* mannitol, 10 m*M* Tricine, 1 m*M* EGTA, pH 7.2) at 5–20 mg mitochondrial protein/ml.

B. Percoll Gradients

The versatility of a colloidal silica sol such as Percoll ensures the formation of isoosmotic gradients and facilitates a range of methods for the density purification of mitochondria through isopycnic centrifugation. The most common method is the sigmoidal, self-generating gradient obtained by centrifugation of a Percoll solution in a fixed-angle rotor. The density gradient is formed during centrifugation at >10,000*g* due to the sedimentation of the poly-dispersed colloid (average particle size 29 nm diameter, average

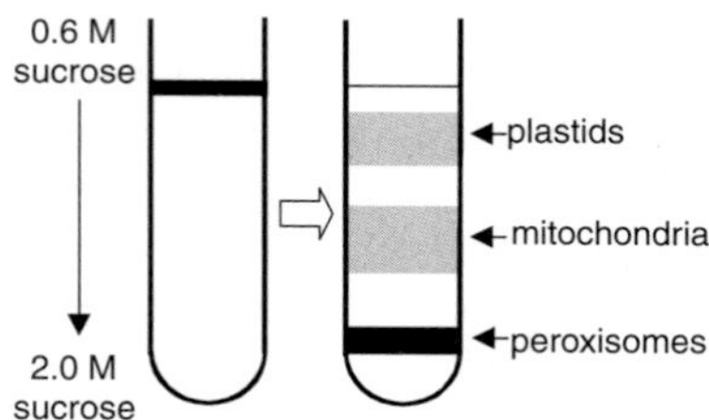

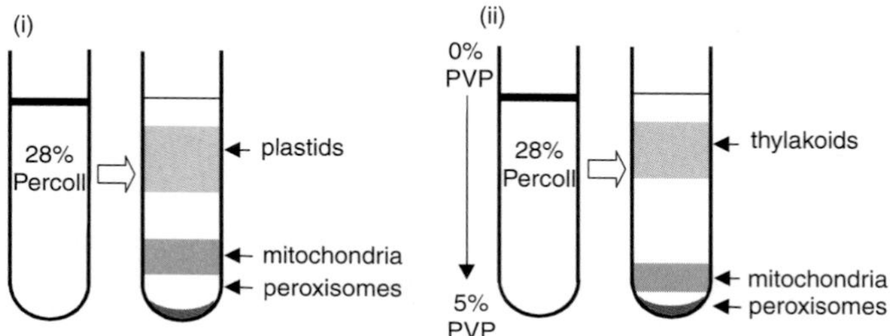

Fig. 1 Sucrose and Percoll gradients for purification of plant mitochondria. (A) A crude mitochondria sample loaded onto a 0.6–1.8 *M* sucrose gradient is separated by centrifugation at $75{,}000g_{max}$ for 60 min, yielding plastid, mitochondrial, and peroxisome fractions. (B) Crude mitochondria samples loaded on either (i) 28% (v/v) Percoll medium or (ii) 28% (v/v) Percoll medium with a 0–5% (w/v) PVP^{-40} gradient, separated by centrifugation at $40{,}000g_{max}$ for 45 min, yielding mitochondrial fractions separated from plastid and peroxisome contamination.

density of 2.2 g/ml). The concentration of Percoll in the starting solution and the time of centrifugation can be varied to optimize a particular separation. A solution of 28% (v/v) Percoll in 0.4 *M* mannitol or 28% (v/v) Percoll in 0.3 *M* sucrose supplemented with 10 m*M* MOPS, pH 7.2, 0.1% (w/v) BSA is prepared on the day of use. Washed mitochondria from up to 80 g of etiolated plant tissue are layered over 35 ml of the gradient solution in a 50-ml centrifuge tube. Gradients are centrifuged at $27{,}000g_{max}$ for 45 min in an angle rotor of a Sorvall or Beckman preparative centrifuge. After centrifugation, mitochondria form a buff-colored band below yellow-orange plastid membranes (Fig. 1Bi). Mitochondria are aspirated with a Pasteur pipette, avoiding collection of the yellow or green plastid fractions. The suspension is diluted with at least 4 volumes of standard wash medium and is centrifuged at $12{,}000g_{max}$ for 15–20 min in 50-ml centrifuge tubes. The resultant loose pellets are resuspended in wash medium and centrifuged again at $12{,}000g_{max}$ for 15–20 min, and the mitochondrial pellet is resuspended in resuspension medium at a concentration of 5–20 mg mitochondrial protein/ml.

Separation of mitochondria from green plant tissues is aided greatly by the inclusion of a 0–5% (w/v) PVP-40 preformed gradient in the Percoll solution (Day *et al.*, 1985). Such gradients can be formed using aliquots of the Percoll wash buffer solution with or without the addition of 5% (w/v) PVP-40 and a standard linear gradient former. This technique allows much greater separation of mitochondria and thylakoid membranes (Fig. 1Bii). Percoll step gradients can also be used; these aid the concentration of mitochondrial fractions on a gradient at an interface between Percoll concentrations. Step gradients can be formed easily by setting up a series of inverted 20-ml syringes (fitted with 19-gauge needles) strapped to a flat block of wood that has been clamped to a retort stand at an angle of 45° over a rack containing the centrifuge tubes. The needles are lowered to touch the bevels against the inside, lower edge of the tubes. Step gradient solutions are then added (from bottom to top) to the empty inverted syringe bodies and each allowed to drain through in turn before the addition of the next step solution. Alternatively, the gradients can be generated by underlaying the less dense solutions with the more dense solutions. Step gradients should, however, only be used once the density of mitochondria from a particular tissue and the density of contaminating components have been determined using continuous gradients. A range of density marker beads is available from Amersham-Pharmacia Biotech for standardizing centrifugation conditions and establishing new protocols.

C. Mitochondrial Minipreparations for Screening Transgenic Lines

Rapid, high-percentage-yield preparations of mitochondria from a large number of small samples are sometimes required. This is the case in the screening of transgenic plants with mitochondrial-targeted changes in protein expression or in experimentation on small amounts of transgenic material. In potato lines expressing antisense constructs to tricarboxylic acid cycle enzyme subunits, we have used a Percoll step gradient protocol that yields 25–35% of the mitochondria present in 3- to 5-g FW tuber samples. This procedure allows for the purification of eight samples at once, using a standard fixed angle rotor housing eight 50-ml tubes. Tuber samples are cooled to 4°C, diced finely (2- to 3-mm squares), and ground in a cooled mortar and pestle in 8–10 ml of standard grinding medium. After filtering through muslin, a $1000g_{max}$ centrifugation for 5 min removes starch and cell debris, and the 10- to 12-ml supernatant is then layered directly onto a Percoll step gradient made in a standard mannitol wash buffer (2 ml 40%, 15 ml 28%, 5 ml 20% Percoll in a 50-ml tube). After centrifugation at $40{,}000g_{max}$ for 45 min, mitochondria are removed from the 28/40% interface and washed twice in 50-ml tubes diluted with wash medium (Fig. 2A). Approximately 0.2–0.5 mg of mitochondrial protein is obtained per sample, and these preparations have specific activities, which are 90–95% of those from samples prepared by conventional methods for preparing potato tuber mitochondria. Similar protocols can be developed for screening transgenic tissue culture or callus samples.

D. Preparation of Cell Culture/Callus Mitochondria

Purification of mitochondria from plant cell culture and callus material can be notoriously difficult. This is due to two main factors: (1) the difficulties in breaking open cells and (2) the densities of mitochondria from nondifferentiated cells tend to be lower than

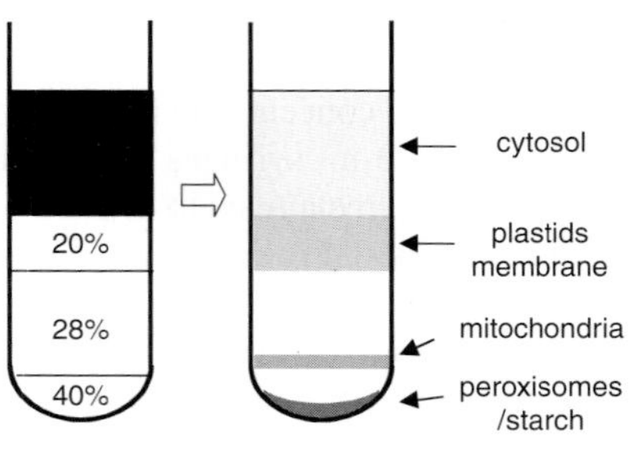

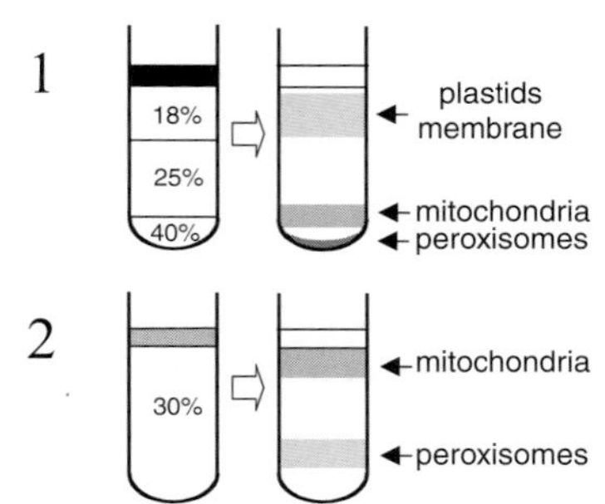

Fig. 2 Percoll density gradients for the purification of potato tuber mitochondria and *Arabidopsis* cell culture mitochondria. (A) Minipreparations of mitochondria from 5 to 10 g FW of potato tuber using 20/28/40% step gradients. (B) Two-stage gradient separation of *Arabidopsis* cell culture mitochondria using 18/25/40% Percoll step gradient followed by layering the mitochondrial fraction onto a 30% self-forming Percoll gradient to maximize removal of peroxisomal material.

those of mitochondria from whole plant tissues, and thus sediment to different regions in density gradients. Cells from cultures are not broken easily by homogenizers or blenders due to the strong cell wall structure. We have found a mortar and pestle with added sand or glass beads (0.4 mm diameter) is reasonably effective in breaking cells, whereas other groups have used enzymatic digestion of the cell wall to release protoplasts, from which mitochondria are then isolated (Vanemmerik *et al.,* 1992; Zhang *et al.,* 1999). The advantage of the mortar and pestle method is the speed of isolation; however, yields are not high (5–8 mg mitochondrial protein/100 g FW cells). The disadvantage of the enzymatic approach is the need to incubate cells for more than 1 hr in digestion enzymes, which may alter the mitochondrial functions of interest, but this approach does tend to give higher yields of purified mitochondria (15–25 mg mitochondrial protein/100 g FW cells).

We have used the following method to isolate plant mitochondria from an *Arabidopsis thaliana* cell culture. Approximately 100 g FW cells are harvested by collecting material from 10 flasks of 100 ml culture by filtration through a 20 × 20-cm square of muslin. In batches of 20 g FW, cells are ground in standard homogenization buffer in the presence of 1–2 g glass beads (0.4 mm diameter). Grinding is initiated upon addition of 10 ml of homogenization buffer and is continued with subsequent additions to a total of 50 ml. Once all the batches are ground to ensure breakage of the majority of cells, the homogenate is filtered and centrifuged in 50-ml tubes at $2500g_{max}$ for 5 min to remove cell debris. The resultant supernatant is then centrifuged at $15{,}000g_{max}$ for 15 min. Following resuspension of the pellets in standard mannitol wash buffer, the two centrifugation steps are repeated. The final pellet is resuspended in several milliliters of mannitol wash buffer and layered over a Percoll step gradient in a 50-ml tube comprising the following steps, bottom to top: 5 ml 40% Percoll, 15 ml 25% Percoll, and 15 ml 20% Percoll, all in mannitol wash buffer. After centrifugation at $40{,}000g_{max}$ for 45 min, mitochondria and peroxisomes are found at the 25/40% interface, whereas the plastid membrane remains in the 20% Percoll step (Fig. 2B). After washing in five times the volume of wash buffer, mitochondria and the peroxisome pellet are resuspended in mannitol wash buffer and layered over 30 ml of 28% Percoll in 0.3 *M* sucrose wash buffer. After centrifugation at $40{,}000g_{max}$ for 45 min, mitochondria band just below the start of the 28% Percoll gradient, whereas peroxisomes are found toward the bottom of this gradient (Fig. 2B). If sufficient material is present, the peroxisomal layer will have a translucent yellow color whereas mitochondria will be a white-buff, opaque colored band. The preparation of mitochondria from other cell cultures may require even lower Percoll concentrations on the first gradient. In the case of some plant cell cultures, gradients of 5 ml 25%, 15 ml 20%, and 15 ml 15% Percoll have been used with some success, with mitochondria banding at the 20–25% Percoll interface.

V. Mitochondrial Yield, Purity, Integrity, Storage, and Function

A. Yield Calculation

The yield of mitochondria from plant tissues is decreased successively through the purification process due to incomplete grinding of plant tissue, loss in discarded supernatant and pellet fractions, and removal of damaged mitochondria on a density basis in the final gradient steps. Typically, 45–75% of total mitochondria membrane marker enzyme activity in a tissue is found in the first high-speed pellet, 15–30% is loaded onto the density gradient after washing, and only 5–15% is found in the washed mitochondrial sample at the end of the purification procedure. In nongreen tissues, cytochrome *c* oxidase can be used as a marker enzyme, but in green tissues a variety of oxidases can prevent the use of this assay. Cytochrome *c* oxidase activity can be measured using an O_2 electrode, as 0.1 m*M* KCN sensitive O_2 consumption in the presence of 10 m*M* ascorbate, 50 μM cytochrome *c*, 0.05% (v/v) Triton X-100 in 10 m*M* TES–KOH (pH 7.2), or as 0.1 m*M* KCN sensitive reduced cytochrome *c* oxidation at 550 nm ($\varepsilon = 28{,}000\ M^{-1}$). In cases where this assay does not work, fumarase is the best candidate for a mitochondria-specific enzyme, but its low activity can hamper the use of this marker in some tissues. Fumarase can be assayed as fumarate formation at 240 nm ($\varepsilon = 4880\ M^{-1}$) by the addition of 50 m*M* malate to samples in 0.1 *M* K_2PO_4–KOH (pH 7.7). In potato tuber tissue we have successfully used assays for mitochondrial citrate synthase, NAD-dependent malic enzyme, fumarase, and cytochrome *c* oxidase to provide an estimate of the mitochondrial content, which averages at 0.26 ± 0.04 mg mitochondrial protein per gram FW potato tuber.

B. Determination of Purity

Marker enzymes for contaminants commonly found in plant mitochondrial samples can be used to assess the purity of preparations. Peroxisomes can be identified by catalase, hydroxy-pyruvate reductase, or glycolate oxidase activities. Chloroplasts can be identified by chlorophyll content, etioplasts by carotenoid content and/or alkaline pyrophosphatase activity, and glyoxysomes by isocitrate lyase activity. The endoplasmic reticulum can be identified by antimycin A-insensitive cytochrome *c* reductase activity and plasma membranes by K^+-ATPase activity. Cytosolic contamination is rare if density centrifugation is performed properly and care is taken in the removal of mitochondrial fractions, but such contamination can be measured easily as alcohol dehydrogenase or lactate dehydrogenase activities. References to and methods for the assay of these enzymes are summarized by Quail (1979) with several additions by Neuburger (1985).

C. Determination of Membrane Integrity

A variety of assays can be used to determine the integrity of the outer and inner membranes of plant mitochondrial samples, thus providing information on the structural damage caused by the isolation procedure. Two latency tests are commonly used to estimate the proportion of broken mitochondria in a sample; both rely on the impermeable nature of an intact outer mitochondrial membrane to added cytochrome *c*. The ratio of cytochrome *c* oxidase activity (measured as in Section V,A) before and after addition of a nonionic detergent [0.05% (v/v) Triton X-100] or before and after osmotic shock (induced by dilution of mitochondria in H_2O followed by the addition of 2X reaction buffer) gives an estimate of the proportion of ruptured mitochondria in a sample. Alternatively, succinate:cytochrome *c* oxidoreductase can be assayed as cytochrome *c* reduction at 550 nm ($\varepsilon = 21000\ M^{-1}$) in wash medium supplemented with 0.5 m*M* ATP, 50 μ*M* cytochrome *c*, 1 m*M* KCN, and 10 m*M* succinate. The ratio of the activity of a control sample over the activity of a sample diluted in H_2O, followed by addition of the 2X reaction buffer, gives an estimate of the proportion of ruptured mitochondria in the sample.

Inner membrane integrity can be assayed by the latency of matrix enzyme activities with or without detergent or osmotic shock. Transport of organic acids and cofactors across the inner mitochondrial membrane does hamper accurate measurements of matrix enzyme activity latency; however, malate dehydrogenase activity is so limited by transport in intact mitochondria that assaying the latency of its activity gives a good indication of inner membrane integrity (Douce, 1985). Finally, the ability of mitochondria samples to maintain a proton-motive force across the inner membrane for ATP synthesis can be assessed by estimating the ratio of respiratory rates in the presence and absence of added ADP and by measurement of the ADP consumed/O_2 consumed ratio. Both these latter assays are affected by the presence and operation of the nonproton pumping plant alternative oxidase (AOX), which should be inhibited by 0.1 m*M* *n*-propylgallate or 1 m*M* salicylhydroxamic acid during experiments to assess coupling of O_2 consumption and ATP synthesis.

D. Storage

Once isolated by density gradient purification, mitochondria from most plant tissues can be kept on ice for 5–6 h without significant losses of membrane integrity and respiratory function. Pure preparations of some storage tuber mitochondria can be kept at

4°C for several days without loss of function, although cofactors such as NAD, TPP, and CoA often need to be added back for full tricarboxylic acid cycle function. Longer term storage of functional mitochondria can be achieved by the addition of dimethyl sulfoxide to 5% (v/v) or ethylene glycol to 7.5% (v/v) followed by rapid freezing of small volumes of mitochondrial samples (<0.5 ml) in liquid N_2 (Schieber *et al.*, 1994; Zhang *et al.*, 1999). Frozen samples can then be stored at −80°C. In our hands, potato mitochondria samples prepared in this way can be stored for over a year and, following thawing of these samples on ice, no significant losses can be detected in respiratory rate, coupling to ATP production, or outer membrane integrity.

E. Assays of Mitochondrial Function

The Clark-type polargraphic O_2 electrode with platinum-silver electrodes has been the routine method of assaying the function of isolated plant mitochondria for many years and has formed the basis of much of the characterization of the novel mitochondrial functions found in plants. Commercial designs of such dissolved O_2 electrodes are available from Hansatech (Kings Lynn, Norfolk, UK) and Rank Brothers Ltd (Bottisham, Cambridgeshire, UK) and allow for rapid measurements of the O_2 consumption rate in 1- to 2-ml volumes containing 0.1–0.3 mg mitochondrial protein. Using a basic reaction medium [0.3 *M* sucrose, 10 m*M* NaCl, 5 m*M* KH_2PO_4, 3 m*M* $MgCl_2$, 0.1% (w/v) BSA, 10 m*M* TES–NaOH, pH 7.2] and by the addition of substrates, cofactors, and effectors of O_2 consumption rate, a large amount of information can be obtained.

In addition to the assays used for studying mitochondrial function from other sources, the following assays can be undertaken in plant mitochondria. NADH (1 m*M*) or NADPH (1 m*M*) can be supplied exogenously as respiratory substrates via the function of cytosolic-facing NADH and NADPH rotenone-insensitive nonproton pumping dehydrogenases that bypass the function of complex 1 and that directly reduce ubiquinone in the inner mitochondrial membrane. Pyruvate (10 m*M*) rarely provides a rapid respiratory rate due to the lack of an endogenous malate/OAA pool in isolated plant mitochondria, and thus a "sparker" addition of malate (50 μ*M*) is required. Malate (10 m*M*) itself is a good respiratory substrate due to the presence of an NAD-dependent malic enzyme in the plant mitochondrial matrix that catalyzes the decarboxylation of malate to pyruvate, thereby providing, along with malate dehydrogenase, the substrates for further TCA cycle operations via citrate synthase. Large amounts of glycine decarboxylase are found in mitochondria from photosynthetic sources, allowing glycine (10 m*M*) to support the highest rates of O_2 consumption in mitochondria from these tissues.

Rotenone does not fully inhibit O_2 consumption mediated via the generation of NADH in the matrix of mitochondria due to the additional presence of matrix-facing, rotenone-insensitive NADH dehydrogenases in plant mitochondria. Cytochrome pathway inhibitors (antimycin A, myxothiazol, KCN, CO) do not fully inhibit O_2 consumption due to the presence of an alternative oxidase (AOX) that oxidizes the reduced ubiquinone pool directly and reduces O_2 to H_2O. The AOX can be inhibited by SHAM (~250 μ*M*) or by gallates such as *n*-propyl gallate (~50 μ*M*) and octylgallate (~5 μ*M*). The AOX is also activated by short 2-oxo acids (such as pyruvate and glyoxylate) and by reduction

Table I
Methods Optimized for the Assessment of Plant Mitochondrial Enzyme Activities and for Partial or Complete Purification of Enzymes from Plant Mitochondria

Enzyme	Reference
Pyruvate dehydrogenase complex	Randall and Miernyk (1990), Millar *et al.* (1998)
Citrate synthase	Stevens *et al.* (1997), Stitt (1984)
NAD(P)-isocitrate dehydrogenase	Rasmusson and Moller (1990), McIntosh and Oliver (1992)
Aconitase	Verniquet *et al.* (1991)
2-Oxoglutarate dehydrogenase complex	Poulsen and Wedding (1970), Millar *et al.* (1999)
Succinyl-coA synthase	Palmer and Wedding (1966)
Fumarase	Behal *et al.* (1996), Behal and Oliver (1997)
Malate dehydrogenase	Hayes *et al.* (1991), Walk *et al.* (1977)
NAD-malic enzyme	Grover *et al.* (1981), Hatch *et al.* (1982), Grover and Wedding (1984)
Glycine decarboxylase	Oliver (1994), Neuburger *et al.* (1986), Walker and Oliver (1986)
Complex I	Leterme and Boutry (1993), Herz *et al.* (1994), Rasmusson *et al.* (1998)
Non-CI NADH dehydrogenases	Moller *et al.* (1993), Soole and Menz (1995)
Complex II	Hattori and Asahi (1982), Igamberdiev and Falaleeva (1994)
Complex III	Braun and Schmitz (1992), Emmermann *et al.* (1993), Braun and Schmitz (1995)
Complex IV	Maeshima and Asahi (1978), Devirville *et al.* (1994)
Alternative oxidase	Hoefnagel *et al.* (1995), Zhang *et al.* (1996), Vanlerberghe and McIntosh (1997)
Complex V	Glaser and Norling (1983), Hamasur and Glaser (1992), Hamasur *et al.* (1992)

of a disulfide bond, which can be catalyzed via reduction of the NADPH pool in the mitochondrial matrix (Vanlerberghe and McIntosh, 1997). Thus assessment of the activity of AOX can vary depending on the substrate used, and the relevant literature should be studied carefully if investigation of this activity is to be undertaken. Specific spectrophotometric assays optimized for enzymes of plant mitochondria have been published in a variety of reviews and research papers; a list in Table I provides the reader with a ready entry to this material.

VI. Subfractionation of Mitochondrial Compartments

A. Separation of Inner and Outer Mitochondrial Membranes, Matrix, and Intermembrane Space

Using a combination of osmotic shock and differential centrifugation, plant mitochondrial samples can be fractionated into four components comprising the two aqueous compartments, the matrix (MA) and the intermembrane space (IMS) and the two membrane compartments, the inner mitochondrial membrane (IM) and the outer mitochondrial membrane (OM). While separation of the IM and MA fractions from each other is relatively trivial, separation of the IMS and OM fractions from each other is a greater challenge and much is owed in the development of these techniques to the work of Mannella and Bonner (1985, 1975) on the plant mitochondrial outer membrane.

To perform a subfractionation, a purified mitochondrial sample that has not been frozen is separated into two aliquots and each added to 49 times the volume of the two separate solutions. One solution contains 10 m*M* sucrose and the other 86 m*M* sucrose, both in 10 m*M* MOPS, pH 7.2 (depending on the source of the mitochondria to subfractionated, these osmotic strengths can be altered to improve yields and purity of final fractions significantly). The two solutions are slowly stirred on ice for 15 min. The two "shocked" mitochondrial samples (10 and 86 m*M*) are then kept separate for the following procedures, except for the OM pellets, which are combined at the end. Using a solution of 2 *M* sucrose in 100 m*M* Tris–HCl, pH 7.4, each solution is raised to a final sucrose concentration of 0.3 *M* and stirring is continued for 15 min. Mitoplasts (IM + MA) are removed from both samples by centrifugation at 15,000g_{max} for 15 min.

The supernatant (OM + IMS) is then transferred to clean tubes and centrifuged at 200,000g_{max} for 90 min. The supernatants of this high-speed centrifugation contain IMS-10 and IMS-86 and can be frozen separately for later analysis. The pellets are OM, but due to the small amount of this membrane obtained, further purification is required to remove contaminating IM and mitoplasts. This is achieved by combining the OM-10 and OM-86 pellets, resuspended in a minimum volume of 100 m*M* Tris-HCl, pH 7.4, and layering the combined sample over a 4-ml 0.6/0.9 *M* step sucrose gradient. This gradient is centrifuged at 50,000g_{max} for 60 min. The OM is collected from the 0.6/0.9 interface and is only visible when large amounts of mitochondria (>100 mg protein) are being fractionated. Visible bands at other locations are probably contaminants and should not be collected.

The mitoplast pellet of the first centrifugation, which may be frozen before this step, is then diluted to 15 mg protein/ml and can then either be ruptured by three rounds of freeze thawing in liquid nitrogen or alternatively sonicated at full power for 3 × 5-s bursts with 20-s rest periods on ice. Centrifugation at 80,000g_{max} for 60 min pellets the IM-10 and IM-86 fractions. The supernatants are retained separately as MA-10 and MA-86 and the remaining fractions are frozen.

Depending on the tissue and the preparation of mitochondria, the 10 or 86 m*M* samples will give higher purity separations of some fractions. The 86 m*M* sample will yield a higher purity IMS fraction due to lower contamination of MA, whereas the 10 m*M* sample will yield a higher purity IM and MA fraction by ensuring a more complete rupture of the mitochondria sample yielding mitoplasts. Sample purity can be assayed by measuring cytochrome *c* oxidase for IM, fumarase for MA, and antimycin A-insensitive NADH : cytochrome *c* oxidoreductase for OM. In mammalian mitochondria, adenylate kinase activity is used as a marker for IMS, but in plants this enzyme activity is found most abundantly associated with the outer face of the IM. In Fig. 3, SDS–PAGE separations of the proteins of these four mitochondrial subcompartments are presented from subfractionation experiments performed on mitochondria isolated from a monocotyledonous plant (maize) and a dicotyledonous plant (potato). Typically, the enzymatic analysis of cross-contamination shows that less than 10% of the protein of any compartment is derived from contaminating proteins from the three other compartments.

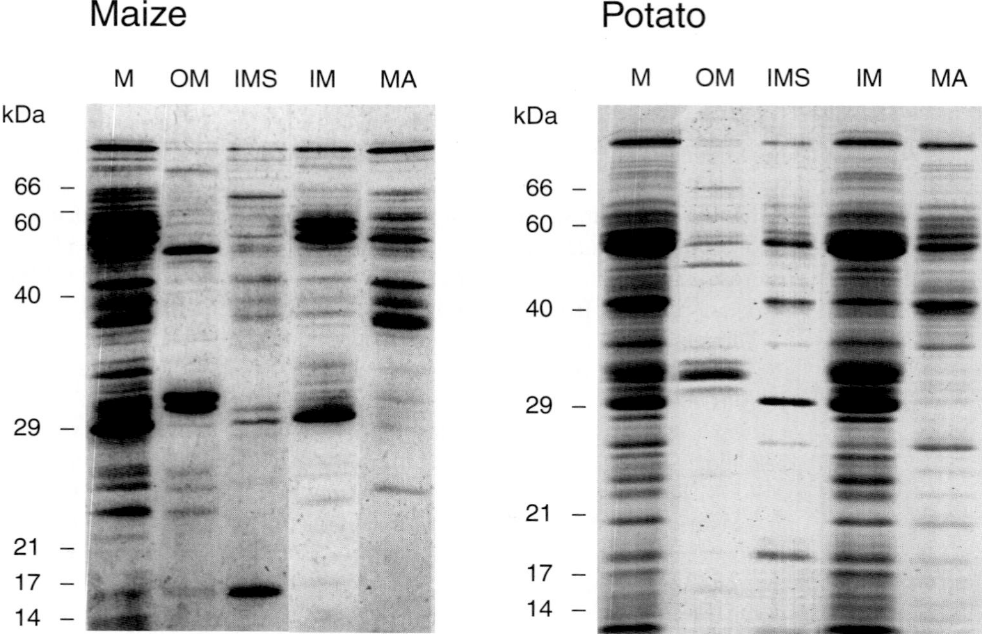

Fig. 3 SDS–PAGE separations of the four compartments of plant mitochondria isolated from maize coleoptiles and potato tubers visualized by Coomassie staining. M, intact mitochondria; OM, outer membrane; IMS, intermembrane space; IM, inner membrane; MA, matrix space.

B. Separation of Mitochondrial Inner Membrane Components by Anion Exchange Chromatography

Samples of IM can be resolved on two simple chromatography columns to provide relatively high purity fractions containing functional components of the electron transport chain and the mitochondrial superfamily of transporters. Samples of mitochondrial membrane (up to 25 mg of protein for a 1-ml Mono Q column) are dissolved in 1% (w/v) *n*-dodecyl-β-D-maltoside and 25 m*M* MOPS (pH 7.5). Approximately 1.5 ml of solution is required for each 10 mg of mitochondrial membrane. The dissolved membrane is then diluted to 0.1% (w/v) *n*-dodecyl-β-D-maltoside and 25 m*M* MOPS (pH 7.5), loaded on a 1-ml Mono Q anion-exchange column (Amersham-Pharmacia Biotech) preequilibrated with the same solution, and eluted using a 0–0.6 *M* NaCl gradient over 30 ml at 1 ml/min with 1-ml fractions collected. The inner mitochondrial transporter superfamily (IMTS) (which all have P_i values of greater than 9.0) are eluted in the void and appear as proteins of 30–35 kDa. During the gradient elution, ETC components appear in the following order: complex V, ATP synthetase (peak fractions 6–8); complex I, NADH-Q oxidoreductase (peak fractions 9–10); complex IV, cytochrome *c* oxidase (peak fractions 11–12, green colored from bound Cu), and

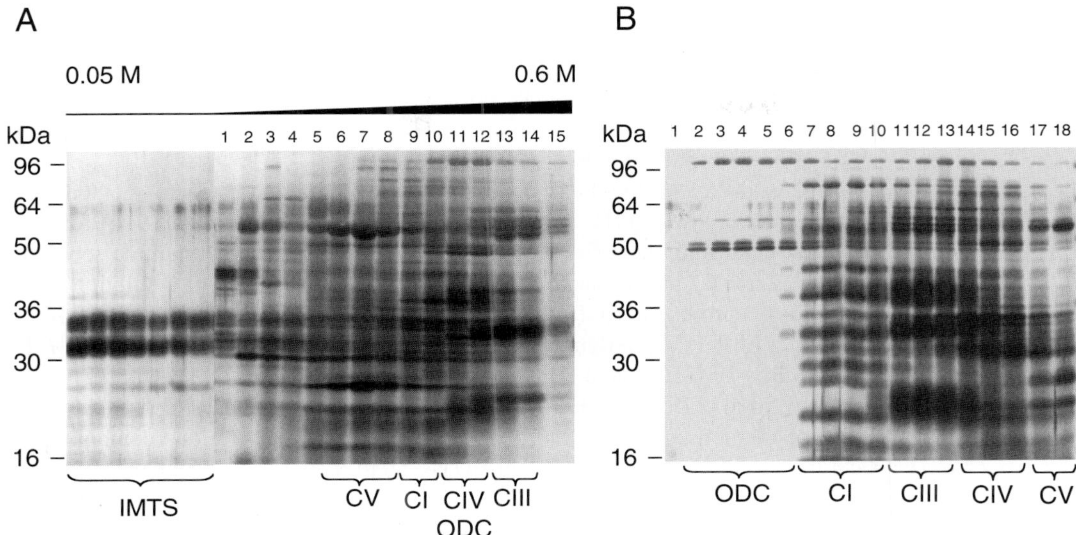

Fig. 4 Separation of potato inner mitochondrial membrane protein complexes using anion-exchange (Mono Q) and gel-filtration (Superose 6) chromatography. (A) Twenty-five micrograms of IM protein dissolved in *n*-dodecyl-β-maltoside was absorbed on a Mono Q column and protein complexes were eluted using a linear NaCl gradient (0.05–0.6 *M*). (B) Fractions 10–12 from the Mono Q run were separated by size on a Superose 6 column, revealing partial purifications of OGDC (2-oxoglutarate dehydrogenase complex), and the respiratory complexes I, III, IV, and V. Aliquots of protein from each fraction were separated by SDS–PAGE and visualized by silver staining.

complex III, Q-cytochrome *c* oxidoreductase (peak fractions 13–14, red colored from bound Fe). The OGDC (2-oxoglutarate dehydrogenase complex) which is bound to the IM, is eluted in fractions 10–12 and can be identified by its prominent 100-kDa subunit (Millar *et al.,* 1999). The ETC components can be further separated by size using a gel-filtration column. In this case, fractions 10–12 were concentrated to 250 μl and loaded on a Superose 6 gel-filtration column (Amersham Pharmacia Biotech) preequilibrated with 0.1% (w/v) *n*-dodecyl-β-D-maltoside, 100 m*M* NaCl, and 25 m*M* MOPS-NaOH (pH 7.5). Complexes are eluted in the same buffer by size: OGDC ~2500 kDa (peak fractions 2–6), complex I ~1000 kDa (peak fractions 7–10), complex III ~650 kDa (peak fractions 11–13), complex IV ~300 kDa (peak fractions 14–16), and some complex V fragments ~200–300 kDa (fractions 17 and 18). A typical chromatographic separation of potato inner mitochondrial membrane protein complexes is shown in Fig. 4.

C. Blue Native–PAGE of Intact Electron Transport Components

Blue-native (BN) PAGE has been used very successfully for the separation of mammalian and plant mitochondrial electron transport protein complexes. This electrophoresis technique utilizes Coomassie Serva Blue G-250 to bind and provide a net negative charge to protein complexes under nondenaturing conditions and can resolve protein complexes in the 200- to 1000-kDa range. A combination of first dimension BN-PAGE and second dimension SDS-PAGE and Western blotting also allows the immuno-detection of proteins in mitochondria and provides evidence of their native state incorporation or association with mitochondrial complexes.

We have used a protocol based largely on the modifications of the method of Schagger and von Jagow (1991), which were adopted by Jansch *et al.* (1996) for the separation of plant mitochondrial proteins. Gels consisted of an acrylamide separating gel (12% T, 2.6% C) and a stacking gel (4% T, 2.6% C) formed in a solution of 0.25 *M* ε-amino-*n*-caproic acid and 25 m*M* Bis–Tris–HCl (pH 7.0) using ammonium persulfate and TEMED for polymerization. The anode buffer consists of 50 m*M* Tricine, 15 m*M* Bis–Tris–HCl, and 0.02% (w/v) Serva Blue G250 (pH 7.0) and the cathode buffer of 50 m*M* Bis–Tris–HCl (pH 7.0). Sample preparation involves resuspension of 1 mg of membrane protein in 75 μl of an ACA buffer solution (containing 0.75 *M* ε-amino-*n*-caproic acid, 0.5 m*M* Na_2EDTA, 50 m*M* Bis–Tris–HCl, pH 7.0) followed by the addition of 15 μl of a freshly prepared solution of 10% (w/v) *n*-dodecylmaltoside. Following centrifugation for 10 min at 20,000g_{max}, supernatants are transferred to microtubes containing 15 μl of solution comprising 5% (w/v) Serva Blue G250 dissolved in ACA buffer solution. Unused protein tracks in gels should be filled with a solution containing ACA buffer, 10% (w/v) *n*-dodecylmaltoside, and 5% Serva Blue solutions in a ratio of 0.7 : 0.15 : 0.15, respectively.

Gels are run at 4°C in a precooled apparatus with all samples and buffers precooled to 4°C. Approximately 30 min before the run, 0.03% (w/v) *n*-dodecylmaltoside is added to the cathode buffer only. Electrophoresis is commenced at 100 V constant voltage. After 45 min the current is limited to 15 mA and the voltage limit is removed. Voltage will rise from 100 V to approximately 500 V during the 5 h of the electrophoresis run time. See Jansch *et al.* (1996) for detailed examples of the use of this technique on plant mitochondria.

VII. Proteome Analysis

Advances in two-dimensional gel electrophoresis utilizing fixed pH gradients for the first isoelectric focusing (IEF) dimension have enhanced the ease and reproducibility of such gels for proteome analysis greatly (Link, 1999). Using this approach, a plant mitochondrial sample can be separated to reveal 400–550 protein features with P_i values ranging between 3.5 and 10 and apparent molecular masses between 150 and 10 kDa. Transfer of proteins from the gel to nitrocellulose and probing with antibodies to mitochondrial proteins can be used to identify proteins on this two-dimensional map. N-terminal sequencing and MS analysis of protein features can also be used for identification.

Samples ranging from 150 μg to 1 mg of mitochondrial protein are acetone extracted for two-dimensional analysis. In a 1.5-ml tube, 400 μl of absolute acetone cooled to −80°C is added to a sample of mitochondria made up to a volume of 100 μl with dH_2O, stored at $-20^\circ C$ for 1 h and centrifuged at $20{,}000g_{max}$ for 15 min. The supernatant is removed and the sample is air dried by placing the open tube in a heating block set at 30°C for 10 min. A total of 200 μl of an IEF sample buffer [6 *M* urea, 2 *M* thiourea, 2% (w/v) CHAPS, 0.3% (w/v) dithiothreitol (DTT), and 0.5% (v/v) Amersham-Pharmacia Biochem IPG buffer, 0.005% w/v] is added and the pellet is resuspended by pipette, vortexing and heating to 30°C. After a further centrifugation to remove insoluble material, the 200-μl sample in IEF buffer is added to a reswell chamber, an 11-cm Immobiline strip is placed over the solution, and mineral oil is layered over the top to allow for reswelling overnight. Reswollen IEF gels are run using a Pharmacia Multiphor II flat-bed electrophoresis tank at 20°C for 17 h using a program of ramped voltages according to the manufacturer's instructions. IEF strips are then slotted into a central single well of an acrylamide (4% T, 2.6% C) stacking gel, formed using a single well comb, above a 0.1 × 16 × 16-cm (12% T, 2.6% C) 0.1% (w/v) SDS–polyacrylamide gel. Gel electrophoresis is performed at 30 mA and is completed in 6 h. Figure 5 shows a typical separation of *Arabidopsis* cell culture mitochondria using a commercial 3–10 pH nonlinear 180-mm IPG strip from Amersham-Pharmacia Biotech and a 12% (12% T,

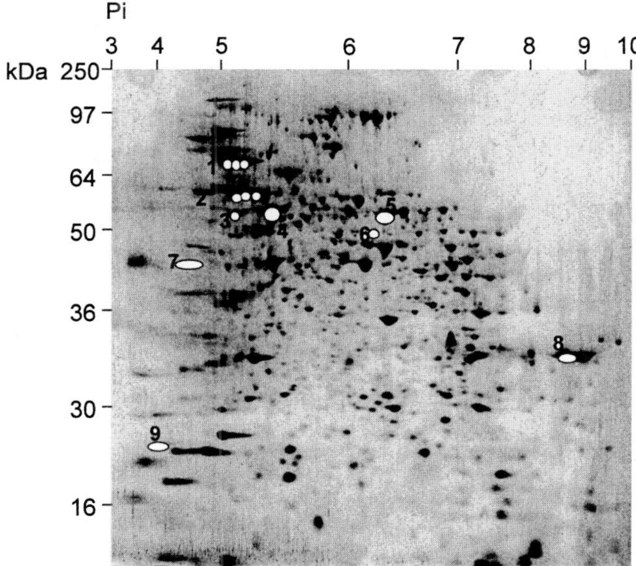

Fig. 5 Two-dimensional gel electrophoresis of *Arabidopsis* cell culture mitochondrial proteins. Isoelectric focusing (left to right) on a pH 3–10 nonlinear immobilized pH gradient strip was followed by SDS–PAGE (top to bottom) on a 12% polyacrylamide gel and silver staining. Approximate P_i and MW of proteins are shown on each axis. Numbered proteins have been identified by the use of antibodies on duplicate gels transferred to nitrocellulose. 1, HSP70s; 2, HSP60s; 3, dihydrolipoamide acetyltransferase (£2); 4, ATP synthetase F1 β subunit; 5, ATP synthetase F1 α subunit; 6, dihydrolipoamide dehydrogenase (E3); 7, pyruvate dehydrogenase (E1α); 8, adenine nucleotide translocator; 9, acyl carrier protein.

2.6% C) SDS–polyacrylamide gel. Following transfer onto nitrocellulose, a variety of protein spots were identified using antibodies to specific enzymes. We are currently also undertaking analysis of a variety of protein spots from these gels by MALDI-TOF and MS *de novo* sequencing analysis.

VIII. Analysis of *in Organello* Translation Products

The mitochondrial genome encodes a small percentage of the proteins found in mitochondria, and the synthesis of these proteins in isolated plant mitochondrial can be analyzed directly by *in organello* translation assays coupled to two-dimensional profiling of the translated products. A method for analysis of mitochondrial translation products from yeast is described in Chapter 25. Mitochondria must be essentially free of contaminating bacteria for successful *in organello* translation assays. To aid this requirement, sucrose gradients are often used for the purification of mitochondria. However, Percoll can be used in purification from sterile plant material (e.g., tissue or cell culture) or easily sterilized plant material. All the buffers used in the preparation of mitochondria can be sterilized by autoclaving or passage through 0.45-μm filters. Percoll must be sterilized on its own and not in the presence of salts and sugars. Following the completion of isolation, mitochondria are suspended at 2 mg ml^{-1} in a solution containing 0.4 *M* mannitol, 10 m*M* Tricine, and 1 m*M* EGTA, pH 7.5 (KOH).

A set of three incubations is performed for each mitochondrial sample in order to assess bacterial contamination and to maximize [^{35}S] methionine incorporation using extramitochondrial or intramitochondrial energy generation systems. Sodium acetate is used in the first incubation as an energy source that is utilized by bacteria but not by mitochondria; ATP is used in the second to provide energy directly to mitochondria and bypass oxidative phosphorylation; and succinate with ADP is used in the third to generate ATP by oxidative phosphorylation.

A reaction mix containing GTP (0.83 m*M*), [^{35}S] methionine (0.2 mCi ml^{-1}), salts [0.3 *M* mannitol, 150 m*M* KCl, 16.8 m*M* Tricine, 20 m*M* KPO_4 (pH 7.2), 16.6 m*M* $MgCl_2$, 1.32 m*M* EGTA] and amino acids (41.67 μ*M* each of aspartic acid, glutamine, lysine, serine, valine, asparagine hydrate, glutamic acid, leucine, proline, tyrosine, arginine·HCl, glycine, isoleucine, phenylalanine, tryptophan, alanine, cysteine, histidine·HCl, and threonine) is prepared. The salts and amino acids can be prepared as 1.25 and 12X stocks, respectively, and stored at −80°C in aliquots. Energy mixes are also prepared in sterilized H_2O containing the following: (1) 100 m*M* sodium acetate, (2) 65 m*M* ATP, and (3) 50 m*M* succinate and 5.4 m*M* ADP. A single incubation will utilize a 150-μl aliquot of the reaction mix, 50 μl of an energy mix, and 50 μl of a 2-mg ml^{-1} mitochondrial suspension.

Aliquots of reaction mix and energy mix are combined for each incubation, maintained at 20°C, and the reaction is started by the addition of the mitochondrial suspension. Following incubation for 60–90 min, reactions are stopped by the addition of 4 volumes (1 ml) of an ice-cold stop solution (0.4 *M* mannitol, 10 m*M* Tricine, 1 m*M* EGTA, and 12 mM L-methionine, pH 7.2). To estimate trichloroacetic acid (TCA) precipitable counts, duplicate 5-μl aliquots are removed from each reaction prior to dilution with the stop solution and spotted onto 2-cm Whatman 3MM chromatography paper disks. The disks are allowed to air dry for a few minutes and are then dropped into ice-cold 10% (w/v) TCA and left on ice for 15 min or overnight. The disks are then transferred to 5% (w/v) TCA preheated to boiling and incubated for 10 min. The disks are then washed four times in 5% (w/v) TCA at room temperature for 5 min per wash, followed by a 15-min wash in absolute ethanol and a final 5-min wash in acetone. The disks are allowed to dry completely prior to scintillation counting, using 4 ml of scintillant (e.g., HiSafe3, Fisher) per disk.

For analysis of protein translation products, mitochondria are pelleted from the reaction solutions by centrifugation at 20,000g_{max} for 5 min. These pelleted samples may be frozen for later analysis or solubilized in SDS–PAGE sample loading buffer or IEF sample buffer. Following separation by one- or two-dimensional electrophoresis, gels are dried for subsequent autoradiography to visualize [^{35}S]methionine incorporation into polypeptides. Note that DTT or mercaptoethanol should only be added after heating in sample loading buffer, to avoid aggregation of subunit 1 of cytochrome *c* oxidase. In Fig. 6, a sample from an *in organello* translation by maize mitochondria was separated by one- and two-dimensional electrophoresis and total proteins and [^{35}S]methionine-labeled proteins are compared. Translated products from the ATP synthase F1 α subunit gene can be seen prominently at 55 kDa.

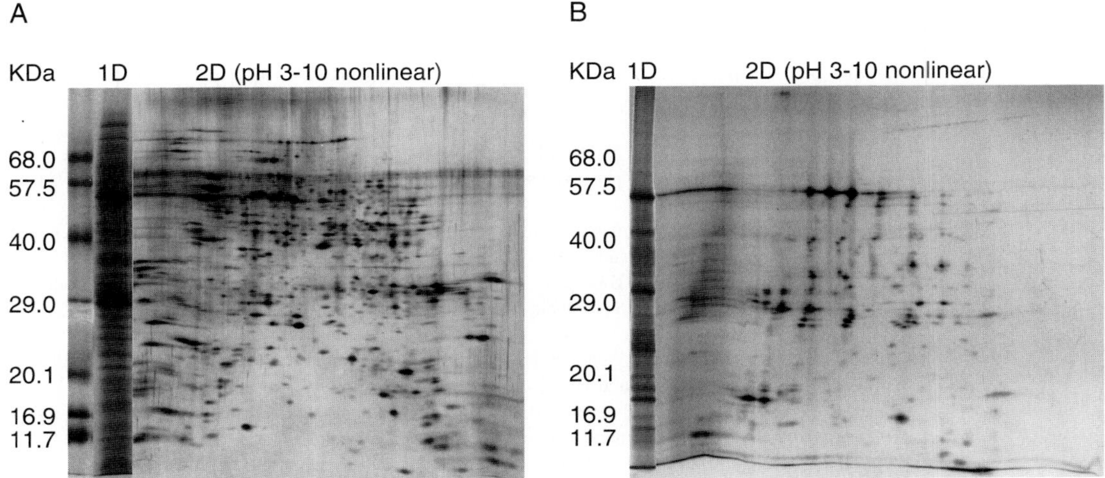

Fig. 6 One- and two-dimension electrophoresis of a protein sample from an *in organello* translation assay performed on maize coleoptile mitochondria. (A) Silver-stained gel of total maize mitochondrial proteome and (B) [^{35}S]methionine-labeled translation products.

IX. Conclusion

These techniques outline a basis for studying plant mitochondria at the level of overall metabolic function, at the level of individual enzyme and protein complex composition and operation, and at the level of each polypeptide chain in the organelle proteome. Integration of these approaches will be essential tools in future identification and characterization of the mitochondrially targeted products of nuclear-encoded genes sequenced in plant genome projects. The recent discoveries of cytochrome *c* release from mitochondria in the early stages of programmed cell death in plants (Balk *et al.,* 1999) and the identification of a mitochondrial ribonuclease specific for polyadenylated RNAs associated with cytoplasmic male sterility in plants (Gagliardi and Leaver, 1999) are examples of the way in which plant research will continue to turn to the study of mitochondrial function and the mitochondrial proteome for answers.

References

Balk, J., Leaver, C. J., and McCabe, P. F. (1999). Translocation of cytochrome *c* from the mitochondria to the cytosol occurs during heat-induced programmed cell death in cucumber plants. *FEBS Lett.* **463,** 151–154.

Behal, R. H., and Oliver, D. J. (1997). Biochemical and molecular characterization of fumarase from plants: Purification and characterization of the enzyme—cloning, sequencing, and expression of the gene. *Arch. Biochem. Biophys.* **348,** 65–74.

Braun, H. P., and Schmitz, U. K. (1992). Affinity purification of cytochrome-*c* reductase from potato mitochondria. *Eur. J. Biochem* **208,** 761–767.

Braun, and Schmitz, U. K. (1995). Cytochrome-*c* reductase/processing peptidase complex from potato mitochondria. *Methods Enzymol.* **260,** 70–82.

Day, D. A., Neuburger, M., and Douce, R. (1985). Biochemical characterization of chlorophyll-free mitochondria from pea leaves. *Aust. J. Plant Physiol.* **12,** 219–228.

Devirville, J. D., Moreau, F., and Denis, M. (1994). Proton to electron stoichiometry of the cytochrome-*c*-oxidase proton pump in plant mitoplasts. *Plant Physiol. Biochem* **32,** 85–92.

Douce, R. (1985). "Mitochondria in Higher Plants: Structure, Function and Biogenesis." Academic Press, London.

Emmermann, M., Braun, H. P., Arretz, M., and Schmitz, U. K. (1993). Characterization of the bifunctional cytochrome-c reductase processing-peptidase complex from potato mitochondria. *J. Biol. Chem.* **268,** 18936–18942.

Gagliardi, D., and Leaver, C. J. (1999). Polyadenylation accelerates the degradation of the mitochondrial mRNA associated with cytoplasmic male sterility in sunflower. *EMBO J.* **18,** 3757–3766.

Glaser, E., and Norling, B. (1983). Kinetics of interaction between the H^+-translocating component of the mitochondrial ATPase complex and oligomycin or dicyclohexylcarbodiimide. *Biochem. Biophys. Res. Commun.* **111,** 333–339.

Grover, S. D., Canellas, P. F., and Wedding, R. T. (1981). Purification of NAD-malic enzyme from potato and investigation of some physical and kinetic properties. *Arch. Biochem. Biophys.* **209,** 396–407.

Grover, S. D., and Wedding, R. T. (1984). Modulation of the activity of NAD-malic enzyme from *Solanum tuberosum* by changes in oligomeric state. *Arch. Biochem. Biophys.* **234,** 418–425.

Hamasur, B., and Glaser, E. (1992). Plant mitochondrial F_0F_1 ATP synthase: Identification of the individual subunits and properties of the purified spinach leaf mitochondrial ATP synthase. *Eur. J. Biochem* **205,** 409–416.

Hamasur, B., Guerrieri, F., Zanotti, F., and Glaser, E. (1992). F_0F_1-ATP synthase of potato tuber mitochondria: Structural and functional characterization by resolution and reconstitution studies. *Biochim. Biophys. Acta* **1101,** 339–344.

Hatch, M. D., Tsuzuki, M., and Edwards, G. E. (1982). Determination of NAD-malic enzyme in leaves of C_4 plants: Effects of malate dehydrogenase and other factors. *Plant Physiol.* **69,** 483–491.

Hattori, T., and Asahi, T. (1982). The presence of two forms of succinate dehydrogenase in sweet potato root mitochondria. *Plant Cell Physiol.* **23,** 515–523.

Hayes, M. K., Luethy, M. H., and Elthon, T. E. (1991). Mitochondrial malate dehydrogenase from corn: Purification of multiple forms. *Plant Physiol.* **97,** 1381–1387.

Herz, U., Schroder, W., Liddell, A., Leaver, C. J., Brennicke, A., and Grohmann, L. (1994). Purification of the NADH ubiquinone oxidoreductase (complex-I) of the respiratory chain from the inner mitochondrial membrane of *Solanum tuberosum*. *J. Biol. Chem.* **269,** 2263–2269.

Hoefnagel, M. H. N., Wiskich, J. T., Madgwick, S. A., Patterson, Z., Oettmeier, W., and Rich, P. R. (1995). New inhibitors of the ubiquinol oxidase of higher plant mitochondria. *Eur. J. Biochem.* **233,** 531–537.

Igamberdiev, A. U., and Falaleeva, M. I. (1994). Isolation and characterization of the succinate-dehydrogenase complex from plant mitochondria. *Biochem, (Moscow)* **59,** 895–900.

Jansch, L., Kruft, V., Schmitz, U. K., and Braun, H. P. (1996). New insights into the composition, molecular-mass and stoichiometry of the protein complexes of plant mitochondria. *Plant J.* **9,** 357–368.

Leaver, C. J., Hack, E., and Forde, B. G. (1983). Protein synthesis by isolated plant mitochondria. *Methods Enzymol.* **97,** 476–484.

Leterme, S., and Boutry, M. (1993). Purification and preliminary characterization of mitochondrial complex I (NADH-ubiquinone reductase) from broad bean (*Vicia faba* L). *Plant Physiol.* **102,** 435–443.

Link, A. J. (1999). "2D Proteome Analysis Protocols." Humana Press, Totowa, New Jersey.

Maeshima, M., and Asahi, T. (1978). Purification and characterisation of sweet potato cytochrome c oxidase. *Arch. Biochem. Biophys.* **187,** 423–430.

Mannella, C. A. (1985). The outer membrane of plant mitochondria. *In* "Higher Plant Cell Respiration" (R. Douce and D. A. Day, eds.), Vol. 18, pp. 106–133. Springer-Verlag, Berlin.

Mannella, C. A., and Bonner, W. D. (1975). Biochemical characteristics of the outer membranes of plant mitochondria. *Biochim. Biophys. Acta* **413,** 213–225.

McIntosh, C. A., and Oliver, D. J. (1992). NAD-linked isocitrate dehydrogenase: Isolation, purification, and characterization of the protein from pea mitochondria. *Plant Physiol.* **100,** 69–75.

Millar, A. H., Hill, S. A., and Leaver, C. J. (1999). The plant mitochondrial 2-oxoglutarate dehydrogenase complex: Purification and characterisation in potato. *Biochem. J.* **343,** 327–334.

Millar, A. H., Knorpp, C., Leaver, C. J., and Hill, S. A. (1998). Plant mitochondrial pyruvate dehydrogenase complex: Purification and identification of catalytic components in potato. *Biochem. J.* **334,** 571–576.

Moller, I. M., Rasmusson, A. G., and Fredlund, K. M. (1993). NAD(P)H-ubiquinone oxidoreductases in plant mitochondria. *J. Bioenerg. Biomembr.* **25,** 377–384.

Neuburger, M. (1985). Preparation of plant mitochondria, criteria for assessment of mitochondrial integrity, and purity, survival in vitro. *In* "Higher Plant Cell Respiration" (R. Douce and D. A. Day, eds.), Vol. 18, pp. 7–24. Springer-Verlag, Berlin.

Neuburger, M., Bourguignon, J., and Douce, R. (1986). Isolation of a large complex from the matrix of pea leaf mitochondria involved in the rapid transformation of glycine into serine. *FEBS Lett.* **207,** 18–22.

Neuburger, M., Journet, E., Bligny, R., Carde, J., and Douce, R. (1982). Purification of plant mitochondria by isopycnic centrifugation in density gradients of Percoll. *Arch. Biochem. Biophys.* **217,** 312–323.

Oliver, D. J. (1994). The glycine decarboxylase complex from plant mitochondria. *Annu. Rev. Plant Physiol. Plant Mol. Biol.* **45,** 323–337.

Palmer, J. M., and Wedding, R. T. (1966). Purification and properties of succinyl-CoA synthetase from Jerusalem artichoke mitochondria. *Biochim. Biophys. Acta* **113,** 167–174.

Poulsen, L. L., and Wedding, R. T. (1970). Purification and properties of the α-ketoglutarate dehydrogenase complex of cauliflower mitochondria. *J. Biol. Chem.* **245,** 5709–5717.

Quail, P. H. (1979). Plant cell fractionation. *Annu. Rev. Plant Physiol.* **30,** 425–484.

Randall, D. D., and Miernyk, J. A. (1990). The mitochondrial pyruvate dehydrogenase comples. *In* "Methods in Plant Biochemistry" (P. J. Lea, ed.), Vol. 3, pp. 175–192. Academic Press, New York.

Rasmusson, A. G., Heiser, V., Zabaleta, E., Brennicke, A., and Grohmann, L. (1998). Physiological, biochemical and molecular aspects of mitochondrial complex I in plants. *Biochim. Biophys. Acta* **1364,** 101–111.

Rasmusson, A. G., and Moller, I. M. (1990). NADP-utilizing enzymes in the matrix of plant-mitochondria. *Plant Physiol.* **94,** 1012–1018.

Schieber, O., Dietrich, A., and Marechal-Drouard, L. (1994). Cryopreservation of plant mitochondria as a tool for protein import or in organello protein synthesis studies. *Plant Physiol.* **106,** 159–164.

Soole, K. L., and Menz, R. I. (1995). Functional molecular aspects of the NADH dehydrogenases of plant mitochondria. *J. Bioenerg. Biomembr.* **27,** 397–406.

Stevens, F. J., Li, A. D., Lateef, S. S., and Anderson, L. E. (1997). Identification of potential inter-domain disulfides in three higher plant mitochondrial citrate synthases: Paradoxical differences in redox-sensitivity as compared with the animal enzyme. *Photosynth. Res.* **54,** 185–197.

Stitt, M. (1984). Citrate synthase. *Methods Enzym. Anal.* **4,** 353–358.

van Emmerik, W. A. M., Wagner, A. M., and van der Plas, L. H. W. (1992). A quantitative comparison of respiration in cells and isolated mitochondria from *Petunia hybrida* suspension cultures: A high yield isolation procedure. *J. Plant Physiol.* **139,** 390–396.

Vanlerberghe, G. C., and McIntosh, L. (1997). Alternative oxidase: From gene to function. *Annu. Rev. Plant Physiol. Plant Mol. Biol.* **48,** 703–734.

Verniquet, F., Gallard, J., Neuburger, M., and Douce, R. (1991). Rapid inactivation of plant aconitase by hydrogen peroxide. *Biochem. J.* **276,** 643–648.

Walk, R. A., Michaeli, S., and Hock, B. (1977). Glyoxysomal and mitochondrial malate dehydrogease of watermelon (*Citrullus vulgaris*) cotyldons. I. Molecular properties of the purified isoenzymes. *Planta* **136,** 211–220.

Walker, J. L., and Oliver, D. J. (1986). Glycine decarboxylase multienzyme complex: Purification and partial characterization from pea leaf mitochondria. *J. Biol. Chem.* **261,** 2214–2221.

Zhang, Q., Wiskich, J. T., and Soole, K. L. (1999). Respiratory activities in chloramphenicol-treated tobacco cells. *Physiol. Plant.* **105,** 224–232.

Zhang, Q. S., Hoefnagel, M. H. N., and Wiskich, J. T. (1996). Alternative oxidase from *Arum* and soybean: Its stabilization during purification. *Physiol. Plant.* **96,** 551–558.

CHAPTER 4

Assessing Functional Integrity of Mitochondria *in Vitro* and *in Vivo*

Mauro Degli Esposti

Cancer Research Campaign Molecular Pharmacology Group
School of Biological Sciences
University of Manchester
Manchester, M13 9PT, United Kingdom

I. Introduction

This chapter presents a compendium of assay procedures and guidelines for studying mitochondrial function either in isolated mitochondrial preparations (*in vitro*) or within live cells. Especially considered are techniques that can be applied successfully both to isolated mitochondrial preparations and to mitochondria within living cells. Whenever possible, the author has provided a critical appraisal of assay techniques that will hopefully guide scientists, including those who are interested in the involvement of mitochondria in apoptosis and disease.

II. *In Vitro* Assays with Mitochondrial Preparations

A. Redox Assay of Respiratory Enzymes

The electron transport activity of mitochondrial respiratory complexes can be measured by either polarographic, e.g., oxygen electrode, or spectrophotometric techniques. Spectrophotometric techniques are used for performing *in vitro* redox assays to functionally dissect a single respiratory enzyme within the biological preparation. These assays require simple instrumentation, namely spectrophotometers, fluorimeters, or plate readers, and provide straightforward, quantitative estimates of specific enzyme activities that are not obtained readily with other techniques. However, the choice of the assay for a given respiratory enzyme may be critical for obtaining meaningful data. The author discusses the principal problems of each technique and provides guidelines that may be particularly valuable for scientists interested in mitochondrial cell physiology but who have limited experience in the redox assays of mitochondrial function.

1. Complex I Assays

The assay of mitochondrial complex I is complicated by several factors that normally do not affect the activity of other respiratory complexes. First of all, NADH is impermeable to the inner mitochondrial membrane unless its integrity is disrupted. This is why the classical respiratory assay of complex I in intact (coupled) mitochondria uses substrates such as glutamate or malate that are carried across the inner membrane and indirectly produce intramitochondrial NADH via matrix enzymes. The disadvantage inherent to this classical assay is that the activity of complex I limits only partially the respiratory rate of glutamate, pyruvate, or malate, a fact that is often overlooked in studies on complex I deficiency and cellular respiration (Davey *et al.,* 1998; Barrientos and Moraes, 1999). Thus, it is preferable to study complex I activity with uncoupled preparations in which the permeability barrier for NADH is disrupted by freezing and thawing, by hypoosmotic shock, or by sonication; detergents should be avoided as they inhibit complex I activity.

Once the preparations are made permeable to NADH, the next problem is to choose an appropriate assay. NADH-ferricyanide reductase only monitors the electron transport of the low potential cofactors of complex I, but remains useful for providing a reasonable estimate of the content of active complex I (cf. Degli Esposti *et al.,* 1996). The NADH-ubiquinone (Q) reductase activity is the recommended assay for complex I; its ideal substrates are decyl-Q (DB, available commercially) or undecyl-Q (UBQ) (Degli Esposti *et al.,* 1994, 1996). Dichlorophenolindophenol (DCIP) and naphtoquinones such as menadione have also been used as artificial electron acceptors for complex I activity (Degli Esposti, 1998), but their physiological significance is limited. Often the NADH-Q reductase activity is measured with the hydrophilic analog Q-1, which produces fast rates of NADH oxidation, but it reacts incompletely with complex I (Degli Esposti *et al.,* 1996; Helfenbaum *et al.,* 1997). Moreover, the rates of NADH-Q reductase are less sensitive to rotenone with Q-1 than with DB or UBQ, especially in mitochondrial preparations from human tissues. Rotenone is the classical potent inhibitor of complex I and is used routinely to distinguish the specific activity of complex I from other NADH oxidation reactions that may be present in tissue fractions. Except for pure heart mitochondria, the rotenone-insensitive oxidation of NADH measured in most mitochondrial preparations derives in part from the incomplete inhibition of complex I by rotenone and in part from microsomal cytochrome P450 and cytochrome b_5 (lipid desaturase). The interference by microsomal redox reactions can be particularly serious in fractions from liver and cell cultures (cf. Rustin *et al.,* 1993).

Submitochondrial particles (SMP or ETP_H; see Chapter 1) are better suited than mitochondria for the study of complex I activity (Degli Esposti, 1998). Beef heart SMP have the highest content of complex I in mammalian tissues, usually around 0.04 nmol/mg of protein. This content approximates to one-fifth of that of complex III and one-tenth of that of cytochrome oxidase in heart mitochondria. However, the content of complex I varies greatly in other tissues. For instance, mitochondria from brain, liver, platelets, and lymphoid cells have a content of complex I that is well below 0.01 nmol/mg of protein, which explains the low activity values that are usually measured in these tissues.

a. NADH-Q Reductase

The buffer is 50 m*M* potassium phosphate, 1 m*M* EDTA, and 2 m*M* KCN, usually at pH 7.6 (but rates are essentially similar between pH 7 and 8). The permeabilized mitochondrial preparation, e.g., freeze-thawed EPT_H particles, is diluted to 4–10 mg/ml with a buffer containing 0.25 *M* sucrose and 10 m*M* Tris–Cl, pH 7.8, and is treated with 1 nmol/mg of antimycin. To reduce the antimycin leak, it is desirable to treat the mitochondria also with a center o inhibitor of complex III that does not inhibit complex I, namely MOA-stilbene or a strobilurin fungicide other than myxothiazol (Degli Esposti *et al.,* 1994). However, in most instances the contribution of the residual complex III activity at center o is no more than 10% of the overall Q reductase rate, and thus it may be ignored. An aliquot of the mitochondrial preparations should be further treated with 2 nmol/mg of rotenone, incubated for 10 min at room temperature and then for 2 h in ice before measuring the rotenone-insensitive rate. The reaction mixture is prepared in disposable 4-ml plastic cuvettes with 1.95 ml of buffer, 0.02–0.1 mg/ml of mitochondrial protein, and 100 μ*M* NADH (10 μl of a 20 m*M* aqueous solution). The mixture is allowed to equilibrate at 30°C on a spectrophotometer and NADH oxidation is measured at either 340 nm ($\varepsilon = 6.2\ \text{m}M^{-1}\text{cm}^{-1}$) or 350 nm ($\varepsilon = 5.5\ \text{m}M^{-1}\text{cm}^{-1}$). The reaction is started with the addition of the quinone substrate (dissolved in ethanol, usually at final concentrations around 20 μ*M*). Note that the addition of DB always produces a small increase in turbidity due to the poor water solubility of the quinone. The assay can also be performed in 96-well microplates using plate readers and 0.15–0.2 ml of the complete reaction mixture per well.

b. NADH-Ferricyanide Reductase Assay

This assay is more useful for determining the content of functional complex I in a preparation than its physiological activity. Indeed, ferricyanide reductase is completely insensitive to rotenone and other potent inhibitors. The buffer for this assay is identical to that used for the NADH-Q reductase assay, and the permeabilized mitochondrial preparation is diluted to approximately 4 mg/ml with 0.25 *M* sucrose, 10 m*M* Tris–Cl, pH 7.8, and treated with 1 nmol/mg antimycin and 2 nmol/mg rotenone for at least 1 h on ice (Degli Esposti *et al.,* 1996). The reaction mixture consists of 1.95 ml of buffer, 1 m*M* K-ferricyanide (10 μl of 0.2 *M* stock solution), and 0.01–0.02 mg/ml of mitochondrial protein. Once equilibrated at 30°C in a spectrophotometer set at 420 nm ($\varepsilon = 1\ mM^{-1}cm^{-1}$), the reaction is started by the addition of 100 μ*M* NADH with a quick mix to enable detection of the rapid initial rate of absorbance decay.

c. NADH-Cytochrome c Reductase

This assay measures the integrated activity of complex I and complex III and usually offers the advantage of a superior sensitivity than the NADH-Q reductase assay. However, the cytochrome *c* reductase activity is much more affected by interference from microsomal reactions than is the Q reductase activity, to the extent that it is not recommended for crude mitochondrial preparations from cell cultures or from tissues such as liver and white blood cells. This assay is sensitive to both rotenone and complex III inhibitors, such as antimycin, but no inhibitor completely blocks the reduction of cytochrome *c,* which is partially driven by superoxide and other reactive oxygen species (ROS) (Degli Esposti *et al.,* 1996). The assay protocol is very similar to that of the NADH-Q reductase, except that mitochondrial preparations are treated with complex III inhibitors only to determine the nonenzymatic reduction of cytochrome *c*. The buffer is 50 m*M* K-phosphate, 1 m*M* EDTA, and 2 m*M* KCN (freshly prepared the same day), pH 7.4–7.6. The reaction mixture consists of 1.95 ml of buffer, 20–30 μ*M* cytochrome *c* (e.g., 20 μl of a 1.5 m*M* stock solution), and 0.01–0.02 mg/ml of mitochondrial protein. Once equilibrated at 30°C in a spectrophotometer set at 550 nm ($\varepsilon = 18.7\ mM^{-1}cm^{-1}$), the reaction is started by the addition of 100 μ*M* NADH with a rapid mix to enable detection of the initial rate.

2. Complex II Assays

Complex II, the most concentrated flavin dehydrogenase in mitochondria, has the slowest turnover among the respiratory complexes. Moreover, the activity of complex II is frequently inhibited by endogenous oxalacetate; this inhibition can be removed effectively by preincubation of the concentrated mitochondrial preparation with 1 m*M* malonate at room temperature. As in the case for complex I, complex II activity can be measured with different methods. The commonly used assay of succinate-cytochrome *c* reductase suffers from the limited turnover of complex II and the reduction of cytochrome by nonmitochondrial redox reactions. The recommended assay of complex II is the reduction of DCIP mediated by quinones. The high extinction coefficient of DCIP in the visible region also enables measurements in rather turbid samples and in very crude cell fractions. Another, and unique, advantage of this assay is that its reagents have no problems of membrane permeability. The buffer required for the DCIP reductase assay is identical to that used for the NADH-Q reductase assay, namely 50 m*M* K-phosphate, 1 m*M* EDTA, and 2 m*M* KCN, pH 7.6. Before the assays, the mitochondrial preparation is dissolved to 10–20 mg/ml with 250 m*M* sucrose, 10 m*M* Tris–Cl, pH 7.8, treated with

1 mM malonate and 1 nmol/mg of antimycin for 30 min at room temperature, and then diluted to 4–5 mg/ml with the same buffer and kept on ice. The reaction mixture, placed in 4-ml plastic cuvettes, consists of 1.9 ml of buffer, 0.02–0.05 mg/ml of the malonate-treated preparation, 20 mM K-succinate (40 μl of a 1 M stock solution, pH 7.6), and 40 μM DCIP (from a 10 mM aqueous solution). The reaction is monitored at 600 nm ($\varepsilon = 21$ mM^{-1}cm^{-1}) at 30°C for a few minutes. The slow absorbance drift, mainly due to DCIP reduction via endogenous Q, is stimulated by the addition of 40 μM Q-1 or 20 μM DB. Contrary to complexes I and III, complex II has a similar affinity for most quinones, including idebenone (Degli Esposti *et al.,* 1996). Usually, the time course shows an induction phase before following a linear rate of absorbance decrease. To verify the specificity of the DCIP reductase to complex II, the reaction can be inhibited by 0.5 mM TTFA.

3. Complex III Assay

The activity of complex III is usually the fastest among those of mitochondrial respiratory enzymes, in isolated mitochondria as well as in cell fractions. There is only one assay for measuring the activity of complex III: the ubiquinol-cytochrome *c* reductase assay. The major problem of this assay is that the (ubi)quinol substrate is not available commercially and needs to be prapared chemically beforehand. The best method is that described originally by Rieske (1967) and yields quinols stable for months when stored under nitrogen at −70°C. Another problem is that quinols directly reduce cytochrome *c*. This nonenzymatic reaction needs to be subtracted, but with hydrophobic quinols such as Q-2H_2 or DBH, its relative proportion is minimal, i.e., 2–5% of the overall rate. To appropriately evaluate the nonenzymatic reaction, an aliquot of the concentrated mitochondrial preparation is set aside and treated with 2 nmol/mg of both antimycin and myxothiazol. One additional problem in assays using cytochrome *c* as a substrate is that the outer mitochondrial membrane (OM) is fundamentally impermeable to the cytochrome in intact mitochondria and in fresh cell fractions. This can be overcome easily by the treatment of the preparation with a small quantity (ca. 0.2%) of detergents, (e.g., Na-deoxycholate or Tween 20) to disrupt membrane integrity and facilitate the accessibility of cytochrome *c* to its site in complex III. Alternatively, the interference of the OM barrier with the assay of complex III can be exploited to measure the integrity of the OM, evaluated from the latency in the rate of cytochrome *c* reduction with respect to detergent-treated mitochondria (Kluck *et al.,* 1999). The integrity of OM is disrupted during many pathways of apoptosis, which induce cytochrome *c* release into the cytosol and consequent activation of the caspase cascade (Kluck *et al.,* 1999).

Ubiquinol-Cytochrome c Reductase

The buffer used routinely for measuring the specific activity of ubiquinol-cytochrome *c* reductase is 25 mM K-phosphate, 1 mM EDTA, 1 mM KCN (freshly prepared), pH 7.4 (Degli Esposti *et al.,* 1994). Note that although the pH optimum is around 8, the absolute rate of the nonenzymatic reaction is much more pronounced at pH 8 than at PH 7.4. The reaction mixture consists of 1.95 ml of buffer, 20 μM horse heart cytochrome *c*, and 0.01–0.02 mg/ml of mitochondrial protein. After equilibration at room temperature in a spectrophotometer set at 550 nm ($\varepsilon = 18.7\ mM^{-1}cm^{-1}$) or 412–450 nm ($\varepsilon = 40\ mM^{-1}cm^{-1}$), the reaction is started by a few microliters of an ethanolic solution of DBH or Q-$2H_2$ to a final concentration of 20 μM quinol. The cuvette is mixed rapidly and the rate is recorded immediately. Because the activity is generally very high, the appropriate amount of mitochondrial protein needs to be adjusted so as to obtain a good resolution of the linear initial rate of the reaction. The nonenzymatic rates are measured at the end of the experiment to avoid contaminations with the inhibitors; complex III inhibitors are so potent that even a slight contamination can affect the measured activity significantly.

4. Complex IV Assay

In many respects, the assay of the redox activity of complex IV, cytochrome *c* oxidase, appears to be the simplest of all the respiratory enzyme assays. Only a single reagent is sufficient, reduced cytochrome *c*. However, this reagent is not available commercially and needs to be prepared in advance by reduction with either dithionite and ascorbate, usually followed by desalting with column chromatography to remove excess reductant. This preparation is only of limited inconvenience due the stability of the reduced preparation in the freezer (Trounce *et al.*, 1996). The other problem of this assay regards the accessibility of cytochrome *c* to the catalytic site that, as discussed earlier, can be impaired by an intact OM. To overcome this, the mitochondrial preparations should be treated with detergents such as dodecyl-maltoside or Triton X-100 before the assay. The basic protocol of cytochrome *c* oxidase can also be used for latency assays of OM permeability in apoptosis studies (Kluck *et al.*, 1999). In these assays, the rate of cytochrome *c* oxidation in the presence of detergent is considered to represent 100% of the activity with no latency. See Chapters 5 and 6 for a detailed discussion on the assay of cytochrome oxidase.

B. Reactive Oxygen Species Assays

It has been known since the early 1970s that mitochondrial respiration is associated with the production of ROS at various sites (Loschen *et al.*, 1971). Complexes I, II, and III contribute extensively to ROS production via their low-potential cofactors and semiquinone radicals. The relative contribution of a single respiratory complex may vary, depending on the method used for assaying ROS production and the coupling state of the preparation; as a rule, uncoupled mitochondria produce a limited amount of ROS (Loschen *et al.*, 1971; Khorsunov *et al.*, 1997). Because of the increasingly recognized importance of mitochondrial ROS in cell death signaling and disease pathogenesis, updated guidelines are required for suitable measurements. Basically, all procedures using optical or luminescent probes (for a review, see Boveris, 1984) have a limited sensitivity and specificity and are generally not applicable to cellular studies.

So far, scopoletin has been the fluorescent probe used most frequently in studies of ROS production in isolated mitochondria. Scopoletin is a fluorescent substrate for peroxidases that, upon reaction with hydrogen peroxide, loses its blue fluorescence

(Loschen *et al.,* 1971; Boveris, 1984; Korshunov *et al.,* 1997). Scopoletin does not affect the redox activity of respiratory complexes and therefore can be safely applied to ROS assays with mitochondrial preparations. However, assays based on scopoletin have four major limitations: (i) requirement of exogenous peroxidases; (ii) spontaneous decrease of fluorescence with time; (iii) major interference from endogenous NAD(P)H fluorescence; and (iv) lack of applicability to ROS assays in intact cells.

ROS production in live cells is measured most frequently with fluorogenic probes, namely compounds that become fluorescent on interaction with hydrogen peroxide, e.g., dichlorodihydrofluorescein diacetate (DCFDA) (Garcia-Ruiz *et al.,* 1997), or superoxide, e.g., dihydroethidine (Budd *et al.,* 1997). Because DCFDA does not suffer from any of the limitations of scopoletin and provides superior sensitivity, it is recommended for ROS assay with either mitochondria or cells.

DCFDA Assay

The basic protocol of DCFDA measurements is suitable for both spectrofluorometers and fluorescence plate readers. With our protocol for automated microplate measurements (Degli Esposti and Mclennan, 1998), the production of hydrogen peroxide driven by respiratory substrates responds linearly to concentrations of mitochondria or SMP ranging from 0.05 to 1 mg/ml. Conversely, the signals depend hyperbolically on the probe concentration at a fixed protein level, with half-maximal values at 1 μ*M* DCFDA, a concentration that does not affect complex I activity.

Each well of a 96-well microtiter plate contains 50 m*M* K-phosphate or phosphate-buffered saline (PBS) buffer, pH 7.6, containing 1 μ*M* of DCFDA [diluted from a 5 m*M* stock solution in dimethyl sulfoxide (DMSO)] and 0.2–0.5 mg/ml of mitochondrial protein to in a final volume of 0.15 ml. Once the microplate is equilibrated in the instrument at 30°C for 10 min, the reaction is started by the addition of a respiratory substrate, e.g., 10 m*M* succinate, and is usually followed for 30 to 60 min. Fluorescence readings (typically with excitation at 485 nm and emission at 520 nm, and 5-nm bandwiths) are accumulated automatically every 60 s for each well, and data files are subsequently manipulated with programs such as Excel. Usually, the rate of DCFDA oxidation shows an induction phase lasting about 10 min before becoming linear with time (Fig. 1). The induction phase is reduced if horseradish peroxidase (HRP) is present in the reaction mixture. The final concentration of HRP should not exceed 1 unit per well because the presence of significant amounts of hydrogen peroxide in any aqueous medium drives relatively high rates of DCFDA oxidation in blanks without any mitochondria. Note that in previous studies using fluorescent substrates of HRP, such as DCFDA or scopoletin, the basal rate produced by HRP alone was not taken into proper consideration (Loschen *et al.,* 1971; Korshunov *et al.,* 1997).

Because the fluorescence emission of DCFDA is very intense, modern instruments do not require the HRP supplement to obtain sensitive measurements above background (Fig.1). Moreover, the omission of HRP allows for the use of KCN to block respiration. This mimics anoxia while also inhibiting the action of endogenous peroxidases such as catalase. Catalase, which is present in all cells and tissues, and abundant in heart, often contaminates mitochondrial preparations. If fluorescence plate readers are

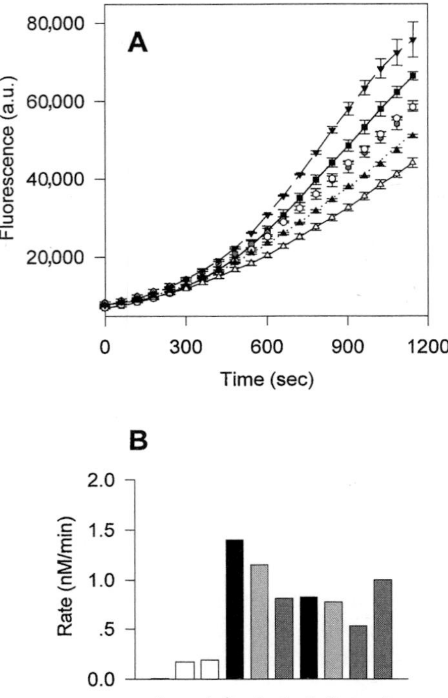

Fig. 1 Measurements of hydrogen peroxide production in submitochondrial particles. ETP_H were suspended in phosphate buffer at 0.5 mg/ml in the presence of 0.1 units of antibody-conjugated HRP. Except for an aliquot run in parallel (nonactivated), the particles were treated with malonate to activate complex II activity. The inhibitors were added with the phosphate buffer at the following concentrations: 0.5 μM for rotenone and antimycin 10 mM for malonate, and 25 μM for carboxin. Incubation of the particles at 30°C for 5 min was followed by the addition of 1 μM DCFDA and 10 mM K-succinate to initiate the plate reader measurements. Readings were accumulated every 60 s in triplicate. (A) Time course of the initial production of hydrogen peroxide. Except for antimycin, which enhanced the rate and extent of fluorescence increase, all other inhibitors showed a time course intermediate between the control and that with the combination of carboxin and rotenone. The fluorescence is expressed in arbitrary units (a.u.). (B) Rates of hydrogen peroxide production measured as the tangent to the linear part of the time course of fluorescence increase that was attained after 15–30 min from time 0, depending on the sample. A coefficient of 4000 a.u. per nM DCF was used in the calculation. Note that the rate of the HRP blank, i.e., without mitochondrial particles, is comparable to that obtained with nonactivated particles.

not available, DCFDA measurements can be carried out in spectrofluorimeters with excitation set at 485 or 503 nm (the latter corresponding to the excitation maximum) and emission at 522 nm. The quantitative evaluation of the DCF fluorescence measured in either a plate reader or a spectrofluorometer should be undertaken with a serial dilution of a commercial DCF solution (Black and Brandt, 1974; Degli Esposti and McLennan, 1998). The concentration range giving linear responses depends on the technical characteristics of the instrument, which in some plate readers extends to picomolar values. All measurements with DCFDA need to be compared with appropriate blanks run in parallel. The rate of spontaneous DCFDA oxidation in the blanks is generally below 5% of the overall rate with mitochondria, and it can be diminished further by adding bovine serum albumin (BSA) at concentrations equivalent to those of the mitochondrial preparation. DCFDA is mainly sensitive to hydrogen peroxide and lipid hydroperoxides. DCFDA oxidation due to hydroxyl radicals usually accounts for less than 20% of that due to hydrogen peroxide. Figure 1 shows results obtained in a microplate experiment in which the oxidation of DCFDA by succinate-respiring ETP_H is also stimulated by HRP, which enhances the rate of ROS production 10-fold (Fig. 1B). These rates range between 3 and 8 pmol/min per milligram of protein, in good agreement with the very low rates of hydrogen peroxide produced by uncoupled mitochondria, which are barely measurable with other methods (e.g., Korshunov *et al.*, 1997).

Using mitochondrial preparations and fractions thereof, the recently introduced OxyBurst probe can be used as an alternative to DCFDA for measuring ROS production in the aqueous phase. OxyBurst is a modified dihydrofluorescein conjugated to BSA that is more stable than DCFDA (www.probes.com) and does not affect the activity of complex I or other respiratory complexes. Measurements should be carried out with 10 μg/ml of the OxyBurst probe, excitation at 485 nm and emission at 527 nm.

C. Membrane Potential Measurements

Membrane potential measurements provide the direct assessment of mitochondrial functional integrity and are conveniently obtained with optical or fluorescent probes. However, it should be kept in mind that they generally provide indirect and semiquantitative evaluations of the transmembrane potential because of complications inherent to the nature of the probe response. Accurate calibrations allow for more quantitative estimates when tightly coupled preparations are available (Rottenberg, 1989; Ghelli *et al.*, 1997). Under normal respiratory conditions, complex III is the predominant generator of membrane potential, mainly due to the electrogenic (i.e., membrane charging) electron transfer between the two *b* hemes embedded in the membrane. The contribution of complexes I and IV to mitochondrial membrane potential varies in different tissues or cell line, depending often on the relative content of the enzymes of the preparation. In principle, membrane potential ($\Delta\psi$) can be measured with either positively charged probes (in mitochondria) or negatively charged probes (in inside-out SMP) that accumulate electrophoretically in the membrane and the internal compartment of the preparation.

a. Mitochondria

Safranine is the classical probe for membrane potential measurements in mitochondria (Akerman and Wikstrom, 1976; Smith, 1990). Its accumulation inside mitochondria is accompanied by a blue shift in absorbance that can be recorded at different wavelength couples, e.g., 511 nm minus 533 nm (Akerman and Wikstrom, 1976) or 520 nm minus 554 nm (Luvisetto *et al.,* 1991), and also from fluorescence changes. Safranine fluorescence increases with membrane potential when measured with excitation at 520 nm and emission at 580–590 nm and probe concentrations around 10 μM (Lapiana *et al.,* 1998).

Safranine signals are of limited sensitivity as compared to those produced by cationic rhodamine derivatives such as rhodamine 123 (Rho123). The mitochondrial accumulation of these probes in response to membrane potential can induce large fluorescence changes, which can be advantageously exploited to monitor $\Delta\psi$ (Emaus *et al.,* 1986; Scaduto *et al.,* 1999). Both optical and fluorescence spectra of Rho123 show a marked (about 10 nm) red shift on accumulation and binding to coupled mitochondria when the probe concentration is 0.2–1 μM in the medium and mitochondrial protein is equal or more than 0.3 mg/ml (Emaus *et al.,* 1986; Fontaine *et al.,* 1998; Scaduto *et al.,* 1999). As a result of this spectral shift, the intensity of Rho123 fluorescence at 525–530 nm with fixed excitation at around 500 nm is largely attenuated by mitochondrial uptake [Emaus *et al.,* 1986]. The resulting decrease in fluorescence reinforces the concomitant quenching due to the aggregation of the probe accumulated inside mitochondria (Metivier *et al.,* 1998). The spectral changes of Rho123 in mitochondria also derive from binding to matrix proteins, especially the F1 domain of ATP synthase (Emaus *et al.,* 1986). Despite its complexity, Rho123 fluorescence quenching has been used frequently not only to monitor $\Delta\psi$ changes but also to evaluate permeability changes in the inner mitochondrial membrane, the so-called PT transition (Fontaine *et al.,* 1998). Conversely, the large increase in the intramitochondrial concentration of Rho123 due to $\Delta\psi$ accounts for the intense red fluorescence exhibited by mitochondria when visualized directly with microscopic techniques (Chen, 1988; Farkas *et al.,* 1989; Huser *et al.,* 1999) or when measured with fluorescence-activated cell sorting (FACS) analysis (Finucane *et al.,* 1999). With these techniques, and especially using low concentrations of both probe and mitochondria, the fluorescence intensity of Rho123 is proportional to the magnitude of $\Delta\psi$, and its depolarization normally induces an attenuation of fluorescence intensity (Chen, 1988).

With respect to the popular Rho123, the tetramethyl rhodamine derivatives TMRM and TMRE are more membrane soluble and exhibit less spectral changes on accumulation in coupled mitochondria (Ehrenberg *et al.,* 1988; Farkas *et al.,* 1989; Scaduto *et al.,* 1999). These probes offer the additional advantage of longer wavelength fluorescence that can be monitored conveniently through the >580-nm band-pass filters normally used in confocal microscopy and FACS analysis. Indeed, using excitation wavelengths near the 488-nm laser line, most of the spectral shift of TMRE accumulated inside mitochondria (also within cells) is canceled out when the emitted fluorescence is monitored at wavelengths equal to or higher than 580 nm (Degli Esposti *et al.,* 1999; Huser *et al.,* 1999). Using these settings and TMRE (or TMRM) concentrations equal or less than 100 nM, the build-up of membrane potential is associated with a fluorescence increase rather than quenching (Huser *et al.,* 1999; Degli Esposti *et al.,* 1999).

Ideally, $\Delta\psi$ measurements should be performed with probes and techniques that are applicable equally to isolated mitochondria and to live cells, and to give a consistent decrease in fluorescence upon depolarization (as in microscopic studies, cf. Chen, 1988; Farkas *et al.,* 1989; Huser *et al.,* 1999; Scorrano *et al.,* 1999). The most sensitive and reliable technique is FACS analysis with either Rho123 (80–100 n*M*) or TMRE (40–50 n*M*) as described by Finucane *et al.* (1999). The two major advantages of FACS analysis are (i) the fluorescence intensity of coupled mitochondria is at least two orders of magnitude greater than that of uncoupler-treated mitochondria and (ii) only a few micrograms of mitochondria are required to obtain reliable results.

In the absence of a cytofluorimeter, the alternative is to use spectrofluorimeters or fluorescence plate readers with a protocol such as the one described here. Mitochondria are suspended at 2 mg/ml in sucrose 200 m*M*, 20 m*M* K-HEPES, pH 7.2, 2 m*M* $MgCl_2$, 1 m*M* EDTA, 5 m*M* succinate, and loaded with 100 n*M* TMRE (from a stock ethanolic solution) for 10 min at room temperature. Mitochondria are then pelleted by centrifugation (e.g., 12,000*g* for 10 min) and resuspended at 0.2 mg/ml in the same buffer without succinate. The mixture is placed in a fluorimeter cuvette or in fluorescence-transparent microplate wells, and fluorescence emission is monitored at room temperature using 580 nm as the emission wavelength (or >580-nm band-pass filters) and 500 nm as the excitation wavelength (or <500-nm broad-pass excitation filters). $\Delta\psi$ build-up is stimulated by the addition of respiratory substrates, e.g., glutamate/malate, and $\Delta\psi$ depolarization is obtained after the addition of either FCCP or a mixture of valinomycin and nigericin. Note that $\Delta\psi$ depolarization produces only a partial decrease in TMRE fluorescence, as the probe released in the medium also fluoresces. Because fluorescent probes such as rhodamines tend to produce radicals on continuous illumination (Bunting, 1992; Huser *et al.,* 1999), it is advisable to use fluorescence excitation in pulsed excitation mode; this is easily achieved in fluorescence plate readers.

b. Submitochondrial Particles

Coupled submitochondrial particles can be obtained from various tissues, including platelets (Baracca *et al.,* 1997). A most convenient probe for measuring $\Delta\psi$ in submitochondrial particles is oxonol VI, which undergoes a large optical red shift on accumulation in the positively charged internal volume of the particles (Bashford and Smith, 1979). $\Delta\psi$ generated by respiratory substrates (NADH, succinate or ubiquinols) is normally measured at 22–24°C by following the absorbance changes at 630 nm minus 601 nm (Degli Esposti *et al.,* 1996; Ghelli *et al.,* 1997). The dual-wavelength mode is not a strict requisite, as large potential-sensitive signals are also obtained with single wavelength measurements at 629 nm. SMP are diluted to 6–12 mg/ml in isoosmotic sucrose buffer, and preferably treated with 2 nmol of oligomycin to limit proton leaks through ATP synthase. Subsequently, the particles are adjusted to a final protein concentration of 0.15–0.3 mg/ml in the assay medium, consisting of 125 m*M* sucrose, 0.05 *M* tricine–OH, 2.5 m*M* $MgCl_2$, and 0.04 *M* KCl, pH 8 (STMK buffer), containing 2–3 μ*M* oxonol-VI. Uncoupler-sensitive absorbance increases of about 0.1 units can be obtained with coupled preparations respiring NADH.

D. Proton Pumping Measurements

Direct assays of proton pumping require sensitive pH electrodes and large amounts of well-coupled mitochondria (Nicholls and Ferguson, 1992). Indirect measurements of proton pumping can be obtained with pH-sensitive probes that display either optical or fluorescence changes. Generally, these probes measure small variations in the proton concentration of the medium, which necessarily needs to be minimally buffered. The most useful optical probes of this kind are phenol red and bromocresol purple (BCP), which allow relatively sensitive measurements in a variety of preparations of coupled mitochondria (Degli Esposti *et al.,* 1982).

Coupled mitochondria are diluted to 1 mg/ml in isotonic buffer containing 200 m*M* sucrose, 30 m*M* KCl, 0.5 m*M* EDTA, 0.5 μg/ml valinomycin, and 40 μ*M* N-ethylmaleimide, adjusted to pH 6.9–7.0 with a minimal quantity of HEPES or MOPS, and incubated for a few minutes at room temperature with 50 μ*M* BCP (from an ethanolic stock solution). The mixture is placed in a cuvette and the absorbance changes are measured at either 590 nm (in single-wavelength mode) or 584 nm minus 629 nm (in dual-wavelength mode). External acidification due to substrate respiration results in a transient decrease in absorbance, which can be calibrated directly by the addition of known concentrations of HCl (Degli Esposti *et al.,* 1982). BCP measurements suffer from smaller interference from endogenous cytochromes than those with phenol red, but produce large signals only at pH below or equal to 7.0. As for most probes, the optimal concentration of BCP/mg of protein needs to be evaluated empirically for the specific system under study.

The proton pumping activity of mitochondria can also be measured with membrane-permeable probes that monitor pH changes inside the aqueous volume of the internal matrix. Neutral red (suitable absorbance changes at 520 nm minus 580 nm) and the acetoxymethyl ester of 2′,7′-bis(carboxyethyl)-5(6)carboxyfluorescein (BCECF, a fluorescent probe with maximal excitation at 450 nm and emission at 530 nm) are used most frequently for measuring pH changes inside coupled mitochondria. However, procedures based on matrix-localized probes tend to be more complex and less sensitive than the BCP assay; the optimal response of BCECF is obtained at 2 mg/ml of mitochondrial protein (Luvisetto *et al.,* 1991).

Coupled submitochondrial particles can also be used for measurements of proton pumping. The most sensitive and reliable procedure with SMP is based on the fluorescent pH-sensitive dye 9-amino-6-chloro-2-methoxyacridine (ACMA) (Helfenbaum *et al.,* 1997). ACMA is a weak base that accumulates inside the particle volume proportionally to the acidic ΔpH inside. This accumulation produces an intermolecular stacking that quenches the strong fluorescence of ACMA extensively, with nearly complete quenching at ΔpH values around 2. The basic reaction mixture of ACMA measurements consists of 2.4 ml of STMK buffer, pH 8, 1.5–2 μ*M* ACMA, and 0.4–0.5 mg/ml of submitochondrial particles. The mixture is prepared directly in a 4-ml fluorimeter cuvette and is allowed to equilibrate in the dark for at least 15 min at room temperature. The cuvette is then placed in a spectrofluorometer under constant stirring (normally at room temperature of 22–24°C), and the fluorescence emission is monitored until it remains stable with time. The spectrofluorimeter is set with excitation at 412 nm (bandwidth of 2.5 nm) and emission at 510 nm (bandwidth of 5 nm). The quenching of the ACMA signal can be calibrated with direct measurements of pH (Helfenbaum *et al.,* 1997) and depends almost linearly on ΔpH values of 0.2 to 2. To obtain maximal signals of ΔpH, the membrane potential is collapsed by the introduction of 0.6–1 μM valinomycin in the assay mixture. The quenching signals are totally abolished by uncouplers or nigericin.

The ACMA assay is particularly suited for measuring the proton pumping activity of complex I and ATP synthase (Helfenbaum *et al.,* 1997; Baracca *et al.,* 1997). An advantage of this assay is that it monitors only the particles with a sealed inside-out configuration, as only these particles actively accumulate ACMA, resulting in the quenching of its fluorescence. The major disadvantage is that ACMA measurements are slow in comparison with membrane potential measurements with oxonol VI. However, ACMA measurements can be adapted easily to fluorescence plate readers.

III. Mitochondrial Functions within Living Cells

This section focuses on some advanced methods for measuring key mitochondrial functions within live cells. Detailed discussions about these and other aspects of mitochondrial function in live cells are presented elsewhere in this book.

A. Fluorescence Measurements of Membrane Potential

So far, Rho123 and carbocyanine dyes such as JC-1 and $DiOC_6(3)$ have been the probes used most frequently for evaluating mitochondrial membrane potential in live cells. Although these probes can provide useful qualitative information, they suffer from several problems, which are summarized in Table I. The rhodamine derivatives TMRM and TMRE, together with the recently introduced CM-MRos (MitoTracker orange) and CM-XRos (MitoTracker red) (Haugland, 1996), are becoming increasingly popular for measuring active mitochondria in cells. Chloromethyl rosamines such as MitoTracker red accumulate in mitochondria in response to membrane potential like rhodamines, but not in a reversible way, because they bind covalently to mitochondrial proteins. Hence, these probes are more suitable for microscopic applications, e.g., subcellular localization (Wolter *et al.,* 1997; Degli Esposti *et al.,* 1999; Krohn *et al.,* 1999), than for functional evaluations of membrane potential. MitoTracker orange and MitoTracker red also present side effects, as they bind to the adenine nucleotide translocator, thereby affecting PT in mitochondria (Macho *et al.,* 1996; Scorrano *et al.,* 1999), and partially inhibit the activity of complex I (M.Degli Esposti, unpublished observations, cf. Scorrano *et al.,* 1999). However, these side effects can be reduced to a minimum by using probe concentrations equal or less than 100 n*M* in serum-supplemented medium at room temperature.

TMRM and TMRE do not present the major problems that affect Rho123, carbocyanine, and MitoTracker probes (Table I). As discussed earlier for their application to

Table I
Properties and Problems of Major Probes for Membrane Potential

Probe	Properties and problems
JC-1 (red aggregate)	Non-Nernstian distribution Staining of a limited proportion of cellular mitochondria Altered response to ionophores No response to potentials lower than 100 mV
$DioC_6(3)$	Staining of nonmitochondrial cell membranes Inhibition of complex I Extensive light-induced radical production[a]
Rhodamine 123	Cytotoxicity (mainly due to ATPase inhibition) Large red shift on binding to mitochondria Self-quenching on mitochondrial uptake
Mito Tracker red or orange	Non-Nernstian distribution (covalent binding) Limited self-quenching on mitochondrial uptake Limited red shift on binding to mitochondria Light-induced radical production[a]
TMRE or TMRM	Approximate Nernstian distribution Red shift on binding to mitochondria No inhibition of complex I Very low cytotoxicity

[a]Note that all probes produce photo damage to cells due to the autooxidation of their long-lived triplet state with molecular oxygen, which produces ROS (Bunting, 1992).

mitochondria, the spectral changes of TMRE or TMRM due to mitochondrial accumulation and binding can be largely reduced by using appropriate wavelengths or filters for excitation and emission (Huser *et al.,* 1998). Because these probes approach a Nernstian equilibrium across the membranes of live cells (Ehrenberg *et al.,* 1988; Farkas *et al.,* 1989; Fink *et al.,* 1999; Scaduto *et al.,* 1999), they generally produce the best quantitative evaluations of $\Delta\psi$ when suitably applied to cells in solution (Fig. 2). The following standard protocol for FACS measurements of cellular potential with TMRE has been developed in collaboration with the laboratory of Doug Green (LaJolla Institue for Allergy and Immunology, San Diego, CA).

a. *FACS Measurements with TMRE*

Cells are removed from culture flasks and suspended at 0.5–1 $\times$ 10^6/ml in standard growth medium, e.g., RPMI containing 10% fetal calf serum, and supplemented with 40–50 n*M* TMRE. The cell suspension is allowed to equilibrate at room temperature in the dark for at least 15 min, and then cellular fluorescence is measured directly in a FACS apparatus using the 488-nm line for excitation and the FL-2 band-pass filter for emission. Ten thousand counting events are generally accumulated and displayed in a histogram form with logarithmic scale (Fig. 2). Separate aliquots of the cell suspension

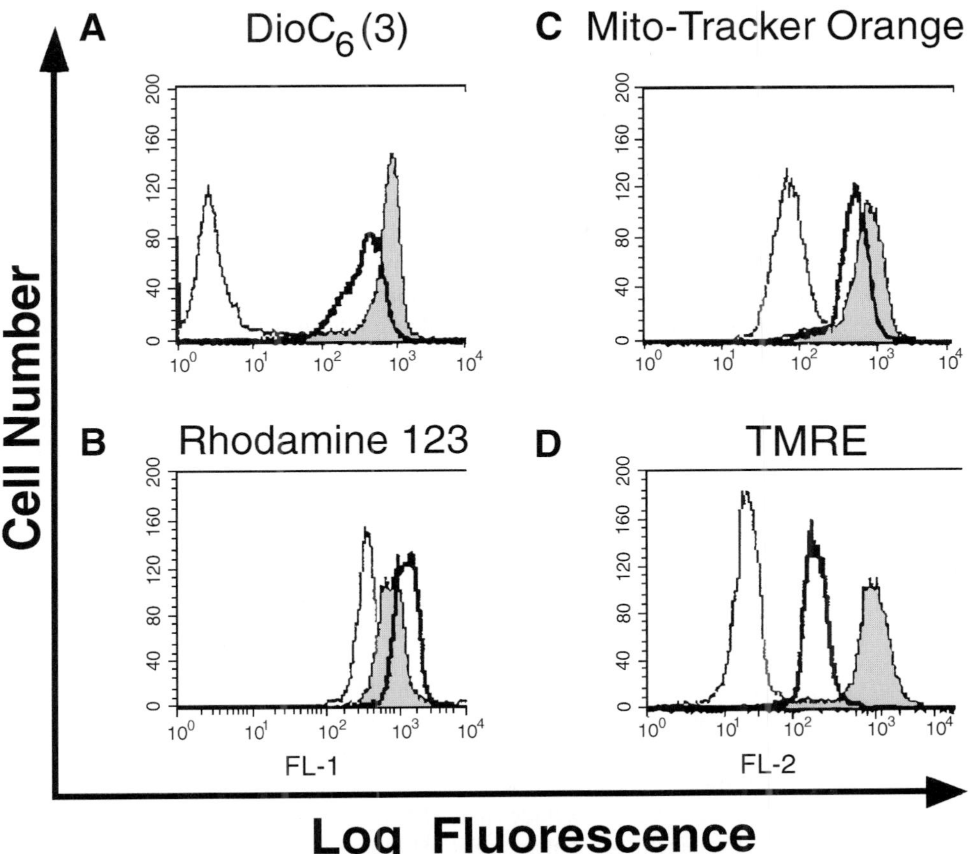

Fig. 2 Membrane potential measurements in cells with different mitochondrial probes. Results were obtained in collaboration with Nigel Waterhouse and Doug Green at the LaJolla Institute for Allergy and Immunology (San Diego, CA) under conditions similar to those reported by Finucane *et al.* (1999). Jurkat T cells were suspended in RPMI medium (supplemented with fetal calf serum) containing the probes, kept at room temperature for 20 min, and subsequently analyzed with 10,000 events per sample in a Beckton-Dickinson Facscalibur cytometer. The thin line histogram on the left of each panel represents the signal obtained after treating the cells with the detergent NP-40 for approximately 10 min. Gray-filled histograms are control samples, and thick-line histograms are samples treated with 5 μM FCCP to depolarize the major component of the mitochondrial membrane potential. (A) Cells were stained with 40 n*M* DioC$_6$(3) (cf. Zamzami *et al.*, 1995; Macho *et al.*, 1997a,b). (B) Cells were stained with 150 n*M* rhodamine 123; note the apparent increase in fluorescence on depolarization, mainly due to dequenching of the probe accumulated in mitochondria (Emaus *et al.*, 1986; Metivier *et al.*, 1998). (C) Cells were stained with 150 n*M* MitoTracker orange (cf. Bossy-Wetzel *et al.*, 1998). (D) Cells were stained with 40 n*M* TMRE.

are measured after a 15-min treatment with 2–5 μM FCCP, or 2 μM nigericin, alone or together with 2 μM valinomycin, for evaluating the overall range of mitochondrial potential, or a nonionic detergent such as NP-40, to assess the nonspecific fluorescence of TMRE due to interaction with cell membranes (Fig. 2).

One important aspect of this protocol is that cells are suspended in growth medium before and during FACS measurements, in contrast to many previous protocols in which cells were washed after staining and resuspended in PBS (Macho *et al.,* 1997a,b; Zamzami *et al.,* 1995; Cossarizza *et al.,* 1993; Salvioli *et al.,* 1997). Washing of TMRE-stained cells is superfluous. Moreover, cells resuspended in PBS rapidly run out of the nutrients present in growth medium and thus lose their glycolytic capacity to sustain mitochondrial membrane potential, either directly by providing pyruvate or indirectly by compensating oxphos deficiency with cytosolic ATP. This fact is partially responsible for the large drops in mitochondrial membrane potential that have been observed after several hours of apoptosis activation using cells washed and maintained in PBS for FACS analysis (Zamzami *et al.,* 1997; Macho *et al.,* 1997a,b; Salvioli *et al.,* 1997). If cells are supplemented with glucose, limited changes in membrane potential are observed in the first few hours after apoptosis induction.

b. Quantitative Estimation of Cellular Potentials with TMRE

Several technical considerations favor TMRE as the most suitable probe for studying $\Delta\psi$ in many cell types. First, the standard 488-nm excitation wavelength eliminates most of the spectral changes in TMRE emission that are due to accumulation into mitochondria in response to $\Delta\psi$. Second, the FL-2 band-pass filters in FACS instruments have a maximum that is very close to the maximum of TMRE emission [around 580 nm in energized mitochondria (Farkas *et al.,* 1989)], thereby facilitating high fluorescence outputs at low probe concentrations. Third, nigericin significantly increases basal TMRE fluorescence, which excludes that TMRE, at concentrations below 100 n*M*, suffers from the self-quenching problems of Rho123 (cf. Metivier *et al.,* 1998). Fourth, the cytotoxicity of TMRE, especially without continuous illumination, is very limited at concentrations below 100 n*M* (Farkas *et al.,* 1989; Huser *et al.,* 1998). Another advantage is that TMRE fluorescence, in view of the almost ideal Nernstian distribution of the probe, can be used to quantify intracellular membrane potentials (Ehrenberg *et al.,* 1988; Fink *et al.,* 1998). One way of performing this quantitation is to apply the Nernst equation to the ratio of maximal signals of TMRE fluorescence in control and detergent-treated cells (Fig. 2; cf. Khron *et al.,* 1999; N. Waterhouse, unpublished results). Because of the indirect nature of this quantitation (Rottenberg, 1979), a term like $\Delta\Psi_F$ is introduced to define a phenomenological membrane potential, derived from the equation:

$$\Delta\Psi_F(\text{mV}) = \Delta\Psi_m \times F = -59 \times \log[\text{FL-2 max control}]/[\text{FL-2 max with NP-40}],$$

where F represents a factor (<1) that accounts for membrane-binding phenomena deviating the distribution of TMRE from a perfect Nernstian behavor. Using this equation, $\Delta\Psi_F$ values in various lymphoid cells generally lie between 110 and 125 mV, which are not far from the generally accepted values of 130–150 mV for coupled mitochondria within cells (Farkas *et al.,* 1989; Chen, 1988).

B. Reactive Oxygen Species Measurements in Living Cells

Because mitochondria are the predominant generators of ROS in animal cells, the measurement of cellular ROS chiefly reflects mitochondrial function or dysfunction. The two major radical species that can be measured within cells are superoxide and hydrogen peroxide. Superoxide is the first species to be produced by the autooxidation of mitochondrial redox groups, such as iron–sulfur clusters and semiquinone radicals. However, the detection of superoxide in live cells can be affected heavily by the expression level of Mn-SOD, which is a highly regulated antioxidant scavenger present in mitochondria. The expression level of Mn-SOD varies greatly among different cell lines and also within a given cell line, depending on stress and apoptosis conditions (Esposito *et al.,* 1999). The reduced form of the DNA stain ethidium bromide, often labeled dihydroethidine (DHE) or hydroethidine, is specifically oxidized by superoxide anions. Within cells, the oxidized ethidium binds strongly to mitochondrial and nuclear DNA with extensive enhancement of its red fluorescence, whereas the reduced DHE has only a weak blue fluorescence. These properties of DHE have been exploited to monitor changes in ROS production during cell death activation in a number of cell systems (Macho *et al.,* 1996, 1997b). The major complication in the use of DHE derives from the positive charge of the probe that favors its electrophoretic accumulation into mitochondria. When the concentration exceeds the binding capacity of mtDNA, the oxidized form of DHE produced inside mitochondria shows attenuated fluorescence, mostly due to self-quenching (Budd *et al.,* 1997). Hence, a collapse of mitochondrial membrane potential releases the accumulated oxidized ethidium into the cytoplasm, with immediate dequenching of its fluorescence, which is followed by additional fluorescence enhancement on binding to nuclear DNA. Consequently, the red fluorescence of DHE is increased by both an increase in intracellular superoxide and a decrease in $\Delta\psi$ (Budd *et al.,* 1997). The concentration of the probe is critical for limiting $\Delta\psi$ interference, which appears to be minimal when DHE is kept around 1 μM in the assay medium (Budd *et al.,* 1997). The only valuable alternative to DHE in cells studies is the chemiluminescent probe lucigenin (bis-*N*-methylacridinium nitrate), which needs to be used at concentrations at or above 50 μM for sensitive signals with live cells (Pitkanen and Robinson, 1996; Imlay, 1995; Hennet *et al.,* 1993). Fortunately, lucigenin does not significantly affect complex I activity, usually the most delicate respiratory enzyme, up to concentrations around 100 μM (H. McLennan and M. Degli Esposti, unpublished observations). The disadvantages of lucigenin derive from the limited sensitivity of its signals and the technical limitations of chemiluminescence, which, for example, cannot be applied conveniently to FACS analysis.

In comparison with the limited procedures available for measuring superoxide, there is ample choice of probes for measuring the cellular production of hydrogen peroxide. Hydrogen peroxide is constantly produced and, due to its high stability, is maintained at detectable steady-state levels within metabolically active cells and tissues. Several probes can be satisfactorily employed to measure hydrogen peroxide production in live cells (see www.probes.com/ for an updated list). Many of these probes have been developed for measuring the oxygen burst reactions, but are not necessarily appropriate for measuring mitochondrial ROS production, as they also inhibit complex I activity. Inhibition of complex I should be avoided because it enhances the normal ROS production of mitochondria, mainly via autooxidation of the overreduced iron–sulfur cofactors of the dehydrogenases (Boveris, 1984). Here only the recently introduced ROS-specific staining of mitochondria is discussed.

c. *Mitochondrial Staining and Microscopy Analysis*

Using the reduced Mito Tracker red probe (CM-H_2XRos), it is possible to obtain a ROS-specific staining of mitochondria within cells that is amenable to both qualitative and quantitative measurements of mitochondrial ROS production (Degli Esposti *et al.,* 1999). Note that CM-H_2XRos does not significantly affect complex I activity even at high concentrations, contrary to its oxidized form Mito Tracker red (CM-XRos) (M. Degli Esposti and H. McLennan, unpublished results). CM-H_2XRos does not fluoresce until it enters an actively respiring cell, where it is oxidized predominantly by reactions involving hydrogen peroxide production. If unbound to proteins, CM-H_2XRos specifically accumulates inside mitochondria due to the positive charge it acquires on oxidation by intracellular ROS. The probe then covalently binds to mitochondrial proteins and forms a permanent red stain of these organelles. If mitochondrial membrane potential is collapsed during oxidation of the probe, red staining can diffuse to other cellular compartment, especially the nucleus (cf. Scorrano *et al.,* 1999). For obtaining an effective ROS-sensitive staining, cells are suspended at $0.4–1 \times 10^6$/ml in fresh growth medium containing 0.5 μM of CM-H_2Xros, which has been prepared just before the experiment by dissolving the content of a commercial vial in 0.1 ml of pure DMSO. Cells are incubated in this staining medium for 15 min at room temperature, washed twice with PBS, and then collected on a slide using a Cytospin apparatus. Cells are then fixed with a fresh 3.7% formaldehyde solution in PBS, followed by washing with PBS containing 30 mM NH_4Cl and then with distilled water. Ideally, confocal microscopy should be undertaken with an intermediate photomultiplier voltage and a 590-nm band-pass filter for fluorescence emission. The basal fluorescence intensity may vary significantly in different types of cells and under different conditions of stimulating ROS production. However, 1 h of apoptosis induction is usually sufficient to detect large change in the ROS-specific staining of mitochondria (Degli Esposti *et al.,* 1999). Slides can also be analyzed with epifluorescence, using long-wavelength emission filters such as the Texas Red filter (note that standard rhodamine filters are inadequate).

C. Cellular Autofluorescence Measurements

The blue autofluorescence of cells can be exploited for evaluating the state of mitochondrial NAD(P)H as described originally in isolated mitochondria (Chance, 1976). Reduced NADH and NADPH, but not their oxidized forms, emit fluorescence with maximum around 450 nm when excited at wavelengths between 350 and 370 nm (the excitation maximum varies in different cellular systems, but usually is near 360 nm). Protein-bound NADPH is the predominant fluorescent species in mitochondria and accounts for about 80% of the cellular autofluorescence intensity around 450 nm, whereas cytosolic NAD(P)H contributes about 10% of the same autofluorescence. Hence, the intrinsic fluorescence exhibited by cells at 450 nm can be used to directly evaluate the redox state of mitochondrial NAD(P)H (Duchen and Biscoe, 1992; Eng *et al.,* 1989; Vlessis, 1990; Degli Esposti *et al.,* 1999; Poot *et al.,* 1999). These measurements are relatively delicate, and care should be exercised to limit photo-damage by the excitation light and the spontaneous fading with time of the autofluorescence. In practice, it is advisable to limit the detection of autofluorescence to periods of less than 20 min of continuous illumination, or at time intervals.

In the basic protocol for measuring NAD(P)H fluorescence in lymphoid cells (Degli Esposti *et al.,* 1999), cells are resuspended in PBS at a concentration of 10^6/ml in PBS containing 10 m*M* glucose (or other nutrients, if required) and autofluorescence is measured at 37°C under conditions of gentle automatic stirring in a fluorimeter set at the excitation maximum of 358 nm and emission maximum (normally around 443 nm) with 10-nm bandwidths. Measurements are carried out in triplicate for less than 15 min of continuous illumination. Alternatively, measurements can been carried out every 5 min in a fluorescence plate reader equipped with appropriate filters and using microplates giving minimal background (this depends on manufacturer and instrument performance). In order to verify the contribution of mitochondrial NAD(P)H to the autofluorescence signals, 2 μ*M* of FCCP is added at the beginning of some experiments to induce a rapid stimulation of NAD(P)H oxidation by the respiratory chain (Duchen and Biscoe, 1992), followed by the addition of rotenone or another complex I inhibitor to completely block uncoupled respiration, and thus restore the maximal level of intramitochondrial NAD(P)H (Vlessis, 1990).

Measurements of NAD(P)H fluorescence have been described using confocal microscopy (Nieminen *et al.,* 1997) and FACS (Poot *et al.,* 1999). These applications require relatively sophisticated, yet commercially available, instruments and may inspire further studies in the future. Indeed, endogenous NAD(P)H offers a powerful intrinsic tool for monitoring the metabolic state of mitochondria within cells, devoid of the problems associated with exogenous probes.

References

Akerman, K. E. O., and Wikstrom, M. (1976). Safranine as a probe of the mitochondrial membrane potential. *FEBS Lett.* **68,** 191–197.

Baracca, A., Bucchi, L., Ghelli, A., and Lenaz, G. (1997). Protonophoric activity of NADH coenzyme Q reductase and ATP synthase in coupled submitochondrial particles from horse platelets. *Biochem. Biophys. Res. Commun.* **235,** 469–473.

Barrientos, A., and Moraes, C. T. (1999). Titrating the effects of complex I impairment in the cell physiology. *J. Biol. Chem.* **274,** 16188–16197.

Bashford, C. L., and Smith, J. C. (1979). The use of optical probes to monitor membrane potential. *Methods Enzymol.* **55,** 569–582.

Black, M. J., and Brandt, R. B. (1974). Spectrofluorometric analysis of hydrogen peroxide. *Anal. Biochem.* **58,** 246–254.

Bossy-Wetzel, E., Newmeyer, D. D., and Green, D. R. (1998). Mitochondrial cytochrome c release in apoptosis occurs upstream of DEDV-specific caspase activation and independently of mitochondrial transmembrane depolarization. *EMBO J.* **17,** 37–49.

Boveris, A. (1984). Determination of the production of superoxide radicals and hydrogen peroxide in mitochondria. *Methods Enzymol.* **105,** 429–435.

Budd, S., Castilho, R. F., and Nicholls, D. G. (1997). Mitochondrial membrane potential and hydroethidine-monitored superoxide generation in cultured cerebellar granule cells. *FEBS. Lett.* **415,** 21–24.

Bunting, J. R. (1992). A test for the singlet oxygen mechanism of cationic dye photosensitization of mitochondrial damage. *Phochem. Photobiol.* **55,** 81–87.

Chance, B. (1976). Pyridine nucleotide as an indicator of the oxygen requirements for energy-linked functions of mitochondria. *Circ. Res.* **38,** I 31–38.

Chen, L. B. (1988). Mitochondrial membrane potential in living cells. *Annu. Rev. Cell Biol.* **4,** 155–181.

Cossarizza, A., Baccarani-Contri, M., Kalashnikova, G., and Franceschi, C. (1993). A new method for the cytofluorimetric analysis of mitochondrial membrane potential using the J-aggregate forming lipophilic cation 5,5',6,6'-tetrachloro-1,1',3,3'tetramethylbenzimidazol-carbocianine iodide (JC-1). *Biochem. Biophys. Res. Commun.* **197,** 40–45.

Davey, G. P., Peuchen, S., and Clark, J. B. (1998). Energy thresholds in brain mitochondria: Potential involvement in neurodegeneration. *J. Biol. Chem.* **273,** 12753–12757.

Degli Esposti, M. (1998). Inhibitors of NADH-ubiquinone reductase: An overview. *Biochim. Biophys. Acta* **1364,** 222–235.

Degli Esposti, M., Saus, J. B., Timoneda, J., Bertoli, E., and Lenaz, G. (1982). The inhibition of proton translocation in the mitochondrial bc_1 region by dicyclohexylcarbodiimide. *FEBS Lett.* **147,** 101–105.

Degli Esposti, M., Crimi, M., and Ghelli, A. (1994). Natural variation in the potency and binding sites of mitochondrial quinone-like inhibitors. *Biochem. Soc. Trans.* **22,** 209–213.

Degli Esposti, M., Ngo, A., Ghelli, A., Benelli, B., Carelli, V., McLennan, H., and Linnane, A. W (1996). The interaction of Q analogs, particularly hydroxyldecyl benzoquinone (idebenone), with the respiratory complexes of heart mitochondria. *Arch. Biochem. Biophys.* **330,** 395–400.

Degli Esposti, M., Hatzinisiriou, I., McLennan, H., and Ralph, S. (1999). Bcl-2 and mitochondrial oxygen radicals: New approaches with ROS-sensitive probes. *J. Biol. Chem.* **274,** 29831–29837.

Degli Esposti, M., and McLennan, H. (1998). Mitochondria and cells produce reactive oxygen species in virtual anaerobiosis: Relevance to ceramide-induced apoptosis. *FEBS Lett.* **430,** 338–342.

Duchen, M. R., and Biscoe, T. J. (1992). Mitochondrial function in type I cells isolated from rabbit arterial chemoreceptors. *J. Physiol. (Lond.)* **450,** 13–31.

Ehrenberg, B., Montana, V., Wei, M., Wuskel, J. P., and Loew, L. M. (1988). Membrane potential can be determined in individual cells from the nernstian distribution of cationic dyes. *Biophys. J.* **53,** 785–794.

Emaus, R. K., Grunwald, J. J., and Lemasters, J. J. (1986). Rhodamine 123 as a probe of transmembrane potential in isolated rat-liver mitochondria: spectral and metabolic properties. *Biochim. Biophys. Acta* **850,** 436–441.

Eng, J., Lynch, R. M., and Balaban, R. S. (1989). Nicotinamide adenine dinucleotide fluorescence spectroscopy and imaging of isolated cardiac myocytes. *Biophys. J.* **55,** 621–630.

Esposito, L. A., Melov, S., Panov, A., Cottrell, B. A., and Wallace, D. C. (1999). Mitochondrial disease in mouse results in increased oxidative stress. *Proc. Natl. Acad. Sci. USA* **96,** 4820–4825.

Farkas, D. L., Wei, M., Febbroriello, P., Carson, J. H., and Loew, L. M. (1989). Simultaneous imaging of cell and mitochondrial membrane potentials. *Biophys. J.* **56,** 1053–1069.

Fink, C., Morgan, F., and Loew, L. M. (1998). Intracellular probes concentrations by confocal microscopy. *Biophys. J.* **75,** 1648–1658.

Finucane, D., Bossy-Wetzel, E., Waterhouse, N., Cotter, T. G., and Green, D. R. (1999). Bax-induced caspase activation and apoptosis via cytochrome c release from mitochondria is inhibitable by Bcl-xL. *J. Biol. Chem.* **274,** 2225–2233.

Fontaine, E., Eriksson, O., Ichas, F., and Bernardi, P. (1998). Regulation of the permeability transition pore in skeletal muscle mitochondria. *J. Biol. Chem.* **273,** 12662–12668.

Garcia-Ruiz, C., Colell, A., Mari, M., Morales, A., and Fernadez-Checa, J. C. (1997). Direct effect of ceramide on the mitochondrial electron transport chain leads to generation of reactive oxygen species. *J. Biol. Chem.* **272,** 11369–11377.

Ghelli, A., Benelli, B., and Degli Esposti, M. (1997). Measurement of the membrane potential generated by complex I in submitochondrial particles. *J. Biochem. (Tokyo)* **121,** 746–755.

Haugland, R. P (1996). Introduction to potentiometric probes. *In* "Handbook of Fluorescent Probes and Research Chemicals," 6th Ed., pp. 586–592. Molecular Probes Inc., Eugene.

Helfenbaum, L., Ngo, A., Ghelli, A., Linnane, A. W., and Degli Esposti, M. (1997). Proton pumping of mitochondrial complex I: Differential activation by analogs of ubiquinone. *J. Bioenerg. Biomembr.* **29,** 71–80.

Hennet, T., Richter, C., and Peterhans, E. (1993). Tumor necrosis factor-α induces superoxide anion generation in mitochondria of L929 cells. *Biochem. J.* **289,** 585–592.

Huser, J., Rechenmacher, C. E., and Blatter, L. A. (1998). Imaging the permeability pore transitions in single mitochondria. *Biophys. J.* **74,** 2129–2137.

Imlay, J. A. (1995). A metabolic enzyme that rapidly produces superoxide: Fumarate reductase of *Eschericia coli. J. Biol. Chem.* **270,** 19767–19773.

Kluck, R., Degli Esposti, M., Perkins, G., Renken, C., Kuwana, T., Bossy-Wetzel, E., Goldberg, M., Allen, T., Barber, M. J., Green, D. R., and Newmeyer, D. D. (1999). The pro-apoptotic proteins, Bid and Bax, cause a limited permeabilization of the mitochondrial outer membrane that is enhanced by cytosol. *J. Cell Biol.* **147,** 809–822.

Korshunov, S. S., Skulachev, V. P., and Starkov, A. A. (1997). High protonic potential actuates a mechanism of production of reactive oxygen species in mitochondria. *FEBS Lett.* **416,** 15–18.

Krohn, A. J., Wahlbrink, T., and Prehn, J. H. (1999). Mitochondrial depolarization is not required for neuronal apoptosis. *J. Neurosci.* **19,** 7394–7404.

La Piana, G., Fransvea, E., Marzulli, D., and Lofrumento, N. E. (1998). Mitochondrial membrane potential supported by exogenous cytochrome *c* oxidaton mimics the early stages of apoptosis. *Biochem. Biophys. Res. Commun.* **246,** 556–561.

Loschen, G., Flohe, L., and Chance, B. (1971). Respiratory chain linked H_2O_2 production in pigeon heart mitochondria. *FEBS Lett.* **18,** 261–264.

Luvisetto, S., Schmehl, I., Cola, C., and Azzone, G. F. (1991). Tracking the proton flow during transition from anaerobiosis to steady state. *Eur. J. Biochem.* **202,** 113–120.

Macho, A., Castedo, M., Marchetti, P., Aguilar, J. J., Decaudin, D., Zamzami, N, Girard, P. M., Uriel, J., and Kroemer, G. (1997a). Mitochondrial dysfunctions in circulating T lymphocytes from human immunodeficiency virus-1 carriers. *Blood* **86,** 2481–2487.

Macho, A., Decaudin, D., Castedo, M., Hirsch, T., Susin, S. A., Zamzami, N., and Kroemer, G. (1996). Chloromethyl-X-rosamine is an aldehyde-fixable potential-sensitive fluorochrome for the detection of early apoptosis. *Cytometry* **25,** 333–340.

Macho, A., Hirsch, T., Marzo, I., Marchetti, P., Dallaporta, B., Susin, S. A., Zamzami, N., and Kroemer, G. (1997b). Glutathione depletion is an early and calcium elevation is a late event of thymocite apoptosis. *J. Immunol.* **158,** 4612–4619.

Metivier, D., Dallaporta, B., Zamzami, N., Larochette, N., Susin, S. A., Marzo, I., and Kroemer, G. (1998). Cytofluorometric detection of mitochondrial alterations in early CD95/Fas/APO-1-triggered apoptosis of Jurkat T lymphoma cells: Comparison of seven mitochondrial-specific fluorochromes. *Immunol. Lett.* **61,** 157–163.

Nicholls, D. G., and Ferguson, S. J. (1992). *"Bioenergetics 2."* Academic Press, London.

Nieminen, A. L., Byrne, A. M., Herman, B., and Leemasters, J. J. (1997). Mitochondrial permeability transition in hepatocytes induced by *t*-BuOOH: NAD(P)H and reactive oxygen species. *Am. J. Physiol.* **272,** C1286–C1294.

Pitkanen, S., and Robinson, B. H. (1996). Mitochondrial complex I deficiency leads to increased production of superoxide radicals and induction of superoxide dismutase. *J. Clin. Invest.* **98,** 345–351.

Poot, M., and Pierce, R. H. (1999). Detection of changes in mitochondrial function during apoptosis by simultaneous staining with multiple fluorescent dyes and correlated multiparameter flow cytometry. *Cytometry* **35,** 311–317.

Rottenberg, H. (1989). Proton electrochemical potential gradients in vesicles, organelles and prokaryotic cells. *Methods Enzymol.* **172,** 63–84.

Rieske, J. S. (1967). Preparation and properties of reduced coenzyme Q-cytochrome c reductase (complex III) of the respiratory chain. *Methods Enzymol.* **10,** 239–245.

Royall, J. A., and Ischiropoulos, H. (1993). Evaluation of 2′,7′-dichlorofluorescin and dihydrorhodamine 123 as fluorescent probes for intracellular H_2O_2 in cultured endothelial cells. *Arch. Biochem. Biophys.* **302,** 348–355.

Rustin, P., Chretien, D., Bourgeron, T., LeBidois, J., Sidi, D., Rotig, A., and Munnich, A. (1993). Investigation of respiratory chain activity in human heart. *Biochem. Med. Metab. Biol.* **50,** 120–126.

Salvioli, S., Ardizoni, A., Franceschi, C., and Cossarizza, A. (1997). JC-1, but not $DiOC_6(3)$ or rhodamine 123, is a reliable fluorescent probe to assess $\Delta\Psi$ changes in intact cells: Implications for studies on mitochondrial functionality during apoptosis. *FEBS Lett.* **411,** 77–82.

Scaduto, R. C., and Grotyohann, L. W. (1999). Measurement of mitochondrial membrane potential using fluorescent rhodamine derivatives. *Biophys. J.* **76,** 469–477.

Scorrano, L., Petronilli, V., Colonna, R., Di Lisa, F., and Bernardi, P. (1999). Chloromethyltetramethylrosamine (Mitotracker Orange) induces the mitochondrial permeability transition and inhibits respiratory complex I. *J. Biol. Chem.* **274,** 24657–24663.

Smith, J. C. (1990). Potential-sensitive molecular probes in membranes of bioenergetic relevance. *Biochim. Biophys. Acta* **1016,** 1–28.

Trounce, I. A., Kim, Y. L., Jun, A. S., and Wallace, D. C. (1996). Assessment of mitochondrial oxidative phosphorylation in patient muscle biopsies, lymphoblasts, and transmitochondrial cell lines. *Methods Enzymol.* **264,** 484–509.

Vlessis, A. A. (1990). NADH-linked substrate dependence of peroxide-induced respiratory inhibition and calcium efflux in isolated renal mitochondria. *J. Biol. Chem.* **265,** 1448–1453.

Wolter, K. G., Hsu, Y. T., Smith, C. L., Nechushtan, A., Xi, X.-G., and Youle, R. J. (1997). Movement of Bax from the cytosol to mitochondria during apoptosis. *J. Cell Biol.* **139,** 1281–1292.

Zamzami, N., Marchetti, P., Castedo, M., Zanin, C., Vayssiere, J. L., Petit, P. X., and Kroemer, G. (1995). Reduction in mitochondrial potential constitutes an early irreversible step of programmed lymphocite death *in vivo*. *J. Exp. Med.* **181,** 1661–1672.

CHAPTER 5

Assaying Mitochondrial Respiratory Complex Activity in Mitochondria Isolated from Human Cells and Tissues

Mark A. Birch-Machin* and Douglass M. Turnbull†

Departments of *Dermatology and †Neurology
Medical School, University of Newcastle upon Tyne
Newcastle upon Tyne, NE2 4HH, United Kingdom

I. Introduction

Mitochondrial cytopathies are a heterogeneous group of multisystem disorders predominantly affecting skeletal and cardiac muscle and the central nervous system, but are associated with a broad spectrum of other clinical phenotypes, including neurodegenerative disease (Chinnery and Turnbull, 1997; Wallace, 1999). In many patients, the impairment of mitochondrial respiratory chain function is due to a mutation in mitochondrial DNA (mtDNA), which may take the form of a rearrangement or a point mutation in a tRNA, rRNA, or protein-encoding gene (Chinnery and Turnbull, 1999). Despite major advances in the molecular investigation of patients with disease due to respiratory chain abnormalities, there are still many unresolved problems in the biochemical investigation of patients, and there is frequently a lack of correlation between biochemical and molecular genetic abnormalities. We believe that this is due, in part, to the limitation of some biochemical assays used in the diagnosis of these conditions to detect partial defects of the respiratory chain complexes. Therefore, accurate biochemical investigation to determine the site and severity of the defect is essential, together with an understanding of the control that each individual complex has on overall substrate oxidation.

II. Preparation of Mitochondrial Fractions

All of the following steps are performed at 0–4°C. Centrifugation conditions are reproduced exactly by the use of an ω^2t-integrator. The recoveries of mitochondria are about 30% of theoretical, and mitochondria obtained are intact and well coupled, as shown by the respiratory control and ADP:O ratios.

A. Tissue (Muscle)

Mitochondrial fractions can be prepared by differential centrifugation of homogenates from a variety of tissues. Whereas some tissues, such as liver, are easily dispersed to release subcellular organelles, the tough myofibrillar framework of striated muscle requires more vigorous disruption. A large number of homogenization techniques have been developed to prepare mitochondria from skeletal muscle, and the different methods have been reviewed (Sherratt *et al.,* 1988). The methodology used must reflect a compromise between a high yield of mitochondria, which requires thorough disruption of the tissue, and the need to preserve mitochondrial integrity. Some workers have used proteolytic enzymes, either before or after mechanical disruption, to improve the purity and yield of the mitochondrial fraction. Despite the possibility of obtaining better mitochondrial yields by the use of proteolytic enzymes, we use vigorous homogenization and differential centrifugation alone because of our concern about the proteases digesting potentially important enzymes on or in the outer mitochondrial membrane.

1. Reagents

Medium A: KCl, 120 m*M* 4-(2-hydroxyethyl)piperazine-1-(2-ethanesulfonic acid) (HEPES), 20 m*M,* pH 7.4 (at 20°C); $MgCl_2$, 2 m*M*; ethyleneglycolbis(β-aminoethyl ether)-N,N,N^1,N^1-tetraacetic acid (EGTA), 1 m*M*; bovine serum albumin (BSA; defatted), 5 mg/ml

Medium B: Sucrose, 300 m*M*; HEPES, 2 m*M,* pH 7.4 (at 20°C); EGTA, 0.1 m*M*

2. Protocol

Human muscle samples are obtained by biopsy from vastus lateralis of patients who have muscle pain but are subsequently shown to have no histochemical or biochemical evidence of muscle disease. In most cases a needle biopsy is taken, which provides approximately 50–150 mg of muscle tissue. In these circumstances, and when only milligram amounts of muscle (i.e., <150 mg) are available (such as from an open biopsy of young children), a mitochondrial homogenate (or muscle post-600-g supernatant) rather than a mitochondrial fraction is prepared.

Preparation of a Muscle Post-600 g Supernatant

The muscle sample is dissected out, weighed, and placed in ice-cold medium A. It is then trimmed of fat and connective tissues, chopped finely with a pair of scissors, and rinsed in medium A to remove any blood. The muscle is chopped further using a scalpel blade and forceps to ensure minimum loss of tissue. The disrupted muscle is then homogenized on ice using an Ultraturrex homogenizer (Scientific Instrument Centre Ltd., Liverpool, UL) (Ystral Y/20, setting 9 for 3.5 s) in a total volume of 1.0 ml of medium A. The homogenate is centrifuged at 4°C ($600g_{av}$ for 10 min) to remove nuclear debris, and the supernatant is respun ($600g_{av}$ for 10 min). Enzyme measurements are performed on this second supernatant.

Preparation of a Muscle Mitochondrial Fraction

For muscle samples >150 mg and for samples obtained by open biopsy of adults (typically 0.5–3 g of tissue may be acquired), we may proceed beyond the post-600*g* stage to prepare a mitochondrial fraction. Muscle is chopped finely as described earlier; the disrupted muscle is then made up to 20 volumes with respect to the original wet weight of tissue, with medium A, and homogenized with the Ultraturrex homogenizer (Ystral Y/20, setting 9 for 3.5 s). The homogenate is centrifuged at $600g_{av}$ for 10 min and the supernatant is filtered through four layers of cheesecloth to remove fat and fibrous tissue. The pellet is resuspended in 8 volumes of medium A with a hand-held Teflon/glass homogenizer (Potter–Elvejhem) and recentrifuged ($600g_{av}$ for 10 min). The second supernatant is filtered through four layers of cheesecloth and combined with the first. The pellet obtained at this stage is retained; it contains the nuclei and some mitochondria and is suitable for molecular genetic studies (Southern blotting, etc.). The combined supernatants are centrifuged ($17000g_{av}$ for 10 min), and pellets containing the mitochondria are resuspended in 10 volmes of medium A and then centrifuged at $7000g_{av}$ for 10 min. The pellets are resuspended in 10 volumes of medium B and centrifuged ($3500g_{av}$ for 10 min), and the mitochondrial fraction is finally suspended in a small volume of medium B (~25–50 mg protein/ml). If large amounts of animal mitochondria are being prepared, this resuspension is performed using a hand-held Teflon/glass homogenizer. However, if human muscle is used, resuspension is performed by gently detaching the pellet and mixing mitochondria using a plastic Pasteur pipette. This procedure minimizes any loss of mitochondria. In addition, we further wash the centrifuge tube with 1.25 ml of medium B and transfer this suspension to an Eppendorf centrifuge tube and centrifuge at $11{,}000g_{av}$

for 10 min. Any mitochondria pelleted during this procedure are kept but are used for enzyme assays only if the original mitochondrial pellet is fully used. The mitochondrial suspension is kept at 0°C for the studies of mitochondrial substrate oxidation. The rest of the suspension is divided into aliquots and is stored at −85°C for further biochemical and molecular studies.

3. Comments

Bovine serum albumin is included in the homogenization medium to prevent uncoupling of mitochondria by long-chain fatty acids and lysolecithins that are released during homogenization.

B. Cultured Cells (Myoblasts and Fibroblasts)

Human myoblasts and skin fibroblasts are cultured under standard conditions as described by Singh-Kler *et al.* (1991) and Clark *et al.* (1999), respectively. Cultured cells are harvested from six to eight confluent culture flasks (80 cm^2) using trypsin. Cells are used between passages 6 and 15, as the percentage recovery of mitochondria in the final fraction decreases with increasing passage number (e.g., in skin fibroblasts there was 37.5% recovery from cells in passage 12 but only 21.3% in passage 26).

1. Reagents

Medium A: Sucrose, 250 m*M*; HEPES, 2 m*M*, pH 7.4 (at 20°C); EGTA, 0.1 m*M*
Medium B: Potassium phosphate, 25 m*M* (pH 7.2) and $MgCl_2$, 5 m*M*

2. Protocol

All procedures are carried out at 0–4°C. The fibroblasts are suspended in 1 ml of medium A and pelleted by centrifugation at $327g_{av}$ for 10 min. The washed cell pellet is homogenized by 10 passes in a tight-fitting glass/Teflon power-driven Potter–Elvejhem homogenizer in approximately 4 ml of medium A. The homogenate is separated by centrifugation at $571g_{av}$ for 10 min, yielding a mitochondria-rich supernatant and a cell debris pellet that is resuspended in medium A and rehomogenized and spun as before. The supernatants are combined, and mitochondria are pelleted by centrifugation ($14{,}290g_{av}$ for 10 min). The resulting mitochondrial pellet is resuspended in 50–100 μl of medium A, giving a protein concentration of about 5 mg/ml. The mitochondrial fraction can be stored at −85°C. Prior to measurement of complex I and complex II activities, the mitochondrial fractions are purified further by suspension in hypotonic medium B and centrifugation at $15{,}339g_{av}$ for 10 min in a microcentrifuge. The resulting washed pellets are resuspended in fresh buffer to a protein concentration of approximately 1 mg/ml, using a microfuge tube homogenizer followed by vortexing. The samples are then freeze-thawed three times using liquid nitrogen.

The sensitivity of the enzyme assays allows for the use of small amounts of mitochondrial material. Enzyme activities are measured using a range of 5–20 μg of protein per assay. Because the preparation of the mitochondrial fraction from six to eight 80-cm^2 flasks yields a total of approximately 300–500 μg of protein, there is enough for the measurement of all the enzyme activities described later.

3. Comments

This mitochondrial preparation method is based on that described by Singh Kler *et al.* (1991), which involved much larger quantities of fibroblasts and the use of a hand-held ground glass homogenizer and a glass/Teflon power-driven Potter–Elvejhem homogenizer. In order to reduce the preparation time and variability, we evaluated the effects on mitochondrial recovery of removing the hand-held ground glass homogenization stage. The percentage recovery of mitochondria in the final pellet, compared to the initial homogenate, was $33.8 \pm 4.4\%$ [mean $\pm$ SD ($n = 3$)] using the glass/Teflon homogenization only, which was comparable to that obtained by the original method [$33.6 \pm 6.3\%$; mean $\pm$ SD ($n = 5$)].

III. Spectrophotometric Measurement of the Activities of Individual Complexes I–IV

All assays are performed at 30°C in a final volume of 1 ml using a Hitachi 557, an SLM Aminco DW2000, or a Perkin-Elmer Lambda 2 spectrophotometer.

A. Complex I: NADH:Ubiquinone Oxidoreductase Activity

1. Reagents

Assay medium: Potassium phosphate, 25 m*M*, pH 7.2 at 20°C; $MgCl_2$, 5 m*M*; KCN 2 m*M*; bovine serum albumin (fraction V), 2.5 mg/ml; NADH, 13 m*M*

$Ubiquinone_1$, 65 m*M* in ethanol

Antimycin A, 1 mg/ml in ethanol

Rotenone, 1 mg/ml in ethanol

2. Protocol

Complex I specific activity is measured by following the decrease in absorbance due to the oxidation of NADH at 340 nm, with 425 nm as the reference wavelength ($\varepsilon = 6.81$ $mM^{-1}\ cm^{-1}$). NADH (0.13 m*M*), $ubiquinone_1$ (65 μ*M*), and antimycin A (2 μg/ml) are added to the assay medium and the absorbance change is recorded for 1–2 min. Mitochondria (20–50 μg of protein) are added, and the NADH:ubiquinone oxidoreductase activity is measured for 3–5 min before the addition of rotenone (2 μg/ml), after which the activity is measured for an additional 3 min. Complex I activity is the rotenone-sensitive NADH:ubiquinone oxidoreductase activity.

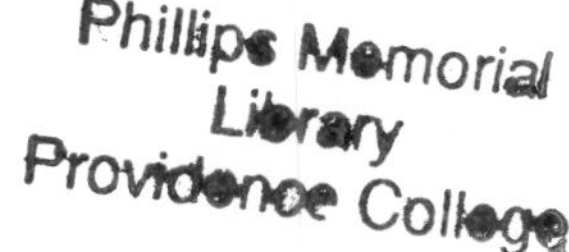

Table I
Effect of Membrane Disruption on Complex I Activity in Human Skeletal Muscle Mitochondria

Mitochondrial treatment	Complex I activity[a]
Freeze-thaw in hypotonic media	100 (90)
Freeze-thaw in hypotonic media + dodecyl maltoside	94.8 ± 14.8 (91)
Sonication in isotonic media + dodecyl maltoside	92.0 ± 9.0 (87)
Sonication in isotonic media	71.5 ± 17.1 (80)
Hypotonic shock	60.0 ± 253 (83)
Freeze-thaw in isotonic media	44 ± 13.4 (64)

[a]Expressed as percentage values, those obtained by freeze-thawing in hypotonic solution representing 100%. Results represent the mean ± SD for five different muscle samples. Figures in parentheses represent the average percentage of total NADH–ubiquinone oxidoreductase activity that was rotenone sensitive. The samples were freeze-thawed three times in either isotonic media (250 m*M* sucrose, 1 m*M* EGTA, 2 m*M* HEPES, pH 7.4) or hypotonic media (20 m*M* potassium phosphate, pH 7.2). Dodecyl maltoside was added to the samples at a concentration of 0.5:1 detergent to protein ratio and incubated for 10 min at 4°C prior to measurement.

3. Comments

The accurate measurement of complex I activity in a mitochondrial fraction depends on access of NADH to its binding site on the inner aspect of the inner mitochondrial membrane. To establish the optimum conditions for the release of complex I activity, we compared several established methods of membrane disruption, such as freeze-thawing mitochondria in hypotonic or isotonic media, sonication (Zheng *et al.*,1990), hypotonic shock (Chretien *et al.*, 1990), or the addition of detergent (Table I). The nonionic detergent, dodecyl maltoside (*n*-dodecyl β-D-maltoside), was selected for the experiment following the result of a comparative study of the ability of several detergents (deoxycholate, CHAPS, and Lubrol) to solubilize complex I. Maximum complex I activity was obtained by freeze-thawing the mitochondria three times in *hypotonic* media [25 m*M* potassium phosphate (pH 7.2), 5 m*M* $MgCl_2$]. In addition, the freeze-thaw technique is simpler and easier to perform with small samples, whereas sonication requires extra equipment and has the added risk of microdroplet formation, which is always a concern with human samples. In light of these experiments, all further studies on complex I were performed using hypotonically freeze-thawed mitochondrial preparations.

Complex I activity is linear with protein over a range of at least 15–70 μg of human muscle mitochondria and up to 82 and 520 μg of rat liver and rat muscle mitochondrial protein, respectively. The concentrations of NADH and ubiquinone used are in excess of the K_m for both, i.e., 2 μM (Friedrich *et al.*, 1989; Majander *et al.*, 1991) and 8 μM (Friedrich *et al.*,1989), respectively. Both $ubiquinone_1$ and $ubiquinone_2$ can be used as electron acceptors, but the rate with $ubiquinone_1$ is faster than that obtained using $ubiquinone_2$ [complex I activity was 195 ± 34 and 82 ± 26 nmol/min per mg protein when measured with $ubiquinone_1$ and $ubiquinone_2$, respectively (mean ± SD for the

Table II
Rotenone Sensitivity of Total NADH–Ubiquinone Oxidoreductase Activity in Mitochondria Fractions from Different Tissues[a]

Tissue	% of total enzyme activity that is rotenone sensitive
Human skeletal muscle	83.8 ± 5.0 (23)
Human liver	55 ± 9.0 (4)
Human platelets	52.8 ± 9.7 (10)
Human fibroblasts	51.8 ± 12.4 (5)

[a]Mitochondrial fractions were isolated from different tissues and NADH–ubiquinone oxidoreductase was measured in the absence and presence of rotenone. The percentage of the total activity that was inhibited by rotenone is shown for each mitochondrial fraction; the values are means ± SD, with the number of experiments in parentheses.

same three human muscle samples)]. These results are similar to the findings reported for the purified enzyme from bovine heart (Hatefi and Rieske, 1967).

As complex I activity is measured as the rotenone-sensitive NADH:ubiquinone oxidoreductase activity, it is important to establish that maximum rotenone binding is achieved. When NADH:ubiquinone oxidoreductase activity is measured in human muscle mitochondrial fractions in the absence of bovine serum albumin, only 52.1 ± 5% (mean ± SD for 10 samples) of total activity is rotenone sensitive. The addition of 2.5 mg ml of bovine serum albumin to the assay medium increased rotenone sensitivity to 83.8 ± 5.0%. No increase in rotenone sensitivity is achieved by increasing further the concentration of bovine serum albumin or by the addition of phospholipid. The proportion of the total NADH:ubiquinone oxidoreductase activity that is rotenone sensitive varies considerably in different tissues (Table II). This, in part, is related to the relative purity of the mitochondrial preparation, and therefore to the degree of contamination by other enzymes with NADH:ubiquinone oxidoreductase activity. A potentially important enzyme contributing to the rotenone-insensitive NADH:ubiquinone oxidoreductase is NADH:cytochrome b_5 reductase (EC 1.6.2.2). This enzyme is involved in the oxidation of NADH generated in the cytosol and is situated on the outer mitochondrial membrane and on the endoplasmic reticulum (Sottocasa *et al.,* 1967). Using an antibody raised against purified NADH:cytochrome b_5 reductase, we have shown that this enzyme makes an important contribution to the rotenone-insensitive NADH:-ubiquinone oxidoreductase activity (Birch-Machin *et al.,* 1994). This has important implications for mitochondria prepared from tissues, other than muscle, that have a high percentage of rotenone-insensitive activity, as under these circumstances it is difficult to detect a partial decrease in rotenone-sensitive NADH oxidation [e.g., lymphocytes (Rustin *et al.,* 1994)]. There is considerable rotenone-insensitive NADH:ubiquinone oxidoreductase activity in fibroblast mitochondrial fractions. Washing in hypotonic buffer appears to partially remove this contaminating enzyme activity and results in an increase in the percentage *of activity* due to complex I (the rotenone-sensitive component) from 48.2 ± 12.4% (n = 50) to 68.1 ± 7.0% (n = 13) (mean ± SD). Additionally, this washing procedure increases the specific activities of both complex I and complex II from 40 ± 17 (n = 9) to 63 ± 15 nmol·min^{-1}·mgprotein^{-1} (n = 10) (mean ± SD).

B. Complex II: Succinate:Ubiquinone Oxidoreductase Activity

1. Reagents

Assay medium: Potassium phosphate, 25 m*M*, pH 7.2 at 20°C; $MgCl_2$, 5 m*M*

Sodium succinate, 1 *M*

KCN, 1 *M*

2,6-Dichlorophenolindophenol, 5 m*M*

$Ubiquinone_1$ or $ubiquinone_2$, 56 m*M* in ethanol

Antimycin A, 1 mg/ml in ethanol

Rotenone, 1 mg/ml in ethanol

2. Protocol

Complex II specific activity is measured by following the reduction of 2,6-dichlorophenolindophenol at 600 nm (ε = 19.1 mM^{-1} cm^{-1}). Mitochondria (10–50 μg of protein) are preincubated in the assay medium plus succinate (20 m*M*) at 30°C for 10 min. Antimycin A (2 μg/ml), rotenone (2 μg/ml), KCN (2 m*M*), and dichlorophenolindophenol (50 μ*M*) are added, and a baseline rate is recorded for 3 min. The reaction is started with ubiquinone (65 μ*M*), and the enzyme-catalyzed reduction of dichlorophenolindophenol is measured for 3–5 min.

3. Comments

In isolated mitochondria, a variable proportion of complex II is deactivated due to the tight binding of the competitive inhibitor oxaloacetate. To measure total complex II activity it is essential to ensure that the enzyme is fully activated, which can be achieved by preincubation with succinate. Comparison of the rate after activation (incubating at 30°C for 10 min with 20 m*M* succinate) with that achieved when the mitochondrial fraction is not activated showed that the latter is only 60% of the fully activated rate. There is no increase in activity by preincubation at a higher temperature (38°C), by incubation for a longer period of time (30°C for 20 min.), or with a higher concentration of succinate (40 m*M*).

Complex II activity is also dependent on the disruption of the inner mitochondrial membrane. The activity in mitochondrial fractions freeze-thawed in hypotonic medium (25 m*M* potassium phosphate, pH 7.2, 5 m*M* $MgCl_2$) is 293 ± 55 nmol/min/mg protein (mean ± SD for five human skeletal muscle mitochondrial fractions), compared with

147 ± 41 nmol/min/mg protein for the same fractions freeze-thawed in isotonic medium. No further increase in activity is found when dodecyl maltoside or Triton X-100, are added to the hypotonically freeze-thawed mitochondria at detergent:protein ratios used to investigate complex I.

Complex II activity is linear with mitochondrial protein over a range of at least 5–50 μg for human muscle, rat muscle, and rat liver mitochondria. The rate of succinate:ubiquinone oxidoreductase activity is greater when dichlorophenolindophenol is used as a terminal electron acceptor [316 ± 82 nmol/min/mg protein (mean ± SD for nine skeletal muscle mitochondrial fractions)] as compared with the rate (100 ± 30 nmol/min per/mg protein) measured with ubiquinone$_1$ (measured by following reduction of ubiquinone$_1$ at 280 nm, with 465 nm as a reference wavelength, $\varepsilon = 13\ \text{m}M^{-1}\ \text{cm}^{-1}$). In contrast to complex I, the activity of complex II, when measured with dicholorophenol as the electron acceptor, is the same whether ubiquinone$_1$ or ubiquinone$_2$ is used in the assay with human skeletal muscle mitochondrial fractions [278 ± 46 nmol/min/mg protein with ubiquinone$_1$, 284 ± 50 nmol/min/mg protein with ubiquinone$_2$ (mean ± SD for the same four samples)].

C. Complex III

In mammals, complex III contains 11 subunits (Gonzalez-Halphen *et al.*, 1988), which includes three electron transfer proteins: cytochrome c_1, the Rieske iron–sulfur protein, and cytochrome *b*, which contains two *b* heme groups. These two heme groups are designated b_H and b_L to denote their approximate E_m values of +50 and −50 mV, respectively. In the mammalian system, the spectral peaks of the b_H and b_L heme groups occur at different wavelengths and therefore may also be designated as b_{562} and b_{566}, respectively.

The proposed mechanism by which electron transport through the three electron transfer proteins is linked to proton translocation is the proton-motive Q cycle. The essential features of this model are that there must be two sites or centers that react with quinones and that electrons are transferred within a transmembranous cytochrome *b* (Fig. 1). The redox potential of the b_{562} heme is approximately 100 mV more positive than that of the b_{566} heme and is found at the high potential center (Q_i) on the proton input side of the membrane, whereas the b_{566} heme is at the low potential center (Q_o), which is on the proton output side of the inner membrane. Electrons derived from the oxidation of quinol are not only transferred within the transmembranous cytochrome *b*, but also to the iron–sulfur protein, then to the heme portion of cytochrome c_1, and finally to cytochrome *c*. As a result of the net oxidation of one molecule of ubiquinol, complex III transfers two electrons to two molecules of cytochrome *c*, four protons are released at the proton output side of the membrane, while two protons are taken up at the proton input side (Fig. 1).

There appears to be at least three groups of complex III inhibitors. Class I inhibitors, such as myxothiazol (Von Jagow and Link, 1986) and stigmatellin, block electron transport to the b_{566} heme and the iron–sulfur protein through binding at the low potential Q_o center (Fig. 1). Electron transport to the b_{566} heme is also blocked by class II inhibitors,

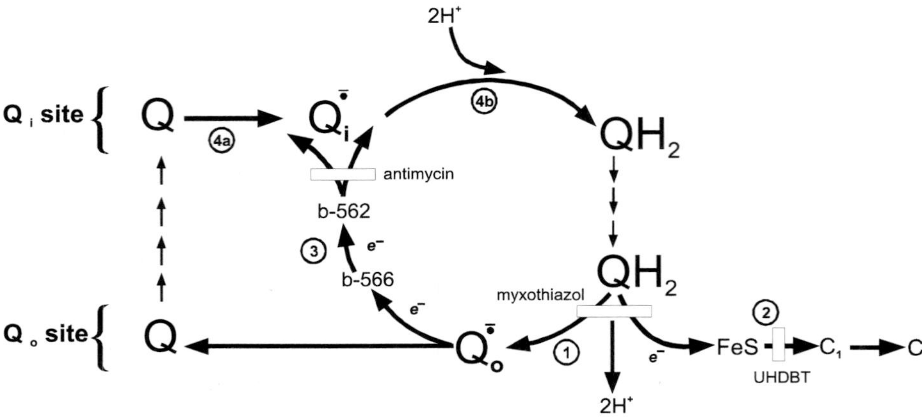

Fig. 1 Protonmotive Q cycle mechanism of electron transfer and proton translocation at complex III. The proposed branched cyclic pathway of electron transfer from ubiquinol (QH_2) to cytochrome *c* (C) is shown. The numbered circles are electron transfer reactions. The key feature of the Q cycle is the two centers of quinol oxidation/quinone reduction. In the Q_o center, at the proton output side of the membrane, step 1 describes the transfer of one electron from a ubiquinol molecule to the high potential iron–sulfur protein (FeS), with the resulting low potential and highly unstable ubisemiquinone anion ($Q_o\bar{\bullet}$) immediately transferring its electron to the b_{566} heme group. Coincident with these events is the release of two protons at the Q_o surface of the membrane. In step 2 the electron on the FeS is transferred to cytochrome c_1 and then to cytochrome *c*. In a probably simultaneous event in step 3 the electron on the b_{566} heme is transferred to the b_{562} heme. In step 4a the latter heme reduces ubiquinone to the relatively stable ubisemiquinone anion ($Q_i\bar{\bullet}$). When heme b_{562} is again reduced by a repeat of the series of reactions just described, the heme transfers its electron ($Q_i\bar{\bullet}$) to form ubiquinol (step 4b). Coincident with this event is the uptake of two protons at the Q_i surface of the membrane. Therefore, two molecules of ubiquinol are oxidized during one complete Q cycle. Open boxes show the sites at which myxothiazol, UHDBT (5-*n*-undecyl-6hydroxy-4,7-dioxobenzoxythiazole), and antimycin A inhibit electron transfer.

such as hydroxyquinone analogues, but electron flow between the iron–sulfur protein and cytochrome c_1 is prevented. Class III inhibitors, such as antimycin (Von Jagow and Link, 1986), block electron flow between quinone and the b_{562} heme by binding at or near the Q_i center. The following methods allow investigators to assess mitochondrial dysfunction at complex III induced either by acute titration experiments or by human disease.

i. Ubiquinol:ferricytochrome *c* oxidoreductase

1. Reagents

Assay medium: Potassium phosphate, 25 m*M,* pH 7.2 at 20°C; $MgCl_2$, 5 m*M*; bovine serum albumin (fraction V), 2.5 mg/ml; KCN, 2 m*M*

Cytochrome *c* (III), 2.5 m*M*

n-Dodecyl-β-D-maltoside, 30 m*M* (15 mg/ml)

Ascorbic acid, solid
Ubiquinol$_2$, 35 m*M* in ethanol
Rotenone, 1 mg/ml in ethanol

2. Protocol

Complex III specific activity is measured by monitoring the reduction of cytochrome *c* (III) at 550 nm, with 580 nm as the reference wavelength ($\varepsilon = 19\ mM^{-1}\ cm^{-1}$). All assays are performed at 30°C in a final volume of 1 ml. Cytochrome *c* (III) (15 μ*M*), rotenone (2 μg/ml), dodecyl-β-D-maltoside (0.6 m*M*), and ubiquinol$_2$ (35 μ*M*) are added to the assay medium, and the nonenzymatic rate is recorded for 1 min. Mitochondria (5–20 μg of protein) are added, and the increase in absorbance is measured. The increase in absorbance rapidly becomes nonlinear, and activity is expressed as an apparent first-order rate constant after reduction of the remaining cytochrome *c* (III) with a few grains of ascorbate. Ubiquinol is prepared by dissolving ubiquinone (10 μmol) in 1 ml of ethanol, acidified to pH 2 with 6 *M* HCl. The quinone is reduced with excess solid sodium borohydride. Ubiquinol is extracted into diethylether:cyclohexane (2:1, v/v) and evaporated to dryness under nitrogen gas, dissolved in 1 ml of ethanol acidified to pH 2 with HCl; this solution is stable at −70°C for 1 year.

3. Comments

The inclusion of KCN in the assay media prevents the reoxidation of the product, cytochrome *c*(II), by inhibition of complex IV of the respiratory chain (cytochrome *c* oxidase). The assay medium also contains rotenone, which prevents any nonspecific changes in the ubiquinol concentration through inhibition of complex I activity. The measured complex III activity is linear, with a protein range of at least 2–30 μg for rat muscle and liver mitochondria and 5–25 μg for human skeletal muscle mitochondria. Using these protein concentrations the chemical rate of cytochrome *c* (III) reduction is less than 10% of the enzyme rate.

Because the respiratory chain enzymes are embedded in the inner mitochondrial membrane, it is important to ensure access of both ubiquinol and cytochrome *c* to the enzyme. The hypotonic freeze-thawing technique used to measure complex I and complex II activities is not appropriate for the measurement of complex III. Indeed, this technique consistently lowers complex III values by 40% compared to nonfreeze-thawed mitochondria. We have evaluated the effect of dodecyl maltoside on complex III activity in human skeletal muscle mitochondria and found that the activity of complex III increases twofold, but only with concentrations of dodecyl maltoside greater than 0.44 m*M* (Fig. 2). This increase in complex III activity by dodecyl maltoside is observed in mitochondrial fractions from pig heart and human muscle, and the effect is independent of the protein concentration (over the range of 4–24 μg human muscle mitochondria). Preincubation of mitochondria with detergent is not required and the detergent is simply added to the assay buffer. A further interesting observation is that the addition of dodecyl maltoside

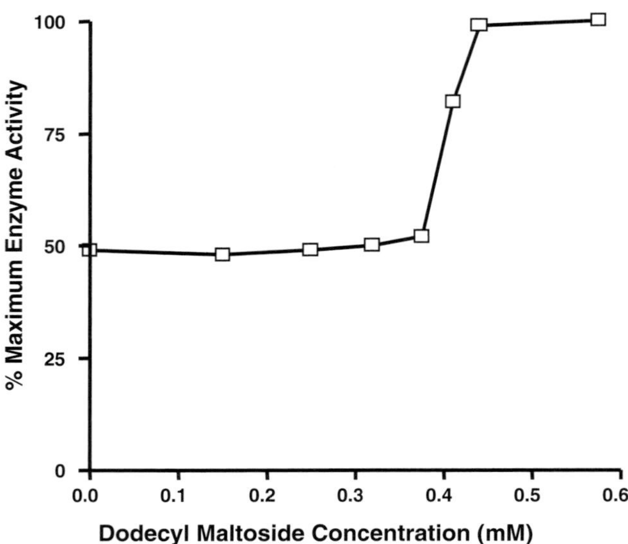

Fig. 2 The effect of dodecyl maltoside on complex III activity in human skeletal muscle mitochondrial fractions. Enzyme activities are expressed as a percentage of the maximum values measured at the appropriate detergent concentration.

(>0.44 m*M*) to mitochondria that had previously been hypotonically freeze-thawed stimulates complex III activity to a greater extent so that the final value is identical to that of the nonfreeze-thawed mitochondria treated with dodecyl maltoside. Detergent concentrations that give maximal enzyme activity are in excess of the critical micellar concentration (0.16 m*M*) of dodecyl maltoside and suggest that the BSA present in the medium is interfering with the effect of the detergent (possibly through its liganding with detergents). However, the inclusion of BSA is necessary because in its absence the reliability of the assay decreases. The mechanism of the effect of dodecyl maltoside on complex III is uncertain; the detergent may affect the solubility of the ubiquinol, alter the conformation of the enzyme, causing a monomer to dimer transition (Nalecz *et al.,* 1985), or allow greater access of cytochrome *c*.

The natural substrate for complex III is ubiquinol_{10}, but this compound is insoluble in aqueous solution, and the use of low molecular weight ubiquinols such as ubiquinol_1 and ubiquinol_2 is a reasonable compromise. Some workers (Zheng *et al.,* 1990) have used custom-synthesized *n*-decyl coenzyme Q-ol. Complex III activity is dependent on the length of the isoprenoid chain of the ubiquinol homologue, and we have measured complex III activity in rat and human muscle mitochondrial fractions using duroquinol, ubiquinol_1, and ubiquinol_2. The activity using duroquinol and ubiquinol_1 was 12 and 48%, respectively, of the activity using ubiquinol_2 in human mitochondria and was 1% (duroquinol) and 12% (ubiquinol_1) of the activity using ubiquinol_2 in rat muscle mitochondria. The finding that activity increases with the length of the isoprenoid chain agrees with those studies on bovine heart submitochondrial particles (Fato *et al.,* 1988) and the purified enzyme (Yu *et al.,* 1985). Complex III activity is inhibited 95% by antimycin A (1 μg/ml) and the residual 5% electron leak through the Q_o center can be prevented by the addition of myxothiazol (1 μg/ml).

ii. Duroquinol-ubiquinone transhydrogenase activity

In addition to measuring flux through the whole complex, it is possible to measure electron flow via the b_{562} heme at the Q_i center (Fig. 1) using the duroquinol-ubiquinone transhydrogenase assay (Boveris *et al.,* 1971). The assay involves the reduction of heme b_{562} by duroquinol, which then reduces ubiquinone.

1. Reagents

Assay medium: HEPES, 50 m*M,* pH 7.0 at 20°C; KCl, 100 m*M*; EGTA, 1 m*M*; KCN, 2 m*M*

All of the following reagents are dissolved in absolute ethanol:

Duroquinol, 35 m*M*, which is prepared by the same method as that described earlier for ubiquinol

Ubiquinone, 4 m*M*

Myxothiazol, 1 mg/ml

Antimycin A, 1 mg/ml

2. Protocol

Duroquinol-ubiquinone transhydrogenase activity is measured by following the formation of ubiquinol from ubiquinone at 284/305 nm ($\varepsilon = 7.1$ mM^{-1}cm^{-1}). Duroquinol (0.1 m*M*) and mitochondria are added to the assay media (final volume of 1 ml at 30°C) 1 min prior to the addition of ubiquinone (8 μ*M*). The nonlinear decrease in absorbance is measured and the activity is expressed as an apparent first-order rate constant. This includes both the enzyme-catalyzed rate, which is inhibited by antimycin A, and a spontaneous chemical reaction.

3. Comments

The enzyme activity is unaffected by myxothiazol (2 μg/ml), which prevents electron flow to heme b_{566} at the Q_o center (Fig. 1). A high mitochondrial protein concentration is required to detect an enzyme rate, even with the advantages of dual-wavelength spectrophotometry. The reaction is linear for 140–300 μg of human heart and skeletal muscle and for 200–400 μg of human liver mitochondrial fraction. The assay has been used successfully in the diagnosis of a patient with a specific defect in complex III (Birch-Machin *et al.,* 1989).

D. Complex IV: Cytochrome *c* Oxidase

1. Reagents

Assay medium: Potassium phosphate, 20 m*M*, pH 7.0 at 20°C
Cytochrome *c* (II), 3 m*M*
Potassium hexacyanoferrate, solid
n-Dodecyl-β-D-maltoside, 30 m*M* (15 mg/ml)

2. Protocol

Complex IV specific activity is measured by following the oxidation of cytochrome *c* (II) at 550 nm with 580 nm as the reference wavelength ($\varepsilon = 19.1\ mM^{-1}\ cm^{-1}$). All assays are performed at 30°C in a final volume of 1 ml. Cytochrome c (II) (15 μ*M*) and dodecyl maltoside (0.45 m*M*) are added to the assay medium, and the nonenzymatic rate is recorded. Mitochondria (5–15 μg of protein) are added, and complex IV activity is measured either as the initial rate or as the apparent first-order rate constant after fully oxidizing cytochrome *c* (II) by the addition of a few grains of potassium hexacyanoferrate.

Cytochrome *c* (II) is prepared by the addition of ascorbate to cytochrome *c* (III), and cytochrome *c* (II) is then purified by Sephadex G-25 chromatography. Cytochrome *c* (II) is stored in aliquots at −85°C, and its concentration is determined spectrophotometrically before each batch of assays.

3. Comments

The kinetics of cytochrome *c* oxidase activity are complex, but the rate can be determined either as an initial rate or as a first-order rate constant using the spectrophotometric assay. The assay is linear with human skeletal muscle mitochondrial protein in the range of 0.5–10 μg when calculating as an initial rate and 2–35 μg using a first-order rate constant.

Similar to complex III, disruption of the inner mitochondrial membrane by freeze-thawing in hypotonic solution also lowers complex IV activity, but only by 15–20%. We have evaluated the effect of dodecyl maltoside on complex IV activity in human skeletal muscle mitochondria and have shown an approximate fourfold increase in activity (Fig. 3) when the concentration of dodecyl maltoside in the assay buffer was greater than the critical micellar concentration. Note that the complex IV values in Fig. 3 represent two different concentrations of mitochondrial protein; the effect of dodecyl maltoside on complex IV activity, like its effect on complex III, is independent of protein concentration. Studies on purified cytochrome *c* oxidase have shown that dodecyl maltoside stimulates activity by conferring an active conformation of the enzyme (Rosevear *et al.,* 1980; Mahapatro and Robinson, 1990). We assume that similar effects occur in the mitochondrial fraction, rather than simply allowing better access of cytocrome *c,* as other detergents, such as Triton X–100 (at concentrations of 0.07–3 m*M*), did not increase activity.

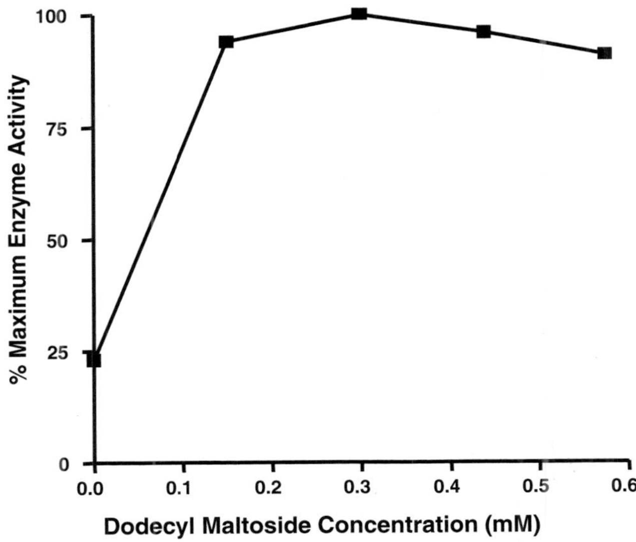

Fig. 3 The effect of dodecyl maltoside on complex IV activity in human skeletal muscle mitochondrial fractions. Enzyme activities are expressed as a percentage of the maximum values measured at the appropriate detergent concentration. The complex IV values are an average of two experiments at 3 and 5 μg of mitochondrial protein. The critical micellar concentration of dodecyl maltoside is 0.16 m*M*.

E. General Comments

It is important to realize that a single disruption technique cannot be applied uniformly in the measurement of individual respiratory chain complexes. We have compared the effects of different methods of disruption, and their combinations, on the activities of complexes I–IV. For complexes I and II we found that the most reproducible and easiest method to obtain maximum activity is to freeze-thaw the mitochondrial fraction in hypotonic media. While we found that dodecyl maltoside is an extremely effective detergent, we did not, however, observe an increase in complex I and II activity over and above that obtained by freeze-thawing.

In contrast, we found that hypotonic freeze-thawing is not the method of choice for the measurement of complexes III and IV activity. Such treatment resulted in a loss of activity, whereas there was a marked increase in the activity of both complexes III and IV by the addition of dodecyl maltoside that is independent of the mitochondrial protein concentration. The lability of complex IV observed by other groups (e.g., Zheng *et al.*, 1990) may well reflect the disruption technique used.

Our measurements of complexes I, II, and III have involved the use of the oxidized or reduced forms of duroquinone, ubiquinone$_1$, and ubiquinone$_2$. Complex I and III activities depend on the quinone analogue used, and so for comparative experiments it is important to maintain the same substrate.

IV. Linked Assays

Electron flow through complex III can be assessed by two linked assays that measure the segment of the respiratory chain linking the activity of complex II with that of complex III. In these linked assays, succinate is the reductant and the terminal electron acceptor is either hexacyanoferrate ([Fe(CN)6]3-) or cytochrome *c* (III).

A. Succinate-Hexacyanoferrate Reductase

1. Reagents

Assay medium: Potassium phosphate, 0.1 m*M*; HEPES, 10 m*M*, pH 7.2 at 20°C; KCl, 130 m*M*; ethylenediaminetetraacetate (EDTA), 1 m*M*; ADP, 0.25 m*M*; KCN, 2 mM; BSA (fraction V), 1.5 mg/ml

Rotenone, 1 mg/ml in ethanol

Potassium hexacyanoferrate, 50 m*M*

Sodium succinate, 1 M

Antimycin A, 1mg/ml in ethanol

2. Protocol

Enzyme activity is measured by following the reduction of the hexacyanoferrate (III) at 420 nm with 475 nm as the reference wavelength ($\varepsilon = 1.05\ \mathrm{m}M^{-1}\,\mathrm{cm}^{-1}$). Intact mitochondria (30–200 μg of protein), succinate (20 m*M*), and rotenone (2 μg/ml) are added to the assay medium (final volume 1 ml at 30°C) and incubated for 10 min at 30°C. The reaction is started by the addition of potassium hexacyanoferrate (0.5 m*M*), and the linear decrease in absorbance is recorded.

3. Comments

a. Principle of the Assay

The artificial electron acceptor hexacyanoferrate (III) is able to accept electrons derived from the oxidation of succinate in the mitochondrial matrix by reacting directly with the respiratory chain at the level of cytochrome *c*, which is located on the outer face of the mitochondrial inner membrane. The hexacyanoferrate ion has a higher midpoint potential, which enables it to accept electrons from cytochrome *c*. Although its molecular weight is small enough to allow it to cross the outer mitochondrial membrane, hexacyanoferrate does not cross the inner membrane and, is therefore, unable to react nonspecifically with reducing equivalents in the matrix. The reaction buffer is isotonic in order to ensure that the mitochondrial membranes remain intact. In this intact membrane state, the transfer of electrons from succinate is almost completely blocked by the addition of antimycin (2 μg/ml), as the succinate dehydrogenase component of complex II is located on the inner face of the inner membrane and electrons derived from succinate can only be conducted to an exogenous acceptor via complex III. ADP is included in the reaction buffer to maintain mitochondria in state 3 respiration, although mitochondria do become progressively uncoupled in the presence of hexacyanoferrate (III). Inclusion of KCN and rotenone ensures inhibition of complex IV and complex I activities, respectively. The succinate-hexacyanoferrate reductase activity is linear, with a protein range of at least 0.03–0.3 mg of liver or skeletal muscle mitochondria from rat or human.

b. Optimization of Activity

In isolated mitochondria, a variable proportion of complex II is deactivated due to the tight binding of the competitive inhibitor oxaloacetate. It is important to measure total complex II activity, as some evidence suggests that complex II activity is the rate-limiting step in the succinate-hexacyanoferrate reductase assay (see comments for the succinate-cytocrome *c* reductase assay for more details). Preincubation of mitochondria at 30°C for 10 min with 20 m*M* succinate ensures that complex II is fully activated, as it routinely increases enzyme activity by 1.7-fold (see section on measurement of complex II activity).

B. Succinate-Cytochrome *c* Reductase

1. Reagents

Assay medium: Potassium phosphate, 25 m*M*, pH 7.2 at 20°C; KCN, 2 m*M*

Rotenone, 1 mg/ml in ethanol

Cytochrome *c* (III), 5 m*M*

Antimycin A, 1 mg/ml in ethanol

Sodium succinate, 1 *M*

2. Protocol

Enzyme activity is measured by following the reduction of cytochrome *c* (III) at 550 nm with 580 nm as the reference wavelength ($\varepsilon = 19\ \text{m}M^{-1}\ \text{cm}^{-1}$). Mitochondria (10–50 μg of protein), succinate (20 m*M*), and rotenone (2 μg/ml) are added to the assay media (final volume 1 ml) and incubated for 10 min at 30°C. The reaction is started by the addition of 37.5 μ*M* cytochrome *c* (III), which gives a linear increase in absorbance.

3. Comments

The succinate-cytochrome *c* reductase (SCR) assay is used widely, and all variations are based on the method of Sottocasa *et al.* (1967). Some groups (e.g., Chretien *et al.*, 1998) include BSA (1 mg/ml) in the reaction, but we have not found this to be necessary. The concentrations of cytochrome *c* (III) used vary from 34 μ*M* (Sottocasa *et al.*, 1967) to 50 μ*M* (Zheng *et al.*, 1990). These concentrations are all in excess of the K_m for complex III and as such have no effect on enzyme activity, apart from increasing the duration over which the absorbance change is linear.

Optimization of Activity

To allow for access of the exogenously added cytochrome *c* to complex III, the mitochondrial fraction is disrupted by freeze-thawing in hypotonic media (25 m*M* potassium phosphate, pH 7.2). In contrast to its effect on complex III activity, the addition of dodecyl maltoside abolishes SCR activity; furthermore, sonication of mitochondria also decreases activity (Zheng *et al.,* 1990). Presumably both of these treatments result in separation of the two respiratory chain components in the linked assay.

C. Limitations of Linked Assays for the Diagnosis of Complex III Deficiency

One of the most commonly used assays in the investigation of patients with defects of the respiratory chain is the measurement of SCR. However, as with all linked or multicomponent assays, it measures several components of the respiratory chain, and the ability to detect a partial defect in one enzyme complex will depend on the amount of control exerted by that enzyme step on overall electron flux.

Therefore, in the SCR assay, it is uncertain to what extent a decrease in complex III activity would slow electron flux through the pathway. To answer this question we first showed that complex III exerts a low degree of control on electron flux through SCR in rat muscle mitochondria (Taylor *et al.,* 1994). Second, in human muscle mitochondria, titration of electron flow through complex III with specific inhibitors (myxothiazol and antimycin A) has shown that significant decreases in individual complex III activity (measured as ubiquinol-cytochrome *c* reductase) are not reflected in the SCR assay (Taylor *et al.,* 1993). *For example, inhibition of complex III by 50% (with either inhibitor) gives no change in flux through SCR.* This has important diagnostic implications as it suggests that a partial decrease in complex III activity would not be detected by measuring SCR activity. Indeed, further work in our laboratory has shown this to be the case in patients with partial defects of complex III (Taylor *et al.,* 1993). This disparity between complex III activity and the corresponding SCR measurement is highlighted by data from a typical complex III patient. This patient data clearly showed a partial defect of complex III (42% of control values), but the SCR activity (80% of control values) was within the normal range. The results for SCR in patients were confirmed by our findings using the succinate-hexacyanoferrate reductase (SHFR) assay. The specific activities of SCR and SHFR are similar, and this latter assay has the advantage of using an inorganic electron acceptor that is unlikely to be influenced by steric factors. This confirms that the measurement of electron flow from succinate to cytochrome c does not reliably detect partial defects of complex III.

In addition, we have investigated the control exerted by complex II on SCR activity by titrating complex II activity with malonate and have found that this complex represents a major point of control of flux through SCR. Thus it is likely that measurement of SCR activity will detect defects of complex II (Bourgeron *et al.,* 1995).

An additional problem with both succinate hexacyanoferrate reductase and succinate-cytochrome c reductase is that the standard deviation from the control mean is much larger than that obtained from the complex III assay.

V. Applications of Techniques to Diagnostic Investigations

The investigation of a patient with a suspected respiratory chain defect requires a combination of clinical, morphological, histochemical, biochemical, and genetic studies. An important component of the investigation of skeletal muscle mitochondrial function is the morphological and histochemical appearance of the muscle (Johnson, 1983). The presence of structurally abnormal mitochondria in increased concentration around the periphery of the muscle fiber is the classic morphological abnormality in patients with defects of the mitochondrial respiratory chain. Cytochemical techniques enable the measurement of succinate dehydrogenase by using a tetrazolium salt as an electron acceptor to yield a coloured formazan, whereas cytochrome *c* oxidase is demonstrated cytochemically by using 3,3′-diaminobenzidine tetrahydrochloride as an electron acceptor to the respiratory chain at the level of cytochrome *c* (Brierley *et al.,* 1998; Andrews *et al.,* 1999). Cytochemical staining for cytochrome *c* oxidase is particularly important in assessing abnormal mitochondrial function in patients with partial defects of this enzyme, which affect only a small proportion of the muscle fibers. Such partial defects of cytochrome *c* oxidase may not necessarily be found by enzyme assay of muscle homogenates or by oxidations of mitochondrial fractions (Birch-Machin *et al.,* 1993), as the proportion of negative fibers may be small.

In the light of this histochemical information, we believe that the optimization, particularly of the individual respiratory chain complex assays, forms an essential part of the diagnostic studies, as the detection of a partial defect of one respiratory chain complex is dependent on the measurement of maximal activity. This involves different methods of disruption of the mitochondrial membranes. Using several established methods of membrane disruption, we have found that optimal activities of complexes I and II are obtained using a combination of freeze-thawing mitochondria in hypotonic potassium phosphate buffer, whereas complex III and IV activities are increased markedly by the addition of the detergent *n*-dodecyl-β-D-maltoside. Therefore it is important to realize that a single disruption technique cannot be applied uniformly in the measurement of individual respiratory chain complexes. The lability of complex IV observed by some groups (Zheng *et al.,* 1990) may well reflect the disruption technique used.

The extent of the measurements made on the mitochondrial fractions will vary with the availability of equipment and quantity of mitochondria isolated from the tissue. Therefore, it is important to realize the limitations and benefits of each technique. Spectrophotometric measurement of substrate oxidation using linked assays, such as SCR and SHFR, ultimately reflect electron flux through the respiratory chain, including complex III, but require the action of several other systems. Therefore, these assays are not, in diagnostic terms, entirely specific for respiratory chain activity. Thus, the ability of these techniques to detect a defect of complex III will depend on the effect that changes in activity of this enzyme have on flux through several pathways, i.e., its control strength. Spectrophotometric measurement of the activities of individual complexes activity avoids these problems. In addition, it requires only microgram quantities of mitochondrial protein, enabling studies to be performed on mitochondria prepared from very small samples of muscle or other tissues. This is of increasing importance because of the numbers of young children now being investigated and because many investigators, including our own group, use renewable sources of tissue, such as cultured skin fibroblasts or blood platelets. The mitochondrial yield from these tissues is small and biochemical studies are limited.

References

Andrews, R. M., Griffiths, P. G., Johnson, M. A., and Turnbull, D. M. (1999). Histochemical localisation of mitochondrial enzyme activity in human optic nerve and retina. *Br. J. Ophthalmol.* **83**(2), 231–235.

Birch-Machin, M. A., Briggs, H. L., Saborido, A. A., Bindoff, L. A., and Turnbull, D. M. (1994). An evaluation of the measurement of the activities of complexes I-IV in the respiratory chain of human skeletal muscle mitochondria. *Biochem. Med. Metab. Biol.* **51**(1), 35–42.

Birch-Machin, M. A., Shepherd, I. M., Watmough, N. J., Sherratt, H. S. A., Bartlett, K., Darley-Usmar, V., Aynsley-Green, A., and Turnbull, D. M. (1989). Fatal lactic acidosis in infancy with a defect of complex III of the respiratory chain. *Pediat. Res.* **25**(5), 553–559.

Birch-Machin, M. A., Singh-Kler, R., and Turnbull, D. M. (1993). Study of skeletal muscle dysfunction *In* "Mitochondrial Dysfunction" (L. H. Lash and D. P. Jones, eds.), pp. 51–69. Academic Press, San Diego.

Bourgeron, T., Rustin, P., Chretien, D., Birch-Machin, M., Viegas-Pequignot, E., Munnich, A., and Rotig, A. (1995). Mutation of a nuclear succinate dehydrogenase gene results in mitochondrial respiratory chain deficiency. *Nature Genet.* **11**(2), 144–149.

Boveris, A., Oshino, R., Erecinska, M., and Chance, B. (1971). Reduction of mitochondrial components by durohydroquinone. *Biochim. Biophys. Acta* **245**(1), 1–16.

Brierley, E. J., Johnson, M. A., Lightowlers, R. N., James, O. J. W., and Turnbull, D. M. (1998). Role of mitochondrial DNA mutations in human aging: Implications for the central nervous system and muscle. *Ann. Neurol.* **43**(2), 217–223.

Chinnery, P. F., and Turnbull, D. M. (1997). Clinical features, investigation, and management of patients with defects of mitochondrial DNA. *J. Neurol. Neurosurg. Psychiat.* **63,** 559–563.

Chinnery, P. F., and Turnbull, D. M. (1999). Mitochondrial DNA and disease. *Lancet* **354**(Suppl. 1), 17–21.

Chretien, D., Bourgeron, T., Rotig, A., Munnich, A., and Rustin, P (1990). The measurement of rotenone-sensitive NADH-cytochrome c reductase activity in mitochondria isolated from minute amounts of human skeletal muscle. *Biochem. Biophys. Res. Commun.* **173,** 26–33.

Chretien, D., Gallego Barrientos, A., Casademont, J., Cardellach Munnich, A., Rotig, A., and Rustin, P. (1998). Biochemical parameters for the diagnosis of mitochondrial respiratory chain deficiency in humans, and their lack of age-related changes. *Biochem. J.* **329**(Pt 2), 249–254.

Clark, K. M., Taylor, R. W., Johnson, M. A., Chinnery, P. F., Chrzanowska Lightowlers, Z. M., Andrews, R. M., Nelson, I. P., Wood, N. W., Lamont, P. J., Hanna, M. G., Lightowlers, R. N., and Turnbull, D. M. (1999). A mtDNA mutation in the initiation codon of the cytochrome *c* oxidase subunit II gene results in lower levels of the protein and a mitochondrial encephalomyopathy. *Am. J. Hum. Genet.* **64,** 1330–1339.

Fato, R., Castelluccio, C., Palmer, G., and Lenaz, G (1988). A simplified method for the determination of the kinetic constants of membrane enzymes utilising hydrophobic substrates: Ubiquinol cytochrome c reductase. *Biochim. Biophys. Acta* **932,** 216–222.

Friedrich, T., Hofhaus, G., Ise, W., Nehls, U., Schmitz, B., and Weiss, H (1989). A small isoform of NADH:ubiquinone oxidoreductase (complex I) without mitochondrially encoded subunits is made in chloramphenicol-treated *Neurospora crassa*. *Eur. J. Biochem.* **180,** 173–180.

Gonzalez-Halphen, D., Lindorfer, M. A., and Capaldi, R. A. (1988). Subunit arrangement in beef heart complex III. *Biochemistry* **27**(18), 7021–7031.

Hatefi, Y., and Rieske, J. S. (1967). Preparation and properties of DPNH-Coenzyme Q reductase (complex I of the respiratory chain). *Methods Enzymol.* **10,** 235–239.

Johnson, M. A. (1983). *In* "Histochemistry in Pathology" (M. I. Filipe and B. D. Lake, eds.), p. 89. Churchill Livingstone, Edinburgh.

Mahapatro, S. N., and Robinson, N. C. (1990). Effect of changing the detergent bound to bovine cytochrome c oxidase upon its individual electron transfer steps. *Biochemistry* **29,** 764–770.

Majander, A., Huoponen, K., Savontaus, M.-L., Nikoskelainen, E., and Wikstrom, M. (1991). Electron transfer properties of NADH-ubiquinone reductase in the ND1/3460 and ND4/11778 mutations of LHON. *FEBS. Lett.* **292,** 289–292.

Nalecz, M. J., Bolli, R., and Azzi, A. (1985). Molecular conversion between monomeric and dimeric states of the mitochondrial cytochrome bc1 complex: Islolation of active monomers. *Arch. Biochem. Biophys.* **236,** 619–628.

Rosevear, P., Van Aken, T., Baxter, J., and Ferguson–Miller, S. (1980). Alkyl glycoside detergents: A simpler synthesis and their effects on kinetic and physical properties of cytochrome c oxidase. *Biochemistry* **19,** 4108–4115.

Rustin, P., Chretien, D., Bourgeron, T., Gerard, B., Rotig, A., Saudubray, J., and Munnich, A. (1994). Biochemical and molecular investigations in respiratory chain deficiencies. *Clin. Chim. Acta* **228**(1), 35–51.

Sherratt, H. S. A., Watmough, N. J., Johnson, M. A., and Turnbull, D. M. (1988). Methods for study of normal and abnormal skeletal muscle mitochondria. *Methods Biochem. Anal.* **33,** 243–335.

SinghKler, R., Jackson, S., Bartlett, K., Bindoff, L. A., Eaton, S., Pourfarzam, M., Frerman, F., Goodman, S. I., Watmough, N., and Turnbull, D. M. (1991). Quantitation of acyl-CoA and acylcarnitine esters accumulated during abnormal mitochondrial fatty acid oxidation. *J. Biol. Chem.* **266,** 22932–22938.

Sottocasa, G. L., Kuylenstierna, B., Ernster, L., and Bergstrand, A (1967). An electron transport system associated with the outer membrane of liver mitochondria. *J. Cell Biol.* **32,** 415–438.

Taylor, R. W., Birch-Machin, M. A., Bartlett, K., Lowerson, S. A., and Turnbull, D. M. (1994). The control of mitochondrial oxidations by complex III in rat muscle and liver mitochondria: Implications for our understanding of mitochondrial cytopathies in man. *J. Biol. Chem.* **269**(5), 3523–3528.

Taylor, R. W., Birch-Machin, M. A., Bartlett, K., and Turnbull, D. M. (1993). Succinate-cytochrome *c* reductase: Assesssment of its value in the investigation of defects of the respira tory chain. *Biochim. Biophys. Acta* **1181,** 261–265.

Von Jagow, G., and Link, T. A. (1986). Use of specific inhibitors on the mitochondrial bc1 complex. *Methods Enzymol.* **126,** 253–271.

Wallace, D. C. (1999). Mitochondrial diseases in man and mouse. *Science* **283,** 1482–1488.

Yu, C.-A., Gu, L., Lin, Y., and Yu, L. (1985). Effect of alkyl side chain variation on the electron transfer activity of ubiquinone derivatives. *Biochemistry* **24,** 3897–3902.

Zheng, X., Shoffner, J. M., Voljavec, A. S., and Wallace, D. C. (1990). Evaluation of procedures for assaying oxidative phosphorylation: Enzyme activities in mitochondrial myopathy muscle biopsies. *Biochim. Biophys. Acta.* **1019,** 1–10.

CHAPTER 6

In Vivo Measurements of Respiration Control by Cytochrome *c* Oxidase and *in Situ* Analysis of Oxidative Phosphorylation*

Gaetano Villani† and Giuseppe Attardi

Division of Biology
California Institute of Technology
Pasadena, California 91125

* Dedicated to the memory of Ferruccio Guerrieri

† Current address: Department of Medical Biochemistry and Biology, University of Bari, Bari, Italy.

I. Introduction

Traditionally, studies on oxidative phosphorylation (OXPHOS) have been carried out on isolated mitochondria. In recent years, however, the increasing evidence that interactions of these organelles with the cytoskeleton, other organelles, and the cytosol play a significant role in the various OXPHOS reactions has created the need for new methods that can probe these reactions under conditions approximating more closely the *in vivo* situation. This need has been accentuated by the discovery since the late 1980s of a large variety of mitochondrial DNA mutations that cause diseases in humans and by the increasing evidence that a reduction in the rate of assembly and in the activities of the OXPHOS apparatus is associated with aging or neurodegenerative diseases (Wallace, 1992; Schon *et al.*, 1997). In particular, the recognition of threshold effects in the capacity of a mitochondrial DNA (mtDNA) mutation to produce a dysfunction in the presence of varying amounts of wild-type mtDNA has stimulated a strong interest in determining the degree of control that a particular OXPHOS step exerts on the rate of mitochondrial respiration and/or ATP production *in vivo*. This chapter describes an approach that has been developed to determine the control that cytochrome *c* oxidase (COX) exerts on the rate of endogenous respiration in intact cells (Villani and Attardi, 1997; Villani *et al.*, 1998), as well as a method for analysis of OXPHOS in permeabilized cells.

The experimental procedures described are based on polarographic measurements of oxygen consumption by means of Clark-type oxygen electrodes. The equipment and its utilization have been described in detail (Hofhaus *et al.*, 1996; Trounce *et al.*, 1996), and new generations of oxygen electrodes and oxygraphic chambers, as well as of software for the automatic analysis of respiratory kinetics, are now available. With any chosen reaction chamber, it is important to measure the exact mixing volume. This corresponds to the minimal volume required to fill the chamber up to the bottom of the channel port of the stopper. This value is important for subsequent calculations. The volume introduced into the chamber has to exceed slightly the mixing volume in order to avoid formation of air bubbles from the vortex created by stirring.

II. Measurements of Endogenous Respiration in Intact Cells

A. Media and Reagents

Dulbecco's modified Eagle's medium (DMEM) lacking glucose, supplemented with 5% fetal bovine serum (FBS)

Tris-based, Mg^{2+}-, Ca^{2+}-deficient (TD) buffer: 0.137 *M* NaCl, 5 m*M* KCl, 0.7 m*M* Na_2HPO_4, 25 m*M* Tris–HCl, pH 7.4, at 25°C

B. Procedure

The measurements are performed at 37°C on exponentially growing cell cultures change the media in cultured cells the day before measurements are made. Allow the temperature of the measurement medium to equilibrate under stirring in the reaction chamber: this will result in a constant horizontal tracing on the recorder paper. Wash with TD buffer and collect cells by centrifugation (for cells growing in suspension) or by trypsinization and centrifugation (for cells growing on plates). The final cell pellet is resuspended in a portion of the equilibrated medium from the chamber by gently pipetting. Cells are then transferred into the oxygraphic chamber. Close the chamber and record the endogenous oxygen consumption rate. After obtaining a reliable rate, open the chamber and remove one sample for cell counting and four 20- to 50-μl samples for total protein determination.

C. Comments

It has been reported (Villani and Attardi, 1997) that the osteosarcoma-derived 143B. TK$^-$ cell line respires in TD buffer at the same rate as in DMEM lacking glucose (Fig. 1a). When the effects of inhibitors or uncouplers of OXPHOS have to be analyzed, the use of TD buffer is recommended.

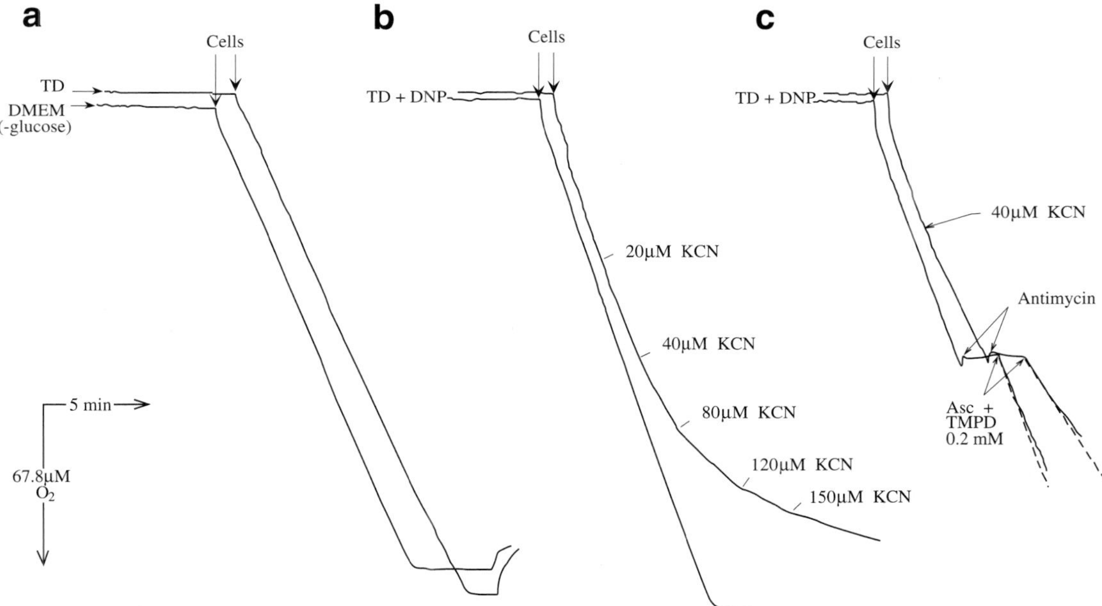

Fig. 1 Polarographic tracings of oxygen consumption in 143B.TK$^-$ cells. (a) Endogenous respiration in TD buffer or DMEM-glucose medium. (b) Cyanide titration of uncoupled endogenous respiration; KCN was added sequentially at different concentrations in one of the two chambers, as indicated. (c) Inhibitory effect of 40 μ*M* KCN on the uncoupled endogenous respiration rate and on the initial rate (dashed lines) of 10 m*M* ascorbate (Asc) + 0.2 m*M* TMPD-supported oxygen consumption.

The oxygen consumption rate, expressed relative to the cell number, tends to decrease with increasing cell density in the culture. This is due, at least in part, to a decrease in the size of the cells, as shown by the fact that the respiratory rate, normalized per milligram of protein, remains fairly constant at different cell densities (Villani *et al.,* 1998).

The mitochondrial respiration rate should be corrected for KCN (2–6 m*M*)-resistant oxygen consumption rate (see next section for preparation of KCN solution), which reflects oxygen metabolism from other pathways. This includes superoxide radical formation that normally does not exceed 1–2% of total oxygen consumption (Ernster, 1986). An increase beyond such a value could be taken as indicative of cellular oxidative stress.

Additional information can be obtained by the analysis of the endogenous respiration rate of intact cells in the presence of membrane-permeant inhibitors or uncouplers. As an example, the degree of coupling of endogenous respiration to ATP production can be estimated by inhibiting the mitochondrial ATPase with oligomycin. This will result in a decrease in the respiratory rate due to the establishment of a maximal steady-state electrochemical membrane potential. The respiration rate can then be restored by the addition of mitochondrial uncouplers. *Note:* Oligomycin should be used at the minimal concentration that results in the maximal ratio between uncoupled and oligomycin-inhibited respiration rate.

III. KCN Titration of COX Activity in Intact Cells

A. Principle

The application of the metabolic control theory (Kacser and Burns, 1973; Heinrich and Rapoport, 1974) to the study of the mitochondrial metabolism can be a valuable approach for determining the degree of control exerted by different OXPHOS steps on the rate of mitochondrial respiration (Groen *et al.,* 1982) and for identifying and quantifying enzymatic defects in the OXPHOS machinery caused by mitochondrial or nuclear DNA mutations. The method used most often to analyze the control that a given enzymatic step exerts on the overall metabolic flux is the inhibitor titration technique (Groen *et al.,* 1982). In this approach, a specific inhibitor is utilized to titrate an enzymatic activity both as an isolated step and as a step integrated in the metabolic pathway. In this way, one can measure, for each concentration of inhibitor used, the percentage of inhibition of the isolated step and to what degree this specific inhibition affects the overall pathway.

This method can be applied to intact cells to study the control exerted by cytochrome *c* oxidase on endogenous respiration. In particular, the isolated COX activity is titrated with KCN after inhibiting the upstream segment of the respiratory chain with antimycin A and reducing the endogenous cytochrome *c* pool with tetramethyl-*p*-phenylene diamine (TMPD), a membrane-permeant, one-equivalent electron donor, in the presence of an excess of ascorbate as the primary reducing agent. The titration curve thus determined is then compared to the KCN titration curve of the endogenous respiration, where the COX activity participates as a respiratory chain-integrated step (Villani and Attardi, 1997).

To avoid any influence of proton cycling and OXPHOS on the respiratory flux, the mitochondrial uncoupler dinitrophenol (DNP) is routinely utilized in these experiments at the concentration that produces the highest stimulation of respiration rate. For DNP titration, after recording the coupled endogenous respiration rate, as described earlier, DNP is added at increasing concentrations, and, at each concentration, the uncoupled respiratory rate is measured and expressed as a percentage of the starting coupled endogenous respiration rate (Villani *et al.,* 1998).

B. Media and Reagents

TD buffer

50 m*M* DNP (Sigma) in absolute ethanol; the stock solution can be stored at −20°C

4 μ*M* antimycin A (Sigma) in absolute ethanol (stored at −20°C)

1 *M* ascorbate: adjust to pH 7.4 by gradual addition of NaOH tablets. Distribute in microfuge tubes and store at −80°C.

300 m*M* TMPD: distribute in tubes and store at −80°C. Note that the solution can turn blue due to the oxidized form of the TMPD, but this will not affect the results, as $TMPD^+$ will be reduced by ascorbate rapidly.

1 *M* KCN: this solution is adjusted to pH 8 by adding concentrated HCl. The stock solution is then distributed in tubes and stored for long periods at −80°C. *Note:* Preparation of the KCN stock must be carried out in a fume hood, as highly toxic volatile HCN may be produced, especially when an excess of added HCl rapidly drops the pH to acidic values. The pH adjustment can lower the effective titer of the solution without affecting the following experimental results, as relative inhibition rates have to be measured. However, if the solution is exposed to excess acid during preparation, it should be discarded.

C. Procedure

For this type of experiment, the parallel use of two oxygraphic chambers equipped with electrodes is recommended. These should be connected to a double-channel oxygraph and recorder. Calibrate the instrument to produce the same 0–100% oxygen range in the two channels. This allows one to correct for small decreases in respiratory rates occurring during the oxygraphic run.

1. KCN Titration of Integrated COX Activity

1. Allow the temperature of the measurement medium to equilibrate under stirring in the reaction chambers.
2. At the same time, wash with TD and collect exponentially growing cell cultures in which the medium was changed the day before.

3. Resuspend the final cell pellet in a volume of measured medium (one-fifth of the chamber volume), which is made by combining two equal samples taken from the two chambers. Gently pipette up and down, and split the resulting suspension between the two chambers to which DNP has been added. As a rule, DNP should be used at the minimal concentration needed to produce maximal stimulation of the endogenous respiration. In the case of 143B.TK$^-$ cells, DNP is added to final concentration of 25–30 μ*M*. The concentration of cells in the chamber is adjusted to ensure that the uninhibited endogenous respiration rate remains fairly constant during the titration experiment, and the TMPD autooxidation rate represents a small percentage of the overall isolated COX activity (see later).

4. Close the chambers and measure the initial oxygen consumption rate. You should obtain two parallel lines on the chart recorder (see Fig. 1b).

5. Sequentially introduce small amounts of a 10 m*M* KCN solution at appropriate intervals near the bottom of one of the chambers using a Hamilton syringe fitted with a 2.5-in. (6.3 cm)-long needle (Hamilton). The intervals of addition are chosen to allow adequate time for reaching constant slopes in the tracing and for constructing detailed inhibition curves (see Fig. 2a).

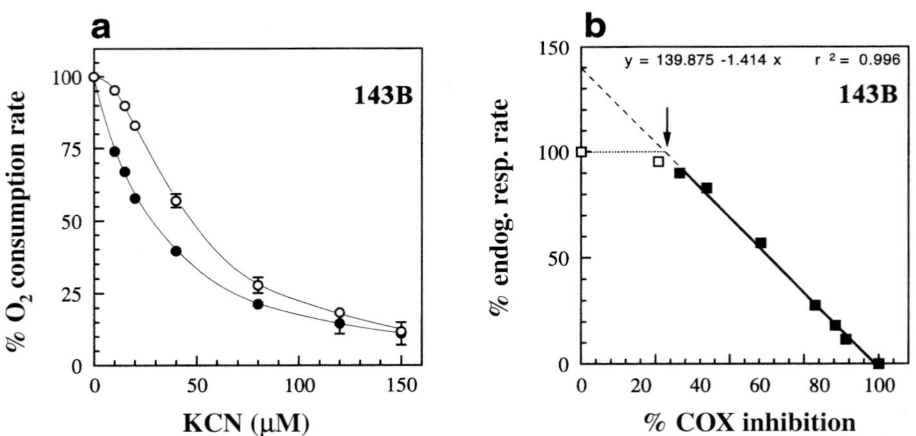

Fig. 2 Inhibition by KCN of endogenous respiration rate in TD buffer in the presence of DNP (○) or ascorbate + TMPD-dependent respiration in the presence of DNP and antimycin A (●) in intact 143B.TK$^-$ cells (a), and percentage of endogenous respiration rate as a function of percentage of isolated COX inhibition (threshold plot) in 143B.TK$^-$ cells, with determination of relative maximum COX capacity (COX_{Rmax}) (b). Adapted from Villani and Attardi (1997). (a) Data shown represent the means ±SE (the error bars that fall within the individual data symbols are not shown) obtained in six or seven determinations on 143B.TK$^-$ cells. The TMPD concentration used for KCN titration of the O_2 consumption rate of antimycin-treated cells was 0.2 m*M*. (b) Percentage rates of endogenous respiration at different KCN concentrations in the experiments illustrated in (a) are plotted against percentages of COX inhibition by the same KCN concentrations, and the least-square regression line through the filled symbols beyond the inflection point (threshold) in the curve (arrow) is extended to zero COX inhibition. The equation describing the extrapolated line is shown. For further details, see text and Villani and Attardi (1997).

2. KCN Titration of Isolated COX Activity

1. Same as in Section III.C1, step 1.
2. Same as in Section III.C1, step 2.
3. Split the cell suspension, prepared as in Section III.C1, step 3, between the two chambers, to which DNP and antimycin A have been added. Antimycin should be used at the minimal concentration needed to produce maximal inhibition of respiration (20 n*M* for 143B.TK$^-$ cells).
4. After closing the chambers, add ascorbate and TMPD to both chambers and measure the initial oxygen consumption rate. You should obtain two parallel lines on the chart recorder. Ascorbate is used at 10 m*M*. For 143B.TK$^-$ cells, addition of ascorbate alone does not result in any significant oxygen consumption. TMPD, which starts oxygen consumption, should be used at a concentration that produces oxygen consumption rates comparable to the endogenous respiration rate. Under these conditions, cytochrome *c* oxidase can be titrated both as an integrated step and as an isolated step under similar electron fluxes.
5. Same as in Section III.C1, step 5.

D. Analysis of Data

In the titration of both integrated and isolated COX activity, the KCN-inhibited activity measured after each addition of KCN is expressed as a percentage of the uninhibited activity measured [in parallel (i.e., at the same O_2 level] in the other chamber. If the starting oxygen consumption rates are not the same, the percentage difference should be used to correct the subsequent calculation of the inhibited activities.

For the measurement of isolated COX activity, the oxygen consumption rates must be corrected for the KCN-insensitive oxygen consumption due to TMPD autooxidation. For this purpose, in a separate experiment, DNP, antimycin, ascorbate, and TMPD are added to TD buffer at the same concentrations used in the assays performed with cells, and the oxygen consumption rate is measured in the oxygen concentration range in which the cyanide titrations are performed. This is important because the rate of oxygen consumption depends on the O_2 concentration in the chamber. In a similar experiment performed in the presence of mtDNA-less cells, we have observed no difference in TMPD oxidation rate. Practically, when measuring the slopes corresponding to the reference and KCN-inhibited isolated COX activities, data obtained should be corrected by substracting the slope corresponding to the TMPD autooxidation rate in the same O_2 concentration range.

An alternative way to perform the titration at each KCN concentration is presented in Fig. 1c. The experiment is set up as before, with cells resuspended in TD plus DNP. After recording the uncoupled respiration rate in both chambers, KCN is added to one of the chambers at a given concentration and the oxygen consumption is recorded until a constant slope is obtained. Then, antimycin is added to both chambers to stop the respiration completely. Finally, ascorbate and TMPD are added to both chambers, and the two isolated COX activities (control and KCN-inhibited) are measured as initial rates and then corrected for the TMPD autooxidation rate as determined in a separate experiment. This method is preferable in the case of unstable oxygen consumption rates.

Two main parameters can be determined using the KCN titration technique: (i) the inhibition "threshold" value, defined as the percentage inhibition of the isolated COX activity where COX activity becomes rate limiting for the overall flux, and (ii) the "maximum relative capacity," which indicates the percentage excess of COX activity over that needed to support the maximal rate of endogenous respiration.

As shown in Fig. 2a, the KCN titration of the O_2 consumption rate, in the presence of ascorbate and TMPD, in antimycin A-inhibited 143B.TK$^-$ cells (isolated step), produces a curve that is quasilinear at nonsaturating concentrations of the inhibitor. However, the variation of the endogenous respiration rate (overall flux) over the same range of KCN concentrations produces a curve that is sigmoidal. This difference is consistent with the occurrence in 143B.TK$^-$ cells of an excess of COX activity over that required to support a normal respiratory rate.

In Fig. 2b, the difference in relative KCN sensitivity between the endogenous respiration rate and the isolated COX activity in 143.TK$^-$ cells is illustrated in the form of a threshold plot; i.e., a plot of the relative endogenous respiration rates against percentage inhibition of isolated COX activity by the same KCN concentrations. In this plot, the endogenous respiration rate remains fairly constant with increasing inhibition of the isolated enzyme, up to the percentage inhibition threshold (COX_T), i.e., ~28%, at which a further decrease in enzyme activity has a marked effect on the rate of endogenous respiration. When the least-square regression line through the filled symbols (beyond the inflexion point in the curve) is extrapolated to zero COX inhibition, the *y* intercept gives an estimate of the maximum COX capacity as an integrated step, expressed relative to the uninhibited endogenous respiration rate (Taylor *et al.,* 1994; Villani and Attardi, 1997).

This method has been applied to a variety of human cell types and has revealed generally low values of reserve COX capacity with interesting differences, which can be helpful in understanding the tissue specificity of mitochondrial diseases (Villani *et al.,* 1998). Furthermore, by means of the same strategy, the impact of the apoptotic release of cytochrome *c* on cell respiration has been investigated (P. Hajek *et al.,* 2000).

IV. *In Situ* Analysis of Mitochondrial OXPHOS

One of the most important bioenergetic parameters to be determined when dealing with mitochondrial metabolism is the efficiency of oxidative phosphorylation. This is usually measured as coupling of mitochondrial respiration to ATP production, namely as the P/O ratio [defined as nanomoles of ATP produced per natoms of oxygen consumed during ADP-stimulated (state III) respiration (Chance and Williams, 1955)]. This classic methodology has been used on isolated mitochondria [for an application to mitochondria isolated from cultured cells, see Trounce *et al.,* (1996)].

A. Media and Reagents

Buffer A: 75 m*M* sucrose, 5 m*M* potassium phosphate (monobasic), 40 m*M* KCl, 0.5 m*M* EDTA, 3 m*M* $MgCl_2$, 30 m*M* Tris–HCl (pH 7.4 at 25°C)

Permeabilization buffer: buffer A + 1 m*M* phenylmethylsulfonyl fluoride (PMSF)

Measurement buffer: buffer A + 0.35% bovine serum albumin (BSA; Sigma)

Substrate stock solutions: 1 *M* glutamate, 1 *M* malate, 1 *M* succinate. Adjust to pH 7.0–7.4 with KOH or NaOH, distribute in tubes, and store at −20°C.

10% digitonin (Calbiochem) in dimethyl sulfoxide

2 m*M* rotenone in absolute ethanol (stored at −20°C)

50 m*M* ADP (Sigma): adjust pH to 6.5–6.8 with diluted KOH and add $MgCl_2$ to 20 m*M*. Distribute in tubes and store at −80°C.

50 m*M* p^1, p^5-di(adenosine-5′) pentaphosphate (Ap_5A) (Sigma) in H_2O (stored at −80°C)

B. Digitonin Titration

The amount of digitonin needed for optimal cell permeabilization must be titrated for each cell type analyzed. A simple functional test of optimal permeabilization can be performed using succinate as the respiratory substrate. Succinate is normally not utilized for endogenous respiration and can enter intact cells only poorly, as tested in a variety of human cell types (Kunz *et al.,* 1995; Villani *et al.,* 1998). Alternatively, a previously described method based on the trypan blue exclusion test (Hofhaus *et al.,* 1996) can be utilized to determine the extent of permeabilization. The exclusion test is particularly useful when working with cells showing a high contribution of rotenone-insensitive endogenous respiration.

Collect cells, resuspend the final pellet, and transfer it to the oxygraphic chamber, as described in the previous sections, by using buffer A equilibrated at 37°C and supplemented with 0.5 m*M* ADP just before adding the cell suspension. Take a sample for cell counting. After recording endogenous respiration, add rotenone to a final concentration of 100 n*M* and wait until maximal inhibition is reached. Then, add digitonin and, after 2 min, add 5 m*M* succinate and measure respiration. The optimal amount of digitonin gives the highest succinate-dependent respiration. This rate should not increase on the addition of exogenous cytochrome *c* if the integrity of the mitochondrial outer membrane is preserved.

C. P/O Ratio Assay

A calibration of the oxygraph is needed at the beginning and at the end of the experiments. If many samples are analyzed, additional dithionite calibrations (Hofhaus *et al.,* 1996) are recommended during the course of the experiment. The experiments are run at 37°C. At this temperature the concentration of O_2 in distilled water under saturating conditions is 217 nmol/ml (Cooper, 1977). All the buffers must be kept at the measurement temperature during the experiment.

Collect cells, wash once in buffer A, and remove a sample for cell counting. Resuspend the final pellet in permeabilization buffer at a cell concentration similar to the one utilized for digitonin titration. Add the optimal amount of digitonin and wait for 1–2 min, occasionally shaking the tube. Then, add 10 volumes of measurement buffer and collect permeabilized cells by centrifugation. Resuspend the final pellet in a sample of measurement buffer previously equilibrated at 37°C, by gentle pipetting and transfer the suspension into the chamber under stirring. After supplying the mixture with 0.3 m*M* Ap_5A, close the chamber and record the baseline. At this point, the oxygen consumption rate is undetectable or very low and produced by endogenous substrates within mitochondria. If the starting oxygen consumption rate is too high, it indicates that the permeabilization procedure has probably not worked properly, and a higher digitonin concentration is usually required. Add glutamate and malate at a concentration of 5 m*M* each and record state IV respiration. When a stable rate is obtained, add ADP (100– 200 nmol for a 1.5-ml chamber); this will result in a transient increase in respiration rate (state III), which will return to state IV respiration when all the ADP has been phosphorylated. After obtaining a linear state IV respiration, a second addition of ADP can be made to obtain a second measurement of the P/O ratio. Thereafter, block respiration by adding rotenone to 100 n*M*, add 5 m*M* succinate, and, after obtaining a stable state IV respiration, add ADP (50–100 nmol); and record the transient state III respiratory rate and the subsequent return to state IV.

D. Analysis of Data

Figure 3 shows the oxygraphic tracing obtained for $143B.TK^-$-permeabilized cells. Values of respiratory control ratio [defined as ratio between respiration rate in the presence of added ADP (state III) and state IV respiration rate obtained following its consumption (Chance, 1959; see also Estabrook, 1967)] of 3–4 have been measured for these cells, with glutamate and malate as respiratory substrates. These results are very similar to previous results obtained on mitochondria isolated from the same cells (Trounce *et al.,* 1996). These values are also very close to the ratio between the endogenous respiration rate [>95% rotenone sensitive (Villani and Attardi, 1997)] and the oligomycin-inhibited respiration rate, as measured in intact cells (data not shown). Furthermore, the glutamate/malate-dependent respiration rate in the presence of ADP is very similar to the coupled endogenous respiration rate of intact cells (data not shown), thus allowing one to measure the ATP production efficiency under physiological respiratory fluxes. This is especially important when the dependence of the H^+/e^- stoichiometry of cytochrome *c* oxidase (Capitanio *et al.,* 1991, 1996) or, in general, of the efficiency of ATP production on the electron transfer rate (Fitton *et al.,* 1994) is considered. P/O ratios can be calculated as ratios between the amount (nanomoles) of added ADP and the oxygen (natoms) consumed during the ADP-induced state III respiration (Chance and Williams, 1955; Estabrook, 1967). In this way, P/O ratios up to about 2.3 and 1.3 have been measured

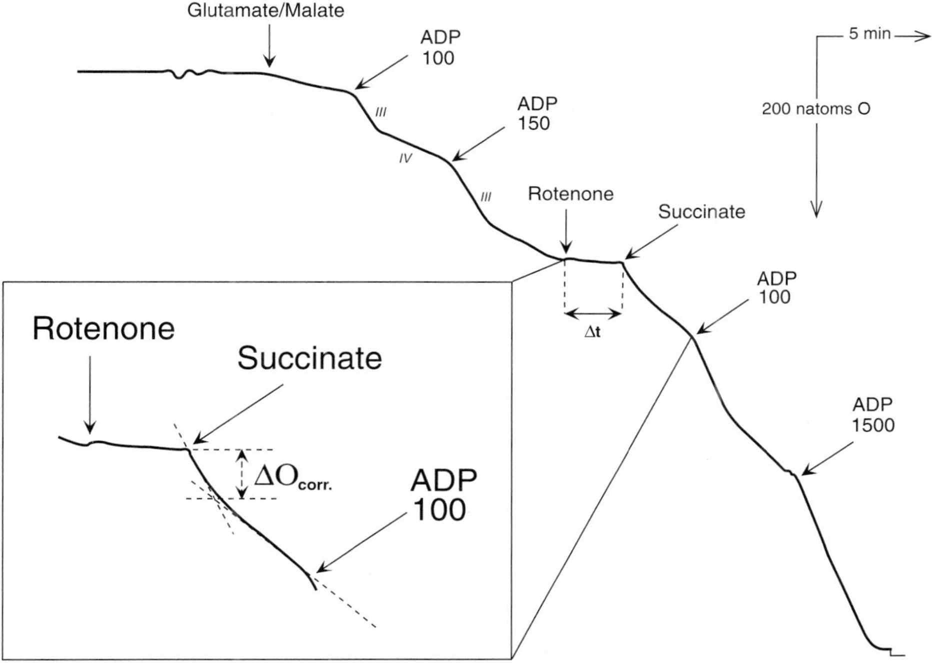

Fig. 3 Polarographic tracings of an experiment for P/O ratio measurements in digitonin-permeabilized 143B.TK$^-$ cells. The additions of ADP are expressed in nanomoles. *III*, ADP-stimulated state III respiration; *IV*, state IV respiration. (Inset) An enlargement of the tracing corresponding to the transient state III respiration obtained on succinate addition. See text for further details.

in 143B.TK$^-$ digitonin-permeabilized cells respiring, respectively, in the presence of glutamate/malate and succinate.

The first addition of succinate after inhibition with rotenone induces a transient stimulation of respiration. This may be due to the ADP that has accumulated as a result of ATP-consuming systems operating in permeabilized cells during the time of rotenone inhibition (Δt in Fig. 3). In fact, the ADP produced under these conditions cannot be rephosphorylated to ATP because of the block of respiration and consequent collapse of the electrochemical transmembrane potential that drives the ATP synthesis by mitochondrial ATP synthase. The rate of ADP production by ATP-consuming processes can be estimated by measuring the amount of oxygen consumed during the transient state III cited earlier (ΔO_{corr} in the inset of Fig. 3), multiplying this amount by the P/O ratio, as measured for succinate respiration, and then dividing the value thus obtained by the time of rotenone inhibition (Δt). The final value thus calculated (ADP/s) can then be used to correct the P/O ratio measurements for state III respiration for the additional amount of ADP produced and then converted to ATP. In the experiment shown in Fig. 3, for instance, ATP consumption has been estimated to account for the cycling of 0.23 nmol ADP/sec.

The final P/O ratio values, corrected by this factor, would then be 2.78 and 2.52 (instead of 2.33 and 2.16) for the first and the second addition of ADP to glutamate/malate-oxidizing cells, respectively, and 1.37 (instead of 1.1) for succinate-driven respiration.

In the case of 143B.TK$^-$ cells, no P/O ratio could be measured if glucose was present in the measurement buffer due to ADP production by membrane-associated hexokinase activity. Therefore, the presence of traces of glucose could result in a lower P/O ratio. Furthermore, it was observed that the presence of PMSF in the measurement buffer resulted in the formation of aggregates in the oxygraphic chamber.

The procedure described in this section is very useful when dealing with a limited amount of sample, as much information can be collected within a single experiment. In fact, substrate-dependent respiration rates, respiratory control ratios, and P/O ratios can be measured at different energy conservation sites with as few as 2–4 $\times$ 10^6 cells/ml, as has been done with 143B.TK$^-$ cells and its derivative cell lines.

Acknowledgments

This work was supported by National Institutes of Health Grant GM-11726 (to G.A.). We are very grateful to M. Greco for help in some of the experiments, to the late F. Guerrieri for helpful discussions and A. Chomyn for her critical reading of the manuscript.

References

Capitanio, N., Capitanio, G., Demarinis, D. A., De Nitto, E., Massari, S., and Papa, S. (1996). Factors affecting the H^+/e^- stoichiometry in mitochondrial cytochrome *c* oxidase: Influence of the rate of electron flow and transmembrane delta pH. *Biochemistry* **35,** 10800–10806.

Capitanio, N., Capitanio, G., De Nitto, E., Villani, G., and Papa, S. (1991). H^+/e^- stoichiometry of mitochondrial cytochrome complexes reconstituted in liposomes. Rate-dependent changes of the stoichiometry in the cytochrome *c* oxidase vesicles. *FEBS Lett.* **288,** 179–82.

Chance, B. (1959). "Ciba Found. Symp. Regulation Cell Metabolism," p. 91. Little Brown, Boston, MA.

Chance, B., and Williams, G. R. (1955). Respiratory enzymes in oxidative phosphorylation. I. Kinetics of oxygen utilization. *J. Biol. Chem.* **217,** 383–393.

Cooper, T. G. (1977). "The tools of Biochemistry." Wiley, New York.

Ernster, L. (1986). Oxygen as an environmental poison. *Chem. Scripta* **26,** 525–534.

Estabrook, R. W. (1967). Mitochondrial respiratory control and the polarographic measurement of ADP:O ratios. *Methods Enzymol.* **10,** 41–47.

Fitton, V., Rigoulet, M., Ouhabi, R., and Guerin, B. (1994). Mechanistic stoichiometry of yeast mitochondrial oxidative phosphorylation. *Biochemistry* **33,** 9692–9698.

Groen, A. K., Wanders, R. J., Westerhoff, H. V., van der Meer, R., and Tager, J. M. (1982). Quantification of the contribution of various steps to the control of mitochondrial respiration. *J. Biol. Chem.* **257,** 2754–2757.

Hájek, P., Villani, G., and Attardi, G. (2001). Rate-limiting step preceding cytochrome *c* release in cells primed for Fas-mediated apoptosis is revealed by cell sorting and respiration analysis. *J. Biol. Chem.* **276,** 606–615.

Heinrich, R., and Rapoport, T. A. (1974). A linear steady-state treatment of enzymatic chains: General properties, control and effector strength. *Eur. J. Biochem.* **42,** 89–95.

Hofhaus, G., Shakeley, R. M., and Attardi, G. (1996). Use of polarography to detect respiration defects in cell cultures. *Methods Enzymol.* **264,** 476–483.

Kacser, H., and Burns, J. A. (1973). *In* "Rate Control of Biological Processes" (D. D. Davies, ed.), pp. 65–104. Cambridge Univ. Press, London.

Kunz, D., Luley, C., Fritz, S., Bohnensack, R., Winkler, K., Kunz, W. S., and Wallesch, C. W. (1995). Oxygraphic evaluation of mitochondrial function in digitonin-permeabilized mononuclear cells and cultured skin fibroblasts of patients with chronic progressive external ophthalmoplegia. *Biochem. Mol. Med.* **54,** 105–111.

Schon, E. A., Bonilla, E., and Di Mauro, S. (1997). Mitochondrial DNA mutations and pathogenesis. *J. Bioenerg. Biomembr.* **29,** 131–149.

Taylor, R. W., Birch-Machin, M. A., Bartlett, K., Lowerson, S. A., and Turnbull, D. M. (1994). The control of mitochondrial oxidations by complex III in rat muscle and liver mitochondria: Implications for our understanding of mitochondrial cytopathies in man. *J. Biol. Chem.* **269,** 3523–3528.

Trounce, I. A., Kim, Y. L., Jun, A. S., and Wallace, D. C. (1996). Assessment of mitochondrial oxidative phosphorylation in patient muscle biopsies, lymphoblasts, and transmitochondrial cell lines. *Methods Enzymol.* **264,** 484–509.

Villani, G., and Attardi, G. (1997). *In vivo* control of respiration by cytochrome *c* oxidase in wild-type and mitochondrial DNA mutation-carrying human cells. *Proc. Natl. Acad. Sci. USA* **94,** 1166–1171.

Villani, G., Greco, M., Papa, S., and Attardi, G. (1998). Low reserve of cytochrome *c* oxidase capacity *in vivo* in the respiratory chain of a variety of human cell types. *J. Biol. Chem.* **273,** 31829–31836.

Wallace, D. C. (1992). Diseases of the mitochondrial DNA. *Annu. Rev. Biochem.* **61,** 1175–1212.

CHAPTER 7

Assay of Mitochondrial ATP Synthesis in Animal Cells

Giovanni Manfredi,* Antonella Spinazzola,† Nicoletta Checcarelli,† and Ali Naini†

* Department of Neurology and Neuroscience
Weill Medical College of Cornell University
New York, New York

† H. Houston Merritt Clinical Research Center for Muscular Dystrophy and Related Disorders
Department of Neurology
Columbia University, College of Physicians and Surgeons
New York, New York

I. Introduction

Mitochondria play a major role in cellular ATP synthesis, as mitochondrial oxidative phosphorylation generates 36 mol of ATP per mole of glucose, as opposed to only 2 mol of ATP generated by glycolysis in the cytoplasm. Therefore, the measurement of mitochondrial ATP synthesis can be considered a pivotal tool in understanding many important characteristics of cellular energy metabolism, both in the normal physiological state and in pathological conditions, such as mitochondrial disorders.

The first issue to be addressed in setting up an experimental procedure to measure mitochondrial ATP synthesis is how to deliver reaction substrates to the mitochondria. Because of the low permeability of the plasma membrane to some hydrophilic substrates, such as ADP, some investigators choose to perform their assays on isolated mitochondria (Tuena de Gómez-Puyou *et al.,* 1984; Tatuch and Robinson 1993; Vazquez-Memije *et al.,* 1996, 1998). For ADP phosphorylation to take place, mitochondria need to be maintained in a coupled state. Therefore, ATP synthesis can only be measured on freshly isolated mitochondria. This involves delicate and time-consuming procedures. Moreover, coupling of isolated mitochondria, especially from cultured cells, is very inconsistent, leading to potential lack of reproducibility.

Measurement of ATP synthesis on whole cells eliminates the need to isolate mitochondria. However, a membrane permeabilization step is required to allow the hydrophilic substrates to penetrate into the cells. Membrane permeabilization can been achieved by treatment with detergents, such as saponin (Kuntz *et al.,* 1993), or by treatment with digitonin (Wanders *et al.,* 1993, 1994, 1996; Houstek *et al.,* 1995). The use of whole permeabilized cells instead of isolated mitochondria also reduces the amount of cell material needed for the assay.

The second issue is ATP detection and quantification. One of the methods commonly employed to quantify ATP produced by isolated mitochondria is based on fluorimetry (Tatuch and Robinson 1993; Wanders *et al.,* 1993, 1994, 1996; Houstek *et al.,* 1995). Another method to detect ATP from isolated mitochondria employs ^{32}Pi incorporation into ADP and subsequent transfer into glucose-6-phosphate by hexokinase, followed by the extraction of unincorporated ^{32}Pi and measurement of radioactivity in a scintillation counter (Tuena de Gomez-Puyou *et al.,* 1984; Vazquez-Memije *et al.,* 1997, 1998).

A rapid and reliable alternative for the measurement of ATP on both intact cells and isolated mitochondria is based on the luciferin-luciferase system. Firefly luciferase is widely employed as a reporter gene in mammalian cells, but its bioluminescence properties have also been used to measure ATP content in isolated mitochondria (Strehler and Totter, 1952; Le Masters and Hackenbrock, 1973, 1976; Wibom *et al.,* 1990, 1991) and in permeabilized whole cells (Maechler *et al.,* 1998; Ouhabi *et al.,* 1998; James *et al.,* 1999). The reaction catalyzed by luciferase is

$$\text{luciferase} + \text{luciferin} + \text{ATP} \rightarrow \text{luciferase-luciferyl-AMP} + \text{PPi}$$

$$\text{luciferase-luciferyl-AMP} + O_2 \rightarrow \text{luciferase} + \text{oxyluciferin} + \text{AMP} + CO_2 + h\nu$$

The reaction produces a flash of yellow-green light, with a peak emission at 560 nm, whose intensity is proportional to the amount of substrates in the reaction mixture (DeLuca, 1979).

Regardless of the detection system, isolated mitochondria or permeabilized cells are incubated with appropriate substrates followed by extraction and measurement of the steady-state level of ATP. In order to obtain a kinetic measurement of ATP synthesis, replicate tests have to be run at different time intervals. Measurement of ATP synthesis with the luciferase-luciferin system in permeabilized cells, does not require repeated sampling and allows kinetic measurements to be performed on a single sample. We have applied these methods to monitor ATP synthesis in mamalian cultured cells.

II. Methodological Considerations

A. Cell Permeabilization (Titration of Digitonin)

Although more convenient than isolating coupled mitochondria, permeabilization procedures also require standardization. An insufficient degree of permeabilization could result in an underestimation of ATP synthesis due to lack of available substrates. In addition, excessive permeabilization could lead to mitochondrial membrane damage and uncoupling. For these reasons, it is important to establish the optimal amount of digitonin needed per unit of cell protein. In our experience, the digitonin concentration at which cybrid cells and fibroblasts become permeable to substrates and still maintain an intact mitochondrial membrane potential is 50 μg/ml with 1–5 mg/ml cell protein after a 1-min incubation followed by a wash step (Fig. 1). Higher concentrations of digitonin result in decreased ATP synthesis. However, because of the variations in membrane cholesterol content, the parameters for digitonin treatment may need to be optimized for each cell type.

B. ATP Detection by Luciferase-Luciferin

A number of factors need to be taken into consideration in setting up a luminescence assay. First, depending on the ATP concentration, firefly luciferase shows two different time courses of light production, possibly due to the binding of substrates at two different

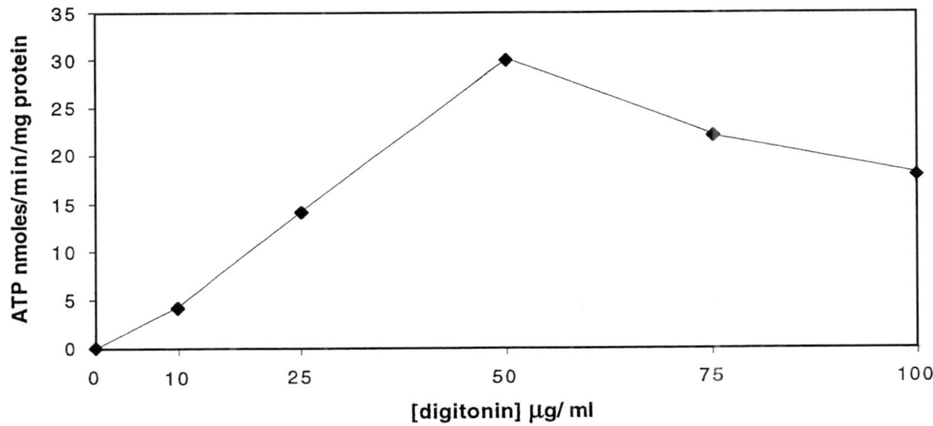

Fig. 1 Titration of digitonin. Wild-type cybrids were incubated with increasing concentrations of digitonin. ATP synthesis was assayed with malate/pyruvate as substrates. The maximum ATP synthesis rate was obtained with 50 μg/ml digitonin.

catalytic sites. At high concentrations of ATP, a short flash of light is produced followed by enzyme inactivation, whereas at low concentrations of ATP there is a less intense but more prolonged production of light (De Luca and McElroy, 1974). In the range of ATP concentration normally present in cells, we find that the initial burst of light and the ensuing inactivation of luciferase are prevented by preincubating luciferase with luciferin for 30 min (see buffer B in Table I). Second, the kinetics of luciferase is unstable at low concentrations of ATP and at high concentrations of luciferin (Lembert and Idahl, 1995). Therefore, the concentration of luciferin in the reaction mixture has to be appropriate for the levels of ATP present in cells.

C. Specificity of the Assay

ATP is formed not only through mitochondrial oxidative phosphorylation, but also through other metabolic pathways, such as glycolysis and adenylate kinase. Therefore, to obtain a specific measurement of mitochondrial ATP synthesis, these other sources of ATP must be either accounted for or, preferably, eliminated. Adenylate kinase catalyzes the following reversible reaction:

$$2\,\text{ADP} \longleftrightarrow \text{ATP} + \text{AMP}.$$

When ADP is added to the cells, the luminescence response shows an initial rapid phase, due to adenylate kinase activity, and a second slow phase deriving from oxidative phosphorylation. For this reason, P^1,P^5-di(adenosine)pentaphosphate, an inhibitor of adenylate kinase (Kurebayashi *et al.,* 1980), is added to the reaction mixture to eliminate the initial rapid phase of light emission. By assaying luminescence with and without inhibitors of mitochondrial ATP synthesis, such as atractyloside or oligomycin, it is possible to determine how much luminescence derives from ATP not produced by oxidative phosphorylation; this is subtracted from the total ATP measured. Moreover, prior to the measurement the cells are rinsed and maintained in glucose-free buffer to deplete most residual intracellular glucose. It is noteworthy that the residual ATP values obtained with oligomycin are slightly higher than with atractyloside (Fig. 3). This difference is likely due to the concomitant inhibition of both mitochondrial ATP synthesis and hydrolysis by oligomycin. Therefore, consumption of residual ATP is slower.

III. Experimental Procedures

This chapter describes the assay of ATP synthesis in three types of cells: fibroblasts from normal individuals; fibroblasts from patients affected by a mitochondrial encephalopathy (Leigh's syndrome) associated with cytochrome *c* oxidase deficiency due to mutations in the SURF-1 gene (Zhu *et al.,* 1998; Tiranti *et al.,* 1998), and cytoplasmic hybrids (cybrids; King and Attardi, 1989) harboring either wild-type mitochondrial DNA (mtDNA) or the T8993G mtDNA mutation in the ATPase 6 gene, which is responsible for a mitochondrial disorder characterized by neuropathy ataxia and retinitis pigmentosa (NARP; Holt *et al.,* 1990).

Table I
Media and Reagents[a]

	Volumes (final concentrations)	Stock solutions	Notes
Buffer A (cell suspension)	160 μl	150 m*M* KCl, 25 m*M* Tris–HCl, 2 m*M* EDTA, 0.1% bovine serum albumin, 10 m*M* K-phosphate, 0.1 m*M* MgCl, pH 7.4	Store at 4 °C; at room temperature for the assay
Digitonin	5 μl (50 μg/ml)	2 mg/ml in buffer A	Prepare the same day; keep on ice
Di(adenosine) pentaphosphate	5 μl (0.15 m*M*)	6 m*M* in water	Store at −20 °C; thaw and keep on ice
Malate[b]	2.5 μl (1 m*M*)	80 m*M* in buffer A	1 *M* stock in water; store at −20 °C; thaw and dilute
Pyruvate[b]	2.5 μl (1 m*M*)	80 m*M* in buffer A	1 *M* stock in water; store at −20 °C; thaw and dilute
Succinate[b]	5 μl (5 m*M*)	200 m*M* in buffer A	1 *M* stock in water; store at −20 °C; thaw and dilute
Rotenone	2 μg/ml	400 μg/ml in ethanol	Store at −20 °C; keep on ice
Atractyloside[b]	10 μl (0.78 m*M*)	12.5 m*M* in water	Prepare the same day; keep on ice
Oligomycin[b]	2 μl (2 μg/ml)	0.2 mg/ml in ethanol	Store at −20 °C; keep on ice
ADP	5 μl (0.1 m*M*)	4 m*M* in buffer A	Prepare the same day; keep on ice
Luciferin	See buffer B	100 m*M* in water	Store at −20 °C; thaw and keep on ice
Luciferase	See buffer B	1 mg/ml in 0.5 *M* Tris–acetate, pH 7.75	Store at −20 °C; thaw and keep on ice
Buffer B	10 μl	0.5 *M* Tris−acetate, pH 7.75, 0.8 m*M* luciferin 20 μg/ml luciferase	Prepare the same day; keep at room temperature (25 °C)
Total volume	To 200 μl with buffer A		

[a]All reagents are from Sigma Biochemicals except digitonin, *Photinus pyralis* luciferase (Roche Molecular Biochemicals), and luciferin (Promega Life Science).

[b]Malate+pyruvate or succinate is used alternatively as substrate; atractyloside or oligomycin is used alternatively as inhibitor.

A. Cell Culture

Human skin fibroblasts were grown in 150-mm culture dishes in Dulbecco's modified Eagle's medium (DMEM) containing high glucose (4.5 mg/ml), 2 m*M* L-glutamine, and 110 mg/liter sodium pyruvate supplemented with 15% fetal bovine serum (FBS). Wild-type and mutant cybrid cells were grown in 150-mm culture dishes in DMEM, supplemented with 5% FBS. When cells had reached approximately 80% confluence, the medium was removed by suction and cells were harvested by trypsinization. Cells were then concentrated by centrifugation at 800*g* for 3 min in a swinging bucket rotor centrifuge at room temperature and the cell pellet was washed twice in phosphate-buffered saline (PBS). Cells were counted in a hemocytometer or in an automated cell counter.

B. Measurement of ATP Synthesis on Samples Extracted at Fixed Intervals

Skin fibroblasts from control individuals prepared as described earlier were resuspended in buffer A [25 m*M* Tris, 150 m*M* KCl, 2 m*M* EDTA, 10 m*M* KPO_4, 0.1 m*M* $MgCl_2$, 50 μg/ml digitonin, 0.1% BSA, 0.15 m*M* P^1,P^5-di(adenosine)pentaphosphate, pH adjusted to 7.75, see Table I] to the concentration of 0.75×10^6 cells/ml (approximately 1 mg/ml cell protein). For each assay, two aliquots of 0.25 ml (250 μg protein) of the cell suspension were transferred into two 1.5-ml microcentrifuge tubes, and 0.75 ml of substrate solution containing 10 m*M* malate, 10 m*M* pyruvate, and 1 m*M* ADP in buffer A was added. One tube was processed for ATP content immediately (time zero). The other tube was incubated at 37 °C for 5 min. An aliquot of 50 μl of the sample from each tube was mixed with 450 μl of boiling extraction buffer (100 m*M* Tris, 4 m*M* EDTA, pH 7.75) and incubated for 2 min at 100 °C. After cooling on ice, tubes were centrifuged at 1000*g* for 1 min and the supernatant was collected and kept on ice for ATP measurement. For the luciferin/luciferase assay, we diluted the supernatant 50 times with cold extraction buffer. An aliquot of 100 μl of the diluted supernatant was mixed with 100 μl of a luciferase reagent (Wame; Boehringer Mannheim GmbH, Germany). Immediately after the addition of the reagent, the tube was placed in a luminometer and the light output was integrated for 1 s. A reagent blank was prepared by mixing 100 μl of incubation mixture minus cells with 100 μl of luciferase reagent. Standard solutions containing 10^{-11} to 10^{-5} *M* ATP were used to construct a standard curve to calculate ATP content. We measured ATP-derived luminescence at different time points and observed that the maximum rate of ATP synthesis was attained between 0 and 10 min, followed by a progressive decrease between 10 and 30 min, after which the luminescence curve became flat (Fig. 2A). Therefore, in order to measure maximum ATP synthesis in a linear range, we decided to use 0- and 5-min incubation times for all our assays. The rate of ATP synthesis in skin fibroblasts obtained from a normal individual is shown in Fig. 2B. The addition of 0.2 m*M* dinitrophenol (a strong mitochondrial uncoupler) reduced the rate of ATP formation by approximately 80%. The residual ATP synthesis observed in the presence of dinitrophenol and substrates could be attributed to residual adenylate kinase activity and glycolysis.

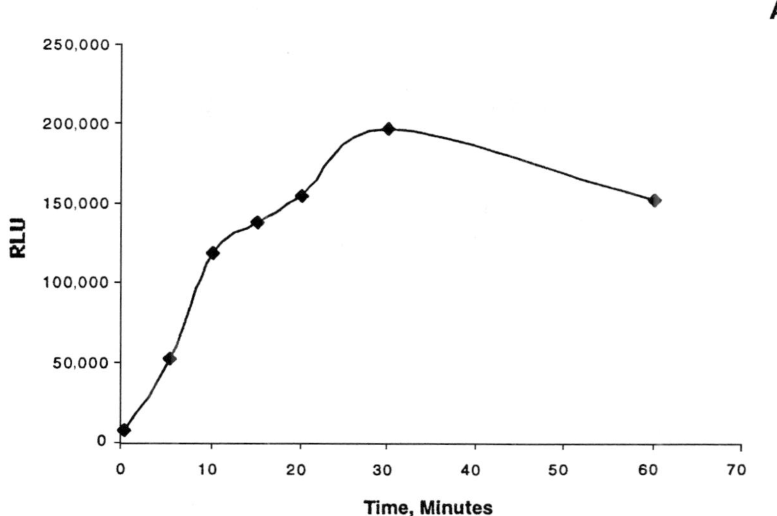

Fig. 2 (A) Luminescence curve in normal permeabilized human skin fibroblasts in the presence of pyruvate plus malate and ADP.

C. Measurement of ATP Synthesis with a Continuous Assay

Cells were resuspended at 1×10^7 cells/ml in buffer A (see Table I) at room temperature and an aliquot of the suspension was set aside for protein measurements. The cell suspension (160 μl) was incubated for 1 min at room temperature with 50 μg/ml digitonin with gentle agitation. Cells were then washed by adding 1 ml of buffer A and then concentrated by centrifugation at 800*g* as described earlier. The cell pellet was resuspended in 160 μl of buffer A and 0.15 m*M* P^1,P^5-di(adenosine)pentaphosphate. Ten microliters of buffer B (containing luciferin and luciferase), 0.1 m*M* ADP, and either 1 m*M* malate plus 1 m*M* pyruvate or 5 m*M* succinate plus 2 μg/ml rotenone (Table I) were added. For each reaction, one replicate tube was prepared containing the components just mentioned plus either 1 m*M* atractyloside or 10 μg/ml oligomycin. Atractyloside inhibits the ADP-ATP translocase (Klingenberg *et al.*, 1985). Oligomycin inhibits ATP synthase. The inhibitors were added to obtain the baseline luminescence corresponding to nonmitochondrial ATP production. We have determined that atractyloside and oligomycin have no effect on luciferase activity. Cells were placed in a counting luminometer, and the light emission was recorded. We used an Optocomp I luminometer (MGM Instruments, Inc.), which allows for multiple recordings in the kinetic measurement setting. The integration time for each reading was set at 1 s and the interval between readings at was set 20 s for a total recording time of 10 min (30 readings). Other luminometers can be used, but those that allow kinetic measurements are preferable because they simplify the experimental procedure. It is also

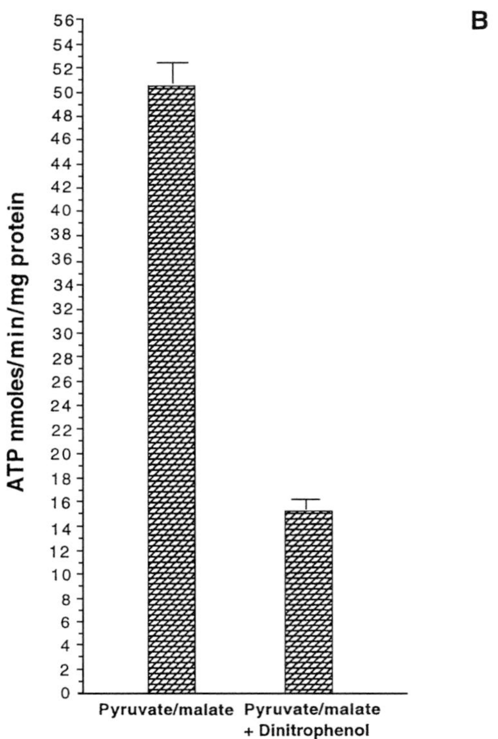

Fig. 2 (B) ATP synthesis rates in normal human skin fibroblast in the presence or absence of an inhibitor. Fibroblasts were incubated at 37 °C for 5 min in buffer containing digitonin, with malate/pyruvate as substrates. Dinitrophenol (0.2 m*M*) was added to the incubation mixture to inhibit mitochondrial ATP synthesis.

useful to interface the instrument with a computer to facilitate data collection and analysis.

Total cellular protein content was measured on an aliquot of cells using the Bio-Rad DC protein assay kit (Bio-Rad Laboratories), a modification of the Lowry protein assay. Serial dilutions of bovine serum albumin were used as standards.

Figure 3 shows a representative assay on wild-type cybrids. The luminescence curves describe the kinetics of ATP synthesis in cells energized with malate plus pyruvate in the presence or absence of mitochondrial inhibitors. In wild-type cells, luminescence increases linearly for approximately 3 min. It reaches a peak at 4 to 5 min, followed by a slow, progressive decrease. The decrease in luminescence is probably due to the chemical properties of luciferin, which is progressively transformed into the inactive derivative deoxyluciferin (Lembert and Idahl, 1995). The linear portion of the curve is used to extrapolate the variation in luminescence per unit of time (change in relative light units or ΔRLU). The ΔRLU measured in the presence of atractyloside is subtracted from the total ΔRLU in order to obtain the proportion of ΔRLU derived from mitochondrial

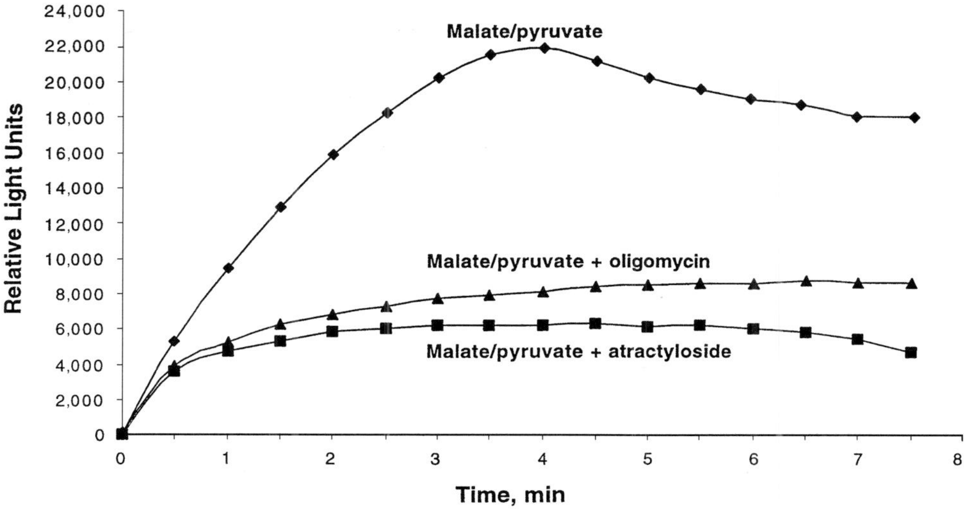

Fig. 3 Luminescence curve in wild-type cybrids. In a typical experiment, cells were assayed in three different conditions: with malate/pyruvate as substrates or with malate/pyruvate after the addition of either oligomycin or atractyloside as inhibitors of mitochondrial ATP synthesis. Luminescence (expressed as relative light units) was recorded at 30-s intervals with a 1-s integration time.

ATP synthesis. The change in luminescence is then converted in ATP concentration based on the values obtained from an ATP standard curve [luminescence obtained with luciferase is directly proportional to ATP concentration in the physiological concentration range (Strehler, 1968)]. A standard ATP/luminescence curve is constructed by measuring luminescence derived from ATP solutions containing 0, 0.01, 0.1, 0.25, 0.5, and 1 m*M* ATP in buffer A plus 10 μl of buffer B (Fig. 4).

ATP synthesis rates in wild-type and mutant cybrids and in control and Leigh's fibroblasts are shown in Fig. 5.

IV. ATP Synthesis in Cellular Models of Mitochondrial Diseases

We applied our continuous assay to measure ATP synthesis in wild-type and homoplasmic T8993G mutant cybrid cells (Manfredi *et al.,* 1999). We considered this a suitable disease model to test, as ATP synthesis is impaired in fibroblasts and lymphoblastoid cells harboring high levels of this mutation (Tatuch *et al.,* 1993; Trounce *et al.,* 1994, Houstek *et al.,* 1995; Vazquez-Memije *et al.,* 1996). Furthermore, because all other respiratory chain complexes and the mitochondrial membrane potential are unaltered in NARP cells, this model allowed us to study a form of "pure" ATP synthesis defect. Mutant cells showed a marked reduction of mitochondrial ATP synthesis as compared to wild type, with all substrates tested (Fig. 5). Residual ATP synthesis in

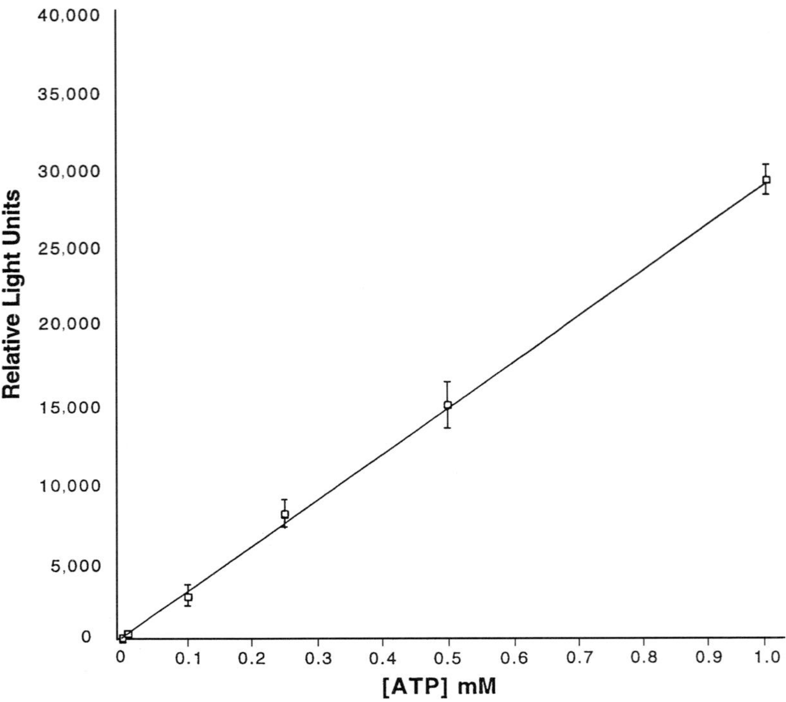

Fig. 4 ATP standard assay. Linear regression analysis for the correlation between ATP concentration and luciferin /luciferase-dependent luminescence (relative light units). Each point indicates the value corresponding to the mean luminescence $\pm$ SD ($n = 4$); $r = 0.998$.

mutant cells was 18% of wild type with malate plus pyruvate and 27% of wild type with succinate.

Our results are in line with those obtained in permeabilized fibroblasts by Houstek and colleagues (1995). In that study, residual ATP synthesis in virtually homoplasmic mutant cells was 33% of control with malate/pyruvate and 50% with succinate. The relatively smaller decrease in ATP synthesis in fibroblasts than in cybrids could be ascribed to the presence of low amounts of wild-type mtDNA in the fibroblasts. However, results obtained in isolated mitochondria were much more variable. For example, ATP synthesis with malate/pyruvate in T8993G mutant cells ranged from 0% of controls in isolated mitochondria from fibroblasts (Vazquez-Memije *et al.,* 1996) to 67% in isolated mitochondria from lymphoblasts (Tatuch *et al.,* 1993). With succinate, ATP synthesis was 58% of controls in both studies. Again, the reason for this discrepancy is not clear, but it could be partly due to intrinsic differences in the metabolic characteristics in the cell types and partly due to the differences in the methods utilized.

We have also measured ATP synthesis in fibroblasts from a patient harboring a defect of cytochrome *c* oxidase due to a mutation in a nuclear gene encoding for SURF-1 (Tiranti

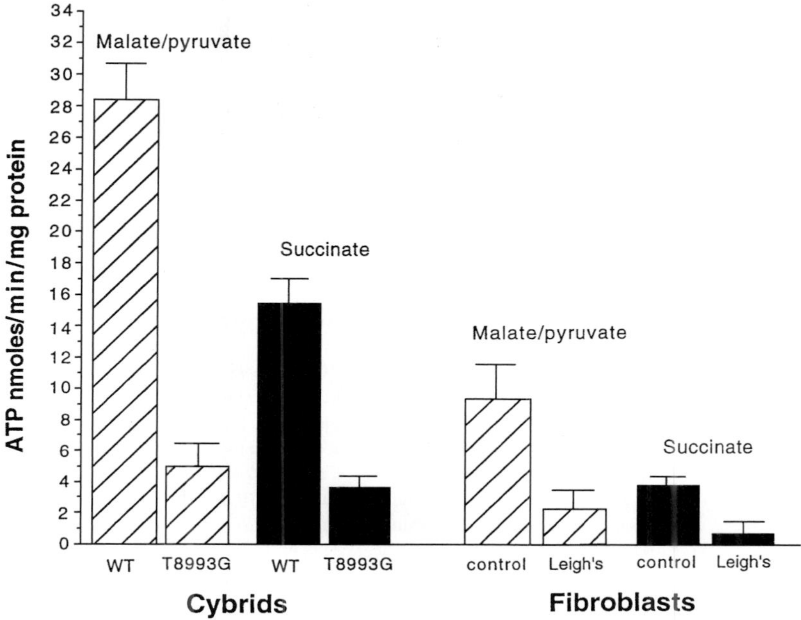

Fig. 5 ATP synthesis rates in wild-type and T8993G mutant cybrids and in control and cytochrome *c* oxidase-deficient Leigh's fibroblasts using either malate/pyruvate or succinate as substrates. Bars indicate the values corresponding to the mean activity $\pm$SD ($n = 3$).

et al., 1998; Zhu *et al.,* 1998). ATP synthesis in cytochrome *c* oxidase-deficient cells was reduced to 27 and 50% of controls with malate plus pyruvate and succinate, respectively.

References

DeLuca, M., and McElroy, W. D. (1974). Kinetics of the firefly luciferase catalyzed reactions. *Biochemistry* **13,** 921–925.

Deluca, M. (1976). Firefly luciferase. *Adv. Enzymol. Relat. Areas Mol. Biol.* **44,** 37–68.

Holt, I. J., Harding, A. E., Petty, R. K., and Morgan-Hughes, J. A. (1990). A new mitochondrial disease associated with mitochondrial DNA heteroplasmy. *Am. J. Hum. Genet.* **46,** 428–433.

Houstek, J., Klement, P., Hermanska, J., Houstkova, H., Hansikova, H., Van den Bogert, C., and Zeman, J. (1995). Altered properties of mitochondrial ATP-synthase in patients with a T→G mutation in the ATPase 6 (subunit a) gene at position 8993 of mtDNA. *Biochim. Biophys. Acta* **1271,** 349–357.

James, A. M., Sheard, P. W., Wei, Y. H., and Murphy, M. P. (1999). Decreased ATP synthesis is phenotypically expressed during increased energy demand in fibroblasts containing mitochondrial tRNA mutations. *Eur. J. Biochem.* **259,** 462–469.

King, M. P., and Attardi, G. (1989). Human cells lacking mtDNA: Repopulation with exogenous mitochondria by complementation. *Science* **246,** 500–503.

Klingenberg, M. (1985). Principles of carrier catalysis elucidated by comparing two similar membrane translocators from mitochondria, the ADP/ATP carrier and the uncoupling protein. *Ann. N.Y. Acad. Sci.* **456,** 279–288.

Kunz, W.S., Kuznetsov, A. V., Schulze, W., Eichhorn, K., Schild, L., Striggow, F., Bohnensack, R., Neuhof, S., Grasshoff, H., Neumann, H. W., *et al.* (1993). Functional characterization of mitochondrial oxidative phosphorylation in saponin-skinned human muscle fibers. *Biochim. Biophys. Acta* **1144,** 46–53.

Kurebayashi, N., Kodama, T., and Ogawa, Y. (1980). P1,P5-Di(adenosine-5′)pentaphosphate(Ap5A) as an inhibitor of adenylate kinase in studies of fragmented sarcoplasmic reticulum from bullfrog skeletal muscle. *J. Biochem.* **88,** 871–876.

Lemasters, J. J., and Hackenbrock, C. E. (1973). Adenosine triphosphate: Continuous measurement in mitochondrial suspension by firefly luciferase luminescence. *Biochem. Biophys. Res. Commun.* **55,** 1262–1270.

Lemasters, J. J., and Hackenbrock, C. R. (1976). Continuous measurement and rapid kinetics of ATP synthesis in rat liver mitochondria, mitoplasts and inner membrane vesicles determined by firefly-luciferase luminescence. *Eur. J. Biochem.* **67,** 1–10.

Lembert, N., and Idahl, L. A. (1995). Regulatory effects of ATP and luciferin on firefly luciferase activity. *Biochem. J.* **305,** 929–933.

Maechler, P., Wang, H., and Wollheim, C. B. (1998). Continuous monitoring of ATP levels in living insulin secreting cells expressing cytosolic firefly luciferase. *FEBS Lett.* **422,** 328–332.

Manfredi, G., Gupta, N., Vazquez-Memije, M. E., Sadlock, J. E., Spinazzola, A., De Vivo, D. C., and Schon, E. A. (1999). Oligomycin induces a decrease in the cellular content of a pathogenic mutation in the human mitochondrial ATPase 6 gene. *J. Biol. Chem.* **274,** 9386–9391.

Ouhabi, R., Boue-Grabot, M., and Mazat, J. P. (1998). Mitochondrial ATP synthesis in permeabilized cells: Assessment of the ATP/O values in situ. *Anal. Biochem.* **263,** 169–175.

Strehler, B. L. (1968). Bioluminescence assay: Principles and practice. *Methods Biochem. Anal.* **16,** 99–181.

Strehler, B. L., and Totter, J. R. (1952). Firefly luminescence in the study of energy transfer mechanism. I. Substrates and enzyme determinations. *Arch. Biochem. Byophys.* **40,** 28–41.

Tatuch, Y., and Robinson, B. H. (1993). The mitochondrial DNA mutation at 8993 associated with NARP slows the rate of ATP synthesis in isolated lymphoblast mitochondria. *Biochem. Biophys. Res. Commun.* **192,** 124–128.

Tiranti, V., Hoertnagel, K., Carrozzo, R., Galimberti, C., Munaro, M., Granatiero, M., Zelante, L., Gasparini, P., Marzella, R., Rocchi, M., Bayona-Bafaluy, M. P., Enriquez, J. A., Uziel, G., Bertini, E., Dionisi-Vici, C., Franco, B., Meitinger, T., and Zeviani, M. (1998). Mutations of SURF-1 in Leigh disease associated with cytochrome c oxidase deficiency. *Am. J. Hum. Genet.* **63,** 1609–1621.

Trounce, I., Neill, S., and Wallace, D. C. (1994). Cytoplasmic transfer of the mtDNA nt 8993 T→G (ATP6) point mutation associated with Leigh syndrome into mtDNA-less cells demonstrates cosegregation with a decrease in state III respiration and ADP/O ratio. *Proc. Natl. Acad. Sci. USA* **91,** 8334–8338.

Tuena de Gomez-Puyou, M., Ayala, G., Darszon, A., and Gomez-Puyou, A. (1984). Oxidative phosphorylation and the Pi-ATP exchange reaction of submitochondrial particles under the influence of organic solvents. *J. Biol. Chem.* **259,** 9472–9478.

Vazquez-Memije, M. E., Shanske, S., Santorelli, F. M., Kranz-Eble, P., Davidson, E., DeVivo, D. C., and DiMauro, S. (1996). Comparative biochemical studies in fibroblasts from patients with different forms of Leigh syndrome. *J. Inherit. Metab. Dis.* **19,** 43–50.

Vazquez-Memije, M. E., Shanske, S., Santorelli, F. M., Kranz-Eble, P., DeVivo, D. C., and DiMauro, S. (1998). Comparative biochemical studies of ATPases in cells from patients with the T8993G or T8993C mitochondrial DNA mutations. *J. Inherit. Metab. Dis.* **21,** 829–836.

Wanders, R. J., Ruiter, J. P., and Wijburg, F. A. (1993). Studies on mitochondrial oxidative phosphorylation in permeabilized human skin fibroblasts: Application to mitochondrial encephalomyopathies. *Biochim. Biophys. Acta* **1181,** 219–222.

Wanders, R. J., Ruiter, J. P., and Wijburg, F. A. (1994). Mitochondrial oxidative phosphorylation in digitonin-permeabilized chorionic villus fibroblasts: A new method with potential for prenatal diagnosis. *J. Inherit. Metab. Dis.* **17,** 304–306.

Wanders, R. J., Ruiter, J. P., Wijburg, F. A., Zeman, J., Klement, P., and Houstek, J. (1996). Prenatal diagnosis of systemic disorders of the respiratory chain in cultured chorionic villus fibroblasts by study of ATP-synthesis in digitonin-permeabilized cells. *J. Inherit. Metab. Dis.* **19,** 133–136.

Wibom, R., Lundin, A., and Hultman, E. (1990). A sensitive method for measuring ATP-formation in rat muscle mitochondria. *Scand. J. Clin. Lab. Invest.* **50,** 143–152.

Wibom, R., Soderlund, K., Lundin, A., and Hultman, E. (1991). A luminometric method for the determination of ATP and phosphocreatine in single human skeletal muscle fibres. *J. Biolumin. Chemilumin.* **6,** 123–129.

Zhu, Z., Yao, J., Johns, T., Fu, K., De Bie, I., Macmillan, C., Cuthbert, A. P., Newbold, R. F., Wang, J., Chevrette, M., Brown, G. K., Brown, R. M., and Shoubridge, E. A. (1998). SURF1, encoding a factor involved in the biogenesis of cytochrome c oxidase, is mutated in Leigh syndrome. *Nature Genet.* **20,** 337–343.

CHAPTER 8

Measurement of Membrane Permeability and Permeability Transition of Mitochondria

Naoufal Zamzami, Carine Maisse, Didier Métivier, and Guido Kroemer

Centre National de la Recherche Scientifique
F-94801 Villejuif, France

I. Introduction

Mitochondria are essential for the evolution of complex animals in aerobic conditions. They carry out most cellular oxidations and produce the bulk of an animal cell's ATP. This organelle has also been implicated in the regulation of cell death processes. Indeed, a major site of the activity of (pro/anti)apoptotic proteins is the mitochondrion. Following a variety of death signals, mitochondria show early alterations of their function,

METHODS IN CELL BIOLOGY, VOL. 65

which, at least in some instances, may be explained by the opening of the permeability transition (PT) pores.

The PT pore, also called the mitochondrial megachannel or multiple conductance channel (Bernardi, 1996; Kinnally *et al.,* 1996; Zoratti and Szabò, 1995), is a dynamic multiprotein complex located at the contact site between the inner and the outer mitochondrial membranes, one of the critical sites of metabolic coordination between the cytosol and the mitochondrial intermembrane and matrix spaces. The PT pore participates in the regulation of matrix Ca^{2+}, pH, $\Delta\Psi_m$, and volume. It also functions as a Ca^{2+}-, voltage-, pH-, and redox-gated channel, with several levels of conductance and little if any ion selectivity (Bernardi, 1996; Beutner *et al.,* 1996; Ichas *et al.,* 1997; Kinnally *et al.,* 1996; Marzo *et al.,* 1998; Zoratti and Szabò, 1995).

In isolated mitochondria, opening of the PT pore causes matrix swelling with consequent distension and local disruption of the mitochondrial outer membrane (whose surface is smaller than that of the inner membrane), release of soluble products from the intermembrane space, dissipation of the mitochondrial inner transmembrane potential, and release of small molecules up to 1500 Da from the matrix, through the inner membrane (Bernardi, 1996; Kantrow and Piantadosi, 1997; Petit *et al.,* 1998; Zoratti and Szabò, 1995). Similar changes are found in apoptotic cells, perhaps with the exception of matrix swelling (which is only observed in a transient fashion), before cells shrink (Kroemer *et al.,* 1995, 1997, 1998; Liu *et al.,* 1996; vander Heiden *et al.,* 1997). In several models of apoptosis, pharmacological inhibition of the PT pore is cytoprotective, suggesting that opening of the PT pore can be rate limiting for the death process (Kroemer *et al.,* 1998; Marchetti *et al.,* 1996; Zamzami *et al.,* 1996). Moreover, the cytoprotective oncoprotein Bcl-2 has been shown to function as an endogenous inhibitor of the PT pore (Kroemer, 1997; Susin *et al.,* 1996; Zamzami *et al.,* 1996). Some authors have suggested that other mechanisms not involving the PT pore may account for mitochondrial membrane permeabilization during apoptosis (Bossy-Wetzel *et al.,* 1998; Kluck *et al.,* 1997). Irrespective of this possibility, mitochondrial membrane permeabilization is a general feature of apoptosis. This chapter details experimental procedures designed to measure apoptosis-associated mitochondrial membrane permeabilization, both in intact cells and in isolated mitochondria.

II. Procedures

A. Mitochondrial Alterations in Intact Cells

Mitochondrial inner membrane depolarization is one of the major alterations observed in apoptotic cells. Accurate quantitation of the mitochondrial inner transmembrane potential ($\Delta\Psi_m$), using appropriate potentiometric fluorochromes, reveals a $\Delta\Psi_m$ decrease in most models of apoptosis. This $\Delta\Psi_m$ decrease is an early event and thus mostly precedes the activation of downstream caspases and nucleases. Our current data are compatible with the notion that a $\Delta\Psi_m$ decrease mediated by the opening of the mitochondrial PT pore constitutes an irreversible event in the apoptotic process (Susin *et al.,* 1998; Zamzami *et al.,* 1998). Therefore, the determination of the $\Delta\Psi_m$ allows for the detection of early apoptosis, both *in vitro* and *ex vivo.* This section proposes several protocols for the quantitation of $\Delta\Psi_m$ and other mitochondrial alterations in purified mitochondria as well as in intact cells.

B. Cytometric Analysis: Lipophilic Cationic Dyes

Lipophilic cations accumulate in the mitochondrial matrix, driven by the electrochemical gradient, following the Nernst equation, according to which every 61.5-mV increase in membrane potential (usually 120–170 mV) corresponds to a 10-fold increase in cation concentration in mitochondria. Therefore, the concentration of such cations is 2 to 3 logs higher in the mitochondrial matrix than in the cytosol.

Several different cationic fluorochromes can be employed to measure mitochondrial transmembrane potentials. These markers include 3,3′-dihexyloxacarbocyanine iodide [$DiOC_6(3)$] (green fluorescence) (Petit *et al.,* 1990), chloromethyl-X-rosamine (CMXRos) (red fluorescence) (Castedo *et al.,* 1996; Macho *et al.,* 1996), and 5,5′,6,6′-tetrachloro-1,1′,3,3′-tetraethylbenzimidazolcarbocyanine iodide (JC-1) (red and green fluorescence) (Smiley, 1991). As compared to rhodamine 123 (Rh123), which we do not recommend for cytofluorometric analyses (Metivier *et al.,* 1998), $DiOC_6(3)$ offers the important advantage that it does not show major quenching effects. JC-1 incorporates into mitochondria, where it either forms monomers (green fluorescence, 527 nm) or, at high transmembrane potentials, aggregates (red fluorescence, 590 nm) (Smiley, 1991). Thus, the quotient between green and red JC-1 fluorescence provides an estimate of $\Delta\Psi_m$ that is (relatively) independent of mitochondrial mass.

C. Single Staining of Mitochondrial Parameters for Flow Cytometric Analysis

1. Materials

Stock solutions of fluorochromes: $DiOC_6(3)$ should be diluted to 40 μM in dimethyl absolute ethanol (ETOH), CMXRos to 1 mM in DMSO, and JC-1 to 0.76 mM in DMSO. All three fluorochromes can be purchased from Molecular Probes and should be stored, once diluted, at $-20°$C in the dark.

Working solutions: Dilute $DiOC_6(3)$ to 400 nM [10 μl stock solution + 1 ml phosphate-buffered saline (PBS)], CMXRos to 2 μM (2 μl stock solution + 1 ml PBS or mitochondrial resuspension buffer), and JC-1 to 7.6 μM (10 μl stock solution + 1 ml PBS). These solutions should be prepared fresh for each series of stainings.

Carbonyl cyanide *m*-chlorophenylhydrazone (CCCP) diluted in ethanol (stock at 20 mM), a protonophore required for control purposes ($\Delta\Psi_m$ disruption).

Cytofluorometer with appropriate filters.

2. Staining Protocol

1. Cells ($5–10 \times 10^5$ in 0.5 ml culture medium or PBS) should be kept on ice until staining. If necessary, cells can be labeled with specific antibodies conjugated to compatible fluorochromes [e.g., phycoerythrin for $DiOC_6(3)$, fluorescein isothiocyanate for CMXRos] before determination of mitochondrial potential.

2. For staining, add the following amounts of working solutions to 0.5 ml of cell suspension: 25 μl $DiOC_6(3)$ (final concentration: 20 n*M*), 10 μl CMXRos (final concentration: 40 n*M*), or 6.5 μl JC-1 (final concentration 0.1 μ*M*), and transfer tubes to a water bath kept at 37°C. After 15–20 min of incubation, return cells to ice. Do not wash cells. As a negative control, aliquots of cells should be labeled in the presence of the protonophore CCCP (100 μ*M*).

3. Perform cytofluorometric analysis within 10 min, while gating the forward and sideward scatters on viable, normal-sized cells. When large numbers of tubes are to be analyzed (>10 tubes), the interval between labeling and cytofluorometric analysis should be kept constant. When using an Epics Profile cytofluorometer (Coulter), $DiOC_6(3)$ should be monitored in FL1, CMXRos in FL2 (excitation: 488 nm; emission: 599) (Fig. 1), and JC-1 in FL1 versus FL3 (excitation: 488 n*M*; emission at 527 and 590 n*M*). Note that the incorporation of these fluorochromes may be influenced by parameters not determined by mitochondria (cell size, plasma membrane permeability, efficacy of the multiple drug resistance pump, etc.). Results can only be interpreted when the staining profiles obtained in different experimental conditions (e.g., controls versus apoptosis) are identical in the presence of CCCP (Fig. 1).

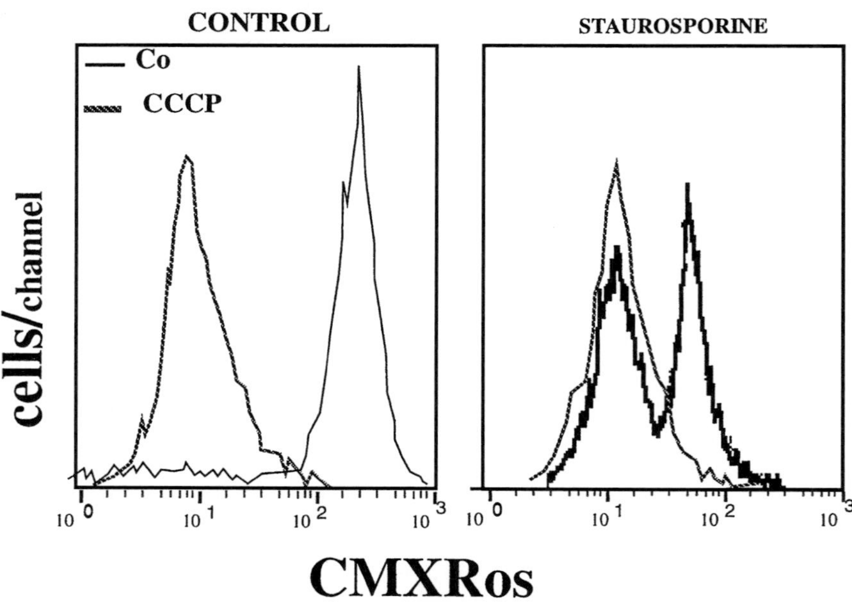

Fig. 1 *Left panel:* $\Delta\Psi_m$ measurement in Rat1 cells. Cells were either left untreated (Co, thin solid) or treated 1/2 hour with 100 μ*M* CCCP to obtain total depolarization (Positive Control, dotted line). *Right panel:* Cells were treated with 0.5 μ*M* staurosporine for 4 h, followed by staining with 40 n*M* CMXROS. Almost 50% of the cells show a decrease of $\Delta\Psi_m$ (*thick solid*) comparable to positive control CCCP (*dotted*) line.

D. Combined Detection of Different Mitochondrial Parameters in Intact Cells

One of the interesting aspects of flow cytometry is the possibility of performing a multiparametric study. It is well known that mitochondrial depolarization is followed by other mitochondrial and nonmitochondrial perturbations. For example, depolarized mitochondria produce elevated levels of reactive oxygen species (ROS), which in turn oxidize mitochondrial cardiolipins. It is possible to determine the production of ROS by dihydroethidine (HE), a substance that is oxidized by superoxide anion to become ethidium bromide (EthBr) and to fluoresce in the red (Rothe and Valet, 1990). HE is more sensitive to superoxide anion than 2′,7′-dichlorofluorescin diacetate, which measures H_2O_2 formation (Carter *et al.,* 1994; Rothe and Valet, 1990). Thus, enhanced HE → EthBr conversion can be observed in cells that cannot be labeled with 2′,7′-dichlorofluorescin diacetate. Alternatively, the damage produced by ROS in mitochondria can be determined indirectly, via assessing the oxidation state of cardiolipin, a molecule restricted to the inner mitochondrial membrane. Nonyl acridine orange (NAO) interacts stoichiometrically with intact, nonoxidized cardiolipin (Petit *et al.,* 1992). As a consequence, a reduction in NAO fluorescence indicates a decrease in cardiolipin content.

1. Materials

Stock solutions of fluorochromes: 4.73 mg/ml (10 m*M*) HE in DMSO should be stored at −20°C; 3.15 mg/ml (10 m*M*) NAO in DMSO should be stored at 4°C. Both fluorochromes are light sensitive. They are available from Molecular Probes.

Working solutions: Dilute 5 μl of either of the stock solution in 0.5 ml PBS (final concentration: 100 μ*M*). Prepare fresh for each series of stainings.

2. Staining Protocol

1. Cells (5–10 × 10^5 in 1 ml culture medium or PBS) should be kept on ice until staining. Before HE staining, cells may be labeled with specific antibodies conjugated to either fluorescein isothiocyanate or phycoerythrine.

2. For staining, add the following amounts of working solutions to 1 ml of cell suspension: 10 μl HE (final concentration: 1 μ*M*), 1 μl NAO (final concentration: 100 n*M*), or 13 μl JC-1 (final concentration 0.1 μ*M*), and transfer tubes to a water bath kept at 37°C. After 15–20 min of incubation, return cells to ice. Do not wash cells.

3. Perform cytofluorometric analysis within 10 min, using viable, normal sized-cells to gate forward and sideward scatter. When using an Epics Profile cytofluorometer (Coulter), HE should be monitored in FL3 and NAO. in FL1 or FL3. For double stainings, compensations have to be adjusted in accord with the apparatus. If HE is combined with NAO, we recommend the following compensations: 27% for FL3–FL1 and 1% for FL1–FL3. In this case, NAO is measured in FL3. For double staining with HE and $DiOC_6(3)$, the recommended compensations are 5% for FL1–FL3 and 2% for FL3–FL1.

E. Detection of Soluble Intermembrane Mitochondrial Proteins by Fluorescence Microscopy

Many apoptotic regulatory proteins are associated with mitochondrial membranes, or translocate from cytosol to mitochondria during the apoptotic process. In addition, mitochondria release apoptogenic factors called soluble intermembrane proteins, or SIMPs (e.g., AIF, cytochrome c, caspases-9 and -2, DDF; Patterson *et al.,* 1999) when they are damaged during apoptosis. Therefore, the detection of the release of such proteins into the cytosol is a complementary indication of the mitochondrial implication in different cell death pathways. Two strategies may be employed to detect the mitochondrial release of such proteins. First, cells may be disrupted and subcellular fractions can be prepared, followed by Western blot analysis of the subcellular redistribution of SIMPs. Second, the localization of SIMPs can be detected by immunofluorescence *in situ,* on fixed intact cells (Fig. 2, see color plate). We recommend to combine both techniques which helps in avoiding artifacts.

F. Immunofluorescence analysis

1. Materials

Adherent cells cultured and treated on coverslips in 12-well-dishes

Slides

1X PBS

Paraformaldehyde, 4% in 1X PBS + picric acid 0.19% (v/v) of a saturated solution

SDS, 0.1% in 1X PBS (v/v) from a 10% SDS stock

Fetal calf serum (FCS) 10% in 1X PBS

PBS-BSA, 1 mg/ml

First antibody (anti-SIMPs)

Corresponding second antibody (anti-isotype to the first antibody), labeled with fluorochrome (phycoerythrine or FITC)

2. Protocol

1. Withdraw the culture medium.
2. Wash once with 1X PBS.
3. Fix with the fixative solution (4% paraformaldehyde + 0.19% picric acid in 1X PBS) for 30–60 min.
4. Wash three times with 1X PBS. Gentle pipetting is recommended to avoid cells detachment.
5. Permeabilize with 0.1% SDS in 1X PBS for 10 min.
6. Wash three times with 1X PBS.
7. Block with 10% FCS in 1X PBS for 20 min.
8. Wash once with 1X PBS.
9. Incubate with the first antibody (anti-SIMPs) in 1 mg/ml PBS-BSA for 60 min.

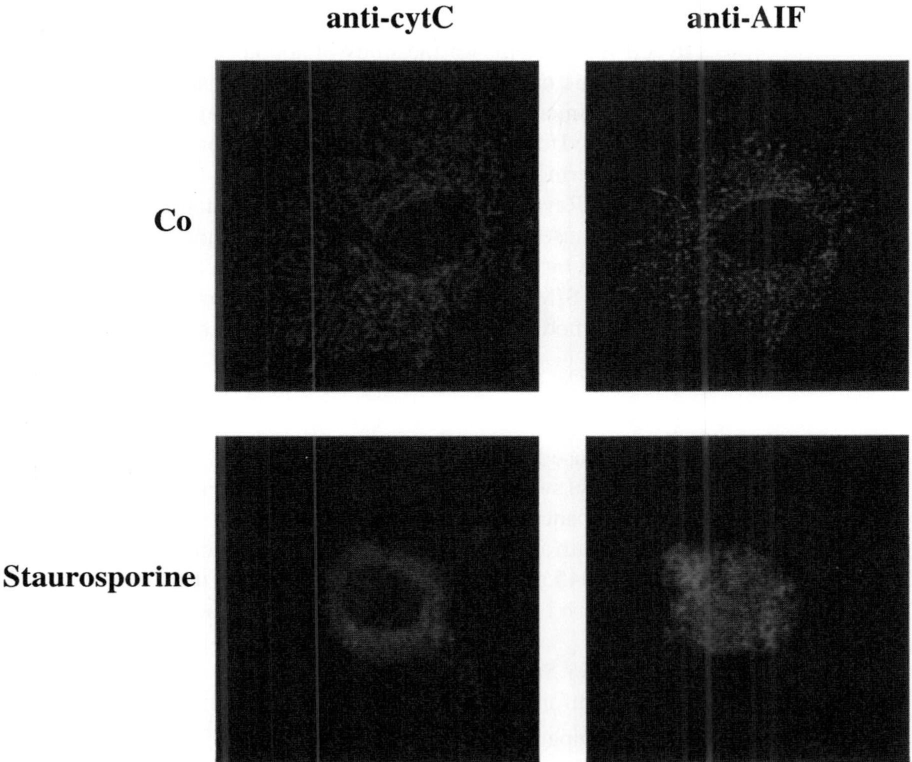

Fig. 2 Analysis of AIF and cytochrome *c* release from mitochondria. Mouse embryonic fibroblasts (MEF) cells, in the absence (Co) and presence of staurosporin (8 h, 2 μM), were stained for AIF (red fluorescence) or cytochrome *c* (green fluorescence), followed by confocal analysis. In these experimental conditions, staurosporin-treated cells have released both AIF and cytochrome *c* from the punctate mitochondrial localization. Note that only AIF translocates to the nucleus, where it induces fragmentation of the DNA into high molecular weight fragments (not shown). Immunostaining protocol: A rabbit antiserum was generated against a mixture of three peptides derived from the mAIF amino acid (aa) sequence (aa 151–170, 166–185, 181–200, coupled to keyhole limpet hemocyanine, generated by Syntem, Nîmes, France). This antiserum (ELISA titer ~10,000) was used (diluted 1/1000) on paraformaldehyde (4%, w/v) and picric acid-fixed (0.19% v/v) cells (cultured on 100-μm coverslips; 18 mm diameter; Superior, Germany), and revealed with a goat antirabbit IgG conjugated to phycoerythrine (Southern Biotechnology, Birmingham, AL). Cells were counterstained for the detection of cytochrome *c* (mAb 6H2.B4 from Pharmingen), revealed by a goat antimouse IgG fluorescein isothiocyanate conjugate (Southern Biotechnology). (See Color Plate.)

10. Wash three times with 1X PBS. From this step onward, samples should be protected from light.
11. Incubate with the fluorescent second antibody (anti-isotype to the first antibody) diluted in 1 mg/ml PBS-BSA for 30 min.
12. Wash three times with 1X PBS.
13. Place the coverslip on a glass slide and examine in a fluorescence microscope.

G. Monitoring Changes in Isolated Mitochondria

In cell-free systems for the study of apoptosis, purified mitochondria can be exposed to apoptogenic molecules such as Ca^{2+}, ganglioside GD3, or recombinant Bax protein, which function as endogenous permeabilizing agents. Similarly, purified mitochondria can be exposed to cytotoxic xenobiotics or viral proteins, which in some cases have direct membrane permeabilizing effects (Jacotot *et al.,* 1999; Larochette *et al.,* 1999; Marchetti *et al.,* 1999; Ravagnan *et al.,* 1999). The quantitation of $\Delta\Psi_m$ in isolated mitochondria is useful because it detects changes in the inner membrane permeability that usually are accompanied by an increase of outer membrane permeability, leading to the release of the so-called SIMPs normally stored in the intermembrane space. This section describes different methods for the measurement of mitochondrial membrane permeability.

1. Volume Changes of Isolated Mitochondria ("Large Amplitude Swelling")

Rat or mouse liver mitochondria are prepared by standard differential centrifugation. Mitochondrial swelling consecutive to PTP opening is followed at 545 nm by the variation of the absorbance (when a spectrophotometer is used) or by the variation of 90° light scattering (with a spectrofluorimeter), in which both excitation and emission wavelengths are fixed at 545 nm. When mitochondria swell, there is a decrease in the absorbance at 545 nm, which is measured by recording the kinetics of absorbance.

a. Reagents for Swelling Assay

Instrument for absorbance analysis (spectrophotometer or spectrofluorimeter)

Freshly prepared mitochondria

Swelling buffer (SB): 0.2 *M* sucrose, 10 m*M* Tris–MOPS, pH 7.4, 5 m*M* succinate, 1 m*M* Pi, 2 μ*M* rotenone, and 10 μ*M* EGTA–Tris. Note that the SB solution should be prepared fresh before each use.

b. Measurement

1. Dilute freshly prepared mitochondria in SB at 0.5 mg/ml and incubate in a thermostated, magnetically stirred cuvette in a final volume of 1 ml.
2. Two minutes after starting to record, the reagents to be tested may be added. A useful positive control consists of the addition of 100 μ*M* Ca^{2+}, which opens the PT pore and causes large amplitude swelling. Cyclosporine A (CsA, 1 μ*M*) added before Ca^{2+} prevents this swelling, and CsA-mediated inhibition of mitochondrial volume change is generally interpreted to mean that the CsA-sensitive PTP mediates this reaction (Fig. 3).

2. Mitochondrial Transmembrane Potential Measurement in Isolated Mitochondria

The $\Delta\Psi_m$ of isolated mitochondria can be quantified by multiple methods. Here we propose two of them. One is based on the cytofluorometric analysis of purified mitochondria on a per mitochondrion basis. In this case, the incorporation of the dye CMXRos is

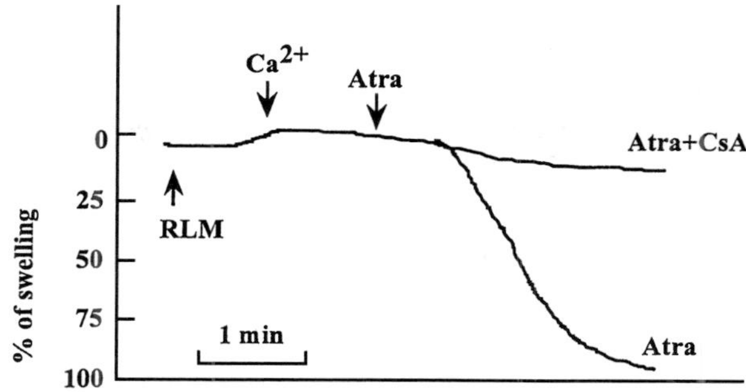

Fig. 3 Induction of PT pore opening in purified mitochondria by atractyloside. The incubation medium contained 0.2 *M* sucrose, 10 m*M* Tris–MOPS, pH 7.4, 5 m*M* succinate–Tris, 1 m*M* Pi, 2 m*M* rotenone, and 10 m*M* EGTA–Tris. Final volume 1 ml, 25°C. All experiments were started by the addition, with stirring, of 1 mg of mouse or rat liver mitochondria (RLM; not shown); 25 μM Ca^{2+} and 100 μM actratyloside were added where indicated Mitochondria were preincubated with 1 μM cyclosporin A.

measured: low levels of CMXRos incorporation indicate a low $\Delta\Psi_m$. The second protocol is performed as a bulk measurement based on the quantitation of Rh123 quenching. At a high $\Delta\Psi_m$ level, most of the rhodamine 123 is concentrated in the mitochondrial matrix and is quenched. At lower $\Delta\Psi_m$ levels, Rh123 is released, causing dequenching and an increase in rhodamine 123 fluorescence. Thus, a low $\Delta\Psi_m$ level corresponds to a higher value of Rh123 fluorescence.

a. Required Materials

M buffer: 220 m*M* sucrose, 68 m*M* mannitol, 10 m*M* KCl, 5 m*M* KH_2PO_4, 2 m*M* $MgCl_2$, 500 μM EGTA, 5 m*M* succinate, 2 μM rotenone, and 10 m*M* HEPES, pH 7.2

Mitochondria are purified as described earlier, kept on ice for a maximum of 4 h, and resuspended in M buffer (or similar buffers)

CMXRos stock solution is prepared as described earlier

Rhodamine 123 stock solution (10 m*M* in ethanol, kept at −20°C and protected against light)

Advanced cytofluorometer capable of detecting isolated mitochondria

Spectrofluorometer

b. Protocol

Mitochondria are incubated for 30 min at 20°C in the presence of the indicated reagent. Use 100 μM CCCP of the protonophore as a control. Determine the $\Delta\Psi_m$ using the potential-sensitive fluorochrome chloromethyl-X-rosamine (100 n*M,* 15 min, room temperature) and analyze in a FACS Vantage cytofluorometer (Becton Dickinson) or similar instrument while gating on single mitochondrion events in forward and side scatters (Fig. 4).

Alternatively, mitochondria (1 mg protein/ml) are incubated in a buffer supplemented with 5 μM rhodamine 123 for 5 min, and the $\Delta\Psi_m$-dependent quenching of rhodamine

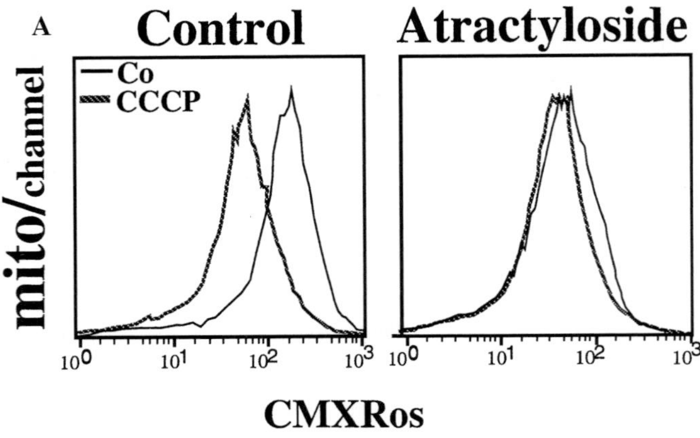

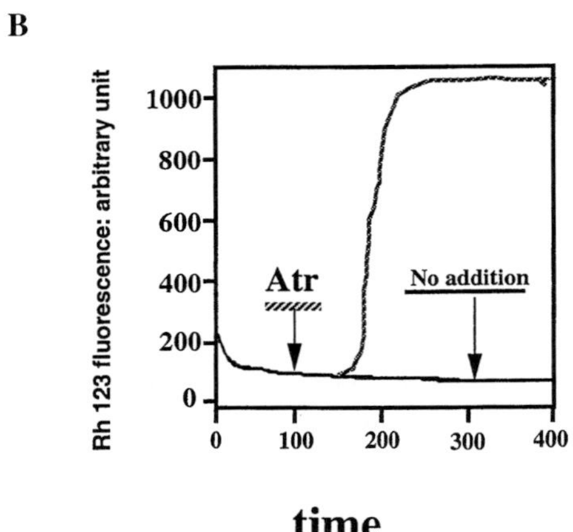

Fig. 4 (A) $\Delta\Psi_m$ measurements in isolated mouse liver mitochondria treated with 100 μM atractyloside, followed by staining with CMXRos in the presence (cccp) or absence (Co) 100 μM of CCCP followed by cytofluorometric evaluation of the CMXRos-dependent fluorescence. Note that higher fluorescence values imply a higher $\Delta\Psi_m$. (B) $\Delta\Psi_m$ measurement using Rh123. The same samples in A were stained with Rh123 and the kinetics of rhodamine dequenching was evaluated in a fluorimeter.

fluorescence (excitation 490 nm, emission 535 nm) is measured continuously in a fluorometer (Shimizu *et al.,* 1998), as shown in Fig. 4.

III. Anticipated Results and Pitfalls

The cytofluorometric determination of the $\Delta\Psi_m$ in purified mitochondria is a delicate procedure and requires an advanced cytofluorometer capable of detecting isolated mitochondria. Note that the quality of mitochondrial preparation is very important and that for each experiment controls must be performed to assess the background incorporation of fluorochromes in the presence of CCCP, a protonophore causing a complete disruption of the $\Delta\Psi_m$. Working with isolated mitochondria requires that the mitochondrial preparations are optimal and fresh (<4 h). Rhodamine 123 fluorescence measurements in a spectrofluorometer generate less problems than cytofluorometric measurements of isolated mitochondria.

References

Bernardi, P. (1996). The permeability transition pore: Control points of a cyclosporin A-sensitive mitochondrial channel involved in cell death. *Biochim. Biophy. Acta (Bioenergetics)* **1275,** 5–9.

Beutner, G., Rück, A., Riede, B., Welte, W., and Brdiczka, D. (1996). Complexes between kinases, mitochondrial porin, and adenylate translocator in rat brain resemble the permeability transition pore. *FEBS Lett.* **396,** 189–195.

Bossy-Wetzel, E., Newmeyer, D. D., and Green, D. R. (1998). Mitochondrial cytochrome *c* release in apoptosis occurs upstream of DEVD-specific caspase activation and independently of mitochondrial transmembrane depolarization. *EMBO J.* **17,** 37–49.

Carter, W. O., Narayanan, P. K., and Robinson, J. P. (1994). Intracellular hydrogen peroxide and superoxide anion detection in endothelial cells. *J. Leukocyte Biol.* **55,** 253–258.

Castedo, M., Hirsch, T., Susin, S. A., Zamzami, N., Marchetti, P., Macho, A., and Kroemer, G. (1996). Sequential acquisition of mitochondrial and plasma membrane alterations during early lymphocyte apoptosis. *J. Immunol.* **157,** 512–521.

Ichas, F., Jouavill, L. S., and Mazat, J.-P. (1997). Mitochondria are excitable organelles capable of generating and conveying electric and calcium currents. *Cell* **89,** 1145–1153.

Jacotot, E., Ravagnan, L., Loeffler, M., Ferri, K., Vieira, H. L. A., Zamzami, N., Costantini, P., Druillennec, S., Hoebeke, J., Brian, J. P., Irinopoulos, T., Daugas, E., Susin, S. A., Cointe, D., Xie, Z. H., Reed, J. C., Roques, B. P., and Kroemer, G. (1999). The HIV-1 viral protein R induces apoptosis via a direct effect on the mitochondrial permeability transition pore. *J. Exp. Med.* **191,** 33–46.

Kantrow, S. P., and Piantadosi, C. A. (1997). Release of cytochrome *c* from liver mitochondria during permeability transition. *Biochem. Biophys. Res. Comm.* **232,** 669–671.

Kinnally, K. W., Lohret, T. A., Campo, M. L., and Mannella, C. A. (1996). Perspectives on the mitochondrial multiple conductance channel. *J. Bioenerg. Biomembr.* **28,** 115–123.

Kluck, R. M., Bossy-Wetzel, E., Green, D. R., and Newmeyer, D. D. (1997). The release of cytochrome *c* from mitochondria: A primary site for Bcl-2 regulation of apoptosis. *Science* **275,** 1132–1136.

Kroemer, G. (1997). The proto-oncogene Bcl-2 and its role in regulating apoptosis. *Nature Medicine* **3,** 614–620.

Kroemer, G., Dallaporta, B., and Resche-Rigon, M. (1998). The mitochondrial death/ life regulator in apoptosis and necrosis. *Annu. Rev. Physiol.* **60,** 619–642.

Kroemer, G., Petit, P. X., Zamzami, N., Vayssière, J.-L., and Mignotte, B. (1995). The biochemistry of apoptosis. *FASEB J.* **9,** 1277–1287.

Kroemer, G., Zamzami, N., and Susin, S. A. (1997). Mitochondrial control of apoptosis. *Immunol. Today* **18,** 44–51.

Larochette, N., Decaudin, D., Jacotot, E., Brenner, C., Marzo, I., Susin, S. A., Zamzami, N., Xie, Z., Reed, J. C., and Kroemer, G. (1999). Arsenite induces apoptosis via a direct effect on the mitochondrial permeability transition pore. *Exp. Cell Res.* **249,** 413–421.

Liu, X. S., Kim, C. N., Yang, J., Jemmerson, R., and Wang, X. (1996). Induction of apoptotic program in cell-free extracts: Requirement for dATP and cytochrome C. *Cell* **86,** 147–157.

Macho, A., Decaudin, D., Castedo, M., Hirsch, T., Susin, S. A., Zamzami, N., and Kroemer, G. (1996). Chloromethyl-X-rosamine is an aldehyde-fixable potential-sensitive fluorochrome for the detection of early apoptosis. *Cytometry* **25,** 333–340.

Marchetti, P., Castedo, M., Susin, S. A., Zamzami, N., Hirsch, T., Haeffner, A., Hirsch, F., Geuskens, M., and Kroemer, G. (1996). Mitochondrial permeability transition is a central coordinating event of apoptosis. *J. Exp. Med.* **184,** 1155–1160.

Marchetti, P., Zamzami, N., Josph, B., Schraen-Maschke, S., Mereau-Richard, C., Costantini, P., Metivier, D., Susin, S. A., Kroemer, G., and Formstecher, P. (1999). The novel retinoid AHPN/CD437 triggers apoptosis through a mitochondrial pathway independent of the nucleus. *Cancer Res.* 15;59(24): 6257–66.

Marzo, I., Brenner, C., Zamzami, N., Susin, S. A., Beutner, G., Brdiczka, D., Rémy, R., Xie, Z.-H., Reed, J. C., and Kroemer, G. (1998). The permeability transition pore complex: A target for apoptosis regulation by caspases and Bcl-2 related proteins. *J. Exp. Med.* **187,** 1261–1271.

Metivier, D., Dallaporta, B., Zamzami, N., Larochette, N., Susin, S. A., Marzo, I., and Kroemer, G. (1998). Cytofluorometric detection of mitochondrial alterations in early CD95/Fas/APO-1-triggered apoptosis of Jurkat T lymphoma cells: Comparison of seven mitochondrion-specific fluorochromes. *Immunol. Lett.* **61,** 157–63.

Patterson, S., Spahr, C. S., Daugas, E., Susin, S. A., Irinopoulos, T., Koehler, C., and Kroemer, G. (1999). Mass spectrometric identification of proteins released from mitochondria undergoing permeability transition. *Cell Death Differ.* **7,** 134–44.

Petit, J. M., Maftah, A., Ratinaud, M. H., and Julien, R. (1992). 10 N-nonyclacridine orange interacts with cardiolipin and allows for the quantification of phospholipids in isolated mitochondria. *Eur. J. Biochem.* **209,** 267–273.

Petit, P. X., Goubern, M., Diolez, P., Susin, S. A., Zamzami, N., and Kroemer, G. (1998). Disruption of the outer mitochondrial membrane as a result of mitochondrial swelling: The impact of irreversible permeability transition. *FEBS Lett.* **426,** 111–116.

Petit, P. X., O'Connor, J. E., Grunwald, D., and Brown, S. C. (1990). Analysis of the membrane potential of rat- and mouse-liver mitochondria by flow cytometry and possible applications. *Eur. J. Biochem.* **220,** 389–397.

Ravagnan, L., Marzo, I., Costantini, P., Susin, S. A., Zamzami, N., Petit, P. X., Hirsch, F., Poupon, M.-F., Miccoli, L., Xie, Z., Reed, J. C., and Kroemer, G. (1999). Lonidamine triggers apoptosis via a direct, Bcl-2-inhibited effect on the mitochondrial permeability transition pore. *Oncogene* **18,** 2537–2546.

Rothe, G., and Valet, G. (1990). Flow cytometric analysis of respiratory burst activity in phagocytes with hydroethidine and 2′,7′-dichlorofluorescin. *J. Leukocyte Biol.* **47,** 440–448.

Shimizu, S., Eguchi, Y., Kamiike, W., Funahashi, Y., Mignon, A., Lacronique, V., Matsuda, H., and Tsujimoto, Y. (1998). Bcl-2 prevents apoptotic mitochondrial dysfunction by regulating proton flux. *Proc. Natl. Acad. Sci. USA* **95,** 1455–1459.

Smiley, S. T. (1991). Intracellular heterogeneity in mitochondrial membrane potential revealed by a J-aggregate-forming lipophilic cation JC-1. *Proc. Natl. Acad. Sci. USA* **88,** 3671–3675.

Susin, S. A., Zamzami, N., Castedo, M., Hirsch, T., Marchetti, P., Macho, A., Daugas, E., Geuskens, M., and Kroemer, G. (1996). Bcl-2 inhibits the mitochondrial release of an apoptogenic protease. *J. Exp. Med.* **184,** 1331–1342.

Susin, S. A., Zamzami, N., and Kroemer, G. (1998). Mitochondrial regulation of apoptosis: Doubt no more. *Biochim. Biophys. Acta (Bioenergetics)* **1366,** 151–165.

vander Heiden, M. G., Chandal, N. S., Williamson, E. K., Schumacker, P. T., and Thompson, C. B. (1997). Bcl-XL regulates the membrane potential and volume homeostasis of mitochondria. *Cell* **91,** 627–637.

Zamzami, N., Brenner, C., Marzo, I., Susin, S. A., and Kroemer, G. (1998). Subcellular and submitochondrial mechanisms of apoptosis inhibition by Bcl-2-related proteins. *Oncogene* **16,** 2265–2282.

Zamzami, N., Susin, S. A., Marchetti, P, Hirsch, T, Gómez-Monterrey, I, Castedo, M, and Kroemer, G. (1996). Mitochondrial control of nuclear apoptosis. *J. Exp. Med.* **183,** 1533–1544.

Zoratti, M., and Szabò, I. (1995). The mitochondrial permeability transition. *Biochem. Biophys. Acta-Rev. Biomembranes* **1241,** 139–176.

CHAPTER 9

Assaying Actin-Binding Activity of Mitochondria in Yeast

Istvan R. Boldogh and Liza A. Pon

Columbia University College of Physicians and Surgeons
New York, New York 10032

I. Introduction

Mitochondrial motility, distribution, and transfer from the mother cell to the bud in vegetatively growing yeast cells are essential processes of the cell cycle. Several observations support the idea that these processes are controlled by the actin cytoskeleton. For example, yeast mitochondria often coalign with actin cables, one of the major actin structures (Drubin *et al.,* 1993; Lazzarino *et al.,* 1994). Also, mutations in the actin gene or in genes that destabilize actin cables result in reduction of mitochondrial motility or in the transfer of mitochondria between mother and daughter cells (Simon *et al.,* 1995, 1997; Herman *et al.,* 1997).

This chapter describes a sedimentation assay to explore the interaction between yeast mitochondria and the actin cytoskeleton. The mitochondria–actin-binding assay is carried out by incubation of isolated mitochondria with purified F-actin. This assay was used to demonstrate the presence of actin-binding activity on the surface of mitochondria (Lazzarino *et al.,* 1994). In addition, comparing actin-binding activities of wild-type mitochondria with those of mutant mitochondria revealed some of the molecular participants in mitochondria–actin interactions (Boldogh *et al.,* 1998).

II. Purification of Yeast Actin

We use DNase I affinity chromatography to isolate actin for the binding assay. This approach has been a useful tool for obtaining actin from nonmuscle cells, where the amount of actin is limited. The basis for this method is that monomeric actin (G-actin) and DNase I form a tight, specific 1:1 complex (Mannherz *et al.,* 1975). Guanidine hydrochloride has been used to dissociate actin from DNase I (Lazarides and Lindberg, 1974), but because guanidine hydrochloride causes denaturation of actin (Water *et al.,* 1980), formamide is used more widely to elute actin from DNase I columns, as it enables recovery of higher amounts of polymerization-competent actin (Zechel, 1980a,b). This latter method was used successfully by Kilmartin and Adams (1984), Drubin *et al.* (1988), and Kron *et al.* (1992) for purifying actin from yeast.

We use fresh wet blocks of commercial baker's yeast, which can be purchased from a local wholesale food supplier. The process is simple: the high-speed supernatant of disrupted cells is applied onto a DNase I column; then actin is eluted from the column with 40% formamide and further purified and concentrated by anion-exchange chromatography. Finally, the actin is subjected to two polymerization–depolymerization cycles, yielding polymerization-competent, highly pure protein. Our yield is 2–4 mg actin per 100 g of baker's yeast.

A. Solutions

Formamide: To remove impurities, formamide is incubated with a mixed bed cation/anion-exchange resin AG 501-×8 (Bio-Rad, Hercules, California) for 4 h at room temperature with gentle rotation (10 ml formamide : 1 g resin). The resin is removed by filtration.

PI-1 and PI-2 protease inhibitor cocktails (all reagents are from Sigma, St. Louis, MO). PI-1: pepstatin A (P4265) and chymostatin (C7268) are dissolved in dimethyl sulfoxide (D8418) at a concentration of 10 mg/ml; antipain (A6191), leupeptin (L2884), and aprotinin (A1153) are dissolved in water at a concentration of 10 mg/ml. To make the cocktail, equal volumes of all five solutions are mixed and water is added so that the final concentration of each protease inhibitor is 0.5 mg/ml.

PI-2: 10 m*M* benzamidine–HCl (B6506) and 1 mg/ml 1,10-phenanthroline (P9375) dissolved in ethanol. Both PI-1 and PI-2 solutions are used at 1000× dilution. They are aliquoted and can be stored for several months at −20°C.

100 m*M* ATP: Dissolved in water. Adjusted to pH 7.0 with KOH. Stored at −20°C.

200 m*M* phenylmethylsulfonyl fluoride (PMSF; Roche, Indianapolis, IN): Dissolved in ethanol. This stock is stable for several months upon storage at −20°C.

1 *M* imidazole, pH 7.4; pH is adjusted with HCl. Autoclaved. Stored at 4°C.

0.5 *M* $CaCl_2$: Autoclaved. Stored at 4°C.

1 *M* $MgCl_2$: Autoclaved. Stored at 4°C.

0.5 *M* EGTA: Dissolved in water and adjusted to pH 8.0 with KOH. Autoclaved. Stored at 4°C.

1 *M* dithiothreitol (DTT): Dissolved in water. Stored at −20°C.

G buffer (G-actin extraction buffer): 2 m*M* imidazole, pH 7.4, 0.2 m*M* ATP, 0.2 m*M* $CaCl_2$, 1× PI-1, 1× PI-2, 1 m*M* PMSF.

DN buffer (DNase I column buffer): 2 m*M* imidazole, pH 7.4, 0.2 m*M* ATP, 0.2 m*M* $CaCl_2$, 0.1 m*M* DTT, 1× PI-1, 1× PI-2, 1 m*M* PMSF.

DE buffer (DE52 column buffer): 10 m*M* imidazole, pH 7.4, 0.2 m*M* ATP, 0.2 m*M* $CaCl_2$, 0.5 m*M* DTT, 0.5× PI-1, 0.5× PI-2, 0.2 m*M* PMSF.

D buffer (depolymerization buffer): 2.5 m*M* imidazole, pH 7.4, 0.2 m*M* ATP, 0.2 m*M* $CaCl_2$, 0.5 m*M* DTT, 0.5× PI-1, 0.5× PI-2, 0.2 m*M* PMSF.

Note: In aqueous solution PMSF undergoes hydrolysis and loss of activity. Therefore it is added to buffers with stirring just prior to use, and all buffers except G buffer are filtered through Whatman 4 filter paper.

B. Preparation of DNase I Column

The DNase I column is made by covalent coupling of bovine pancreatic DNase I, grade II (Roche), to CNBr-activated Sepharose 4B resin (Amersham Pharmacia, Uppsala, Sweden) according to the manufacturer's instructions. Briefly, 10 g of dry resin is swollen and washed with 2 liters of 1 m*M* HCl on a sintered glass filter. The resin, still on the filter, is washed quickly with 50 ml coupling buffer consisting of 0.5 *M* NaCl, 0.1 *M* $NaHCO_3$, pH 8.3. Confirm that the pH of the buffer is basic after the wash. If not, wash quickly with an additional 20 ml of coupling buffer. Thirty-five milliliters of the pretreated resin is incubated with 270 mg of DNase I dissolved in 70 ml coupling buffer containing 1 m*M* PMSF overnight with gentle mixing on a rotating platform at 4°C. PMSF in the coupling buffer inhibits potential proteolytic degradation of DNase I by contaminating proteases. Under these conditions, DNase I is quantitatively coupled to the resin. To confirm this, remove a 30- to 50-μl aliquot of the coupling reaction mixture at the beginning and end of the coupling reaction. Separate the resin from the uncoupled material by centrifugation at 10,000*g* for 20 s. Supernatants are analyzed by SDS–PAGE.

After coupling, the resin is washed on a sintered glass filter with 150 ml coupling buffer and incubated in 100 ml 1 *M* ethanolamine, pH 8.0, at room temperature for 2 h with gentle mixing on a rotating platform. After this incubation the resin is washed on a sintered glass funnel with 150 ml of 0.1 *M* K–acetate, pH 4.0, containing 0.5 *M* KCl, followed by 150 ml 0.1 *M* Tris–HCl, pH 8.0, containing 0.5 *M* KCl. This low- and high-pH wash is repeated two times. Afterward, the resin is transferred to a 26 × 65-mm column (30 ml) and washed in successive steps with at least 3 volumes of each of the following solutions: D buffer, D buffer containing 2 mg/ml bovine serum albumin, D buffer containing 0.5 *M* NaCl, and finally D buffer. If the column is not used immediately, store at 4°C in D buffer with 0.02% sodium azide to prevent bacterial growth. We use a single column at least four times over a 1 year period with approximately a 10–12% decrease in actin recovery after each usage. After each usage the column is washed with 3 volumes of D buffer containing 0.5 *M* KCl followed by 3 volumes of D buffer containing 0.02% sodium azide.

C. Preparation of Yeast Cell Extract

Yeast cell disruption is carried out by agitation with glass beads using a bead-beating device (Biospec Products, Bartlesville, OK). Because actin is extracted in monomer (G-actin) form, it is important to promote depolymerization during the disruption step. Increasing concentrations of mono- and divalent cations induce G-actin polymerization to the filamentous (F-actin) form. For example, measurable yeast actin polymerization occurs in the presence of Mg^{2+} below 1 m*M* and in the presence of K^{+} below 50 m*M* (Greer and Schekman, 1982). Therefore, disruption is carried out using low ionic strength buffer. Temperature also influences the state of actin polymerization: elevated temperatures promote polymer formation by decreasing the critical concentration for actin polymerization (Gordon *et al.,* 1977). For this reason, the bead-beater chamber is kept in ice-water slush during the homogenization process.

To prepare yeast cell extract, 300 g of baker's yeast is suspended in ice-cold G buffer to a final volume of 500 ml. This cell suspension is homogenized in three aliquots; one-third of the suspension is transferred to a 350-ml bead-beater chamber filled to one-half capacity with prechilled 0.5-mm glass beads and homogenized for 6 × 0.5-min pulses with 1-min cooling intervals between each pulse. Eighty to 90% of the cells are disrupted, as determined by visualization of the extract with bright-field microscopy. The bead beater chamber is always filled to full capacity to avoid foaming, which denatures proteins. Also, yeast vacuoles release substantial amounts of proteases on cell disruption. Therefore, in each round of cell homogenization we add an additional 300 μl of 200 m*M* PMSF to the chamber. After homogenization the cell extract is subjected to centrifugation at 120,000*g* at 4°C for 90 min. Before centrifugation each tube is overlaid with filtered G buffer. This layer, which can be removed by aspiration, will collect lipids as a cloudy white top layer. Thus, overlaying the extract with fresh buffer minimizes the loss of extracted proteins. The supernatant is filtered through several layers of cheesecloth to remove particulate matter. Finally, DTT is added to 1 m*M* final concentration. The volume of this clarified extract is 250–280 ml. This solution, referred to as the high-speed supernatant, is loaded immediately onto the DNase I column (Fig. 1, lane 1).

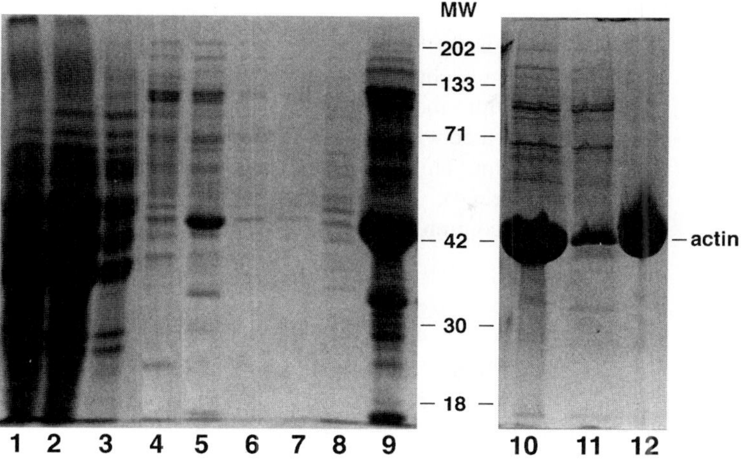

Fig. 1 Purification of yeast actin. 12% SDS–PAGE showing the Coomassie-stained protein pattern of steps during the purification of yeast actin. Lane 1, high-speed supernatant of whole cell extracts; lane 2, DNase I column flow-through; lane 3, DNase I column wash; lane 4, DNase I column 10% formamide wash; lane 5, DNase I column 40% formamide eluate; lane 6, DE52 column flow-through; lane 7, DE52 column wash; lane 8, DE52 column 50 m*M* KCl wash; lane 9, DE52 column 300 m*M* KCl eluate; lane 10, G-actin solution before polymerization in the presence of 0.5 *M* KCl; lane 11, supernatant of high-speed centrifugation of F-actin; and lane 12, pellet of high-speed centrifugation of F-actin (the pellet was resuspended to the same volume as before centrifugation). Amounts loaded: lanes 1–3, 2.5 μl of each fraction; lanes 4–9, 16 μl of each fraction; and lanes 10–14, 15 μl of each fraction.

D. DNase I Affinity Chromatography

Note: all steps of actin purification are carried out at 4°C.

1. The high-speed supernatant of homogenized cells is loaded onto the DNase I column at a flow rate of 2 ml/min. The flow-through is collected and reloaded onto the DNase I column to increase actin binding to the column (Fig. 2, lane 2).

2. The column is washed with 200 ml DN buffer, followed by a wash with 100 ml DN buffer containing 10% formamide (Fig. 1, lanes 3 and 4). These and subsequent washing steps are carried out at a flow rate of 1.6 ml/min.

3. Dissociation of actin from DNase I is facilitated by wash with 120 ml DN buffer containing 40% formamide. This eluate is collected directly into a dialysis membrane (Spectra/Por 2, Spectrum, Houston, TX) containing 20 ml DE buffer and immersed in 2 liters of DE buffer. Because prolonged exposure to 40% formamide leads to increased denaturation of actin, the direct collection of actin into the dialysis membrane ensures rapid dilution and removal of formamide.

4. The column is washed with 200 ml DN buffer. The first 30–40 ml of this wash is collected into the dialysis membrane.

5. The actin-containing eluate is dialyzed overnight against 2 liters of DE buffer with at least one change of the dialysis buffer (Fig. 1, lane 5).

E. DE52 Chromatography

The actin is further purified and concentrated using a diethylaminoethyl cellulose (DE52) anion-exchange resin (Whatman, Maidstone, England). The resin is prepared according to the manufacturer's instructions. Four grams of preswollen ion-exchange resin is incubated with 50 ml 0.2 *M* imidazole, pH 7.4, for 15 min. The buffer is decanted from the resin. This incubation and wash step is repeated three times. The resin is poured into a 20 × 17-mm column (5 ml). The column is washed with 20 volumes of DE buffer. This column is always made fresh, 1 day before use.

Protocol

1. The dialysate is filtered through glass wool to remove particulate matter and loaded onto a DE52 column. The flow rate during loading and subsequent washing steps is 1 ml/min. See Fig. 1, lane 6 for flow-through.
2. The column is washed with 15 ml DE buffer, followed by a wash with 15 ml DE buffer containing 50 m*M* KCl (Fig. 1, lanes 7 and 8).
3. Actin is eluted into one fraction with 20 ml DE buffer containing 300 m*M* KCl at a flow rate of 0.5 ml/min (Fig. 1, lane 9).

Note: As an alternative, the dialysis step between the DNase I and DE52 chromatographies can be omitted. In this case, loading of the high-speed supernatant from the cell extract onto the DNase I column and the subsequent wash with DN buffer are done overnight. The 40% formamide eluate from the DNase I column is loaded immediately onto the DE52 column by connecting the outlet of the DNase I column directly to the DE52 column. The overnight loading and washing steps require a chromatography system with an automated buffer selection device. Although the purity of the final actin product prepared this way is similar to that obtained by the dialysis method, the yield is generally 10–20% lower.

F. Polymerization and Depolymerization of Actin

The actin recovered after DNase I and DE52 chromatography steps is contaminated with other proteins and denatured actin. Therefore, it is purified further by two cycles of polymerization and depolymerization. Actin polymerization is initiated by incubation in buffers containing high salt, ATP, and $MgCl_2$. F-actin is separated from soluble material by ultracentrifugation. Thereafter, F-actin is depolymerized by dialysis against buffers containing low salt. All steps are carried out at 4°C. This procedure separates polymerization-competent, active actin from polymerization-incompetent actin and actin aggregates. In addition, because actin polymerization is carried out in the presence of high salt, this method removes protein contaminants, including salt-extractable proteins that bind F-actin.

Protocol

1. The actin eluate from the DE52 column is polymerized by the addition of $MgCl_2$ to 4 m*M* and ATP to 1 m*M*, and incubation overnight at 4°C. The actin solution at this point is viscous and somewhat cloudy.

2. F-actin is separated from soluble material by centrifugation at 150,000*g* for 90 min at 4°C. The pellet is rinsed once very quickly with D buffer. The actin pellet is hard and glossy; it is difficult to see and resuspend. To avoid losses during resuspension, first the pellet is "softened" by incubation with a small volume (1 ml) of D buffer for 1 h. The pellet is then homogenized with a glass or Teflon-coated pestle. Because this actin solution is still viscous, a disposable pipette tip with a cut end should be used for transfer of the resuspended F-actin to the dialysis membrane. To maximize actin recovery, wash the centrifuge tube and homogenizer several times with a small volume of depolymerization buffer and pool this material with the resuspended F-actin. The total volume of D buffer added should be 4 ml.

3. Actin is dialyzed against 0.5 l D buffer for 36 h with at least three changes of dialysis buffer. After dialysis, G-actin is clarified by centrifugation at 150,000*g* for 60 min at 4°C (Fig. 1, lane 10). The pellet, which contains actin aggregates and depolymerization-incompetent actin, is discarded.

4. The clarified G-actin is polymerized by the addition of imidazole, pH 7.4, to 10 m*M*, KCl to 0.5 *M*, $MgCl_2$ to 4 m*M*, EGTA, pH 8.0, to 1 m*M*, and ATP to 1 m*M* and incubation at 4°C for at least 4 h. F-actin is sedimented at 150,000*g* for 90 min at 4°C and resuspended in 4 ml of D buffer into a dialysis membrane as in step 2 (Fig. 1, lane 12). The supernatant of this centrifugation step (Fig. 1, lane 11) is discarded.

5. Actin is dialyzed in 0.5 liter in D buffer for 36 h with at least three changes of dialysis buffer. After dialysis, G-actin is clarified by centrifugation at 150,000*g* for 60 min at 4°C.

6. Purified actin is stored as F-actin, as this form is more stable than G-actin. This final polymerization is initiated by the addition of imidazole, pH 7.4, to 10 m*M*, KCl to 50 m*M*, $MgCl_2$ to 4 m*M*, EGTA, pH 8.0, to 1 m*M*, and ATP to 0.5 m*M*. This F-actin solution is stored on ice at 2 mg/ml protein concentration and is used for mitochondria-actin binding assays. If this solution is stored more than 2 months, carry out one cycle of depolymerization–repolymerization process prior to use.

III. Sedimentation Assay for Binding of Actin to Mitochondria

The sedimentation assay described in this section measures interaction between isolated mitochondria and F-actin. In a typical assay, isolated yeast mitochondria are incubated with phalloidin-stabilized yeast actin filaments in the presence or absence of ATP for 10 min at 30°C. Thereafter, mitochondria are separated from the reaction mixture by low-speed centrifugation through a sucrose cushion. Proteins recovered in low-speed pellets are identified by Western blot analysis using antibodies raised against actin and mitochondrial marker proteins. Under these low-speed centrifugation conditions, all unbound F-actin is recovered in the supernatant. Therefore, any actin recovered in the mitochondrial pellet is bound to the organelle. A schematic outlining this assay is shown in Fig. 2A.

A. Reagents

1. Mitochondria

For the mitochondria-actin binding assay we use Nycodenz-purified mitochondria stored at 5–20 mg/ml concentration in liquid nitrogen in a buffer containing 0.6 *M* sorbitol, 20 m*M* HEPES–KOH, pH 7.4, 2 m*M* $MgCl_2$, 1 m*M* EGTA, pH 8.0, 200 m*M* PMSF, 1× PI-1, 1× PI-2, and 15–17% Nycodenz. A detailed description of purifying mitochondria on a Nycodenz step gradient is provided by Glick and Pon (1995) and Diekert *et al.* (see Chapter 2). This method yields highly purified organelles.

The frozen mitochondrial suspension is thawed at 30°C and diluted with 5 volumes of RM0 buffer. The organelles are isolated from this suspension by centrifugation at 12,500*g* for 5 min. The supernatant is removed, and the mitochondrial pellet is resuspended by pipetting gently in RM buffer to a final protein concentration of 10 mg/ml. Only freshly thawed and reisolated mitochondria should be used in the sedimentation assay, as prolonged storage of the organelle on ice leads to decreased actin-binding activity. With the exception of the mitochondria–actin-binding reaction, all steps are carried out at 4°C. All protein concentration measurements are done by the BCA protein assay (Pierce, Rockford, IL).

2. Actin

F-actin is purified from yeast as described earlier. In this assay actin is stabilized in the filamentous form with phalloidin. Without the addition of phalloidin we have observed polymerization of G-actin and apparent depolymerization of F-actin. These variations in the state of actin polymerization can prevent reproducible quantitation of actin filaments bound to mitochondria. Therefore, treatment with phalloidin is critical to maintain all added actin in the filamentous form. Prior to the assay we add 1/50 volume of 0.5 mg/ml phalloidin to 2 mg/ml F-actin solution (see Section II). The F-actin phalloidin mixture is incubated on ice for 30 min and diluted with 1 volume of 2× RM buffer. This solution is subjected to centrifugation at 12,500*g* for 10 min before the assay in order to remove possible F-actin aggregates and to minimize nonspecific cosedimentation of F-actin with mitochondria.

3. Other Reagents

100 m*M* ATP: Dissolved in water. Adjusted to pH 7.0 with KOH. Stored at −20°C.

500 unit/ml apyrase (Sigma, A6410): Dissolved in 20 m*M* HEPES, pH 7.4. Aliquoted, stored at −20°C.

5 mg/ml creatine kinase (Sigma, C3755): Dissolved in 50% glycerol. Stored at −20°C.

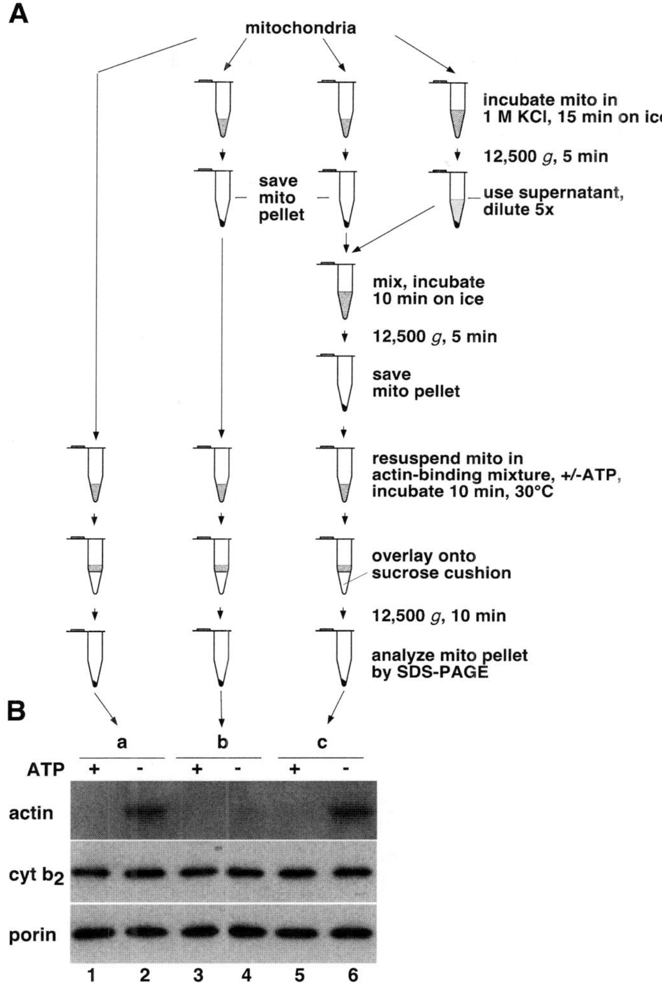

Fig. 2 Sedimentation assays of isolated mitochondria from D273-10B yeast with phalloidin-stabilized F-actin in the presence or absence of ATP. (A) Schematic illustration of sedimentation assay using untreated and salt-treated mitochondria. (B) SDS–PAGE and Western blot analysis of levels of actin and of two mitochondrial marker proteins (cytochrome b_2 and porin) in mitochondrial pellets of the sedimented reaction mixtures. Actin was detected using a monoclonal antiactin antibody, c4d6 (Lessard, 1988). Recovery and intactness of mitochondria after sedimentation were determined with polyclonal anticytochrome b_2 and antiporin antibodies (gift from G. Schatz). Lanes 1 and 2, untreated mitochondria; lanes 3 and 4, salt-washed mitochondria; and lanes 5 and 6, salt-washed mitochondria pretreated with mitochondrial salt extract. Sample a, which is a control for salt washing, contains mitochondria that were incubated in buffer without KCl for 15 min on ice. Mitochondria in samples b and c were incubated in 1 *M* KCl. In all cases, 170 μg mitochondria was loaded for actin detection and 16 μg mitochondria was loaded for cytochrome b_2 and porin detection.

0.5 *M* creatine phosphate: Dissolved in water. Stored at −20°C.

2.4 *M* sorbitol: Autoclaved. Stored at 4°C.

2.5 *M* KCl. Autoclaved and stored at 4°C.

50% sucrose: Dissolved in water. Stored at −20°C.

1 *M* HEPES, pH 7.4. pH is adjusted with KOH. Autoclaved. Stored at 4°C

0.5 mg/ml phalloidin (Roche): Dissolved in ethanol. Stored at −20°C.

200 m*M* PMSF (Roche): Dissolved in ethanol. This stock is stable for several months upon storage at −20°C.

PI-1 and PI-2 protease inhibitor cocktails are prepared as described in Section II,A.

RM0 buffer: 0.6 *M* sorbitol, 20 m*M* HEPES, pH 7.4, 2 m*M* $MgCl_2$, 1× PI-1, 1× PI-2, 1 m*M* PMSF

RM1 buffer: 0.6 *M* sorbitol, 20 m*M* HEPES, pH 7.4, 2 m*M* $MgCl_2$, 1 *M* KCl, 1× PI-1, 1× PI-2, 1 m*M* PMSF

RM buffer: 0.6 *M* sorbitol, 20 m*M* HEPES, pH 7.4, 2 m*M* $MgCl_2$, 0.1 *M* KCl, 1 m*M* DTT, 1 mg/ml bovine serum albumin (fatty-acid free), 1× PI-1, 1× PI-2, 1 m*M* PMSF

Sucrose cushion solution: 25% (w/v) sucrose, 20 m*M* HEPES, pH 7.4, 1× PI-1, 1× PI-2, 1 m*M* PMSF

Note: All buffers containing PMSF are filtered through Whatman 4 filter paper, as adding PMSF to aqueous solution results in some level of precipitation, which might interfere with the sedimentation assay.

B. Sedimentation Assay Protocol

A standard mitochondria–actin-binding assay is carried out in RM buffer in 80 μl final volume and contains 200 μg mitochondria and 15 μg phalloidin-stabilized F-actin. In ATP-treated samples, ATP levels are maintained at 2 m*M* by the addition of ATP and an ATP-regenerating system consisting of creatine kinase and creatine phosphate. To deplete samples of ATP, apyrase is added to the mixture.

	Final concentration	Sample (+ATP)	Sample (−ATP)
2× RM buffer	1×	22.5 μl	22.5 μl
Water		17.7 μl	20.9 μl
ATP (100 m*M*)	2 m*M*	1.6 μl	—
Creatine kinase (5 mg/ml)	0.1 mg/ml	1.6 μl	—
Creatine phosphate (0.5 *M*)	10 m*M*	1.6 μl	—
Apyrase (500 U/ml)	12.5 U/ml	—	2 μl
F-actin (1 mg/ml)	188 μg/ml	15 μl	15 μl
Mitochondria (10 mg/ml)	2.5 mg/ml	20 μl	20 μl

In all cases, binding reactions are initiated by adding mitochondria to the mixture last. This mitochondria–actin mixture is incubated for 10 min at 30°C. After incubation the reaction mixture is cooled down on ice and overlaid on a 500-μl sucrose cushion. Samples are then subjected to centrifugation at 12,500g for 10 min at 4°C. Sedimenting mitochondria through a sucrose cushion allows for the efficient separation of mitochondria-bound actin from unbound actin, which remains on the top of the cushion. Residual nonspecifically associated actin is removed by gently overlaying 100 μl of RM0 buffer on the surface of mitochondrial pellet. This "wash" is then removed by aspiration without disturbing the integrity of the pellet. The mitochondrial pellet is then resuspended in 20 μl of RM0 buffer and is solubilized by the addition of 10 μl of 3× SDS sample buffer [0.19 *M* Tris–HCl, pH 6.8, 12% (v/w) SDS, 30% (v/v) glycerol, 15% (v/v) 2-mercaptoethanol, 0.015% (v/w) bromphenol blue]. The actin content, as well as the presence of mitochondrial marker proteins in the solubilized mitochondrial pellet, is determined by SDS–PAGE and Western blot analysis (Fig. 2B, lanes 1 and 2). Usually 25 μl of this mixture is used to detect actin and 2.5 μl is used to detect the mitochondrial markers, cytochrome b_2 and porin.

C. Salt Treatment of Mitochondria

High ionic strength (1 *M* KCl) facilitates the dissociation of loosely associated, peripheral membrane proteins from the mitochondrial membrane surface, but does not extract integral membrane proteins or affect the integrity of the organelle. Therefore, the mode of association of actin-binding activity with mitochondria can be tested by measuring the actin binding of mitochondria after incubation of the organelle in 1 *M* KCl. Treatment of a 33-mg/ml mitochondria solution with 1 *M* KCl usually yields 0.5–0.6 mg/ml salt-extracted proteins in the supernatant. After salt treatment, mitochondria are subjected to centrifugation to remove dissociated peripheral membrane proteins, which remain in the supernatant. Using the sedimentation assay, the ATP-sensitive, actin-binding activity of salt-washed mitochondria can be tested. Upon this treatment, mitochondria–actin-binding activity is lost (Fig. 2B, lanes 3 and 4). This activity, however, can be restored by incubation of salt-washed mitochondria with salt-extracted membrane proteins in a buffer containing 200 m*M* KCl. Upon this treatment, with increasing amounts of salt extract, the salt-washed organelle shows increasing restoration of ATP-sensitive, actin-binding activity. To regenerate the actin-binding activity of salt-washed mitochondria to the level of untreated mitochondria, salt-washed mitochondria must be incubated with 6–10 times the amount of salt extract as was removed by salt treatment (Fig. 2B lanes 5 and 6)

The removal and regeneration of actin-binding activity in mitochondria by salt treatment indicate the involvement of a peripheral protein component on the outer membrane in mitochondria–actin interactions. We named this activity mitochondria–actin-binding activity (mABP). The mitochondria–actin-binding assay is also useful in identifying integral protein components that are required for mitochondria–actin interactions. Isolated mitochondria from mutants of two integral outer membrane proteins, Mmm1p and Mdm10p (Burgess *et al.,* 1994; Sogo and Yaffe, 1994), showed no actin-binding activity and no restoration of actin-binding activity to salt-washed mitochondria upon incubation with salt extract from wild-type mitochondria. These properties of Mmm1p and Mdm10p mutants indicate that Mmm1p and Mdm10p might function as docking proteins for mABP (Boldogh *et al.,* 1998).

Protocols

a. Isolation of Salt-Washed Mitochondria

1. Thaw a frozen mitochondrial suspension at 30°C and dilute it with 5 volumes of RM0 buffer.
2. Recover mitochondria from this suspension by centrifugation at 12,500*g* for 5 min.
3. Resuspend mitochondrial pellet in RM1 buffer to a final concentration of 10 mg/ml. Divide the suspension into 20-μl aliquots (200 μg mitochondria per aliquot).
4. Incubate these aliquots for 15 min on ice.
5. Separate mitochondria by centrifugation at 12,500*g* for 5 min. Discard supernatant.
6. Resuspend salt-washed mitochondrial pellet in RM buffer to a final concentration of 10 mg/ml. These samples are used directly for the binding assay. If salt-washed mitochondria are to be regenerated with salt extract before the binding assay, resuspend the salt-washed mitochondrial pellet in RM0 buffer rather than RM buffer to a final protein concentration of 10 mg/ml.

b. Preparation of Mitochondrial Salt-Extractable Membrane Proteins

1. Thaw a frozen mitochondrial suspension at 30°C and dilute it with 5 volumes of RM0 buffer.
2. Recover mitochondria from this suspension by centrifugation at 12,500*g* for 5 min.
3. Resuspend mitochondrial pellet in RM1 buffer to a final concentration of 33 mg/ml.
4. Incubate 15 min on ice.
5. Separate salt extract from salt-washed mitochondria by centrifugation at 12,500*g* for 5 min. Save supernatant (salt extract).
6. Dilute the KCl concentration of supernatant to 0.2 *M* by adding 4 volumes of RM0 buffer. This solution is used for restoration of actin-binding activity of salt-washed mitochondria.

c. Restoration of Actin-Binding Activity of Salt-Washed Mitochondria

1. Mix 300 μl diluted salt extract (see step b.6) with 200 μg salt-washed mitochondria (see step a.6).
2. Incubate this mixture for 15 min on ice.
3. Separate mitochondria by centrifugation at 12,500*g* for 5 min and resuspend the pellet to a final protein concentration of 10 mg/ml in RM buffer. This suspension is used for the binding assay.

IV. Analysis of Mitochondria–Actin Binding

A. Scanning Densitometry

To quantitate the actin and mitochondrial marker bands detected on fluorograms of immunoblots, the images of fluorograms are digitized by a transmission scanner. We use the public domain NIH Image program for analyzing these images: an appropriate selection is made that covers each of the individual bands and computes their mean gray values. A selection with the same area is used to measure the background gray level just above and below the band of interest. The average of these values is subtracted from the gray value of the corresponding band. For quantitation of actin binding, the density of the band of actin recovered in the mitochondrial pellet is compared with the densities of known amounts of actin standards. Our standards range from 50 to 500 ng actin per lane. We find that usually 100–150 ng actin binds to 200 μg of mitochondria. Because

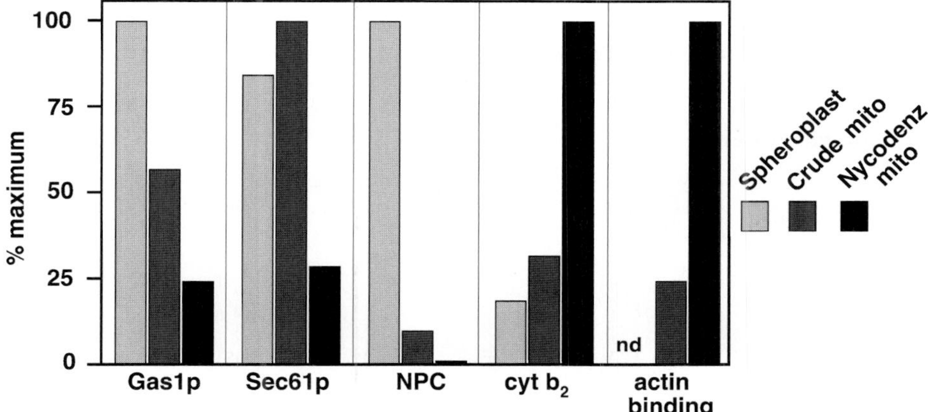

Fig. 3 Relative levels of membrane marker proteins and actin-binding activity in spheroplasts, crude mitochondria, and gradient-purified mitochondria. Spheroplasts, crude mitochondria, and gradient-purified mitochondria fractions were subjected to centrifugation at 12,500*g* for 5 min. The pellets were resuspended in a buffer containing 0.6 *M* sorbitol, 20 m*M* HEPES–KOH, pH 7.4, 2 m*M* $MgCl_2$, 1 m*M* EGTA, pH 8.0, 200 m*M* PMSF, 1× PI-1, and 1× PI-2, and the protein concentrations of the samples were determined with the BCA assay (Pierce) using bovine serum albumin as the standard. Because no purified membrane marker proteins were available for standard curves, only relative levels of marker proteins were determined in the three fractions. To do so, three different amounts of each fraction were analyzed by SDS–PAGE and Western blot. The gray values of the bands that reflected the loading differences in all three fractions on the same blot were used to normalize the relative amount of membrane markers in these fractions. For each marker protein 100% was defined as the maximum amount of marker per milligram of total protein. The marker proteins used were Gas1p, a glycosyphophatidylinositol-anchored protein that is transported through the endoplasmic reticulum (ER), Golgi apparatus, and secretory vesicles prior to insertion into the plasma membrane (Nuoffer *et al.*, 1993); Sec61p, an ER integral membrane protein (Stirling *et al.*, 1992); NPC, nuclear pore complex proteins (Davis *et al.*, 1986); and cytochrome b_2 (cyt b_2), a mitochondrial inner membrane protein. Relative actin-binding activity per milligram protein was determined in crude and gradient-purified mitochondria. One hundred percent was defined as the amount of binding activity in the sample containing the highest specific activity. Membrane contaminants are present in crude mitochondria. Upon further purification using the Nycodenz gradient, we observed an enrichment of the mitochondrial marker and removal of the membrane markers examined. This increase in the purity of mitochondria correlates with an enrichment in the ATP-sensitive actin-binding activity. Thus, ATP-sensitive actin-binding activity detected in isolated organelles resides in mitochondria and not in membrane contaminants within the mitochondrial preparation. nd, not determined.

long exposure of the fluorograms can result in saturated protein bands, it is important to have actin standards with the samples on the same gel and to use only the data points that are in the linear range of the calibration curve. This usually requires loading different amounts of the same samples and multiple exposures of the immunoblots.

B. Establishing Correlation between Actin-Binding Activity and Mitochondrial Purity

We use highly purified mitochondria for our assay. Low levels of membrane contaminants, however, can still be present in the preparation. For this reason it is important to compare the fractionation pattern of mitochondria–actin-binding activity with that of mitochondria and other membrane compartments. In different cellular fractions the levels of selected membrane marker proteins are evaluated by Western blot analysis and are compared with actin-binding activity (see Fig. 3 for details).

Acknowledgments

We thank members of the Pon laboratory for critical evaluation of the manuscript. This work was supported by research grants to L.P. from the National Institutes of Health (GM45735) and the American Cancer Society (RPG-97-163-01-C).

References

Boldogh, I., Vojtov, N., Karmon, S., and Pon, L. A. (1998). Interaction between mitochondria and the actin cytoskeleton in budding yeast requires two integral mitochondrial outer membrane proteins, Mmm1p and Mdm10p. *J. Cell Biol.* **141,** 1371–1381.

Burgess, S. M., Delannoy, M., and Jensen, R. E. (1994). *MMM1* encodes a mitochondrial outer membrane protein essential for establishing and maintaining the structure of yeast mitochondria. *J. Cell Biol.* **126,** 1375–1391.

Davis, L. I., and Blobel, G. (1986). Identification and characterization of a nuclear pore complex protein. *Cell* **45,** 699–709.

Drubin, D. G., Jones, H. D., and Wertman, K. F. (1993). Actin structure and function: Roles in mitochondrial organization and morphogenesis in budding yeast and identification of the phalloidin-binding site. *Mol. Biol. Cell.* **4,** 1277–1294.

Drubin, D. G., Miller, K. G., and Botstein, D. (1988). Yeast actin-binding proteins: Evidence for a role in morphogenesis. *J. Cell Biol.* **107,** 2551–2561.

Glick, B. S., and Pon, L. A. (1995). Isolation of highly purified mitochondria from *Saccharomyces cerevisiae. In* "Methods in Enzymology" (G. M. Attardi and A. Chomyn, eds.), Vol. 260, pp. 213–223. Academic Press, San Diego.

Gordon, D. J., Boyer, J. L., and Korn, E. D. (1997). Comparative biochemistry of non-muscle actins. *J. Biol. Chem.* **252,** 8300–8309.

Greer, C., and Schekman, R. (1982). Actin from *Saccharomyces cerevisiae*. *Mol. Cell. Biol.* **2,** 1270–1278.

Hermann, G. J., King, E. J., and Shaw, J. M. (1997). The yeast gene, MDM20, is necessary for mitochondrial inheritance and organization of the actin cytoskeleton. *J. Cell Biol.* **137,** 141–153.

Kilmartin, J. V., and Adams, A. E. (1984). Structural rearrangements of tubulin and actin during the cell cycle of the yeast *Saccharomyces*. *J. Cell Biol.* **98,** 922–933.

Kron, S. J., Drubin, D. G., Botstein, D., and Spudich, J. A. (1992). Yeast actin filaments display ATP-dependent sliding movement over surfaces coated with rabbit muscle myosin. *Proc. Natl. Acad. Sci. USA* **89,** 4466–4470.

Lazarides, E., and Lindberg, U. (1974). Actin is the naturally occurring inhibitor of deoxyribonuclease I. *Proc. Natl. Acad. Sci. USA* **71,** 4742–4746.

Lazzarino, D. A., Boldogh, I., Smith, M. G., Rosand, J., and Pon, L. A. (1994). Yeast mitochondria contain ATP-sensitive, reversible actin-binding activity. *Mol. Biol. Cell.* **5,** 807–818.

Lessard, J. L. (1988). Two monoclonal antibodies to actin: One muscle selective and one generally reactive. *Cell Motil. Cytoskel.* **10,** 349–362.

Mannherz, H. G., Leigh, J. B., Leberman, R., and Pfrang, H. (1975). A specific 1:1 G-actin:DNAase I complex formed by the action of DNAase I on F-actin. *FEBS Lett.* **60,** 34–38.

Nuoffer, C., Horvath, A., and Riezman, H. (1993). Analysis of the sequence requirements for glycosylphosphatidylinositol anchoring of *Saccharomyces cerevisiae* Gas1 protein. *J. Biol. Chem.* **268,** 10558–10563.

Simon, V. R., Karmon, S. L., and Pon, L. A. (1997). Mitochondrial inheritance: Cell cycle and actin cable dependence of polarized mitochondrial movements in *Saccharomyces cerevisiae*. *Cell Motil. Cytoskel.* **37,** 199–210.

Simon, V. R., Swayne, T. C., and Pon, L. A. (1995). Actin-dependent mitochondrial motility in mitotic yeast and cell-free systems: Identification of a motor activity on the mitochondrial surface. *J. Cell Biol.* **130,** 345–354.

Sogo, L. F., and Yaffe, M. P. (1994). Regulation of mitochondrial morphology and inheritance by Mdm10p, a protein of the mitochondrial outer membrane. *J. Cell Biol.* **126,** 1361–1373.

Stirling, C. J., Rothblatt, J., Hosobuchi, M., Deshaies, R., and Schekman, R. (1992). Protein translocation mutants defective in the insertion of integral membrane proteins into the endoplasmic reticulum. *Mol. Biol. Cell.* **3,** 129–142.

Water, R. D., Pringle, J. R., and Kleinsmith, L. J. (1980). Identification of an actin-like protein and of its messenger ribonucleic acid in *Saccharomyces cerevisiae*. *J. Bacteriol.* **144,** 1143–1151.

Zechel, K. (1980a). Dissociation of the DNAse-I: actin complex by formamide. *Eur. J. Biochem.* **110,** 337–341.

Zechel, K. (1980b). Isolation of polymerization-competent cytoplasmic actin by affinity chromatography on immobilized DNase I using formamide as eluant. *Eur. J. Biochem.* **110,** 343–348.

CHAPTER 10

Analysis and Prediction of Mitochondrial Targeting Peptides

Olof Emanuelsson,* Gunnar von Heijne,* and Gisbert Schneider†

* Stockholm Bioinformatics Center, Stockholm University, S-10691 Stockholm, Sweden

† F. Hoffmann-La Roche Ltd., Pharmaceuticals Division CH-4070 Basel, Switzerland

I. Introduction

Protein import into mitochondria depends on sorting signals present in the nascent polypeptide chain (Neupert, 1997). The targeting signals are collectively referred to as mitochondrial targeting peptides (mTPs). mTPs contain all of the information to target a cytosolically synthesized protein to mitochondria and to the correct submitochondrial compartment. Although both internal and C-terminal mitochondrial import signals have also been identified the best understood mTP is an N-terminal extension (von Heijne, 1994; Diekert *et al.,* 1999; Lee *et al.,* 1999). This chapter reviews the characteristics of mTPs defined by statistical and experimental studies and discusses their use for predicting the presence of mTPs in protein sequences.

II. What Mitochondrial Targeting Peptides (mTPs) Look Like

Two mitochondrial sorting compartments for soluble proteins are known: the matrix and the intermembrane space. Transport of proteins into soluble compartments is mediated by two major mTP classes: the matrix-targeting mTP and a bipartite signal that is composed of an N-terminal mTP followed by an IMS-targeting signal (Fig. 1).

It has long been thought that mTPs can form positively charged, amphiphilic α helices and that this structural element is the main targeting determinant (Waltner and Weiner, 1995; Roise, 1997). This notion has received ample support from mutagenesis studies (e.g., Chu *et al.,* 1987; Thornton *et al.,* 1993; Hammen *et al.,* 1996), from nuclear magnetic resonance, and other spectroscopic structure determinations (Hammen *et al.,* 1994, 1996; McLachlan *et al.,* 1994; Lancelin *et al.,* 1996; Jarvis *et al.,* 1995; Abe *et al.,* 2000), and from statistical analyses (von Heijne, 1986; Gavel *et al.,* 1988; Lemire *et al.,* 1989). It is further believed that mTPs interact initially with the Tom20 receptor in the outer mitochondrial membrane (Schleiff *et al.,* 1999) and that this interaction is mediated primarily by the apolar face of the amphiphilic helix (Pfanner *et al.,* 1997). Subsequently, the mTP is transferred to the Tom22 receptor, which seems to recognize mainly the charged residues (Pfanner *et al.,* 1994; Moczko *et al.,* 1997). Differences in receptor–precursor interaction have been found for cleavable and noncleavable targeting peptides (Brix *et al.,* 1999). The net positive charge of the mTP is also thought to promote its $\Delta\Psi$-driven translocation across the inner membrane (Voos *et al.,* 1999). Finally, the mTP is cleaved off by the matrix mitochondrial processing peptidase (MPP) (Arretz *et al.,* 1991; Brunner *et al.,* 1994). About one-third of all precursor proteins are processed a second time by the matrix-localized mitochondrial intermediate peptidase (MIP) (Kalousek *et al.,* 1988; Isaya and Kalousek, 1994). Details about folding and unfolding of precursor sequences can be found in Schwartz *et al.* (1999).

Even though the role of the amphiphilic α helix is reasonably well understood, the sequence determinants that guide the processing events are more uncertain. Roughly

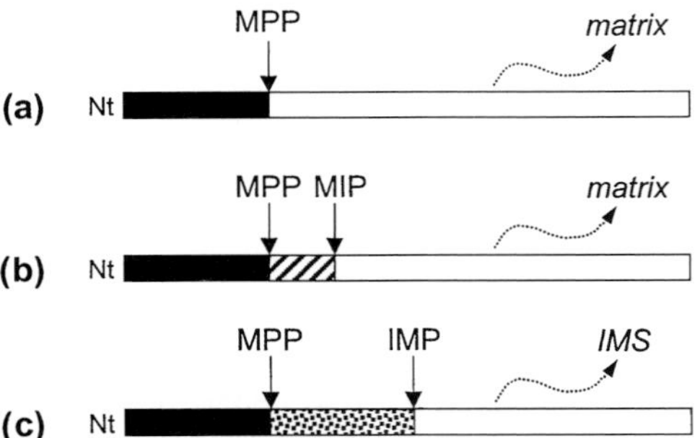

Fig. 1 Classes of mitochondrial targeting peptides (mTP). Arrows indicate proteolytic cleavage sites. MPP, matrix processing peptidase; MIP, mitochondrial intermediate peptidase; IMP, mitochondrial inner membrane peptidase; Nt, N terminus of the precursor sequence.

60% of known mTPs have an Arg residue two or three amino acids N-terminal to the MPP cleavage site (our unpublished data), though this in itself can hardly be enough to specify a unique cleavage site. To complicate matters even further, MIP activity removes an additional eight (or sometimes nine) N-terminal residues from a subset of the MPP/PEP-cleaved precursors (Isaya *et al.*, 1991). Finally, it is not entirely clear how MIP-cleaved proteins differ from those that are not cleaved.

Protein precursors destined for import into the IMS, e.g., cytochrome b_2, cytochrome c_1, and cytochrome peroxidase, have prepeptides with a bipartite structure consisting of two targeting signals (Fig. 1c; Gasser *et al.*, 1982; van Loon *et al.*, 1987). These prepeptides usually contain an N-terminal amphiphilic α helix followed by a downstream IMS targeting signal. Upon precursor import from the cytosol, the amphiphilic α helix is cleaved in reactions catalyzed by MPP within the mitochondrial matrix, thereby bringing the second targeting signal into action. The IMS signal has much in common with bacterial secretion signals (signal peptides) (Behrens *et al.*, 1991; Nunnari *et al.*, 1993; Pratje *et al.*, 1994). It directs the shortened precursor to the inner membrane and enables the process of precursor translocation into the IMS. Inner membrane peptidase (IMP) activity in the IMS removes the IMS targeting signal, releasing the mature protein (Pratje *et al.*, 1994). Some IMS proteins contain internal targeting information, which is not removed proteolytically (Schatz, 1993; Folsch *et al.*, 1996; Diekert *et al.*, 1999), e.g., yeast cytochrome heme lyases (Steiner *et al.*, 1995). Inner membrane proteins such as cytochrome c_1, cytochrome b_1, and D-lactate dehydrogenase contain stop-transfer type targeting signals (Glick *et al.*, 1992; Arnold *et al.*, 1998; Rojo *et al.*, 1998). An uncleaved IMS targeting signal has also been observed in the major adenylate kinase (Aky2p) from yeast (Bandlow *et al.*, 1998). For further details about common features among different amino-terminal targeting signals, see Rusch and Kendall (1995) and Claros *et al.* (1997).

III. Automatic Classification of Known mTPs

Apparently mTPs of different species share common motifs. Although efficient transport of nonnative proteins in general requires a targeting signal from a host cell protein (Rusch and Kendall, 1995), it has been possible to construct fusion proteins containing foreign presequences and observe import into host cell mitochondria. For example, the targeting signal of yeast mitochondrial Trp-tRNA-synthetase was able to target bacterial β-glucuronidase to the mitochondria of transgenic tobacco plants (Schmitz and Lonsdale, 1989). Such experimental observations support attempts to perform a computer-assisted classification and prediction of mTPs.

It is possible to discriminate mTPs reasonably well from the N-terminal parts of cytoplasmic proteins simply by their amino acid composition. mTPs generally contain higher numbers of Arg, Ala, and Ser residues and have a significantly lower content of negatively charged residues (Asp, Glu).

A straightforward approach is to consider the amino acid composition of N-terminal sequence parts and visualize the resulting data distribution by appropriate projection techniques. Encoding each amino acid sequence by its respective composition vector

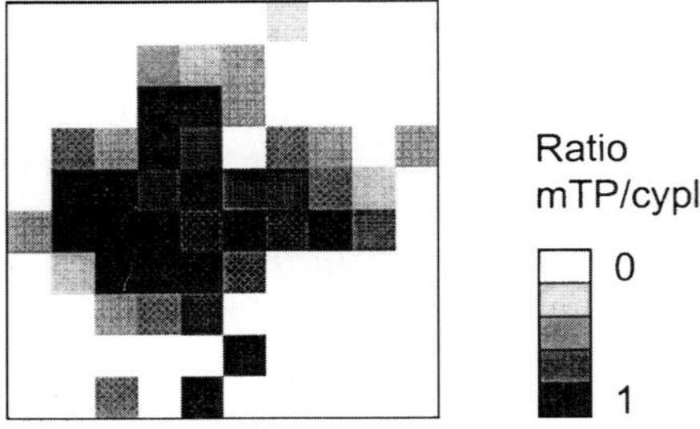

Fig. 2 Distribution of mitochondrial targeting peptides (mTP) and N-terminal sequence stretches of cytoplasmic proteins (N-terminal 30 residues) in sequence space. Sequences were encoded by their amino acid composition. The map was generated using a self-organizing neural network (Schneider, 1999). Dark regions indicate mTP clusters, and white areas are dominated by cytoplasmic sequences. Note that the map forms a torus.

leads to a 20-dimensional sequence space, which can be appropriately displayed by self-organizing maps (SOM) (Schneider, 1999). Many other encoding, classification, and visualization techniques exist that can be applied to sequence classification (for review, see Schneider and Wrede, 1998). There are, however, some characteristic features of SOM that make them particularly useful tools for automatic, exploratory sequence classification: (i) the resulting nonlinear two-dimensional data projection preserves the topology of the original multidimensional space, (ii) SOMs are easy to implement and are reasonably fast, and (iii) the method is scalable to very large data sets. Figure 2 gives an example of a SOM projection that was obtained from a collection of mitochondrial and cytoplasmic sequences encoded by their amino acid composition vectors.

SOMs reveal clustering of most mTPs. Many mTPs that fall outside of the primary cluster are exceptionally long. Other mTPs that fall outside of the primary cluster contain species-specific variations. For example, because mTPs and chloroplast targeting peptides share fundamental features (Huang *et al.,* 1990; De Castro Silva Filho *et al.,* 1996; Akashi *et al.,* 1998), plants have developed more elaborate mechanisms to ensure accurate targeting to mitochondria. Indeed, there is evidence for mTP-independent mechanisms for targeting (Hugosson *et al.,* 1995). In addition, plant mTPs contain higher serine levels than other mTPs and fall into a subclass of mTPs that is distinguished from other mTPs by SOM (Schneider *et al.,* 1998; Sjöling and Glaser, 1998; Chou and Elrod, 1999).

IV. Automatic Identification of Unknown mTPs

Identifying mTPs in newly discovered sequences requires classification of known sequences in a data set. Here, the objective is to generalize the information of known transit peptides into a working prediction algorithm to identify mTPs in unknown proteins or known proteins with unknown subcellular localization. The methods presented in this chapter for predicting the subcellular localization of a protein are based on (i) the use of a number of physicochemical parameters (abundance of certain amino acids, hydrophobicity in certain regions, etc.) or (ii) analysis of the plain residue patterns in (part of) the amino acid sequence.

Predicting subcellular localization based on structure information has not yet been successful in the case of mitochondria (or in other subcellular compartments that require a presequence on the protein for import, e.g., the chloroplast) due to the lack of sufficient structural information on the transit peptide. Although it is widely accepted that many mTPs form an amphiphilic α helix, not enough experimental evidence exists to date to be used in a prediction scheme.

A. Discrimination of Subcellular Locations

MitoProtII (Claros and Vincens, 1996) and PSORT (Nakai and Kanehisa, 1992) are two readily available and well-documented subcellular localization predictors. The former aims at predicting mitochondrial or nonmitochondrial proteins (binary decision), whereas the latter examines a large set of potential subcellular compartments. Another predictor that discriminates between different N-terminal presequences [e.g. mTP, chloroplast transit peptide (cTP), and signal peptide for ER membrane targeting (SP)] has recently been published (Emanuelsson *et al.,* 2000). Two further predictors will also be presented. Only PSORT treats IMS proteins (processed by IMP) as a separate group, whereas the others regard the mitochondrion as one entity.

To compare these predictors, we have tested them on two different data sets, both from SWISS-PROT (Bairoch and Apweiler, 1999). One set consists of a set of human sequences available in SWISS-PROT as of October 19, 1999. This includes 174 mTP, 1370 SP, and 1875 "other" proteins with explicit subcellular location annotated. 2075 proteins in this set have no subcellular localization and were therefore not analyzed. The other set includes 459 *Arabidopsis thaliana* sequences: 113 cTP, 31 mTP, 56 SP, and 259 "other" proteins. Here too, 361 sequences without explicit subcellular localization annotation were not analyzed. The results are shown in Table I.

MitoProt II (Claros and Vincens, 1996) is a program that predicts mitochondrial localization of a sequence by calculating several (11) physicochemical parameters from the sequence and then computing a linear discriminant function (LDF). The parameters considered by MitoProt are in many cases derived from earlier work (e.g., Gavel and von Heijne, 1990; Nakai and Kanehisa, 1992). In total, 47 parameters are used. The parameters that have the greatest value in identifying mTPs are amino acid content in certain regions (e.g., the amino acid composition in the N-terminal part) and the hydrophobicity of regions (e.g., the highest total hydrophobicity for a 17 residue window). The hydrophobicity parameters are measured by four different scales to minimize the influence by any particular scale. Other parameters, including the overall charge of the putative transit peptide region, seem to contribute very little in identifying mTPs.

From the outcome of the LDF, MitoProt II calculates the probability for mitochondrial localization, with no further discrimination as to the different suborganellar compartments present in the mitochondrion. MitoProt II does not explicitly look for the existence of a targeting peptide (an mTP), although such an algorithm, based on a subset of the

Table I

Properties, Performances, and User Interfaces of Tested Predictors

Predictor[b] name	No. of locations	Test set[c]	MCC[d]	Sensitivity[e]	Specificity[f]	% corr.[g]	CS prediction[h]	CS performance[i]	Availability of the predictor[j]			Internet resources[k]
									www	e-mail	down load	
MitoProtII	2	Human	0.46	0.92	0.27	86.7	Yes	36.1	—	S	M	ftp://ftp.ebi.ac.uk/pub/software/ ftp://ftp.ens.fr/pub/molbio/ mitoprot@biologie.ens.fr
PSORT II	10	Human	0.46	0.81	0.31	90.0(68.6)	No	—	S	—	M	http://psort.nibb.ac.jp
PSORT I	17	Plant	0.47	0.61	0.42	91.5(74.2)	No	—	S	S	M	http://psort.nibb.ac.jp psort@nibb.ac.jp
TargetP	3	Human	0.66	0.95	0.49	94.7(86.3)	Yes	65.3	M	—	—	http://genome.cbs.dtu.dk/services/
TargetP (plant)	4	Plant	0.52	0.65	0.48	92.8(86.1)	Yes	Not tested	M	—	—	http://genome.cbs.dtu.dk/services/
NNPSL	4	Human	0.31	0.76	0.18	81.0(57.0)	No	—	M	—	—	http://predict.sanger.ac.uk/nnpsl/
MTS	2	Human	0.37	0.94	0.19	79.5	Yes	12.5	M	M	—	http://202.247.6.235/

[a]All performance figures are calculated for a two-category prediction, taking into account only the ability to predict mitochondrial/nonmitochondrial localization (if not otherwise stated).

[b]For references, see text.

[c]The human (*H. sapiens*) test set consisted of 3419 sequences with clearly annotated subcellular localization from SWISS-PROT release 38 (Bairoch and Apweiler, 1999). The plant test set contained 459 *A. thaliana* sequences extracted from SWISS-PROT under identical conditions.

[d]Mathews correlation coefficient (Mathews, 1975), calculated according to the formula: $\text{MCC} = (PN{-}OU)/\text{sqrt}[(N+U)(N+O)(P+U)(P+O)]$, where P is the number of correctly predicted positive examples, N is the number of correctly predicted negative examples, O is the number of false positives (overpredicted), and U is the number of false negatives (underpredicted). For a perfect predictor, MCC = 1.

[e]$P/(P+U)$.

[f]$P/(P+O)$.

[g]Percentage correctly predicted sequences. For those predictors that have several (>2) potential subcellular locations, the value considering all its available locations has been included in parentheses.

[h]Does the predictor provide a cleavage site (CS) prediction?

[i]Percentage predicted cleavage sites within ±2 residues of correct CS. The prediction performance is tested on the 72 sequences of the human mTP test set that were unambiguously annotated as mTP containing (without annotation such as "PROBABLE," "POTENTIAL," "BY SIMILARITY," or question marks in the FT field of the SWISS-PROT entry). With these restrictions applied, the *A. thaliana* set contained too few mTP sequences to allow testing.

[j]On the www, via e-mail, and possibility to download the program for local runs. S, one sequence at a time only; M, multiple sequences (e.g., in a file); —, not available.

[k]URL/e-mail/ftp.

47 parameters, has been described in the paper as a possible alternative. Still, many of the erroneously predicted true mitochondrial proteins (false negatives) are reported to lack an mTP. Prediction accuracy on its original data yields a sensitivity of 0.76 and a Matthews correlation coefficient (Matthews, 1975) of 0.31 (Claros and Vincens, 1996). In our performance test on the human data set from SWISS-PROT (see earlier discussion), MitoProt II scored a correlation coefficient (computed as mitochondrial vs all other types) of 0.46 (Table I).

PSORT is a knowledge-based, multicategory subcellular localization program. Predictions are made by calculating a set of sequence-derived parameters and comparing them to a representation of a number of "localization rules" that have been collected from the literature and stored in the knowledge base. PSORT is actually two different programs. The original PSORT (Nakai and Kanehisa, 1992), which will be referred to as PSORT I, and PSORT II (Horton and Nakai, 1997), where the latter is an evolution of the former, written in a different programming language and with the possibility for the users to train the predictor on their own data sets. The web version of PSORT II was trained on a set of 1080 yeast proteins and cannot be used to identify plant (or prokaryotic) sequences. PSORT I was trained on a set of 295 proteins from various species. As a result, PSORT I is the best version for identifying plant mTPs. PSORT I deals with 17 subcellular compartments, including all four subcompartments within mitochondria. PSORT II comprises 10 localizations and does not make predictions regarding submitochondrial localization. Perhaps the greatest difference between the two versions is that the reasoning algorithms are different. PSORT I represents the knowledge base as a set of "if-then" rules and calculates a "certainty factor" for each localization, whereas PSORT II stores the entire training set and predicts a new sequence by choosing the majority localization among a certain number of nearest training examples (Euclidean distance in physicochemical space). Apart from resulting in better performance (Horton and Nakai, 1997), this approach also facilitates the incorporation of new training data in the prediction. Testing PSORT on our two benchmarking data sets yielded a Matthews correlation coefficient of 0.47 for the plant set (PSORT I) and 0.46 for the human set (PSORT II) (Table I).

A novel subcellular predictor (TargetP) that relies on the detection of an N-terminal transit peptide has been developed in our laboratory (Emanuelsson *et al.*, 2000). It uses neural networks for pattern recognition and exists in two versions: one for plant proteins (including cTPs, SPs, and mTPs) and one for other eukaryotic proteins (including SPs and mTPs). Both have in common the fact that they are made of networks in two layers, where the first layer consists of two (three for the plant version) parallel networks, each of which assigns a value of "signal peptideness" or "mTP-ness" (or "cTP-ness" for the plant version), respectively, to each residue. The top layer network then integrates these values for the 100 N-terminal residues in the sequence and outputs one score per potential subcellular location, i.e., three scores for the nonplant version (mTP, SP, other) and four scores for the plant version (cTP, mTP, SP, other). In the default implementation, the highest output score simply determines the prediction ("winner takes all"), but, it is possible to demand the output also to be above a certain threshold to be valid as a prediction, thus altering the expected sensitivity/specificity balance.

The training sets consisted of clearly annotated and redundancy reduced (using Hobohm algorithm II; Hobohm *et al.,* 1992) SWISS-PROT sequences of the just-mentioned three (plant: four) types, where the "other" set included known nuclear and cytosolic proteins. The plant version was trained on plant proteins and the nonplant version without such proteins. However, as the number of mitochondrial sequences is quite limited, mTPs of all types were included in the mTP training sets. Although earlier studies (Schneider *et al.,* 1998) show that plant mTPs deviate somewhat from other mTPs in their amino acid composition and may be considered as a subgroup of mTPs, as discussed earlier, the common features were deemed to be strong enough to justify the gathering of all mTPs into one group. In our tests, the predictor scored a Matthews correlation coefficient of 0.52 and 0.66 on the plant and human set, respectively (Table I).

NNPSL (Reinhardt and Hubbard, 1998) is another neural network-based predictor, but unlike the one described earlier, it uses only the amino acid composition of the sequence to assign one of four subcellular localization (cytosolic, extracellular, nuclear, and mitochondrial) to a query sequence. It does not deal with plant sequences. The advantage of only looking at amino acid composition as opposed to the actual amino acid sequence is that the prediction is less sensitive to incomplete or partly faulty sequences. In the training of this predictor, some 3000 sequences (from SWISS-PROT) were used. Redundancy reduction was performed so that no sequence was more than 90% identical to any other sequence. NNPSL scored 0.31 on our human set (Table I).

A hidden Markov model method for predicting mitochondrial transit peptides (MTS) was developed by Fujiwara and co-workers (1997). The development of this predictor used SWISS-PROT entries annotated to contain an mTP as the positive set, whereas the negative set consisted of yeast sequences with annotations indicating other subcellular localization. Accordingly, it does not predict plant sequences. The prediction service is free, but requires a one-time member registration. On our human set, a correlation coefficient of 0.37 was obtained.

B. Cleavage Site Prediction

In contrast to recognition of potential targeting peptides, the quality of current MPP and MIP cleavage site prediction systems is rather poor. The main problem is the large fraction of false-positive predictions ("overprediction"), i.e., several potential cleavage sites are indicated in a precursor sequence. This makes an unambiguous identification of the N terminus of the mature protein difficult, although in most cases the experimentally observed precursor processing site is among the cleavage site candidates. Two related methods have been developed: linear weight matrices as in MitoProt (Claros and Vincens, 1996) and conventional linear and nonlinear neural networks (Schneider *et al.,* 1995, 1998). Both methods use a sliding window approach to detect the border between the mTP and the mature part of the protein precursor. MitoProt computes a "CleavSite" parameter, which is the number of the first residue in the mature protein. Our novel predictor TargetP assigns a potential cleavage site by applying three weight matrices (one for each class of mTP with Arg in −2, −3, or −10) and choosing the site corresponding to the highest score; 59.9% of the cleavage sites of our set of redundancy-reduced mTPs were correctly predicted within ±2 residues. Straightforward enough, this approach obviously still needs to be refined and improved. A comparison of cleavage site prediction performance on a subset of 72 unambiguously annotated sequences from the human mTP data set is presented in Table I.

While there exists no simple residue pattern defining the MPP target site (Arretz *et al.,* 1994; Schneider *et al.,* 1998), an octapeptide motif located C-terminal to the dipeptide RX was identified that seems to be a necessary prerequisite for MIP processing: (F,L,I)XX(T,S,G)XXXX (Hendrick *et al.,* 1989; Gavel and von Heijne, 1990; Schneider *et al.,* 1998). Experimental and theoretical studies support the idea that specific features of the N-terminal mature sequence (in the case of single-step processing by MPP) or the octapeptide (in the case of MPP and subsequent MIP processing) define the precise MPP cleavage site and direct MPP to its appropriate target position (Gavel and von Heijne, 1990; Isaya *et al.,* 1991; Isaya and Kalousek, 1994). The distribution and clustering of positive charges along the respective sequence stretches seem to play a major role in this process (Ou *et al.,* 1994).

V. Conclusions

Several computational methods have been developed to predict mTPs and their MPP/MIP cleavage sites. Although satisfactory results can be obtained for mTP recognition, no sufficiently accurate technique exists for cleavage site prediction. The current lack of validated sequence data and poor understanding of the structural aspects of targeting, translocation, and processing make the development of more successful algorithms difficult.

Most methods for mTP prediction focus on the amino terminus of possible mitochondrial proteins. However, it is clear that other portions of proteins also contribute to the import process. For example, there is evidence supporting the existence of internal functional topogenic signals (Folsch *et al.,* 1996; Diekert *et al.,* 1999). It has also been shown that C-terminal features of the small subunit of Rubisco modulate import of its precursor into chloroplasts (Dabney-Smith *et al.,* 1999). A surprising observation is C- to N-terminal translocation of preproteins into mitochondria (Folsch *et al.,* 1998), and in some studies an influence of the mature part of the protein on translocation efficiency has been observed (e. g., Waltner *et al.,* 1996). However, the small number of such known examples prohibits a thorough analysis of underlying sequence features. Furthermore, some experimental evidence has been provided for the existence of a cotranslational pathway for protein import into mitochondria (Verner, 1993; Crowley and Payne, 1998). It was demonstrated that the mTP is involved in the regulation of ribosome binding to the mitochondrial outer membrane. This import pathway may require additional, hitherto unknown mTP features. Thus, improved mTP prediction tools will have to account for features of the mature protein, rather than focusing only on the N-terminal parts of the precursor.

References

Abe, Y., Shodai, T., Muto, T., Mihara, K., Torii, H., Nishikawa, S., Endo, T., and Kohda, D. (2000). Structural basis of presequence recognition by the mitochondrial protein import receptor Tom20. *Cell* **100,** 551–560.

Akashi, K., Grandjean, O., and Small, I. (1998). Potential dual targeting of an *Arabidopsis* archaebacterial-like histidyl-tRNA synthetase to mitochondria and chloroplasts. *FEBS Lett.* **431,** 39–44.

Arnold, I., Folsch, H., Neupert, W., and Stuart, R. A. (1998). Two distinct and independent mitochondrial targeting signals function in the sorting of an inner membrane protein, cytochrome c1. *J. Biol. Chem.* **273,** 1469–1476.

Arretz, M., Schneider, H., Guiard, B., Brunner, M., and Neupert, W. (1994). Characterization of the mitochondrial processing peptidase of *Neurospora crassa. J. Biol. Chem.* **269,** 4959–4967.

Arretz, M., Schneider, H., Wienhues, U., and Neupert, W. (1991). Processing of mitochondrial precursor proteins. *Biomed. Biochim. Acta* **50,** 403–412.

Bairoch, A., and Apweiler, R. (1999). The SWISS-PROT protein sequence data bank and its supplement TrEMBL in 1999. *Nucleic Acids Res.* **27,** 49–54.

Bandlow, W., Strobel, G., and Schricker, R. (1998). Influence of N-terminal sequence variation on the sorting of major adenylate kinase to the mitochondrial intermembrane space in yeast. *Biochem. J.* **329,** 359–367.

Behrens, M., Michaelis, G., and Pratje, E. (1991). Mitochondrial inner membrane protease 1 of *Saccharomyces cerevisiae* shows sequence similarity to the *Escherichia coli* leader peptidase. *Mol. Gen. Genet.* **228,** 167–176.

Brix, J., Rudiger, S., Bukau, B., Schneider-Mergener, J., and Pfanner, N. (1999). Distribution of binding sequences for the mitochondrial import receptors Tom20, Tom22, and Tom70 in a presequence-carrying preprotein and a non-cleavable preprotein. *J. Biol. Chem.* **274,** 16522–16530.

Brunner, M., Klaus, C., and Neupert, W. (1994). The mitochondrial processing peptidase. *In* "Signal Peptidases" (G. Von Heijne, ed.), pp. 73–86. R. G. Landes Company, Austin.

Chou, K. C., and Elrod, D. W. (1999). Protein subcellular location prediction. *Protein Eng.* **12,** 107–118.

Chu, T. W., Grant, P. M., and Strauss, A. W. (1987). Mutation of a neutral amino acid in the transit peptide of rat mitochondrial malate dehydrogenase abolishes binding and import. *J. Biol. Chem.* **262,** 15759–15764.

Claros, M. G. (1995). MitoProt, a Macintosh application for studying mitochondrial proteins. *Comput. Appl. Biosci.* **11,** 441–447.

Claros, M. G., Brunak, S., and von Heijne, G. (1997). Prediction of N-terminal targeting signals. *Curr. Opin. Struct. Biol.* **7,** 394–398.

Claros, M. G., and Vincens, P. (1996). Computational method to predict mitochondrially imported proteins and their targeting sequences. *Eur. J. Biochem.* **241,** 779–786.

Crowley, K. S., and Payne, R. M. (1998). Ribosome binding to mitochondria is regulated by GTP and the transit peptide. *J. Biol. Chem.* **273,** 17278–17285.

Dabney-Smith, C., van Den Wijngaard, P. W., Treece, Y., Vredenberg, W. J., and Bruce, B. D. (1999). The C terminus of a chloroplast precursor modulates its interaction with the translocation apparatus and PIRAC. *J. Biol. Chem.* **274,** 32351–32359.

De Castro Silva Filho, M., Chaumont, F., Leterme, S., and Boutry, M. (1996). Mitochondrial and chloroplast targeting sequences in tandem modify protein import specificity in plant organelles. *Plant Mol. Biol.* **30,** 769–780.

Diekert, K., Kispal, G., Guiard, B., and Lill, R. (1999). An internal targeting signal directing proteins into the mitochondrial intermembrane space. *Proc. Natl. Acad. Sci. USA* **96,** 11752–11757.

Emanuelsson, O., Nielsen, H., Brunak, S., von Heijne, G. (2000). Predicting subcellular localization of proteins based on their N-terminal amino acid sequence. *J. Mol. Biol.* **300,** 1005–1016.

Folsch, H., Gaume, B., Brunner, M., Neupert, W., and Stuart, R. A. (1998). C- to N-terminal translocation of preproteins into mitochondria. *EMBO J.* **17,** 6508–6515.

Folsch, H., Guiard, B., Neupert, W., and Stuart, R. A. (1996). Internal targeting signal of the BCS1 protein: A novel mechanism of import into mitochondria. *EMBO J.* **15,** 479–487.

Fujiwara, Y., Asogawa, M., and Nakai, K. (1997). Prediction of mitochondrial targeting signals using Hidden Markov Models. *In* "Genome Informatics," pp. 53–60.

Gasser, S. M., Ohashi, A., Daum, G., Bohni, P. C., Gibson, J., Reid, G. A., Yonetani, T., and Schatz, G. (1982). Imported mitochondrial proteins cytochrome b2 and cytochrome c1 are processed in two steps. *Proc. Natl. Acad. Sci. USA* **79,** 267–271.

Gavel, Y., Nilsson, L., and von Heijne, G. (1988). Mitochondrial targeting sequences: Why "non-amphiphilic" peptides may still be amphiphilic. *FEBS Lett.* **235,** 173–177.

Gavel, Y., and von Heijne, G. (1990). Cleavage-site motifs in mitochondrial targeting peptides. *Protein Eng.* **4,** 33–37.

Glaser, E., Sjöling, S., Tanudji, M., and Whelan, J. (1998). Mitochondrial protein import in plants: Signals, sorting, targeting, processing and regulation. *Plant Mol. Biol.* **38,** 311–338.

Glick, B. S., Brandt, A., Cunningham, K., Muller, S., Hallberg, R. L., and Schatz, G. (1992). Cytochromes c1 and b2 are sorted to the intermembrane space of yeast mitochondria by a stop-transfer mechanism. *Cell* **69,** 809–822.

Hammen, P. K., Gorenstein, D. G., and Weiner, H. (1994). Structure of the signal sequences for two mitochondrial matrix proteins that are not proteolytically processed upon import. *Biochemistry* **33,** 8610–8617.

Hammen, P. K., Gorenstein, D. G., and Weiner, H. (1996). Amphiphilicity determines binding properties of three mitochondrial presequences to lipid surfaces. *Biochemistry* **35,** 3772–3781.

Hammen, P. K., Waltner, M., Hahnemann, B., Heard, T. S., and Weiner, H. (1996). The role of positive charges and structural segments in the presequence of rat liver aldehyde dehydrogenase in import into mitochondria. *J. Biol. Chem.* **271,** 21041–21048.

Hendrick, J. P., Hodges, P. E., and Rosenberg, L. E. (1989). Survey of amino-terminal proteolytic cleavage sites in mitochondrial precursor proteins: Leader peptides cleaved by two matrix proteases share a three-amino acid motif. *Proc. Natl. Acad. Sci. USA* **86,** 4056–4060.

Hobohm, U., Scharf, M., Schneider, R., and Sander, C. (1992). Selection of representative protein data sets. *Protein Sci.* **1,** 409–417.

Horton, P., and Nakai, K. (1997). Better prediction of protein cellular localization sites with the *k* nearest neighbors classifier. *ISMB* **5,** 147–152.

Huang, J., Hack, E., Thornburg, R. W., and Myers, A. M. (1990). A yeast mitochondrial leader peptide functions in vivo as a dual targeting signal for both chloroplasts and mitochondria. *Plant Cell* **2,** 1249–1260.

Hugosson, M., Nurani, G., Glaser, E., and Franzén, L. G. (1995). Peculiar properties of the PsaF photosystem I protein from the green alga *Chlamydomonas reinhardtii*: Presequence independent import of the PsaF protein into both chloroplasts and mitochondria. *Plant Mol. Biol.* **28,** 525–535.

Isaya, G., and Kalousek, F. (1994). Mitochondrial intermediate peptidase. *In* "Signal Peptidases" (G. Von Heijne, ed.), pp 87–103. R. G. Landes Company, Austin.

Isaya, G., Kalousek, F., Fenton, W. A., and Rosenberg, L. E. (1991). Cleavage of precursors by the mitochondrial processing peptidase requires a compatible mature protein or an intermediate octapeptide. *J. Cell Biol.* **113,** 65–76.

Jarvis, J. A., Ryan, M. T., Hoogenraad, N. J., Craik, D. J., and Hoj, P. B. (1995). Solution structure of the acylated and noncleavable mitochondrial targeting signal of rat chaperonin 10. *J. Biol. Chem.* **270,** 1323–1331.

Kalousek, F., Hendrick, J. P., and Rosenberg, L. E. (1988). Two mitochondrial matrix proteases act sequentially in the processing of mammalian matrix enzymes. *Proc. Natl. Acad. Sci. USA* **85,** 7536–7540.

Lancelin, J. M., Gans, P., Bouchayer, E., Bally, I., Arlaud, G. J., and Jacquot, J. P. (1996). NMR structures of a mitochondrial transit peptide from the green alga *Chlamydomonas reinhardtii. FEBS Lett.* **391,** 203–208.

Lee, C. M., Sedman, J., Neupert, W., and Stuart, R. A. (1999). The DNA helicase, Hmi1p, is transported into mitochondria by a C-terminal cleavable targeting signal. *J. Biol. Chem.* **274,** 20937–20942.

Lemire, B. D., Fankhauser, C., Baker, A., and Schatz, G. (1989). The mitochondrial targeting function of randomly generated peptide sequences correlates with predicted helical amphiphilicity. *J. Biol. Chem.* **264,** 20206–20215.

Matthews, B. W. (1975). Comparison of the predicted and observed secondary structure of T4 phage lysozyme. *Biochim. Biophys. Acta* **405,** 442–451.

McLachlan, L. K., Haris, P. I., Reid, D. G., White, J., Chapman, D., Lucy, J. A., and Austen, B. M. (1994). A spectroscopic study of the mitochondrial transit peptide of rat malate dehydrogenase. *Biochem. J.* **303,** 657–662.

Moczko, M., Bomer, U., Kubrich, M., Zufall, N., Honlinger, A., and Pfanner, N. (1997). The intermembrane space domain of mitochondrial Tom22 functions as a trans binding site for preproteins with N-terminal targeting sequences. *Mol. Cell Biol.* **17,** 6574–6584.

Nakai, K., and Kanehisa, M. (1992). A knowledge base for predicting protein localization sites in eukaryotic cells. *Genomics* **14,** 897–911.

Neupert, W. (1997). Protein import into mitochondria. *Annu. Rev. Biochem.* **66,** 863–917.

Nunnari, J., Fox, T. D., and Walter, P. (1993). A mitochondrial protease with two catalytic subunits of nonoverlapping specificities. *Science* **262,** 1997–2004.

Ou, W., Kumamoto, T., Mihara, K., Kitada, S., Niidome, T., Ito, A., and Omura, T. (1994). Structural requirements for recognition of the precursor proteins by the mitochondrial processing peptidase. *J. Biol. Chem.* **269,** 24673–24678.

Pfanner, N., Craig, E. A., and Meijer, M. (1994). The protein import machinery of the mitochondrial inner membrane. *Trends Biochem. Sci.* **19,** 368–372.

Pfanner, N., Craig, E. A., and Honlinger, A. (1997). Mitochondrial preprotein translocase. *Annu. Rev. Cell Dev. Biol.* **13,** 25–51.

Pratje, E., Esser, K., and Michaelis, G. (1994). The mitochondrial inner membrane peptidase. *In* "Signal Peptidases" (G. Von Heijne, ed.), pp. 105–112. R. G. Landes Company. Austin.

Reinhardt, A., and Hubbard, T. (1998). Using neural networks for prediction of the subcellular location of proteins. *Nucleic Acids Res.* **26,** 2230–2236.

Roise, D. (1997). Recognition and binding of mitochondrial presequences during the import of proteins into mitochondria. *J. Bioenerg. Biomembr.* **29,** 19–27.

Rojo, E. E., Guiard, B., Neupert, W., and Stuart, R. A. (1998). Sorting of D-lactate dehydrogenase to the inner membrane of mitochondria: Analysis of topogenic signal and energetic requirements. *J. Biol. Chem.* **273,** 8040–8047.

Rusch, S. J., and Kendall, D. A. (1995). Protein transport via amino-terminal targeting sequences: Common themes and diverse systems. *Mol. Membr. Biol.* **12,** 295–307.

Schatz, G. (1993). The protein import machinery of mitochondria. *Protein Sci.* **2,** 141–146.

Schleiff, E., Heard, T. S., and Weiner, H. (1999). Positively charged residues, the helical conformation and the structural flexibility of the leader sequence of pALDH are important for recognition by hTom20. *FEBS Lett.* **461,** 9–12.

Schmitz, U. K., and Lonsdale, D. M. (1989). A yeast mitochondrial presequence functions as a signal for targeting to plant mitochondria in vivo. *Plant Cell* **1,** 783–791.

Schneider, G. (1999). How many potentially secreted proteins are contained in a bacterial genome? *Gene* **237,** 113–121.

Schneider, G., Schuchhardt, J., and Wrede, P. (1995). Peptide design in machina: Development of artificial mitochondrial protein precursor cleavage sites by simulated molecular evolution. *Biophys. J.* **68,** 434–447.

Schneider, G., Sjöling, S., Wallin, E., Wrede, P., Glaser, E., and von Heijne, G. (1998). Feature-extraction from endopeptidase cleavage sites in mitochondrial targeting peptides. *Proteins* **30,** 49–60.

Schneider, G., and Wrede, P. (1998). Artificial neural networks for computer-based molecular design. *Prog. Biophys. Mol. Biol.* **70,** 175–222.

Schwartz, M. P., Huang, S., and Matouschek, A. (1999). The structure of precursor proteins during import into mitochondria. *J. Biol. Chem.* **274,** 12759–12764.

Sjöling, S., and Glaser, E. (1998). Mitochondrial targeting peptides in plants. *Trends Plant Sci.* **3,** 136–140.

Steiner, H., Zollner, A., Haid, A., Neupert, W., and Lill, R. (1995). Biogenesis of mitochondrial heme lyases in yeast: Import and folding in the intermembrane space. *J. Biol. Chem.* **270,** 22842–22849.

Thornton, K., Wang, Y., Weiner, H., and Gorenstein, D. G. (1993). Import, processing, and two-dimensional NMR structure of a linker-deleted signal peptide of rat liver mitochondrial aldehyde dehydrogenase. *J. Biol. Chem.* **268,** 19906–19914.

Van Loon, A. P., Brandli, A. W., Pesold-Hurt, B., Blank, D., and Schatz, G. (1987). Transport of proteins to the mitochondrial intermembrane space: The 'matrix-targeting' and the 'sorting' domains in the cytochrome c1 presequence. *EMBO J.* **6,** 2433–2439.

Verner, K. (1993). Co-translational import into mitochondria: An alternative view. *Trends Biochem. Sci.* **18,** 366–371.

Von Heijne, G. (1994). Design of protein targeting signals and membrane protein engineering. *In* "Concepts in Protein Engineering and Design" (P. Wrede and G. Schneider, eds.), pp. 263–279. Walter-de-Gruyter, Berlin.

Von Heijne, G. (1986). Mitochondrial targeting sequences may form amphiphilic helices. *EMBO J.* **5,** 1335–1342.

Von Heijne, G., Steppuhn, J., and Herrmann, R. G. (1989). Domain structure of mitochondrial and chloroplast targeting peptides. *Eur. J. Biochem.* **180,** 535–545.

Voos, W., Martin, H., Krimmer, T., and Pfanner, N. (1999). Mechanisms of protein translocation into mitochondria. *Biochim. Biophys. Acta* **1422,** 235–254.

Waltner, M., Hammen, P. K., and Weiner, H. (1996). Influence of the mature portion of a precursor protein on the mitochondrial signal sequence. *J. Biol. Chem.* **271,** 21226–21230.

Waltner, M., and Weiner, H. (1995). Conversion of a nonprocessed mitochondrial precursor protein into one that is processed by the mitochondrial processing peptidase. *J. Biol. Chem.* **270,** 26311–26317.

CHAPTER 11

Assaying Protein Import into Mitochondria

Michael T. Ryan, Wolfgang Voos, and Nikolaus Pfanner

Institut für Biochemie und Molekularbiologie
Universität Freiburg
D-79104 Freiburg, Germany

I. Introduction

Most proteins found in the mitochondria originate in the cytosol and therefore must be imported into this organelle following their synthesis. Over the past decade, most of the translocase components of the outer and inner mitochondrial membranes (TOM and TIM, respectively) have been identified in *Saccharomyces cerevisiae* and *Neurospora crassa* (Neupert *et al.,* 1997; Koehler *et al.,* 1999; Voos *et al.,* 1999). The role of each of these components in protein import has been characterized through a combination of both genetic and biochemical means.

A preprotein is directed to the mitochondria by virtue of its targeting signal, which is most often situated at the N terminus (see Chapter 10). Preproteins are first bound to receptors of the mitochondrial membrane—Tom70, Tom22, and Tom20—before being translocated across the outer membrane channel, which is formed by Tom40 (Hill *et al.,* 1998). Many preproteins destined to reside in the outer membrane transverse laterally out of this channel upon contact of their hydrophobic anchors with the TOM machinery. For some proteins that are targeted to the intermembrane space (IMS), their engagement with the TOM machinery and subsequent translocation across the outer membrane is sufficient for their import (Lill *et al.,* 1992; Kurz *et al.,* 1999). For preproteins that are targeted to the inner membrane or matrix, the mitochondrial membrane potential ($\Delta\Psi$) is required as a driving force along with, in many cases, matrix-located heat shock protein 70 (mtHsp70). Preproteins containing matrix targeting signals are translocated partially or completely across the inner membrane through the channel formed by the TIM23 complex. The N-terminal targeting signal is then often removed by the matrix processing peptidase (MPP). Preproteins destined for the inner membrane move laterally out of the TIM23 complex, whereas those headed for the matrix may require the assistance of mtHsp70 and cofactors for their complete import. Some preproteins utilizing this pathway contain stop-transfer signals that result in their arrest in the inner membrane and an IMS-located protease then cleaves the hydrophobic anchor from the protein, thereby releasing it into the IMS as a soluble protein. Another group of inner membrane proteins contain internal targeting signals that direct them from the inner face of the TOM machinery to a different TIM complex (the TIM22 complex) via their association with small intermembrane space TIM proteins (Koehler *et al.,* 1999; Truscott and Pfanner, 1999). The characterization of these different import pathways has been achieved mainly through the use of mitochondrial *in vitro* import studies. In such cases, isolated mitochondria are incubated with *in vitro*-translated ^{35}S-labeled preproteins under varying conditions, and their import characteristics and requirements are analyzed. The following sections describe the various methods employed to characterize both the mitochondrial import of preproteins and the protein import machinery *in vitro.*

II. Synthesis of Preproteins

In most cases, preproteins are synthesized *in vitro* using rabbit reticulocyte lysate systems. A number of kits are available commercially (e.g., from Amersham Pharmacia

Biotech and Promega) that either require the addition of RNA that is synthesized separately or are coupled transcription/translation systems. Because the coupled systems are relatively straightforward to use, we will mention briefly the procedures that employ separate transcription and translation reactions.

A. Template DNA

We routinely clone open reading frames (ORFs) into the pGEM-4Z Vector (Promega), whereby the initiation codon of the ORF is cloned downstream of the SP6 RNA polymerase transcription initiation site. Generally, we have found that SP6 RNA polymerase-generated transcripts are produced more efficiently than those synthesized by T7 or T3 RNA polymerases (nevertheless, both of these polymerases can produce sufficient amounts of RNA). While it is often recommended to linearize the vector DNA downstream of the ORF for optimal transcription, we find that in most cases it is not necessary. One point that may be important before cloning ORFs into vectors to be used for transcription is to limit the size of the 5′-untranslated region so that ATG codons are not present upstream of the initiation codon, as they may otherwise result in false priming of protein synthesis during the translation reaction.

An alternative procedure to cloning ORFs into vectors is to generate polymerase chain reaction (PCR) products of the ORF that includes an SP6 (or other) RNA polymerase promoter and a transcription initiation site within the 5′ primer so that the resultant PCR product can be used directly in the transcription reaction (Fig. 1). Furthermore, this approach can often result in higher transcript yields compared to an ORF cloned into a vector, as the PCR product is both short and linear (and thus has reduced secondary structure). In addition, one can use a modification of this method to perform very quick 5′ and 3′ deletions of the ORF to determine regions within the translated preprotein that are important for its mitochondrial biogenesis, e.g., to identify the presence of an N-terminal mitochondrial targeting signal (Fig. 1; Kurz *et al.,* 1999). The following section outlines a general approach for generating PCR products for their following use in *in vitro* transcription reactions.

1. Polymerase Chain Reaction Approach for Template Generation

a. Reagents

5′ primer including RNA polymerase promoter region upstream of the initiation codon (in bold), e.g., for SP6 5′ GGATT TAGGT GACAC TATAG AATAC **ATG** N_{15-18} and for T7 5′ TCTAA TACGA CTCAC TATAG GGAGA **ATG** N_{15-18}, where N stands for nucleotides complementary to the template DNA

3′ primer complementary to template either downstream of or spanning the stop codon

DNA template (cloned gene, cDNA, or total yeast genomic DNA)

Proof-reading thermostable DNA polymerase, e.g., Vent polymerase (New England Biolabs) or *Pfu* polymerase (Stratagene) and 10× PCR buffer

2.5 m*M* dNTP stock

A. Generation of template DNA by PCR

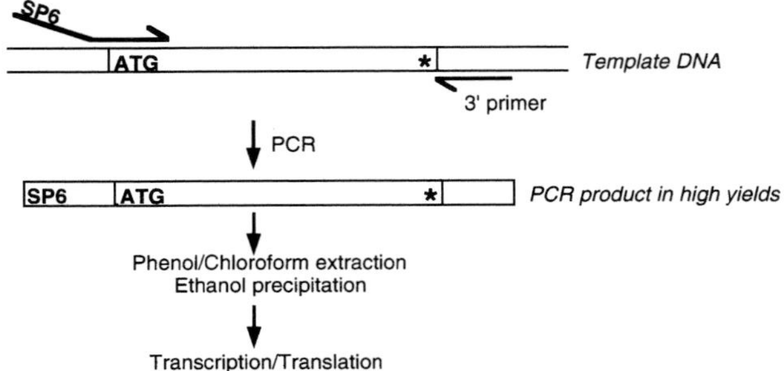

B. Generation of template DNA by PCR for C-terminal preprotein deletions

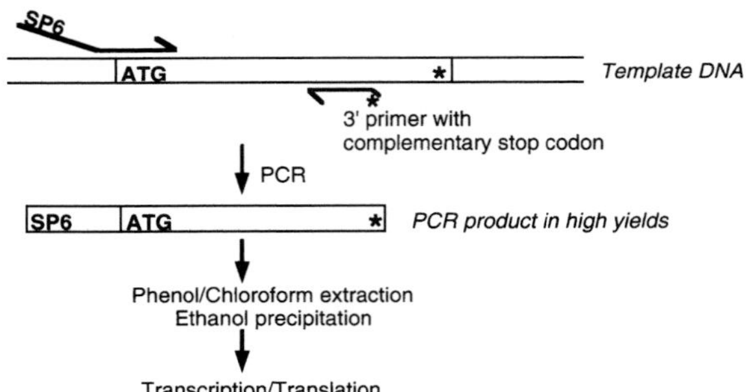

C. Generation of template DNA by PCR for N-terminal preprotein deletions

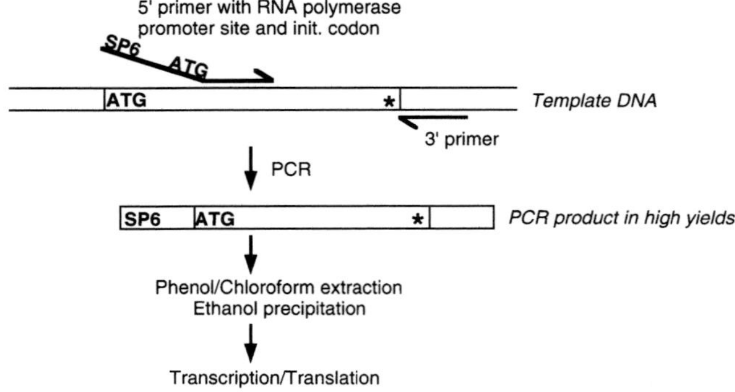

Sterile water
Phenol/chloroform

b. Procedure

The following reagents are added to a sterile PCR tube: 10 μl 10× PCR buffer, 10 μl 10 pmol/μl 5′ primer, 10 μl 10 pmol/μl 3′ primer, 5 μl 2.5 m*M* dNTPs, ~50 ng DNA template, H_2O to 100 μl, and 1–2 μl DNA polymerase (10 U/μl)

Overlay with mineral oil or equivalent and program the PCR cycling program with the following:

initial denaturation step:	94°C	2 min
1. denaturation step:	94°C	1 min
2. annealing step:	48–55°C	1 min
3. extension step:	72°C	2 min/1000-bp product (for Vent polymerase)
	steps 1–3 are cycled 25–30 times	
final extension:	94°C	5 min

An aliquot (5 μl) of the PCR product is tested for purity and abundance by agarose-gel electrophoresis prior to phenol/chloroform extraction and ethanol precipitation of the remainder. The PCR product is typically resuspended in 50 μl of sterile H_2O and half is used in a standard transcription reaction (see later).

B. *In Vitro* Transcription

Transcription reactions may be performed according to the instructions included with the appropriate RNA polymerase from the manufacturer. For the large-scale synthesis of RNA that we routinely perform in the laboratory, sufficient RNA quantities are obtained to synthesize approximately 1 ml of preprotein-containing lysate.

Fig. 1 Generation of PCR products for use as template in *in vitro* transcription reactions. (A) For the generation of full-length ORFs, a 5′ primer containing an RNA polymerase promoter (in this case SP6) upstream of a region complementary to the template DNA [encompassing or upstream of the initiation codon (ATG) of the ORF] is used in PCR along with a 3′ primer complementary to the template DNA [encompassing or downstream of the stop codon (*)]. The PCR product contains the complete ORF and includes the RNA polymerase promoter site for use in *in vitro* transcription. (B) The method in A is modified to include a 3′ primer that is instead complementary to a region upstream of the authentic stop codon. The primer contains an in-frame stop codon so that the PCR product used in *in vitro* transcription/translation results in a preprotein lacking the desired C-terminal region. (C) The 5′ primer containing the RNA polymerase promoter site is made complementary to a region downstream of the initiation codon. The primer therefore must contain an in-frame initiation codon in order to generate a PCR product that, following *in vitro* transcription and translation, will result in a preprotein lacking the desired N-terminal region.

1. Transcription Reaction for Large-Scale Synthesis of RNA *in Vitro*

a. Reagents

10× transcription buffer: 400 m*M* HEPES–KOH, pH 7.4, 60 m*M* $Mg(OAc)_2$, and 20 m*M* spermidine; store in 1-ml aliquots at −20°C

Transcription premix: 1 ml 10× transcription buffer, 20 μl 20 mg/ml fatty acid-free bovine serum albumin (BSA), 100 μl 1 *M* dithiothreitol (DTT), 50 μl 0.1 *M* ATP, 50 μl 0.1 *M* CTP, 50 μl 0.1 *M* UTP, 4 μl 0.1 *M* GTP, and 8.7 ml sterile H_2O; filter sterilize and store in 120-μl aliquots at −20°C

m^7(5′)ppp(5′)G (Amersham Pharmacia Biotech) made as a 1 m*M* stock concentration in sterile H_2O (it should be noted that many RNA transcripts do not require a cap structure in order to be translated efficiently, although we routinely add cap to our reactions)

Template DNA, e.g., purified plasmid or purified PCR product containing upstream RNA polymerase transcription initiation site

Appropriate RNA polymerase

RNasin ribonuclease inhibitor (Promega)

b. Procedure

20 μg plasmid DNA or ∼2 μg PCR product DNA template, 120 μl premix, 10 μl m^7GpppG, 5 μl RNasin (40 U/μl), and 2 μl RNA polymerase (50 U/μl); total volume of 200 μl adjusted with sterile H_2O

After incubation at 37°C for 1–2 h, RNA is precipitated by the addition of 20 μl of 10 *M* LiCl and 600 μl ethanol. The sample is placed at −20°C for at least 15 min prior to pelleting the RNA in a table-top centrifuge at maximum speed at 4°C for 30 min. The supernatant is discarded and residual ethanol is removed by a quick centrifugation before resuspending the pellet in 100 μl ribonuclease-free water containing 40 U RNasin. The RNA is stored in 25-μl aliquots at −80°C. RNA prepared in this way is in most cases stable for more than 1 year. Typically one RNA aliquot is used in a 250-μl translation reaction.

C. *In Vitro* Translation

1. General Considerations

Rabbit reticulocyte lysates are generally employed for the production of radiolabeled preproteins. Other translation systems that may be used, however, are yeast (Garcia *et al.,* 1991; Hönlinger *et al.,* 1995; Fünfschilling and Rospert, 1999) and wheat germ lysates (from Promega). The yeast translation system is not available commercially and is quite laborious to prepare, whereas the wheat germ lysates may lack the necessary chaperones to maintain preproteins in an import competent form (Hachiya *et al.,* 1993). Rabbit reticulocyte lysates are prepared by the manufacturer according to Pelham and Jackson (1976). Briefly, rabbits are injected with phenylhydrazine, which results in reticulocyte build-up. Reticulocytes are purified, lysed, and treated with micrococcal nuclease to destroy endogenous mRNA (of which hemoglobin-encoding transcripts are by far the majority). Other components (e.g., an ATP-regenerating system and additional tRNAs) are added to optimize translational efficiency.

2. Procedure

Translation reactions are performed in the presence of [^{35}S]methionine and cysteine (Pro-mix L-[^{35}S] *in vitro* cell-labeling mix from Amersham Pharmacia Biotech, specific activity >1000 Ci/mmol) at 30°C for 60–90 min prior to the addition of cold methionine to a final concentration of 5 m*M*. Ribosomes are pelleted (108,000*g*, 15 min, 2°C), and the supernatants are adjusted with sucrose to a final concentration of 250 m*M* (to make the lysate isotonic for mitochondria) prior to aliquoting and storage at −80°C. In some cases, it may be necessary to optimize the RNA and salt concentrations used in the translation reaction in order to obtain suitable amounts of translation product. Often, lower molecular weight translation products are observed due to the initiation of protein synthesis from internal methionine residues. In most cases this is not a problem as most of these N-terminally truncated preproteins lack the targeting information necessary for mitochondrial import. Increasing the concentration of $MgCl_2$ and KOAc in the translation reaction can help suppress internal initiation. Following the *in vitro* translation reaction, 1 μl of lysate is normally analyzed by SDS–PAGE and digital autoradiography (phosphorimaging or equivalent) to ascertain the success of the reaction. Radioactive marker proteins or the use of radioactive ink laid on top of conventional stained marker proteins serve as standards.

3. Ammonium Sulfate Precipitation

In some cases, the presence of the large amount of hemoglobin in the lysate, which runs at the 15- to 8-kDa range on SDS gels, may pose a hindrance to detecting radiolabeled proteins within this area, especially given that hemoglobin can be radiolabeled nonspecifically. In most cases, this is not a problem when an import reaction is performed, as hemoglobin does not bind to mitochondria whereas the preprotein should. However, there is no way to confirm that a preprotein has been synthesized in the translation reaction before performing import, and it is difficult to determine the efficiency of preprotein binding/import into mitochondria if one cannot quantitate the original amount of preprotein added to the reaction. Thus it may be necessary to perform an ammonium sulfate precipitation of the lysate following the translation reaction for those preproteins that are obscured by hemoglobin.

Generally, most preproteins will precipitate in ammonium sulfate solutions of 66% (v/v) saturation while the majority of hemoglobin remains in the supernatant. Two volumes of saturated ammonium sulfate are added to the preprotein-containing lysate. The sample is left on ice for 20 min prior to pelleting the precipitate in a table-top centrifuge at maximum speed for 15 min at 4°C. An additional advantage of this technique is that unincorporated radiolabel is also removed, resulting in a safer working environment. The pellet is then resuspended in its original volume in buffer prior to performing the import assay. Presumably most cytosolic factors that are important for promoting import are also precipitated with the preprotein, as no noticeable differences are observed in the import of most preproteins tested.

Ammonium sulfate precipitation of preproteins is also employed when more detailed import studies, such as analyzing the contribution of protein unfolding in import, are undertaken (Ostermann *et al.*, 1989). In this case, the ammonium sulfate pellet containing the radiolabeled preprotein is resuspended in buffer containing 8 *M* urea and the import characteristics are compared with the import of the preprotein that was resuspended in buffer containing no urea.

III. Standard Protocol for Protein Import into Isolated Mitochondria

Because most preproteins destined for mitochondria are translocated across the outer membrane, they become protected from externally added protease. In addition, most preproteins are translocated into the inner membrane or matrix and therefore require the presence of a membrane potential for their import. Matrix-targeted preproteins often contain an N-terminal presequence that is proteolytically removed upon import (in one or two steps), giving rise to a processed mature preprotein that can be separated readily from the precursor form by SDS–PAGE.

A. Isolation of Mitochondria

Mitochondria are prepared according to published procedures (see Chapters 1–3) and are used fresh in the case of mammalian or *Neurospora crassa* mitochondria. Yeast mitochondria are aliquoted at 5–10 mg/ml in SEM buffer (250 m*M* sucrose, 1 m*M* EDTA, 10 m*M* MOPS–KOH, pH 7.2) and stored at −80°C without deleterious effects to preprotein import.

B. Import Time Course Experiment

When starting with a new preprotein, it is always convenient to perform an import time course study to ascertain the appropriate kinetic range (i.e., a linear import rate), which is important for further characterization as described in the following sections. The assay described here involves the incubation of a radiolabeled preprotein with mitochondria and removing time point samples prior to splitting of each sample in two and treating one-half with protease. A typical additional control that is employed is to incubate the preprotein with mitochondria lacking a membrane potential.

1. Reagents and Methodology

a. Reagents

BSA buffer: 3% (w/v) fatty acid-free bovine serum albumin, 250 m*M* sucrose, 80 m*M* KCl, 5 m*M* $MgCl_2$, 2 m*M* KH_2PO_4, 5 m*M* methionine, 10 m*M* MOPS–KOH, pH 7.2 (stored frozen in aliquots)

ATP: 0.2 *M* stock in H_2O, titrated to pH 7.0 (stored frozen in aliquots)

AVO: 0.8 m*M* antimycin A, 0.1 m*M* valinomycin, 2 m*M* oligomycin in ethanol (stored at −20°C). Antimycin serves to block complex III of the respiratory chain; oligomycin blocks the F_0/F_1-ATPase; valinomycin is a potassium ionophore.

Valinomycin: 0.1 m*M* valinomycin in ethanol (stored at −20°C)

Proteinase K: 2.5 mg/ml freshly prepared in SEM buffer

Phenylmethylsulfonyl fluoride (PMSF): 200 m*M* stock in isopropanol (PMSF is stable in isopropanol for >6 months)

SEM buffer

NADH: 0.2 *M* NADH freshly prepared in H_2O. For protein import into mammalian mitochondria, NADH should be replaced with sodium succinate (pH 7.5) to a final concentration of 10 m*M*. Both of these reagents are used as respiratory substrates, which act to maintain the mitochondrial membrane potential. However, yeast mitochondria contain an intermembrane space-facing NADH dehydrogenase that transfers electrons from NADH to ubiquinone, whereas mammalian mitochondria lack such an enzyme. Because NADH cannot be transported across the inner membrane, succinate is used as the substrate for the reduction of ubiquinone by succinate:ubiquinone oxidoreductase.

b. Procedure

The logic behind the following experiment is to minimize pipetting errors involved with working with small volumes. Tube I contains enough sample to take four time point samples while still leaving a remainder. The amount of reticulocyte lysate containing ^{35}S-preprotein added to the import reaction ranges between 1 and 20% (v/v), although we typically use 2.5% (v/v).

	Tube I	Tube II
BSA buffer	828 μl	186 μl
0.2 *M* NADH	9 μl	—
0.2 *M* ATP	9 μl	2 μl
AVO (in ethanol)	—	2 μl
Ethanol	9 μl	—
Mitochondria (10 mg/ml)	22.5 μl	5 μl
Mix samples on ice		
Reticulocyte lysate containing ^{35}S-preprotein	22.5 μl	5 μl
Mix samples well and incubate at 25–30°C		

At time points 2, 5, 10, and 15 min, remove 200 μl of the import reaction from tube I and place each into a 1.5-ml tube containing 1 μl of valinomycin on ice and quickly mix.

It is important to pipette the reaction contents up and down at least once prior to removing each sample time point as mitochondria tend to sediment in suspension. In the case of most preproteins (i.e., those requiring a membrane potential), import is stopped by the addition to valinomycin. Even for those preproteins not requiring a membrane potential, import is slowed considerably at low temperatures. At 15 min, tube II is also placed on ice.

At this time there are 5 tubes on ice. Set up another 10 tubes on ice and split each of the original 200-μl samples in half (after mixing). To one set of tubes, add 2 μl of 2.5 mg/ml proteinase K (final concentration of 50 μg/ml). Mix the samples and incubate on ice for 10 min. Following protease treatment, add 1 μl of 200 m*M* PMSF and incubate for a further 10 min. Alternatively, trypsin can be substituted for proteinase K; it is used at concentration (20–50 μg/ml). Trypsin is inactivated by the addition of a

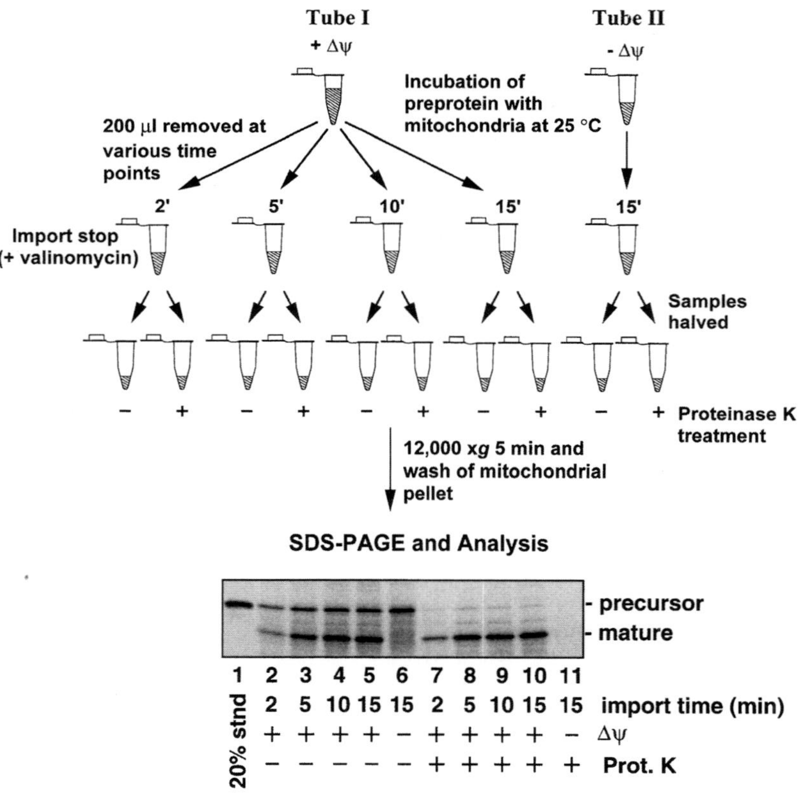

Fig. 2 *In vitro* assay for protein import into mitochondria. The preprotein pSu9-DHFR was incubated with yeast mitochondria at 25°C according to the protocol in Section III,b and as outlined in the schematic diagram. Mitochondria were isolated and analyzed by SDS–PAGE and phosphorimaging. Lane 1: 20% of total added preprotein to each import reaction is shown. The precursor and the proteolytically processed mature forms of Su9-DHFR are indicated.

30-fold excess of soybean trypsin inhibitor with thorough mixing. Following centrifugation (12,000*g*, 5 min, 4°C), the supernatants are removed carefully, taking care not to disrupt the very small mitochondrial pellets. The pellets are washed once in SEM buffer (without resuspension) and centrifuged as described earlier. Mitochondrial pellets are resuspended in SDS–PAGE loading dye (Laemmli, 1970) prior to heating at 95°C for 5 min and are subjected to SDS–PAGE. An aliquot of the original preprotein used in the import experiment (1–2 μl reticulocyte lysate) is loaded on the gel in a lane adjacent to the import reactions, to be used both as a standard and to quantitate the efficiency of import (most preproteins usually have a maximal import efficiency between 20 and 40% of total added). An example of such an import experiment is shown in Fig. 2 using a preprotein destined for the matrix compartment of mitochondria.

IV. Assaying Mitochondrial Localization of Imported Preproteins

The experiment just described may give a number of clues regarding the mitochondrial localization of the preprotein being tested (Table I). However, in order to narrow down the location of the preprotein more precisely, further experiments need to be performed. For example, suppose a preprotein is not processed and is degraded by protease treatment independent of a mitochondrial membrane potential. This preprotein may be inserted into the outer face of the outer membrane or it may be just nonspecifically associated with the mitochondrial fraction. Even if the preprotein is protected from proteases, indicating that it may be translocated into the mitochondrial intermembrane space or be buried

Table I
Experimental Variables Used to Determine the Location of a Preprotein Following Mitochondrial Import/Binding[a]

Proteolytically processed?	Resistant to external protease?	Requires $\Delta\Psi$?	Possible location	Further experiments
Yes	Yes	Yes	Matrix Inner membrane IMS	Sonication Carbonate extraction Swelling + protease
No	Yes	Yes	Matrix Inner membrane	Sonication Carbonate extraction Swelling + protease
No	Yes	No	OM (buried or facing IMS) IMS Protease-stable aggregate	Carbonate extraction Swelling + protease Detergent + protease
No	No	No	Outer face of OM Unspecific binding	Carbonate extraction

[a] $\Delta\Psi$, membrane potential; IMS, intermembrane space; OM, outer membrane.

within the outer membrane, it is equally likely that it has pelleted with the mitochondria as a protease-stable aggregate.

The following experiments assess the specificity of preprotein association with mitochondria and also give clues as to their final location within mitochondria. To perform these tests, the preprotein is imported into mitochondria for a length of time that ensures maximal import to the final location (e.g., 20 min). If the preprotein is protected from external protease following its import, then the nonimported species is first degraded with protease prior to subsequent analysis.

A. Extraction of Soluble Proteins

Most matrix and intermembrane space soluble proteins can simply be separated from membrane-bound proteins by sonicating mitochondrial extracts. A microtip sonifier is sufficient for such purposes. Following import, mitochondria are resuspended in 10 m*M* Tris–Cl, pH 7.6, 100 m*M* NaCl. Thereafter, mitochondria are subjected to sonication on ice in a sonicating water bath for 5 min or using a microtip sonifier (3×6 pulses; 40% duty cycle; microtip output setting 5). The membranes are centrifuged at 100,000*g* for 30 min at 4°C before removal of the supernatant containing soluble proteins. The pellet is resuspended in SDS–PAGE loading dye while the supernatant is subjected to TCA precipitation. Briefly, sodium deoxycholate is added to the supernatant to a final concentration of 0.0125% (w/v) before the addition of one-fifth volume of 72% (w/v) TCA. The sample is left on ice for 30 min and the precipitated protein is pelleted at >18,000*g* for 30 min prior to washing with ice-cold acetone and recentrifugation. The often invisible pellet is dried briefly, resuspended in SDS–PAGE loading dye, and incubated for ~10 min allow for solubilization of proteins prior to heating of the samples at 95°C for 5 min and subsequent SDS–PAGE analysis.

B. Determination of Membrane Integration

This technique involves the extraction of soluble and peripheral membrane proteins from integral membrane proteins through the use of carbonate extraction (Fujiki *et al.,* 1992). Following import, samples are split in half and mitochondrial pellets are washed and reisolated as described in Section III,b. One pellet is solubilized in SDS–PAGE loading dye while the other pellet is resuspended in 200 μl of freshly prepared 0.1 *M* sodium carbonate, pH 11.5. Following incubation on ice for 30 min, mitochondrial membranes are reisolated by ultracentrifugation at 100,000*g* for 30 min at 4°C. The pellet is resuspended in SDS–PAGE loading dye while the supernatant is subjected to TCA precipitation (Section IV,A). Samples are subjected to SDS–PAGE analysis, and the results of carbonate extraction are confirmed by immunodecorating with marker proteins (Fig. 3). Antibodies against known integral membrane (and hence carbonate resistant) mitochondrial proteins should be employed along with antibodies against known peripheral and soluble (and hence carbonate extractable) mitochondrial proteins.

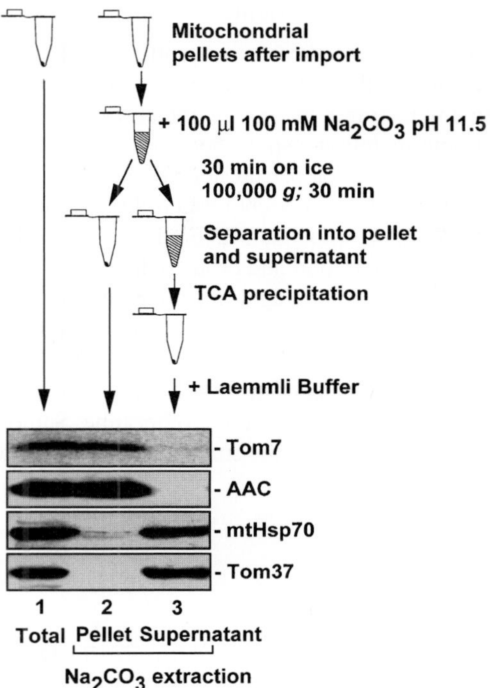

Fig. 3 Extraction of soluble and peripheral membrane proteins by carbonate extraction. *In vitro*-translated ^{35}S-labeled Tom7 was incubated with isolated mitochondria prior to performing the carbonate extraction (top). Samples were subjected to SDS–PAGE and Western blotting. Following phosphorimage analysis to identify Tom7, the blot was immunodecorated for the inner membrane protein, the ADP/ATP carrier (AAC), the soluble matrix protein mtHsp7O, and the peripheral outer membrane protein Tom37. Proteins that remain in the pellet following carbonate extraction are most likely to be integral membrane proteins.

C. Protease Degradation of Preproteins within the Mitochondria

In order to confirm that a preprotein is indeed protease protected due to its import across the outer mitochondrial membrane, it is important to show that, under some situations, it can be degraded. Detergent solubilization of mitochondrial membranes makes proteins accessible to externally added protease. Following the import assay, a mitochondrial pellet is resuspended in an appropriate volume of SEM buffer (e.g., 200 μl) and the sample is halved. To one sample, Triton X-100 [from a 10% (v/v) stock] is added to a final concentration of 0.5% (v/v) prior to the addition of proteinase K to both samples (10–50 μg/ml; a high concentration of proteinase K is not normally necessary to degrade the preprotein as the reticulocyte lysate and the BSA from the import buffer are absent from the solution, making the preprotein more accessible). If, however, the protein is still resistant to protease, it may be that it cosediments nonspecifically with the mitochondrial pellet as an aggregate. It is therefore important to perform an import experiment in the absence of mitochondria to exclude this possibility.

D. Mitochondrial Swelling Experiments

Lowering the sucrose concentration within a mitochondrial suspension results in the swelling of the mitochondrial matrix, leading to rupturing of the outer membrane but leaving the inner membrane intact. Following such treatment, proteins within or facing the intermembrane space become accessible to externally added protease, resulting in their complete or partial degradation (Fig. 4). Some soluble IMS proteins may even be released from mitochondria following such treatment and can therefore be separated from swollen mitochondria (or mitoplasts) following centrifugation (Martin *et al.*, 1999).

Following import and protease treatment, mitochondria are centrifuged at 10,000*g* for 5 min (higher centrifugal force may rupture some mitochondria) and are washed in SEM buffer and recentrifuged. Mitochondrial pellets are resuspended gently but thoroughly in 40 μl of SM buffer (250 m*M* sucrose, 10 m*M* MOPS–KOH, pH 7.2). The sample is split in half and 180 μl of 10 m*M* MOPS–KOH, pH 7.2, is added to one tube while 180 μl of SM buffer is added to the other tube. Samples are mixed and left on ice for 15 min. Following this time, the contents of each tube may be pipetted up and down 10 times in

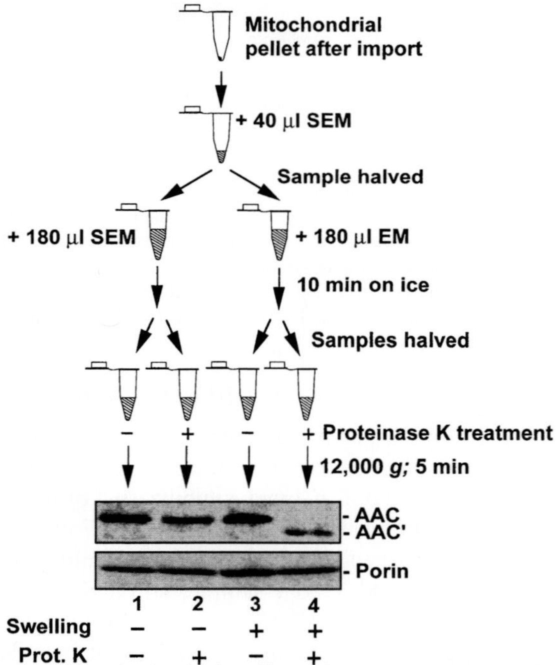

Fig. 4 Procedure employed for mitochondrial swelling and outer membrane disruption followed by protease treatment. *In vitro*-translated ^{35}S-labeled AAC and ^{35}S-labeled porin were incubated with mitochondria and then treated to mitochondrial swelling as outlined in the schematic diagram. Swelling and protease treatment results in the cleavage of a small portion of the inner membrane-embedded AAC, giving rise to a fragment (AAC'). The imported form of porin is embedded totally within the outer membrane and is therefore resistant to protease.

order to facilitate the rupturing of the outer membrane from the swollen mitochondria. Once again each of the samples are split in two and one of each sample is treated with proteinase K (20–50 μg/ml).

Following protease inactivation with 2 m*M* PMSF, mitochondria and mitoplasts are reisolated, washed, and analyzed by SDS–PAGE. Swelling efficiency can be measured by immunodecorating with mitochondrial proteins that are only accessible to protease treatment following swelling [e.g., the ADP/ATP carrier (Kübrich *et al.,* 1998) or IMS proteins such as cytochrome b_2 or the small Tim proteins (Kurz *et al.,* 1999)]. It is also important to test that swelling has not ruptured the inner membrane and so immunodecorating for the protease resistance of a matrix protein is recommended.

E. Cofractionation Studies

1. Digitonin Solubilization

This method utilizes differences in the solubility of the outer and inner mitochondrial membranes to the detergent digitonin (Hartl *et al.,* 1986). The outer membrane is solubilized at low digitonin concentrations compared to the inner membrane, thereby resulting in the progressive susceptibility of proteins within the different compartments to protease treatment.

a. Reagents

SEMK buffer: SEM buffer containing 100 m*M* KCl

Digitonin: Most preparations of digitonin contain impurities and it is therefore necessary to perform a recrystallization step. This is performed by solubilizing digitonin powder in boiling ethanol and then placing the solution at −20°C for 20 min. The digitonin precipitate is centrifuged at 5000*g* and the pellet is dried in a vacuum dessicator. The powder is kept in an air-tight container or made up into a 2% (v/v) solution in H_2O. Even following this procedure, impurities are often present and will precipitate in solution over time and when exposed to cooler temperatures.

b. Procedure

Following mitochondrial import and reisolation, mitochondrial pellets (50 μg) are resuspended well in 10 μl of SEMK buffer and are placed on ice. An equal volume of SEMK containing increasing concentrations of digitonin [0–1% (w/v)] is added to each of the samples and mixed well. Following incubation on ice for 1 min, samples are diluted with 80 μl SEMK buffer and treated with 25 μg/ml proteinase K as described earlier. Samples are precipitated with TCA and subjected to SDS–PAGE and immunoblotting (for monitoring success with antibodies against marker proteins) along with autoradiography.

2. Subfractionation

Following import and reisolation, mitochondria are swollen and homogenized in order to separate mitoplasts (matrix and inner membrane) from the intermembrane space fraction and also from a fraction of the outer membrane (a large fraction of outer membrane is retained in the mitoplast fraction). The sample is subjected to sucrose gradient centrifugation [see Chapters 2 and 3 and Pon *et al.* (1989)], and fractions are eluted from the gradient, precipitated by TCA, and subjected to SDS–PAGE. The gel is transferred to PVDF membranes for immunodecoration with antibodies against marker proteins from different mitochondrial subcompartments for comparison with the location of the radiolabeled, imported preprotein. The limitation of this technique is that a large amount of preprotein would be needed in order to detect it within the diluted protein fractions, and the submitochondrial fractions may often overlap. Mixing additional mitochondria with mitochondria used in the original import reaction may lead to better detection of the submitochondrial fractions.

V. Energetic Requirements for Preprotein Import

The disruption of the mitochondrial membrane potential ($\Delta\Psi$) prevents preproteins from engaging productively with the TIM machinery. However, some preproteins can still traverse the mitochondrial outer membrane in the absence of a $\Delta\Psi$, thereby protecting them from externally added protease. These preproteins include members of the carrier protein family, such as the ADP/ATP carrier (AAC), which utilizes an import pathway that differs from preproteins containing N-terminal presequences. In the absence of a $\Delta\Psi$, AAC is associated with the TOM machinery at the inner face of the outer membrane, where it can be degraded following mitochondrial swelling and protease treatment. In the presence of a $\Delta\Psi$, AAC is inserted into the inner membrane, whereby it is largely protected from protease following mitochondrial swelling (Pfanner and Neupert, 1987; Kübrich *et al.,* 1998).

Many preproteins require not only a membrane potential for their import into mitochondria, but also the presence of nucleotide triphosphates. External ATP is required for a subset of preproteins that may utilize cytosolic (or reticulocyte lysate) cofactors for their import, whereas internal (or matrix) ATP is utilized by mtHsp70, which acts to drive the forward movement of the preprotein across the outer and inner mitochondrial membranes. To maintain high ATP levels during import, an ATP regenerating system, consisting of 0.5 m*M* ATP, 5 m*M* creatine phosphate, and 100 μg/ml creatine kinase, is added. This system can also contribute to maintaining the mitochondrial membrane potential through electrogenic exchange of ATP and ADP by the inner membrane AAC (Rowley *et al.,* 1994). An elaborate assay that manipulates either or both of the internal and external mitochondrial ATP levels has been developed by Glick (1995). More simpler methods to deplete ATP levels are to add 25 U/ml apyrase to the mitochondrial and reticulocyte lysate samples separately and preincubate for 10 min at 25°C prior to import. The addition of efrapeptin (2 μg/ml) and/or oligomycin (5 μg/ml) to mitochondria will block further activity by the F_0/F_1-ATPase, whereas the addition of 25 μg/ml atractyloside acts to inhibit AAC from transporting ATP and ADP across the inner membrane.

VI. Assaying Import Pathways

Further characterization of preprotein biogenesis can be performed by analyzing which members of the translocation machinery are utilized for the import of a preprotein into mitochondria.

A. Bypass Import

One approach in the characterization of preprotein import is to analyze whether the TOM receptors are strictly required. To perform this experiment, the receptor domains are cleaved off from their hydrophobic membrane anchors by mild protease digestion of mitochondria prior to performing the import assay [termed bypass import (Pfaller *et al.,* 1989)]. In yeast mitochondria, trypsin treatment (20 μg/ml) removes the cytosolic domains of the receptors Tom70, Tom20, and Tom22, whereas the small receptor protein Tom5 and the import channel protein Tom40 are left intact (Dietmeier *et al.,* 1997; Hill *et al.,* 1998; Fünfschilling and Rospert, 1999). In most cases, the import of preproteins into mitochondria lacking TOM receptors is reduced, although the import of some preproteins such as apocytochrome *c* (Nicholson *et al.,* 1988) seems to be unaffected.

1. Methodology

a. Reagents

SEM buffer

Soybean trypsin inhibitor (SBTI) solubilized in SEM buffer at 30 mg/ml

Trypsin solubilized in SEM buffer at 1 mg/ml

b. Procedure

Two tubes, each containing 100 μg of mitochondria in 500 μl SEM buffer, are set up on ice, and 10 μl of 30 mg/ml SBTI is added to the first tube (serving as the negative control) and mixed thoroughly before the addition of 10 μl of 1 mg/ml trypsin to both tubes. Samples are incubated on ice for 10 min prior to the addition of 10 μl of SBTI to the second tube and both tubes are incubated for a further 10 min on ice. Mitochondria are pelleted (10,000*g*, 5 min at 4°C) and washed with SEM buffer in the presence of 1 mg/ml SBTI to remove traces of trypsin prior to performing import analysis (Section III).

B. Inhibition of Receptor Activity by Antibody Preincubation

Preincubation of mitochondria with antibodies raised against the different TOM components prior to import analysis has been a traditional means to determine both receptor functions and preprotein import pathways (Ohba and Schatz, 1987; Vestweber *et al.,*

1989; Hines *et al.,* 1990; Söllner *et al.,* 1989, 1990, 1991; Kiebler *et al.,* 1993; Dietmeier *et al.,* 1997). Large amounts of antibodies are typically required to inhibit preprotein import reactions. Isolated mitochondria are mixed with a 5- to 15-fold weight excess of purified IgGs (resuspended in SEM buffer) and are incubated on ice for 30 min. Import assays can be performed directly or following reisolation of mitochondria and removal of excess IgG. Due to their divalent nature, the IgGs may sterically hinder import indirectly of their binding to TOM receptors. The use of Fab fragments instead of antibodies may circumvent such a problem (Kiebler *et al.,* 1993; Alconada *et al.,* 1995). Incubation of mitochondria with preimmune antibodies or with antibodies against outer membrane proteins not involved in protein import (e.g., porin/VDAC) serve as negative controls.

C. Protein Import into Yeast Mutants with Defects in Protein Import

1. Defects in the TOM Machinery

Perhaps the most effective means of analyzing the receptor requirement of a preprotein in question is to perform protein import studies in mitochondria isolated from yeast strains lacking various receptors (as a result of genetic manipulations). Yeast mutants lacking the receptors Tom70, Tom20, and Tom5 are viable, although the expression of Tom22 is reduced in *tom20*Δ cells, leading to a strong growth phenotype [which can be overcome by restoring the wild-type levels of Tom22 through overexpression (Dekker *et al.,* 1998)]. In some yeast strains, deletion of the *TOM22* gene is possible (van Wilpe *et al.,* 1999). Other mutants have been used to analyze the requirement of the general import pore complex in protein import by using strains lacking the small TOM proteins or by using temperature-sensitive mutations in the channel protein Tom40 (Kassenbrock *et al.,* 1993; Schleiff *et al.,* 1999). In performing such studies, it is important to carry out a time course import study within the linear import range predetermined for the preprotein and to use mitochondria isolated from the corresponding wild-type strain as a control. Mitochondrial preparations to be compared should always be isolated at the same time, and the same concentrations of mitochondria must be used in order for such comparisons to be valid. Finally, it is recommended to include control mitochondrial preproteins with previously determined receptor dependencies in such studies (Kurz *et al.,* 1999). Preproteins usually show a reduced import in mitochondria lacking its preferred receptor, but because the TOM receptors possess overlapping specificities for preproteins, import is not totally inhibited. One must also bear in mind that the use of yeast mutants in these studies is only truly valid for studying the import pathway of those preproteins that are encoded by yeast genes, and not those of other species. While TOM homologs are found in other organisms, they may differ in their preprotein-binding specificity.

2. Defects in the TIM Machinery

Almost all known components of the TIM machinery are encoded by essential genes. Most studies have therefore utilized yeast TIM mutants that have been screened for temperature-sensitive growth defects. A defective import of preproteins *in vitro* is still often observed when mitochondria are isolated from such mutants, grown under temperatures permissive for growth (Dekker *et al.,* 1997; Bömer *et al.,* 1998; Koehler *et al.,* 1998). In another approach, the gene encoding the desired TIM protein is placed under the control of a galactose-inducible promoter. Shifting of yeast cells to growth on glucose represses the synthesis of the TIM protein, resulting in its eventual depletion in mitochondria. Mitochondria harvested during this time may have a defective protein import machinery (Sirrenberg *et al.,* 1996, 1998), although there may be pleiotropic defects associated with such depletion.

3. Defects in mtHsp70

Temperature-sensitive yeast strains containing mutations in the *SSC1* gene, which encodes mtHsp70, have been generated (Gambill *et al.,* 1993; Voos *et al.,* 1993). Mitochondria isolated from these strains and the corresponding wild-type strains, when grown at 23°C, have no obvious defects in the *in vitro* import of preproteins. However, when these mitochondria are pretreated with a 37°C heat shock for 15 min prior to import studies at 25°C, the import of many preproteins that require mtHsp70 for their import becomes inhibited. These mutants are therefore ideal to study the importance of preprotein unfolding prior to their import and also to analyze the import pathway of a preprotein into the matrix or inner membrane (Gambill *et al.,* 1993; Voos *et al.,* 1993; Voisine *et al.,* 1999; Kurz *et al.,* 1999).

D. Formation of Import Intermediates

1. Low-Temperature Intermediates

When a preprotein is incubated with mitochondria for short times and at low temperatures (0–10°C), it may engage with the import machinery but be trapped at some step within the import pathway. These import intermediates may be either bound to components of the TOM machinery, spanning the TOM and TIM machinery, or partially inserted/assembled within the appropriate membrane (Schleyer and Neupert, 1985; Söllner *et al.,* 1991; Rapaport and Neupert, 1999). The characterization of import intermediates can be performed through techniques such as protease digestion studies, carbonate extraction, blue-native PAGE (see Section VII), and chase assays. It is possible to test for the association of import intermediates with mitochondrial import machinery by performing cross-linking analysis and immunoprecipitations (see Chapter 13).

2. Nucleotide-Dependent Intermediates

Some preproteins require external ATP for their translocation across the outer membrane, in some cases for the release of cytosolic chaperones from the preprotein. For the ADP/ATP carrier, depletion of external ATP from the lysate (Section V) results in its stable association with Tom70 that can be observed by blue-native PAGE (Ryan *et al.,* 1999).

The addition of ATP and incubation at 25°C can chase the preprotein across the outer membrane and into a protease-protected location. For preproteins requiring matrix ATP, depletion of internal ATP can result in a preprotein that has partially inserted into the matrix, leading to the removal of its presequence. However, due to the inability of mtHsp70 to pull the remainder of the preprotein across the mitochondrial membranes, the preprotein can be found arrested between the TOM and TIM machinery, nested in the TIM machinery, or it can even slip out of the TIM channel and be trapped in the intermembrane space.

3. Fusion Protein Intermediates

Most preproteins must be in a relatively extended state in order to be imported into mitochondria. Thus, the addition of a protein domain capable of being tightly folded to the C-terminal end of a preprotein can result in its arrest in the mitochondrial TOM and TIM channels. The most commonly employed fusion domain is dihydrofolate reductase (DHFR), which is maintained in a tightly folded state by the addition of methotrexate, a folate analogue. Fusion of a mitochondrial presequence to the N terminus of DHFR leads to its targeting and import into mitochondria. In the presence of methotrexate, however, translocation of the fusion protein across the outer membrane is prevented (Eilers and Schatz, 1986). When the preprotein region of the fusion construct is long enough to traverse both mitochondrial membranes, then processing of the mitochondrial presequence can occur by matrix proteases, and the fusion construct can be prevented from slipping out of mitochondria due to its binding to Tim23 and/or mtHsp70 (Dekker *et al.,* 1997).

A recombinantly expressed preprotein–DHFR construct can be added to mitochondria in saturating amounts (Dekker *et al.,* 1997). In this case, the preprotein blocks all Tim23-containing translocation channels, but because the TOM channels are in excess, a subset of preproteins are still capable of being imported across the outer membrane (in most cases, preproteins directed to the matrix are not found stably associated with the TOM machinery without a functional TIM machinery, and can slip out). Preproteins destined for the outer membrane or IMS, or those which utilize the TIM22 machinery, are still import competent under these conditions (Dekker *et al.,* 1997; Kurz *et al.,* 1999).

VII. Assaying Protein Assembly by Blue-Native Electrophoresis

An additional means of testing import of proteins into mitochondria, especially those destined for mitochondrial membranes, is to analyze the assembly of the preprotein into its macromolecular complex. This is achieved most readily by performing blue-native electrophoresis following the import assay (Schägger and von Jagow, 1991) (see also Chapter 14). It is important to use a detergent that is gentle in solubilizing mitochondrial membranes but that does not disrupt the complexes to a great degree. Typically, following the import assay, we solubilize mitochondrial pellets (25–100 μg protein) in 50 μl ice-cold digitonin buffer [1% (w/v) digitonin, 20 m*M* Tris–C1, pH 7.4, 0.1 m*M* EDTA, 50 m*M*

NaCl, 10% (v/v) glycerol, 1 m*M* PMSF]. After a clarifying spin (12,000*g* for 5 min at 4°C), 5 μl of sample buffer [5% (w/v) Coomassie brilliant blue G-250, 100 m*M* Bis–Tris, pH 7.0, 500 m*M* 6-aminocaproic acid] is added and the samples are electrophoresed at 4°C through a 6–16 or 4–16% polyacrylamide gradient gel (molecular mass range of 10–1000 kDa).

For immunoblotting, the cathode buffer is exchanged midway during electrophoresis with buffer lacking Coomassie dye, as the removal of excess dye aids in protein transfer. Following electrophoresis, the native gel is soaked in blotting buffer [20 m*M* Tris–base, 150 m*M* glycine, 20% (v/v) methanol, 0.08% (w/v) SDS] for 20 min prior to electroblotting onto a PVDF membrane.

In some cases, it may be necessary to perform a second-dimensional SDS–PAGE step in order to test for other subunits found in complex with the imported protein (Dietmeier *et al.,* 1997; Dekker *et al.,* 1998). The desired lane is excised from the blue-native gel, and the blue migration front at the bottom (containing dye and detergent) is removed. The gel strip is sandwiched between two SDS–PAGE glass plates and the separation gel is poured, leaving space for the stacking gel. Following polymerization, the stacking gel is poured carefully with the gel tilted in order to prevent bubbles from being trapped under the blue-native gel strip. A comb containing two to three lanes is also placed in the stacking gel at the side of the gel strip in order to load molecular weight markers and a mitochondrial import reaction in the first dimension for comparison.

It must be emphasized that many preproteins do not form large complexes and/or may not be resolved using this system. However, blue-native PAGE has been a particularly useful technique to analyze the assembly of preproteins of members of the TOM (Dekker *et al.,* 1998; Dietmeier *et al.,* 1997; Rapaport and Neupert, 1999) and TIM (Dekker *et al.,* 1998; Bömer *et al.,* 1997; Kurz *et al.,* 1999) machinery, and also for the outer membrane protein porin (unpublished observations), as well as inner membrane carrier proteins, such as the ADP/ATP carrier (Ryan *et al.,* 1999). Typically, such preproteins lack cleavable targeting signals and therefore additional approaches, such as BN–PAGE, enable a better understanding into their import and biogenesis.

VIII. Import of Purified Recombinant Proteins into Mitochondria

The amount of radiolabeled preprotein used in standard import experiments is certainly very low. A rough assessment indicates that the concentration of preproteins in reticulocyte lysate is ~2 pmol/ml (Rassow *et al.,* 1989). This indicates that the maximum amount reached in a standard import reaction is about 1 pmol/mg mitochondria. Although the synthesis of preproteins by *in vitro* translation is a fast and convenient method, it is limited to the analytical approach. It remains unclear how far the conclusions obtained by the substoichiometric amounts of preprotein, produced by *in vitro* translation, are applicable to the *in vivo* situation.

However, the use of a recombinantly expressed preprotein purified from *Escherichia coli* results in unlimited amounts of preprotein. Indeed, our studies indicate that the import machinery of the inner membrane can be saturated under these conditions (Dekker *et al.,* 1997). As was determined by titration of the amount of accumulated translocation intermediates, the total amount of import sites in the inner membrane is about 15–20 pmol/mg mitochondria (Bömer *et al.,* 1998). Routinely, we use a preprotein concentration of about 650 pmol/mg mitochondria to reach saturation conditions for import. On average, import rates *in vitro* reach about 10–15 pmol imported protein/min/mg mitochondria.

Import of recombinant preproteins can be performed using radioactive-labeled preproteins or with unlabeled preproteins that are detected immunologically. Usually, the second approach is chosen due to its experimental convenience. An additional advantage is the possibility of obtaining reasonably good quantitative data by using standardized immunoblots in comparison with the imported preprotein (see later). However, the consequence of using the nonradioactive approach is that no naturally occurring mitochondrial precursor proteins can be used, as the imported preprotein must be identified in the background of total mitochondrial protein. For preproteins, we use artificial fusion proteins containing the mitochondrial targeting signal of a preprotein (cytochrome b_2) fused to the complete mouse dihydrofolate reductase protein. The fusion protein is imported and acquires protease resistance just like any mitochondrial preprotein. Because no DHFR-like proteins exist in mitochondria, the imported DHFR can be identified easily.

A. Preparation of Preprotein

The structure of a representative fusion protein is shown in Fig. 5A. The amino-terminal cytochrome b_2 portion supplies the presequence. In this case, the presequence carries a 19 residue deletion that results in the deactivation of the second part of the presequence that usually directs cytochrome b_2 into the intermembrane space. All fusion proteins with this deletion behave as matrix-targeted mitochondrial precursors. In the matrix, these preproteins are proteolytically processed twice: one is performed by the standard peptidase MPP and the other represents the removal of an octapeptide by another specific peptidase.

The coding sequences of these fusion proteins are cloned into the *E. coli* expression vector pUHE, which places the open reading frame under the control of an IPTG-inducible promoter.

1. Expression in *E. coli*

E. coli cells containing the relevant plasmids are grown in LB medium (1% tryptone, 0.5% yeast extract, 0.5% NaCl) at 37°C to an OD_{600} of 1. IPTG is added to a final concentration of 1 m*M* to induce expression of the fusion protein. After 2 h of shaking at 37°C, cells are harvested and washed in wash buffer (30% sucrose, 20 m*M* KP_i, pH 8.0, 1 m*M* EDTA, 10 m*M* DTT, 1 m*M* PMSF). After resuspension in buffer A (20 m*M* MOPS–KOH, pH 8.0, 1 m*M* EDTA), lysozyme is added to a concentration of 1 mg/ml and cells are incubated on ice for 10 min prior to the addition of 0.1% Triton X-100.

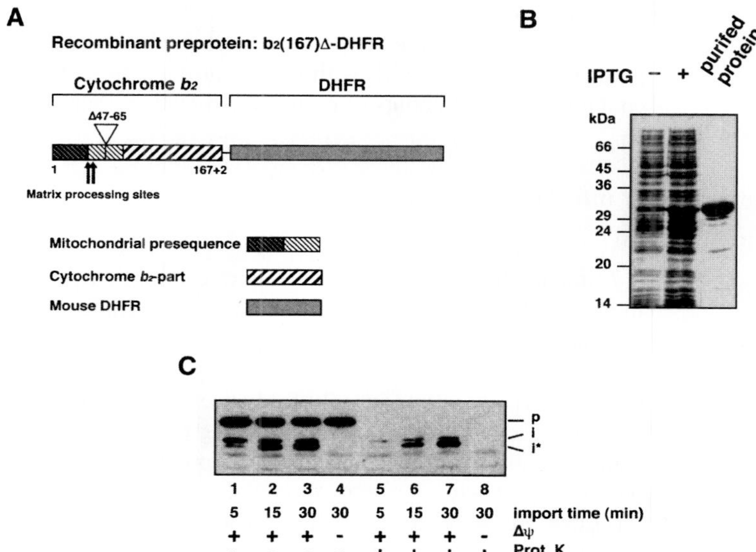

Fig. 5 Import of a purified recombinant preprotein into mitochondria. (A) Schematic diagram showing the fusion construct of the preprotein $b_2(167)\Delta$-DHFR used in import. (B) Coomassie-stained SDS–PAGE showing expression and purification of $b_2(167)$ Δ-DHFR. (C) *In vitro* import of $b_2(167)$ Δ-DHFR into mitochondria in the presence or absence of a membrane potential (ΔΨ) followed by protease treatment. Samples were subjected to SDS–PAGE and were subsequently immunodecorated with antibodies against DHFR. p, precursor; i, i^* processed protein.

The suspension is incubated for a further 10 min on ice prior to sonification (Branson Sonifier with 3 × 20 pulses; 40% duty cycle; microtip setting 7). Cell debris is pelleted by centrifugation at 15,000*g* for 20 min at 4°C.

2. Purification

Cyt b_2–DHFR fusion proteins have a high p*I* and therefore have an overall positive charge, making them ideal substrates for the purification via cation-exchange chromatography. Supernatants are filtered through a 0.2-μm filter and then loaded onto an FPLC Mono S column (Amersham Pharmacia Biotech) equilibrated previously in buffer A. Most of the contaminating proteins are removed already in the flow-through and by washing with 3 column volumes of buffer A; bound proteins are then eluted by a NaCl gradient in buffer A. The fusion protein elutes at about 300 m*M* NaCl. The average purity of the recombinant preprotein is greater than 95%, as judged by SDS–PAGE analysis (Fig. 5B) The most prominent biochemical feature is that the DHFR moiety is folded and is resistant to treatment with proteinase K, similar to DHFR alone. The purified proteins are concentrated and stored in small aliquots at −80°C.

Protein amounts are measured by the Bradford protein assay (Bio-Rad) or, more precisely, by comparing the Coomassie-stained profile of a dilution series of the purified preprotein on SDS–PAGE with marker proteins from an electrophoresis calibration kit containing exact amounts of proteins (Amersham Pharmacia).

B. Protein Import into Mitochondria

1. Import Reaction

Mitochondria are prepared and import reactions are performed essentially as described in Section III. Recombinant preproteins can be diluted to the appropriate concentration in water just before the import reaction. To obtain saturating conditions (see earlier discussion), about 600 pmol recombinant preprotein is used per import assay with 25 μg mitochondria, resulting in a 30-fold excess of precursor over inner membrane import sites. In cases where membrane-spanning translocation intermediates are accumulated using a tightly folded DHFR domain, preproteins are preincubated for 5 min ice directly before import with 4 μ*M* methotrexate to stabilize the DHFR conformation. Import reactions are stopped by cooling on ice and by the addition of 1 μ*M* valinomycin, destroying the $\Delta\Psi$. Complete import is assessed by proteinase K treatment.

2. Detection

After separation of mitochondrial proteins by SDS–PAGE; proteins are blotted on nitrocellulose or PVDF membranes. Imported DHFR fusion proteins are detected by immunodecoration with antibodies against DHFR. The use of affinity-purified antibodies substantially improves the quality and sensitivity of the immunoblot reaction.

A typical result of an import reaction is shown in Fig. 5C. In our hands, recombinant preproteins are imported *in vitro* with high efficiency and show all typical characteristics of a successfully imported preprotein, such as binding to the mitochondrial surface, membrane potential-dependent specific processing (in this case to the intermediate forms i and i^*), and protease resistance of the processed forms.

a. Affinity Purification of Antibodies

To prepare affinity-purified antibodies, the antigen is first coupled to CNBR–Sepharose (Amersham Pharmacia). The dry material is resuspended and washed three times with a large volume of 1 m*M* HCl. The wet Sepharose is transferred immediately into 10 volumes of coupling buffer (0.1 *M* $NaHCO_3$, pH 8.3, 0.5 *M* NaCl) before the addition of 5–10 mg purified antigen per gram dry Sepharose. The slurry is incubated with gentle agitation for 12 h at 4°C. The Sepharose is washed once in coupling buffer, and free activated groups are blocked by a 2-h incubation with 0.1 *M* Tris–HCl, pH 8.0. The Sepharose is then washed three times with 0.5 *M* NaCl, 0.1 *M* sodium acetate, pH 4.0, and three times with 0.5 *M* NaCl, 0.1 *M* Tris–HCl, pH 8.0. The slurry is packed into small gravity-flow columns, washed with Tris-buffered saline (TBS), and stored with 0.01% NaN_3 at 4°C.

Depending on the titer of the antiserum, between 5 and 20 ml of antiserum is diluted in TBS and passed over the antigen column at room temperature. After washing with 5–10 column volumes, the bound antibodies are eluted by adding 1 column volume of 100 m*M* glycine–HCl, pH 2.5. Fractions of approximately 1 ml are collected and immediately neutralized with 100 μl 1 *M* Tris–HCl, pH 7.4. The fractions are-tested in diluted form for reactivity on immunoblots and are stored together with 0.1 mg/ml BSA and 0.01% NaN_3 at −20°C.

b. Standardized Western Blot

To obtain data for the absolute amounts of imported preproteins, a dilution series of pure preprotein (1–10% of the added amount) is run on the SDS–PAGE gel and used as a standard value to be compared with the signal after import. After immunodecoration with affinity-purified anti-DHFR antibodies (1:500 diluted), proteins are detected by chemiluminescence (ECL; Amersham Pharmacia). Several exposure times are required to avoid overexposure and saturation of the signal. Signals are quantified using an ImageMaster scanner system (Amersham Pharmacia).

Acknowledgments

We thank A. Geissler, B. Guiard, F. Martin, and K. Model for helpful suggestions and materials. Work in the authors' laboratory was supported by grants from the Deutsche Forschungsgemeinschaft, the Sonderforschungsbereich 388, and the Fonds der Chemischen Industrie.

References

Alconada, A., Gärtner, F., Hönlinger, A., Kübrich, M., and Pfanner, N. (1995). Mitochondrial receptor complex from *Neurospora crassa* and *Saccharomyces cerevisiae. In* "Methods of Enzymology" (G. M. Attardi and A. Chomyn, eds), Vol. 260, pp. 263–286. Academic Press, San Diego.

Bömer, U., Maarse, A. C., Martin, F., Geissler, A., Merlin, A., Schönfisch, B., Meijer, M., Pfanner, N., and Rassow, J. (1998). Separation of structural and dynamic functions of the mitochondrial translocase: Tim44 is crucial for the inner membrane import sites in translocation of tightly folded domains, but not of loosely folded preproteins. *EMBO J.* **17,** 4226–4237.

Bömer, U., Meijer, M., Maarse, A. C., Dekker, P. J. T., Pfanner, N., and Rassow, J. (1997). Multiple interactions of components mediating preprotein translocation across the inner mitochondrial membrane. *EMBO J.* **16,** 2205–2216.

Dekker, P. J. T., Martin, F., Maarse, A. C., Bömer, U., Müller, H., Guiard, B., Meijer, M., Rassow, J., and Pfanner, N. (1997). The Tim core complex defines the number of mitochondrial translocation contact sites and can hold arrested preproteins in the absence of matrix Hsp70-Tim44. *EMBO J.* **16,** 5408–5419.

Dekker, P. J. T., Ryan, M. T., Brix, J., Müller, H., Hönlinger, A., and Pfanner, N. (1998). The preprotein translocase of the outer mitochondrial membrane: Molecular dissection and assembly of the general import pore complex. *Mol. Cell. Biol.* **18,** 6515–6524.

Dietmeier, K., Hönlinger, A., Bömer, U., Dekker, P. J. T., Eckerskorn, C., Lottspeich, F., Kübrich, M., and Pfanner, N. (1997). Tom5 functionally links mitochondrial preprotein receptors to the general import pore. *Nature* **388,** 195–200.

Eilers, M., and Schatz, G. (1986). Binding of a specific ligand inhibits import of a purified precursor protein into mitochondria. *Nature* **322,** 228–232.

Fujiki, Y., Fowler, S., Shio, H., Hubbard, A. L., and Lazarow, P. B. (1982). Polypeptide and phospholipid composition of the membrane of rat liver peroxisomes: Comparison with endoplasmic reticulum and mitochondrial membranes. *J. Cell Biol.* **93,** 103–110.

Fünfschilling, U., and Rospert, S. (1999). Nascent polypeptide-associated complex stimulates protein import into yeast mitochondria. *Mol. Biol. Cell* **10,** 3289–3299.

Gambill, D., Voos, W., Kang, P. J., Miao, B., Langer, T., Craig, E. A., and Pfanner, N. (1993). A dual role for mitochondrial heat shock protein 70 in membrane translocation of preproteins. *J. Cell Biol.* **123,** 109–117.

Garcia, P. D., Hansen, W., and Walter, P. (1991). In vitro protein translocation across microsomal membranes of *Saccharomyces cerevisiae. In* "Methods of Enzymology" (C. Guthrie and G. R. Fink, eds.), Vol. 194, pp. 675–682. Academic Press, San Diego.

Glick, B. S. (1995). Pathways and energetics of mitochondrial import in *Saccharomyces cerevisiae. In* "Methods of Enzymology" (G. M. Attardi and A. Chomyn, eds.), Vol. 260, pp. 224–231. Academic Press, San Diego.

Hachiya, N., Alam, R., Sakasegawa, Y., Sakaguchi, M., Mihara, K., and Omura, T. (1993). A mitochondrial import factor purified from rat liver cytosol is an ATP-dependent conformational modulator for precursor proteins. *EMBO J.* **12,** 1579–1586.

Hartl, F. U., Schmidt, B., Wachter, E., Weiss, H., and Neupert, W. (1986). Transport into mitochondria and intramitochondrial sorting of the Fe/S protein of ubiquinol-cytochrome c reductase. *Cell* **47,** 939–951.

Hill, K., Model, K., Ryan, M. T., Dietmeier, K., Martin, F., Wagner, R., and Pfanner, N. (1998). Tom40 forms the hydrophilic channel of the mitochondrial import pore for preproteins. *Nature* **395,** 516–521.

Hines, V., Brandt, A., Griffiths, G., Horstmann, H., Brütsch, H., and Schatz, G. (1990). Protein import into yeast mitochondria is accelerated by the outer membrane protein MAS 70. *EMBO J.* **9,** 3191–3200.

Hönlinger, A., Keil, P., Nelson, R. J., Craig, E. A., and Pfanner, N. (1995). Posttranslational mitochondrial protein import in a homologous yeast in vitro system. *Biol. Chem.* **376,** 515–519.

Kassenbrock, C. K., Cao, W., and Douglas, M. G. (1993). Genetic and biochemical characterization of ISP6, a small mitochondrial outer membrane protein associated with the protein translocation complex. *EMBO J.* **8,** 3023–3034.

Kiebler, M., Keil, P., Schneider, H., van der Klei, I. J., Pfanner, N., and Neupert, W. (1993). The mitochondrial receptor complex: A central role of MOM22 in mediating preprotein transfer from receptors to the general insertion pore. *Cell* **74,** 483–492.

Koehler, C. M., Jarosch, E., Tokatlidis, K., Scmid, K., Schweyen, R. J., and Schatz, G. (1998). Import of mitochondrial carriers mediated by essential proteins of the imtermembrane space. *Science* **279,** 369–373.

Koehler, C. M., Merchant, S., and Schatz, G. (1999). How membrane proteins travel across the intermembrane space. *Trends Biochem. Sci.* **24,** 428–432.

Kübrich, M., Rassow, J., Voos, W., Pfanner, N., and Hönlinger, A. (1998). The import route of ADP/ATP carrier into mitochondria separates from the general import pathway of cleavable preproteins at the *trans* side of the outer membrane. *J. Biol. Chem.* **273,** 16374–16381.

Kurz, M., Martin, H., Rassow, J., Pfanner, N., and Ryan, M. T. (1999). Biogenesis of Tim proteins of the mitochondrial carrier import pathway: Differential targeting mechanisms and crossing-over with the main import pathway. *Mol. Biol. Cell* **10,** 2461–2474.

Laemmli, U. K. (1970). Cleavage of structural proteins during the assembly of the head bacteriophage T4. *Nature* **227,** 680–685.

Lill, R., Stuart, R. A., Drygas, M. E., Nargang, F. E., and Neupert, W. (1992). Import of cytochrome c heme lyase into mitochondria: A novel pathway into the intermembrane space. *EMBO J.* **11,** 449–456.

Martin, H., Eckerskorn, C., Gärtner, F., Rassow, J., and Pfanner, N. (1999). The yeast mitochondrial intermembrane space: Purification and analysis of two distinct fractions. *Anal. Biochem.* **265,** 123–128.

Neupert, W. (1997). Protein import into mitochondria. *Annu. Rev. Biochem.* **66,** 863–917.

Nicholson, D. W., Hergesberg, C., and Neupert, W. (1988). Role of cytochrome c heme lyase in the import of cytochrome c into mitochondria. *J. Biol Chem.* **263,** 19034–19042.

Ohba, M., and Schatz, G. (1987). Protein import into yeast mitochondria is inhibited by antibodies raised against 45-kd proteins of the outer membrane. *EMBO J.* **6,** 2109–2115.

Ostermann, J., Horwich, A. L., Neupert, W., and Hartl, F. U. (1989). Protein folding in mitochondria requires complex formation with hsp60 and ATP hydrolysis. *Nature* **341,** 125–130.

Pelham, H. R., and Jackson, R. J. (1976). An efficient mRNA-dependent translation system from rabbit reticulocyte lysate. *Eur. J. Biochem.* **67,** 247–256.

Pfaller, R., Pfanner, N., and Neupert, W. (1989). Mitochondrial protein import: Bypass of proteinaceous surface receptors can occur with low specificity and efficiency. *J. Biol. Chem.* **264,** 34–39.

Pfanner, N., and Neupert, W. (1987). Distinct steps in the import of ADP/ATP carrier into mitochondria. *J. Biol. Chem.* **262,** 7528–7536.

Pon, L., Moll, T., Vestweber, D., Marshallsay, B., and Schatz, G. (1989). Protein import into mitochondria: ATP dependent protein translocation activity in a submitochondrial fraction enriched in membrane contact sites and specific proteins. *J. Cell Biol.* **109,** 2603–2616.

Rapaport, D., and Neupert, W. (1999). Biogenesis of Tom40, core component of the TOM complex of mitochondria. *J. Cell Biol* **146,** 321–331.

Rassow, J., Guiard, B., Wienhues, U., Herzog, V., Hartl, F. U., and Neupert, W. (1989). Translocation arrest by reversible folding of a precursor protein imported into mitochondria: A means to quantitate translocation contact sites. *J. Cell Biol.* **109,** 1421–1428.

Rowley, N., Prip-Buus, C., Westermann, B., Brown, C., Schwarz, E., Barrel, B., and Neupert, W. (1994). Mdj1p, a novel chaperone of the DnaJ family, is involved in mitochondrial biogenesis and protein folding. *Cell* **22,** 249–259.

Ryan, M. T., Müller, H., and Pfanner, N. (1999). Functional staging of the ADP/ATP carrier translocation across the outer mitochondrial membrane. *J. Biol. Chem.* **274,** 14851–14854.

Schägger, H., and von Jagow, G. (1991). Blue native electrophoresis for isolation of membrane protein complexes in enzymatically active form. *Anal. Biochem.* **199,** 223–231.

Schleiff, E., Silvius, J. R., and Shore, G. C. (1999). Direct membrane insertion of voltage-dependent anion-selective channel protein catalyzed by mitochondrial Tom20. *J. Cell Biol.* **31,** 973–978.

Schleyer, M., and Neupert, W. (1985). Transport of proteins into mitochondria: Translocational intermediates spanning contact sites between outer and inner membranes. *Cell* **43,** 339–350.

Sirrenberg, C., Bauer, M. F., Guiard, B., Neupert, W., and Brunner, M. (1996). Import of carrier proteins into the mitochondrial inner membrane mediated by Tim22. *Nature* **384,** 582–585.

Sirrenberg, C., Endres, M., Fölsch, H., Stuart, R. A., Neupert, W., and Brunner, M. (1998). Carrier protein import into mitochondria mediated by the intermembrane proteins Tim10/Mrs11 and Tim12/Mrs5. *Nature* **391,** 912–915.

Söllner, T., Griffiths, G., Pfaller, R., Pfanner, N., and Neupert, W. (1989). MOM19, an import receptor for mitochondrial precursor proteins. *Cell* **59,** 1061–1070.

Söllner, T., Pfaller, R., Griffiths, G., Pfanner, N., and Neupert, W. (1990). A mitochondrial import receptor for the ADP/ATP carrier. *Cell* **62,** 107–115.

Söllner, T., Rassow, J., and Pfanner, N. (1991). Analysis of mitochondrial protein import using translocation intermediates and specific antibodies. *In* "Methods in Cell Biology" (A. M. Tartakoff, ed.), Vol. 34, pp. 345–358. Academic Press, San Diego.

Truscott, K. N., and Pfanner, N. (1999). Import of carrier proteins into mitochondria. *Biol. Chem.* **380,** 1151–1156.

van Wilpe, S., Ryan, M. T., Hill, K., Maarse, A. C., Meisinger, C., Brix, J., Dekker, P. J. T., Moczko, M., Wagner, R., Meijer, M., Guiard, B., Hönlinger, A., and Pfanner, N. (1999). Tom22 is a multifunctional organizer of the mitochondrial preprotein translocase. *Nature* **401,** 485–489.

Vestweber, D., Brunner, J., Baker, A., and Schatz, G. (1989). A 42K outer-membrane protein is a component of the yeast mitochondrial protein import site. *Nature* **341,** 205–209.

Voisine, C., Craig, E. A., Zufall, N., von Ahsen, O., Pfanner, N., and Voos, W. (1999). The protein import motor of mitochondria: Unfolding and trapping of preproteins are distinct and separable functions of matrix Hsp70. *Cell* **97,** 565–574.

Voos, W., Gambill, D., Guiard, B., Pfanner, N., and Craig, E. A. (1993). Presequence and mature part of preproteins strongly influence the dependence of mitochondrial protein import in heat shock protein 70 in the matrix. *J. Cell Biol.* **123,** 119–126.

Voos, W., Martin, H., Krimmer, T., and Pfanner, N. (1999). Mechanisms of protein translocation into mitochondria. *Biochim. Biophys. Acta* **1422,** 235–254.

CHAPTER 12

Analysis of Protein–Protein Interactions in Mitochondria by Coimmunoprecipitation and Chemical Cross-Linking

Johannes M. Herrmann, Benedikt Westermann, and Walter Neupert

Institut für Physiologische Chemie
Universtität München
80336 München, Germany

I. Introduction

Because many mitochondrial proteins are present in multisubunit structures the analysis of protein–protein interactions is of particular importance for the understanding of mitochondrial architecture and biogenesis. Among these oligomers are the respiratory chain complexes and the F_0F_1-ATPase in the inner membrane, the mitochondrial

METHODS IN CELL BIOLOGY, VOL. 65

ribosome in the matrix, and the protein translocation complexes. The elucidation of the stochiometry and organization of these complexes requires a detailed analysis of the interactions among their various protein subunits. In addition to the study of these permanent steady-state interactions, various techniques have been employed to monitor transient protein–protein interactions. These are of special interest for the investigation of various aspects of mitochondrial biogenesis, such as protein translocation across mitochondrial membranes, folding and assembly of proteins, protein synthesis in the mitochondrial matrix, and protein degradation.

To understand the nature of protein–protein interactions and their impact on the biogenesis of mitochondria, the different interactions have to be identified and their consequences determined. This chapter discusses two classical approaches to monitor protein–protein interactions: copurification and cross-linking. Typical protocols for each technique will be presented.

II. General Considerations

Several points have to be considered before starting an analysis of protein–protein interactions. (i) Protein–protein interactions are often rather weak or unstable. For example, in the case of membrane protein complexes, solubilization with detergents often destabilizes oligomers so that they cannot be analyzed in extracts. (ii) Interactions can be transient and possibly persist only for short periods of time. This may prevent coisolation of the interaction partners, as the binding might not last during the experiment. (iii) Proteins often can interact with several different binding partners. For example, during import into mitochondria, a precursor protein contacts several proteins of the translocation machinery in a sequential manner. (iv) Many proteins bind to other proteins after extraction. In particular, membrane proteins and nonassembled subunits of complexes often expose hydrophobic domains, causing them to stick to other proteins. These artifactual interactions sometimes appear rather specific.

For these reasons it is crucial that experiments addressing protein–protein interactions are well controlled and supported by complementary approaches that reveal the same interactions consistingly. Various methods have been employed in the past to monitor interactions of mitochondrial proteins. Copurification procedures are the standard assays to detect protein–protein interactions, as they do not require purified proteins and thus allow for the isolation of previously unknown binding partners. Chemical cross-linkers can stabilize weak or short-lived interactions, and therefore are used extensively to monitor transient interactions. However, cross-linking often is rather inefficient and the isolation and identification of yet unknown interaction partners are therefore difficult. This low efficiency prevents quantitative predictions of interactions. Advantages and limitations of both techniques are discussed later.

In addition to copurification and chemical cross-linking, several other methods are commonly employed to assess protein–protein interactions. These include (i) blot overlay assays (also referred to as affinity blotting). Here, the binding of molecules to bait proteins immobilized on nitrocellulose or comparable membranes is visualized. This technique has hardly been used for the study of mitochondrial proteins. (ii) The composition of mitochondrial protein complexes has been studied extensively after separation of these complexes by size-exclusion chromatography, gradient centrifugation, or native gel electrophoresis (for the latter, see Chapter 14). (iii) To study the interaction of purified proteins, several biophysical methods can be used. For example, plasmon resonance was used to measure the binding affinities of preproteins to receptors of the mitochondrial outer membrane (Iwata and Nakai, 1998). (iv) Interactions of proteins can also be tested by two-hybrid analysis in the yeast *Saccharomyces cerevisiae* (Fields, 1993). This method has been used in several cases to track down interacting domains of mitochondrial proteins (for examples see Armstrong *et al.,* 1999; Brown *et al.,* 1994; Haucke *et al.,* 1996). Approaches for the analysis of protein–protein interactions have been reviewed by Phizicky and Fields (1995).

In copurification or pull-down assays, a known protein serves as a bait to coisolate binding partners. Innumerable versions of copurification approaches have been used to identify protein–protein interactions. Two different experimental setups are possible. First, the bait protein or peptide can be immobilized on a resin to which a protein extract is applied. Interacting proteins will then stick to the bait and unbound proteins can be washed off. Here, the interaction occurs after generation of the extracts, e.g., after lysis of mitochondria. This method is typically employed to assess binding parameters such as affinities. It can also be used for the isolation of larger quantities of binding partners. This type of copurification is normally referred to as affinity chromatography.

Alternatively, the bait can be pulled down from an extract and interacting proteins are thereby coisolated. Here, the protein complexes are formed before preparation of the extracts and therefore reflect the authentic interactions that are present in the cell. This type of assay is typically used to monitor *in vivo* interactions. The most commonly used procedure to isolate protein complexes is coimmunoprecipitation (see later). Alternatively, affinity tags on bait proteins can be used for isolation (for an overview see Jarvik and Telmer, 1998). This chapter illustrates the techniques of immuno- and coimmunoprecipitation, as well as chemical cross-linking, and presents typical protocols and examples of each approach.

III. Immunoprecipitation

Immunoprecipitation does not directly assess protein–protein interactions. Rather it is used to identify proteins in more or less complex protein mixtures. Here we describe the technique as the basis for coimmunoprecipitation of mitochondrial proteins.

We use yeast as a system but the method can be directly applied to other organisms. Preferably, isolated mitochondria should be used as starting material in order to reduce the background, which might be rather high when starting from whole cells (see Chapter 2 for isolation of mitochondria). Mitochondria can be used directly or following various manipulations, e.g., after import of radiolabeled precursor proteins. In the example described here, mitochondrial translation products are first radiolabeled according to the protocol described in Chapter 24. Mitochondria are lysed under denaturing conditions to break protein–protein interactions and the extract is cleared by centrifugation. Protein A–Sepharose CL-4B beads (Amersham Pharmacia, Sweden) are incubated with the antiserum raised against the protein of interest in our example cytochrome *c*-oxidase subunit II (CoxII). In addition, a control serum obtained from the same rabbit prior to the first immunization should be used (preimmune serum). The beads are washed to remove unbound proteins, incubated with the extract, washed again, and bound proteins are resolved by SDS–PAGE. Boiling of the samples is omitted at all steps as several mitochondrial translation products are very hydrophobic and tend to form insoluble aggregates at high temperature.

A. Immunoprecipitation of CoxII Following *in Organello* Translation

1. Pretreatment of Isolated Mitochondria to Radiolabel Translation Products

Mitochondrial translation products are radiolabeled in 200 μg yeast mitochondria for 30 min at 25°C as described in Chapter 25. Mitochondria are reisolated by centrifugation for 10 min at 14,000*g* at 2°C and washed in 0.6 *M* sorbitol, 250 m*M* KCl, 10 m*M* HEPES, pH 7.4.

2. Lysis

Mitochondria are lysed by vigorous shaking for 10 min at 4°C in 30 μl 1% SDS. After the addition of 1 ml lysis buffer [1% Triton X-100 (w/v), 300 m*M* NaCl, 5 m*M* EDTA, 1 m*M* phenylmethylsulfonyl fluoride, 10 m*M* Tris–HCl, pH7.4], the sample is mixed again for 5 min at 4°C and centrifuged for 10 min at 30,000*g* at 2°C. The resulting supernatant is the extract used for immunoprecipitation.

3. Coupling of Antibodies to Protein A–Sepharose Beads

For coupling of the antibodies to the beads, 20 μl protein A–Sepharose CL-4B slurry (50% suspension) is washed twice in lysis buffer and incubated either with 30 μl rabbit antiserum against CoxII or with 30 μl preimmune serum for 30 min at room temperature. The beads are washed three times in lysis buffer.

4. Immunoprecipitation

Five hundred microliters of the extract is added to the beads loaded with antibodies against CoxII or preimmune antibodies. After shaking for 2 h at 4°C, the beads are washed three times with 1 ml lysis buffer, once with 10 m*M* Tris–HCl, pH 7.4, and resuspended in gel-loading buffer containing SDS. Isolated proteins are separated by SDS–PAGE and visualized by autoradiography as described in Chapter 25. The result of this experiment is shown in Fig. 1.

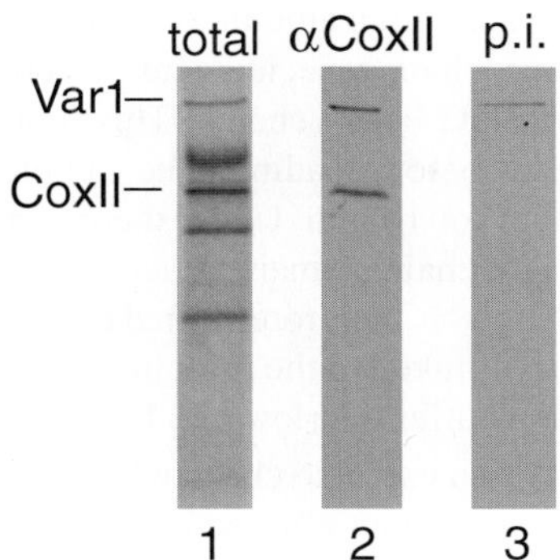

Fig. 1 Immunoprecipitation of cytochrome *c*-oxidase subunit II (CoxII) after radiolabeling of mitochondrial translation products. Mitochondrial translation products were labeled *in organello*, mitochondria were lysed, and the resulting extract was applied directly to the gel (lane 1, total) or used for immunoprecipitations with serum against CoxII (αCoxII, lane 2) or preimmune serum (p.i., lane 3). For details, see Section III,A. For comparison, 10% of the extract used for the immunoprecipitations was incubated with 12% trichloroacetic acid for 30 min on ice and centrifuged for 30 min at 30,000*g* at 2°C. The resulting pellet was washed with ice-cold acetone, dried for 5 min at 37°C, and dissolved in gel-loading buffer (total).

Comments

1. The example shows the importance of a comparison to unrelated or preimmune sera. Var1 is a mitochondrially encoded protein that assembles into the small subunit of the mitochondrial ribosome. Unassembled Var1 is very "sticky" and sediments with the beads even if unrelated sera or no antibodies are coupled to the protein A–Sepharose (Fig. 1, lane 3).

2. For immunoprecipitations of proteins that do not tend to aggregate upon boiling, mitochondria should be heated for 3 min at 96°C after solublilization in 1% SDS and also finally before the samples are loaded on the gels.

3. Affinity purification of the antibodies often significantly reduces the background obtained with antiserum (for protocols, see Harlow and Lane, 1988).

4. Instead of prebinding the antibodies to the protein A–Sepharose, the serum (or purified antibodies) and the protein A–Sepharose can be added directly to the extract. However, in this case the amount of serum required has to be determined carefully, because it is important that all antibodies will be bound by the protein A–Sepharose during the incubation.

5. The efficiency of immunoprecipitation of epitope-tagged proteins by monoclonals is often quite low. This can be improved significantly by the addition of several copies of the tag onto the protein, thereby increasing the avidity of antibody binding to the antigen (e.g., three or more myc epitopes in tandem).

6. Instead of protein A–Sepharose, protein G–Sepharose can be used, which has a higher affinity for some IgG subtypes. This has to be tested, especially if one uses monoclonals.

7. The rather high amount of IgG often leads to a bad quality of the SDS–PAGE. This can be a problem, especially for proteins running at about 50 to 60 kDa, which is the size of the IgG heavy chains. This problem may be solved by omitting the boiling of the samples before loading. The samples are mixed instead in loading buffer at room temperature for 10 min. Under these conditions the disulfide bridges between the heavy and the light chains remain intact even in the presence of 2.5% β-mercaptoethanol, and the IgG "blob" is therefore shifted to about 90 kDa. If this is not sufficient, the antibodies can be cross-linked to the protein A–Sepharose so that they will not be released into the gel loading buffer (Harlow and Lane,1988) (see Section III,B).

8. It is often useful to check whether the immunoprecipitation was quantitative. To do this, the supernatant after sedimentation of the protein A–Sepharose/antibody/antigen conjugates can be analyzed either by Western blotting or by reimmunoprecipitation. After quantitative immunoprecipitation the supernatant will be completely depleted of the antigen.

B. Cross-Linking of Antibodies to Protein A–Sepharose

1. Binding of Antibodies to Protein A–Sepharose Beads

Antiserum (250 μl) is incubated with 500 PBS (8 g/liter NaCl, 0.2 g/liter KCl, 1.44 g/liter Na_2HPO_4, 0.24 g/liter NaH_2PO_4) and 200 μl protein A–Sepharose slurry (50% in PBS) for 1 h at 25°C.

2. Cross-linking

The beads are washed three times in 0.2 *M* $NaBO_4$, pH 9.0, and resuspended in 1.8 ml of this buffer with 10.4 mg dimethyl pimelidate (DMP; Pierce).

3. Quenching

After incubation for 30 min at 30°C, the beads are washed and then incubated in 0.2 *M* ethanolamine for 2 h at 25°C to quench the cross-linker. Finally, they are resuspended in 150 μl PBS/20 m*M* NaN_3. This solution can be stored at 4°C for several weeks.

In some cases, cross-linking can interfere with the binding efficiency of the antibodies to the antigen. This has to be tested for each serum. The extent of the immobilization of antibodies on the protein A–Sepharose varies significantly. At least in most cases, however, cross-linking of the IgGs to the protein A–Sepharose clearly improves the quality of the gels.

IV. Coimmunoprecipitation

In contrast to immunoprecipitation experiments, nondenaturing conditions are used for coimmunoprecipitation so that protein complexes, instead of single proteins, are pulled down. Thus, the critical step in this protocol is the solubilization of the protein complex. This is normally done by the addition of nonionic detergents, such as Triton X-100, octyl glucopyranoside, dodecyl maltoside, or digitonin (all can be obtained from Sigma). Triton X-100 is easy to handle, highly soluble in water, stable (in the dark), inexpensive and the solubilization is very reproducible. However, many complexes are not stable in Triton X-100. In the case of fragile complexes, digitonin is often the detergent of choice. The quality of digitonin varies greatly between different batches and suppliers. It is essential to recrystallize digitonin before use (see Section IV,A). In addition to the type of detergent, the ionic strength and the pH of the solubilization buffer can be critical. Both have to be tested and adapted for each complex. We normally use buffers with 50 to 150 m*M* NaCl and pH 7.0–7.5.

A. Recrystalization of Digitonin

One gram of digitonin is dissolved in 20 ml ethanol p.A. and boiled carefully in a water bath using a stir bar. The solution is then cooled very slowly (within several hours) down to −20°C. After incubation for 16 h at −20°C the solution is centrifuged, the supernatant removed, and the pelleted digitonin crystals are dried under vacuum for several hours. The resulting powder can be stored at 4°C.

B. Coimmunoprecipitation of ATPase Oligomers after Protein Synthesis in Isolated Mitochondria

As an example of coimmunoprecipitation, we present an experiment assessing the oligomerization of mitochondrial translation products. Under appropriate conditions, newly synthesized ATPase9, a subunit of the F_0 part of the F_0F_1 ATPase, assembles into a 48-kDa complex that can be bound by mt-Hsp70. This complex is converted into a larger 54-kDa complex. Both ATPase9 complexes can be found in association with the F_1 part of the ATPase, and can be coimmunoprecipitated with F_1 subunit α ($F_1\alpha$) [for details, see Herrmann *et al.* (1994) and Chapter 24].

1. Pretreatment of Isolated Mitochondria to Radiolabel Translation Products

Mitochondrial translation products are radiolabeled in 140 μg yeast mitochondria [isolated from chloramphenicol-pretreated cells, see Herrmann *et al.* (1994)] for 20 min at 30°C, as described in Chapter 24 with the exception that phosphoenolpyruvate, α-ketoglutarate, and pyruvate kinase are omitted from the translation buffer. Then apyrase is added (0.8 U/μl final concentration) to deplete mitochondrial ATP. After another incubation for 8 min at 30°C mitochondria are reisolated by centrifugation at 2°C for 10 min at 14,000*g* and resuspended in 900 μl 0.6 *M* sorbitol, 250 m*M* KCl, 10 m*M* HEPES, pH 7.4.

2. Lysis

Mitochondria are reisolated and lysed by gentle agitation for 10 min at 4°C in 1 ml ice-cold lysis buffer [0.1% Triton X-100 (w/v), 150 m*M* NaCl, 5 m*M* EDTA, 1 m*M* phenylmethylsulfonyl fluoride, and 10 m*M* Tris–HCl, pH 7.4]. The extract is cleared by centrifugation for 10 min at 30,000*g* at 2°C.

3. Coupling of Antibodies to Protein A–Sepharose Beads

For coupling of antibodies to the beads, 20 μl protein A–Sepharose CL-4B slurry (50% suspension) is washed twice in lysis buffer and incubated with 30 μl rabbit serum against mt-Hsp70 (Ssc1p) or $F_1\alpha$, or with preimmune serum, for 30 min at room temperature. The beads are washed three times in lysis buffer.

4. Coimmunoprecipitation

The extract (320 μl) is added to the beads loaded with antiserum against mt-Hsp70 (Ssc1p), $F_1\alpha$, or preimmune serum, respectively. The samples are inverted for 1 h at 4°C. The beads are then washed twice in lysis buffer, once in 10 m*M* Tris–HCl, pH 7.4, and are then resuspended in gel-loading buffer. Isolated proteins are separated by SDS–PAGE and visualized by autoradiography as described in Chapter 24. The result of this experiment is shown in Fig. 2.

Comments

1. If the same experiment is performed without ATP depletion, the ATPase9 oligomer is not coimmunoprecipitated by the mt-Hsp70 antiserum because the ATP-bound state of mt-Hsp70 does not interact stably with its substrates. Thus, ATP depletion arrests mt-Hsp70 in a substrate-bound state. Coimmunoprecipitation of mt-Hsp70/substrate

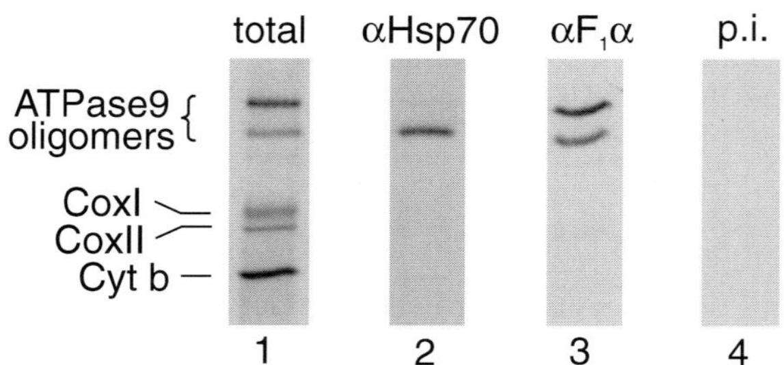

Fig. 2 Coimmunoprecipitation of ATPase9 oligomers after radiolabeling of mitochondrial translation products. Mitochondrial translation products were labeled *in organello,* mitochondria were lysed, and the resulting extract was applied directly to the gel (lane 1, total) or used for immunoprecipitations with serum against mt-Hsp70 (αHsp70, lane 2), ATPase subunit α ($\alpha F_1\alpha$, lane 3), or preimmune serum (p.i., lane 4). For details, see Section IV,B. For total (lane 1), 10% of the extract used for the coimmunoprecipitations (lanes 2 to 4) was TCA precipitated as described in Fig. 1. CoxI, cytochrome *c*-oxidase subunit I; Cyt b, cytochrome *b*.

complexes is possible under these conditions, although this is normally a transient interaction. The coimmunoprecipitation with the $F_1\alpha$ antiserum, however, is not affected by ATP. This is an example to demonstrate the importance of the lysis conditions used.

2. A useful control for the specificity of a coimmunoprecipitation is the use of an extract that does not contain the antigen, e.g., extracts obtained from a deletion strain.

3. The mild buffer conditions may cause a problem known as "postlysis binding," meaning that the interaction of the bait with the coisolated proteins occurs only in the extract and therefore does not reflect an *in vivo* interaction. This can be a serious problem, and coimmunoprecipitations therefore have to be controlled carefully. Three typical controls are used. (i) In competition experiments, an excess of potential binding partners of the bait is added after lysis. This should not influence the result of the coimmunoprecipitation. Potential binding partners might be identical to the coisolated protein. Alternatively, other substrates for the bait can be added. (ii) Dilution experiments. If the binding of the bait to its binding partner occurred before lysis, the protein concentration in the lysate should play a minimal role. In contrast, postlysis interactions should be more pronounced at higher protein concentrations in the lysate. (iii) In the case of mixing experiments, bait and potential substrate proteins are combined after lysis. No coimmunoprecipitation should be observed in this case.

V. Cross-Linking

Chemical cross-linking introduces covalent bonds within or between proteins. Thus, proteins that are in proximity to each other can be connected stably. This makes cross-linking a powerful method, especially for the detection of transient or weak interactions that might not be detected by coimmunoprecipitation. Cross-linking has been used extensively to assess interactions among mitochondrial proteins. A limitation of cross-linking is the dependence on specific side chains of amino acids in the proteins that are involved in the coupling reaction. Obviously, these side chains need to be in a certain distance and orientation to allow cross-linking. Cross-linking is usually not very efficient: in the case of stable interactions, cross-linking yields might exceed 10%; however, in the case of transient interactions, it is typically less than 1%. Therefore, a sensitive method to detect the cross-links is required. For stable interactions (such as the interactions of the subunits of the respiratory chain complexes) cross-links can be detected by Western blotting. Transient interactions are usually assessed by the use of radiolabeled proteins. Synchronization of the interactions is often required to increase the yield of specific cross-links. In the case of experiments addressing protein translocation across mitochondrial membranes, radiolabeled precursor proteins containing a C-terminal fusion to mouse dihydrofolate reductase (DHFR) can be used. DHFR can be stably folded by addition of the substrate analog methotrexate, and thereby allows the arrest of translocation intermediates of defined length. This has been used in many examples to cross-link precursor proteins to components of the mitochondrial protein import machineries (e.g., Berthold *et al.,* 1995; Vestweber *et al.,* 1989a).

There are two different approaches to use cross-linking reagents: (i) [photoreactive] cross-linkers can be incorporated into proteins co- or posttranslationally. Then these proteins are allowed to interact with other proteins and the cross-linking reaction is performed (by a light flash) (Brunner, 1993, 1996). This type of cross-linking led to the identification of Tom40/Isp42 as a component of the translocation machinery in the mitochondrial outer membrane (Vestweber *et al.,* 1989a,b). (ii) The second type uses bifunctional cross-linking reagents to connect preexisting complexes. This method is rather versatile and can be used to investigate many different processes. For example, cross-linking of subunits of the respiratory chain complexes has been used extensively to identify nearest neighbors within these multisubunit oligomers (e.g., Briggs and Capaldi, 1977; Smith *et al.,*1978; Todd and Douglas, 1981). It also has been used in many cases to screen for components in the proximity to a precursor protein during mitochondrial import or folding (Leuenberger *et al.,* 1999; Rapaport *et al.,* 1998; Sirrenberg *et al.,* 1996, 1998; Söllner *et al.,* 1992).

A large variety of bifunctional cross-linkers is available commercially (e.g., from Pierce or Molecular Probes). Cross-linking reagents differ in several properties.

1. Cross-linkers differ in their specificity, depending on their reactive groups. Targets can be primary amines (the amino terminus and lysine residues), sulfhydryls (cysteine residues), carboxy groups (aspartate and glutamate residues and the carboxy terminus), or carbohydrates. Some photoreactive reagents such as phenyl azides are more or less nonselective. Both homo- and heterobifunctional cross-linkers are available.
2. Some cross-linkers have to be activated by a light flash, which allows the trapping of protein–protein interactions at a specific time point.
3. Cross-linking reagents differ in the length of the spacer arm, which is usually in the range of 0.4–1.2 nm.
4. Some cross-linkers allow the cleavage of the cross-bridge to release the adduct from the bait.
5. Some cross-linkers can be radiolabeled. This can be used, for example, to iodinate binding partners specifically so that they can be identified after cleavage of the cross-bridge.
6. Several cross-linkers are available in membrane-permeable and -impermeable versions.

Because the yields of cross-links between interacting proteins are not predictable, several cross-linkers (e.g., differing in the length of the spacer arm) have to be titrated and the conditions for each interaction have to be optimized. Cross-linking reagents that have been used to detect interactions of mitochondrial proteins include *m*-maleimidobenzoyl-*N*-hydroxysuccinimide ester (MBS), 1,5-difluoro-2,4-dinitrobenzene (DFDNB), dithiobis (succinimidyl propionate) (DSP), disuccinimidyl glutarate (DSG), disuccinimidyl suberate (DSS), and 1,4-di(3′-2′-pyridyldithio-propionamido)butane (DPDPB).

In the following protocol, a radiolabeled precursor protein was arrested during an *in vitro* import reaction and treated with the cross-linker DSS. By immunoprecipitation, one cross-linking adduct was identified as Tim17, a component of the translocation machinery in the inner membrane.

A. Cross-Linking of a Radiolabeled Precursor Protein to a Component of the Import Machinery

1. Pretreatment of Mitochondria to Arrest a Radiolabeled Precursor Protein in the Import Channel

Mitochondria (150 μg) are preincubated in 600 μl import buffer (220 m*M* sucrose, 80 m*M* KCl, 10 m*M* MOPS–KOH, pH 7.2, 3% bovine serum albumin, 25 m*M* potassium phosphate, 5 m*M* MgOAc, and 1 m*M* $MnCl_2$) in the presence of 1 μ*M* methotrexate for 5 min at 0°C. Then 20 μl reticulocyte lysate containing ^{35}S-radiolabeled pSu9(1-112)-DHFR is added. After another incubation for 5 min on ice to allow binding of the methotrexate to DHFR, the sample is incubated for 15 min at 25°C.

2. Cross-Linking

The sample is split into a 50-μl aliquot for mock treatment (total, without cross-linker) and a 550-μl aliquot. To the first aliquot, 0.5 μl dimethyl sulfoxide (DMSO) and to the latter 5.5 μl DSS (30 m*M* stock in DMSO) are added. After incubation for 30 min at 25°C, Tris–HCl, pH 8.0, is added to 100 m*M* final concentration to quench unreacted cross-linker.

3. Immunoprecipitation

Fifty-five microliters of the cross-linked sample is removed (total, with cross-linker). Mitochondria are reisolated from all samples by centrifugation, washed in 0.6 *M* sorbitol, 250 m*M* KCl, 10 m*M* HEPES, pH 7.4, and either dissolved in gel-loading buffer (for totals) or in 80 μl 1% SDS,100 m*M* Tris–HCl, pH 7.4 (for immunoprecipitation). Material used for immunoprecipitation is vortexed, boiled for 3 min, diluted with 1 ml lysis buffer [1% Triton X-100 (w/v), 300 m*M* NaCl, 5 m*M* EDTA, 1 m*M* phenylmethylsulfonyl fluoride, 10 m*M* Tris–HCl, pH 7.4], and centrifuged for 10 min at 30,000*g* at 2°C. The resulting supernatant is divided and used for immunoprecipitation as described in protocol 1, either with a serum against Tim17 or with a preimmune serum as control. The result of this experiment is shown in Fig. 3.

Comments

1. DSS belongs to the *N*-hydroxysuccinimide esters, (NHS esters), which are by far the most commonly used cross-linking reagents. They react with deprotonated primary amines, which excludes using Tris as a buffer during cross-linking. Tris is often used as a quenching reagent following cross-linking. Alternatively, glycine (pH 7.5) can be used for quenching. For NHS esters it is important to keep the pH above 7–7.5 to reduce protonation of the amino groups. These reagents are rather unstable in water, especially

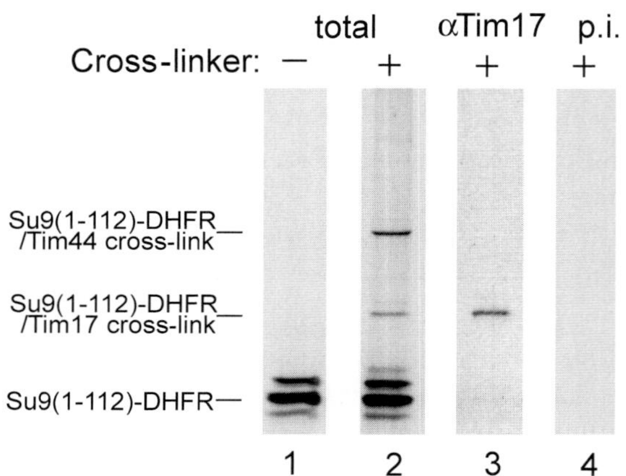

Fig. 3 Cross-linking of Su9(1-112)-DHFR to Tim17. ^{35}S-radiolabeled Su9(1-112)-DHFR was arrested in the mitochondrial import channel prior to incubation in the absence (lane 1) or presence (lanes 2-4) of the cross-linking reagent DSS. Mitochondria were lysed and either loaded directly on the gel (lanes 1 and 2) or the resulting extract was used for immunoprecipitation with serum against Tim17 (αTim17, lane 3) or preimmune serum (p.i., lane 4). Total samples correspond to 20% of the material used for immunoprecipitation. For details, see Section V,A.

at a pH above 8. To prevent hydrolysis of the cross-linkers during storage, they have to be kept dry. It is very important to warm the vials containing the cross-linkers to room temperature before opening to protect them against condensing water. The cross-linker solutions in DMSO should be prepared fresh before use.

2. Quenching reagents should be present in all steps after the cross-linking reaction.

3. We did not observe any effects of the presence of up to 2–4% DMSO on protein import into isolated mitochondria or *in organello* synthesis of mitochondrial translation products.

4. ATP depletion, dissipation of the membrane potential, or low temperature can be used to arrest translocating chains instead of using DHFR/methotrexate.

VI. Summary

Many different techniques have been employed to analyze protein–protein interactions. Coimmunoprecipitation and chemical cross-linking have been used extensively to study mitochondrial biogenesis. Both techniques have proven to be powerful methods to investigate the sequential interactions of precursor proteins with the various components of the translocation machineries in the mitochondrial membranes. Similarly, protein–protein interactions during processes such as protein synthesis, folding, and degradation can be studied. Moreover, the composition of the oligomeric protein complexes of mitochondria, such as respiratory chain complexes or protein translocation machineries, can be determined. The general principles and protocols of these methods are described and illustrated with typical examples.

Acknowledgment

The authors thank Dr. Albrecht Gruhler for providing the experiment shown in Fig. 3.

References

Armstrong, L. C., Saenz, A. J., and Bornstein, P. (1999). Metaxin 1 interacts with metaxin 2, a novel related protein associated with the mammalian mitochondrial outer membrane. *J. Cell Biochem.* **74,** 11–22.

Berthold, J., Bauer, M. F., Schneider, H.-C., Klaus, C., Dietmeier, K., Neupert, W., and Brunner, M. (1995). The MIM complex mediates preprotein translocation across the mitochondrial inner membrane and couples it to the mt-Hsp70/ATP driving system. *Cell* **81,** 1085–1093.

Briggs, M. M., and Capaldi, R. A. (1977). Near-neighbor relationships of the subunits of cytochrome c oxidase. *Biochemistry* **16,** 73–77.

Brown, N. G., Constanzo, M. C., and Fox, T. D. (1994). Interactions among three proteins that specifically activate translation of the mitochondrial COX3 mRNA in *Saccharomyces cerevisiae*. *Mol. Cell. Biol.* **14,** 1045–1053.

Brunner, J. (1993). New photolabeling and crosslinking methods. *Annu. Rev. Biochem.* **62,** 483–514.

Brunner, J. (1996). Use of photocrosslinkers in cell biology. *Trends Cell Biol.* **6,** 154–157.

Fields, S. (1993). The two-hybrid system to detect protein-protein interactions. *Methods Enzymol.* **5,** 116–124.

Harlow, E., and Lane, D. (1988). "Antibodies: A Laboratory Manual." Cold Spring Harbor Laboratory Press, Cold Spring Harbor, NY.

Haucke, V., Horst, M., Schatz, G., and Lithgow, T. (1996). The Mas20p and Mas70p subunits of the protein import receptor of yeast mitochondria interact via the tetratricopeptide repeat motif in Mas20p: Evidence for a single hetero-oligomeric receptor. *EMBO J.* **15,** 1231–1237.

Herrmann, J. M., Stuart, R. A., Craig, E. A., and Neupert, W. (1994). Mitochondrial heat shock protein 70, a molecular chaperone for proteins encoded by mitochondrial DNA. *J. Cell Biol.* **127,** 893–902.

Iwata, K., and Nakai, M (1998). Interaction between mitochondrial precursor proteins and cytosolic soluble domains of mitochondrial import receptors, Tom20 and Tom70, measured by surface plasmon resonance. *Biochem. Biophys. Res. Commun.* **253,** 648–652.

Jarvik, J. W., and Telmer, C. A. (1998). Epitope tagging. *Annu. Rev. Genet.* **32,** 601–618.

Leuenberger, D., Bally, N. A., Schatz, G., and Koehler, C. M. (1999). Different import pathways through the mitochondrial intermembrane space for inner membrane proteins. *EMBO J.* **18,** 4816–4822.

Phizicky, E. M., and Fields, S. (1995). Protein-protein interactions: Methods for detection and analysis. *Microbiol. Rev.* **59,** 94–123.

Rapaport, D., Künkele, K.-P., Dembowski, M., Ahting, U., Nargang, F., Neupert, W., and Lill, R. (1998). Dynamics of the TOM complex of mitochondria during binding and translocation of preproteins. *Mol. Cell Biol.* **18,** 5256–5262.

Sirrenberg, C., Bauer, M. F., Guiard, B., Neupert, W., and Brunner, M. (1996). Import of carrier proteins into the mitochondrial inner membrane mediated by Tim22. *Nature* **384,** 582–585.

Sirrenberg, C., Endres, M., Fölsch, H., Stuart, R. A., Neupert, W., and Brunner, M. (1998). Carrier protein import into mitochondria mediated by the intermembrane proteins Tim10/Mrs11p and Tim12/Mrs5p. *Nature* **391,** 912–915.

Smith, R. J., Capaldi, R. A., Muchmore, D., and Dahlquist, F. (1978). Cross-linking of ubiquinone cytochrome c reductase (complex III) with periodate-cleavable bifunctional reagents. *Biochemistry* **17,** 3719.

Söllner, T., Rassow, J., Wiedmann, M., Schlossmann, J., Keil, P., Neupert, W., and Pfanner, N. (1992). Mapping of the protein import machinery in the mitochondrial outer membrane by crosslinking of translocation intermediates. *Nature* **355,** 84–87.

Todd, R. D., and Douglas, M. G. (1981). A model for the structure of the yeast mitochondrial adenosine triphosphatase complex. *J. Biol. Chem.* **256,** 6984–6989.

Vestweber, D., Brunner, J., Baker, A., and Schatz, G. (1989a). A 42K outer-membrane protein is a component of the yeast mitochondrial protein import site. *Nature* **341,** 205–209.

Vestweber, D., Brunner, J., and Schatz, G. (1989b). Modified precursor proteins as tools to study protein import into mitochondria. *Biochem. Soc. Trans.* **17,** 827–828.

CHAPTER 13

Blue-Native Gels to Isolate Protein Complexes from Mitochondria

Hermann Schägger

Institut für Biochemie I
Universitätsklinikum Frankfurt
60590 Frankfurt am Main, Germany

I. Introduction

Since the first description of blue-native polyacrylamide gel electrophoresis (BN–PAGE; Schägger and Von Jagow, 1991), the original protocol and related techniques have been improved and expanded considerably. These modifications and supplementary protocols have been published, but are scattered in the literature. This chapter summarizes these developments, presents a new buffer system for BN–PAGE, which, in contrast to the previous protocols, does not disturb protein determinations, and presents the author's suggestions for choice of specific detergent and of the appropriate detergent/protein ratio for preservation of native protein–protein interactions. These techniques led to findings including demonstration that the yeast F_1F_0-ATP-synthase exists as a dimer (Arnold *et al.*, 1998), yeast and mammalian respiratory chains exist as a network of supercomplexes (Schägger and Pfeiffer, 2000), and detergent-sensitive multiprotein complexes could be isolated without dissociation of detergent-labile subunits.

The basic principles of BN–PAGE, however, are unchanged. Mild neutral detergents are used for solubilization of biological membranes, similar to the solubilization step usually applied for chromatographic isolation protocols. After solubilization, an anionic dye (Coomassie blue G-250) is added that binds to the surface of all membrane proteins and to many water-soluble proteins. This binding of a large number of negatively charged dye molecules to protein has several useful effects. (1) It shifts the isolectric point of the proteins to more negative values. As a result all proteins, even basic ones, migrate to the anode irrespective of their original isoelectric points upon electrophoresis at pH 7.5. The separating principle of BN–PAGE is not the charge/mass ratio, but the decreasing pore size of the acrylamide-gradient gel, which reduces the protein migration velocity according to the mass of the protein, and can completely stop protein mobility at a mass-specific pore-size limit. (2) The excess negative charges on the surfaces of individual dye-associated proteins repel each other. As a result, the tendency of aggregation of membrane proteins is reduced considerably. (3) Dye-associated membrane proteins are negatively charged and therefore soluble in detergent-free solution. As a result, detergent can be omitted from the gel, and the risk of denaturation of detergent-sensitive membrane proteins is minimized. (4) Dye-associated proteins are visible during electrophoresis and migrate as blue bands. This facilitates excision of selected bands and recovery of native proteins by electroelution. Finally the subunit composition of multiprotein complexes can be determined by excision of selected bands followed by SDS–PAGE for resolution in a second dimension.

II. Materials and Methods

A. Chemicals

Dodecyl-ß-D-maltoside, Triton X-100, and 6-aminohexanoic acid are from Fluka, acrylamide, bis-acrylamide (the commercial, twice-crystallized products), Serva Blue G (Coomassie blue G-250) from Serva, and digitonin from Roth. All other chemicals are from Sigma.

B. First-Dimension: Blue-Native Electrophoresis

1. Detergents, Stock Solutions, and Buffers

In principle, any neutral detergent, and also mild anionic detergents such as cholic acid derivatives, can be used for the solubilization of biological membranes for BN–PAGE if the detergent can solubilize the desired protein and keep it in the native state. We preferentially use dodecyl-ß-D-maltoside, Triton X-100, and digitonin, all of which are stored as 10% stock solutions. Stock solutions of Tricine (1 *M*), 6-aminohexanoic acid

Table I
Buffers for BN–PAGE[a]

Solution	Composition
Deep blue cathode buffer B[b]	50 m*M* Tricine, 7.5 m*M* imidazole (the resulting pH is around 7.0), plus 0.02% Coomassie blue G-250
Slightly blue cathode buffer B/10[c]	Similar to cathode buffer B, but lower dye concentration (0.002%)
Gel buffer (3×); triple concentrated	75 m*M* imidazole/HCl (pH 7.0), 1.5 *M* 6-aminohexanoic acid[d]
Anode buffer	25 m*M* imidazole/HCl (pH 7.0)
5% Coomassie blue	Suspended in 500 m*M* 6-aminohexanoic acid
AB mix (acrylamide–bisacrylamide mixture; 49.5% T, 3% C)[e]	48 g acrylamide and 1.5 g bisacrylamide per 100 ml

[a]Imidazole is introduced instead of previously used bis–Tris because bis–Tris interferes with many commonly used protein determinations, e.g., the Lowry method. The imidazole concentrations required are half the previous bis–Tris concentrations due to slight differences in p*K* values.

[b]Cathode buffer B is stirred for several hours before use and stored at room temperature. During prolonged storage at low temperatures Coomassie dye may form aggregates, which can prevent proteins from entering the gel.

[c]Cathode buffer B/10 and all other solutions can be stored at 7°C.

[d] 6-Aminohexanoic acid is not essential for BN–PAGE, but it improves protein solubility and is an efficient and inexpensive serine protease inhibitor. After BN–PAGE gels can be stored at 4°C for several days. We did not observe protease degradation under these conditions.

[e]%T, total concentration of acrylamide and bisacrylamide monomers; %C, percentage of cross-linker to total monomer.

(2 *M*), imidazole (1 *M*), and imidazole/HCl (1 *M*; pH 7.0) are stored at 7°C. Buffers for BN–PAGE are listed in Table I.

2. Gel Types

Acrylamide gradient gels are used commonly in BN–PAGE. A selection of gel types to cover specific molecular mass ranges is given in Table II. Each of these gel types is used either for analytical purposes ("analytical gel" with dimensions: 14 × 14 × 0.16 cm; sample wells 0.5 or 1.0 cm) or for preparative purposes ("preparative gel" with dimensions: 14 × 14 × 0.3 cm; one 14-cm sample well). Gradient gel preparation is exemplified in Table III.

Table II
Gel Types

Molecular mass range (kDa)	Sample gel (% T)	Gradient separation gel (% T)
100–10,000	3.0	3 → 13
100–3,000	3.5	4 → 13
100–1,000	4.0	5 → 13
20–500	4.0	6 → 18

Table III
Gradient Gel Preparation[a]

	Sample gel 4% T	Gradient separation gel 5% T	13% T
AB mix	0.5 ml	1.88 ml	3.9 ml
Gel buffer (3×)	2 ml	6 ml	5 ml
Glycerol	—	—	3 g
Water	3.5 ml	10 ml	3 ml
10% APS[b]	50 μl	100 μl	75 μl
TEMED	5 μl	10 μl	7.5 μl
Total volume	6 ml	18 ml	15 ml

[a]Volumes for one analytical gel (14 × 14 × 0.16 cm). Linear gradient separation gels are cast at 4°C and maintained at room temperature for polymerization. The volume of the 5% T solution is greater than that of the 13% T solution containing glycerol. This assures that the two solutions initially are not mixed when the connecting tube is opened. The sample gel is cast at room temperature. After removal of the combs, gels are overlaid with gel buffer (1×) and are stored at 4°C.

[b]10% aqueous ammonium persulfate solution, freshly prepared.

3. Sample Preparation

a. How Much Detergent Is Required for Complete Solubilization of Membrane Proteins from Biological Membranes?

As a first rule of thumb, a Triton X-100 or dodecylmaltoside/protein ratio of 1 g/g solubilizes bacterial membranes, but 2–3 g/g is required for mitochondrial membranes. Using digitonin, the detergent/ protein ratio is doubled (2 and 4–6 g/g, respectively).

Second general rule: For first trials do not use a single detergent/protein ratio, but a series of detergent to protein ratios, e.g., 0.5, 1.0, 1.5, and 2.0 Triton X-100 per gram of protein (for bacterial membranes). This is a test for the efficiency of solubilization, and may also detect whether the physiological state of the desired (multi)protein complex is sensitive to this detergent. For example, the protein might be dimeric at a low detergent/protein ratio, but monomeric at higher detergent/protein ratios, or some protein subunits might be lost using higher detergent/protein ratios (detectable only after resolution by SDS–PAGE in a second dimension).

b. What Is the Maximal and Minimal Protein Load for BN–PAGE?

As a general rule the maximal protein load seems to depend on the amount of DNA present in the solubilized sample. The maximal protein load for a 1 × 0.16-cm sample well, starting from isolated mitochondria (low DNA), is 200–400 μg of total protein. Using bacterial membranes directly, the protein load should be reduced to 50–100 μg.

DNA seems to fill the pores of the sample gel and thus prevents proteins from entering the gel. Use of DNases does not help much; deep blue and ragged artifact bands will appear in the gel. The best way to remove DNA seems to be mild disruption of cells and differential centrifugation.

The minimal protein load to BN–PAGE depends on the sensitivity of the protein detection method. When specific antibodies are available there is essentially no minimal protein load for BN–PAGE, provided that the same detergent/protein ratio is adjusted as for solubilization of larger protein amounts, and the detergent concentration is clearly above the critical micelle concentration (cmc). Use low final volumes for solubilization of low protein amounts, e.g., 5 μl, if the detergent concentration would approach the cmc when adjusting to a standard final volume of 40 μl.

c. General Scheme for Solubilization of Biological Membranes

Use aliquots of suspended biological membranes (50–200 μg of total protein). These samples usually can be shock-frozen in liquid nitrogen and stored at −80°C if suspended in carbohydrate solution (e.g., >250 m*M* sucrose or 400 m*M* sorbitol) or 15% glycerol. Ideally, the salt concentration in these suspensions should be low (around 50 m*M* NaCl; potassium and divalent cations can cause aggregation of Coomassie dye and thus precipitate proteins together with the dye). The preferred buffer is 50 m*M* imidazole/HCl, pH 7.0. Concentrated protein suspensions (>10 mg/ml) can be used directly for solubilization. Diluted suspensions have to be pelleted by centrifugation.

Sedimented 50- to 200-μg protein samples are usually solubilized by adding 20–40 μl 50 m*M* NaCl, 5 m*M* 6-aminohexanoic acid, 50 m*M* imidazole/HCl, pH 7.0, and detergent from 10% stock solutions (detergent/protein ratios from 0.5 to 6 g/g; see earlier discussion). This scheme of solubilization usually yields better results than previous protocols, which used 500 m*M* 6-aminohexanoic acid as a dielectric instead of sodium chloride. However, the NaCl concentration should not exceed 50 m*M*. Higher salt concentrations can lead to extreme stacking of proteins (the extremely high protein concentration within these sharp protein bands causes protein aggregregation).

Solubilization is complete within several minutes. The sample is then centrifuged for 30 min at 100,000*g* (if the molecular mass of the desired protein does not exceed 1000 kDa) or for 20 min at 20,000*g* for giant multiprotein complexes up to 10,000 kDa.

The density of the supernatants is usually higher than that of the cathode buffer. Addition of 5–15% glycerol or 200–500 m*M* 6-aminocaproic acid can facilitate sample loading. Shortly before application of the sample to BN–PAGE, Coomassie dye from a 5% suspension in 500 m*M* 6-aminohexanoic acid is added. The amount of Coomassie dye depends on the amount of detergent used. The optimal Coomassie dye/detergent ratio is in the range from 1 : 4 to 1 : 10 g/g. This addition of dye is especially helpful with samples containing high protein and lipid concentrations, which require high detergent concentrations, e.g., 2–4%, for solubilization.

Excess lipid/detergent micelles, which would impede protein resolution, receive a lot of negative charges when incorporating the anionic dye and are removed rapidly as they migrate to the running front. The mixture of a neutral detergent and an anionic dye mimics a mild anionic detergent. For most multiprotein complexes this "anionic detergent" is not critical because a potential denaturing effect is compensated for by the presence of lipids extracted from the membranes.

There are two special situations when the addition of Coomassie dye to the sample is not recommended, and a noncolored sample (or slightly colored by addition of the dye Ponceau S) is applied to the gel for BN–PAGE. The first group includes protein complexes with easily dissociating protein subunits, e.g., protein complexes of the mitochondrial protein import machinery. The second group consists of all samples that are not generated by the direct solubilization of biological membranes, but are prepurified, e.g., by chromatographic protocols. The stabilizing effect of lipids is missing in these samples, as they are usually isolated in a partially or completely delipidated state. In both cases the source of the Coomassie dye is the cathode buffer B or B/10.

We speculate that the beneficial effect of omitting the dye from the sample is due to a preseparation of the desired protein from the bulk of the detergent before Coomassie dye from the cathode buffer can bind to the protein. Proteins with isoelectric points below p*I* 7 migrate to the anode while the bulk of the neutral detergent stays in the sample well. Coomassie dye from the cathode buffer is retarded initially by these detergent micelles, but it catches up with the protein later, when the protein already has entered the gel.

4. Running Conditions

BN–PAGE usually is performed at 4–7°C, as broadening of bands was observed at room temperature; and cathode buffer B is commonly used. Initially, 100 V is applied until the sample has entered the gel. The run is continued with current limited to 15 mA and voltage limited to 500 V. After about one-third of the total running distance, cathode buffer B is removed, and the run is continued using cathode buffer B/10. This buffer exchange improves detection of faint protein bands, and the performance of SDS electrophoresis in the second dimension. Run times are 3–5 h.

5. Electroelution of Native Proteins

Two points are essential for the electroelution of native proteins. First, a band of the desired protein or a band of a marker protein in the vicinity should be detectable during the run. This reduces the number of gel strips to be excised and electroeluted. Second with respect to the recovery of the protein from the gel, it is also essential that BN–PAGE is stopped as soon as the desired complex is resolved as a detectable band, e.g., after half of the normal running distance. At that point the protein still has some mobility and therefore can be efficiently extracted. In contrast, protein complex mobility and the extraction efficiency drop considerably when approaching the pore size limit of the acrylamide gradient gel.

Blue protein bands are excised from the gel and squeezed through a syringe into the cathodic arm of the H-shaped elutor vessel built according to Hunkapiller *et al.* (1983) and Schägger (1994a), and both lower ends are sealed with dialysis membranes with a low cutoff value, e.g., 2 kDa (the mechanical stability of thick low cutoff dialysis membranes is usually much higher than that of normal dialysis membranes). Both arms of the chamber and the horizontal connecting tube are filled with electrode buffer (25 m*M* Tricine, 3.75 m*M* imidazole, pH 7.0, and 5 m*M* 6-aminohexanoic acid are added for protease inhibition). Extraction is performed for several hours at 500 V with the current limited to 2 mA per elutor vessel (to prevent damage if a wrong high salt buffer were used erroneously). Partially aggregated proteins collect as a thin blue layer on the anodic dialysis membrane.

6. Semidry Electroblotting of Native Proteins

As with native electroelution, native electroblotting also requires short runs of BN–PAGE for the efficient transfer of proteins (see earlier discussion). Use of cathode buffer B/10 during BN–PAGE is recommended, because excess Coomassie dye binds strongly to PVDF membranes and reduces the protein-binding capacity. Use of PVDF membranes, not nitrocellulose membranes, is recommended because the latter bind Coomassie dye rather strongly and cannot be destained under the conditions described below.

PVDF membranes, e.g., Immobilon P, are wetted with methanol and soaked in electrode buffer (25 m*M* Tricine, 7.5 m*M* imidazole; the resulting pH is around 7.0). The gel and then the PVDF membrane are placed on a 3-mm stack of papers soaked with electrode buffer (the lower electrode is the cathode in this arrangement). Finally another 3-mm stack of papers soaked with electrode buffer is added, the anode is mounted, and a 5-kg load is put on top before the transfer is started. The transfer at 4°C is completed after about 1 h at 20 V. For background destaining, 25% methanol, 10% acetic acid is used.

C. Second Dimension: SDS–PAGE

The preferred SDS–PAGE system for the resolution of protein subunits in the second dimension is Tricine–SDS–PAGE (Schägger, 1994a; Schägger and Von Jagow, 1987) because protein bands are sharper, 2–3% lower acrylamide concentrations compared to the Tris–glycine system, Laemmli (1970), can be used, and the electrotransfer of proteins to PVDF membranes is facilitated.

Individual lanes from BN–PAGE (0.5 cm) are placed on a glass plate at the usual position for stacking gels and are usually soaked with 1% SDS, 1% mercaptoethanol for up to 2 h. Mercaptoethanol is an efficient inhibitor of acrylamide polymerization. Therefore, it must be removed prior to pouring gels. If cleavage of disulfide bonds is not essential, and if there is no extremely strong protein band in BN–PAGE, pre-incubation in SDS/Mercaptoethanol solutions is not required. SDS originating from the cathode buffer is sufficient to dissociate low amounts of multiprotein complexes within the native gel.

Mount spacers, which can be considerably thinner than the native gel (e.g., 0.7 mm for 1.6-mm native gels), and put the second glass plate on top. The 0.5-cm lanes of the native gel are squeezed to a width of about 1 cm and will not move when the glass plates are brought to a vertical position. Pour the separating gel mixture between the glass plates, usually a 16% T, 3% C mixture, leaving a gap of about 1.5 cm below the native gel strip, and overlay with water. After polymerization, a 10% T, 3% C acrylamide mixture is poured in and overlayed with water, which may contact the native gel. After polymerization the native gel strip is pushed down to the 10% acrylamide gel by using the smooth side of an inverted sample application comb. The empty areas to the left and to the right of the native gel strip finally are filled by a 10% acrylamide native gel mixture (see Table III).

Second-dimension SDS–PAGE of this special gel type, with dimensions 0.07 × 14 × 14 cm, are started at room temperature, with a maximal voltage of 200 V and current limited to maximally 50 mA. When the current drops below 50 mA, the voltage can be increased to 300 V, with the current still limited to maximally 50 mA. The run time is 5–6 h.

III. Applications

There are several ways to proceed after BN–PAGE. For example, individual lanes can be cut out and processed by second-dimension SDS–PAGE. This type of two-dimensional (2D) resolution is especially valuable for analytical purposes, e.g., quantification of OXPHOS protein amounts and analysis of subunit composition after Coomassie or silver staining (Schägger, 1994a,b, 1996). These 2D gels can also be electroblotted to PVDF or nitrocellulose membranes for detection of proteins by specific antibodies. In our opinion, immunodetection after 2D resolution is preferable to immunodetection after electroblotting of blue-native gels.

There are two major ways to use BN–PAGE for preparative purposes. The blue bands of individual proteins or protein complexes can be cut out and electroeluted as described earlier (Schägger and Von Jagow, 1991; Schägger,1995b). These purified blue proteins can be used for immunization and antibody production and for protein chemical work, such as chemical or enzymatic fragmentation and subsequent amino-terminal protein sequencing.

For the latter purpose, an alternative shorter method is preferable. Blue protein bands from preparative native gels (0.3 × 14 × 14 cm) are excised to obtain a 14-cm gel strip with dimensions of about 0.2 × 0.3 × 14 cm. This strip is cut into five pieces, assembled to form a 1.5-cm stack of gel pieces, incubated with SDS/mercaptoethanol as described earlier, and resolved using a 1.6-mm SDS gel. After electroblotting to PVDF membranes, a sufficient amount of protein is usually available to obtain amino-terminal protein sequences by automated Edman degradation.

Electroeluted proteins can be used for enzymatic analyses, with some restrictions (Schägger and Von Jagow, 1991). Removal of the dye, incubation with lipid/detergent solutions, or incorporation into lipid vesicles is often required for this application. However, see also Zerbetto *et al.* (1997) for direct histochemical stain of blue-native polyacrylamide gels.

A. Isolation of Protein Complexes from Mitochondria and Chloroplasts

BN–PAGE originally was developed as a means for isolation of the five membrane multiprotein complexes of the oxidative phosphorylation system (OXPHOS complexes I–V) from bovine mitochondria (Schägger and Von Jagow, 1991; Schägger, 1995b; Zerbetto *et al.* 1997) and for the analysis of the development of cytochrome oxidase isoforms during the perinatal period of the rat (Schägger *et al.,* 1995b). It has also been used for isolation of OXPHOS complexes from yeast mitochondria (Arnold *et al.,* 1998), plant mitochondria (Jänsch *et al.,* 1994), and chloroplasts (Kügler *et al.,* 1997; Burrows *et al.,* 1998), and for the analysis of protein complexes of the mitochondrial protein import machinery of yeast (Dietmeir *et al.,* 1997; Ryan *et al.,* 1999) and plants (Jänsch *et al.,* 1998; Caliebe *et al.,* 1997), including translocase of the outer membrane (Tom), and translocase of the inner membrane (Tim). Especially interesting results have been obtained using BN–PAGE to detect a supercomplex of preprotein translocase complexes Tim and Tom linked by imported preproteins (Dekker *et al.,* 1997).

BN–PAGE is also useful for the isolation of complexes from other biological membranes, e.g., the vacuolar ATPase holoenzyme and its membrane sector from chromaffin granula (Ludwig *et al.,* 1998). All of the complexes dicussed are in the molecular mass range of 100 to 1000 kDa, and one standard gel type (4% sample gel; 5 → 13% acrylamide gradient gel; (see Table II) was usually applied. Here we show that even complexes as large as 10,000 kDa can be isolated when starting the gradients at a lower acrylamide concentration.

1. Giant Protein Complexes from Isolated Mitochondria

Bovine heart mitochondria (330 μg) were solubilized as described earlier using Triton X-100 for solubilization at a detergent/protein ratio of 3 g/g. After a 10-min centrifugation at 15,000*g*, Coomassie blue G-250 dye was added to a dye:detergent ratio of 1 : 4. The solubilized multiprotein complexes were separated by BN–PAGE and their subunits were resolved by SDS–PAGE in the second dimension (Fig. 1). Use of dodecylmaltoside gave similar results.

2. Determination of Molecular Masses of Native Protein Complexes

From the separation of multiprotein complexes in Fig. 1 it is apparent that BN–PAGE separates proteins according to their molecular masses. Many commercially available water-soluble standard proteins and the membrane protein complexes from bovine heart mitochondria have been shown to fit a general calibration curve (Schägger, 1994b; Schägger *et al.,* 1994) with maximal deviations of 20% (with the exception of some extremely basic membrane proteins and some water-soluble proteins that do not bind Coomassie dye). Making use of this fast high-resolution technique (compared to gel filtration and analytical ultracentrigation), it was possible to discriminate between dimeric and oligomeric states of enzymes, using microgram amounts of partially purified protein (Schägger *et al.,* 1994). The technique was also applied for control of purity and homogeneity of column eluates, for the detection of subcomplexes generated by dissociation of individual subunits, and for analysis of possible protein–protein interactions of two proteins found in the same fraction (Schägger *et al.,* 1994).

Comparing different forms of a specific multiprotein complex, e.g., dimeric and monomeric forms, or sub complexes generated by dissociation of one or several subunits, the apparent molecular mass differences fit ideally to the calculated differences. Dissociation of a single subunit representing, e.g., 5% of the total molecular mass of a complex, can be detected.

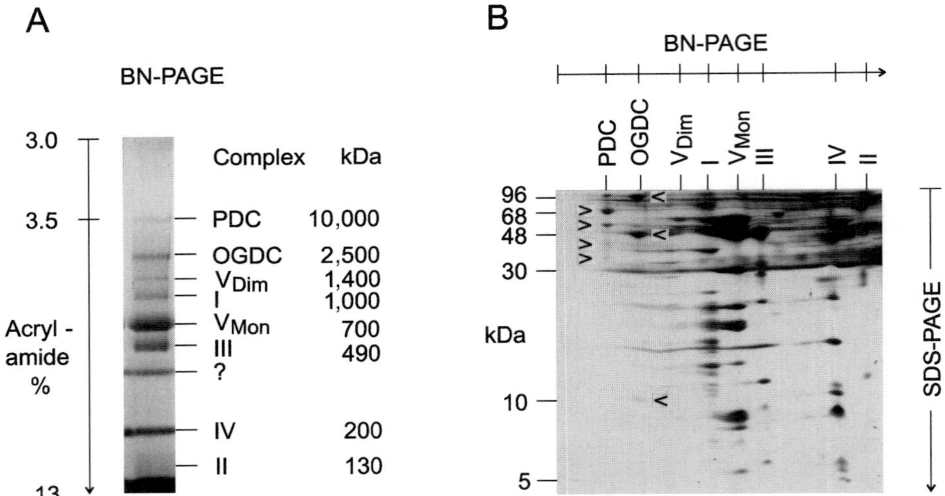

Fig. 1 Isolation of giant multiprotein complexes from bovine heart mitochondria. (A) Multiprotein complexes from bovine heart mitochondria were solubilized by Triton X-100 and separated by BN–PAGE using a linear acrylamide gradient gel from 3.5 to 13%, overlaid by a 3% sample gel. The F_1F_0-ATP synthase (also called complex V) was identified in a major monomeric form (V_{Mon}) and a minor dimeric form V_{Dim}). OGDC, oxoglutarate decarboxylase complex; PDC, pyruvate decarboxylase complex. (B) Subunits of native complexes resolved by BN–PAGE were separated by Tricine–SDS–PAGE using a 16.5% T, 3% C gel type. The identity of PDC and OGDC was confirmed by amino-terminal sequencing of the 68-kDa subunit E2 of PDC and the 96- and 48-kDa subunits E1 and E2, respectively, of OGDC.

However, the apparent masses determined by BN–PAGE potentially vary with glycosylation. We could not discriminate unambiguously between the monomeric or a dimeric form of the glycosylated membrane sector of V-ATPase (Ludwig *et al.,* 1998). We also noticed that the apparent masses of specific membrane protein complexes can differ by 20% using different detergents, although the subunit composition is identical (e.g., 600 kDa using Triton X-100 or dodecylmaltoside compared to 750 kDa using digitonin). We assume that this shift of apparent molecular masses is due to a different degree of delipidation (less delipidation using digitonin leads to higher apparent masses). This is in line with the observation that use of low Triton X-100 or dodecylmaltoside/protein ratios for solubilization also causes a shift to higher apparent masses.

B. Isolation of Protein Complexes from Total Tissue Homogenates and Cell Lines

Mitochondrial protein may be equal to or more than 5–10% of the total cellular protein in skeletal muscle or heart muscle. Therefore, we attempted isolation of OXPHOS complexes directly from total tissue homogenates without a separate preparation of mitochondria. Our ultimate goal was to use this technique for the localization of deficiencies of OXPHOS complexes in human mitochondrial encephalomyopathies. Because only

10–20 mg of total skeletal muscle can be obtained by needle biopsy from the patients, the starting material was not sufficient for a careful subcellular fractionation. The two-dimensional technique using BN–PAGE in the first dimension and SDS–PAGE in the second dimension has been applied successfully for the localization of OXPHOS deficiencies (loss of assembly of individual or multiple OXPHOS complexes) in mitochondrial encephalomyopathies (Bentlage *et al.,* 1995; Schägger *et al.,* 1996) and in other disorders with deficiencies in OXPHOS complexes (Schägger and Ohm, 1995; Kretzler *et al.,* 1999). The protocols for processing different tissues (Schägger, 1996) and cell lines (Schägger *et al.,* 1996) has been described elsewhere and will not be repeated here.

C. Final Purification by BN–PAGE of Chromatographically Prepurified Complexes

1. Preparative Aspects

Purification of membrane proteins and complexes is often difficult, and only partial purification may be achieved by conventional isolation protocols. BN–PAGE is a valuable means for final purification of these problematic proteins and has been used as a final purification step for preparation of protein for antibody production and for protein chemical work. An example for such a final purification is described in (Schägger, 1995b). About 1–2 mg of a partially purified OXPHOS complex could be loaded per preparative gel and recovered from each gel by native electroelution.

First rule: the salt concentration of the sample to be applied for preparative BN–PAGE should be as low as possible, but high enough to keep proteins in solution e.g., 25–50 m*M* NaCl. 6-Aminohexanoic acid (500 m*M*) is very effective for the solubilization of biological membranes by neutral detergents, but it has only a negligible effect for the solubility of partially purified proteins and cannot replace NaCl during gel filtration. If a protein, for example, is eluted from an anion exchanger by 300 m*M* NaCl, salt reduction by gel filtration or dialysis is absolutely required. When using this approach it is important to test the eluate or dialysate for protein aggregation (e.g., by centrifugation).

Second rule: Avoid an excess of detergent for elution of the desired protein from a chromatograpohic column because the combination with the anionic Coomassie dye used for BN–PAGE generates a mild anionic detergent. However, too low detergent concentrations may cause oligomerization. The minimal detergent concentrations are 0.05–0.1% for Triton X-100 and 0.1–0.2% for dodecylmaltoside.

Third rule: Do not add Coomassie dye to the sample if the desired protein complex contains labile subunits. The Coomassie dye required is supplied with cathode buffer B. Cathode buffer B/10 often can be used directly when the detergent concentration in the sample is as low as 0.2%.

2. Analysis of Neurotransmittor Receptor Assembly

The assembly of distinct combinations of nicotinic acetylcholine receptor subunits has been analyzed by BN–PAGE after prepurification of digitonin extracts from oocytes by Ni^{2+} NTA agarose chromatography (Nicke *et al.,* 1999). Using a similar protocol the oligomeric states and the assembly of the $P2X_1$, $P2X_3$, and glycine receptors have been analyzed (Nicke *et al.,* 1998; Griffon *et al.,* 1999). Coomassie dye was added to the sample before application to the native gel, which is not in line with the "third rule" (see earlier discussion). The observation of distinct oligomeric structures in these cases indicates that rather stable assemblies exist.

IV. Outlook

The application range of BN–PAGE gradually expanded from the analysis of mitochondrial OXPHOS complexes from bovine and human tissue (mitochondrial encephaolomyopathies) to the analysis of OXPHOS systems from plant mitochondria, yeast OXPHOS mutants, and chloroplasts. BN–PAGE has been applied with great success to the analysis of mitochondrial protein import complexes and of receptors in signal transduction.

Some publications demonstrate that BN–PAGE is also a valuable technique for the identification of physiological protein–protein interactions: preprotein translocase complexes linked together by their substrates, the preproteins (Dekker *et al.,* 1997), yeast ATP synthase existing in a dimeric state (Arnold *et al.,* 1998), and respiratory chain complexes from yeast and mammalian mitochondria organized to form a network of supercomplexes (Schägger and Pfeiffer, 2000). This feature of BN–PAGE seems to be an essential one, as only two other techniques for the detection of physiological protein–protein interactions are available: immunoprecepitation and the two-hybrid system, which often has been used successfully, but often produces false-positive results.

We expect that the application of BN–PAGE will expand to the analysis of further receptors in the plasma membrane and in the membranes of cellular organelles, to the analysis of protein–chaperone interactions, and to the detection and analysis of antigen–antibody interactions. Protein–antibody interactions are stable under the conditions of BN–PAGE, as deduced from a shift of the apparent molecular mass of the protein antigen, and detection of IgG heavy and light chains in 2D SDS–PAGE at the position of the protein antigen (Schägger, unpublished).

Acknowledgment

This work was supported by grants of the Deutsche Forschungsgemeinschaft Sonderforschungsbereich 472, Frankfurt, and the Fonds der Chemischen Industrie.

References

Arnold, I., Pfeiffer, K., Neupert, W., Stuart, R. A., and Schägger, H. (1998). Yeast F_1F_0-ATP synthase exists as a dimer: Identification of three dimer specific subunits. *EMBO J.* **17,** 7170–7178.

Bentlage, H., De Coo, R., Ter Laak, H., Sengers, R., Trijbels, F., Ruitenbeek, W., Schlote, W., Pfeiffer, K., Gencic, S., von Jagow, G., and Schägger, H. (1995). Human diseases with defects in oxidative phosphorylation. I. Decreased amounts of assembled oxidative phosphorylation complexes in mitochondrial encephalomyopathies. *Eur. J. Biochem.* **227,** 909–915.

Burrows, P. A., Sazanov, L. A., Svab, Z., Maliga, P., and Nixon, P. J. (1998). Identification of a functional respiratory complex in chloroplasts through analysis of tobacco mutants containing disrupted plastid Ndh genes. *EMBO J.* **17,** 868–876.

Caliebe, A., Grimm, R., Kaiser, G., Lubeck, J., Soll, J., and Heins, L. (1997). The chloroplastic protein import machinery contains a Rieske-type iron cluster and a mononuclear iron-binding protein. *EMBO J.* **16,** 7342–7350.

Dekker, P. J. T., Martin, F., Maarse, A. C., Bömer, U., Müller, H., Guiard, B., Meijer, M., Rassow, J., and Pfanner, N. (1997). The Tim core complex defines the number of mitochondrial translocation contact sites and can hold arrested preproteins in th absence of matrix Hsp70-Tim44. *EMBO J.* **16,** 54048–5419.

Dekker, P. J. T., Ryan, M. T., Brix, J., Müller, H., Hönlinger, A., and Pfanner, N. (1998). Preprotein translocase of the outer mitochondrial membrane: Molecular dissection and assembly of the general import pore complex. *Mol. Cell. Biol.* **18,** 6515–6524.

Dietmeier, K., Hönlinger, A., Bömer, U., Dekker, P. J. T., Eckerskorn, C., Lottspeich, F., Kübrich, M., and Pfanner, N. (1997). Tom5 functionally links mitochondrial preprotein receptors to the general import pore. *Nature* **388,** 195–200.

Dionisi-Vici, C., Ruitenbeek, W., Fariello, G., Bentlage, H., Wanders, R. J. A., Schägger, H., Piantadosi, C., Sabetta, G., and Bertini, E. (1997). New familial mitochondrial encephalopathy with macrocephaly, cardiomyopathy and complex I deficiency. *Ann. Neurol.* **42,** 661–665.

Griffon, N., Büttner, C., Nicke, A., Kuhse, J., Schmalzing, G., and Betz, G. (1999). Molecular determinants of glycine receptor subunit assembly. *EMBO J.* **18,** 4711–4721.

Hill, K., Model, K., Ryan, M. T., Dietmeier, K., Martin, F., Wagner, R., and Pfanner, N. (1998). Tom40 forms the hydrophilic channel of the mitochondrial import pore for preproteins. *Nature* **395,** 516–521.

Hunkapiller, M. W., Lujan, E., Ostrander, F., and Hood, L. E. (1983). Isolation of microgram quantities of proteins from polyacrylamide gels for amino acid sequence analysis. *Methods Enzymol.* **91,** 227–236.

Jänsch, L., Kruft, V., Schmitz, U., and Braun, H. P. (1996). New insights into the composition, molecular mass and stoichiometry of the protein complexes of plant mitochondria. *Plant J.* **9,** 357–368.

Jänsch, L., Kruft, V., Schmitz, U. K., and Braun, H. P. (1998). Unique composition of the preprotein translocase of the outer mitochondrial membrane from plants. *J. Biol. Chem.* **273,** 17251–17257.

Kretzler, M., Haltia, A., Solin, M.-L., Schägger, H., Taanman, J.-W., Kritz, W., Kerjaschki, D., Schlödorff, D., and Holthöfer, H. (1999). Altered gene expression and functions of mitochondria in human nephrotic syndrome. *FASEB J.* **13,** 523–532.

Kügler, M., Jänsch, L., Kruft, V., Schmitz, U. K., and Braun, H. P. (1997). Analysis of the chloroplast protein complexes by blue-native polyacrylamide gel electrophoresis. *Photosynth. Res.* **53,** 35–44.

Kurz, M., Martin, H., Rassow, J., Pfanner, N., and Ryan, M. T. (1999). Biogenesis of Tim proteins of the mitochondrial carrier import pathway: Differential targeting mechanisms and crossing over with the main import pathway. *Mol. Cell. Biol.* **10,** 2461–2474.

Laemmli, U. K. (1970). Cleavage of structural proteins during the assembly of the head of bacteriophage T4. *Nature* **227,** 680–685.

Ludwig, J., Kerscher, S., Brandt, U., Pfeiffer, K., Getlawi, F., Apps, D. K., and Schägger, H. (1998). Identification and characterization of a novel 9.2 kDa membrane sector associated protein of vacuolar proton-ATPase from chromaffin granules. *J. Biol. Chem.* **273,** 10939–10947.

Nicke, A., Baumert, H. G., Rettinger, J., Eichele, A., Lambrecht, G., Mutschler, E., and Schmalzing, G. (1998). $P2X_1$ and $P2X_3$ receptors form stable trimers: A novel structural motif of ligand-gated ion channels. *EMBO J.* **17,** 3016–3028.

Nicke, A., Rettinger, J., Mutschler, E., and Schmalzing, G. (1999). Blue native PAGE as a useful method for the analysis of the assembly of distinct combinations of nicotinic acatylcholine receptor subunits. *J. Receptor Signal Transd. Res.* **19,** 493–507.

Ryan, M. T., Müller, H., and Pfanner, N. (1999). Functional staging of ADP/ATP carrier translocation across the outer mitochondrial membrane. *J. Biol. Chem.* **274,** 20619–20627.

Schägger, H. (1994a). Denaturing electrophoretic techniques. *In* "A Practical Guide to Membrane Protein Purification" (G. Von Jagow and H. Schägger, eds.). Academic Press, San Diego.

Schägger, H. (1994b). Native gel electrophoresis. *In* "A Practical Guide to Membrane Protein Purification" (G. Von Jagow and H. Schägger, eds.). Academic Press, San Diego.

Schägger, H. (1995a). Quantification of oxidative phosphorylation enzymes after blue native electrophoresis and two-dimensional resolution: Normal complex I protein amounts in Parkinson's disease conflict with reduced catalytic activities. *Electrophoresis* **16,** 763–770.

Schägger, H. (1995b). Native electrophoresis for isolation of mitochondrial oxidative phosphorylation protein complexes. *Methods Enzymol.* **260,** 190–202.

Schägger, H. (1996). Electrophoretic techniques for isolation and quantitation of oxidative phosphorylation complexes from human tissues. *Methods Enzymol.* **264,** 555–566.

Schägger, H., Bentlage, H., Ruitenbeek, W., Pfeiffer, K., Rotter, S., Rother, C., Böttcher-Purkl, A., and Lodemann, E. (1996). Electrophoretic separation of multiprotein complexes from blood platelets and cell lines: Technique for the analysis of diseases with defects in oxidative phosphorylation. *Electrophoresis* **17,** 709–714.

Schägger, H., Brandt, U., Gencic, S., and von Jagow, G. (1995a). Ubiquinol-cytochrome c-reductase from human and bovine mitochondria. *Methods Enzymol.* **260,** 82–96.

Schägger, H., Cramer, W. A., and Von Jagow, G. (1994). Analysis of molecular masses and oligomeric states of protein complexes by blue native electrophoresis and isolation of membrane protein complexes by two-dimensional native electrophoresis. *Anal. Biochem.* **217,** 220–230.

Schägger, H., Noack, H., Halangk, W., Brandt, U., and von Jagow, G. (1995b). Cytochrome c oxidase in developing rat heart: Enzymatic properties and aminoterminal sequences suggest identity of the fetal heart and the adult liver isoform. *Eur. J. Biochem.* **230,** 235–241.

Schägger, H., and Ohm, T. (1995). Human diseases with defects in oxidative phosphorylation. II. F_1F_0 ATP-synthase defects in Alzheimer's disease revealed by blue native polyacrylamide gel electrophoresis. *Eur. J. Biochem.* **227,** 916–921.

Schägger, H., and Von Jagow, G. (1987). Tricine-sodium dodecyl sulfate polyacrylamide gel electrophoresis for the separation of proteins in the range from 1-100 kDalton. *Anal. Biochem.* **166,** 368–379.

Schägger, H., and Von Jagow, G. (1991). Blue native electrophoresis for isolation of membrane protein complexes in enzymatically active form. *Anal. Biochem.* **199,** 223–231.

Schägger, H., and Pfeiffer, K. (2000). Super complexes in the respiratory chains of yeast and mammalian mitochondria. *EMBO J.* **19,** 1777–1783.

Van Wilpe, S., Ryan, M. T., Hill, K., Maarse, A. C., Meisinger, C., Brix, J., Dekker, P. J. T., Mocko, M., Wagner, R., Meijer, M., Guiard, B., Hönlinger, A., and Pfanner, N. (1999). Tom22 is a multifunctional organizer of the mitochondrial preprotein translocase. *Nature* **401,** 485–489.

Zerbetto, E., Vergani, L., and Dabbeni-Sala, F. (1997). Quantification of muscle mitochondrial oxidative phosphorylation enzymes via histochemical staining of blue native polyacrylamide gels. *Electrophoresis* **18,** 2059–2064.

CHAPTER 14

Application of Electron Tomography to Mitochondrial Research

Carmen A. Mannella

Resource for the Visualization of Biological Complexity
Wadsworth Center
Albany, New York 12201

I. Introduction

Some of the first, and at times most controversial, biological applications of transmission electron microscopy (TEM) were investigations into the internal organization of the mitochondrion [see review of the pioneering work of Palade, Sjostrand, and others by Rasmussen (1995)]. Electron micrographs of thin sections cut from plastic blocks and stained with heavy metals show mitochondria in profile, with a smooth outer membrane enveloping a more complicated looking inner membrane. The images in Munn's (1974) "The Structure of Mitochondria" graphically depict the wide variety of morphologies displayed by the mitochondrial inner membrane in different organisms, in different tissues of the same organism, and in different metabolic states of the same mitochondria. However, thin-section electron micrographs are inherently two dimensional and their interpretation is not unambiguous. This limitation has led, over time, to the formulation of several competing models for the three-dimensional (3D) structure of the inner membrane, despite the common perception of a consensus model (e.g., Mannella, 1994).

The two dimensionality of conventional TEM imaging is imposed by the physics of the interaction of electrons with the specimen. Most structural information in electron micrographs is provided by electrons scattered elastically (i.e., without energy loss to the specimen) only once. Heavy-metal staining of specimens is used because it increases the ratio of elastic to inelastic scattering events, which improves contrast and reduces the extent of beam damage (resulting from transfer of energy to the specimen). However, for incident electrons of a given energy, as specimen thickness increases, the fraction of singly elastically scattered electrons drops and the fractions of singly inelastically and multiply elastically and inelastically scattered electrons increase, reducing image contrast greatly. Electron microscopes in most laboratories operate at an accelerating voltage of 100 kV, for which optimum specimen thickness is below 100 nm. So-called thin sections of plastic-embedded specimens are typically 80 nm thick. Mitochondria have dimensions on the order of hundreds or thousands of nanometers, with discernible detail in micrographs on the order of 5–10 nm. Thus, TEM projection images through 80-nm-thick slices at the same time underrepresent the organelle as a whole and undersample (in the third or z direction) features of interest.

TEM imaging of serially cut thin sections can be a useful strategy for obtaining 3D information (e.g., McEwen and Marko, 1999). When applied to mitochondria by Daems and Wisse (1966), the results indicated that the standard, textbook model for mitochondria, depicting the inner membrane with accordion-like folds or baffles, is inaccurate. However, conventional serial sectioning cannot be used to reconstruct the complicated internal structure of mitochondria, as the z resolution is limited by the thickness at which plastic blocks can be sliced uniformly (Winslow *et al.,* 1991). Even if it were possible to routinely serially cut 5- to 10-nm sections, it would take on the order of 100 such sections to reconstruct a significant portion of a mitochondrion.

An important advance toward achieving 3D TEM imaging was the introduction in the 1960s of electron microscopes that operate at significantly higher accelerating voltages, up to and exceeding 1000 kV. The more energetic electrons produced by these instruments have a longer mean free path for elastic scattering, allowing useful images to be obtained from specimens hundreds or a few thousands of nanometers thick. By the late 1970s, there were three high-voltage (1000 kV) electron microscopes (HVEMs) dedicated for biological applications in the United States alone. However, the ability to record electron micrographs from thick specimens was, by itself, insufficient for meaningful 3D imaging. The next challenge was how to interpret the tremendous overlap of structural information contained in projection images through hundreds of nanometers of biological material (e.g., Fig. 1A). In many cases, stereo viewing of pairs of TEM images recorded from specimens tilted at small angles can help sort out structural detail. However, obtaining the full range of structural information in thick biological specimens was not achieved until the development of electron tomography, the reconstruction of a volume from scores of micrographs recorded as the specimen is tilted around one or more axes (Frank, 1992). The first HVEM tomographic reconstruction of a biological specimen was published by McEwen *et al.* (1986) and the first tomographic study of mitochondria by Mannella *et al.* (1994).

Tomography of thick, plastic section specimens has the potential to become a routine tool for the 3D analysis of biological structure. Furthermore, advances in methodology are leading to the extension of this 3D imaging technique to frozen-hydrated specimens

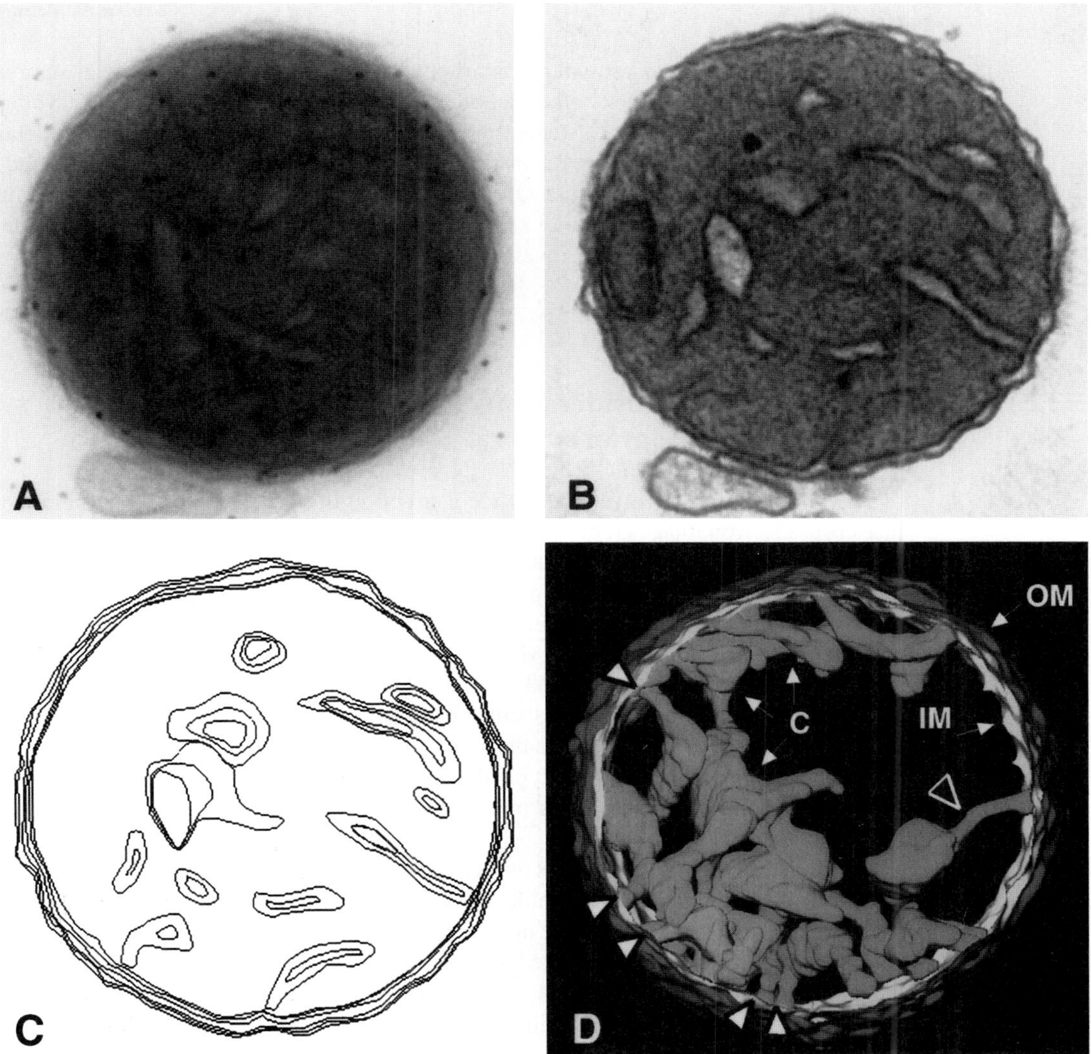

Fig. 1 Tomographic reconstruction of a conventionally fixed and plastic-embedded rat liver mitochondrion. (A) Projection image recorded at 1000 kV of a 0.5-μm-thick section of an isolated, partially "condensed" rat liver mitochondrion 1.5 μm in diameter. (B) Five-nanometer slice through a tomogram reconstructed by weighted back-projection of a double-tilt series of images from the mitochondrion in A by the procedure of Penczek *et al.* (1995). (C) Contours drawn of membrane profiles in three slices from the 3D reconstruction, 15 nm apart. (D) Surface-rendered model of the mitochondrion generated from a full set of membrane contours showing the outer membrane (OM), inner boundary membrane (IM), and selected cristae (C). Arrowheads point to narrow (ca. 30 nm) tubular regions of the cristae that connect the internal compartments to each other and to the inner boundary membrane. Reproduced with permission from Mannella (2000).

(see later). However, at present, tomography is usually undertaken only at national resource centers and a few other laboratories where the necessary higher voltage microscopes and computer systems are available for the biological user. The details of tomographic data collection, reconstruction, and visualization have been presented at length in McEwen and Marko (1999). This chapter explains the advantages and limitations of the technique as applied to mitochondrial research and outlines the important steps in a typical tomographic investigation, with special attention to optimization of specimen preparation.

II. Preliminary Considerations for Application of Electron Tomography

TEM tomography involves the 3D reconstruction of a selected specimen field from numerous electron images representing different angular views. The process of data collection, computation of the reconstruction, and detailed visualization may take several days. Thus, sampling is a serious problem and makes tomography ill-suited to address questions that require statistical answers, e.g., does a particular metabolic change increase or decrease mitochondrial volume. Other techniques like stereology are better suited to answer such questions as they involve random sampling of large numbers of 2D profiles, from which 3D conclusions may be legitimately drawn (e.g., Weibel, 1973). In contrast, tomography can provide valuable information about the 3D structure of a few typical (or interesting atypical) mitochondria in a particular specimen. This information can be valuable in its own right and can be used to guide other kinds of analyses.

The first 3D tomographic images of rat liver mitochondria produced by Mannella *et al.* (1994, 1997) demonstrated that, unlike the standard model shown in textbooks, the inner membrane involutions (cristae) are not simply accordion-like baffles. Instead, the cristae are pleiomorphic and are connected by narrow, sometimes very long, tubular regions to each other and to the peripheral region of the inner membrane (Fig. 1D). Restricted connections of the cristae to the inner membrane of rat liver mitochondria were first inferred by Daems and Wisse (1966), who called them *pediculi.* Perkins *et al.* (1997) confirmed their presence in other kinds of mammalian mitochondria and determined the average diameter of the features, which they called *crista junctions,* to be 28 nm by careful measurements from multiple tomographic reconstructions. Another structural feature of interest in electron micrographs of thin sections of conventionally prepared mitochondria are the sites of close apposition or contact between outer and inner membranes, first described by Hackenbrock (1966). Tomographic analysis has shown the contact regions to be punctate not linear (cf. Van Venetie and Verkleij, 1982), ranging in size from 14 to 40 nm (Mannella *et al.,* 1996; Perkins *et al.,* 1997). By knowing the shape and average dimensions of crista junctions and contact sites, their frequency/surface area can be reliably determined in cells or isolated mitochondrial fractions by analysis of conventional thin-section micrographs.

Answering specific questions about possible effects of osmotic, metabolic, or genetic changes on mitochondrial structure, and of mitochondrial structure on function, will often involve the use of tomography in conjunction with other experimental or computational techniques. Correlative approaches are possible with tissue culture cells, in which an event is observed by light microscopy, followed by fixation and 3D electron microscopy of the field of interest (Rieder and Cassels, 1999). For mitochondrial suspensions, techniques such as polarography, spectroscopy, or light scattering can be used to monitor events prior to fixation (e.g., Mannella *et al.*, 2000). Once obtained, tomographic data can be the input for modeling studies of cellular processes. For example, the virtual cell program (Schaff *et al.*, 1997) is being used to map the distribution of important metabolites such as ADP within mitochondrial compartments whose 3D shapes have been provided by tomography (Moraru *et al.*, 2000).

Finally, tomography is a team effort, with optimum results accruing from close collaboration between microscopists and biologists. Critical steps, such as the selection of the field to be reconstructed and the appropriate modes of visualizing the resulting 3D volumes, will usually require insights about a system that can only be provided by the experimentalist asking the biological questions. Electron microscopy facilities that provide tomography on a collaborative or service basis for biological users have internet-based capabilities for real-time remote viewing of specimens and specialized, interactive computer visualization tools that can be used on-site or distributed. Examples are described on the websites of NIH-supported national resource centers such as those in Albany, New York (www.wadsworth.org/rvbc), San Diego, California (www-ncmir.ucsd.edu), and Boulder, Colorado (bio3d.colorado.edu).

III. Methodology

A. Specimen Preparation

1. Conventional Plastic-Embedded Specimens

In general, for a given biological specimen, tomographic data collection can be done using whichever fixation and embedding protocol yields optimum results for conventional thin-section TEM imaging of that specimen. Conventional procedures typically involve glutaraldehyde fixation, osmication, dehydration, infiltration by the plastic monomer and catalyst, and polymerization. These blocks are cut with standard ultramicrotomes into sections that can range from 250 nm to over 1000 nm. The thick sections are poststained with uranyl and lead salts, using protocols that involve longer washes than typically used for thin sections. Colloidal gold is deposited on one or both surfaces of the specimen to serve as fiducial markers for alignment programs.

Of the two most commonly used embedding media, "Epon" is preferred over Spurr's resin for tomography because of better contrast (apparently due to easier infiltration of the stain into the thick sections) and less beam-induced distortion of the sections during data collection (see later). Age of the blocks is not a factor in data collection. Tomography can be done on thick sections cut from blocks that are many years old, allowing the use of 3D imaging to re-investigate previous experiments.

As a rule, procedures that enhance the contrast of features of interest without disrupting normal ultrastructure should be considered when imaging thick sections in high-voltage electron microscopes. For example, fixation protocols employing ferrocyanide-reduced osmium may improve the contrast of membranes relative to matrix in plastic-embedded isolated mitochondria significantly (Willingham and Rutherford, 1984).

Techniques exist for fixation and embedding of tissue culture cells grown on glass coverslips for light microscopic experiments. Thick sections cut from flat-embedded cells (after dissolving away the glass in hydrofluoric acid) have been used for tomographic analysis of the mitotic apparatus (Khodjakov *et al.,* 1997). However, it should be noted that exposure of tissue culture cells to light in the presence of at least one class of fluorescent dyes commonly used to label mitochondria (MitoTracker Red) leads to disruption of the mitochondrial structure (Minamikawa *et al.,* 1999; Smaili, Buttle, Russell, and Mannella, unpublished observations). Optimum conditions for correlative light/electron microscopy of mitochondria in tissue culture cells are under investigation.

2. Cryosubstitution and Related Techniques

There is considerable evidence that techniques involving rapid freezing and low-temperature replacement of water by organic solvents and plastic provide superior retention of native structure in biological specimens compared to the conventional procedure described earlier. Numerous protocols fall under the general heading of cryosubstitution (noted in McEwen and Marko, 1999). Results from the tomographic reconstruction of a 220-nm-thick section of a mitochondrion in a specimen of cryosubstituted rat liver are shown in Figs. 2A and 2B (see color plate; Mannella *et al.,* 1998). The essential 3D organization of this mitochondrion is very similar to that of mitochondria in rat liver prepared by perfusion-fixation with glutaraldehyde and room-temperature dehydration (not shown). There is less interconnectivity of the cristae, which is typical for mitochondria in the so-called "orthodox" state normally seen *in situ.* The mitochondrion of Fig. 1 has a somewhat contracted matrix and larger intracristal spaces, often seen in isolated mitochondria. It is said to be in a partially "condensed" state. However, the significant feature of restricted junctions of cristae to the peripheral surface of the inner membrane is observed in all cases. Cryosubstitution protocols usually involve extensive osmication of the specimen. This can result in a very dense matrix and thicker membrane profiles than observed in conventionally prepared specimens. There is usually no observable space between outer and inner membranes, which makes contact sites difficult to discern. Whether this is a result of artificial thickening of the membranes and obscuring of detail by osmium or an indication that the inner and outer membranes are normally in very close apposition is currently a matter of discussion.

Cells or tissue samples can be frozen rapidly by several types of procedures, including freeze-slamming and high-pressure freezing (e.g., Sitte, 1996). The latter technique provides optimum preservation of structure throughout many microns of material due to inhibition of ice crystal formation at high pressure. Whichever technique is employed,

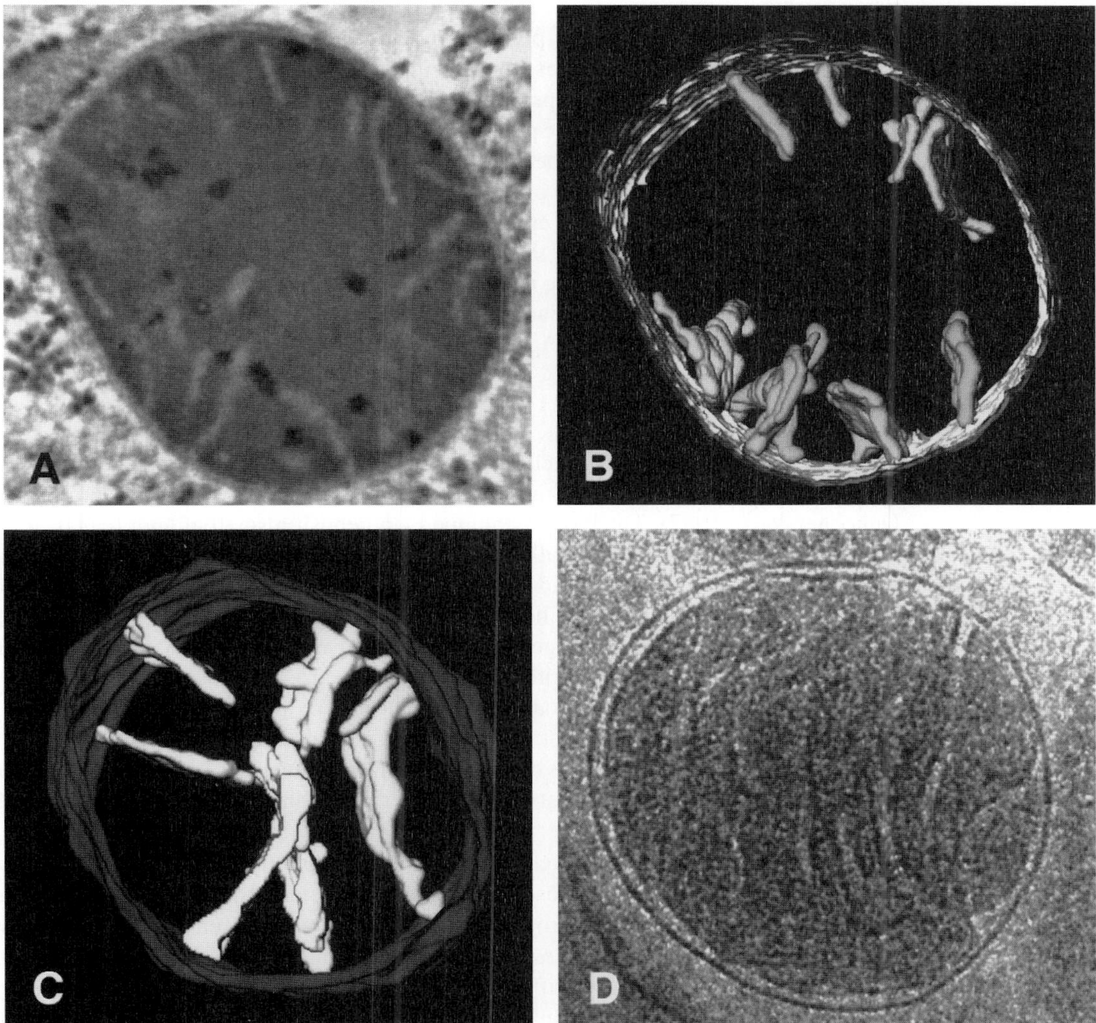

Fig. 2 Application of cryotechniques to mitochondrial imaging. (A) A 5-nm slice from a tomogram of an "orthodox" mitochondrion (700 nm across, 220 nm thick) in a specimen of high-pressure frozen, cryosubstituted rat liver. (B) Surface-rendered model of the mitochondrion in A showing outer and inner boundary membranes and a few of the cristae. A and B reproduced with permission from Mannella *et al.* (1998). (C) Three-dimensional reconstruction of a mitochondrion (900 nm across, 250 nm thick) in a section of rat liver prepared by the procedure of Tokuyasu (1989). Reproduced with permission from Mannella *et al.* (1997). (D) Cryoelectron microscopic image of an unstained, frozen-hydrated intact mitochondrion (500 nm diameter) isolated from *Neurospora crassa.* (see Color Plate.)

when working with tissue, it is essential that the time between dissection of the material and freezing be kept as short as possible. To this end, needle-biopsy type procedures have been developed for high-pressure freezing applications (Hohenberg *et al.,* 1996).

In general, procedures used to produce frozen sections for histological and clinical applications do not provide adequate preservation of ultrastructure for electron microscopy. A cryosectioning procedure has been developed by Tokuyasu (1989) for TEM immunolabeling that involves infiltration of the specimen with a cryoprotectant (a concentrated sucrose solution) and subsequent embedding in a polyalcohol. The procedure is intended to optimize immunoreactivity of the specimen, while sacrificing ultrastructural preservation. Interestingly, tomographic reconstruction indicates that the internal organization of mitochondria in rat liver prepared by this technique is preserved, in particular, the tubular junctions of the cristae (Fig. 2C, see color plate).

3. Frozen-Hydrated Specimens

Suspensions of isolated organelles such as mitochondria can be embedded in vitreous (noncrystalline) ice without any chemical fixation or heavy-metal staining (e.g., Dubochet *et al.,* 1987). The basic technique is to deposit a small aliquot (ca. 10 μl) of mitochondrial suspension (typically in a buffer containing 0.3 *M* sucrose or mannitol) on a standard TEM specimen grid covered with a holey carbon film, then partially blot and rapidly plunge the grid into liquid ethane in thermal contact with liquid nitrogen. (Direct plunging into liquid nitrogen is avoided because of problems with cavitation and resulting poor thermal transfer with the warm specimen.) The thickness of the ice layer in these specimens can be several hundred nanometers, depending on the amount of suspension applied and the extent of evaporation and blotting of the specimen prior to plunging. Flattening of organelles caused by thinning of the water layer prior to plunging improves contrast but also can distort normal membrane structure and usually should be kept to a minimum.

Frozen-hydrated specimens are transferred to and directly imaged (using low dose conditions) in an electron microscope by means of a cryotransfer specimen holder. Figure 2D (see color plate) is a projection image of an intact frozen-hydrated mitochondrion isolated from *Neurospora crassa.* Note that the contrast is not due to stain (because none is used) but to the inherent differences in the mass density of membranes and matrix relative to water. Although the image is noisier than those of conventionally fixed and stained plastic sections, there is considerable detail, e.g., contact sites between outer and inner membrane are visible in the upper segment of the organelle. The details of data collection for cryotomography are described in the following section.

B. Data Collection and Analysis

1. Electron Microscopy

At the Albany microscopy resource, two electron microscopes are available for tomographic data collection from thick specimens: a 1200-kV AEI EM7 high-voltage electron microscope (HVEM) and a 400-kV JEOL JEM 4000FX intermediate-voltage instrument (IVEM). The performance of both TEMs with plastic sections less than 500 nm thick is comparable. The HVEM offers important advantages for thicker plastic sections: shorter exposure times and greater depth of focus (2600 nm vs 1300 nm under routine imaging conditions). This is especially important with tilted sections because the effective thickness doubles at 60° tilt. The newer IVEM has the advantage of computer control, allowing fully automated data collection, important for minimizing dose during imaging of beam-sensitive specimens (see later).

In principle, images should be obtained from a specimen over a full 180° tilt range so that all possible angular views are represented. The increment at which these views are recorded defines the resolution of the final reconstruction. In practice, images are collected in a series of tilts around one axis, spanning (minimally) 120° (+/−60°) in 2° increments, followed (if possible) by rotation of the specimen by 90° and collection of another tilt series (Penczek *et al.,* 1995). The resulting "missing wedge" (one tilt axis) or "missing pyramid" (two axes) of information leads to a loss of resolution by a factor of approximately 2–3 in the z direction, i.e., if the resolution in the plane of the section is 5 nm, it is 10–15 nm in z. In the course of tracking, focusing and recording 100+ images with normal manual operation, specimens receive radiation doses of several hundreds to thousands of electrons per square angstrom. The irradiation initially induces a rapid, apparently uniform shrinkage of thick plastic sections in the z direction by approximately 20% (Deng *et al.,* 1999). During data collection, irradiation can occasionally cause sections to buckle or bend, effectively ending useful data collection from a given field. Procedures to minimize beam exposure (e.g., using image intensifier cameras for scanning fields and focussing) are available on the HVEM that effectively reduce the frequency of the latter problem with plastic sections. In addition, cooling sections to liquid nitrogen temperature (using cryo stages) may also reduce beam-induced specimen shrinkage (e.g., Lamvik, 1981).

The IVEM is equipped with a 1024 × 1024 pixel, cooled-charge-coupled device (CCD) camera and computer-controlled image acquisition system that allows automated data collection and off-axis focusing (Rath *et al.,* 1997). Using this system, total specimen irradiation can be kept to less than 100 electrons per square angstrom for a single-axis tilt series of a field 1000 × 1000 nm^2. This is sufficient for successful tomographic reconstruction of mitochondria embedded in vitreous ice over 500 nm thick at a resolution of 6–12 nm (Mannella *et al.,* 1999). Despite their thickness, these frozen-hydrated specimens have considerable phase contrast. This allows contrast to be enhanced in images like Fig. 2D by defocusing the objective lens, i.e., tuning the contrast transfer function of the IVEM to the features of interest. A defocus of −15 to −20 μm is typically used when imaging mitochondria in the IVEM, which enhances detail around 10 nm and cuts off information below 5 nm. Another potentially important mechanism of contrast enhancement is energy filtration, available on some microscopes, which removes inelastically scattered electrons that contribute to background (Koster *et al.,* 1997; Nicastro *et al.,* 2000).

2. Computer Analysis

The first step in generating a 3D reconstruction from tilt-series images collected on photographic emulsions is digitization. This can be done with a variety of commercially available, high-precision digital scanners. At the Albany resource, a high-resolution video camera is also available, which is fast and offers considerable flexibility in its operation (see McEwen and Marko, 1999). Images are digitized with pixels whose dimension (in nanometers on the specimen) is one-third to one-fourth of the anticipated resolution in the final reconstruction. Typically, a pixel size of 2 nm is used for mitochondrial studies of either thick plastic sections or frozen hydrated specimens, as the final resolution is not expected to be better than 6 nm. In the case of CCD imaging, the 1024×1024 pixel array is usually binned to 512×512, with a final pixel size of 2 nm. Note that, at this sampling, a data set of 100 images of a field $1000 \times 1000\ nm^2$ contains 2.6×10^7 pixels. (A cubic volume enclosing a 1000-nm-diameter spherical mitochondrion at 1 nm resolution would contain 10^9 voxels.) Suffice it to say that computational procedures with data sets this large require fast, multiprocessor workstations.

At the Albany resource, alignment, normalization, and tomographic reconstruction steps are done using protocols incorporated in the Spider image processing system (Frank *et al.,* 1996). Standard procedures involve a program that aligns the two independent tilt series in three dimensions using several colloidal gold particles distributed across the field as fiducial markers and a weighted back-projection reconstruction algorithm (Penczek *et al.,* 1995). The result is a volume containing a continuous 3D map of the specimen density. In some cases, this 3D image can be visualized by volume rendering using commercial programs such as VoxelView (Vital Images; e.g., Mannella *et al.,* 1994). The more usual analytical procedures involve examination of the individual slices of the 3D volume, like that shown in Fig. 1B. Simple sequential displays of the slices (a so-called walk through of the volume) can convey considerable information about the internal structure of the specimen. Generating surface-rendered 3D models like those shown in Fig. 1D, 2B and 2C involves interactive contour tracing of the features of interest (e.g., membrane profiles) in individual slices through the reconstruction, as shown in Fig. 1C. Tracing can be done using Sterecon (Marko and Leith, 1996) or equivalent programs. This step in the analysis of the tomograms is generally the most time-consuming part of the entire process and requires that careful attention be paid to tracking the continuity of features between slices. Depending on the expected resolution in the z direction, every two or three consecutive x–y slices can be summed through the volume, which reduces the number of slices to be analyzed and improves the signal-to-noise ratio. These contour images can then be displayed (as is or filled) in Sterecon or ported to commercial or public domain programs such as Iris Explorer or Movie.BYU for tilting and generation of continuous surfaces.

Acknowledgments

Michael Marko is thanked for his assistance with the preparation of figures for this manuscript and for his many helpful suggestions. He and Karolyn Buttle, Bruce McEwen, Pawel Penczek, Dean Leith, Bimal Rath,

Joachim Frank, and Conly Rieder are gratefully recognized for their contributions to the development and application of procedures at the Wadsworth Center's Resource for the Visualization of Biological Complexity (formerly the Biological Microscopy and Image Reconstruction Resource). This facility is supported as a national resource center by NIH/NCRR Grant RR01219, awarded by the Biomedical Research Technology Program. Some of this work was also supported by NSF Grant MCB 9506113.

References

Daems, W. T., and Wisse, E. (1966). Shape and attachment of the cristae mitochondriales in mouse hepatic cell mitochondria. *J. Ultrastruct. Res.* **16,** 123–140.

Deng, Y., Marko, M., Buttle, K., Leith, A., Mieczkowski, M., and Mannella, C. A. (1999). Cubic membrane structure in amoeba (*Chaos carolinensis*) mitochondria determined by electron microscopic tomography. *J. Struct. Biol.* **127,** 231–239.

Dubochet, J., Adrian, M., Chang, J.-J., LePault, J., and McDowall, A. W. (1987). Cryoelectron microscopy of vitrified specimens. *In* "Cryotechniques in Biological Electron Microscopy" (R. A. Steinbrecht and K. Zieold, eds.), pp. 114–131, Springer-Verlag, Berlin.

Frank, J., Radermacher, M., Penczek, P., Zhu, J., Li, Y., Ladjadj, M., and Leith, A. (1996). SPIDER and WEB: Processing and visualization of images in 3D electron microscopy and related fields. *J. Struct. Biol.* **116,** 190–199.

Frank, J. (ed.) (1992). "Electron Tomography." Plenum, New York.

Hackenbrock, C. R. (1966). Ultrastructural bases for metabolically linked mechanical activity in mitochondria. I.Reversible ultrastructural changes with change in metabolic steady state in isolated liver mitochondria. *J. Cell. Biol.* **30,** 269–297.

Hohenberg, H., Tobler, M., and Muller, M. (1996). High-pressure freezing of tissue obtained by fine-needle biopsy. *J. Microsc.* **183,** 133–139.

Khodjakov, A., Cole, R., McEwen, B. F., Buttle, K., and Rieder, C. L. (1997). Chromosome fragments possessing only one kinetochore can congress to the spindle equator in PtK1 cells. *J. Cell Biol.* **136,** 229–241.

Koster, A. J., Grimm, R., Typke, D., Hegerl, R., Stoschek, A., Walz, J., and Baumeister, W. (1997). Perspectives of molecular and cellular electron tomography. *J. Struct. Biol.* **120,** 276-308.

Lamvik, M. K. (199). Radiation damage in dry and frozen-hydrated organic material. *J. Microsc.* **161,** 171–181.

Mannella, C. A. (2000). Our changing views of mitochondria. *J. Bioenerg. Biomembr.* **32,** 1–4.

Mannella, C. A., Buttle, K., Bradshaw, P., and Pfeiffer, D. R. (2000). Changes in internal compartmentation of yeast mitochondria after onset of the permeability transition and osmotic recontraction. *Biophys. J.* **78,** 140A.

Mannella, C. A., Buttle, K., and Marko, M. (1997). Reconsidering mitochondrial structure: New views of an old organelle. *Trends Biochem. Sci.* **22,** 37–38.

Mannella, C. A., Buttle, K. F., O'Farrell, K. A., Leith, A., and Marko, M. (1996). Structure of contact sites between the outer and inner mitochondrial membranes investigated by HVEM tomography. *In* "Microscopy and Microanalysis" (G. W. Bailey *et al.,* eds.), Vol. 2, pp. 966–967, San Francisco Press, San Francisco.

Mannella, C. A., Buttle, K. F., Tessitore, K., Rath, B. K., Hsieh, C., D'Arcangelis, D., and Marko, M. (1998). Electron microscopic tomography of cellular organelles: Chemical fixation vs. cryo-substitution of rat-liver mitochondria. *In* "Microscopy and Microanalysis" (G. W. Bailey *et al.,* eds.), Vol. 4, pp. 430–431, Springer, Baton Rouge.

Mannella, C. A., Hsieh, C., and Marko, M. (1999). Electron microscopic tomography of whole, frozen-hydrated rat-liver mitochondria at 400 kV. *In* "Microscopy and Microanalysis" (G. W. Bailey *et al.,* eds.), Vol. 5, pp. 416–417, Springer, Baton Rouge.

Mannella, C. A., Marko, M., Penczek, P., Barnard, D., and Frank, J. (1994). The internal compartmentation of rat-liver mitochondria: Tomographic study using the high-voltage transmission electron microscope. *Microsc. Res. Tech.* **27,** 278–283.

Marko, M., and Leith, A. (1996). Sterecon-Three-dimensional reconstructions from stereoscopic contouring. *J. Struct. Biol.* **116,** 93–98.

McEwen, B., and Marko, M. (1999). Three-dimensional transmission electron microscopy and its application to mitosis research. *In* "Methods in Cell Biology" (C. L. Rieder ed.), Vol. 61, pp. 81–111, Academic Press, San Diego.

McEwen, B. F., Radermacher, M., Rieder, C. L., and Frank, J. (1986). Tomographic three-dimensional reconstruction of cilia ultrastructure from thick sections. *Proc. Natl. Acad. Sci. USA* **83,** 9040–9044.

Minamikawa, T., Sriratana, A., Williams, D. B., Bowser, D. N., Hill, J. S., and Nagley, P. (1999). Chloromethyl-X-rosamine (MitoTracker Red) photosensitizes mitochondria and induces apoptosis in intact human cells. *J. Cell Sci.* **112,** 2419–2430.

Moraru, I., Slepchenko, B., Mannella, C. A., and Loew, L. M. (2000). Role of cristae morphology in regulating mitochondrial nucleotide and H^+ dynamics. *Biophys. J.* **78,** 194A.

Munn, E. A. (1974). "The Structure of Mitochondria." Academic Press, London.

Nicastro, D., Frangakis, A. S., Typke, D., and Baumeister, W. (2000). Cryo-electron microscopy of Neurospora mitochondria. *J. Struct. Biol.* **129,** 48–56.

Penczek, P., Marko, M., Buttle, K., and Frank, J. (1995). Double-tilt electron tomography. *Ultramicroscopy* **60,** 393–410.

Perkins, G., Renken, C., Martone, M. E., Young, S. J., Ellisman, M., and Frey, T. (1997). Electron tomography of neuronal mitochondria: Three-imensional structure and organization of cristae and membrane contacts. *J. Struct. Biol.* **119,** 260–272.

Rasmussen, N. (1995). Mitochondrial structure and the practice of cell biology in the 1950s. *J. His. Bio.* **28,** 381–429.

Rath, B. K., Marko, M., Radermacher, M., and Frank, J. (1997). Low-dose automated electron tomography: A recent implementation. *J. Struct. Biol.* **120,** 210–218.

Rieder, C. L., and Cassels, G. (1999). Correlative light and electron microscopy of mitotic cells in monolayer cultures. *In* "Methods in Cell Biology" (C. L. Rieder, ed.), Vol. 61, pp. 297–315, Plenum, New York.

Schaff, J., Fink, C. C., Slepchenko, B., Carson, J. H., and Loew, L. M. (1997). A general computational framework for modeling cellular structure and function. *Biophys. J.* **73,** 1135–1146.

Sitte, H. (1996). Advanced instrumentation and methodology related to cryoultramicrotomy: A review. *Scan. Electron Micros. Suppl.* **10,** 387–466.

Tokuyasu, K. T. (1989). Use of poly(vinylpyrrolidone) and poly(vinyl alcohol) for cryoultramicrotomy. *Histochem. J.* **21,** 163–171.

Van Venetie, R., and Verkleij, A. J. (1982). Possible role of non-bilayer lipids in the structure of mitochondria: A freeze-fracture electron microscopy study. *Biochim. Biophys. Acta* **692,** 379–405.

Weibel, E. R. (1973). Stereological techniques for electron microscopic morphometry. *In* "Principles and Techniques of Electron Microscopy, Biological Applications"(M. A. Hayat, ed.), Vol. 3, pp. 239–296. Van Nostrand Reinhold, New York.

Willingham, M. C., and Rutherford, A. V. (1984). The use of osmium-thiocarbohydrazide-osmium (OTO) and ferrocyanide-reduced osmium methods to enhance membrane contrast and preservation in cultured cells. *J. Histochem. Cytochem.* **32,** 455–460.

Winslow, J. L., Hollenberg, M. J., and Lea, P. J. (1991). Resolution limit of serial sections for 3D reconstruction of tubular cristae in rat liver mitochondria. *J. Electron Microsc. Tech.* **18,** 241–248.

CHAPTER 15

Epitope Tagging and Visualization of Nuclear-Encoded Mitochondrial Proteins in Yeast

Dan W. Nowakowski, Theresa C. Swayne, and Liza A. Pon

Department of Anatomy and Cell Biology
Columbia University
New York, New York 10032

I. Introduction

Probing mitochondrial function has never been more exciting. Many nuclear-encoded proteins that are targeted to mitochondria remain to be identified, and the function of many mitochondrial proteins is unknown. The tagging of proteins with epitopes, fluorochromes, or affinity labels is an effective tool to study protein localization, interaction dynamics, and function. The first part of this chapter describes a method for C- or N-terminal epitope tagging, detection, and initial characterization of nuclear-encoded proteins in mitochondria. The second part addresses specialized methods for visualizing tagged as well as untagged mitochondrial proteins in yeast by indirect immunofluorescence.

II. Single-step Modification of Nuclear-Encoded Mitochondrial Genes in Yeast: An Overview

A technique has been developed that enables single-step tagging of chromosomal genes at their 5′ or 3′ ends by homologous recombination in the budding yeast *Saccharomyces cerevisiae* and the fission yeast *Schizosaccharomyces pombe* (McElver and Weber, 1992; Wach *et. al.,* 1997; Bähler *et al.,* 1998; Longtine *et al.,* 1998). This single-step polymerase chain reaction (PCR)-based method is rapid, economical, and does not require plasmid clones of the genes of interest. The first step is PCR amplification from a plasmid template of an insertion cassette consisting of the desired tag and a selectable marker for growth selection (Fig 1.). The PCR primers have 3′ends (20 nucleotides) that correspond to, and enable amplification of, sequences encoding the epitope tag and a selectable marker and 5′ends (40 or more nucleotides) with perfect homology to the desired *target gene* sequences. The amplified DNA is transformed directly into yeast using a standard transformation protocol (Gietz and Woods, 1998), and homologous recombinants that carry the modified target gene are identified and characterized.

This method is versatile as various tags, including multiple copies of HA, Myc, or glutathione *S*-transferase (GST) (detectable with commercially available antibodies), can be inserted into a target gene using a single set of primers. Because cassette modules carrying the *Aequorea victoria* green fluorescent protein (GFP) and new GFP variants with altered emission spectra are available, this application can be adapted for single and double labeling in living cells (Heim and Tsien, 1996; Niedenthal *et al.* 1996; Prasher, 1995; Yeast Resource Center, University of Washington, Seattle). In addition, spectral variants of GFP can be adapted to study *in vivo* protein complexes using fluorescence resonance energy transfer (Heim and Tsien, 1996; GFP-variant plasmid templates available from YRC, University of Washington, Seattle). Finally, using this method, tagged genes can be expressed either from their endogenous promoter or by introducing the *GAL1*-regulatable promoter in its place. The following section illustrates an example of a nuclear-encoded mitochondrial gene product that is tagged with 13-Myc (Evan *et al.,* 1985; Longtine *et al.,* 1998) and expressed under its normal promoter.

III. Generation and Analysis of the Modified Gene Product: Technical Considerations

A. Preparation of the Cassette for Transformation

1. Primer Design

Mdm10p is a nuclear-encoded protein that is localized to the mitochondrial outer membrane (Sogo and Yaffe, 1994). Phenotypic analysis of cells bearing a deletion in

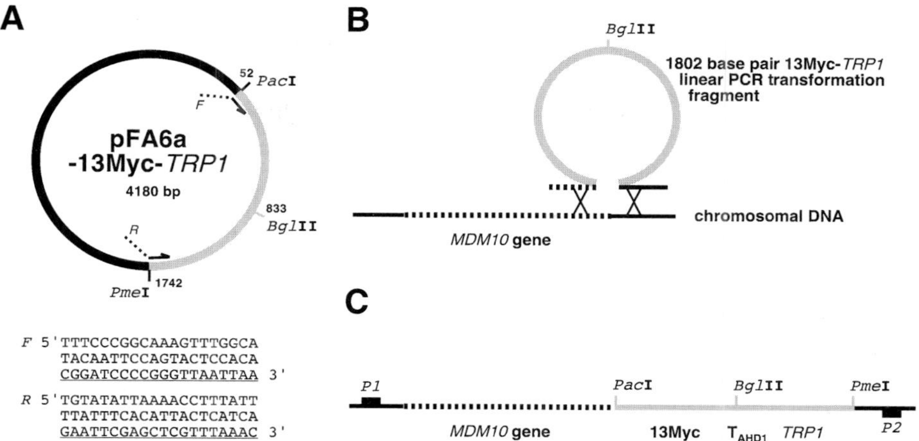

Fig. 1 One-step PCR-based gene modification in yeast. (A) To make a double-stranded transformation fragment for C-terminal tagging, DNA encoding the tag-selectable marker module is PCR amplified using ~60 base hybrid primers [primers *F* and *R* each contain 20 bases, underlined, corresponding to the pFA6a-13Myc-*TRP1* plasmid template (solid arrow segments) and 40 or more bases to the gene being tagged (dotted arrow segments); in this example for *MDM10* C-terminal tagging with 13Myc, the *F* primer includes 40 bases of the *MDM10* sequence just prior to but not including the stop codon, whereas the *R* primer sequence includes the stop codon as well as the 3′UTR sequence which follows]. (B) Integration of the PCR-generated tag-marker transformation fragment into yeast chromosomal DNA occurs by homologous recombination; following a standard yeast transformation procedure, transformed cells are isolated based on growth on selectable media, in this example on media lacking tryptophan. (C) Integration into the targeted locus can be characterized initially by agarose DNA electrophoresis via diagnostic PCR done on isolated genomic DNA using 20 base primers that are homologous to regions outside the target gene (primers *P1* and *P2*, shown, and/or primer *P3*, not shown, with unique homology to the inside of the tag-marker transformation fragment). T_{ADH1} marks the site of an *ADH1* terminator. Not drawn to scale.

MDM10 supports a role for this protein in linking mitochondria to the actin cytoskeleton and in control of mitochondrial movement and inheritance in yeast (Boldogh *et al.*, 1998). We describe a method to tag Mdm10p at its C terminus by homologous recombination using DNA amplified via PCR from plasmid pFA6a-13Myc-*TRP1*, and oligonucleotide primers purified by SDS–PAGE. The primers used for *MDM10* tagging are:

Forward-5′TTTCCCGGCAAAGTTTGGCATACAATTCCAGTACTCCACACGGATCCCCGGGTTAATTAA 3′
Reverse-5′TGTATATTAAAACCTTTATTTTATTTCACATTACTCATCAGAATTCGAGCTCGTTTAAAC 3′

Primers for the transformation of *S. cerevisiae* yeast should typically be 60 or more bases long and composed of two parts. Twenty bases at the 3′ends of the primers (underlined) are complementary to, and enable amplification of, sequences encoding the epitope tag and the selectable marker from the plasmid template (see also Fig 1.). These 20 bases are invariant and work to amplify any tag-marker combination from the pFA6a-derived plasmids (for presently available combinations, see Fig 2). For the C-terminal tag-forward primer, the 5′end should include at least 40 bases of perfectly homologous

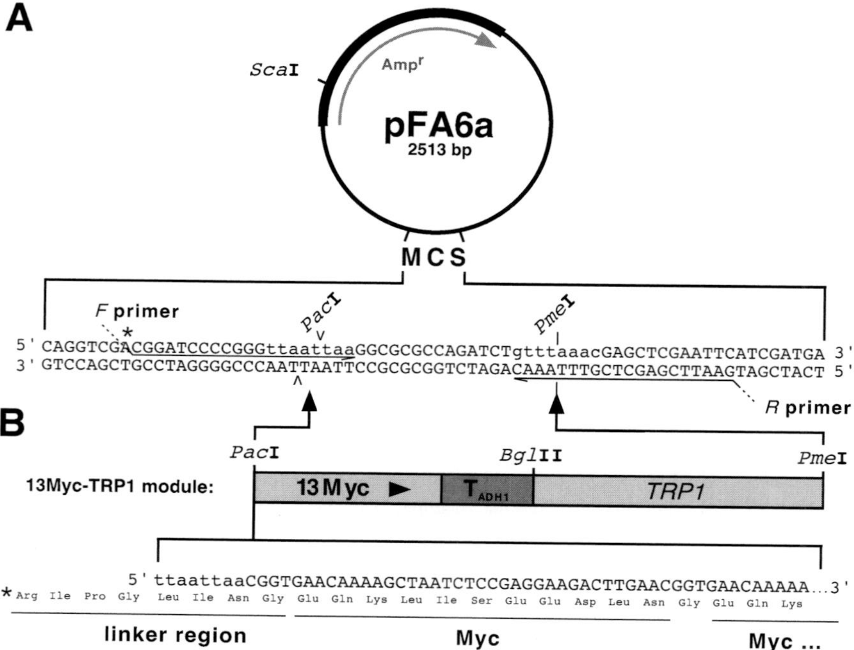

Fig. 2 Schematic depiction of pFA6a-derived plasmids. (A) pFA6a and DNA sequence with a region of its multiple cloning site (MCS) showing underlined 20 base stretches (common to all pFA6a-derived plasmids) used in primer design for single step-PCR amplification of the transformation cassette. In this example used for C-terminal tagging, primer regions marked *F* and *R* (indicated by the broken line) represent sequence homology (40 bases or more recommended) to the gene being tagged. (B) Abbreviated map of the 13Myc-*TRP1* module inserted into pFA6a and its initial sequence showing a translated region that introduces a linker between the 3′end of the tagged gene and the 13 repeats of the Myc tag sequence (an asterisk marks the site where additional amino acids can be inserted into the linker region during primer design, see text for discussion). A library of plasmids made by inserting other combinations of tags and selectable marker modules into pFA6a *Pac*I and *Pme*I sites is available (i) for C-terminal tagging: GFP(S65T)-*kanMX6,* GFP(S65T)-*TRP1,* GFP(S65T)-*HIS3MX6,* 3HA-*kanMX6,* 3HA-TRP1, 3HA-*HIS3MX6,* 13Myc-*kanMX6,* 13Myc-*TRPI,* 13Myc-*HIS3MX6,* GST-*kanMX6,* GST-*TRPI,* GST-*HIS3MX6,* and (ii) for N-terminal tagging: *kanMX6*-PGAL1-3HA, *TRPI*-PGAL1-3HA, *HIS3MX6*-PGAL1-3HA, *kanMX6*-PGAL1-GST, *TRPI*-PGAL1-GST, *HIS3MX6*-PGAL1-GST, *kanMX6*-PGAL1-GFP, *TRPI*-PGAL1-GFP, *HIS3MX6*-PGAL1-GFP; T_{ADH1} marks the site of an *ADH1* terminator (Wach *et al.,* 1994; Longtine *et al.,* 1998). One set of primers can thus serve to amplify all C-terminal or all N-terminal tag-selectable marker modules for use in one-step modification of the gene of interest.

sequence just 3′ to the stop codon of the target gene. The reverse primer sequence can encompass the region of the stop codon as well as at least 37 bases of the 3′-UTR sequence that follows.

Special note on primer design for GFP–fusion constructs. Maximum fluorescence output is especially important for mitochondrial fusion proteins used in live imaging applications. We find that GFP fluorescence can often be optimized by varying the length of the linker region between the target gene and the GFP molecule. For C-terminal tagging, this is accomplished during initial design of the forward primer by introducing a sequence encoding extra amino acids between the 40 bases of DNA homologous to the target gene and the 20 bases corresponding to the plasmid template (see Fig. 2B*); we suggest starting with five alanines.

2. PCR Amplification and Cassette Purification

We find that the same PCR conditions amplify different tag-marker cassettes with different efficiencies. However, optimum amplification conditions can often be found by titrating the $Mg(OAc)_2$ concentration in the PCR reaction. As a starting point, we suggest the following PCR conditions:

Reagent	Volume (μl)
PCR buffer (3.3 × stock, contains no Mg^{2+})	30.3
$Mg(OAc)_2$ (25 m*M* stock)[1]	4
dNTP mix (10 m*M* each dNTP)	8
Forward primer (2×10^{-5} *M* stock)[2]	1
Reverse primer (2×10^{-5} *M* stock)[2]	1
pFA6a-13Myc-*TRP1* template (~50 ng/μ stock)	1
rTth DNA polymerase (P.E. high-fidelity enzyme)	1
dH_20	53.7

[1]For different plasmid templates, we suggest first titrating the $Mg(OAc)_2$ concentration; e.g., try four reactions with final $Mg(OAc)_2$ concentrations of 0.5, 1.0, 2.0, and 3.0 m*M*.

[2]To assure PCR specificity for new primers, we suggest three control reactions: (i) no primers in the reaction, (ii) forward primer only, and (iii) reverse primer only.

Thermocycler Conditions:

1 cycle of 94°C (5 min)

30 cycles of 94°C (1 min), 55°C (1 min), and 72°C (2 min)

1 cycle of 72°C (10 min)

We often pool up to ten 100 μl PCR reactions to obtain DNA sufficient for a single yeast transformation (>1 μg of DNA per 1×10^8 cells in a single transformation reaction). Furthermore, we suggest agarose gel electrophoresis purification of the PCR-amplified DNA prior to transformation (a 1% agarose gel is effective to separate the amplified DNA from template and other contaminants in the PCR reaction). After cutting out the PCR fragment from the gel, we isolate and concentrate it using a bead Qiaex II gel extraction kit (Qiagen, Valencia, CA).

3. Yeast Transformation

If the ultimate purpose of tagging the target protein is visualization in live or fixed cells, it is important to select a strain with a large cell diameter (>4 μm) for optimal imaging resolution. *S. cerevisiae* yeast strains from the D273-10B genetic background have been used extensively to study mitochondrial function and should serve as a good starting point. PCR-amplified DNA (for our example, the 1802-bp 13Myc-*TRP1* fragment) is transformed into yeast using a lithium acetate transformation protocol (Gietz and Woods, 1998). This protocol is also available at the following website: http://www.umanitoba.ca/faculties/medicine/human_genetics/gietz/method.html.

The following is an example of a yeast transformation reaction. Grow cells to midlog phase [in yeast peptone dextrose (YPD) media; OD_{600} measurements should fall between 0.1 and 1.0 depending on strain; OD_{600} of 1.0 is approximately equal to 1×10^7 cells]. Cells are then washed with 100 m*M* lithium acetate and resuspended to a final volume of 50 μl in 100 m*M* lithium acetate (total of 1×10^8 of cells per transformation reaction). The lithium acetate wash should be as brief as possible, as this reagent is toxic to cells.

Add the following reagents to cells in lithium acetate: 240 μl polyethylene glycol [50% (w/v) stock (PEG 3350)], 36 μl 1.0 *M* lithium acetate, 25 μl 2.0 mg/ml stock single-stranded calf thymus DNA, and 50 μl dH_20 and transformation cassette DNA (1.0 μg minimum)

Following thorough mixing of reagents, the reaction is incubated in a water bath at 30°C for 30 min, followed by heat shock at 42°C for 15–25 min. Cells are concentrated by mild centrifugation (30 sec, 7000*g*). The lithium acetate-containing supernatant is decanted, and cells are resuspended in 200 μl of sterile water.

B. Isolation and Characterization of Modified Strains

1. Strain Selection

For transformations using the 13Myc-*TRP1* cassette, clones are isolated by spreading all cells from a single transformation reaction onto an SC-Trp plate (synthetic complete plate lacking tryptophan) and growing for 3–6 days. A single reaction (1.0 μg transformation fragment DNA per 1×10^8 of cells) typically yields 10–20 transformed clones for further characterization.

For transformations using *kanMX6*-based cassettes, which specify resistance to the drug geneticin (G418), clone selection is slightly different due to the added toxicity of the selection drug. Briefly, all cells from a single transformation reaction are spread onto a YPD plate. This plate is incubated for 1–2 days and then replica plated onto a YPD+G418 (200 mg/liter) plate. To select stable transformants, the YPD+G418 plate is incubated for 3–4 days and then replica plated onto a fresh YPD+G418 plate. *Note:* The salt concentration of SC plates interferes with G418 drug activity. As a result, large numbers of false positives are isolated when using SC+G418 plates for selection. Thus, we recommend using YPD+G418 plates for isolation of *kanMX6*-transformed cells.

2. PCR Screening, Protein Expression, and Functional Analysis

Transformants isolated on selective plates are further screened by extracting genomic DNA and testing via PCR for homologous recombination at the target locus. Using a unique pair of 20 base primers that anneal to the genomic sequence outside the coding region of the gene (see Fig 1C; also Longtine *et al.,* 1998), this reaction can distinguish between wild-type or nonintegrated false clones from clones that have been correctly modified by targeted integration (Fig 3A). Optional PCR can be carried out using a third primer that anneals to a region within the integration cassette that is not found elsewhere in the genome (not shown). If further manipulation is desired, any of the resulting PCR fragments can be subcloned into a plasmid vector or sequenced, provided that desired restriction sites have been engineered into the primers used for PCR. A rapid colony PCR method is described in which genomic DNA isolated from one-half of a colony (2–5 mm in size) directly from a selection plate serves as a template for diagnostic PCR reactions (modified from Hoffman and Winston, 1987); the other half of a colony is used to streak clones onto fresh selective plates for further growth and characterization.

Fifteen-minute colony genomic DNA extraction:

1. Pour ~0.3 g of 0.5-mm glass beads (Biospec Products Inc., Bartlesville, OK) into an Eppendorf tube. Mix with 0.2 ml of buffer A (2% Triton X-100, 1% SDS, 100 m*M* NaCl, 10 m*M* Tris–HCl pH 8.0, 1 m*M* Na_2EDTA) and 0.2 ml of phenol:chloroform:isoamyl alcohol (25 : 24 : 1; Boehringer Mannheim).
2. Resuspend one-half colony (approximately 2–5 mm in size) directly from the selection plate into the bead mix.
3. Vortex vigorously for 3–4 min.
4. Add 0.2 ml of Tris–EDTA (TE), pH 8.0.
5. Centrifuge for 5 min at ~12,500*g*. Transfer the aqueous layer to a fresh Eppendorf tube. Add 1.0 ml of 100% ethanol. Mix by inversion.
6. Centrifuge for 2 min at ~12,500*g*. Discard the supernatant. Resuspend the pellet in 0.4 ml of TE (pH 8.0) plus 3 μl of a 10-mg/ml solution of RNase A. Incubate for 5 min at 37°C. Add 10 μl of 4 *M* ammonium acetate, plus 1.0 ml of 100% ethanol. Mix by inversion.
7. Centrifuge for 2 min at ~12,500*g*. Discard the supernatant. Air-dry the pellet and resuspend in 5.0 μl of TE (pH 8.0).
8. Of the genomic DNA-TE suspension, 2.5μl can be directly used as a template in a 25-μl diagnostic PCR reaction (see Fig. 3A).[3]

Transformants that are positive for integration at the target locus can be validated further by analysis of protein expression via Western blot (Fig. 3B) and by visualization of the tagged construct in cells via immunofluorescence staining as described later. Finally, tagged constructs can often be characterized for functionality given knowledge about the phenotype of the gene of interest. For instance, deletion of the open reading frame of the *MDM10* gene results in yeast that are unable to grow on nonfermentable

[3]For each of the 11 mitochondria-specific constructs tested by us in *S. cerevisiae,* only 10–30% of the cells isolated on selective plates showed integration into the target locus. Bähler *et al.* (1998) reported that using 80–100 nucleotide primers for the modification of nine genes in *S. pombe,* 20–100% of the transformants obtained showed correct integration.

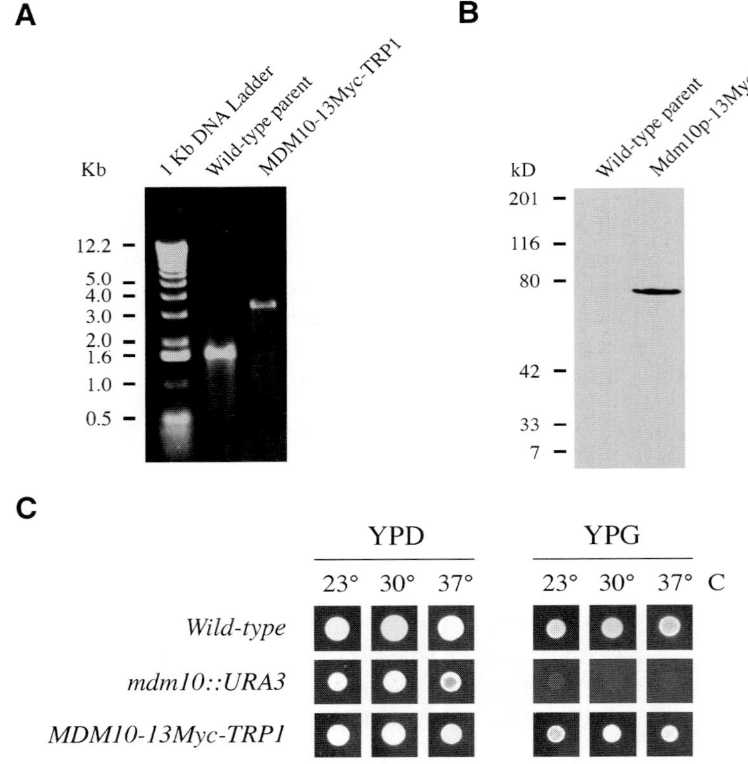

Fig. 3 After isolation of transformants on selective plates, clones are validated via PCR, protein expression, and functionality assays. (A) Integration of the cassette into the target locus is initially verified via PCR on isolated genomic DNA. In the 1% agarose gel, PCR products from an untransformed parent wild-type haploid yeast, and yeast transformed with the 13Myc-*TRP1* cassette, are distinguished readily by size in wild-type *MDM10* and *MDM10-13Myc-TRP1* strains (*MDM10* was amplified using 20 base primers, which correspond to chromosomal DNA 55 nucleotides upstream and downstream of the gene.) (B) Clones positive for integration are verified for protein expression via Western blot analysis. The 10% SDS–PAGE gel of whole cell extracts from wild-type and tagged strains was probed with a 9E10 αMyc monoclonal antibody. The Mdm10p-13Myc lane, but not the wild-type parent strain lane, shows expression of a single tagged protein of the expected molecular weight. (C) Tagged constructs can often be characterized for functionality given knowledge about the phenotype of the gene of interest. While deletion of the open reading frame of the *MDM10* gene results in yeast that is unable to grow on nonfermentable carbon sources (YPGlycerol), yeast bearing Mdm10p-13Myc behaves just like the wild type in that it retains the ability to grow on nonfermentable carbon sources at all temperatures.

carbon sources (Fig. 3C), and accumulate large spherical mitochondria (Sogo and Yaffe, 1994). In the case of the Mdm10p-13Myc construct, the modified strain retains the ability to grow on nonfermentable carbon sources (YPGlycerol, Fig. 3C) and exhibits wild-type tubular mitochondrial morphology (not shown). This suggests that tagging Mdm10p at the C terminus does not compromise the function of the protein.

Table I
Useful Marker Antigens for Yeast Mitochondria

Protein	Location[a]	References
Porin	OM	Roeder *et al.* (1998), Mihara *et al.* (1985)
Cytochrome oxidase subunit III	IM	Haugland *et al.* (1998), Taanman *et al.* (1993)
Citrate synthase I	MAT	Azpiroz *et al.* (1993)
OM14	OM	Riezman *et al.* (1983), McConnell *et al.* (1990)
ABF2	mtDNA	Diffley *et al.* (1991)

[a] OM, mitochondrial outer membrane; IM, mitochondrial inner membrane; MAT, mitochondrial matrix; mtDNA, mitochondrial DNA.

IV. Visualizing Yeast Mitochondria by Immunostaining

Several well-characterized proteins can serve as markers for immunofluorescence visualization of mitochondria in *S. cerevisiae.* These include proteins targeted to each of the submitochondrial compartments (Table I). In addition to these antibodies to specific proteins, a polyclonal antibody raised against outer mitochondrial membranes has been used successfully for immunofluorescence (Riezman *et al.,* 1983; Smith *et al.,* 1995).

A. Growth of *S. cerevisiae* for Immunofluorescent Staining

The protocol given here is modified from the method of Pringle and colleagues (1989, 1991). While it is sometimes possible to perform immunofluorescent staining on cells picked from colonies on solid media, we prefer midlog phase yeast grown in liquid media. Whether cells are observed alive or fixed, the internal structures, including mitochondria, of rapidly growing yeast are better preserved and the cell population is more reproducible.

The choice of carbon source for growth affects the abundance and morphology of mitochondria (Damsky, 1976; Stevens, 1977; Visser *et al.,* 1995). For localization of mitochondrial antigens or analysis of mitochondrial motility and morphology, we usually use lactate, a nonfermentable carbon source. Lactate medium selects for cells with mtDNA and mitochondrial metabolic potential; it is also inexpensive and easy to prepare (see recipe later). For strains that cannot grow on nonfermentable carbon sources, we use a raffinose-based rather than the traditional glucose-based medium because glucose represses mitochondrial biogenesis and consequently makes the organelles more difficult to visualize.

Another consideration in choice of medium is autofluorescence. Lactate and raffinose media as described later produce little general autofluorescence. However, in any carbon source, the red material that accumulates in vacuoles of *ade2*$^{-}$ cells is fluorescent and can interfere with microscopy in the red or green emission channels. Provision of extra adenine (two- to fivefold more than in normal media) can prevent this problem.

For more complete information on growth and maintenance of *S. cerevisiae,* see Sherman (1991).

1. Solutions for Growth of *S. cerevisiae* for Immunofluorescent Staining

Lactate growth medium (For 1 liter): 3 g yeast extract, 0.5 g glucose, 0.5 g $CaCl_2$, 0.5 g NaCl, 0.6 g $MgCl_2$, 1 g KH_2PO_4, 1 g NH_4Cl, 22 ml 90% lactic acid, and 7.5 g NaOH

(optional: for growth of *ade2*$^-$ cells, add 0.1 mg/ml adenine)

Dissolve ingredients in 900 ml distilled H_2O. Adjust pH to 5.5 with NaOH. Bring volume to 1 liter with distilled H_2O. Dispense into flasks so that the volume of media is one-fifth the volume of the flask (e.g., place 50 ml media in a 250-ml flask). Stopper the flasks with a cotton plug wrapped in cheesecloth and cover top of flask with aluminum foil. Autoclave.

Raffinose growth medium (for 1 liter): 10 g yeast extract, 20 g Bacto-peptone, and 20 g raffinose

(optional: for growth of *ade2*$^-$ cells, add 0.1 mg/ml adenine)

Dissolve ingredients in 900 ml distilled H_2O. Bring volume to 1 liter with distilled H_2O. Dispense into flasks so that the volume of media is one-fifth the volume of the flask. Stopper the flasks with a cotton plug wrapped in cheesecloth and cover top of flask with aluminum foil. Autoclave.

2. Protocol for Growth of *S. cerevisiae* for Immunofluorescent Staining

1. Prepare media as described earlier.

2. Inoculate flasks with a small amount of *S. cerevisiae* from a plate or preculture. (Growth rates vary with genotype; to grow wild-type cells to midlog phase overnight, we typically inoculate with between 1/100 and 1/250 volume from a stationary-phase preculture.)

3. Grow cells at 30°C with shaking at 250 rpm.

4. Cells are ready to fix when they are in the midlog phase of growth. A good way to check the log phase status and general health of the culture is to count budding cells on a standard bright-field microscope with a 40x or higher magnification objective. For best results, at least 70–90% of cells should have a bud.

B. Fixation of *S. cerevisiae* with Paraformaldehyde

There are two main classes of fixatives: cross-linkers and precipitants. This protocol uses paraformaldehyde, a cross-linking fixative that forms hydroxymethylene bridges between spatially adjacent amino acid residues. The most important variables in this type of fixation are the concentration of paraformaldehyde, the duration of fixation, and the pH of the fixative solution. High paraformaldehyde concentrations, low pH, and a long fixation period increase the number of cross-links formed and thereby improve structural preservation. Excessive cross-linking, however, will make the antigens inaccessible to antibody binding. The optimal protocol strikes a balance between structural preservation and antigen accessibility.

The conditions under which cells are fixed can strongly affect the quality of the immunofluorescence results. Common manipulations, including centrifugation and increasing or decreasing temperature, can alter internal structures in *S. cerevisiae* (Lillie and Brown, 1994). Therefore, we add the fixative directly to a liquid culture under growth conditions to minimize such disruptions. However, prolonged fixation in growth media causes high levels of autofluorescence. Consequently, after a brief initial fixation under growth conditions, the cells are transferred to media-free fixative for the remainder of the fixation period (Fig. 4, see color plate).

A mixture of methanol and acetone is suitable for some immunofluorescence staining, including staining of the actin cytoskeleton. However, we find it to be a poor choice for mitochondria. It solubilizes many membranes and can cause redistribution of some mitochondrial antigens from subcellular membranes to other locations (I. R. Boldogh, W. D. Nowakowski, S. L. Karmon, and L. A. Pon, unpublished data).

1. Solutions for Fixation of *S. cerevisiae* with Paraformaldehyde

a. First Fix

For 100 ml:

1. Add 16.5 ml 1 *M* K_2HPO_4 (dibasic potassium phosphate) to 40 ml distilled water and heat to 60°C in a fume hood.
2. Add 18.5 g paraformaldehyde (electron microscopy grade).
3. Stir until dissolved (typically 1 h), keeping temperature between 55° and 60°C.
4. Remove from heat. Add 33.5 ml 1 *M* KH_2PO_4 (monobasic potassium phosphate).
5. Bring to 100 ml with distilled water.
6. Filter through Whatman paper.

Note: First Fix solution can be used immediately or stored at 4°C for up to 24 h. To avoid temperature shock, bring the solution to growth temperature before adding it to cells.

b. Second Fix

For 50 ml:

1. Add 1.65 ml 1 *M* KH_2PO_4 (dibasic potassium phosphate) to 40 ml distilled water and heat to 60°C in a fume hood.
2. Add 2.5 g paraformaldehyde (electron microscopy grade).

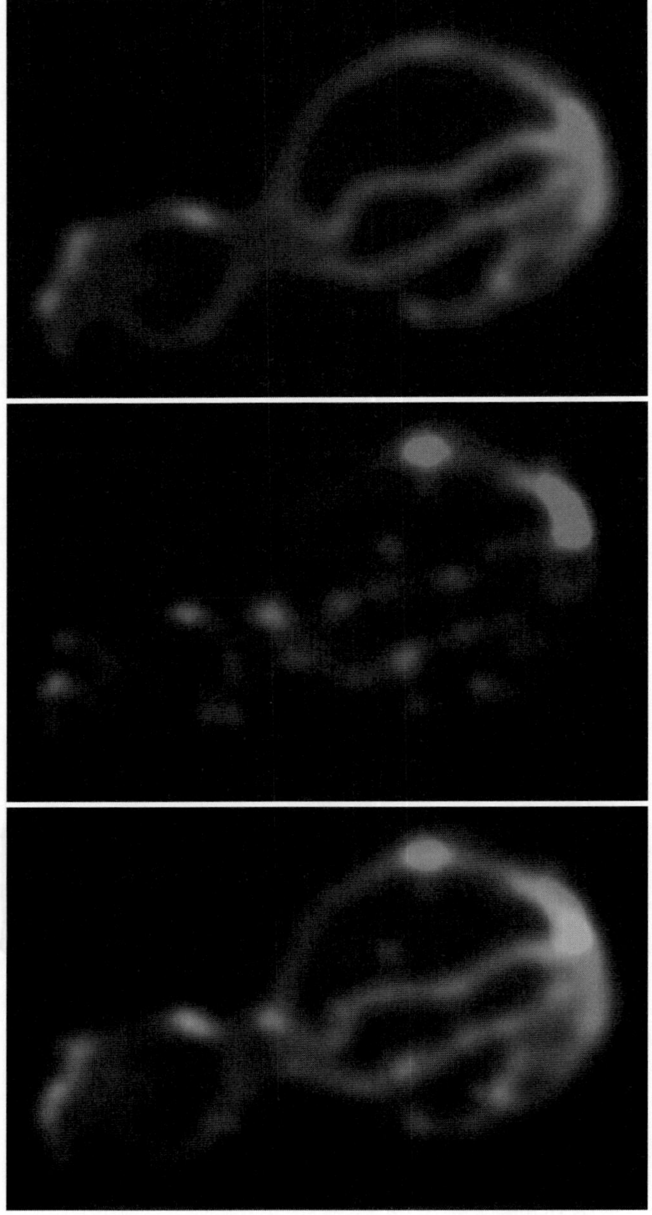

Fig. 4 Indirect immunofluorescence of a *S. cerevisiae* cell fixed with paraformaldehyde showing staining of a cytoplasmic protein, Arp2p, that is targeted to mitochondria. The green image shows fluorescence staining obtained with a CS1-GFP marker, which targets GFP to the mitochondrial matrix via a 52 amino acid citrate synthase leader sequence. The red image shows staining obtained using a primary rabbit antibody raised against a unique peptide within the Arp2 protein sequence, and a secondary rhodamine-conjugated goat–antirabbit antibody. The combined image shows Arp2p punctate structures coalign with mitochondrial tubules. (See Color Plate.)

3. Stir until dissolved (typically 30 min), keeping temperature between 55° and 60°C.
4. Remove from heat. Add 3.3 ml 1 *M* KH_2PO_4 (monobasic potassium phosphate) and 0.1 ml 1 *M* $MgCl_2$.
5. Bring volume to 50 ml with distilled water.
6. Filter through Whatman paper.

Note: Second Fix solution can be used immediately or stored at 4°C for up to 24 h.

1 *M* KP_i buffer, pH 7.5 (for 100 ml): 84 ml 1 *M* K_2HPO_4 and 16 ml 1 *M* KH_2PO_4;
adjust pH to 7.5 with appropriate potassium phosphate solution if necessary and then autoclave

Wash solution (25 m*M* potassium phosphate, pH 7.5) (for 500 ml): 12.5 ml 1 *M* KP_i, pH 7.5, and 400 ml 1 *M* KCl;
bring volume to 500 ml with distilled water and then autoclave

2. Protocol for Fixation of *S. cerevisiae* with Paraformaldehyde

1. Bring First Fix solution to growth temperature.
2. Add 1/4 volume (6.25 ml per 25 ml) First Fix solution to culture flask and continue shaking for 15 min.
3. Centrifuge the suspension at ~8000*g* for about 5 min. Remove supernatant. Add 1 ml of Second Fix solution to pellet and pipette gently up and down to resuspend cells. Transfer suspension to a microcentrifuge tube if desired. Incubate for 1.5 h at room temperature.
4. Centrifuge suspension for 5 s at ~12,500*g*. Remove supernatant and add 1 ml wash solution to pellet. Resuspend cells with gentle pipetting. Repeat this step two more times. If necessary, fixed cells can now be stored overnight in wash solution at 4°C.

C. Immunofluorescent Staining of Fixed Yeast Cells

For indirect immunofluorescence staining, fixed yeast cells are converted to spheroplasts by enzyme-catalyzed removal of the cell wall. The fixed spheroplasts are then exposed to a primary antibody that binds to the antigen of interest (either an endogenous protein for which antibodies are available or an epitope tag). After a wash to remove unbound primary antibody, a secondary antibody is added. The secondary antibody recognizes the invariant (Fc) region of the primary antibody and is tagged with a fluorophore for microscopic visualization.

After the unbound secondary antibody is washed away, the coverslip is mounted on a microscope slide. The mounting solution used here consists mostly of glycerol, which reduces spheroplast movement during imaging and prevents freezing of the samples during storage at −20°C. It also contains *p*-phenylenediamine, an antiphotobleaching agent, which helps prevent destruction of fluorophores by oxygen radicals generated during illumination.

1. Pretreatment of Antibodies with Yeast Cell Walls

A special complication of using animal-derived antisera for immunostaining in yeast arises from the ubiquity of yeast in the environment. Rabbit antisera, even after affinity purification, usually contain antibodies recognizing the yeast cell wall. These contaminating antibodies bind to the residual cell wall on spheroplasts, generating background staining, which may be punctate or distributed uniformly over the surface of the spheroplast. The following simple method is used for removing these contaminating antibodies by preadsorbing them to intact yeast cells. A batch of antibody may be pretreated in this way and stored for later use.

a. Solutions for Pretreatment

10× phosphate-buffered saline (PBS) (for 1 liter): 80 g NaCl, 2 g KCl, 14.4 g Na_2HPO_4, and 2.4 g KH_2PO_4.

Add ingredients to 800 ml distilled H_2O. Adjust pH to 7.2 with HCl. Bring volume to 1 liter with distilled H_2O and autoclave.

PBS^+ (for 50 ml): 5 ml 10 × PBS, 40 ml distilled H_2O, 0.5 g bovine serum albumin, and 0.5 ml 10% NaN_3.

Bring volume to 50 ml with distilled H_2O. Make on day of use.

b. Protocol for Pretreatment of Antibodies with Yeast Cell Walls

1. Grow a liquid culture of yeast cells to stationary phase, typically 24–48 h.
2. Dilute antiserum to 1/25 in PBS^+ supplemented with 1 m*M* phenylmethylsulfonyl fluoride (PMSF).
3. For each milliliter of diluted antibody, remove 500 μl cells from the stationary phase preculture into a centrifuge tube.
4. Concentrate cells by centrifugation for 5 s at ~12,500*g* and remove supernatant. Add a volume of PBS^+ approximately equal to the volume of the cell pellet. Pipette gently to resuspend cells. Repeat this step three more times.
5. For each milliliter of diluted antibody, remove 250 μl of this washed cell suspension to a new centrifuge tube. (Save the remainder of the washed suspension for step 7.) Centrifuge as just described, remove supernatant, and add the diluted antibody. Pipette gently to resuspend cells.
6. Incubate with shaking or rotation at 4°C for 2 h.
7. Take the remaining washed cell suspension from step 5 and centrifuge for 5 s at ~12,500*g*. Discard the supernatant. Centrifuge the antiserum–cell suspension as

described earlier to concentrate cells and cell-associated contaminating antibodies. Transfer the antiserum supernatant to the fresh cell pellet.

8. Incubate again with shaking at 4°C for 2 h.

9. Centrifuge as described previously to concentrate cells; transfer supernatant to a new tube. The supernatant can now be used for immunofluorescent staining. It should be stored in aliquots at −20°C.

2. Spheroplasting

Antibodies will not penetrate the yeast cell wall. For immunostaining, therefore, fixed cells must be converted to spheroplasts: they are treated with an enzyme solution that removes the wall by breaking down cell wall polysaccharides. Under optimal conditions, the spheroplasting treatment results in breakdown of the barrier to antibody penetration but not breakdown of the spheroplast itself.

a. Solutions for Spheroplasting

Tris/DTT (for 100 ml): 10 ml 1 *M* Tris–SO_4, pH 9.4, and 1 ml 1 *M* dithiothretol (DTT)

Bring volume to 100 ml with distilled H_2O. Make on day of use.

NS (for 500 ml): 10 ml 1 *M* Tris–HCl, pH 7.5, 21.4 g sucrose, 1 ml 0.5 *M* EDTA, 0.5 ml 1 *M* $MgCl_2$, 0.05 ml 1 *M* $ZnCl_2$, and 0.05 ml 0.5 *M* $CaCl_2$.

Bring to 500 ml with distilled H_2O. Store in 40-ml aliquots at −20°C. Sterilize by passing through a 0.2-μm filter before use.

NS^+: On day of use, supplement 10 ml NS with the following:

50 μl 200 m*M* PMSF in ethanol and 500 μl 10% NaN_3

b. Protocol for Spheroplasting of Fixed S. cerevisiae

1. Centrifuge cell suspension for 5 s at ~12,500*g*. Remove supernatant and add 1 ml Tris/DTT to pellet. Resuspend cells with gentle pipetting. Incubate for 20 min in a 30°C water bath.

2. Dissolve 0.125 mg/ml zymolyase (Zymolyase 20T from *Arthrobacter luteus,* Seikagaku Inc., Tokyo, Japan) in wash solution. Mix thoroughly until zymolyase is completely dissolved.

3. Centrifuge cell suspension for 5 s at ~12,500*g*. Remove supernatant and add 1 ml of zymolyase solution to pellet. Resuspend cells with gentle pipetting.

4. Incubate for 0.5–2.5 h in a 30°C water bath. Check progress of spheroplasting every ~30 min by phase-contrast or bright-field microscopy. An intact cell has prominently refractile edges, while a spheroplast is less refractile. Incubate until the majority of cells have become spheroplasts (typically 1.5 h), but avoid excessive zymolyase treatment, which can damage cellular structures.

5. Centrifuge spheroplast suspension for 5 s at ~12,500*g*. Remove supernatant and add 1 ml of NS^+ to pellet. Resuspend spheroplasts with gentle pipetting. Repeat this step two more times, resuspending the final pellet in 2 volumes NS^+.

6. Store spheroplasts at 4°C until used. They should be stained within 1 week of fixation.

3. Indirect Immunofluorescent Staining: The Coverslip Method

a. Solutions for Immunofluorescent Staining

Polylysine: Dissolve 0.5 mg/ml polylysine in distilled H_2O. Sterilize by passing through a 0.2-μm filter. Store in aliquots at −20°C.

PBT (1× PBS, 1% bovine serum albumin, 0.1% Triton X-100, 0.1% sodium azide) (for 50 ml): 5 ml 10× PBS, 0.5 g bovine serum albumin, 0.5 ml 10% Triton X-100, and 0.5 ml 10% NaN_3.

Bring volume to 50 ml with distilled H_2O. Make on day of use.

Mounting solution (for 100 ml): 10 ml 1 × PBS, 100 mg *p*-phenylenediamine.

Stir vigorously until dissolved. Adjust pH to 9 with NaOH.

Add 90 ml glycerol and mix thoroughly.

Optional: For DAPI counterstaining of nuclear and mitochondrial DNA, add 100 μl of a stock solution of 1 mg/ml DAPI (4′,6′diamidino-2-phenylindole) in distilled H_2O.

Store at −20°C in aliquots.

b. Protocol for Indirect Immunofluorescent Staining of Fixed Spheroplasts

1. Preparation of staining chamber: For all incubations in this protocol, the coverslip is placed with the spheroplast-coated side down on a drop of incubation solution in a dark, humid chamber. This method requires only a small amount of antibody because the incubation volume is 20–40 μl per coverslip. During incubation, the fluorescent dyes should be protected from light, and the spheroplasts must not be allowed to dry out. The dark, humid incubation chamber serves both of these purposes.To make the staining platform, press together several sheets of Parafilm, about 10 by 15 cm, and crimp at the edges to make a durable platform for multiple coverslips. The Parafilm platform can be reused if it is washed after use with a small amount of soap and rinsed thoroughly with distilled water followed by a squirt of ethanol. To make a dark, humid chamber, set the Parafilm sheet on a 2-cm-high pile of damp paper towels and then invert an opaque tray or pan over the pile.

2. Place a drop (about 40 μl) of polylysine solution on the clean sheet of Parafilm. Lay a 22-mm^2 coverslip on the drop, taking care to avoid creating bubbles at the coverslip–slide interface. (Coverslips should be handled throughout with forceps, grasping near the edges.) The polylysine coating will serve as an adhesive to immobilize spheroplasts on the coverslip. If desired, the coverslip may be labeled with a fine-point marker in one corner of the uncoated side.

3. After at least 10 s, pick up the coverslip by pipetting 200–300 μl distilled water under one edge. This will float the coverslip off the Parafilm so it can be lifted with forceps. Rinse by pipetting about 1 ml over the coated surface of the coverslip, letting the excess drip off into a waste container. Gently remove remaining drops of liquid by tapping the edge of the coverslip on a paper towel or wicking with the edge of a piece of filter paper. Place the coverslip, polylysine side up, on the Parafilm.

4. Place a drop of 100 μl sterile filtered 1× PBS on the coated side of the coverslip. Add 10 μl of spheroplast suspension to the PBS, mixing and gently spreading spheroplasts over the coverslip with the side of the micropipette tip. Incubate for 30 min at room temperature in the dark, humid chamber.

5. Prepare 20–40 μl of primary antibody at the appropriate concentration in PBT. Keep on ice until ready to use.

6. Pick up the coverslip and rinse off unbound spheroplasts by gently pipetting about 1 ml of PBT over the coverslip, letting the excess drip off. Repeat the rinse four more times. Remove stray drops of liquid as described in step 3. The Triton X-100 in PBT permeabilizes spheroplasts and makes antigens more accessible to antibody binding; the BSA blocks nonspecific protein–protein interactions and thereby reduces the level of background staining.

7. Place the 20–40 μl of diluted primary antibody in a drop on the Parafilm platform. Lay the spheroplast-coated side of the coverslip slowly on the drop of antibody, making sure that the entire surface is exposed to antibody and no bubbles are formed. If bubbles form under the coverslip, try gently tapping or pushing down on the coverslip with forceps to nudge the bubbles toward the edge. If this is unsuccessful, lift the coverslip by pipetting 200–300 μl PBT under the edge, remove excess liquid, and lay the coverslip on a fresh drop of diluted antibody. Incubate for 2 h at room temperature in the chamber.

8. Prepare 20–40 μl of secondary antibody at the appropriate concentration in PBT. Keep on ice until ready to use.

9. Lift the coverslip by introducing 200–300 μl PBT under the edge of the coverslip with a micropipette. Gently pipette about 1 ml of PBT over the spheroplasts to rinse, letting the excess drip off. Repeat the rinse four more times. Remove excess liquid.

10. Place the coverslip on a 20- to 40-μl drop of secondary antibody. Incubate for 1 h at room temperature in the chamber.

11. Lift and rinse the coverslip as in step 9. Rinse five more times with PBS. Remove excess liquid.

12. Place 1–2 μl mounting solution (with or without added DAPI) on a clean microscope slide. Too little mounting solution leads to air pockets and poor preservation of staining, whereas too much causes spheroplasts to float around and leads to overstaining if DAPI is used.

13. Slowly lower the coverslip onto the drop, taking care to avoid creating bubbles at the coverslip–slide interface. Spheroplasts can be flattened significantly on a microscope slide, which makes for good conventional micrographs, but will cause obvious distortion

in three-dimensional imaging (e.g., confocal microscopy). If mounting spheroplasts on a slide for three-dimensional microscopy, do not apply pressure to the coverslip. Otherwise, tap lightly with forceps to spread out and flatten the preparation. Dry any residual liquid from the edges of the coverslip. Seal the edges with clear nail polish and let dry.

14. To remove salt deposits on the coverslip that may damage an oil-immersion microscope lens, rinse the coverslip surface with distilled H_2O and dry gently with a Kimwipe or Q-tip.

15. For best results, view samples as soon as possible, and not more than a week after preparation.

Notes: Fluorescently labeled phalloidin can be included to counterstain for actin filaments. Include 5 μl labeled phalloidin (200 U/ml, Molecular Probes, Inc., Eugene, OR) in the 40-μl antibody mix for both primary and secondary antibody incubations.

Detection of two antigens simultaneously can be achieved simply by combining the two primary antibodies in the first step and the two secondary antibodies in the second step. For such an experiment, the primary antibodies must be raised in different species, and the fluorophores used for detection must have sufficiently separated excitation and emission spectra. For example, mouse antiactin and rabbit antiporin primary antibodies could be used, followed by a rhodamine-labeled goat antimouse secondary antibody and a fluorescein-labeled goat antirabbit secondary antibody.

Acknowledgments and Plasmid Requests

The authors thank Mark S. Longtine for plasmid libraries. For plasmid requests and a file of pFA6a-derived plasmid sequences, please write to M. S. Longtine (email: mlunc@isis.unc.edu; fax: 919-962-0320; Department of Biology, CB -3280, Coker Hall, University of North Carolina, Chapel Hill, NC 27599-3280). We also thank the Yeast Resource Center at University of Washington, Seattle, for CFP and YFP plasmids; for these plasmids, please contact Dale Hailey (YRC Microscopy, Box 357350, Department of Biochemistry, University of Washington, Seattle, WA 98195-7350; http://depts.washington.edu/%7Eyeastrc/fm_home1.htm). Finally, we thank Michael Yaffe for yeast strains. Work on these protocols was supported by grants to L. A. P. (National Institutes of Health GM 45735 and American Cancer Society) and to the Confocal Microscopy Core Facility at Columbia University (NIH Grant 5 P30 CA13696 to the Herbert Irving Comprehensive Cancer Center).

References

Bähler, J., Wu, J. Q., Longtine, M. S., Shah, N. G., McKenzie, A., III, Steever, A. B., Wach, A., Philippsen, P., and Pringle, J. R. (1998). Heterologous modules for efficient and versatile PCR-based gene targeting in *Schizosaccharomyces pombe. Yeast* **14,** 934–951.

Boldogh, I., Vojtov, N., Karmon, S., and Pon, L. A. (1998). Interaction between mitochondria and the actin cytoskeleton in budding yeast requires two integral mitochondrial outer membrane proteins, Mmm1p and Mdm10p. *J. Cell Biol.* **141,** 1371–1381.

Damsky, C. H. (1976). Environmentally induced changes in mitochondria and endoplasmic reticulum of *Saccharomyces carlsbergensis yeast. J. Cell Biol.* **71,** 123–135.

Diffley, J. F., and Stillman, B. (1991). A close relative of the nuclear, chromosomal high-mobility group protein HMG1 in yeast mitochondria. *Proc. Natl. Acad. Sci. USA* **88,** 7864–7868.

Evan, G. I., Lewis, G. K., Ramsay, G., and Bishop, J. M. (1985). Isolation of monoclonal antibodies specific for human c-myc proto-oncogene product. *Mol. Cell Biol.* **5,** 3610–3616.

Gietz, R. D., and Woods, R. A. (1998). Transformation of yeast by the lithium acetate/single-stranded carrier DNA/PEG method. *In* "Methods in Microbiology" (A. J. P Brown M. F. Tuite, eds.) Vol. 26. Academic Press, New York.

Hase, T., Riezman, H., Suda, K., and Schatz, G. (1983). Import of proteins into mitochondria: Nucleotide sequence of the gene for a 70-kD protein of the yeast mitochondrial outer membrane. *EMBO J.* **2,** 2169–2172.

Haugland, R. P. (1996). "Handbook of Fluorescent Probes and Research Chemicals." Molecular Probes, Eugene, OR.

Heim, R., and Tsien, R. Y. (1996). Engineering green fluorescent protein for improved brightness, longer wavelength and fluorescence resonance energy transfer. *Curr. Biol.* **6,** 178–182.

Hoffman, C. S., and Winston, F. (1987). A ten-minute DNA preparation from yeast efficiently releases autonomous plasmids for transformation of *Escherichia coli. Gene* **57,** 267–272.

Lillie, S. H., and Brown, S. S. (1994). Immunofluorescence localization of the unconventional myosin, Myo2p, and the putative kinesin-related protein, Smy1p, to the same regions of polarized growth in *Saccharomyces cerevisiae. J. Cell Biol.* **125,** 825–842.

Longtine, M. S., McKenzie, A., III, Demarini, D. J., Shah, N. G., Wach, A., Brachat, A., Philippsen, P., and Pringle, J. R. (1998). Additional modules for versatile and economical PCR-based gene deletion and modification in *Saccharomyces cerevisiae. Yeast* **14,** 953–961.

McConnell, S. J., Stewart, L. C., Talin, A., and Yaffe, M. P. (1990). Temperature-sensitive yeast mutants defective in mitochondrial inheritance. *J. Cell Biol.* **111,** 967–976.

McElver, J., and Weber, S. (1992). Flag N-terminal epitope expresion of bacterial alkaline phosphatase and flag C-terminal epitope tagging by PCR one-step targeted integration. *Yeast* **12,** 773–786.

Mihara, K., and Sato, R. (1985). Molecular cloning and sequencing of cDNA for yeast porin, an outer mitochondrial membrane protein: A search for targeting signal in the primary structure. *EMBO J.* **4,** 769–774.

Niedenthal, R. K., Riles, L., Johnston, M., and Hegemann, J. H. (1996). Green fluorescent protein as a marker for gene expression and subcellular localization in budding yeast. *Yeast* **12,** 773–786.

Prasher, D. C. (1995). Using GFP to see the light. *Trends Genet.* **11,** 320–323.

Pringle, J. R., Adams, A. E. M., Drubin, D. G., and Haarer, B. K. (1991). Immunofluorescence methods for yeast. *In* "Methods in Enzymology" (C. Guthrie G. R. Fink, eds.), Vol. 194, pp. 565–602. Academic Press, San Diego.

Pringle, J. R., Preston, R. A., Adams, A. E., Stearns, T., Drubin, D. G., Haarer, B. K., and Jones, E. W. (1989). Fluorescence microscopy methods for yeast. *Methods Cell Biol.* **31,** 357–435.

Riezman, H., Hase, T., van Loon, A. P., Grivell, L. A., Suda, K., and Schatz, G. (1983). Import of proteins into mitochondria: A 70 kilodalton outer membrane protein with a large carboxy-terminal deletion is still transported to the outer membrane. *EMBO J.* **2,** 2161–2168.

Roeder, A. D., Hermann., G. J., Keegan, B. R., Thatcher, S. A., and Shaw, J. M. (1998). Mitochondrial inheritance is delayed in *Saccharomyces cerevisiae* cells lacking the serine/threonine phosphatase PTC1. *Mol. Biol. Cell* **9,** 917–930.

Sherman, F. (1991). Getting started with yeast. *In* "Methods in Enzymology" (C. Guthrie G. R. Fink, eds.), **194,** p. 1. Academic Press, San Diego.

Simon, V. R., Karmon, S. L., and Pon, L. A. (1997). Mitochondrial inheritance: Cell cycle and actin cable dependence of polarized motochondrial movements in *Saccharomyces cerevisiae. Cell. Motil Cytoskel.* **37,** 199–210.

Simon, V. R., Swayne, T. C., and Pon, L. A. (1995). Actin-dependent mitochondrial motility in mitotic yeast and cell-free systems: Identification of a motor activity on the mitochondrial surface. *J. Cell Biol.* **130,** 345–354.

Smith, M. G., Simon, V. S., O'Sullivan, H., and Pon, L. A. (1995). Organelle-cytoskeletal interactions: Actin mutations inhibit meiosis-dependent mitochondrial rearrangement in the budding yeast *Saccharomyces cerevisiae. Mol. Biol. Cell* **6,** 1381–1396.

Sogo, L. F., and Yaffe, M. P. (1994). Regulation of mitochondrial morphology and inheritance by Mdm10p, a protein of the mitochondrial outer membrane. *J. Cell Biol.* **126,** 1361–1373.

Stevens, B. J. (1977). Variation in number and volume of the mitochondria in yeast according to growth conditions: A study based on serial sectioning and computer graphic reconstruction. *Biol. Cell* **28,** 37–56.

Taanman, J. W., and Capaldi, R. A. (1993). Subunit VIa of yeast cytochrome c oxidase is not necessary for assembly of the enzyme complex but modulates the enzyme activity: Isolation and characterization of the nuclear-coded gene. *J. Biol. Chem.* **268,** 18754–18761.

Visser, W, van Spronsen, E. A., Nanninga, N., Pronk, J. T., Gijs Kuenen, J., and van Dijken, J. P. (1995). Effects of growth conditions on mitochondrial morphology in *Saccharomyces cerevisiae. Antonie van Leeuwenhoek* **67,** 243–253.

Wach, A., Brachat, A., Alberti Segui, C., Rebischung, C., and Philippsen, P. (1997). Heterologous HIS3 marker and GFP reporter modules for PCR-targeting in *Saccharomyces cerevisiae. Yeast* **13,** 1065–1075.

CHAPTER 16

Targeting of Green Fluorescent Protein to Mitochondria

Koji Okamoto, Philip S. Perlman, and Ronald A. Butow

Department of Molecular Biology
University of Texas Southwestern Medical Center
Dallas, Texas 75390-9148

I. Introduction

Although many cell biological studies of mitochondria have been performed by indirect immunofluorescence techniques using antibodies against mitochondrial proteins or to epitope tags fused to mitochondrial proteins, the protocols are laborious and do not allow examination of mitochondria in living cells. Some vital staining dyes, such as MitoTracker and $DiOC_6$, have also been widely used as mitochondrial markers for living cells, but those reagents cannot be selectively targeted to different compartments within mitochondria and may not reflect the behavior of mitochondrial proteins. The green fluorescent protein (GFP) of the jellyfish *Aequorea victoria* has been applied as a visual marker of mitochondria in living cells, making it possible to generate fluorescent fusion proteins specific to the different mitochondrial compartments. This chapter describes GFP marker proteins that are correctly targeted to different mitochondrial compartments: the matrix, the inner and outer membranes, and mitochondrial DNA (mtDNA) in cells of the budding yeast *Saccharomyces cerevisiae*. These marker proteins can be used not only for the study of their sorting patterns both in vegetatively growing cells and in zygotes (Okamoto *et al.*, 1998), but also for the examination of other interesting phenomena, such as dynamic changes in mitochondrial morphology, including fusion and fission events, as well as distribution and motility of mitochondria and mtDNA.

II. Mitochondrial Green Fluorescent Protein (GFP) Fusion Markers

A. Structures of Plasmid-Encoded GFP Fusions

1. Galactose-Inducible GFP Markers

Yeast plasmid expression vectors containing mitochondrial GFP fusion genes are shown in Fig. 1. Construct A is a fusion gene that encodes a protein in which GFP is fused to the C terminus of the first 52 amino acids of the matrix protein, citrate synthase 1 (CS1), which contains a cleavable N-terminal presequence for targeting to the matrix. Construct B codes for a protein where GFP is fused to the C terminus of the inner membrane protease subunit, Yta10p, which also contains a cleavable N-terminal presequence and two transmembrane domains for insertion to the inner mitochondrial membrane (Arlt *et al.*, 1996; Tauer *et al.*, 1994). In construct C, GFP is fused to the

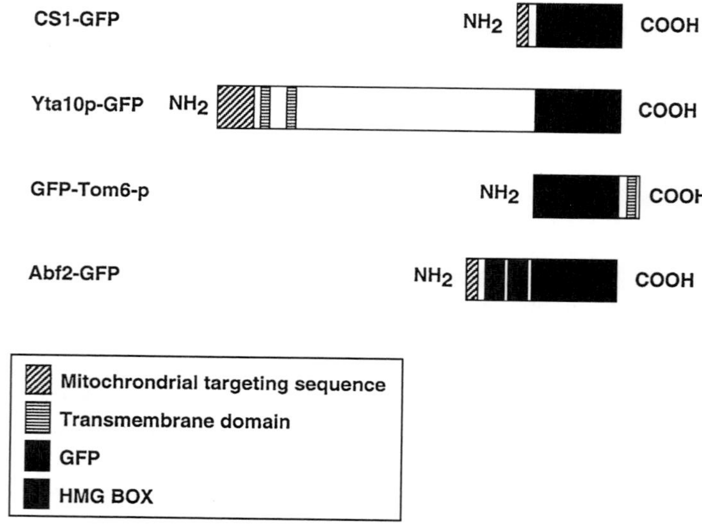

Fig. 1 Diagrammatic representation of fusion proteins between GFP and specific mitochondrial proteins: (A) CS1-GFP, (B) Yta10p-GFP, (C) GFP-Tom6p, and (D) Abf2p-GFP. The various domains of the fusion proteins, including the locations of the GFP tag, are indicated.

N terminus of the integral outer membrane protein, Tom6p, which is a component of the translocase of the outer membrane and has a transmembrane domain (Alconada *et al.,* 1995; Kassenbrock *et al.,* 1993). In construct D, GFP is fused to the C terminus of the mtDNA-binding protein, Abf2p (Diffley and Stillman, 1992; Newman *et al.,* 1996; Zelenaya-Troitskaya *et al.,* 1998), which contains a cleavable N-terminal presequence for the import of the protein to the mitochondrial matrix where mtDNA is located. These fusion genes have been placed in the *CEN-URA3* plasmid, pGAL68, so that their expression is under the control of the *GAL1-10* promoter, allowing induction of these proteins by growing cells on galactose medium (Okamoto *et al.,* 1998). In this way, mitochondria of cells can be labeled transiently with the particular GFP fusion protein, allowing one to follow their sorting pattern, e.g., in zygotes formed with cells of the opposite mating type that do not contain the fusion protein.

2. CS1-GFP and Abf2p-GFP Expressed from Endogenous Promoters

Sometimes it is advantageous to analyze cells in which a GFP fusion proteins is expressed at a level that is comparable to that of the wild-type version of the protein. Such instances include situations where the protein is a stoichiometric component of a complex or, as in the case of Abf2p, is associated with DNA, or when overexpression of a protein has deleterious effects. Two useful constructs for visualizing mitochondria and mtDNA are the fusion genes encoding CS1-GFP and Abf2p-GFP, respectively, whose expression is under the control in their endogenous promoters. The *CEN-URA3* plasmid pRS416/CS1-GFP has 0.8 kb of upstream sequence of the *CIT1* gene, which is sufficient for the full expression of wild-type CS1 (Rosenkrantz *et al.,* 1994). Thus, yeast cells containing pRS416/CS1-GFP grown on nonfermentable carbon sources show strong fluorescence intensity of CS1-GFP localized in the mitochondrial matrix. In contrast, CS1-GFP is strongly repressed in cells grown on glucose medium, as is the endogenous *CIT1* gene. Other nonrepressing fermentable carbon sources, such as raffinose and galactose, allow for relatively strong expression of the CS1-GFP fusion protein suitable for a range of microscopic analyses.

Plasmid pRS416/ABF2-GFP contains the ABF2-GFP gene fusion together with 5′- and 3′-flanking regions of the *ABF2* gene. Abf2p-GFP expressed under the control of the endogenous *ABF2* promoter is functional because it complements the mtDNA instability phenotype of *abf2* null mutants (Zelenaya-Troitskaya *et al.,* 1998). Essentially all of the Abf2p-GFP expressed from the endogenous *ABF2* promoter colocalizes with DAPI-stained mtDNA. A mutant version of Abf2p-GFP, which is compromised in its DNA-binding activity when also expressed in this configuration, only partially colocalizes with mtDNA so that the majority of the GFP fluorescence appears like the matrix protein, CS1-GFP (Zelenaya-Troitskaya *et al.,* 1998).

B. Induction of Galactose-Induced Mitochondrial GFP Markers

The expression of CS1-GFP, Yta10p-GFP, and GFP-Tom6p by continued growth of cells on galactose medium has no detectable effect on cell growth, whereas the over-expression of Abf2p-GFP results in the loss of mtDNA [as does overexpression of wild-type Abf2p (Zelenaya-Troitskaya *et al.*, 1998)]. Expression of Abf2p-GFP should, therefore, be induced by growth on YNB (0.67% yeast nitrogen base) 2% galactose medium for no longer than 45–60 min. Under these conditions, we have found that the amount of the fusion protein is roughly equivalent to that of wild-type Abf2p expressed in the same medium under the control of its endogenous promoter. The expression of Yta10p-GFP and GFP-Tom6p takes 18–24 h to reach levels required for detection by fluorescence microscopy; these levels are some two- to threefold lower than that of CS1-GFP induced for 6 h. Following induction, the synthesis of these fusion proteins can be quickly shut off by transferring the cells to glucose medium. All of these fusion proteins were found to remain detectable for at least 6 h after transfer to glucose medium even though new synthesis was blocked because their mRNAs were no longer detected after only 2 h of incubation in glucose medium.

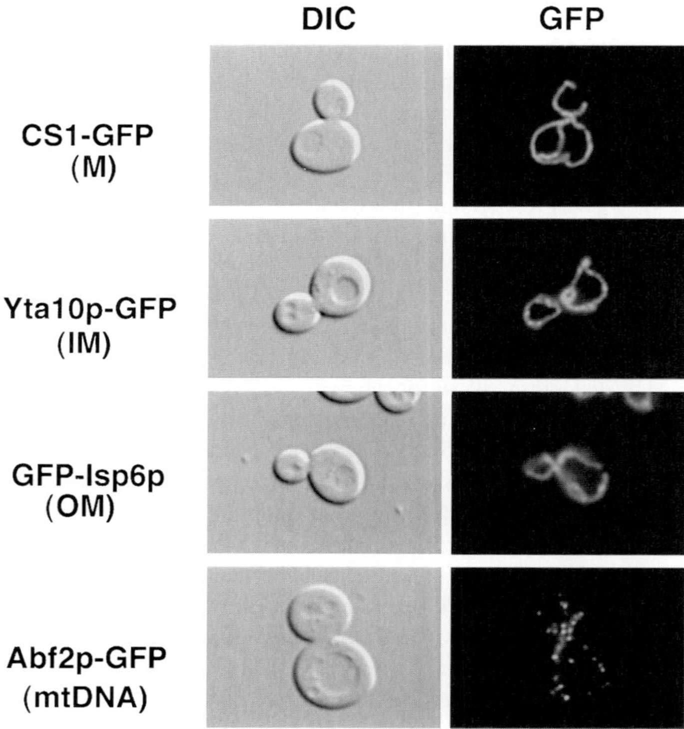

Fig. 2 GFP fluorescence patterns in living cells expression the indicated fusion proteins targeted to the mitochondrial matrix (M), inner membrane (IM), and outer membrane (M) and to mtDNA (taken from Okamoto *et al.*, 1998)

C. Expression and Localization of GFP Fusion Proteins

Yeast cells are transformed with the plasmids encoding mitochondrial GFP fusion proteins by standard procedures and grown to mid-log phase on selective medium (usually YNB) containing 2% raffinose. To induce expression of GFP fusion proteins, cells are transferred into YNB medium containing 2% galactose and 2% raffinose and grown for predetermined times (see earlier discussion), optimized for expression of the particular fusion protein. With the exception of Abf2p-GFP, the expression of each of the GFP fusion proteins results in a fluorescence pattern of a tubular network located near the periphery of the cell, which is a typical morphological characteristic of yeast mitochondria (Fig. 2). In contrast, Abf2p-GFP reveals a punctate pattern that has been shown to be exclusively colocalized with that of DAPI-stained mtDNA (Zelenaya-Troitskaya *et al.*, 1998).

D. Fixation of Cells Expressing Mitochondrial GFP Markers

In some experiments with cells expressing CS1-GFP, Yta10-GFP, GFP-Tom6p marker proteins, cells were analyzed after fixation with 3.7% formaldehyde at 30°C for 1 h. There is no strong effect of the fixation of the fluorescence intensity, pattern, and localization of those GFP markers. Fixed cells can also be stored at 4°C overnight prior to analysis. In contrast, Abf2p-GFP is somehow sensitive to formaldehyde fixation. A portion of the fixed wild-type cells expressing Abf2p-GFP showed patterns similar to those of either CS1-GFP or the mutant form of Abf2p-GFP. It is, therefore, necessary to use Abf2p-GFP without fixation.

III. General Concerns about Targeting of GFP Markers into Mitochondria

The specific targeting of the matrix, inner, and outer membrane GFP markers described here have been confirmed by biochemical analyses that include submitochondrial fractionation, susceptibility to protease digestion, and association with particular membrane fractions (Okamoto *et al.,* 1998). Such experiments are critical to ensure that correct targeting of the fusion protein has been achieved. In addition to possibilities of mistargeting, other problems can arise, such as low levels of expression of the fusion protein due to intrinsic instability, which can often occur if the fusion protein is not functional or if it is not incorporated into the native macromolecular complex in which the wild-type form of the protein is found.

Because GFP is quite a large tag (about 27 kDa), it is possible that some mitochondrial GFP fusion proteins may not be targeted correctly. Mislocalization of GFP fusion markers might be due to masking of targeting signals or prevention of their insertion into the membranes. For example, we found that the fusion protein between GFP and the C terminus of Qcr6p, subunit 6 of the cytochrome bc_1 complex (Van Loon *et al.,*

1984), was mislocalized to the cytosol. The fusion proteins between GFP and the N or C termini of porin, the mitochondrial outer membrane voltage-dependent anion channel, were also mistargeted and found to be distributed throughout the cytosol.

Galactose induction is ideal for transient expression in yeast cells. Some mitochondrial GFP fusion proteins, however, may be toxic or unstable, especially when overproduced. We found, for example, that the fusion protein between GFP and the C terminus of Bcs1p, an integral inner membrane protein required for the assembly of the cytochrome bc_1 complex (Nobrega *et al.,* 1992), although targeted correctly to the inner mitochondrial membrane, was unstable and disappeared within 6 h following galactose induction. Additionally, it should be noted that galactose overexpression of some mitochondrial GFP markers might result in significant damage to mitochondria. As noted earlier, overexpression of Abf2p-GFP leads to loss of mtDNA. Cells overexpressing GFP fused to the C terminus of Tom 70p, an integral outer membrane receptor for import (Hines *et al.,* 1990; Steger *et al.,* 1990), results in aberrant mitochondrial morphology where the reticular mitochondrial network was aggregated rapidly and where mitochondrial transmission of the daughter cell was strongly impaired. Tom70p-GFP appeared to be targeted to mitochondria, but its exact location or topology remains to be determined.

In summary, it is important to check fluorescence intensity, localization, stability, and potential toxic effects of mitochondrial GFP fusion proteins by both microscopic and biochemical analyses. When possible, the functionality of GFP fusion proteins should be confirmed.

IV. Conclusions

GFP fusion markers targeted to the different compartments within mitochondria are useful for a variety of studies. Our laboratory has made extensive use of specific fusion proteins to analyze the sorting of mitochondrial constituents in zygotes of synchronously mated cells (Okamoto *et al.,* 1998). Such studies have not only confirmed that mitochondrial fusion events occur soon after zygote formation, but that mtDNA is preferentially transmitted to the emerging diploid bud. These findings suggest that mtDNA transmission is nonrandom and governed by a segregation machinery that assures its faithful inheritance to daughter cells. Because haploid parental mitochondrial components are initially resolved spatially in young zygotes, this experimental system has provided insights into the dynamics of mitochondrial components that would have been more difficult, if not impossible, to resolve by analysis of vegetatively growing cells.

How eukaryotic cells regulate mitochondrial fusion is largely unknown. A protein involved in this event has been identified from *Drosophila* and yeast (Hales and Fuller, 1997; Hermann *et al.,* 1998; Rapaport *et al.,* 1998). Mitochondrial GFP markers described in this chapter will be especially useful to analyze mutants defective for mitochondrial fusion.

Acknowledgment

Research in the authors' laboratory was supported by Grant GM33510 from the National Institutes of Health.

References

Alconada, A., Kübrich, M., Moczko, M., Hönlinger, A., and Pfanner, N. (1995). The mitochondrial receptors complex: The small subunit Mom8b/Isp6 supports association of receptors with the general insertion pore and transfer of preproteins *Mol. Cell. Biol.* **15,** 6196–6205.

Arlt, H., Tauer, R., Feldmann, H., Neupert, W., and Langer, T. (1996). The YTA10-12 complex, an AAA protease with chaperone-like activity in the inner membrane of mitochondria. *Cell* **85,** 875–885.

Diffley, J. F. X., and Stillman, B. (1992). DNA binding properties of an HMG1-related protein from yeast mitochondria. *J. Biol. Chem.* **267,** 3368–3374.

Hales, K. G., and Fuller, M. T. (1997). Developmentally regulated mitochondrial fusion mediated by a conserved, novel, predicted ATPase. *Cell* **90,** 121–129.

Hermann, G. J., Thatcher, J. W., Mills, J. P., Hales, K. G., Fuller, M. T., Nunnari, J., and Shaw, J. M. (1998). Mitochondrial fusion in yeast requires the transmembrane GTPase Fzo1p. *J Cell Biol.* **143,** 359–373.

Hines, V., Brandt, A., Griffiths, G., Horstmann, H., Brutsch, H., and Schatz, G. (1990). Protein, import into yeast mitochondria is accelerated by the outer membrane protein MAS70. *EMBO J.* **9,** 3191–3200.

Kassenbrock, C. K., Cao, W., and Douglas,, M. G. (1993). Genetic and biochemical characterization of ISP6, a small mitochondrial outer membrane protein associated with the protein translocation complex. *EMBO J.* **12,** 3023–3034.

Newman, S. M., Zelenaya-Troitskaya, O., Perlman, P. S., and Butow, R. A. (1996). Analysis of mitochondrial DNA nucleoids in wild-type and a mutant strain of *Saccharomyces cerevisiae* that lacks the mitochondrial HMG-box protein, Abf2p. *Nucleic Acids Res.* **24,** 386–393.

Nobrega, F. G., Nobrega, M. P., and Tzagoloff, A. (1992). *BCS1,* a novel gene required for the expression of functional Rieske iron-sulfur protein in *Saccharomyces cerevisiae. EMBO J.* **11,** 3821–3829.

Okamoto, K., Perlman, P. S., and Butow, R. A. (1998). The sorting of mitochondrial DNA and mitochodrial proteins in zygotes: Preferential transmission of mitochondrial DNA to the medial bud. *J. Cell. Biol.* **142,** 613–623.

Rapaport, D., Brunner, M., Neupert, W., and Westermann, B. (1998). Fzolp is a mitochondrial outer membrane protein essential for the biogenesis of functional mitochondria in *Saccharomyces cerevisiae. J. Biol. Chem.* **273,** 20150–20155.

Rosenkrantz, M., Kell, C. S., Pennell, E. A., Webster, M., and Devenish, L. J. (1994). Distinct upstream activation regions for glucose-repressed and derepressed expression of the yeast citrate synthase gene, *CIT1. Curr. Genet.* **25,** 185–195.

Steger, F., Sollner, T., Kiebler, M., Dietmeier, K. A., Pfaller, R., Trulzsch, K. S., Tropschug, M., Neupert, W., and Pfanner, N. (1990). Import of ADP/ATP carrier into mitochondria: Two receptors act in parallel. *J. Cell Biol.* **111,** 2353–2363.

Tauer, R., Mannhaupt, G., Schnall, R., Pajic, A., Langer, T., and Feldmann, H. (1994). Ytal0p, a member of a novel ATPase family in yeast, is essential for mitochondrial function *FEBS Lett.* **353,** 197–200.

Van Loon, A. P. G. M., De Groot, R. J., De Haan, M., Dekker, A., and Grivell, L. A. (1984). The DNA sequences of the nuclear gene coding for the 17-kd subunit VI of the yeast ubiquinol-cytochrome c reductase: A protein with an extremely high content of acidic amino acids. *EMBO J.* **3,** 1039–1043.

Zelenaya-Troitskaya, O., Newman, S. M., Okamoto, K., Perlman, P. S., and Butow, R. A. (1998). Functions of the HMG box protein, Abf2p, in mitochondrial DNA segregation, recombination and copy number in *Saccharomyces cerevisiae. Genetics* **148,** 1763–1776.

CHAPTER 17

Assessment of Mitochondrial Membrane Potential *in Situ* Using Single Potentiometric Dyes and a Novel Fluorescence Resonance Energy Transfer Technique

James A. Dykens and Amy K. Stout

MitoKor
San Diego, California 92121

I. Introduction

Assessment of mitochondrial membrane potential ($\Delta\Psi_m$) provides the single most comprehensive reflection of mitochondrial bioenergetic function, primarily because it directly depends on the proper integration of diverse metabolic pathways that converge at the mitochondrion. For example, the functional integrity of electron transfer redox centers of oxidative phosphorylation, the catalytic integrity of the enzymes in β-oxidation pathways and Kreb's cycle, and the diverse transport mechanisms that link the cytosol with the mitochondrial interior, not to mention the structural integrity of mitochondrial architecture, all act in concert to establish and maintain $\Delta\Psi_m$ (Bernardi, 1999; Dykens *et al.,* 1999; Skulachev, 1999; Dykens, 1999; Lemasters *et al.,* 1998). Although it is only one of many factors that act in concert to establish and maintain the protonmotive force (Δp) across the mitochondrial inner membrane, it is the major component of that force, contributing some 150 mV of the total 180–220 mV (the remainder being due to the pH gradient) (reviewed by Nicholls and Ferguson 1992). Thus, it is the predominant driving force responsible for mitochondrial Ca^{2+} uptake, ATP generation, and the production of highly reactive oxygen-centered radicals (Dykens, 1994, 1995, 1997; Nicholls and Budd, 1998; Nicholls and Ward, 2000).

As the most salient index of mitochondrial function, measuring and understanding $\Delta\Psi_m$ have been fundamental to understanding the role of mitochondrial dysfunction in pathology (Beal *et al.,* 1997; Cassarino and Bennett, 1999), including the activation of crucial apoptosis pathways via efflux of mitochondrial cytochrome *c* (Fontaine and Bernardi, 1999; Bernardi *et al.,* 1998; Murphy *et al.,* 1999; Skulachev, 1999). Given the central role of mitochondrial integrity in both apoptotic and necrotic cell death, a therapeutic strategy targeting the improvement of mitochondrial function offers a potentially beneficial approach to pathology, ranging from chronic neurodegenerative diseases with mitochondrial involvement (reviewed by Beal *et al.,* 1997; Moos *et al.,* 1999; Shults *et al.,* 1999; Fiskum *et al.,* 1999) to acute ischemic pathology associated with stroke and myocardial infarction (Reynolds, 1999; Siesjo *et al.,* 1999).

Although assessment of $\Delta\Psi_m$ provides a comprehensive index of mitochondrial function and integrity, it provides little insight into the etiology of dysfunction. As a result, further characterization is generally required to identify whether loss of inner membrane integrity, impairment of electron transfer system (ETS) components, dysfunction of the transmembrane ion or adenylate transport mechanisms, or some other dysfunction underlies the diminution of potential (Andreyev and Fiskum, 1999).

Over the decades, a host of methods for measuring mitochondrial membrane potential have been developed, and they all have distinct advantages and disadvantages (see, e.g., wide-ranging reviews in Mason, 1993). This chapter focuses on several commonly used techniques most likely to yield information that can be put into the context of the extant literature on mitochondrial function. In so doing, we have attempted to identify several potential pitfalls and benefits of each in order to illustrate caveats inherent in their application.

Clearly, the experimental question and availability of instrumentation at hand will dictate which technique is most appropriate. Nevertheless, all of the methods discussed here are based on a common principle: the quantification of the potential difference across any lipid bilayer membrane is determined indirectly by measuring the distribution of a diffusible cationic molecule at equilibrium across that membrane (Johnson *et al.,* 1981; Ehrenberg *et al.,* 1988). This is true regardless of whether it is the plasma membrane or inner mitochondrial membrane potential under scrutiny, and distinguishing between the two in intact cells presents a number of technical problems that limit the use of potentiometric dyes and that therefore prompted development of the fluorescence resonance energy transfer (FRET) assay described in the second section of this chapter.

Positively charged molecules of sufficient lipophilicity will diffuse into cells, and hence into the mitochondrial matrix, because of the potential differences (inside negative) and the imposed concentration gradients across these membranes. In the ideal situation, these charged molecules would cross membranes unimpeded and equilibrate rapidly according only to the electrochemical gradient, without interference by binding to cellular components, or by transport by membrane carriers. Under such ideal conditions, the equilibrium reached is a function of its concentration, diffusion coefficient, and charge, as described by the Nernst equation:

$$V = -RT/ZF \log C_i/C_o \cong -60 \text{ mV} \log C_i/C_o,$$

where V is the potential difference across the membrane, R is the ideal gas law constant, T is the absolute temperature, Z is the charge on the diffusible ion, F is Faraday's constant, and C_i and C_o are the concentrations of the ion inside and outside the membrane, respectively. Therefore, by measuring the cation concentrations on both sides of a membrane, the potential difference across that membrane can be calculated. The accuracy of any such estimate diminishes as circumstances diverge from the ideal situation, and given the complexity of cellular systems, the ideal is rarely, if ever, obtained. In cells, simple application of the Nernst equation is confounded by facilitated diffusion and dye binding, as well as by transmembrane equilibria—and ensuing membrane potentials—that result from interactions of both diffusible and nondiffusible ions and that show elaborate compensatory mechanisms to maintain potential. Despite such complexity, the Nernst equation continues to foster not only our fundamental understanding of how biological membranes function, but also development of new techniques to study membrane potential.

One of the more accurate methods of measuring mitochondrial membrane potential involves measuring the accumulation of radioactive rubidium (or potassium) ions inside mitochondria rendered permeable to the cation by the ionophore valinomycin (Mitchell and Moyle, 1969; Padan and Rottenberg, 1973; Nicholls, 1974; Rottenberg, 1979). This method is appropriate for studying isolated mitochondria or mitochondria in cells where the plasma membrane has been permeabilized. Drawbacks associated with this technique include use of a labile and potentially hazardous radioisotope, the requirement for separating ^{86}Rb taken up into the mitochondria from ^{86}Rb remaining in the media, and the necessity of making some determination of the mitochondrial matrix volume in order to calculate C_i. Furthermore, valinomycin will transport endogenous K^+ ions in addition to the exogenous ions used as the indicator. Consequently, valinomycin itself can have effects on membrane potential and/or mitochondrial function. Finally, although this is a caveat for many biological assessments, another limitation of this method is that it is only truly useful for measuring the average membrane potential from a large pool of isolated mitochondria or permeabilized cells.

Similar to this ^{86}Rb technique, the accumulation of radiolabeled tetraphenylphosphonium ion (TPP$^+$) has also been used to determine mitochondrial membrane potential (Rottenberg, 1984; Jackson and Nicholls, 1986). One advantage of using TPP$^+$ instead of ^{86}Rb is that TPP$^+$ is sufficiently membrane permeable that it is not necessary to add an ionophore to enable it to cross membranes. While the hydrophobicity of TPP$^+$

obviates the need for ionophore-facilitated movement across membranes, it gives rise to other complications, including binding and retention by membranes, which undermine the accuracy of the Nernst equation.

Real-time, kinetic measurement of mitochondrial membrane potential using a TPP^+ electrode is another useful and popular technique (Kamo *et al.*, 1979; Jackson and Nicholls, 1986). After calibration of the electrode in the absence of tissue, the sample is added and the concentration of TPP^+ in the extramitochondrial space is monitored, with the assumption that the remainder has distributed into the mitochondrial matrix (although it also partitions into other membrane-bound compartments with potentials). Typically, in these studies, C_i is not calculated and the C_o measured is not converted to an absolute millivolt measurement. Instead, decreases or increases in C_o, indicative of hyperpolarization or depolarization, respectively, are monitored in real time upon addition of various stimuli.

Use of the TPP^+ electrode also requires either permeabilization of the plasma membrane or isolation of mitochondria prior to assessment. For some experiments, these nonphysiological conditions are unacceptable. For instance, in isolated mitochondria, or even permeabilized cells, it would be impossible to measure changes in mitochondrial membrane potential occurring upon receptor-mediated increases in intracellular Ca^{2+} concentrations. Also, as discussed earlier, this protocol measures the mean membrane potential of the entire population of cells in the experiment. An abiding caveat in any such experiment is that large changes in the membrane potential of a small portion of the mitochondria in the sample are indistinguishable from small changes in membrane potential occurring more homogeneously throughout the sample. If one needs to circumvent this limitation and determine the heterogeneity of responses across a population of cells, researchers should consider measuring mitochondrial membrane potential at the single cell level, using fluorescence microscopy or flow cytometry methods.

II. Single Potentiometric Dyes

Regardless of the method, a number of positively charged, lipophilic fluorescent dyes are commonly used to make qualitative, and even quantitative, measurements of mitochondrial membrane potential (Table I). Of those that are available, commercially, JC-1, tetramethylrhodamine TMR, rhodamine 123, and $DiOC_6(3)$ are used most frequently. Similar to all Nernstian techniques, quantification of the equilibration of these dyes across membranes is used as an indirect indicator of membrane potential. It is beyond the scope of this chapter to review technologies available for the assessment of fluorescent probes, and readers unfamiliar with the requisite instrumentation are encouraged to consult the comprehensive reviews by Johnson (1998) and the earlier volume in this series edited by Matsumoto (1993).

An obvious caveat to bear in mind for all these dyes is that they equilibrate not only across mitochondrial membranes based on charge and concentration, but also across the plasma membrane in a similar manner. For technical reasons, it is generally not feasible to determine mitochondrial membrane potential by measuring dye fluorescence from the interior of the mitochondrion while simultaneously but separately measuring dye

Table I
Potentiometric Dyes for Assessment of $\Delta\Psi_m$

Dye	Full name	Excitation maximum (nm)	Emission maximum (nm)
$DiOC_6(3)$	3,3′-Dihexyloxacarbocyanine iodide	488	501
Rhod 123	Rhodamine 123	507	529
TMRM	Tetramethylrhodamine methyl ester	549	573
TMRE	Tetramethylrhodamine ethyl ester	549	574
JC-1	5,5′,6,6′-Tetrachloro-1,1′,3,3′-tetraethylbenzimidazolylcarbocyanine iodide	514	529 and 590

fluorescence from the surrounding cytosol as indicators of C_i and C_o (but see Farkas *et al.,* 1989; Loew *et al.,* 1993).

Fluorescence microscopy experiments typically measure real-time changes in the total cell fluorescence as a qualitative determination of mitochondrial membrane potential rather than making quantitative calculations. In flow cytometry experiments, the total cell fluorescence at only a single time point is usually determined. In either case, however, any experimental manipulation that alters plasma membrane potential can potentially have an effect on dye localization or on the aggregation state, both of which can alter the resulting fluorescence signal. While these confounding effects are undesirable, they can sometimes be avoided or ignored. For example, digitonin-permeabilized cells can be used to eliminate the contamination of signal by the plasma membrane (Floryk and Houstek, 1999). The FRET assay described in this chapter offers an alternative approach. In addition, because of the higher surface area to volume ratio of the two compartments, cationic fluorescent dyes equilibrate more rapidly across mitochondrial membranes than across the plasma membrane (Nicholls and Ward, 2000). By focusing on only the initial, rapid changes in fluorescence, it is possible to distinguish temporally between changes in mitochondrial and plasma membrane potential (Nicholls and Ward, 2000).

A prevailing assumption made when using these indicators is that dye uptake into a cell or organelle is proportional to its membrane potential. In the absence of any gross non-Nernstian behavior due to dye binding, active transport, or concentration effects (see later), this is generally a safe approximation. However, the inference that the fluorescence signal is directly proportional to membrane potential is unjustified. When many of these dyes, including rhodamine 123, TMRE, and TMRM, are sequestered into the mitochondrial matrix at high concentrations, the close proximity of dye molecules in this low volume space fosters intramolecular energy transfer that results in self-quenching and loss of signal (Emaus *et al.,* 1986; Bunting *et al.,* 1989; Scaduto and Grotyohann, 1999; and see discussion of FRET later). Under these conditions, fluorescence is not proportional to membrane potential. As a result, in experiments where a single measure of the mean or total fluorescence from a cell is made, care should be taken to use these dyes at concentrations where self-quenching does not occur. In contrast, when fluorescence can be monitored continuously in real time, as during imaging experiments, such overloading with dye can be tolerated. Under these conditions, depolarization of the mitochondrial membrane potential is indicated as an increase in the fluorescence signal as the positively charged dye molecules unquench upon exiting the mitochondria.

As a class, the quality of data generated with any of these dyes can be compromised by several factors. One major problem frequently encountered when using these dyes is that many of them are substrates for the P-glycoprotein transporter associated with multidrug resistance (MDR; Gottesman and Pastan, 1993). Transformed cell lines frequently exhibit MDR, and the accumulation of fluorescent dyes, including many of the ones discussed here, is used as an indicator of MDR in screens for inhibitors of the P-glycoprotein-dependent carrier (Efferth *et al.,* 1989; Kessel *et al.,* 1991; Webb *et al.,* 1996). Obviously, the active ejection of dye molecules out of the cell alters the cytosolic concentration, and hence the equilibrium with the mitochondrial matrix. Such active transport invalidates the assumption underlying strict application of the Nernst equation and underscores the need for caution in interpreting such data (Petronilli *et al.,* 1998). Likewise, washing away excess extracellular dye prior to measurement perturbs the established equilibrium and causes intracellular dye to redistribute along the new concentration gradients. To minimize the confounding effects of the MDR transporter and to avoid artifacts associated with the reestablishment of equilibria after washing, whenever permitted by instrumentation, experiments should be conducted in an extracellular media containing the same concentration of dye as used for loading. This strategy worked well for the FRET assay described in Section III.

Another common problem with many of these dyes is their photolability (Bunting, 1992a; Huser *et al.,* 1998), which is usually detected as photobleaching of the dye when cells are overexposed to the excitation light source. This artifactual decrease in signal could be interpreted erroneously as a potential change. Less obviously, many dyes can function as photosensitizers that act synergistically with ambient molecular oxygen to yield reactive species such as singlet oxygen or oxygen-centered radicals. Huser *et al.* (1998) reported laser-induced reactive oxygen species production in cells loaded with tetramethylrhodamine methyl ester that eventually triggered mitochondrial membrane depolarization. Control experiments in which cells are exposed to excitation illumination without the experimental stimulus, or exposed to stimulus without light, should be done for comparison. In any event, for biological systems, the goal should be to minimize both the intensity of illumination and the extent of dye loading, at least as much as practical.

There are other less apparent reasons to avoid excess dye loading. Excessive mitochondrial accumulation of cationic dye at high concentrations can depolarize $\Delta\Psi_m$ and compromise mitochondrial function. Furthermore, it has been suggested that when the ratio of dye molecules to cells is lower, then changes in dye fluorescence predominantly reflect changes in mitochondrial membrane potential rather than plasma membrane potential (Wilson *et al.,* 1985). Similarly, Scaduto and Grotyohann (1998) reported that the sensitivity for detecting changes in mitochondrial membrane potential is increased when the dye-to-cell ratio is decreased.

While the lipophilic nature of these compounds allows them to diffuse through membranes in response to potential and concentration gradients, movement and equilibration of these dyes occur much more slowly than the actual changes in membrane potential being monitored. In fact, all of the dyes considered here are categorized as slow-response dyes. Fast-response dyes undergo a structural change, and consequently a fluorescence change, in response to a change in the electric field surrounding them. These types of probes are capable of detecting potential changes on a millisecond time scale. In contrast, the fluorescence changes monitored when using slow-response dyes are generally a result of increases or decreases in dye concentrations and aggregation states that occur only after the dyes redistribute across membranes in response to potential changes. Because the rate of dye redistribution is dependent on its diffusion coefficient and the imposed concentration gradient, dye redistribution can lag behind the potential changes by seconds to minutes (Bunting, 1992b). Such temporal delays should be considered when monitoring mitochondrial membrane potential in real time.

Such a time lag may be a limitation in some studies, but there are instances when this slow response time is an advantage. For instance, slow permeation can minimize the contribution of plasma membrane potential changes to the overall fluorescence signal observed (Nicholls and Ward, 2000). Furthermore, the lag time for dye efflux from mitochondria is generally greater than that for dye influx (Bunting, 1992b), suggesting that measurements of depolarization may suffer more than measurements of repolarization or hyperpolarization.

A. Dye Selection

1. Rhodamine 123

Historically, rhodamine 123 has been a very popular choice for monitoring mitochondrial membrane potential. However, it illustrates several of the limitations associated with these dyes. For example, its high nonspecific binding to mitochondria compromises its Nernstian distribution (Emaus *et al.,* 1986; Scaduto and Grotyohann, 1999). In addition, not only is rhodamine 123 an established substrate for MDR pumps (Efferth *et al.,* 1989; Kessel *et al.,* 1991; Webb *et al.,* 1996), it also inhibits respiration and uncouples mitochondria significantly at micromolar concentrations (Emaus *et al.,* 1986; Bunting *et al.,* 1989; Scaduto and Grotyohann, 1999). As mentioned earlier, rhodamine 123 exhibits extensive self-quenching (Emaus *et al.,* 1986; Bunting *et al.,* 1989; Scaduto and Grotyohann, 1999), increasing the requirement for judicious dye-loading protocols (see later). In an attempt to overcome some of the limitations of this dye, Juan *et al.* (1994) have developed a modified method for rhodamine 123 use in which they base their measurements of mitochondrial membrane potential on the initial rate of dye uptake into cells. Phototoxicity is another side effect one must take care to avoid with this dye as well (Nicholls and Ward, 2000).

2. $DiOC_6(3)$

An alternative to rhodamine 123 is 3,3′-dihexyloxacarbocyanine iodide [$DiOC_6(3)$], one of a large family of carbocyanine dyes with high fluorescence efficiency. $DiOC_6(3)$

has been used in multiple studies as an indicator of mitochondrial membrane potential as this positively charged species stains polarized mitochondria. However, $DiOC_6(3)$ also stains the endoplasmic reticulum (Terasaki and Reese, 1992; Sabnis *et al.*, 1997). According to Rottenberg and Wu (1998), only at concentrations below 1 n*M* is $DiOC_6(3)$ staining selective for mitochondria. However, the appropriate loading concentration should be determined empirically using the relevant cell type and detection methodology. Like rhodamine 123, $DiOC_6(3)$ may inhibit mitochondrial respiration. Rottenberg and Wu (1998) indicated that $DiOC_6(3)$ inhibits oxygen consumption in intact cells by ~90% at only 40 n*M*, whereas Anderson *et al.* (1993) have shown that $DiOC_6(3)$ inhibits complex I activity with an IC_{50} of $\sim 1 \mu M$.

3. TMR

Tetramethylrhodamine methyl ester (TMRM) and tetramethylrhodamine ethyl ester (TMRE), two dyes that are structurally very similar to rhodamine 123, have been used as indicators of mitochondrial membrane potential in multiple studies since they were first described in 1988 by Ehrenberg *et al.* Predictably, these dyes suffer from many of the same problems as rhodamine 123. Both have been reported to inhibit respiration, and TMRE is more potent and TMRM is less potent than rhodamine 123 in this respect (Scaduto and Grotyohann, 1999). Similar to rhodamine 123, TMRE and TMRM also exhibit significant nonspecific dye binding to membranes (Scaduta and Grotyohann, 1999), and both are substrates for MDR. As described previously, TMRE can generate toxic reactive oxygen species (Huser *et al.*, 1998). Thus, TMRM may be the preferred dye of the pair. Furthermore, these tetramethylrhodamine compounds are more membrane permeable than rhodamine 123 (Bunting, 1992b), making the contribution of plasma membrane potential changes more prominent with these dyes (Nicholls and Ward, 2000).

4. JC-1

While its use is based on similar principles, JC-1 (5,5′,6,6′-tetrachloro-1,1′,3,3′-tetraethylbenzimidazolylcarbocyanineiodide) differs from the other potentiometric dyes discussed earlier because it can exist in two different states with different emission spectra; at low concentrations, JC-1 monomers exhibit green fluorescence, whereas at higher concentrations the dye forms aggregates that exhibit red fluorescence (Reers *et al.*, 1995). This is in contrast to the self-quenching that occurs upon aggregation of the rhodamine dyes at high concentrations. This behavior is useful for measuring mitochondrial membrane potential because the spectral shift that occurs upon aggregate formation within the mitochondrial matrix is dependent on dye concentration, which in turn is dictated by membrane potential. As a result, JC-1 can be used as a ratiometric indicator, with the ratio of red : green fluorescence providing an internal normalizing benchmark useful for comparing data across experiments. One drawback associated with this dye is that more complicated hardware and software are required for ratiometric measurements.

With JC-1, a drop in mitochondrial membrane potential is theoretically reflected by both an increase in the green fluorescence signal and a decrease in the red fluorescence signal. There is at least one report indicating that the aggregate form of the dye is more sensitive to potential changes and that at small depolarizations, only a decrease in red signal is noted (Di Lisa *et al.,* 1995). According to these researchers, only upon more profound depolarization is an increase in green fluorescence observed. However, another group has reported that the primary response to a drop in mitochondrial membrane potential (whether induced by Ca^{2+} loading or uncoupling) is an increase in monomer green fluorescence in the cytosol (Scanlon and Reynolds, 1998). In these studies, a decrease in red fluorescence was only observed after an imposed oxidative stress. It is unclear whether this fluorescence change represents a change in membrane potential or is an artifact of dye oxidation.

There is at least one report of JC-1 being used to assess multidrug resistance (Kuhnel *et al.,* 1997), suggesting that it too may be influenced by active transport mechanisms that would compromise its ability to report membrane potential. Under conditions of active transport of JC-1 or rhodamine 123 by MDR proteins, the addition of a compound that competes for transport would alter dye distribution and therefore fluorescence. These types of responses could mistakenly be interpreted as membrane potential changes. Interestingly, cyclosporin A, a compound that is used frequently in investigations of mitochondrial function, is a substrate for the P-glycoprotein pump. Although it may be possible to pharmacologically inhibit MDR pumps in order to acquire more accurate results, such treatment always invites artifact.

In summary, in addition to idiosyncratic advantages and disadvantages, these dyes share a number of caveats for their use, including self-quenching, cationic dissipation of $\Delta\Psi_m$, photochemistry, and inhibition of respiration. There are a number of published reports in which several of these dyes were compared directly in a number of different cell types (Garner *et al.,* 1997; Salvioli *et al.,* 1997; Mathur *et al.,* 2000). While these studies unanimously conclude that JC-1 is superior, a cautious approach would entail comparison from multiple probes and even multiple techniques before drawing firm conclusions about $\Delta\Psi_m$ changes. Regardless of the dye and the experimental paradigm, because of the Nernst equation, data from studies using potentiometric dyes follow a logarithmic, not linear, relationship.

B. Practical Hints with Potentiometric Dyes

It is clear that optimal dye-loading protocols vary widely and need to be determined empirically for the cell type, cell number, and detection technology at hand. The optimal dye concentration should be sufficient to yield a viable signal-to-noise ratio without compromising mitochondrial function. Because most dyes exhibit some toxicity, and all permeant cations diminish $\Delta\Psi_m$ in proportion to their concentration, initial emphasis should be placed on maximizing detection gain or on increasing biomass in the detector rather than increasing dye concentration. However, given the photochemistry of these dyes, increasing the signal by simply increasing illumination intensity may yield photobleaching or photosensitized radical formation with toxic ramifications. Therefore, before increasing illumination, attempts to maximize detection gain should be made by broadening the band pass of the emission filters or by capitalizing on photomultiplier or CCD technologies.

As a general guideline, most intact immortalized cells containing typical densities of respiring mitochondria will require exposure to 0.2–2 μM dyes such as TMRE for at least 10 min at 37°C to obtain robust mitochondrial staining. A simple technique to determine whether excessive dye loading has yielded self-quenching is to monitor the fluorescence signal as $\Delta\Psi_m$ is collapsed using protonophores such as CCCP or FCCP (ca. 0.5–5 μM), both of which induce virtually instantaneous $\Delta\Psi_m$ collapse. An increase in fluorescence that corresponds with CCCP exposure indicates release of self-quenching, and lower dye concentrations, or shorter loading times, should be attempted until CCCP collapse produces a rapid loss of fluorescence with no lag or upswing in signal.

Clearly, there is a point of diminishing returns in this strategy, and at some low dye concentration, no signal will be detectable. If detection of $\Delta\Psi_m$ collapse is the major parameter being assessed in the experiment, then the increase of fluorescence upon loss of self-quenching can be used as an index of $\Delta\Psi_m$ collapse, bearing in mind that excess dye may have correspondingly undermined mitochondrial function.

When using nonratiometric dyes in a nonquenching mode, the total fluorescence of a cell will still be dependent on factors in addition to membrane potential. For example, the extent of dye loading, and hence fluorescence, will also be proportional to overall cell size and to the mitochondrial mass or volume. Corrections for these types of variables, or paired comparison experimental design, should be considered whenever possible. Similarly, to test whether MDR activity is repressing dye uptake, examine the signal in the presence of 1–10 μM verapamil, FK-506, or some other MDR inhibitor.

Regardless of the dye or methodology chosen, control experiments with compounds that have known effects on $\Delta\Psi_m$ should be done to validate and determine the range of the system. For instance, depolarization with an uncoupler and hyperpolarization with oligomycin and nigericin (which exchanges protons for K^+, thereby decreasing ΔpH while increasing $\Delta\Psi_m$) should be evaluated. However, failure to detect an effect with oligomycin could indicate a lack of sensitivity, as the hyperpolarization induced by this agent is only about 5 mV (Scott and Nicholls, 1980). Similarly, control experiments in which the plasma membrane potential is selectively depolarized should be performed to assess the extent or lack of an effect on $\Delta\Psi_m$.

Finally, all the techniques outlined here are designed to measure relative changes in $\Delta\Psi$m. If an absolute value for $\Delta\Psi_m$ is required, many of these dyes can be calibrated to a voltage scale by clamping $\Delta\Psi_m$ at a known potential using a combination of valinomycin and variable potassium concentrations (method described by Di Lisa *et al.,* 1995). This protocol works only with mitochondria or permeabilized cells, although it is also theoretically possible to calibrate fluorescence measurements obtained under the same rigid conditions for direct conversion into millivolts (Di Lisa *et al.,* 1995).

III. Novel Fluorescence Resonance Energy Transfer Assay for $\Delta\Psi_m$

As discussed earlier, assessing real-time kinetics of $\Delta\Psi_m$ in cells has previously only been possible with membrane-permeant cationic dyes using single-cell imaging systems, curtailing the use of these measurements as a high-throughput screening assay [although see Nuydens *et al.* (1999) for an interesting ratiometric, end-point method]. Although useful, as discussed earlier, these potentiometric dyes for $\Delta\Psi_m$ assessment in intact cells have several inherent limitations, most notably a lack of specificity for mitochondrial potential as opposed to the potential present at the plasma membrane ($\Delta\Psi p$). As noted here and elsewhere (Nicholls and Ward, 2000), care must be taken to use these dyes at concentrations where they neither inordinately deplete $\Delta\Psi_m$ nor self-quench and thereby obscure the kinetics and magnitude of responses. Such low dye concentrations are amenable to microscopy and flow analysis, where intense illumination of single cells and sensitive detection are available (Sureda *et al.,* 1999), but they are not readily amenable to most of the technologies available for high-throughput screening of vast compound libraries that are the basis for contemporary drug discovery programs.

Cell-permeant cationic fluorescent dyes, such as DASPMI or the tetramethylrhodamines, must be used at excessively high concentrations in order to obtain sufficient signal levels for detection by most commercially available plate readers. This is dictated in large measure by the limited intensity of excitation illumination and sensitivity of detection. As these high concentrations of cationic dyes aggregate in the mitochondrial matrix, they dissipate $\Delta\Psi_m$, thereby increasing metabolic burden as compensatory mechanisms attempt to maintain $\Delta\Psi_m$. At some combination of concentration and duration, these dyes become toxic to mitochondria, and hence to cells (Scorrano *et al.,* 1999). For example, between 40 and 50 μM DASPMI is required to detect adequate mitochondrial signals from confluent SH-SY5Y cells in a 96-well plate using a state-of-the-art conventional plate reader optimized for kinetic analysis. Although these dye concentrations in our experiments yielded acceptable signal-to-noise ratios, and the responses of pharmacological agents targeting various mitochondrial functions were in accord with expectations, the use of ionomycin to induce a Ca^{2+} -mediated $\Delta\Psi_m$ collapse caused cells to detach from the plate and exacerbated subsequent toxicity (as assessed via live/dead dye exclusion protocols).

Potentiometric dyes—indeed almost all fluorescent dyes—are sensitive to environmental variables such as pH, solvent polarity, and, in particular, to the proximity of other dyes and paramagnetic molecules such as O_2. For example, the emission efficiency (but not the emission spectra) of many dyes decreases when the excited state species interacts with molecules capable of absorbing this energy, such as proteins with large resonance capacities, or other dyes whose excitation spectra overlaps the emission spectra. When FRET occurs, the energy of emission of one dye is transferred to a second dye molecule with no photon emission, resulting in the quenching of the first (loss of its signal), along with the corresponding excitation and emission of the second dye. The extent of energy transfer in FRET is altered by numerous factors, such as the extent of overlap between emission and excitation spectra of the dye pairs, as well as physical characteristics of the dyes, such as dipole orientations and their individual quantum efficiencies. However, the overriding factor in defining the efficiency of FRET is the proximity of the exchanging dye molecules, with energy transfer decreasing as a function of molecular separation to the sixth power (r^6). The intramolecular separation where 50% of the excitation is quenched by a second dye is defined as the Forster radius, and for most biologically relevant dyes, this is on the order of 10–100 Å. In practice, the relatively small Stokes

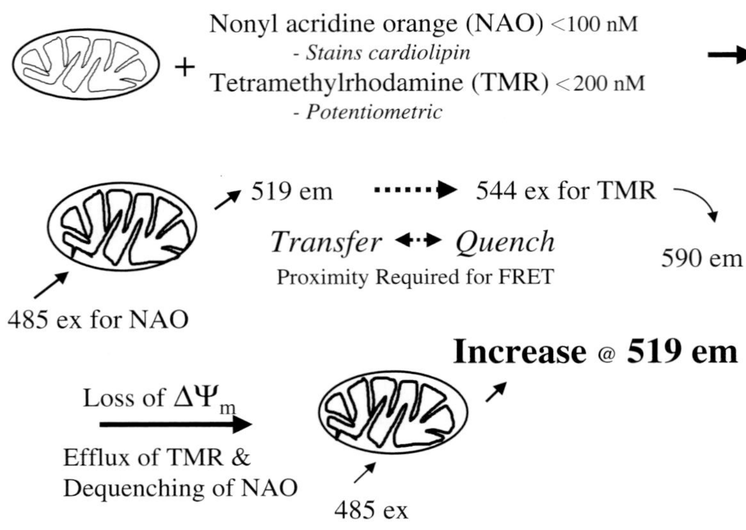

Fig. 1 The FRET-based assay of mitochondrial membrane potential ($\Delta\Psi_m$) depends on close proximity and energy transfer between the cardiolipin stain, nonyl acridine orange (NAO), and the potentiometric dye, tetramethylrhodamine (TMR). Loss of $\Delta\Psi_m$ yields efflux of TMR, and hence dequenching of NAO, which is excited at 485 nm and monitored in real time at 519 nm.

shifts and limited efficiencies of most dyes, as well as the broad band-pass filters found in most nonspecialized plate readers, dictate that either the quenching of the first dye or emission of the second be monitored. Of course, assessing both dyes simultaneously not only is internally corroborating, but such data can also be used to optimize the efficiency of the interactions.

We have developed an assay for mitochondrial $\Delta\Psi_m$ based on FRET between two dyes that are both localized to the mitochondria. In so doing, the FRET assay minimizes, and within specific limitations (outlined later), circumvents the confounding variables of plasma membrane potential and heterogeneous intracellular pH in different microdomains (e.g., between the mitochondrial matrix and the intermembrane space) that have previously made it difficult to isolate $\Delta\Psi_m$ in intact cells in a convenient and reproducible manner using single potentiometric dyes.

In the FRET assay for mitochondrial $\Delta\Psi_m$, the excitation dye is nonyl acridine orange (NAO) (Fig. 1). NAO is predominantly a stain for cardiolipin (diphosphatidyl glycerol), a lipid distributed almost exclusively (>99%) in the mitochondrial inner membrane, with the remainder in the nuclear envelope. The second dye is tetramethylrhodamine (methyl and ethyl esters work equally well), a potentiometric dye that is sequestered into the mitochondrial matrix by both $\Delta\Psi_m$ and the imposed concentration gradients in accord with the Nernstian principles outlined earlier. As such, FRET between these two dyes occurs only when $\Delta\Psi_m$ exists and TMR enters the matrix. It is the specificity of NAO staining

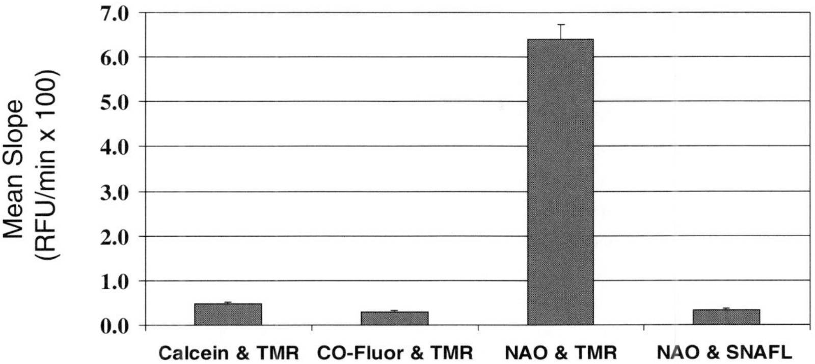

Fig. 2 Resonance energy transfer with NAO only occurs when both dyes are in the mitochondria. In the presence of TMR, substitution of NAO by either calcein AM (λex 494, λem 517) or carboxyfluorescein AM (λex 492, λem 517), both of which have excitation/emission characteristics comparable to NAO (λex 485, λem 517), but both of which are cytosolic, yields no change in signal upon addition of 0.5 μM CCCP, a protonophore that collapses $\Delta\Psi_m$. Likewise, in the presence of NAO, substitution of TMR (λex 544, λem 590) by the cell-permeant diacetate form of SNAFL (λex 514, λem 546), which is also cytoplasmic, yields no energy transfer with NAO, despite a more favorable overlap between NAO emission and SNAFL excitation than with TMR. Only when both dyes are in the mitochondria does NAO quenching and dequenching by CCCP occur, as shown by the robust signal with NAO and TMR.

for cardiolipin, combined with the prerequisite for the close proximity of the two dyes, that allows this FRET assay to report $\Delta\Psi_m$ unconfounded by the signal from the potentiometric dye associated with the plasma membrane potential. Importantly, FRET does not occur when nonmitochondrial fluorescent dyes having suitable excitation/emission spectra are substituted for either the acceptor or the donor dye (Fig. 2). For example, no FRET with TMR is apparent when NAO is replaced by the cytoplasmic dyes calcein AM or carboxyfluorescein, despite their having suitably overlapping excitation spectra. Likewise, no FRET with NAO is detectable when TMR is replaced by the cytoplasmic dye carboxy SNAFL (Fig. 2). Such findings are not surprising in light of the stringent requirement for close proximity in order for FRET to occur.

Because, uptake and retention of the potentiometric dye depend on $\Delta\Psi_m$ at a given concentration, the extent of quenching of NAO by the potentiometric dye reflects the magnitude of $\Delta\Psi_m$. Conversely, dissipation of $\Delta\Psi_m$ results in efflux of the potentiometric dye from the mitochondrion. The corresponding loss of proximity abolishes TMR quenching of NAO so that loss of $\Delta\Psi_m$ is detected as an increase in the NAO signal as it dequenches.

This is not to say that this FRET assay for $\Delta\Psi_m$ is completely independent of events at the plasma membrane. Indeed, any effect that alters the concentration of TMR in the cytoplasm will alter NAO quenching, at least to the extent that the equilibrium of TMR between the cytosol and the mitochondrial matrix depends on the imposed concentration gradient and the diffusion coefficient for the dye (as dictated by the Nernst equation). In this light, ejection of TMR from the cytoplasm, e.g., by the multidrug resistance transporter (see later), will lower cytoplasmic concentration and hence perturb the equilibrium with the mitochondrial matrix. Likewise, changes in the cytoplasmic concentration of the potentiometric dye resulting from depolarization (or hyperpolarization) of the plasma membrane will also alter the cytoplasmic-mitochondrial equilibrium and hence affect NAO quenching. In order to minimize such perturbations in the concentration gradients across the extracellular, cytoplasmic, and mitochondrial compartments, the extracellular potentiometric dye is not removed during the assay, thereby buffering changes. Because the assay monitors changes in NAO, not TMR, emission, retaining excess extracellular TMR does not interfere with readings.

As an additional consideration, loss of $\Delta\Psi_m$ in this FRET assay is detected as NAO dequenching, a result of TMR effluxing from the mitochondria into the cytoplasm. Given the enormous extracellular reservoir of dye present throughout the assay, especially considering the miniscule intracellular and intramitochondrial volumes, the mitochondrial release of TMR should transiently increase only the localized concentration of cytosolic TMR, which would correspondingly increase the gradient driving TMR back into the mitochondria via diffusion. Thus, by retaining extracellular dye and thereby buffering concentration equilibria, the effects on the NAO signal due to changes in potential are augmented, whereas those due to concentration effects alone are minimized.

Through a series of extensive validation and optimization studies, several of which are outlined here, we have established the FRET-based $\Delta\Psi_m$ assay as a sensitive and reliable *in situ* indicator of mitochondrial membrane potential and mitochondrial integrity. The assay was initially designed for use with intact cells, employing Ca^{2+}-ionophores such as ionomycin or A 23187 to model cytosolic Ca^{2+} loading during excitotoxicity. However, the inability of classical mitochondrially active agents such as bongkrekic acid and ruthenium red to exert the expected effects in intact cells prompted us to pursue the assay using cells permeabilized with digitonin. Of course, plasma membrane potential is no longer an issue when cells are permeabilized, and the increased bioavailability at the mitochondrion afforded by removing the plasma membrane as a diffusion barrier broadens the diversity of chemical matter arising from HTS compound discovery. In any event, the FRET assay works in both intact and permeabilized cells, offering a choice of experimental parameters, including use of adherent and nonadherent cell lines. Within the physical constraints of illumination and detection, it could be adapted readily to use on intact tissues and in microscopy.

A. Cell Selection

The assay was developed using a SH-SY5Y neuroblastoma cell line, but the FRET assay can be adapted to a host of cell types, including cell suspensions. Experience with H-460 (lung carcinoma), MCF-7 (breast carcinoma), and cybrids made from SH-SY5Y indicates that dye concentrations and protocols must be optimized for each cell type, a likely consequence of differing mitochondrial masses and differing $\Delta\Psi_m$ and $\Delta\Psi p$, as well as dissimilar cell sizes (resulting in different surface area to volume ratios) and permeabilities.

In the case at hand, SH-SY5Y cells are propagated in high glucose DMEM, 5% FCS, 10 m*M* HEPES. Twenty-four hours prior to assay, cells are trypsinized and plated in clear-bottom, black-walled plates (Costar, Cambridge, MA). Use of the latter is recommended, as cross-talk between adjacent wells in clear-walled plates can erode signal-to-noise ratios when fluorescence is intense. Cells are typically plated at 60,000 cells/well for use in the high throughput screening protocol in FLIPR (Fluorescence Imaging Plate Reader, Molecular Devices, Sunnyvale, CA), although the FRET signal is linear over a wide range of cell densities with a threshold of detection in FLIPR (dictated largely by instrument sensitivity) ca. 3×10^3 SY5Y cells/well in a 96-well plate, and an upper limit of 1.5×10^6 (this cell line is confluent at ca. 1.0×10^5/well).

B. Dyes and Protocol

The FRET $\Delta\Psi_m$ assay depends on the interaction of two dyes: nonyl acridine orange and a suitable potentiometric quenching dye, such as tetramethylrhodamine (both from Molecular Probes, Eugene, OR). Mitochondrial cardiolipin (diphosphatidyl glycerol) is stained with NAO by adding the appropriate volume of a 0.40 μM solution [in Hanks' balanced salt solution (HBSS) buffered by 25 m*M* HEPES] directly to cells in culture media to a final concentration of 85 n*M* and incubated for 6 min. Again, dye concentrations need to be optimized for different cell lines and instrumentation. NAO is prepared every 4 h from a concentrated stock (1.68 m*M* in DMSO, stored in opaque container at 4°C) because it photobleaches even under ambient fluorescent lighting, with a corresponding diminution in the initial signal.

Early reports using NAO indicated that it is sequestered by respiring mitochondria according to $\Delta\Psi_m$ (Maftah *et al.*, 1989), and evidence continues to support this contention (Keij *et al.*, 2000). Indeed, an increase in NAO signal is detected using the FRET-based assay in intact cells stained with NAO alone when $\Delta\Psi_m$ is collapsed by carbonyl cyanide 3-chlorophenylhydrazone (CCCP) or Ca^{2+} when cells are permeabilized by digitonin. However, because comparable NAO staining occurs in cells where $\Delta\Psi_m$ has been collapsed previously via CCCP (0.5 μM), the observed increases are more likely a consequence of secondary effects, such as changes in mitochondrial morphology resulting in changes in the effective surface area for NAO fluorescence and detection, or reorganization of membrane lipids with corresponding spatial aggregation of cardiolipin and increased NAO fluorescence, as suggested by line broadening in electron paramagnetic spin label experiments (Vercesi *et al.*, 1997), and not due to changes in $\Delta\Psi_m$ per se.

Once the cells are stained with NAO, the medium is aspirated, or the plate inverted and shaken, to remove medium and excess dye. Before being transferred to the plate reader, the cells are washed three times with warm HEPES-buffered HBSS for experiments using intact cells or for studies using cells permeabilized by digitonin, with warm KCl media (see later).

In the protocol described here, the quenching dye (tetramethylrhodamine methyl ester, TMR) is added to a final concentration of 0.15 μM while the sample is in the FLIPR plate reader. Although this provides an initial NAO reading useful for HTS calculations (see later), if the available instrumentation does not permit fluid addition during an assay, the quenching dye can be added readily prior to reading. However, to avoid unstable baselines due to TMR efflux from the cytoplasm down the concentration gradient, it is recommended that the quenching dye be present for the duration of the assay. For example, in an early iteration of the protocol, cells were stained as described earlier with NAO, followed by a 10-min incubation with TMR, and then excess dyes were removed via washing. This dye-loading protocol generated baselines of NAO quenching that gradually, but steadily, decreased as TMR effluxed from the mitochondria and repartitioned in the absence of extracellular dye among the mitochondrial, cytosolic, and extracellular compartments along the Nernstian gradient. Keeping the quenching dye present in the media throughout the assay minimizes this problem. Because the concentration gradient between the cytosol and the mitochondrial matrix is dependent on $\Delta\Psi_m$, retaining dye does not diminish the signal-to-noise ratio for $\Delta\Psi_m$.

In the protocol used for HTS at MitoKor, TMR is added in the plate reader, and the mitochondrial uptake of TMR is monitored in real time during the first portion of the assay, thereby ensuring that TMR distribution is at equilibrium before compound responses are evaluated (Fig. 3). This also provides an internal benchmark useful for data normalization; within the error of the assay (and in light of the previously mentioned

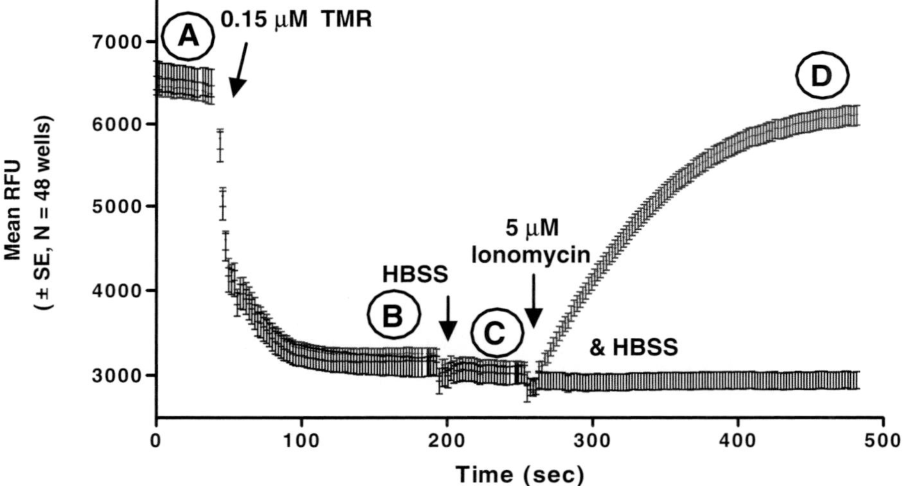

Fig. 3 Collapse of $\Delta\Psi_m$ in intact SY5Y neuroblastoma cells is induced in a dose-dependent manner by acute Ca^{2+} load using the ionophore ionomycin. In the typical assay protocol, quenching of the initial NAO signal (A) is apparent immediately after the addition of TMR (final concentration, 150 n*M*), and steady state is obtained within 1–2 min (B). Monitoring this steady state after the addition of compound (C) can assess potential effects of compound under scrutiny, and potential effects on Ca^{2+}-mediated $\Delta\Psi_m$ collapse can be detected by comparing the response to ionomycin (5 μ*M*) and HBSS buffer alone. CCCP (0.5 μ*M*) also collapses $\Delta\Psi_m$ (data not shown). In this and subsequent figures, SH-SY5Y neuroblastoma cells are plated at 6×10^4 cells/well, 24 h prior to assay, as described in the text.

increases in the NAO signal independent of $\Delta\Psi_m$), the amount of initial NAO quenching, obtained by subtracting the reading at equilibrium (B) from the initial reading of NAO alone (A), should closely approximate the amount of dequenching on complete mitochondrial depolarization, obtained as (D-C) (Fig. 3).

As an additional benchmark for standardization over time, the extent of $\Delta\Psi_m$ collapse (D-C/A-B) can also be related to both positive and negative controls on the same plate exposed to agents that collapse or have no effect on $\Delta\Psi_m$, respectively. For example, in the protocol using permeabilized cells, the ratio of recovered over initial quench is expressed as a percentage of controls exposed to Ca^{2+} alone. In this way, the response by the same compound on different days can be assessed as a percentage of the response to Ca^{2+} on that day, independent of experimental variables that are difficult to precisely regulate, such as cell density, which alters the effective Ca^{2+} concentration per unit cell, or plasma membrane cholesterol content, the primary determinant of digitonin efficiency. Expressing data as a ratio violates an assumption for ANOVA, necessitating data transformation prior to analysis. However, this is not an issue when data are used to generate the nonlinear regressions necessary to appraise compound dose responses.

C. Assay Validation through Mitochondrial Pharmacology

The response in this assay of compounds with well-characterized mitochondrial effects in intact cells is in accord with their pharmacology. For example, collapsing $\Delta\Psi_m$ with uncouplers such as CCCP prior to TMR exposure prevents TMR uptake, and hence abolishes NAO quenching, whereas inhibition of ATP synthase (complex V) by oligomycin yields a dose-dependent hyperpolarization of $\Delta\Psi_m$, but has no effect on Ca^{2+}-mediated $\Delta\Psi_m$ collapse (Table II). In addition, cyclosporin A (CsA) moderates Ca^{2+}-mediated $\Delta\Psi_m$ collapse in a dose-dependent manner, probably by forestalling induction of the permeability transition secondary to Ca^{2+} loading (Table II).

Table II
FRET-Based $\Delta\Psi_m$ Assay in Intact SH-SY5Y Cells Recapitulates Known Mitochondrial Pharmacology

Agent	Effect on TMR uptake (= NAO quenching)	Effect on NAO dequenching (= TMR efflux)
CCCP	Prevents	Collapses $\Delta\Psi_m$ $EC_{50} = 2\ \mu M$
Ionomycin	Prevents	Dissipates $\Delta\Psi_m$ $EC_{50} = 5.6\ \mu M$
Cyclosporin A	Increases (via MDR Inhibition) $EC_{50} = 0.81\ \mu M$	Moderates ionomycin $EC_{50} = 3.8\ \mu M$
Oligomycin	Increases $EC_{50} = 0.90\ \mu M$	No effect on ionomycin

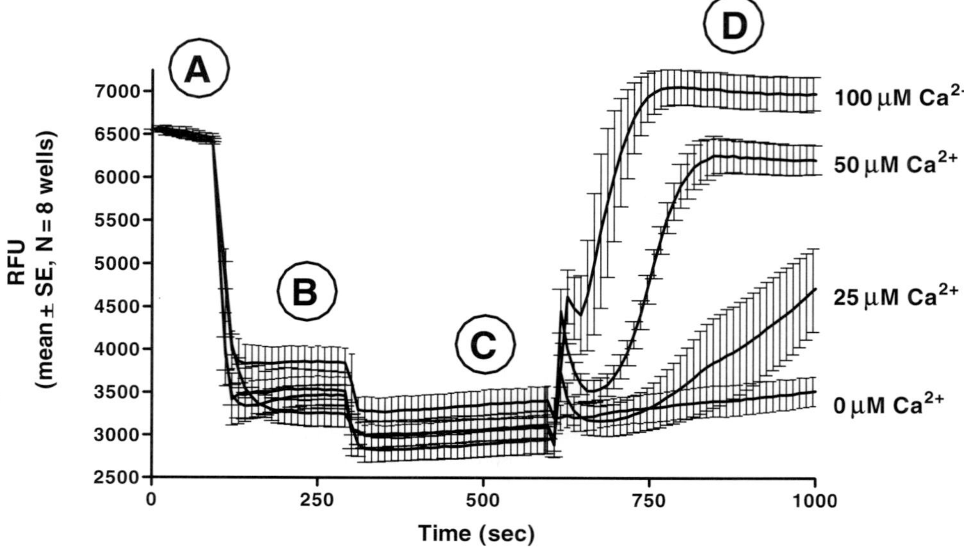

Fig. 4 The FRET $\Delta\Psi_m$ assay can also be performed using cells permeabilized with digitonin where the bioavailability of compounds and Ca^{2+} concentrations are controlled more readily. In this mode, $\Delta\Psi_m$ collapse is induced by the direct addition of Ca^{2+} rather than via ionophore. Excessive Ca^{2+} concentrations (ca. 100 μM) induce rapid $\Delta\Psi_m$ collapse with little or no recovery apparent, whereas intermediate Ca^{2+} concentrations (ca. 50 μM) permit partial recovery of $\Delta\Psi_m$ prior to secondary collapse due to permeability transition.

It should be noted in this context that CsA also induces a discernible hyperpolarization of $\Delta\Psi_m$, i.e., increased TMR uptake and correspondingly increased NAO quenching. However, this apparent hyperpolarization is moderated, although not completely abolished (see Fig. 6) by digitonin permeabilization in the presence of TMR, suggesting that it is an artifact of a plasma membrane effect and not due to an authentic increase in $\Delta\Psi_m$. Indeed, comparable NAO quenching is also induced by FK-506 and verapamil, suggesting that "hyperpolarization" in response to CsA is reasonably ascribed to its well-known inhibition of the p-glycoprotein-dependent multidrug resistant transporter (MDR) in the plasma membrane (Gottesman and Pastan, 1993; Jachez *et al.,* 1993; Hall *et al.,* 1999). As discussed in the previous section dealing with single potentiometric dyes, rhodamines, including TMR, are among the many substrates of MDR, and inhibition of the latter would serve to increase the cytosolic concentration of TMR (Efferth *et al.,* 1989; Kessel *et al.,* 1991; Webb *et al.,* 1996; Nelson *et al.,* 1998). Because TMR follows a Nernstian distribution, the resulting increased cytosolic concentration would translate into increased mitochondrial uptake, and hence increased NAO quenching that is indistinguishable from authentic $\Delta\Psi_m$ hyperpolarization, except that digitonin abolishes it. Such MDR activity is likely a confounding variable in many transformed cell lines, and experiments with fluorescent dyes, especially when used in combination with pleiotropic agents such as CsA, must be interpreted accordingly (Petronilli *et al.*, 1998).

In any event, to avoid these confounding MDR effects, the FRET $\Delta\Psi_m$ assay was also validated using cells permeabilized by digitonin (Fig. 4). In this protocol, which is also amenable to a high throughput format, the cells are stained with NAO as described earlier, but are then washed with 125 m*M* KCl containing 1 m*M* $MgCl_2$, plus 5 m*M* each of succinate, glutamate, and malate as oxidizible substrates, 2 m*M* HEPES, 1 μM EGTA (to remove adventitious Ca^{2+}), and 1 m*M* KH_2PO_4 (pH 7.0). The cells are permeabilized in FLIPR upon addition of TMR containing 0.06% digitonin (final concentration 0.008–0.01%; titration for cell type is essential).

In this format, $\Delta\Psi_m$ collapse is induced not via ionophores, but rather by the direct addition of Ca^{2+} (EC_{50} ca. 30 μM), with some interesting and potentially useful caveats (Fig. 4). At high Ca^{2+} levels (ca. 50–75 μM), $\Delta\Psi_m$ collapses rapidly, restoring the NAO signal to initial levels with simple hyperbolic kinetics reminiscent of CCCP, or high ionomycin concentrations, in intact cells. Below a threshold Ca^{2+} concentration that varies with cell number and type (and extent of permeabilization), a rapid but greatly diminished $\Delta\Psi_m$ collapse is followed by complete recovery to initial baseline. At intermediate Ca^{2+}concentrations (ca. 25 μM), however, rapid initial $\Delta\Psi_m$ dissipation is followed by partial recovery and then a secondary, more gradual loss of $\Delta\Psi_m$ that, because it is moderated by CsA, reflects permeability transition (Figs. 4 and 6).

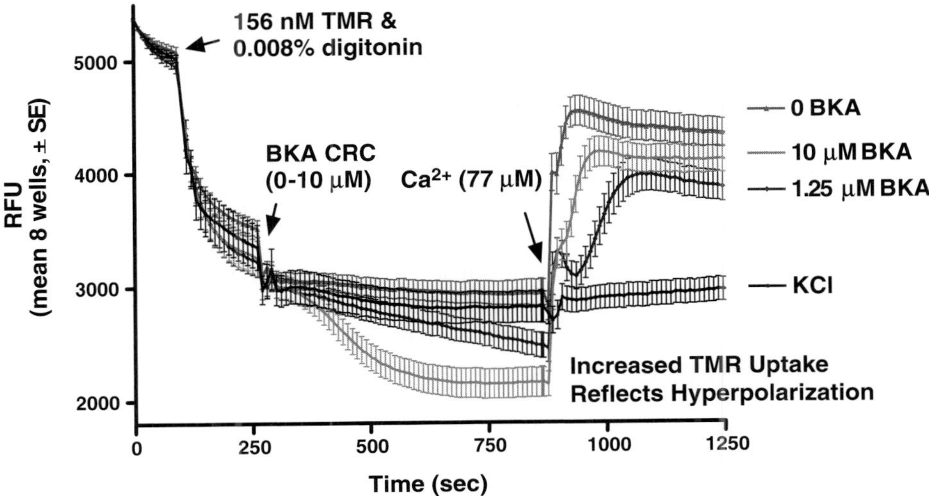

Fig. 5 The FRET $\Delta\Psi_m$ assay indicates that bongkrekic acid (BKA), an inhibitor of adenine nucleotide translocase, moderates secondary loss of $\Delta\Psi_m$ after acute Ca^{2+} exposure. At concentrations above 1.25 μM, BKA apparently hyperpolarizes mitochondria in SH-SY5Y neuroblastoma cells, as shown by dose-dependent increases in NAO quenching due to increased TMR uptake. Such hyperpolarization is reasonably ascribed to the inability to dissipate $\Delta\Psi_m$ via phosphorylation in the absence of ADP as substrate.

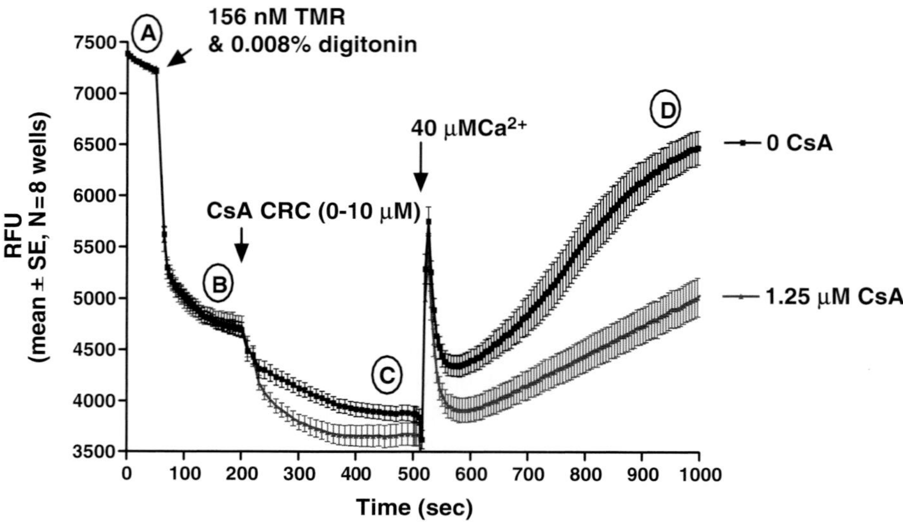

Fig. 6 The FRET $\Delta\Psi_m$ assay indicates that cyclosporin A (CsA) moderates secondary loss of $\Delta\Psi_m$ following acute Ca^{2+} exposure. Note that CsA does not alter the magnitude or time course of the initial collapse of $\Delta\Psi_m$ occurring immediately on Ca^{2+} loading, although it does improve recovery. It remains to be determined whether $\Delta\Psi_m$ hyperpolarization in the "C" interval reflects a real increase in $\Delta\Psi_m$ or whether it is an artifact of incomplete permeabilization and hence due to inhibition of MDR activity by CsA.

As is also the case for intact cells (albeit at higher concentrations), atractyloside and bongkrekic acid, both of which inhibit the adenine nucleotide translocator (ANT), have opposing effects in the $\Delta\Psi_m$ assay; atractyloside, even at 1 μM, collapses $\Delta\Psi_m$ completely in intact cells within 5 min (data not shown), whereas BKA moderates Ca^{2+}-mediated $\Delta\Psi_m$ collapse (Fig. 5). Although both agents inhibit ANT activity, atractyloside locks the protein into a conformation where the adenylate-binding site faces the cytosol (the c-conformation), a position that facilitates permeability transition. Conversely, bongkrekate locks ANT into the m-conformation with the adenine-binding site facing the matrix, a form that imparts resistance to Ca^{2+}-mediated $\Delta\Psi_m$ collapse. In addition to providing some protection against Ca^{2+}-mediated $\Delta\Psi_m$ collapse, bongkrekate also induces $\Delta\Psi_m$ hyperpolarization, as reflected by increased NAO quenching during the "C" interval, no doubt attributable to its preventing ADP entry and consequently preventing dissipation of $\Delta\Psi_m$ via oxidative phosphorylation (Fig. 5).

Given that this assay depends on FRET within the mitochondria and that the cytoplasmic concentration of the potentiometric dye is buffered by an extracellular pool present during the assay, permeabilization of the plasma membrane represents the final step to eliminate artifacts induced by $\Delta\Psi_p$. In addition, removing the plasma membrane as a diffusion barrier substantially increases the bioavailability at the mitochondria of many compounds that have been characterized in isolated mitochondria, but that are generally

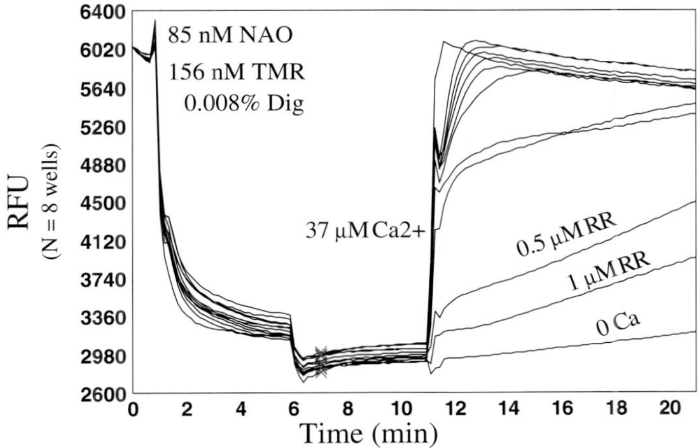

Fig. 7 Using the dye ruthenium red to block mitochondrial Ca^{2+} influx via the uniporter prevents the initial collapse of $\Delta\Psi_m$ on Ca^{2+} loading in a dose-dependent manner and correspondingly forestalls the secondary loss of $\Delta\Psi_m$. This response is apparent at lower concentrations with RU-360, the most active component in the ruthenium red mixture (see text).

impermeant to intact cells. For example, ruthenium red, a well-characterized and specific inhibitor of the mitochondrial Ca^{2+} uniporter, fails to moderate ionomycin-mediated $\Delta\Psi_m$ collapse in intact SH-SY5Y cells, even at millimolar concentrations. However, in cells permeabilized by 0.01% digitonin, nanomolar concentrations of both ruthenium red and its active component RU-360 block the initial rapid $\Delta\Psi_m$ collapse that occurs upon Ca^{2+} exposure, and hence completely abrogate secondary permeability transition (Fig. 7).

In the context of high throughput compound screening, a fair amount of selectivity can be obtained by the judicious selection of conditions. For example, selecting a Ca^{2+} concentration that permits partial $\Delta\Psi_m$ recovery prior to secondary PT will serve to identify compounds that mimic the desirable mitochondrial activity of cyclosporin A, as opposed to agents that moderate Ca^{2+} uptake per se (Fig. 7). Conversely, compounds that dissipate or hyperpolarize $\Delta\Psi_m$ are generally detrimental to long-term cell viability, inducing necrosis and/or apoptosis, depending on the mode of action. In this light, focusing on agents that collapse $\Delta\Psi_m$ may provide novel antitumor therapies that target mitochondrial integrity.

In summary, the FRET assay for $\Delta\Psi_m$ described here circumvents many of the limitations and confounding variables encountered when assessing $\Delta\Psi_m$ in intact cells with single potentiometric dyes. By focusing on the nonpotentiometric dye in the FRET pair, the FRET assay selectively reports $\Delta\Psi_m$ independent of background fluorescence from the potentiometric dye due to the plasma membrane potential. However, the FRET $\Delta\Psi_m$ assay is not entirely independent of plasma membrane effects; because of Nernstian constraints, quenching of mitochondrial NAO by any potentiometric dye is a function of the cytosolic concentration of the potentiometric dye, which is determined by diffusion, active dye transport and changes in $\Delta\psi_p$. Nevertheless, by maintaining extracellular dye concentration during the assay, the response due to potential can be maximally resolved, while minimizing the concentration and diffusion effects. This FRET protocol is adaptable to single cells and cell suspensions, as well as intact and permeabilized adherent cells, using a variety of fluorescence assessment technologies and imaging systems, ranging from microscopes to HTS formats such as FLIPR. The FRET assay has not yet been applied to intact tissues, but given the high permeability of both dyes, it is likely that the physical issues of illumination and detection will pose the major hurdles, not the biology.

Although additional pathologically relevant inducers of mitochondrial dysfunction are currently under development, the use of Ca^{2+} load to induce mitochondrial instability in high throughput FRET-based screens should identify agents capable of improving mitochondrial dysfunction, and hence moderating cellular necrosis and apoptosis, regardless of whether impairment results from acute transient ischemia associated with infarct or stroke or arises from idiopathic etiology of chronic degenerative diseases, such as Alzheimer's, Parkinson's, and osteoarthritis, in which mitochondrial dysfunction is increasingly implicated.

References

Anderson, W. M., Wood, J. M., and Anderson, A. C. (1993). Inhibition of mitochondrial and *Paracoccus denitrificans* NADH-ubiquinone reductase by oxacarbocyanine dyes: A structure-activity study. *Biochem. Pharmacol.* **45,** 2115–2122.

Andreyev, A., and Fiskum, G. (1999). Calcium induced release of mitochondrial cytochrome c by different mechanisms selective for brain versus liver. *Cell Death Differ.* **6,** 825–832.

Andreyev, A. Y., Fahy, B., and Fiskum, G. (1998). Cytochrome c release from brain mitochondria is independent of the mitochondrial permeability transition. *FEBS Lett.* **439,** 373–376.

Antunes, F., Salvador, A., Marinho, H. S., Alves, R., and Pinto, R. E. (1996). Lipid peroxidation in mitochondrial inner membranes. I. An integrative kinetic model. *Free Radic. Biol. Med.* **21,** 917–943.

Beal, M. F., Bodis-Wollner, I., and Howell, N. (1997). "Neurodegenerative Diseases: Mitochondria and Free Radicals in Pathogenesis." Wiley, New York.

Bernardi, P. (1999). Mitochondrial transport of cations: Channels, exchangers, and permeability transition. *Physiol. Rev.* **79,** 1127–1255.

Bernardi, P., Colonna, R., Costantini, P., Eriksson, O., Fontaine, E., Ichas, F., Massari, S., Nicolli, A., Petronilli, V., and Scorrano, L. (1998). The mitochondrial permeability transition. *Biofactors* **8,** 273–281.

Bunting, J. R. (1992a). A test of the singlet oxygen mechanism of cationic dye photosensitization of mitochondrial damage. *Photochem. Photobiol.* **55,** 81–87.

Bunting, J. R. (1992b). Influx and efflux kinetics of cationic dye binding to respiring mitochondria. *Biophys. Chem.* **42,** 163–175.

Bunting, J. R., Phan, T. V., Kamali, E., and Dowben, R. M. (1989). Fluorescent cationic probes of mitochondria: Metrics and mechanism of interaction. *Biophys. J.* **56,** 979–993.

Cassarino, D. S., and Bennett, J. P., Jr. (1999). An evaluation of the role of mitochondria in neurodegenerative diseases: Mitochondrial mutations and oxidative pathology, protective nuclear responses, and cell death in neurodegeneration. *Brain Res. Brain Res. Rev.* **29,** 1–25.

Di Lisa, F., Blank, P. S., Colonna, R., Gambassi, G., Silverman, H. S., Stern, M. D., and Hansford, R. G.

(1995). Mitochondrial membrane potential in single living adult rat cardiac myocytes exposed to anoxia or metabolic inhibition. *J. Physiol.* (*Lond*) **486,** 1–13.

Dykens, J. A. (1994). Isolated cerebellar and cerebral mitochondria produce free radicals when exposed to elevated Ca^{2+} and Na^{+}: Implications for neurodegeneration. *J. Neurochem.* **63,** 584–591.

Dykens, J. A. (1995). Mitochondrial radical production and mechanisms of oxidative excitotoxicity. *In* "The Oxygen Paradox" (K. J. A. Davies and F. Ursini, eds.), pp. 453–467. Cleup Press.

Dykens, J. A. (1997). Mitochondrial free radical production and the etiology of neurodegenerative disease. *In* "Neurodegenerative Diseases: Mitochondria and Free Radicals in Pathogenesis" (M. F. Beal, I. Bodis-Wollner, and N. Howell, eds.), pp. 29–55. Wiley, New York.

Dykens, J. A. (1999). Free radicals and mitochondrial dysfunction in excitotoxicity and neurodegenerative diseases. *In* "Cell Death and Diseases of the Nervous System" (V. E. Koliatos and R. R. Ratan, eds.), pp. 45–68. Humana Press, New Jersey.

Dykens, J. A., Moos, W. H., and Davis, R. E. (1999). An introduction to mitochondrial genetics and physiology. *Drug Dev. Res.* **46,** 2–13.

Efferth, T., Lohrke, H., and Volm, M. (1989). Reciprocal correlation between expression of P-glycoprotein and accumulation of rhodamine 123 in human tumors. *Anticancer Res.* **9,** 1633–1637.

Ehrenberg, B., Montana, V., Wei, M.-D., Wuskell, J. P., and Loew, L. M. (1988). Membrane potential can be determined in individual cells from the nernstian distribution of cationic dyes. *Biophys. J.* **53,** 785–794.

Emaus, R. K., Grunwald, R., and Lemasters, J. J. (1986). Rhodamine 123 as a probe of transmembrane potential in isolated rat-liver mitochondria: Spectral and metabolic properties. *Biochim. Biophys. Acta* **850,** 436–448.

Farkas, D. L., Wei, M. D., Febbroriello, P., Carson, J. H., and Loew, L. M. (1989). Simultaneous imaging of cell and mitochondrial membrane potentials. *Biophys. J.* **56,** 1053–1069.

Fiskum, G., Murphy, A. N., and Beal, M. F. (1999). Mitochondria in neurogeneration: Acute ischemia and chronic neurodegenerative diseases. *J. Cereb. Blood Flow Metab.* **19,** 351–369.

Floryk, D., and Houstek, J. (1999). Tetramethyl rhodamine methyl ester (TMRM) is suitable for cytofluorometric measurements of mitochondrial membrane potential in cells treated with digitonin. *Biosci. Rep.* **19,** 27–34.

Fontaine, E., and Bernardi, P. (1999). Progress on the mitochondrial permeability transition pore: Regulation by complex I and ubiquinone analogs. *J. Bioenerg. Biomembr.* **31,** 335–345.

Garner, D. L., Thomas, C. A., Joerg, H. W., DeJarnette, J. M., and Marshall, C. E. (1997). Fluorometric assessments of mitochondrial function and viability in cryopreserved bovine spermatozoa. *Biol. Reprod.* **57,** 1401–1406.

Gottesman, M. M., and Pastan, I. (1993). Biochemistry of multidrug resistance mediated by the multidrug transporter. *Annu. Rev. Biochem.* **62,** 385–427.

Hall, J. G., Cory, A. H., and Cory, J. G. (1999). Lack of competition of substrates for P-glycoprotein in MCF-7 breast cancer cells overexpressing MDR1. *Adv. Enzyme Regul.* **39,** 113–128.

Huser, J., Rechenmacher, C. E., and Blatter, L. A. (1998). Imaging the permeability pore transition in single mitochondria. *Biophys. J.* **74,** 2129–2137.

Iwase, H., Takatori, T., Nagao, M., Iwadate, K., and Nakajima, M. (1996). Monoepoxide production from linoleic acid by cytochrome c in the presence of cardiolipin. *Biochem Biophys. Res. Commun.* **222,** 83–89.

Jachez, B., Boesch, D., Grassberger, M. A., and Loor, F. (1993). Reversion of the P-glycoprotein-mediated multidrug resistance of cancer cells by FK-506 derivatives. *Anticancer Drugs* **4,** 223–229.

Jackson, J. B., and Nicholls, D. G. (1986). Methods for the determination of membrane potential in bioenergetic systems. *Methods Enzymol.* **127,** 557–577.

Johnson, I. (1998). Fluorescent probes for living cells. *Histochem. J.* **30,** 123–140.

Johnson, L. V., Walsh, M. L., Bockus, B. J., and Chen, L. B. (1981). Monitoring of relative mitochondrial membrane potential in living cells by fluorescence microscopy. *J. Cell Biol.* **88,** 526–535.

Juan, G., Cavazzoni, M., Saez, G. T., and O'Connor, J. E. (1994). A fast kinetic method for assessing mitochondrial membrane potential in isolated hepatocytes with rhodamine 123 and flow cytometry. *Cytometry* **15,** 335–342.

Kamo, N., Muratsugu, M., Hongoh, R., and Kobatake, Y. (1979). Membrane potential of mitochondria measured with an electrode sensitive to tetraphenyl phosphonium and relationship between proton electrochemical potential and phosphorylation potential in steady state. *J. Membr. Biol.* **49,** 105–121.

Keij, J. F., Bell-Prince, C., and Steinkamp, J. A. (2000). Staining of mitochondrial membranes with 10-nonyl acridine orange, MitoFluor Green, and MitoTracker Green is affected by mitochondrial membrane potential altering drugs. *Cytometry* **39,** 203–210.

Kessel, D., Beck, W. T., Kukuruga, D., and Schulz, V. (1991). Characterization of multidrug resistance by fluorescent dyes. *Cancer Res.* **51,** 4665–4670.

Kuhnel, J. M., Perrot, J. Y, Faussat, A. M., Marie, J. P., and Schwaller, M. A. (1997). Functional assay of multidrug resistant cells using JC-1, a carbocyanine fluorescent probe. *Leukemia* **11,** 1147–1155.

Lemasters, J. J., Nieminen, A. L., Qian, T., Trost, L. C., Elmore, S. P., Nishimura, Y., Crowe, R. A., Cascio, W. E., Bradham, C. A., Brenner, D. A., and Herman, B. (1998). The mitochondrial permeability transition in cell death: A common mechanism in necrosis, apoptosis and autophagy. *Biochim. Biophys. Acta* **1366,** 177–196.

Lemasters, J. J., Qian, T., Bradham, C. A., Brenner, D. A., Cascio, W. E., Trost, L. C., Nishimura, Y., Nieminen, A. L., and Herman, B. (1999). Mitochondrial dysfunction in the pathogenesis of necrotic and apoptotic cell death. *J. Bioenerg. Biomembr.* **31,** 305–319.

Loew, L. M., Tuft, R. A., Carrington, W., and Fay, F. S. (1993). Imaging in five dimensions: Time-dependent membrane potentials in individual mitochondria. *Biophys. J.* **65,** 2396–2407.

Maftah, A., Petit, J. M., Ratinaud, M. H., and Julien, R. (1989). 10-N nonyl-acridine orange: A fluorescent probe which stains mitochondria independently of their energetic state. *Biochem. Biophys. Res. Commun.* **164,** 185–190.

Mason, W. T. (1993). "Fluorescent and Luminescent Probes for Biological Activity." Academic Press, San Diego.

Moos, W. H., Dykens, J. A., and Davis, R. E. (1999). Mitochondrial biology and neurodegenerative diseases. *Pharm. News* **6,** 15–29.

Mathur, A., Hong, Y., Kemp, B. K., Barrientos, A. A., and Erusalimsky, J. D. (2000). Evaluation of fluorescent dyes for the detection of mitochondrial membrane potential changes in cultured cardiomyocytes. *Cardiovasc. Res.* **46,** 126–138.

Matsumoto, B. (1992). Cell biological applications of confocal microscopy. *In* "Methods in Cell Biology," Vol. 38, Academic Press, San Diego.

Mitchell, P., and Moyle, J. (1969). Estimation of membrane potential and pH difference across the cristae membrane of rat liver mitochondria. *Eur. J. Biochem.* **7,** 471–184.

Murphy, A. N., Fiskum, G., and Beal, M. F. (1999). Mitochondrial ion neurodegeneration: Bioenergetic function in cell life and death. *J. Cereb. Blood Flow Metab.* **19,** 231–245.

Nelson, E. J., Zinkin, N. T., and Hinkle, P. M. (1998). Fluorescence methods to assess multidrug resistance in individual cells. *Cancer Chemother Pharmacol.* **42,** 292–299.

Nicholls, D. G. (1974). The influence of respiration and ATP hydrolysis on the proton-electrochemical gradient across the inner membrane of rat-liver mitochondria as determined by ion distribution. *Eur. J. Biochem.* **50,** 305–315.

Nicholls, D. G., and Ward, M. W. (2000). Mitochondrial membrane potential and neuronal glutamate excitotoxicity:mortality and millivolts. *Trends Neurosci.* **23,** 166–174.

Nicholls, D. G., and Budd, S. L. (1998). Neuronal excitotoxicity: The role of mitochondria. *Biofactors* **8,** 287–299.

Nicholls, D. G., and Ferguson, S. J. (1992). "Bioenergetics 2," Academic Press, London.

Nuydens, R., Novalbos, J., Dispersyn, G., Weber, C., Borgers, M., and Geerts, H. (1999). A rapid method for the evaluation of compounds with mitochondria-protective properties. *J. Neurosci. Methods* **15,** 153–159.

Padan, E., and Rottenberg, H. (1973). Respiratory control and the proton electrochemical gradient in mitochondria. *Eur. J. Biochem.* **40,** 431–437.

Petronilli, V., Miotto, G., Canton, M., Colonna, R., Bernardi, P., and Di Lisa, F. (1998). Imaging the mitochondrial permeability transition pore in intact cells. *Biofactors* **8,** 263–272.

Reers, M., Smiley, S. T., Mottola-Hartshorn, C., Chen, A., Lin, M., and Chen, L. B. (1995). Mitochondrial membrane potential monitored by JC-1 dye. *Methods Enzymol.* **260,** 406–417.

Reynolds, I. J. (1999). Mitochondrial membrane potential and the permeability transition in excitotoxicity. *Ann. N.Y. Acad. Sci.* **893,** 33–41.

Rottenberg, H. (1979). The measurement of membrane potential and delta pH in cells, organelles, and vesicles. *Methods Enzymol.* **55,** 547–569.

Rottenberg, H. (1984). Membrane potential and surface potential in mitochondria: Uptake and binding of lipophilic cations. *J. Membr. Biol.* **81,** 127–138.

Rottenberg, H., and Wu, S. (1998). Quantitative assay by flow cytometry of the mitochondrial membrane potential in intact cells. *Biochim. Biophys. Acta* **1404,** 393–404.

Sabnis, R. W., Deligeorgiev, T. G., Jachak, M. N., and Dalvi, T. S. (1997). $DiOC_6(3)$: A useful dye for staining the endoplasmic reticulum. *Biotech. Histochem.* **72,** 253–258.

Salvioli, S., Ardizzoni, A., Franceschi, C., and Cossarizza, A. (1997). JC-1, but not $DiOC_6(3)$ or rhodamine 123, is a reliable fluorescent probe to assess delta psi changes in intact cells: Implications for studies on mitochondrial functionality during apoptosis. *FEBS Lett.* **411,** 77–82.

Scaduto, R. C., Jr., and Grotyohann, L. W. (1999). Measurement of mitochondrial membrane potential using fluorescent rhodamine derivatives. *Biophys. J.* **76,** 469–477.

Scanlon, J. M., and Reynolds, I. J. (1998). Effects of oxidants and glutamate receptor activation on mitochondrial membrane potential in rat forebrain neurons. *J. Neurochem.* **71,** 2392–2400.

Scorrano, L., Petronilli, V., Colonna, R., Di Lisa, F., and Bernardi, P. (1999). Chloromethyltetramethylrosamine (Mitotracker Orange) induces the mitochondrial permeability transition and inhibits respiratory complex I: Implications for the mechanism of cytochrome c release. *J. Biol. Chem.* **274,** 24657–24663.

Scott, I. D., and Nicholls, D. G. (1980). Energy transduction in intact synaptosomes: Influence of plasma-membrane depolarization on the respiration and membrane potential of internal mitochondria determined *in situ*. *Biochem. J.* **186,** 21–33.

Skulachev, V. P. (1999). Mitochondrial physiology and pathology; concepts of programmed death of organelles, cells and organisms. *Mol. Aspects Med.* **20,** 139–184.

Siesjo, B. K., Elmer, E., Janelidze, S., Keep, M., Kristian, T., Ouyang, Y. B., and Uchino, H. (1999). Role and mechanisms of secondary mitochondrial failure. *Acta Neurochir. Suppl. (Wien)* **73,** 7–13.

Shults, C. W., Haas, R. H., and Beal, M. F. (1999). A possible role of coenzyme Q10 in the etiology and treatment of Parkinson's disease. *Biofactors* **9,** 267–272.

Sureda, F. X., Gabriel, C., Comas, J., Pallas, M., Escubedo, E., Camarasa, J., and Camins, A. (1999). Evaluation of free radical production, mitochondrial membrane potential and cytoplasmic calcium in mammalian neurons by flow cytometry. *Brain Res. Brain Res. Protoc.* **4,** 280–287.

Terasaki, M., and Reese, T. S. (1992). Characterization of endoplasmic reticulum by colocalization of BiP and dicarbocyanine dyes. *J. Cell Sci.* **101,** 315–322.

Webb, M., Raphael, C. L., Asbahr, H., Erber, W. N., and Meyer, B. F. (1996). The detection of rhodamine 123 efflux at low levels of drug resistance. *Br. J. Haematol.* **93,** 650–655.

Wilson, H. A., Seligmann, B. E., and Chused, T. M. (1985). Voltage-sensitive cyanine dye fluorescence signals in lymphocytes: plasma membrane and mitochondrial components. *J. Cell. Physiol.* **125,** 61–71.

CHAPTER 18

Optical Imaging Techniques (Histochemical, Immunohistochemical, and *in Situ* Hybridization Staining Methods) to Visualize Mitochondria

Kurenai Tanji and Eduardo Bonilla

Department of Neurology
Columbia University,
New York, New York 10032

I. Introduction

Mitochondria, the primary ATP-generating organelles in mammalian cells, produce ATP via oxidative phosphorylation, and the five respiratory complexes located in the inner mitochondrial membrane. Some components of the respiratory chain are encoded by mitochondrial DNA (mtDNA). The human mitochondrial genome is a 16,569-bp double-stranded DNA. It is highly compact and contains only 37 genes: 2 genes encode ribosomal RNAs (rRNAs), 22 encode transfer RNAs (tRNAs), and 13 encode polypeptides. All 13 polypeptides are components of the respiratory chain, including 7 subunits of complex I or NADH dehydrogenase–ubiquinone oxidoreductase, 1 subunit of complex III or ubiquinone–cytochrome *c* oxidoreductase, 3 subunits of complex IV–cytochrome *c* oxidase (COX), and 2 subunits of complex V or ATP synthase (Attardi and Schatz, 1988). The respiratory complexes also contain nuclear DNA (nDNA) encoded sububits, which are imported into the organelle from the cytosol and assembled, together, with the mtDNA-encoded subunits, into the respective holoenzymes in the mitochondrial inner membrane (Attardi and Schatz 1988; Neupert, 1997). Complex II or succinate dehydrogenase–ubiquinone oxidoreductase contains only nDNA-encoded subunits.

Mitochondrial disorders encompass a heterogeneous group of diseases in which mitochondrial dysfunction produces clinical manifestations. Because of the dual genetic makeup of mitochondria (Attardi and Schatz, 1988), these diseases are typically caused by genetic errors in either mtDNA or nDNA. Disorders produced by mtDNA mutations were identified, in part, because they display non-Mendelian inheritance patterns, and are maternally inherited (Anderson *et al.,* 1981; Giles *et al.,* 1980).

In the last decade, pathogenic mtDNA mutations have been identified in three of the "prototypes" of the mitochondrial disorders. First, large-scale mtDNA rearrangements [i.e., deletions (Δ-mtDNA) and/or duplications (dup-mtDNA)] have been associated with sporadic Kearns–Sayre syndrome (KSS) and are often seen in patients with isolated ocular myopathy (OM) (Zeviani *et al.,* 1988; Moraes *et al.,* 1989). Second, myoclonus epilepsy with ragged-red fibers (MERRF) has been frequently associated with two different point mutations, both in the $tRNA^{Lys}$ gene (Shoffner *et al.,* 1990; Silvestri *et al.,* 1992). Third, mitochondrial encephalopathy, lactic acidosis, and stroke-like episodes (MELAS) have been mainly associated with two different point mutations, both in the $tRNA^{Leu(UUR)}$ gene (Goto *et al.,* 1990, 1991). Other point mutations in tRNAs and protein-coding genes of the mitochondrial genome, as well as mendelian-inherited disorders leading to the generation of multiple Δ-mtDNAs or to depletion of mtDNA, have been described (DiMauro and Bonilla, 1997).

Because mitochondria are inherited only from the mother (Giles *et al.,* 1980), pedigrees harboring defects in mtDNA genes should exhibit maternal inheritance. Moreover, because there are hundreds or even thousands of mitochondria in each cell, with an average of five mtDNAs per organelle in somatic tissues, (Bogenhagen *et al.,* 1984; Satoh and Kuroiwa, 1991), mutations in mtDNA result in two populations of mtDNAs, mutated and wild type, a condition known as heteroplasmy. The phenotypic expression of a mtDNA mutation is regulated by the threshold effect, i.e., the mutant phenotype is expressed in heteroplasmic cells only when the relative proportion of mutant mtDNAs exceeds a minimum value (Wallace, 1992; Schon *et al.,* 1997).

In contrast to the rapid progress that has accumulated on mtDNA mutations, the elucidation of nDNA errors responsible for respiratory chain defects has been slow. Only recently have a few nuclear defects been documented at the molecular level, most in patients with Leigh syndrome associated with complex I, complex II, or complex IV deficiency (Bourgeron *et al.,* 1995; Loeffen *et al.,* 1998; Zhu *et al.,* 1998; Tiranti *et al.,* 1998; Papadopoulou *et al.,* 1999).

Brain and muscle, whose function is highly dependent on oxidative metabolism, are the most severely affected tissues in the mitochondrial disorders (Wallace, 1992; Schon *et al.,* 1997). Consequently, genetic as well as morphologic studies of individual muscle fibers or neuronal nuclei from patients with mitochondrial diseases have been instrumental in extending our understanding of the pathogenesis of mitochondrial dysfunction (Mita *et al.,* 1989; Moraes *et al.,* 1992; Sparacco *et al.,* 1995; Tanji *et al.,* 1999).

The purpose of this chapter is to present the histochemical, immunohistochemical, and *in situ* hybridization (ISH) methods that, in our experience, are reliable for identification of mitochondria in frozen or paraffin-embedded tissue sections. We illustrate application of these techniques on specific pathological samples and provide an updated version of the methods. While the described protocols refer to skeletal muscle or brain mitochondria, they can be applied to any cell type (Szabolic *et al.,* 1994; Tanji *et al.,* 1999). It is not our intention to cover every study or method related to morphological aspects of mitochondria, but rather to provide enough information to allow investigators to apply these selected light microscopy tools to a particular scientific or diagnostic question.

II. Histochemistry

The visualization of normal and pathological mitochondria in frozen tissue sections can be carried out using a number of cytochemical techniques. These include the modified Gomori trichrome and hematoxylin–eosin stains and histochemical methods for the demonstration of oxidative enzyme activity.

The most informative histochemical alteration of mitochondria in skeletal muscle is the ragged-red fiber (RRF) observed in frozen sections stained with the trichrome method of Engel and Cunningham (1963). The name derives from the reddish appearance of the trichrome-stained muscle fiber as a result of subsarcolemmal and/or intermyofibrillar proliferation of the mitochondria. Fibers harboring abnormal deposits of mitochondria are most often type I myofibers; they may also contain increased numbers of lipid droplets. Because accumulations of materials other than mitochondria may simulate RRF formation, the identification of deposits suspected of being mitochondrial proliferation should be confirmed histochemically by the application of oxidative enzyme stains.

In our experience, enzyme histochemistry for the activity of succinate dehydrogenase (SDH) and COX has proven to be the most reliable method for the correct visualization of normal mitochondria and for diagnosis and interpretation of some mitochondrial disorders affecting skeletal muscle (DiMauro and Bonilla, 1997; Bonilla and Tanji 1998).

A. Succinate Dehydrogenase

Succinate dehydrogenase is the enzyme that catalyzes the conversion of succinate to fumarate in the tricarboxylic acid cycle. It consists of two large subunits (a 70-kDa flavoprotein and a 30-kDa iron–sulfur-containing protein), which along with two smaller subunits (cybs and cybl) which attach SDH to the inner mitochondrial membrane form complex II of the mitochondrial respiratory chain (Hatefi, 1985; Beinert, 1990). Because complex II is the only component of the respiratory chain whose subunits are all encoded by the nDNA, SDH histochemistry is extremely useful for detecting any variation in the fiber distribution of mitochondria, independently of any alteration affecting the mtDNA.

The histochemical method for the microscopic demonstration of SDH activity in frozen tissue sections is based on the use of a tetrazolium salt (nitro blue tetrazolium, NBT) as an electron acceptor with phenazine methosulfate (PMS) serving as an intermediate electron donor to NBT (Seligman and Rutenburg, 1951; Pette, 1981). The specificity of the method may be tested by performing control experiments in which an SDH inhibitor, sodium malonate (0.01 *M*), is added to the incubation medium.

Using this method for detecting SDH activity in normal muscle sections, two populations of fibers are seen, resulting in a checkerboard pattern. Type II fibers, which rely on glycolytic metabolism, show a light blue network-like stain. Type I fibers, whose metabolism is highly oxidative and therefore contain more mitochondria, show a more elaborate and darker mitochondrial network (Fig. 1A, see color plate).

In samples with pathological proliferation of mitochondria (RRF), the RRF show an intense blue SDH reaction corresponding to the distribution of the mitochondria within the fiber (Fig. 1C). This proliferation of mitochondria is associated with most mtDNA defects (deletions and tRNA point mutations). RRF are also observed in disorders that are thought to be due to defects of nDNA, such as the depletion of muscle mtDNA, and the fatal and benign COX-deficient myopathies of infancy (DiMauro *et al.,* 1994).

SDH histochemistry is also useful for the diagnosis of complex II deficiency. Several patients with myopathy and complex II deficiency have been reported. In agreement with the biochemical observations, SDH histochemistry showed complete lack of reaction in muscle (Haller *et al.,* 1991; Taylor *et al.,* 1996).

B. Cytochrome *c* Oxidase

COX, the last component of the respiratory chain, catalyzes the transfer of reducing equivalents from cytochrome *c* to molecular oxygen (Hatefi, 1985). The holoenzyme

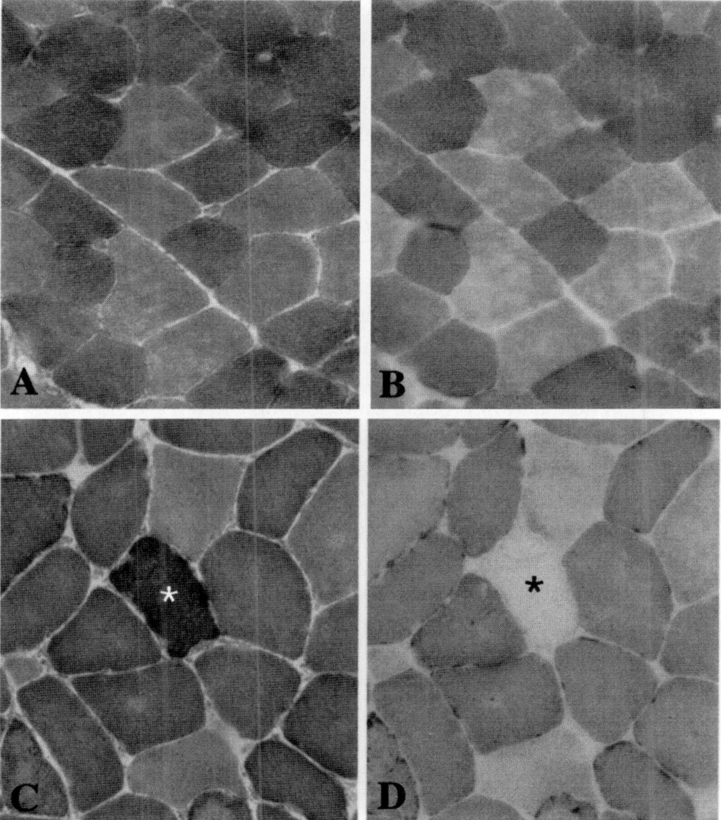

Fig. 1 Histochemical stains for SDH (A) and COX (B) activities on serial muscle sections from a normal and a KSS patient. Material from the unaffected individual shows a checkerboard pattern with both enzymes (A and B). In one section, KSS samples show one RRF (asterisk) by SDH stain (C). The same fiber on the serial section (asterisk) shows reduced COX activity (D). Bar: 50 μm. (See Color Plate.)

contains two heme *a* moieties (*a* and *a*3) and three copper atoms (two in the Cu_A site and one in the and Cu_B site) bound to a multisubunit protein frame embedded in the mitochondrial inner membrane (Cooper *et al.,* 1991; Taanman, 1997). In mammals, the apoprotein is composed of 13 different subunits. The three largest polypeptides (I, II, and III), which are encoded by mtDNA and are synthesized within mitochondria, confer the catalytic and proton pumping activities to the enzyme. The 10 smaller subunits are synthesized in the cytoplasm under the control of nuclear genes and are presumed to confer tissue specificity, thus adjusting the enzymatic activity to the metabolic demands of different tissues (Kadenbach *et al.,* 1987; Capaldi, 1990). Additional nDNA-encoded factors are required for the assembly of COX, including those involved in the synthesis of heme *a* and *a*3, transport and insertion of copper atoms, and proper coassembly of the mtDNA- and nDNA-encoded subunits. Several COX assembly genes have been identified in yeast (Kloeckener-Gruissem *et al.,* 1987; Pel *et al.,* 1992), and pathogenic mutations in the human homologues of three of these genes, SURF1, SCO2 and COX10, have been discovered in patients with COX-deficient Leigh syndrome (Zhu *et al.,* 1998; Tiranti *et al.,* 1998), with a cardioencephalomyopathy characterized by COX deficiency (Papadopoulou *et al.,* 1999), and with leukodystrophy (Valnot *et al.,* 2000).

The dual genetic makeup of COX and the availability of a reliable histochemical method to visualize its activity have made COX one of the ideal tools for investigations of mitochondrial biogenesis, nDNA–mtDNA interactions, and for the study of human mitochondrial disorders at both light and electron microscopic levels (Bonilla *et al.,* 1975; Johnson *et al.,* 1983; Wong-Riley, 1989).

The histochemical method to visualize COX activity is based on the use of 3,3′-diaminobenzidine (DAB) as an electron donor for cytochrome *c* (Seligman *et al.,* 1968). The reaction product on oxidation of DAB occurs in the form of a brown pigmentation corresponding to the distribution of mitochondria in the tissue. The specificity of the method may also be tested by performing control experiments in which the COX inhibitor, potassium cyanide (0.01 *M*), is added to the incubation medium.

As in the case of SDH, staining of normal muscle for COX activity also shows a checkerboard pattern. Type I fibers stain darker due to their mainly oxidative metabolism and more abundant mitochondria content, and type II fibers show a finer and less intensely stained mitochondria network (Fig. 1B).

The application of COX histochemistry to the investigation of KSS, MERRF, and MELAS has revealed one of the most important clues for the study of pathogenesis in these disorders. Muscle from KSS and MERRF patients shows a mosaic expression of COX, consisting of a variable number of COX-deficient and COX-positive fibers (Johnson *et al.,* 1983) (Fig. 1D). Before the advent of molecular genetics, it was difficult to understand the reason for the appearance of this mosaic. However, when it was discovered that these patients harbored mutations of mtDNA in their muscles, it became evident that the mosaic was an indicator of the heteroplasmic nature of the genetic defects. The mosaic pattern of COX expression in mitochondrial disorders is now considered the "histochemical signature" of a heteroplasmic mtDNA mutation affecting the expression of mtDNA-encoded genes in skeletal muscle (Bonilla *et al.,* 1992; Shoubridge, 1993).

Muscle biopsies from patients with MELAS also show COX-deficient fibers, but almost unique to MELAS RRF are COX-positive: the activity is decreased in the center of the fibers and is largely preserved in the subsarcolemmal regions.

It should also be noted that a generalized pattern of COX deficiency and COX-negative RRF is also observed in infants with depletion of muscle mtDNA and with the fatal and benign COX-deficient myopathies of infancy (DiMauro and Bonilla, 1997). In addition, a generalized pattern of COX deficiency in extrafusal muscle fibers, intrafusal fibers, and arterial walls of blood vessels is seen in children affected with COX-deficient Leigh syndrome resulting from mutations in either SURF1 or SCO2 genes (Zhu *et al.,* 1998; Tiranti *et al.,* 1998; Papadopoulou *et al.,* 1999).

These observations indicate that histochemical studies can provide significant information about both the nature and the pathogenesis of mitochondrial disorders. Moreover, they provide useful clues as to which molecular testing is needed to provide a specific diagnosis.

III. Immunohistochemistry

Immunohistochemical detection of specific proteins in single cells is the method of choice to study the expression of both mtDNA and nDNA genes in mitochondria of small and heterogeneous tissue samples. Technical advances have increased the scope of immunohistochemistry greatly and made it accessible to a variety of investigators with minimal expertise in immunology.

Several immunological probes are presently available to perform immunohistochemical studies of mitochondria in frozen tissue sections. These include (i) antibodies directed against mtDNA- and nDNA-encoded subunits of the respiratory chain complexes and other mitochondrial proteins and (ii) antibodies against DNA that allow the detection of mtDNA (Andreetta *et al.,* 1991; Bonilla *et al.,* 1992; Papadopoulou *et al.,* 1999). Because the entire mitochondrial genome has been sequenced, any mtDNA-encoded respiratory chain subunit is potentially available for immunocytochemical studies, and it is anticipated that the same will soon be true for all the nDNA-encoded subunits of the respiratory chain (Taanman *et al.,* 1996).

There are several immunocytochemical methods for the study of mitochondria in tissue sections. These include enzyme-linked methods (peroxidase, alkaline phosphatase, and glucose oxidase) and methods based on the application of fluorochromes. For studies in frozen tissue sections, we favor the use of fluorochromes because they allow for the direct visualization of the antigen–antibody-binding sites and because they are more flexible for double-labeling experiments. For studies of mitochondria on formalin-fixed and paraffin-embedded samples, we routinely employ the avidin–biotin–peroxidase complex (ABC) method (Hsu *et al.,* 1981; Bedetti, 1985).

A. Immunolocalization of Nuclear DNA and Mitochondrial DNA-Encoded Subunits of the Respiratory Chain in Frozen Samples

The main advantage of immunofluorescence to localize proteins in frozen sections is that it allows for the visualization of two different probes in the same mitochondria and in the same plane of section. These methods eliminate the inferences that must be made with studies on serial sections and are particularly useful in investigations of mitochondria in nonsyncitial tissues, such as heart, kidney and brain.

We routinely use a monoclonal antibody against COX IV as a probe for a nDNA-encoded mitochondrial protein and a polyclonal antibody against COX II as a probe for a mtDNA-encoded protein. For these studies, the sections are first incubated with both polyclonal and monoclonal antibodies at optimal dilution (i.e., the lowest concentration of antibody giving a clear particulate immunostain corresponding to the localization of mitochondria in normal muscle fibers). Subsequently, the sections are incubated with goat antirabbit IgG fluorescein (to visualize the mtDNA probe in "green") and goat antimouse IgG Texas red (to visualize the nDNA probe in "red"). We carry out these studies with unfixed frozen sections, but with some antibodies it may be required to permeabilize the mitochondrial membranes to uncover the antigenic sequences or to facilitate the penetration of the probes into the inner mitochondrial compartment. In agreement with Johnson *et al.* (1988), we have also found that fixation of fresh frozen sections with 4% formaldehyde in 0.1 *M* $CaCl_2$, pH 7, followed by dehydration in serial alcohols (outlined in the final section of this chapter), provides the most reproducible and successful results.

Using unfixed serial muscle sections from normal samples, a checkerboard pattern resembling the one described for histochemistry is usually observed; type I fibers appear brighter due to their higher content of mitochondria (Figs. 2A and 2B, see color plate).

In muscle serial sections from patients with KSS harboring a documented deletion of mtDNA (Δ-mtDNA), COX-deficient RRF typically show an absense or marked reduction of COX II, whereas the immunostain is typically normal in both COX-positive and COX-deficient fibers using antibodies against COX IV (Figs. 2C and 2D). Presumably, this is due to the fact that even the smallest deletion eliminates essential tRNAs that are required for the translation of the mitochondrial genome (Nakase *et al.,* 1990; Moraes *et al.,* 1992). Cell culture studies have confirmed this hypothesis: transfer of Δ-mtDNA to mtDNA-free human cell lines produces a severe defect in the synthesis of mtDNA-encoded subunits of the respiratory chain in recipient cells containing predominantly Δ-mtDNAs (Hayashi *et al.,* 1991).

B. Immunolocalization of Mitochondrial DNA on Frozen Samples

Immunohistochemistry using anti-DNA antibodies has been applied as an alternative method to *in situ* hybridization for the studies of localization and distribution of mtDNA in normal and pathological conditions (Andreetta *et al.,* 1991; Tritschler *et al.,* 1992). The advantages of this method are that both mitochondrial and nuclear DNA are detected simultaneously, at the single cell level, and that the nuclear signal can be used as an internal control.

For detection of mtDNA using immunological probes, we use forzen sections and monoclonal antibodies against DNA. In muscle sections from normal controls, these antibodies show an intense staining of nuclei and a particulate stain of the mitochondrial network (Andreetta *et al.,* 1991). When muscle sections from patients with KSS are studied, the mitochondrial network is stained intensely in RRF (Andreetta *et al.,* 1991). Conversely, when muscle biopsies from patients with mtDNA depletion are analyzed, the particulate immunostaining of the mitochondrial network is not detectable or it is only present in a small number fibers, but the intensity of immunostaining in nuclei shows no alteration (Bonilla *et al.,* 1992).

Immunohistochemistry utilizing antibodies against DNA is a useful method for the rapid evaluation of the distribution of mtDNA in normal cells and for the detection of depletion of muscle mtDNA. The method is particularly precise for the diagnosis of mtDNA depletion when it is confined to only a subpopulation of fibers (Vu *et al.,* 1998).

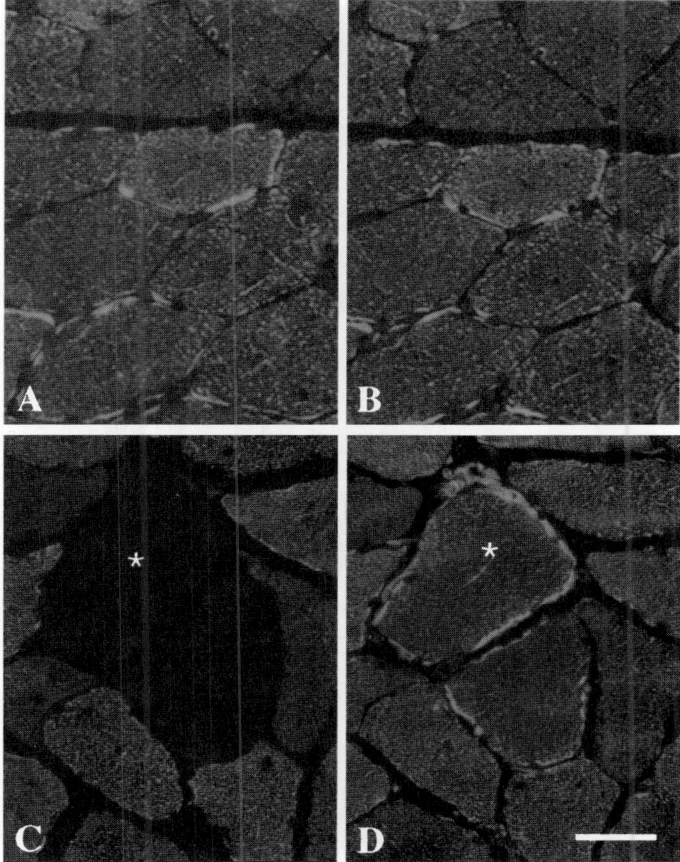

Fig. 2 Immunolocalization of COX II (A) and COX IV (B) on serial muscle sections from a normal and a KSS patient. The normal muscle shows an almost identical mitochondrial network for COX II and for COX IV (A and B). KSS samples show one RRF (asterisk) that lacks COX II immunostain in one section (C), and the same fiber on the serial section (asterisk) shows enhanced stain for COX IV (D). Bar: 50 μm. (See Color Plate.)

C. Immunolocalization of Mitochondrial Proteins on Paraffin-Embedded Samples

As mentioned earlier, we prefer immunoperoxidase for the localization of mitochondria on parafin-embedded brain samples, particularly the ABC method (Hsu *et al.,* 1981; Bedetti, 1985). This method is based on the high affinity of avidin, an egg white protein for the vitamin biotin. In this technique, avidin can be view as an antibody against the biotin-labeled peroxidase. A reliable ABC kit can be obtained from a commercial source (Vector Laboratories); alternatively, the ABC reagents can be prepared according to a previously published method (Hsu *et al.,* 1981).

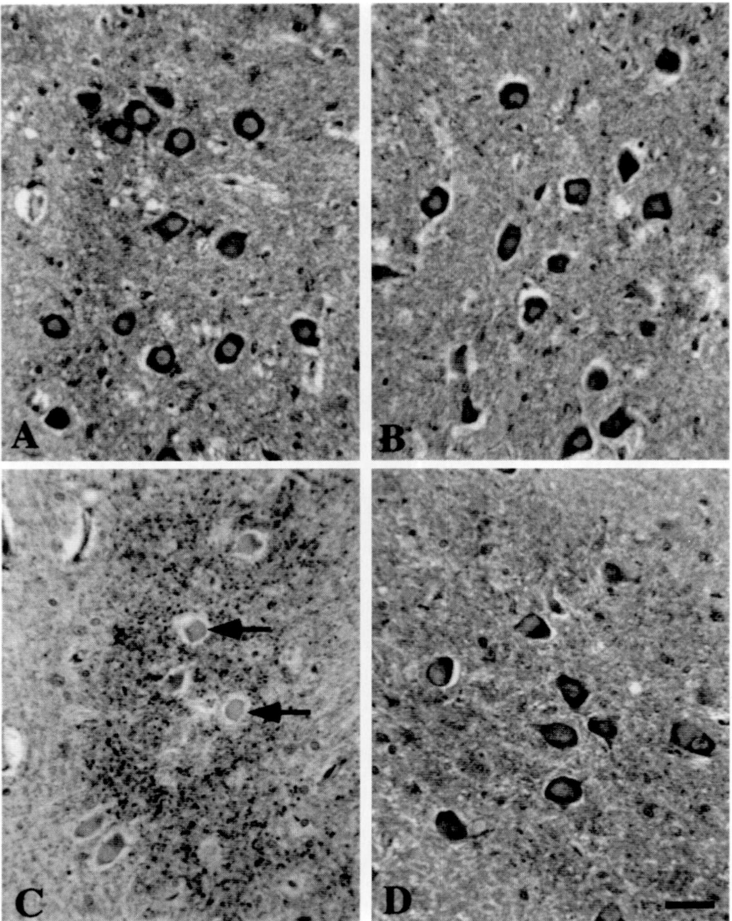

Fig. 3 Immunostaining of sections of the olivary nucleus from a control (A and B) and from a MERRF patient (C and D) for the localization of the mtDNA-encoded COX II subunit of complex IV (A and C) and the nDNA-encoded FeS subunit of complex III (B and D). The patient shows a marked decrease of immunostain in neurons for COX II (arrows), but normal stain for FeS. Bar: 50 μm. (See Color Plate.)

Studies of the mitochondrial respiratory chain on paraffin-embedded samples of brain using the ABC method have provided significant information regarding the pathogenesis of neuronal dysfunction in mitochondrial disorders and of the role of mitochondria in neurodegenerative disorders of the central nervous system, including Parkinson and Alzheimer's disease (Hattori *et al.,* 1990; Sparaco *et al.,* 1995; Tanji *et al.,* 1999; Bonilla *et al.,* 1999). For example, in patients with MERRF, a syndrome characterized clinically by myoclonic epilepsy, cerebellar ataxia, and myopathy, studies of the mitochondrial respiratory chain have shown a severe defect in the expression of the mtDNA-encoded subunit II of COX in neurons of the cerebellar system, including the cerebellar cortex, and the dentate and olivary nuclei (Fig. 3, see color plate). Based on these observations, we propose that in MERRF, mitochondrial dysfunction in neurons of the cerebellar system play a significant role in the genesis of cerebellar ataxia (Sparaco *et al.*, 1995).

IV. *In Situ* Hybridization and Single-Fiber Polymerase Chain Reaction (PCR)

In the last decade two important molecular genetic methods have been applied to study muscle mtDNA at the cellular level. The first is *in situ* hybridization, which is a technique that allows for the precise cellular localization and identification of cells that express a particular nucleic acid sequence. The essence of this method is the hybridization of a nucleic acid probe with a specific nucleic acid sequence found in a tissue section. The second method is the microdissection of histochemically defined single-fiber segments for PCR amplification in order to study heteroplasmic mtDNA mutations.

A. ISH to mtDNA

ISH has been used extensively in human pathologic conditions to correlate mitochondrial abnormalities with the presence of mutated mtDNAs. Because this method relies on hybridization of nucleic acid probes to mtDNA, it has been applied mainly to identify large-scale mtDNA deletions in samples from patients with KSS or isolated OM rather than point mutations and other subtle change, in mtDNA. In these studies, two probes have been used to determine the distribution of wild-type and Δ-mtDNAs: one, in the undeleted region, that hybridizes to both wild-type and Δ-mtDNAs (the "common" probe) and another, contained within the deletion, that hybridizes only to wild-type mtDNAs (the "wild-type"probe). Typical results on serial muscle sections showed an abundant hybridization signal (focal accumulations of mtDNAs) with the common probe, but not with the wild-type probe, in COX-deficient RRF of patients (Mita *et al.*, 1989; Bonilla *et al.*, 1992; Moraes *et al.*, 1992a). These studies reveal that the predominant species of mtDNAs in COX-deficient RRF of KSS and OM patients are Δ-mtDNAs. Furthermore, they showed that the concentration of Δ-mtDNAs in these RRF reached the required threshold level to impair the translation of the mitochondrial genome because they were characterized immunohistochemically by a lack or marked reduction of the mtDNA-encoded COX II polypeptide (Mita *et al.*, 1989; Bonilla *et al.*, 1992; Moraes *et al.*, 1992a).

An important extension of ISH to mtDNA is regional ISH, which has been applied to determine the spatial distribution of multiple Δ-mtDNAs in samples from patients with mendelian-inherited progressive external opthalmoplegia (MI-PEO). These are disorders characterized by progressive external ophthalmoplegia and mitochondrial myopathy. In MI-PEO, hundreds of different deletions coexist within the same muscle in affected family members. Thus, as opposed to the single deletions found in sporadic

KSS and OM, MI-PEO is associated with multiple Δ-mtDNAs, which are apparently generated over the life span of the individual. The genetic defect in these disorders most likely is due to mutations in a nuclear gene affecting the proclivity of the mtDNA to undergo deletions. ISH of serial muscle sections from these patients, in which a different mtDNA regional probe was used on each section, showed a variablity of signal among RRFs (i.e., subsets of RRFs lost hybridization signal with each specific probe, while the remaining RRFs hybridized intensely). These observations have provided the strongest evidence to date that each RRF in MI-PEO contains a different clonal expansion of a single Δ-mtDNA species (Moslemi *et al.,* 1996; Vu *et al.,* 2000).

Although ISH utilizing RNA probes is used more widely in cell and molecular biology applications, most of our experience derives from the use of digoxigenin (DIG)-labeled DNA probes (Manfredi *et al.,* 1997; Vu *et al.,* 2000). We described a method for the identification of RRF (Fig. 4, see color plate) and for the detection of depletion of mtDNA on muscle frozen sections. Please note that the use of RNA probes requires special care, but may provide better results when the levels of the target sequence are low. Although tissue samples can be frozen or fixed and parafin embedded, our procedure is optimized for frozen sections and thoroughly tested on muscle sections from patients with mitochondrial myopathies.

Three factors are critical for the use of ISH, and interpretation of ISH results-obtained. First, because of the focal pattern of distribution of mutant and wild-type mtDNAs in muscle fibers, one must make sure that serial sections above and below the ones used for ISH are characterized histochemically by staining for COX and SDH activities (Fig. 4). Second, the size of the labeled probe is very important and should balance specificity and tissue penetration. We have obtained excellent results with 500 nucleotide probes. However, sizes between 300 and 400 nucleotides are optimum. DNA probes can be prepared in different ways. We routinely use PCR-generated DIG-labeled mtDNA probes. Third, kits and polymerases for DNA labeling are available from most molecular biology companies; however, it is important that the concentration of the probe be determined by dot blot prior to ISH experiments and that the specificity of the probe be tested by Southern blot.

B. Single-Fiber Segments for PCR

Although ISH provides important information on the cellular distribution of Δ-mtDNAS and wild-type mtDNAS, the method does not detect mitochondrial genomes harboring point mutations. Furthermore, only semiquantitative analysis can be obtained by ISH. Because of these limitations, several investigators have utilized a PCR-based method to study mtDNA mutations in single-fiber segments (Shoubridge *et al.,* 1990; Moraes *et al.,* 1992b; Andreu *et al.,* 1999). Briefly, histochemically characterized muscle fiber segments are dissected under the microscope and subjected to PCR amplification. Mutant mtDNAs and wild-type mtDNAs are detected by size or by restriction fragment length polymorphism (RFLP). The observation of higher levels of mutant mtDNAs in muscle fiber segments with an abnormal histochemical phenotype provides strong evidence of pathogenicity for a mtDNA mutation.

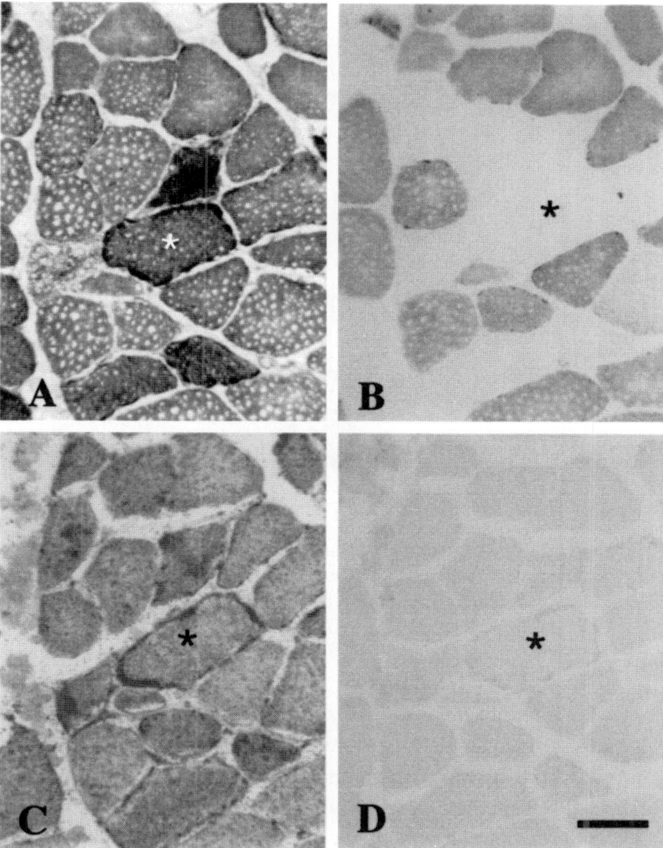

Fig. 4 Cellular localization of mtDNA by ISH. Serial muscle sections from a patient with MERRF were stained for SDH (A) and COX (B) activity, and for mtDNA localized by ISH (C). The ISH signal is seen as red material. Note the strong mtDNA signal in a RRF (asterisk) characterized by increased SDH activity and lack of COX activity. A control section subjected to ISH without the denaturing step (D) shows no hybridization signal. Bar: 50 μm. (See Color Plate.)

The dissection of muscle fiber segments does not require expensive equipment such as micromanipulators or sophisticated microscopes, only skillful hands and a few days of practice. We use a mouth suction apparatus illustrated in Fig. 5. The tubing and the mouthpiece can be found in boxes of microcapillaries, plastic pipettes are found in any laboratory, and an inverted microscope is found in any tissue culture laboratory. Capillaries are prepared by flaming the center of a 5-μl microcapillary and pulling the ends apart when the glass starts to melt. The pulled glass is broken by rubbing it with another microcapillary. After preparation of several pin-pointed microcapillaries and assembling of the dissecting apparatus, microdissection is performed with the aid of an inverted microscope.

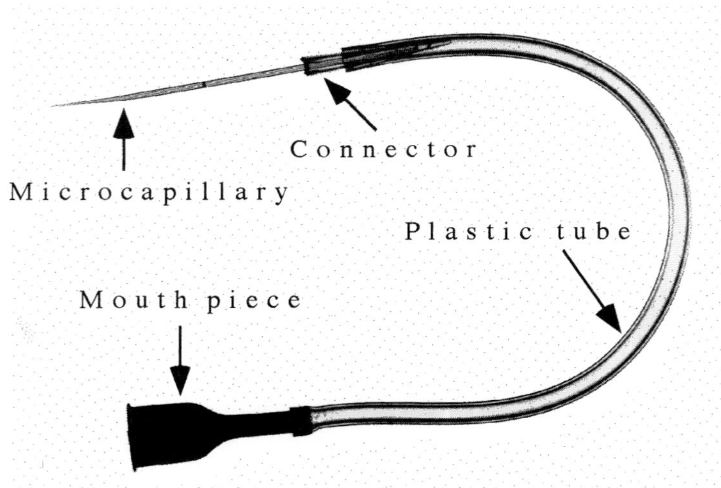

Fig. 5 Illustration of the homemade microdissection device. The tube with the mouthpiece is attached to a connector (sharp end of a plastic pipette). The microcapillaries are inserted to the connector and are changed constantly, one for each single-fiber segment, while the rest of the device can be used until the tube shows signs of air leakage.

The thickness of the sections for microdissection may vary. We find that 20- to 30-μm frozen sections give the most reliable result. Thicker sections stain very strongly with the histochemical methods, and it becomes almost impossible to distinguish the cellular phenotypes. It is important to use slides coated with polylysine (0.1%) in order to avoid movement of the sections during the dissection. Cross sections from test and controls are placed side by side on the same slide and stained for enzyme activity of COX, SDH, or both, as outlined in the histochemical methods. The stained sections provide the dissector a visual phenotype that will direct the dissection. The slides are then immersed in a 100-mm^2 cell culture dish containing 50 ml of 100% ethanol. Single-fiber segments are isolated by gentle mouth suction under the inverted microscope. The fiber segments are placed directly into 50 μl of water in a 250-μl Eppendorf tube. Use a 10× magnifying glass to confirm that the fiber segment has been released into the water-containing Eppendorf tube.

After DNA is released from the single-fiber segments (see methods), the samples are ready for PCR. For the detection of single nucleotide mutations, a standard PCR reaction is performed, and the PCR product is evaluated by agarose gel electrophoresis. If the correct product is obtained, PCR products are subjected to an additional cycle after the addition of 10 μCi of [α-^{32}P]dATP, 100 pmol of each primer, and 2 U of *Taq* polymerase in 1× PCR buffer (last cycle hot PCR). Restriction endonuclease-digested fragments are separated by native 12% polyacrylamide gel electrophoresis and exposed to an X-ray

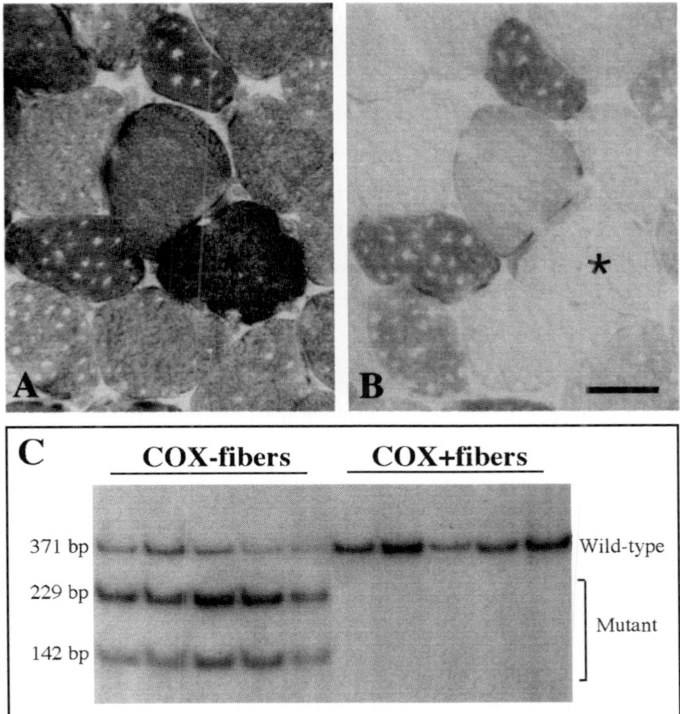

Fig. 6 Histochemistry and PCR analysis of microdissected single-fiber segments from a patient harboring a pathogenic point mutation in the COX 1 gene of mtDNA. This mutation created a novel *HpaI* restriction site that can be detected by RFLP. Histochemical stains for SDH and COX activities on serial sections illustrate the pattern of COX deficiency in the patient. One section shows a RRF (asterisk) by SDH stain (A), and the same fiber on the serial section (asterisk), as well as other fibers, shows lack of COX activity. (B) Bar: 50 μm. For RFLP analysis, COX-positive (COX+) and COX-negative (COX−) muscle fiber segments were microdissected, and a selected region of the COX I gene was amplified by PCR. The PCR fragments were analyzed by RFLP after digestion with the enzyme *Hpa*I. The autoradiogram (C) shows that mutant mtDNAs are confined to COX- single-fiber segments (mean 65%). The level of mutant mtDNAs observed in total muscle DNA was 61%.

film. Figure 6 illustrates results obtained in COX-positive and COX-negative single-fiber segments from a patient with a nonsense mutation in the COX I gene of mtDNA.

Single-fiber PCR has also been used for the quantification of deleted-mtDNAs (Sciacco *et al.,* 1994). In such cases, three or more primers have to be used during the amplification (two located outside the deleted region and the third inside the deletion). In addition, single-fiber PCR has been applied to estimate the absolute amounts of mtDNA, an application that has been particularly useful in the study of mtDNA depletion in muscle (Sciacco *et al.,* 1998).

V. Histochemical Methods

A. Succinate Dehydrogenase

1. Collect 8-μm-thick cryostat sections on poly(L-lysine)-coated (0.1%) coverslips. Dissolve the following in 10 ml of 5 m*M* phosphate buffer, pH 7.4.

5 m*M* ethylenediaminetetraacetic acid (EDTA)
1 m*M* potassium cyanide (KCN)
0.2 m*M* phenazine methosulfate (PMS)
50 m*M* succinic acid
1.5 m*M* NBT
Adjust pH to 7.6

2. Filter solution with filter paper. (Whatman No. 1).
3. Incubate sections for 20 min at 37°C. For control sections, sodium malonate (0.01 *M*) is added to the incubation medium.
4. Rinse 5 min × 3 times in distilled water, at room temperature (RT).
5. Mount on glass slides with warm glycerin gel.

B. COX

1. Collect 8-μm-thick cryostat sections on poly(L-lysine)-coated (0.1%) coverslips. Dissolve the following in 10 ml of 5 m*M* phosphate buffer, pH 7.4.

0.1% 10 mg DAB
0.1% 10 mg cytochrome *c* (from horse heart)
0.02% 2 mg catalase
Adjust pH to 7.4

2. Do not expose solution to light.
3. Filter solution with filter paper. (Whatman No. 1).
4. Incubate sections for 1 h at 37°C. For control sections, potassium cyanide (0.01 *M*) is added to the incubation medium.
5. Rinse with distilled water, 3 × 5 min at RT.
6. Mount on glass slides with warm glycerin gel.

VI. Immunohistochemical Methods

A. Simultaneous Visualization of Mitochondrial DNA and Nuclear DNA-Encoded Subunits of the Respiratory Chain Using Different Fluorochromes

1. Collect 4-mm-thick cryostat sections on poly(L-lysine)-coated (0.1%) coverslips.

2. Incubate the sections for 2 h at RT (in a humid chamber) with anti-COX II polyclonal antibody and with anti-COX IV monoclonal antibody at optimal dilutions (1 : 100 to 1 : 500; Molecular Probes, Eugene, OR) in phosphate buffer saline containing 1% bovine serum albumin (PBS/BSA). Control sections are incubated without the primary antibodies.

3. Rinse the samples with PBS 3 × 5 min at RT.

4. Incubate the sections for 1 h at RT (in a humid chamber) with antirabbit IgG fluorescein and with antimouse IgG Texas red diluted 1 : 100 (Molecular Probes, Eugene, OR) in 1% BSA/PBS.

5. Rinse the samples with PBS, 3 × 5 min at RT.

6. Mount on slides with 50% glycerol in PBS.

B. Localization of Mitochondrial DNA Using Immunofluorescence

1. Collect 4-mm-thick cryostat sections on poly(L-lysine)-coated (0.1%) coverslips.
2. Fix the sections in 4% formaldehyde in 0.1 *M* $CaCl_2$, pH 7, for 1 h at RT.
3. Dehydrate the sections in 70, 80, 90% ethyl alcohol, 5 min each, and in 100% ethyl alcohol for 15 min.
4. Rinse the samples with PBS, 3 × 5 min at RT.
5. Incubate the sections for 2 h at RT (in a humid chamber) with anti-DNA monoclonal antibody (1 : 100–1 : 500; Chemicon, Temecula, CA) in 1% BSA/PBS. Control sections are incubated without the primary antibodies.
6. Rinse the samples with PBS, 3 × 5 min at RT.
7. Incubate the sections for 30 min at RT (wet chamber) with biotynilated antimouse IgG (1 : 100) in 1% BSA/PBS.
8. Rinse the samples with PBS, 3 × 5 min at RT.
9. Incubate the sections for 30 min at RT (in a humid chamber) with streptavidin-fluorescein or streptavidin-Texas red (1 : 250; Molecular Probes, Eugene, OR) in 1% BSA/PBS.
10. Rinse the samples with PBS, 3 × 5 min at RT.
11. Mount on slides with 50% glycerol in PBS.

C. Immunolocalization of Mitochondrial Proteins in Paraffin-Embedded Sections using the ABC Method

1. Collect 4-μm paraffin-embedded sections on poly(L-lysine)-coated (0.1%) slides. Deparaffinize the sections through xylene (concentration) and descending ethanol series (100, 95, 80, and 75%).
2. Incubate the sections in methanol containing 5% H_2O_2 for 30 min at RT.
3. Place the slides in PBS and then incubate the samples with 5% normal serum (from the same species as the host of the second antibody) for 1 h at RT.

4. Incubate the slides with the primary antibody (1 : 1000–1 : 2000) overnight at 4°C.
5. Rinse the slides with PBS, 3 × 5 min at RT.
6. Incubate the slides with the biotynilated second antibody at the optimal conditions (1 : 100–1 : 300) for 1 h at RT.
7. Rinse the slides with PBS, 3 × 5 min at RT.
8. Incubate the slides with ABC complex (prepare 30 min–1 h prior to use).
9. Rinse the slides with PBS, 3 × 5 min at RT.
10. Incubate the slides with DAB-H_2O_2 solution [40 mg of 3,3′-diaminobenzidine tetrahydrochloride dissolved in 100 ml of PBS or 0.05 *M* Tris–HCl buffer (pH 7.6) containing 0.005% H_2O_2] for 1–3 min at RT.
11. Rinse the slides with distilled water (dH_2O) several times.
12. Counterstain the slides briefly with hematoxylin.
13. Rinse the slides with dH_2O, dehydrate through ascending ethanol series, and clear in xylene.
14. Mount the slides with synthetic resin (Permount).

VII. Molecular Genetic Methods

A. *In Situ* Hybridization to Mitochondrial DNA

1. Collect 8-μm frozen sections on poly(L-lysine)-coated (0.1%) slides.
2. Fix the sections with 4% paraformaldehyde for 30 min at RT.
3. Rinse the slides with PBS containing 5 m*M* $MgCl_2$ (PBS-$MgCl_2$, pH 7.4), 3 × 5 min, at RT.
4. Incubate the sections with 5 mg/ml proteinase K for 1 h at 37°C.
5. Place the slides in PBS-$MgCl_2$ at 4°C for 5 min to inactivate proteinase K.
6. Acetylate the sections by incubating in 0.1 *M* triethanolamine containing 0.25% acetic anhydride for 10 min at RT.
7. Treat the slides with 5 mg/ml RNase (DNase-free) in RNase buffer (50 m*M* NaCl and 10 m*M* Tris–HCl, pH 8) for 30 min 37°C.
8. Rinse the slides briefly with PBS-$MgCl_2$.
9. Dehydrate the sections through ascending series of ethanol 70%, 80%, 90%, and 100% (optional).
10. Incubate with hybridization buffer without probe (prehybridization) for 1–2 h at 37°C (in our experience, this step can be eliminated)
11. Denature the probe/hybridization solution (50% deionized formamide, 20 m*M* Tris–HCl, 0.5 m*M* NaCl, 10 m*M* EDTA, 0.02% Ficol, 0.02% polyvynil pyrollidone, 0.12% BSA, 0.05% salmon sperm DNA, 0.05% total yeast RNA, 0.01% yeast tRNA, and 10% dextran sulfate) (20 ng/ml) at 92°C for 10 min and immediately place the solution on ice until it is applied to the sections.

12. Dot off excess prehybridization solution and apply probe/hybridization solution to the sections.
13. Denature the sections covered with the hybridization solution at 92°C for 10–15 min.
14. Hybridize overnight at 42°C.
15. Rinse the slides briefly with 2× SSC (3 *M* NaCl, 0.3 *M* Sodium citrate, pH 7.0) at RT followed by rinsing twice with 1× SSC for 15 min at 45°C and once with 0.2× SSC for 30 min at 45°C.
16. Rinse the slides with PBS for 5 min at RT.
17. Treat the sections with 1% BSA for 1 h at RT.
18. Incubate the slides with alkaline phosphate-conjugated anti-DIG antibody (1 : 1000–1 : 5000) for 1 h at RT.
19. Rinse the slides with PBS, 3 × 5 min at RT.
20. Incubate the slides with color-substrate solution (SIGMA FAST Fast Red TR/Napthol AS-MX alkaline phosphatase substrate tablets) until obtaining the desired strength of the signal.
21. Mount the sections with glycerol–PBS (1 : 1).

B. DNA Release from Single-Fiber Segments for PCR

1. After all single-fiber segments are picked, spin the Eppendorf tubes in a microcentrifuge at maximum speed for 5 min.
2. Remove the supernatant and add 5 μl of alkaline lysis solution (200 m*M* KOH, 50 m*M* dithiothreitol) to each sample.
3. Incubate the samples at 65°C for 30 min and then add 5 μl of a buffered neutralizing solution (900 m*M* Tris–HCl, pH 8.3, 200 m*M* HCl). Aliquots of the resulting 10-μl solution are used for PCR.

Acknowledgments

This work was supported by grants from the National Institutes of Health (NS11766 and PO1HD32062).

References

Anderson, S., *et al.* (1981). Sequence and organization of the human mitochondrial genome. *Nature* **290,** 457–465.

Andreetta, F., *et al.* (1991). Localization of mitochondrial DNA using immunological probes: A new approach for the study of mitochondrial myopathies. *J. Neurol. Sci.* **105,** 88–92.

Andreu, A. L., *et al.* (1999). Isolated myopathy associated with a nonsense mutation in the subunit 4 of the mitochondrial NADH-dehydrogenase gene. *Ann. Neurol.* **45,** 820–823.

Attardi, G., and Schatz, G. (1988). Biogenesis of mitochondria. *Annu. Rev. Cell. Biol.* **4,** 289–333.

Bedetti, C. D. (1985). Immunocytological demonstration of cytochrome *c* oxidase with an immunoperoxidase method. *J. Histochem. Cytochem.* **33,** 446–452.

Beinert, H. (1990). Recent developments in the field of iron-sulfur proteins. *FASEB J.* **4,** 2483–2491.

Bogenhagen, D., and Clayton, D. A. (1984). The number of mitochondrial DNA genomes in mouse and human HeLa cells. *J. Biol. Chem.* **249,** 7991–7995.

Bonilla, E., *et al.* (1975). Electron cytochemistry of crystalline inclusions in human skeletal muscle mitochondria. *J. Ultrastruct. Res.* **51,** 404–408.

Bonilla, E., *et al.* (1992). New morphological approaches for the study of mitochondrial encephalopathies. *Brain Pathol.* **2,** 113–119.

Bonilla, E., and Tanji, K. (1998). Ultrastructural alterations in encephalomyopathies of mitochondrial origin. *BioFactors* **7,** 231–236.

Bonilla, E., *et al.* (1999). Mitochondrial involvement in Alzheimer's disease. *Biochem. Biophys. Acta* **1410,** 171–182.

Bourgeron, T., *et al.* (1995). Mutation of a nuclear succinate dehydrogenase gene results in mitochondrial respiratory chain deficiency. *Nature Genet.* **11,** 144–149.

Capaldi, R. A. (1990). Structure and assembly of cytochrome *c* oxidase. *Arch. Biochem. Biophys.* **280,** 252–262.

Cooper, C. E., *et al.* (1991). Cytochrome *c* oxidase: Structure, function, and membrane topology of the polypeptide subunits. *Biochem. Cell Biol.* **69,** 596–607.

DiMauro, S., *et al.* (1994). Cytochrome oxidase deficiency: Progress and problems. *In* "Mitochondrial Disorders in Neurology" (A. H. V. Shapira and S. DiMauro, eds.), pp. 91–115. Butterworth-Heiemann, Oxford.

DiMauro, S., and Bonilla, E. (1997). Mitochondrial encephalomyopathies. *In* "The Molecular and Genetic Basis of Neurological Disease" (R. N. Rosenberg, S. B. Prusiner, S. DiMauro, and R. L. Barchi, eds.), pp. 201–235. Butterworth-Heinemann, Newton.

Engel, W. K., and Cunningham, G. G. (1963). Rapid examination of muscle tissue: An improved trichrome stain method for fresh frozen biopsy sections. *Neurology* **13,** 919–926.

Giles, R. E., *et al.* (1980). Maternal inheritance of human mitochondrial DNA. *Proc. Natl. Acad. Sci. USA* **77,** 6715–6719.

Goto, Y. I., *et al.* (1990). A mutation in the $tRNA^{Leu(UUR)}$ gene associated with the MELAS subgroup of mitochondrial encephalopathies. *Nature* **348,** 651–653.

Goto, Y. I., *et al.* (1991). A new mtDNA mutation associated with mitochondrial myopathy, encephalopathy, lactic acidosis, and stroke-like episodes. *Biochim. Biophys. Acta* **1097,** 238–240.

Haller, R., *et al.* (1991). Deficiency of skeletal muscle SDH and aconitase. *J. Clin. Invest.* **88,** 1197–1206.

Hatefi, Y. (1985). The mitochondrial electron transport and oxidative phosphorylation system. *Annu. Rev. Biochem.* **54,** 1015–1076.

Hattori, N., *et al.* (1990). Immunohistochemical studies of complexes I, II, III and IV of mitochondria in Parkinson disease. *Ann. Neurol.* **30,** 563–571.

Hayashi, J.-J., *et al.* (1991). Introduction of disease-related mitochondrial DNA deletions into HeLa cells lacking mitochondrial DNA results in mitochondrial dysfunction. *Proc. Natl. Acad. Sci. USA* **88,** 10614–10618.

Hsu, S. M., *et al.* (1981). Use of avidin-peroxidase complex (ABC) in immunoperoxidase techniques: A comparison between ABC and unlabeled antibody (PAP) procedures. *J. Histochem. Cytochem.* **29,** 577–580.

Johnson, M. A., *et al.* (1983). A partial deficiency of cytochrome *c* oxidase in chronic progressive external ophtalmoplegia. *J. Neurol. Sci.* **60,** 31–53.

Johnson, M. A., *et al.* (1988). Immunocytochemical studies of cytochrome *c* oxidase subunits in skeletal muscle of patients with partial cytochrome oxidase deficiency. *J. Neurol. Sci.* **87,** 75–90.

Kadenbach, B., *et al.* (1987). Evolution of a regulatory enzyme: Cytochrome *c* oxidase (complex IV). *Curr. Topics Bioenerg.* **15,** 113–161.

Kloeckener-Gruissem, B., *et al.* (1987). Nuclear functions required for cytochrome *c* oxidase biogenesis in *Saccharomyces cerevisiae*: Multiple trans-acting nuclear genes exert specific effects on expression of each of the cytochrome *c* oxidase subunits encoded on mitochondrial DNA. *Curr. Genet.* **12,** 311–322.

Loeffen, J., *et al.* (1998). The first nuclear-encoded complex I mutation in a patient with Leigh syndrome. *Am. J. Hum. Genet.* **63,** 1598–1608.

Manfredi, G., *et al.* (1997). Association of myopathy with large-scale mitochondrial DNA duplications and deletions: Which is pathogenic? *Ann. Neurol.* **42,** 180–188.

Mita, S., *et al.* (1989). Detection of deleted mitochondrial genomes in cytochrome *c* oxidase-deficient muscle fibers of a patient with Kearns-Sayre syndrome. *Proc. Natl. Acad. Sci. USA* **86,** 9509–9513.

Moraes, C. T., *et al.* (1989). Mitochondrial DNA deletions in progressive external ophthalmoplegia and Kearns-Sayre syndrome. *N. Engl. J. Med.* **320,** 1293–1299.

Moraes, C. T., *et al.* (1992a). Molecular analysis of the muscle pathology associated with mitochondrial DNA deletions. *Nature Genet.* **1,** 359–367.

Moraes, C. T., *et al.* (1992b). The mitochondrial tRNA-Leu(UUR) mutation in MELAS: Genetic, biochemical and morphological correlations. *Am. J. Hum. Genet.* **50,** 934–949.

Moslemi, A. R., *et al.* (1996). Clonal expansion of mitochondrial DNA with multiple deletions in aurosomal dominant progressive external ophthalmoplegia. *Ann. Neurol.* **48,** 707–713.

Nakase, H., *et al.* (1990). Transcription and translation of deleted mitochondrial genomes in Kearns-Sayre syndrome: Implications for pathogenesis. *Am. J. Hum. Genet.* **46,** 418–427.

Neupert, W. (1997). Protein import into mitochondria *Annu. Rev. Biochem.* **66,** 863–917.

Papadopoulou, L. C., *et al.* (1999). Fatal infantile cardioencephalopathy with cytochrome *c* oxidase (COX) deficiency and mutations in *SCO2,* a human COX assembly gene. *Nature Genet.* **23,** 333–337.

Pel, H. J., *et al.* (1992). The identification of 18 nuclear genes required for the expression of the yeast mitochondrial gene encoding cytochrome *c* oxidase subunit I. *Curr. Genet.* **21,** 139–146.

Pette, D. (1981). Microphotometric measurement of initial maximum reaction rates in quantitative enzyme histochemistry in situ. *Histochem. J.* **13,** 319–327.

Satoh, M., and Kuroiwa, T. (1991). Organization of multiple nucleotides of DNA molecules in mitochondria of a human cell. *Exp. Cell Res.* **196,** 137–140.

Schon, E. A., *et al.* (1997). Mitochondrial DNA mutations and pathogenesis. *J. Bioenerg. Biomembr.* **29,** 131–149.

Sciacco, M., *et al.* (1994). Distribution of deleted and wild-type mtDNA in normal and respiration-deficient muscle fibers from patients with mitochondrial myopathy. *Hum. Mol. Genet.* **3,** 13–19.

Sciacco, M., *et al.* (1998). Study of mitochondrial DNA depletion in muscle by single-fiber PCR. *Muscle Nerve* **21,** 1374–1381.

Seligman, A. M., and Rutenburg, A. M. (1951). The histochemical demonstration of succinc dehydrogenase. *Science* **113,** 317–320.

Seligman, A. M., *et al.* (1968). Non-droplet ultrastructural demonstration of cytochrome oxidase activity with a polymerizing osmiophilic reagent, diaminobenzidine (DAB). *J. Cell Biol.* **38,** 1–15.

Shoffner, J. M., *et al.* (1990). Myoclonic epilepsy and ragged-red fiber disease (MERRF) is associated with a mitochondrial DNA $tRNA^{Lys}$ mutation. *Cell* **61,** 931–937.

Shoubridge, E. A., *et al.* (1990). Deletion mutants are functionally dominant over wild-type mitochondrial genomes in skeletal muscle fiber segments in mitochondrial disease. *Cell* **62,** 43–49.

Shoubridge, E. A. (1993). Molecular histology of mitochondrial diseases. *In* "Mitochondrial DNA in Human Pathology" (S. DiMauro D. C. Wallace, eds.), pp. 109–123. Raven Press, New York.

Silvestri, G., *et al.* (1992). A new mtDNA mutation in the $tRNA^{Lys}$ gene associated with myoclonic epilepsy and ragged-red fibers (MERRF). *Am. J. Hum. Genet.* **51,** 1213–1217.

Sparaco, M., *et al.* (1995). Myoclonus epilepsy with ragged-red fibers (MERRF): An immunohistochemical study of the brain. *Brain Pathol.* **5,** 125–133.

Szabolics, M. J., *et al.* (1994). Mitochondrial DNA deletion: A cause of chronic tubulointerstitial nephropathy. *Kidney Int.* **45,** 1388–1396.

Taanman, J.-W. (1997). Human cytochrome *c* oxidase: Structure, function and deficiency. *J. Bioenerg. Biomembr.* **29,** 151–163.

Taanman, J.-W., *et al.* (1996). Subunit specific monoclonal antibodies show different steady-state levels of various cytochrome *c* oxidase subunits in chronic external ophthalmoplegia. *Biochim. Biophys. Acta* **1315,** 199–207.

Tanji, K., *et al.* (1999). Kearns-Sayre syndrome: Unusual pattern of expression of subunits of the respiratory chain in the cerebellar system. *Ann. Neurol.* **45,** 377–383.

Taylor, R. W., *et al.* (1996). Deficiency of complex II of the mitochondrial respiratory chain in late-onset optic atrophy and ataxia. *Ann. Neurol.* **39,** 224–232.

Tiranti, V., *et al.* (1998). Mutations of *SURF-1* in Leigh disease associated with cytochrome *c* oxidase deficiency. *Am. J. Hum. Genet.* **63,** 1609–1621.

Tritschler, H. J., *et al.* (1992). Mitochondrial myopathy of childhood with depletion of mitochondrial DNA. *Neurology* **42,** 209–217.

Valnot, I., *et al.* (2000). A mutation in the human heme A: Farnesyltransferase gene (*COX10*) causes cytochrome *c* oxidase deficiency. *Hum. Mol. Genet.* **9,** 1245–1249.

Vu, T. H., *et al.* (1998). Mitochondrial DNA depletion in a patient with long survival. *Neurology* **51,** 1190–1193.

Vu, T. H., *et al.* (2000). Analysis of muscle from patients with multiple mtDNA deletions by *in situ* hybridization. *Muscle Nerve* **23,** 80–85.

Wallace, D. C. (1992). Diseases of the mitochondrial DNA. *Annu. Rev. Biochem.* **61,** 1175–1212.

Wong-Riley, M. T. T. (1989). Cytochrome oxidase: An endogenous metabolic marker for neuronal activity. *Trends Neurosci.* **12,** 94–101.

Zeviani, M., *et al.* (1988). Deletions of mitochondrial DNA in Kearns-Sayre syndrome. *Neurology* **38,** 1339–1346.

Zhu, Z., *et al.* (1998). *SURF-1,* encoding a factor involved in the biogenesis of cytochrome *c* oxidase, is mutated in Leigh syndrome. *Nature Genet.* **20,** 337–343.

CHAPTER 19

Visualization of Mitochondrial Movement in Yeast

Hyeong-Cheol Yang, Viviana Simon,* Theresa C. Swayne, and Liza Pon

Department of Anatomy and Cell Biology
Columbia University
New York, New York 10032

* Banting and Best Department of Medical Research
University of Toronto
Toronto, Ontario, Canada M5G 1L6

I. Introduction

A great deal of what we know today about mitochondrial morphology, reorganization, and inheritance in budding yeast came from studies in fixed cells or with purified mitochondria. Three-dimensional reconstruction from transmission electron microscopy of serial thin sections of whole yeast cells demonstrated that mitochondria are largely tubular structures, ranging from 0.3 to 0.5 μm in diameter and from 0.5 to 5 μm in length. In addition, these studies showed that mitochondrial volume, morphology, and position during inheritance are controlled by metabolic- and cell division-specific regulators (Hoffmann and Avers, 1973; Stevens, 1977, 1981). Light and electron microscopic analysis in fixed cells revealed colocalization of mitochondria with actin structures and defects in mitochondrial morphology and inheritance upon disruption of the actin cytoskeleton by actin-destabilizing drugs or mutations (Drubin *et al.,* 1993; Lazzarino *et al.,* 1994). Finally, cell-free studies revealed that purified yeast actin filaments cosediment with isolated yeast mitochondria. The mitochondria–actin interaction is mediated by peripheral membrane actin-binding proteins on the mitochondrial surface and by two integral mitochondrial outer membrane proteins, Mdm 10p and Mmm1p (Lazzarino *et al.,* 1994; Burgess *et al.,* 1994; Sogo and Yaffe, 1994; Boldogh *et al.,* 1998).

Building on these observations, visualization of mitochondria in living yeast reveals that mitochondria are highly dynamic organelles that fuse, fragment, and undergo directed movement. For example, we found that mitochondria undergo a series of cell cycle-linked motility events during mitotic cell division (Fig. 1; Simon *et al.,* 1997). In late G_1 phase, subsequent to START, mitochondria orient toward the site of bud emergence and actin polarization. Linear, directed movement of mitochondria from mother cells into developing buds begins in S phase and continues through G_2 and M. Concomitant with this, some mitochondria are immobilized in the tip of the mother cell distal to the site of bud emergence and in the bud tip. These immobilization events contribute to a 50:50 distribution of mitochondria between mother and daughter.

Because mitochondria are essential organelles that must be produced from preexisting mitochondria, inheritance of the organelle is an integral and indispensable part of cell division. Therefore, visualization of mitochondrial movement in living yeast contributes

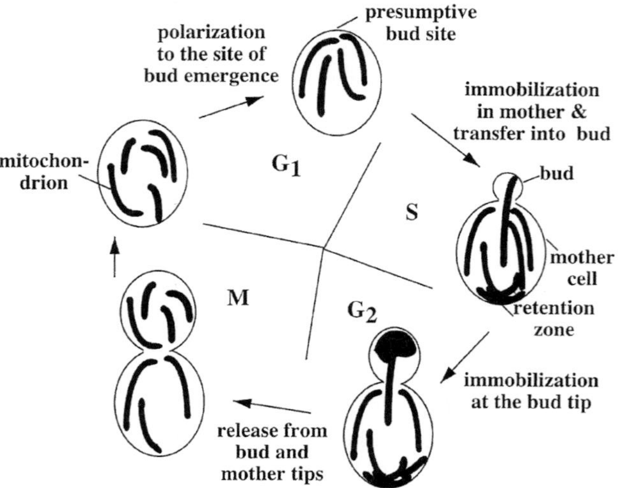

Fig. 1 The mitochondrial inheritance cycle.

to our understanding of the cell division cycle. This type of approach is also applicable to study of the mechanisms underlying enrichment of mitochondria at sites of ATP utilization and changes in the morphology and volume of mitochondria under different environmental conditions. This article describes methods for using mitochondria-specific vital dyes and optical microscopy to study mitochondrial movement and morphology in living cells. Our studies utilize the budding yeast, *Saccharomyces cerevisiae,* as a model system. However, these techniques are readily applicable to the study of mitochondrial dynamics in other eukaryotes.

II. Yeast Cell Growth and Synchronization

A. Growth Media and Conditions

Mitochondrial structure and morphology are known to depend on metabolic and respiratory activity in *S. cerevisiae,* which can perform either fermentation or respiration (or both). Glucose represses mitochondrial biosynthesis, reducing the number of mitochondria up to 90% in some genetic backgrounds. Nonfermentable substrates, such as glycerol, ethanol, or acetone, sometimes cause fragmentation of mitochondria under aerobic conditions (Pon and Schatz, 1991; Visser, 1995). In addition to carbon source, the growth phase of a culture (i.e., cell density) affects mitochondrial morphology. In midlog phase cells, most mitochondria maintain a long, tubular structure (Pon and Schatz, 1991). As the yeast cells stop dividing and enter stationary phase, mitochondria fragment into smaller units.

The long tubular structure in control cells allows us to recognize, for example, the fragmented, clumped, or spherical mitochondria appearing upon treatment with the actin-depolymerizing drug, Lat-A, or in actin cytoskeleton mutants (Sogo and Yaffe, 1994; Boldogh *et al.,* 1998). To maintain long, tubular mitochondria that are abundant and therefore well suited to visualization, we recommend growth of yeast to midlog phase using nonrepressing media. Lactate-based liquid media are the media of choice for yeast cells containing fully functional mitochondria. It is easy to prepare and inexpensive. For cells bearing defects in mitochondrial function that cannot grow on nonfermentable carbon sources, we recommend raffinose-based media. If abundance of the organelle is not an issue, cells can be grown in rich, glucose-based media (YPD).

Media for Yeast Growth

Lactate growth medium (for 1 liter): 3 g yeast extract, 0.5 g Glucose, 0.5 g $CaCl_2$, 0.5 g NaCl, 0.6 g $MgCl_2$, 1 g KH_2PO_4, 1 g NH_4Cl, 22 ml 90% lactic acid, and 7.5 g NaOH

Optional: For growth of *ade2*$^-$ cells, add 0.1 mg/ml adenine.

Dissolve ingredients in 900 ml distilled H_2O. Adjust pH to 5.5 with NaOH. Bring volume to 1 liter with distilled H_2O. Dispense into flasks so that the volume of media is one-fifth the volume of the flask (e.g., place 50 ml media in a 250-ml flask). Stopper the flasks with a cotton plug wrapped in cheesecloth and cover top of flask with aluminum foil. Autoclave.

Raffinose growth medium (for 1 liter): 10 g yeast extract, 20 g Bacto-peptone, and 20 g raffinose

Optional: For growth of *ade2⁻* cells, add 0.1 mg/ml adenine.

Dissolve ingredients in 900 ml distilled H_2O. Bring volume to 1 liter with distilled H_2O. Dispense into flasks so that the volume of media is one-fifth the volume of the flask. Stopper the flasks with a cotton plug wrapped in cheesecloth and cover top of flask with aluminum foil. Autoclave.

Glucose growth medium (YPD) (for 1 liter): 10 g yeast extract, 20 g Bacto-peptone, and 20 g glucose

Optional: For growth of *ade2⁻* cells, add 0.1 mg/ml adenine.

Dissolve ingredients in 900 ml distilled H_2O. Bring volume to 1 liter with distilled H_2O. Dispense into flasks so that the volume of media is one-fifth the volume of the flask. Stopper the flasks with a cotton plug wrapped in cheesecloth and cover top of flask with aluminum foil. Autoclave.

B. Preparation of Synchronized Cells

Synchronization of cells in the G_1 phase is highly recommended for extended time-lapse imaging of mitochondrial movements. Patterns of mitochondrial motility can be characterized easily in a synchronized cell population because any given field will contain multiple cells at the same stage. We use the cell size fractionation method described. In brief, cells are applied to a step gradient. After a short centrifugation time, the cells are distributed through the gradient on the basis of size, with unbudded and small-budded cells near the top. Although the efficiency of synchrony is lower than for other synchronization methods, cell size fractionation using gradient centrifugation can be performed easily and rapidly without the need for specialized equipment and without affecting cellular structure.

Protocol for Cell Size Fractionation

1. Grow yeast cells in a shaking incubator to midlog phase in 25–100 ml of liquid medium.

2. In a 30-ml centrifuge tube, prepare a 60–100% Percoll step gradient as follows, using sterile solutions.

Step 1 (100% Percoll): 2.5 ml Percoll

Step 2 (90% Percoll): 6.75 ml Percoll and 0.75 ml 10X media

Step 3 (75% Percoll): 5.63 ml Percoll, 1.5 ml 5X media, and 0.37 ml distilled H_2O

Step 4 (60% Percoll): 4.5 ml Percoll, 1.5 ml 5X media, and 1.5 ml distilled H_2O

3. Collect cells by centrifugation (13,000*g*) for 10 s at room temperature. Remove all but ~1 ml of the supernatant and store it to be used as conditioned medium after

fractionation. Vortex the concentrated residual cells (in the remaining 1 ml of medium) for 20 s to break up aggregates.

4. Apply the cell suspension to the top of the Percoll gradient. Centrifuge at 1000*g* for 1 min. (Optimal centrifugation speed and time are variable depending on the strains used.)

5. Collect the top 1–10% of the gradient. Resuspend in the conditioned medium and return to the shaking incubator for 30 min to allow the cells to recover.

6. The resulting cell suspension contains up to 70–80% unbudded and small budded cells. Aliquots of this synchronized culture can be stained, if necessary, and used for long-term time-lapse video microscopy in a growth chamber. If desired, the culture may be returned to the incubator and samples removed at various time points to examine mitochondrial dynamics throughout the cell cycle.

Cell size fractionation can also be achieved by elutriation, a method that separates cells by using centrifugal force and counterflow as opposing forces and generates a highly synchronized cell population (Creanor and Mitchison, 1979). Among other techniques for cell synchronization are the use of temperature-sensitive *cdc* mutants (Osley *et al.,* 1986) and treatment with the pheromone α factor (Wilkinson and Pringle, 1974). Hydroxyurea (HU) is also able to induce transient cell cycle arrest (Simchem *et al.,* 1976). Although these methods result in better synchrony than the gradient method, each bears its own limitations. For example, centrifugal elutriation requires specialized equipment that is expensive to set up. Pheromone treatment cannot be used for diploid cells, and shmoos produced during the synchronization are not ideal for the purpose of long-term analysis of mitochondrial motility. Moreover, chemical-induced cell cycle arrest is nonphysiological and potentially toxic. HU inhibits DNA replication and can cause unbalanced cell growth with longer incubation, an effect that may be detrimental to the characterization of mitochondrial movement throughout the cell cycle.

III. Vital Staining of Yeast Mitochondria

Yeast mitochondria are not clearly visible by bright-field, phase-contrast, or differential interference contrast/Nomarski microscopy. Therefore, specific labeling is essential for observing mitochondria in living yeast. Methods for targeting fluorescent proteins such as GFP to yeast mitochondria are described in Chapter 17 of this volume. This section describes an alternative approach: vital staining.

The advantages of vital staining include a minimal time investment and the ability to demonstrate functional properties of the organelle, such as membrane potential or DNA content. This technique also offers a different choice of fluorophores and is independent of the ability of the cell to express and import foreign proteins.

Staining mitochondria in living yeast is a simple and rapid procedure. In most cases, cells can be collected, washed, stained, and mounted on a microscope slide for viewing within 20 min. However, some extra time spent on optimizing staining conditions will yield greatly improved results. The dye concentrations suggested in Table I should be

Table I
Vital Dyes for Yeast Mitochondria

Compound	Target	Maximum excitation/ emmission (nm)	Suggested working concentration	Reference
DAPI[a]	DNA	358/461	0.1 μg/ml (300 n*M*)	McConnell *et al.* (1990)
SYTO18[a,b]	DNA	468/533	7 μg/ml (10 μ*M*)	Haugland (1999)
$DiOC_6(3)$[b]	Negative membrane potential	484/501	0.1 μg/ml (175 n*M*)	Koning *et al.* (1993), Simon *et al.* (1995)
Rhodamine 123[b]	Negative membrane potential	505/534	10–20 μg/ml (30–50 μ*M*)	Haugland (1999)
Rhodamine B hexyl ester	Negative membrane potential	555/579	0.06 μg/ml (100 n*M*)	Haugland (1999)
DASPMI[b,d]	Negative membrane potential	475/605	25 μg/ml (70 μ*M*)	McConnell *et al.* (1990)
MitoTracker[b,c,e]	Negative membrane potential	Several available	25–500 n*M*	Haugland (1999)

[a] Can be used in fixed cells.
[b] Can be viewed with fluorescein filter sets.
[c] Can be viewed with rhodamine filter sets.
[d] Excitation and emission wavelengths are lower under physiological conditions (fluorescein filter set is adequate for viewing).
[e] Stained living cells may be fixed with paraformaldehyde and permeabilized with cold acetone.

considered only as a starting point, as we have found several sources of variability in optimal staining conditions. Different concentrations of dye or duration of staining may be required for some yeast strains or growth conditions; there can also be variation between batches of a particular dye.

There are two main classes of mitochondrial vital dyes for yeast: DNA-binding compounds and lipophilic dyes. These are discussed in detail in the next two sections.

A. DNA-Binding Dyes

1. DAPI

The most common DNA-binding dye used in yeast is the blue-fluorescing DAPI (4′,6′-diamidino-2-phenylindole). DAPI binds in the minor groove in AT-rich regions of double-stranded helical DNA, and also binds to RNA with lower affinity. Upon binding to nucleic acids, DAPI fluorescence increases greatly, and the increase is more pronounced with DNA than with RNA. These characteristics make DAPI a strong nuclear and mitochondrial DNA marker with little cytoplasmic background staining. Another advantage is that it stains mtDNA regardless of the metabolic state of the mitochondria. Consequently, it can be used in cells whose mitochondrial function may be impaired, and can even be used in fixed cells. Because mitochondria are the only extranuclear organelles that contain DNA, DAPI staining structures are diagnostic for mitochondria in cases where the identity of an organelle is not clear from morphology or localization.

The main disadvantage of DAPI is that it stains nuclear as well as mitochondrial DNA so that mitochondria close to the nucleus will not be well resolved. A second disadvantage arises during prolonged visualization. Excitation of DAPI fluorescence requires ultraviolet light, which is toxic to cells; this toxicity is increased in the presence of the dye. In our experience, mitochondrial fragmentation or rupture can occur in DAPI-stained cells after only 1–2 min total illumination time under the microscope. To avoid this artifact, imaging of a particular field should be carried out promptly, without exposing the cells to excitation light for more than a minute. If cells have been illuminated too long, simply move to a new field of view.

Protocol for Staining Live Yeast Cells with DAPI

1. Grow yeast cells to midlog phase in liquid media. Synchronize cells, if desired.
2. Remove an aliquot of cells and concentrate by centrifugation at ~13,000*g* for 5 s. (*Note:* Extended centrifugation can cause breakage of tubular mitochondria.) Remove supernatant.
3. Resuspend cell pellet in 1× phosphate-buffered saline (PBS) or growth media. (As a general rule, PBS can be used unless cells must be observed for more than 30 min.) Centrifuge again as in step 2 and remove supernatant.
4. Add 1× PBS or growth media, as appropriate, to resuspend cells to a density of $\sim 2 \times 10^8$ cells/ml in a microcentrifuge tube. Read the optical density of a diluted sample at 600 nm in order to determine the correct volume. (One milliliter of a cell suspension with OD_{600} 1.0 contains approximately 10^7 cells.)
5. Add 1/1000 volume of the DAPI stock solution. Mix thoroughly. DAPI stock solution (1000×): Dilute commercial DAPI to 0.1 mg/ml (equivalent to 300 μ*M* for dihydrochloride form) in distilled H_2O. Store in aliquots at 20°C in the dark.
6. Wrap the tube in foil and incubate in the dark for 15 min at room temperature.
7. Pipette gently to resuspend cells.
8. Mount cells for short- or long-term observation as described in Section IV.

2. SYTO 18

SYTO 18 is a nucleic acid-binding dye, developed by Molecular Probes, Inc. (Eugene, OR), which is reported by the manufacturer to stain yeast mitochondria. No published references were available at the time of this writing, but it appears that SYTO 18, which can be viewed with fluorescein filter sets, may be a useful alternative to DAPI in cases where ultraviolet excitation is impossible or undesirable. See the manufacturer's web site (http://www.probes.com/) for information on staining protocols.

B. Membrane Potential-Sensing Dyes

Like DNA-binding dyes, membrane potential-sensing mitochondrial dyes are designed to take advantage of a quality unique to mitochondria: in this case, a strong membrane potential. Because functioning mitochondria contain an excess of negative charge inside the inner membrane, positively charged lipophilic molecules will accumulate in the mitochondrial membranes much more readily than in any other compartment. Lipophilic dyes stain the entire mitochondrial membrane and are therefore excellent for investigating mitochondrial distribution and morphology. They can also serve as functional markers to demonstrate the mitochondrial membrane potential.

Many potential-sensing mitochondrial dyes have been developed. This article focuses on several that are known to work well in yeast: $DiOC_6(3)$, DASPMI, rhodamine 123, rhodamine B hexyl ester, and the MitoTracker family. Selection of a suitable dye for a given application should be based on several factors. If double labeling is desired (with a fluorescent protein or another vital dye), the mitochondrial dye chosen must have nonoverlapping excitation and emission spectra. A dye should be tested on the strain of interest to find a concentration that provides sufficient sensitivity and specificity. Finally, the fluorescence must be stable enough to persist for the required observation time without cytotoxicity.

1. $DiOC_6(3)$

$DiOC_6(3)$, a green-fluorescing dye, was the primary tool in early studies of yeast mitochondrial motility (Simon *et al.,* 1995). It is a member of the carbocyanine class, which is characterized by high-intensity fluorescence and photostability (Haugland, 1999). A pair of alkyl chains contribute to the lipophilicity of this dye. $DiOC_6(3)$ concentrations ranging from 10 to 100 ng/ml (in a cell suspension of 1×10^7 cell/ml) do not affect mitochondrial morphology or function. With higher concentrations of $DiOC_6(3)$, however, specific mitochondrial staining is lost and organelle structure is altered. Figure 2 shows the staining pattern of $DiOC_6(3)$ at low and high concentrations in yeast cells. Mitochondria are clearly visible at 20 ng/ml (Fig. 2A). Specific mitochondrial staining was verified by costaining with the DNA-binding dye, DAPI. Under this condition, mitochondria remain detectable for over 30 min without additional $DiOC_6(3)$. Higher $DiOC_6(3)$ concentrations (500–1000 ng/ml) result in staining of additional structures, including the endoplasmic reticulum (ER) (Fig. 2B).

2. DASPMI

DASPMI, also called 4-Di-1-ASP (Molecular Probes) or 2-Di-1-ASP (Sigma Chemical Co., St. Louis, MO), is a lipophilic styryl dye used in studies of yeast mitochondrial morphology, including the screening of mutants (McConnell *et al.,* 1990). DASPMI (Bereiter-Hahn, 1976) produces both green and red fluorescence when excited with blue light from a fluorescein excitation filter or a 488-nm laser line. It is viewed most commonly with a fluorescein filter set, but has also been used in a double-label experiment with fluorescein (Li *et al.,* 1995). When doing such an experiment, it is important to examine single-labeled samples with both filter sets to ascertain the amount of crossover before drawing a conclusion from apparent colocalization.

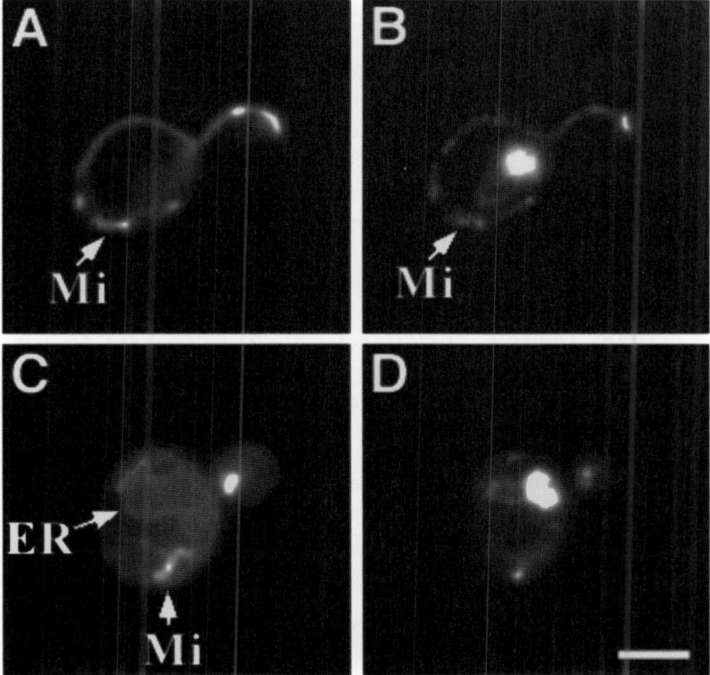

Fig. 2 $DiOC_6(3)$ staining pattern of *S. cerevisiae*. Mitochondria (Mi) in living yeast cells were stained with 20 (A) and 500 (B) ng/ml $DiOC_6(3)$. Chromosomal and mitochondrial DNA of cells in A and B were visualized by DAPI staining (C and D). Nuclear membrane-associated ER (ER) was stained by $DiOC_6(3)$ at high concentration (B and D). Bar: 2 μm.

3. Rhodamine 123

Rhodamine 123 is one of several rhodamine variants that have been used to label mitochondria. Compared to other derivatives, rhodamine 123 is relatively nontoxic. Despite its name, it is best viewed with filters for fluorescein rather than rhodamine.

4. Rhodamine B Hexyl Ester

Another rhodamine derivative, this dye can be viewed with rhodamine filter sets, making it a potential double-label reagent with GFP. It is more lipophilic than rhodamine 123 and consequently will stain the ER at high concentrations (Terasaki and Reese, 1992).

5. MitoTracker Dyes

The chief distinction of these proprietary dyes is that they remain associated with mitochondria after fixation with paraformaldehyde. This can be useful for confirming the mitochondrial localization of the dye by immunofluorescence or for colocalization with antigens of interest in fixed cells. However, MitoTracker-stained cells cannot be permeabilized with the commonly used detergent Triton X-100; cold acetone must be used instead, a somewhat harsher treatment. MitoTracker green is not retained even after acetone permeabilization. Hence, not all immunofluorescence protocols may be compatible with MitoTracker preservation.

MitoTracker dyes are available in a variety of colors (green, orange, and red). Standard fluorescein and rhodamine filter sets can be used to view the green and red variants, respectively; the orange variants can be viewed suboptimally with a rhodamine filter set, but are best viewed through a filter set designed for R-phycoerythrin or DiI.

Standard Protocol for Staining Live Cells with Membrane Potential-Sensing and Lipophilic Dyes

1. Prepare stock solution of the vital dye. All stocks should be stored in the dark.

$DiOC_6$ (5000–20,000× stock): 17.5 m*M* (10 mg/ml) in ethanol (for long-term storage) and (50–200× stock): 175 μ*M* (100 μg/ml) in ethanol (for addition to cell suspension)

DASPMI (100–200× stock): 1 mg/ml in ethanol

Rhodamine 123 (500–1000× stock): 25 m*M* (10 mg/ml) in dimethyl sulfoxide (DMSO)

Rhodamine B hexyl ester (10,000× stock): 1 m*M* (0.63 mg/ml) in DMSO

MitoTracker: Supplied as a solid to be resuspended in DMSO to make a 1 m*M* stock (2000–40,000×)

2. Grow yeast cells to midlog phase in liquid media. Synchronize cells as described earlier, if desired.

3. Remove an aliquot of cells and concentrate by centrifugation at ~13,000*g* for 5 s. (*Note:* Extended centrifugation can cause breakage of tubular mitochondria.) Remove supernatant.

4. Resuspend cell pellet in 1× PBS or growth media. (As a general rule, PBS can be used unless you need to observe the cells for more than 30 min.) Centrifuge again as in step 2 and remove supernatant.

5. Add 1× PBS or growth media (as appropriate) to resuspend cells to a density of $\sim 2 \times 10^8$ cells/ml. Read the optical density of a diluted sample at 600 nm in order to determine the correct volume. (One milliliter of a cell suspension with OD_{600} 1.0 contains approximately 10^7 cells.)

6. Add appropriate volume of dye stock solution. (DMSO is toxic and may also inhibit partitioning of the dye into the aqueous environment of the cells. For dyes dissolved in DMSO, prepare a stock concentration that is at least 100× so that the volume of DMSO added will be minimal.) Mix thoroughly.

7. Wrap the tube in foil and incubate in the dark for 15–30 min at room temperature.

8. Pipette gently to resuspend cells.

9. Mount cells for short-term or long-term observation as described in Section IV.

IV. Preparation of Growth Chamber

To immobilize cells and maintain cell viability during long-term observation, a growth chamber is needed. Cells mounted directly on a slide and observed for longer than ~10 min will either dry out or suffer nutrient starvation. The growth chamber described here prevents these problems. In addition, it immobilizes the cells on an agarose bed, while remaining transparent and thin enough for observation with oil-immersion lenses. The agarose causes minimal autofluorescence in the green channel used for GFP, $DiOC_6(3)$, and other mitochondrial dyes. An oxygen scavenger system (Oxyrase) is added to reduce toxicity due to illumination. This protocol is a modification of the method described by Koning *et al.* (1993).

1. Place a double layer of cellophane tape strips on a glass microscope slide. Using a razor, cut a 1-cm^2 window in the middle of the tape. Peel the remaining tape carefully from the slide and place on a new, clean glass slide. This tape well forms the sides of the chamber.
2. Prepare growth media containing 3% high-melting point agarose (Sigma) and heat until the agarose dissolves. Add 1 μg/ml of $DiOC_6(3)$ (or suitable amount of any other vital dye) and Oxyrase at a concentration recommended by the manufacturer (Oxyrase Inc., Mansfield, OH). Fill the tape well with this heated solution.
3. Place a coverslip over the filled well and apply uniform force to the coverslip to compress the agarose into the well. Once the agarose has solidified, remove the coverslip and excess agarose from the well.
4. Take an aliquot of cells (synchronized if desired) and concentrate by centrifugation for 5 s at 13,000*g* (*Note:* Extended centrifugation can cause breakage of tubular mitochondria.) Resuspend in media to a concentration of 2×10^8 cells/ml. Spread 3 μl of this concentrated cell suspension ($\sim 6 \times 10^5$ cells) across the surface of the solidified agarose and incubate for a few minutes so that the cells and media sink into the agarose.
5. Cover the chamber with a coverslip and seal with clear nail polish. Preparation of fresh chambers prior to each recording is highly recommended to prevent drying of the agarose.

This growth chamber supports cell growth for a period up to 5 h at a rate similar to that of cell growth in liquid culture. Oxyrase-dependent oxygen depletion has little or no effect on cell growth during the period of visualization (up to 150 min); the length of the cell division cycle for wild-type cells in YPD-agarose under these conditions, approximately 90 ± 10 min, is not significantly affected by the addition of Oxyrase.

Use of the growth chamber is recommended to maintain cell growth and mitochondrial function during a long period of observation. However, the growth chamber can be replaced by only a microscope slide glass and a coverslip if mitochondria are observed within 10 min. Such short time-lapse recordings can give useful data on velocity and direction of movement if the time interval between images is also short. Although the analysis of motility during a short period of time cannot reveal patterns of mitochondrial behavior during the entire cell cycle, it is still useful for the investigation of the mechanisms of mitochondrial distribution and movement. Specific characteristics of the movement, such as its dependence on the actin cytoskeleton, were elucidated by the analysis of short-term recordings. Recordings of mitochondrial movements in Latrunculin-A (Lat-A)-treated cells and in cells bearing temperature-sensitive mutations in actin or actin-binding proteins provided evidence for the requirement of an intact actin cytoskeleton for the control of mitochondrial organization and movement (Simon *et al.*, 1995; Smith *et al.*, 1995; Boldogh *et al.*, 1998). A protocol for the preparation of yeast and staining of mitochondria for short-term visualization is described.

1. Grow yeast cells to midlog phase in liquid medium. Concentrate cells by centrifugation (10 s at 13,000*g*) and resuspend in YPD to a cell density of 1×10^7 cells/ml.
2. Incubate the resuspended cells with 20 ng/ml of $DiOC_6(3)$ (for stock solution, see Section III) for 5 min in the dark at room temperature. Wash the stained cells once with media as described in step 1 to remove excess dye. Resuspend to a final cell density of 2×10^8 cells/ml in media.
3. Drop 1.8–2.2 μl of this denser culture on a microscope slide and cover with a 22 × 22-mm coverslip for visualization by fluorescence microscopy. The volume used is important because excess volume can cause cells to float, and too little volume can affect cell structure. Avoid creating bubbles between slide and coverslip. Do not seal; view immediately.

For visualization of GFP-tagged mitochondria, grow yeast cells expressing the CS1-GFP fusion (see Chapter 17) to midlog phase in selective media to prevent loss of the plasmid. Cell concentration is less critical when vital dyes are not used. Therefore, one can simply take 1 ml of a midlog phase culture, concentrate by centrifugation (5 s at 13,000*g*; extended centrifugation can cause breakage of tubular mitochondria), and resuspend in 5–10 μl of the selective media to make a concentrated cell suspension. Place cells in a growth chamber for long-term observation or on a slide as described earlier for short-term viewing.

V. Image Acquisition

A. Equipment

Selection of an appropriate microscope and camera to detect fluorescence is essential for time-lapse recordings of mitochondrial movements. The small size of yeast cells requires a microscope equipped with a high-magnification objective lens and a high-resolution camera. Phototoxicity and photobleaching can be reduced greatly by image acquisition with an ultrasensitive digital camera that can detect fluorescence with relatively weak and short illumination. Data processing speed and memory capacity of the computer linked to the digital camera occasionally limit imaging speed and therefore should be tested thoroughly before committing to a particular imaging system.

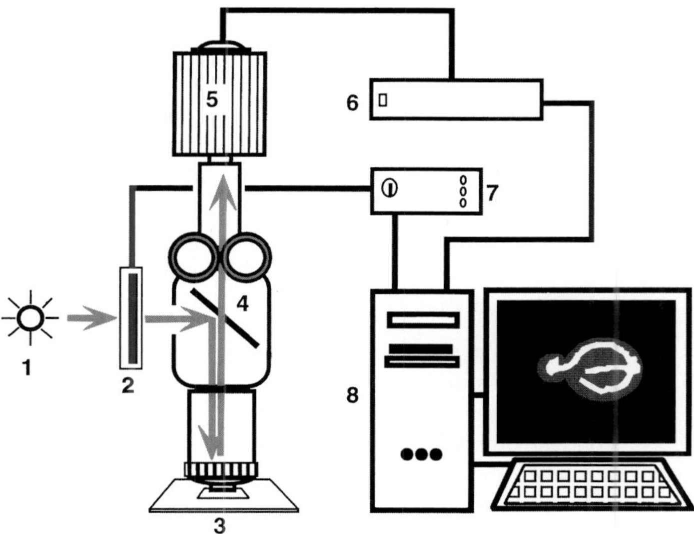

Fig. 3 A schematic diagram of an imaging system for tracking mitochondrial motility. 1, mercury arc lamp for excitation of fluorophore; 2, magnetic shutter controlling excitation light; 3, growth chamber or sample slide; 4, epifluorescence microscope; 5, CCD (charge-coupled device) camera; 6, CCD camera controller; 7, shutter driver; and 8, personal computer connected to camera controller and shutter driver. The shutter for excitation light and the electronic camera shutter are controlled and synchronized by the computer.

Figure 3 shows a schematic diagram of an imaging system used for our mitochondrial motility assay. In this setup, images are taken with a Zeiss Axioplan II microscope (Carl Zeiss, Oberkochen, Germany) using a Plan-Apochromat 100× 1.4 NA objective lens, and a Orca1 digital camera (C4742-95, Hamamatsu, Bridgewater, NJ) with a 2× projection tube (Diagnostic Instruments, Sterling Heights, MI). Fluorescence of $DiOC_6(3)$ or GFP is viewed with excitation and emission wavelengths of 460–500 and 510–560 nm, respectively. A shutter driver (Uniblitz D122, Vincent Associates, Rochester, NY) is synchronized with the camera shutter to control excitation light from the 100-W mercury arc lamp. A software package (IPLab, Scanalytics, Inc., Fairfax, VA) is used to control the camera and shutter, capture images at defined time intervals, and export them to TIFF format for further analysis.

B. Obtaining the Best Image for a Given Application

Phototoxicity of excitation light, photobleaching of fluorescence, and quality of the image are some of the factors that must be considered in a serial image acquisition protocol. These factors can be controlled by adjusting the exposure time and the time interval between acquisitions, as well as the light intensity used for excitation, and the image resolution of the system. Specific criteria can be used to rule out the deleterious effect of any of these variables.

1. Intact mitochondrial structure. Intense and long excitation can cause cleavage or fragmentation of these organelles.

2. Mitochondrial movements. Behavior of photodamaged mitochondria has not been fully investigated; however, photodamage should be suspected if mitochondria show extremely low velocity of movement in a control sample (wild-type cells). Usually, alteration of the organelle structure accompanies changes in dynamics.

3. Mitochondria remain detectable and can be resolved throughout the entire recording time. Severe photobleaching is undesirable and can be avoided by using brief, low-intensity excitation.

We suggest a flow chart (Fig. 4) to help determine the optimum conditions for serial image acquisition.

Sometimes the imaging conditions required by the cells produce images with low contrast or brightness, or with distracting noise. To counteract this, most currently available image acquisition and processing software allows both automatic and manual adjustment of image display properties such as brightness, contrast, and gamma (the relationship between the actual intensity values measured by the camera and the intensity displayed on the monitor or in a print). Such adjustment can increase the ratio of visible signal to noise, which is helpful for restoring photobleached images. However, excessive enhancement should be avoided, because it can introduce artifacts that affect the measurement of mitochondrial motility. In addition, any quantitation of the fluorescence intensity must precede enhancement or take into account all changes made to the images.

VI. Analysis and Quantitation of Movement

Velocities of mitochondrial movement are determined from time-lapse series of images obtained with 10- to 20-s intervals over 10 min of real time. The velocity of mitochondrial movement can be determined by tracking the change in position of individual mitochondria in successive time-lapse frames. This process depends on consistent criteria for identifying both individual mitochondria and meaningful movement. As described earlier, mitochondria usually form an extensive tubular reticulum. Often, different parts of the same organelle are seen to move in different directions or at different times, suggesting that a single mitochondrion can be coupled to more than one motility-generating complex. Consequently, most velocity measurements are made by tracking the leading tip or trailing tail of a mitochondrial tubule because these parts of the mitochondrion are most clearly recognizable.

When measuring movements over nanometer distances, distinguishing between random and active, directed movement is important. For this purpose, we define motile mitochondria as particles (tips, tails, or whole mitochondria) that display linear movement of at least one pixel for three consecutive still frames, or 30–60 s. All other movements are disregarded.

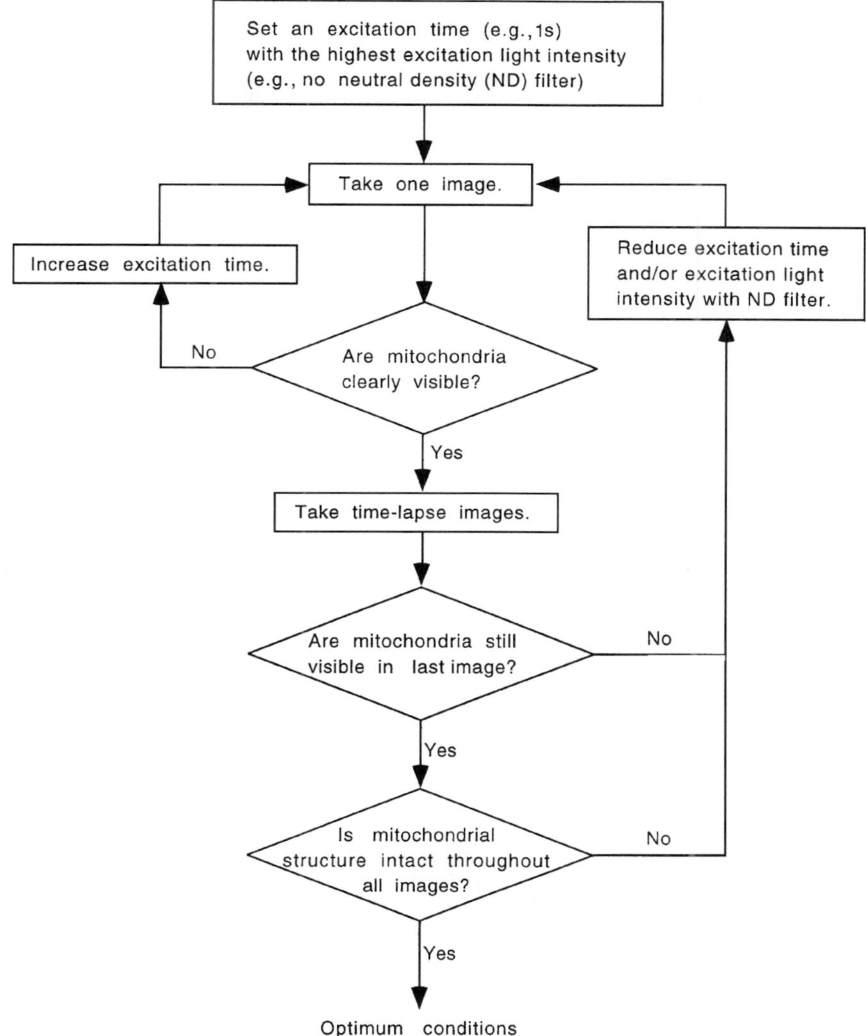

Fig. 4 A flow chart for setting optimum conditions for fluorescence image acquisition.

The free program NIH Image V1.55 can be used to determine the positions (*x–y* coordinates) of moving mitochondria in a series of time-lapse frames. The linear distances between successive positions of mitochondria are calculated using Pythagoras' theorem in a Microsoft Excel spreadsheet, and the instantaneous velocities are averaged to give the mean velocity over the entire period of movement. Measured velocities in wild-type cells vary from 20 to 50 nm/s with a standard deviation of 5–20 nm/s (Simon *et al.,* 1995, 1997; Boldogh *et al.,* 1998).

For analysis of mitochondrial motility as a function of progression through the cell division cycle, time-lapse series are obtained using 2.5- to 5-min intervals for periods up to 120 min. These long-term imaging studies reveal the patterns of mitochondrial motility throughout the entire cell cycle. Organelles move into and out of the plane of focus during 2.5- to 5-min intervals, a condition that precludes tracking of either the direction or the velocity of mitochondrial movement.

Mitochondrial movements in budding cells are documented in Fig. 5. The tip of a mitochondrial tubule moving toward the bud is marked in each time-lapse image. Black dots representing the mitochondrial tip become progressively distant from the original location (the arrow) over time, demonstrating linear and polarized movement.

Figure 6 compares a representative tracing of mitochondrial movement in a control yeast cell to one from a cell treated with Lat-A, a G-actin-binding agent. Lat-A blocks actin polymerization and results in loss of all detectable F-actin in yeast within 2 min of treatment. Mitochondrial tips in the time-lapse images are marked as in Fig. 5, and the dots are connected by a line to track the movement. Mitochondrial movements in untreated cells are linear and directed toward the bud (Fig. 6A). In Lat-A-treated cells in which F-actin disappeared (Fig 6F), mitochondria are fragmented and clumped (Fig. 6D) and show nonlinear and nonpolarized movement with a low velocity (Fig. 6B). The characteristics of mitochondrial movements in Lat-A-treated cells directly support previous findings that mitochondrial movement in budding yeast is dependent on intact actin structure (Simon *et al.*, 1997; Boldogh *et al.*, 1998).

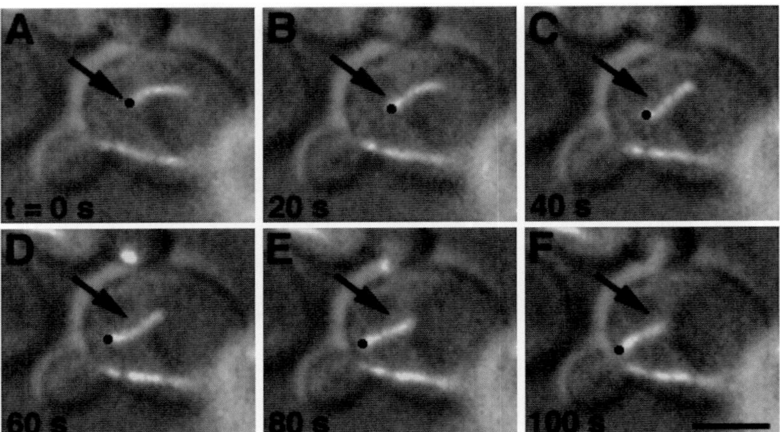

Fig. 5 Time-lapse images of mitochondrial movement in a budding wild-type *S. cerevisiae* cell (DDY 1495). Mitochondria were labeled with GFP using the CS1-GFP fusion; fluorescence images were captured at 20-s intervals. Fluorescence images are superimposed on a DIC image of the cell. One tip of a mitochondrion (shown as a black dot in each frame) is moving toward the bud, and the translocation of the tip from the original site (arrow) is illustrated by the distance between the dot and the arrow (F). Bar: 2 μm.

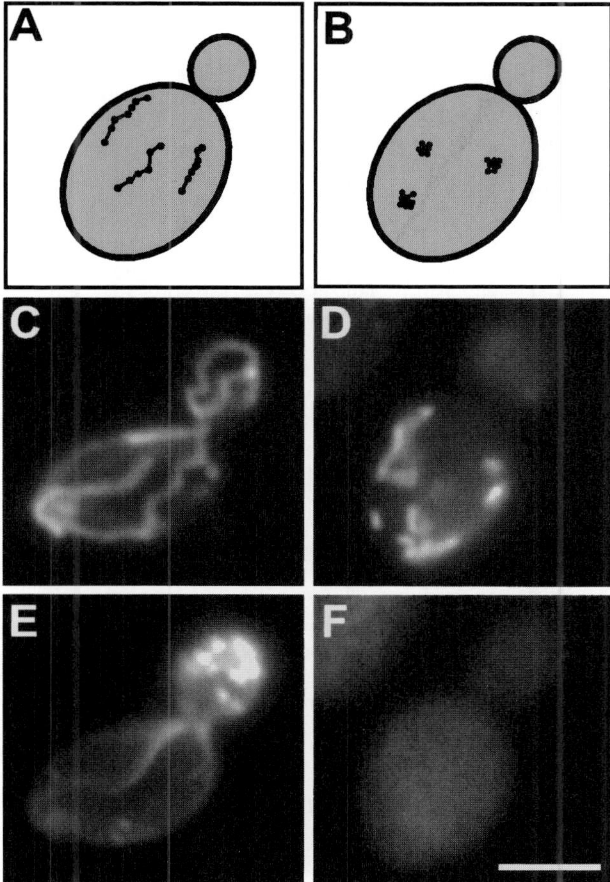

Fig. 6 Representative paths of GFP-labeled mitochondria in control (A) and Lat-A-treated (B) yeast cells (DDY 1495). Cells were incubated in the absence or presence of 0.4 m*M* Lat-A for 2 min before motility analysis. Tracings of the movements of individual mitochondria were made by marking the tip of motile organelles during the time in which they remained in the plane of focus. Points represent the position of the organelle at 20-s intervals. (C and E) Mitochondria and F-actin structure, respectively, in control cells. (D and F) Corresponding structures in Lat-A-treated cells are shown. Bar: 2 μm.

References

Bereiter-Hahn, J. (1976). Dimethylaminostyrylmethylpyridiniumiodine (DASPMI) as a fluorescent probe for mitochondria in situ. *Biochim. Biophys. Acta* **423,** 1–14.

Boldogh, I., Vojtov, N., Karmon, S., and Pon, L. A. (1998). Interaction between mitochondria and the actin cytoskeleton in budding yeast requires two integral mitochondrial outer membrane proteins, Mmm1p and Mdm10p. *J. Cell Biol.* **141,** 1371–1381.

Burgess, S. M., Delannoy, M., and Jensen, R. E. (1994). *MMM1* encodes a mitochondrial outer membrane protein essential for establishing and maintaining the structure of yeast mitochondria. *J. Cell Biol.* **126,** 1375–1391.

Chen, L. B. (1989). Fluorescent labeling of mitochondria. *In* "Methods in Cell Biology" (Y.-L. Wang and D. L. Taylor, eds.), Vol. 29. Academic Press, San Diego.

Creanor, J., and Mitchison, J. M. (1979). Reduction of perturbations in leucine incorporation in synchronous cultures of *Schizosaccharomyces pombe. J. Gen. Microbiol.* **112,** 385–388.

Drubin, D. G., Miller, K. G., and Wertman, K. F. (1993). Actin structure and function: Roles in mitochondrial organization and morphogenesis in budding yeast and identification of the phalloidin binding site. *Mol. Biol. Cell.* **4,** 1277–1294.

Glick, B. S., and Pon., L. A. (1995). Isolation of highly purified mitochondria from *Saccharomyces cerevisiae. Methods Enzymol.* **260,** 213–223.

Haugland, R. P. (1999). "Handbook of Fluorescent Probes and Research Chemicals." Molecular Probes, Eugene, OR.

Hoffmann, H. P., and Avers, C. J. (1973). Mitochondria of yeast: Ultrastructural evidence for one giant, branched organelle per cell. *Science* **181,** 749–751.

Koning, A. J., Lum, P. Y., Williams, J. M., and Wright, R. (1993). $DiOC_6$ staining reveals organelle structure and dynamics in living yeast cells. *Cell Motil. Cytoskel.* **25,** 111–128.

Lazzarino, D. A., Boldogh, I., Smith, M. G., Rosand, J., and Pon, L. A. (1994). ATP-sensitive, reversible actin binding activity in isolated yeast mitochondria. *Mol. Biol. Cell.* **5,** 807–818.

Li, J., Eygensteyn, J., Lock, R. A. C., Verbost, P. M., Van Der Heijden, A. J. H., Bonga, S. E. W., and Flik, G. (1995). Branchial chloride cells in larvae and juveniles of freshwater tilapia *Oreochromis mossambicus. J. Exp. Biol.* **198,** 2177–2184.

McConnell, S. J., Stewart, L. C., Talin, A., and Yaffe, M. P. (1990). Temperature-sensitive yeast mutants defective in mitochondrial inheritance. *J Cell Biol.* **111,** 967–976.

Mitchison, J. M. (1988). Synchronous cultures and age fractionation. *In* "Yeast: A Practical Approach" (I. Campbell and J. H. Duffus, eds.), pp. 51–63. IRL Press, Oxford.

Mulholland, J., Preuss, D., Moon, A., Wong, A., Drubin, D., and Bostein, D. (1994). Ultrastructure of the yeast actin cytoskeleton and its association with the plasma membrane. *J. Cell Biol.* **125,** 381–391.

Osley, M. A., Gould, J., Kim, S., Kane, M., and Hereford, L. (1986). Identification of sequences in a yeast histone promoter involved in periodic transcription. *Cell* **45,** 537–544.

Pon, L., and Schatz, G. (1991). Biogenesis of yeast mitochondria. *In* "The Molecular and Cellular Biology of the Yeast *Saccharomyces:* Genome Dynamics, Protein Synthesis, and Energetics" (J. R. Broach, J. R. Pringle, and E. W. Jones, eds.), pp. 333–406. Cold Spring Harbor Laboratory Press, Cold Spring Harbor, NY.

Simon, V. R., Karmon, S. L., and Pon, L. A. (1997). Mitochondrial inheritance: Cell cycle and actin cable dependence of polarized mitochondrial movements in *Saccharomyces cerevisiae. Cell Motil. Cytoskel.* **37,** 199–210.

Simon, V. R., Swayne, T. C., and Pon, L. A. (1995). Actin-dependent mitochondrial motility in mitotic yeast and cell-free systems: Identification of a motor activity on the mitochondrial surface. *J. Cell Biol.* **130,** 345–354.

Simchen, G., Idar, D., and Kassir, Y. (1976). Recombination of hydroxyurea inhibition of DNA synthesis in yeast meiosis. *Mol. Gen. Genet.* **144,** 21–27.

Smith, M. G., Simon, V. R., O'Sullivan, H., and Pon, L. A. (1995). Organelle-cytoskeletal interactions: Actin mutations inhibit meiosis-dependent mitochondrial rearrangement in the budding yeast *Saccharomyces cerevisiae. Mol. Biol. Cell* **6,** 1381–1396.

Sogo, L. F., and Yaffe, M. P. (1994). Regulation of mitochondrial morphology and inheritance by Mdm 10p, a protein of the mitochondrial outer membrane. *J. Cell. Biol.* **126,** 1361–1373.

Stevens, B. (1977). Variation in number and volume of the mitochondria in yeast according to growth conditions: A study based on serial sectioning and computer graphics reconstitution. *Biol. Cell.* **28,** 37–56.

Stevens, B. (1981). Mitochondrial structure. *In* "The Molecular Biology of the Yeast *Saccharomyces:* Life

Cycle and Inheritance" (J. N. Strathern, E. W. Jones, and J. R. Broach, eds.), pp 471–488. Cold Spring Harbor Laboratory Press, Cold Spring Harbor, NY.

Terasaki, M., and Reese, T. S. (1992). Characterization of endoplasmic reticulum by colocalization of BIP and dicarbocyanine dyes. *J. Cell Sci.* **101,** 315–322.

Visser, W., van Spronsen, E. A, Nanninga, N., Pronk, J., T., Kuenen, J. G., and van Dijken, J. P. (1995). Effects of growth conditions on mitochondrial morphology in *Saccharomyces cerevisiae*. *Antonie van Leeuwenhoek* **67,** 243–253.

Wilkinson, L. E., and Pringle, J. R. (1974). Transient G1 arrest of *S. cerevisiae* of mating type alpha by a factor produced by cell of mating type a. *Exp. Cell Res.* **89,** 175–87.

Yang, H.-C., Palazzo, A., Swayne, T. C., and Pon, L. A. (1999). A retention mechanism for distribution of mitochondria during cell division in budding yeast. *Curr Biol.* **9,** 1111–1114.

CHAPTER 20

Targeting of Reporter Molecules to Mitochondria to Measure Calcium, ATP, and pH

Anna M. Porcelli,[*] Paolo Pinton,[†] Edward K. Ainscow,[‡] Anna Chiesa,[§] Michela Rugolo,[*] Guy A. Rutter,[‡] and Rosario Rizzuto[§]

[*] Department of Biology
University of Bologna
40126 Bologna, Italy

[†] Department of Biomedical Sciences and C.N.R.
Center for the Study of Biomembranes
35121 Padova, Italy

[‡] Department of Biochemistry
University of Bristol
Bristol, BS8 1TD United Kingdom

[§] Department of Experimental and Diagnostic Medicine
Section of General Pathology
44100 Ferrara, Italy

I. Introduction

The study of isolated mitochondria, dating back to the 1960s, provided information on the biochemical routes that couple oxidation of substrates to the production of ATP. In this work, new concepts (such as the chemiosmotic mechanism of energy conservation, the import of proteins into mitochondria, and the existence of a mitochondrial genome with a different genetic code) became established dogmas of modern biology. At the same time, the availability of efficient probes and imaging systems allowed cell biologists to study mitochondria in living cells. These studies revealed that a variety of extracellular stimuli cause a rise in intracellular Ca^{2+} concentration of high spatiotemporal complexity, that in turn is decoded by intracellular effectors. Among these effectors are mitochondria that are endowed with a low-affinity transport system for Ca^{2+} and respond to microdomains of high Ca^{2+} generated in proximity to Ca^{2+} channels.

Recombinant reporter proteins containing mitochondrial targeting sequences are emerging as the tools of choice to study mitochondria in living cells. This chapter describes the development and use of protein chimeras targeted to either the mitochondrial matrix or intermembrane space. Aequorin, the first of the targeted recombinant probes, is a photoprotein of jellyfish of the genus *Aequorea* that emits light upon binding of Ca^{2+} to three high-affinity-binding sites. The second protein probe described in this chapter is luciferase. We discuss how the measurement of luciferase light emission *in situ* allows one to monitor the dynamics of ATP concentration changes in living cells. Finally, we describe use of variants of green fluorescent protein (GFP) as pH probes. GFP fluorescence is intense and resistant to photobleaching. Moreover, spectral variants of GFP have been developed for double or triple label and/or energy transfer experiments. As a result, GFP has emerged as the most widely employed probe for cell biologists and is used routinely to study protein sorting and trafficking, gene expression, and organelle structure and motility. For some applications, the pH sensitivity of GFP is a disadvantage. We describe how this property can be exploited to probe pH of mitochondria in living cells.

A. Targeting Strategy

Targeting a heterologous protein to mitochondria is now fairly straightforward. Indeed, apart from the 13 polypeptides encoded by mitochondrial DNA in humans, all other mitochondrial proteins are encoded by nuclear genes. The signal peptides and routes allowing import of nuclear-encoded proteins into mitochondria have been investigated and clarified extensively in the past years. Therefore, while we describe targeting strategies for construction of our chimeric probes, it is obvious that similar, alternative approaches can be employed for developing novel chimeras.

1. Mitochondrial Matrix

To target heterologous reporter proteins to the mitochondrial matrix, we added at their N terminus the targeting sequence of the smallest subunit of human cytochrome *c* oxidase, COX8. This 44-amino acid (aa) COX subunit is encoded as a larger polypeptide, with a 25-aa-long N-terminal extension. This extension has the hallmarks of "mitochondrial presequences:" it is rich in basic and hydroxylated residues and devoid of acidic ones. In addition, it is recognized by the import machinery, triggering the transfer of the polypeptide to the matrix, and is then cleaved by resident proteases.

Figure 1A shows a generic map of a matrix-targeted recombinant probe. In all cases, an epitope tag is added to the coding region of the reporter protein, which allows for immunolocalization of the transfected protein. Indeed, our experience has been that matrix-targeted constructs are imported to the expected location, but it is always prudent, particularly with photoproteins (that cannot be directly visualized under the microscope), to confirm the appropriate sorting. In the chimeric construct, the cDNA encoding the epitope-tagged heterologous protein is fused in-frame downstream of a 150-bp fragment of the COX8 cDNA that encodes the first 33 aa of the mitochondrial precursor protein. The complete chimeric cDNA encodes, from the N to the C terminus: the 25-aa mitochondrial presequence, 8 aa of the mature COX8 protein, the 9-aa-long HA1 (hemagglutinin) tag, and the reporter protein of interest (aequorin, luciferase, or GFP mutant). Figure 1B shows the immunolocalization of one of these constructs (mitochondrially targeted aequorin, mtAEQ). The typical rod-like appearance of mitochondria can be easily appreciated.

2. Intermembrane Space (IMS)

The other target for novel probes was the outer surface of the ion-impermeable inner mitochondrial membrane. Indeed, although in equilibrium with the cytosol through the ion-permeable outer membrane, this region will contain local domains that differ in ion concentration from the bulk cytosol. In order to obtain the proper sorting and orientation across the inner mitochondrial membrane, we used the targeting sequence of glycerol phosphate dehydrogenase, an integral protein of the inner mitochondrial membrane with a large C-terminal portion protruding into the IMS. In chimeras destined for the intermembrane space, the reporter protein was fused to this C-terminal region (Fig. 2A). The immunolocalization shows that the protein chimeras are properly transported to the mitochondria (Fig. 2B). Localization of the chimera in the IMS cannot be determined by immunofluorescence, but can be deduced from the measurements carried out with the probe. In particular, the [Ca^{2+}] changes revealed by the IMS-targeted chimera are not compatible with a matrix localization of the probe.

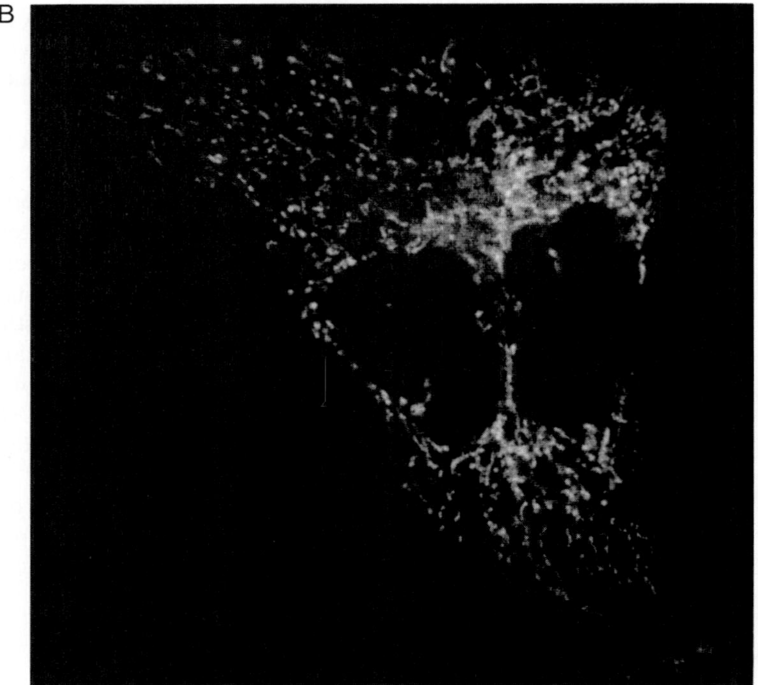

Fig. 1 (A) Generic map of a matrix-targeted recombinant probe. (B) Immunolocalization of mitochondrially targeted aequorin. Staining with a monoclonal antibody recognizing the hemagglutinin epitope (HA1) tag was detected with a TRITC-conjugated secondary antibody. The image, acquired on an inverted epifluorescence microscope, was captured with a back-illuminated CCD camera (Princeton Instruments) using the Metamorph software (Universal Imaging).

Rather, they are the response expected from a region in rapid equilibrium with the cytosol, but also sensing local domains of high [Ca^{2+}] generated at the surface of mitochondria by the release of Ca^{2+} from neighboring endoplasmic reticulum (ER) cisternae.

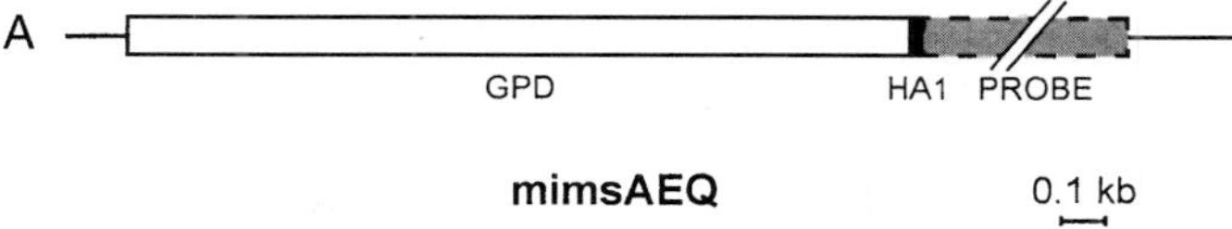

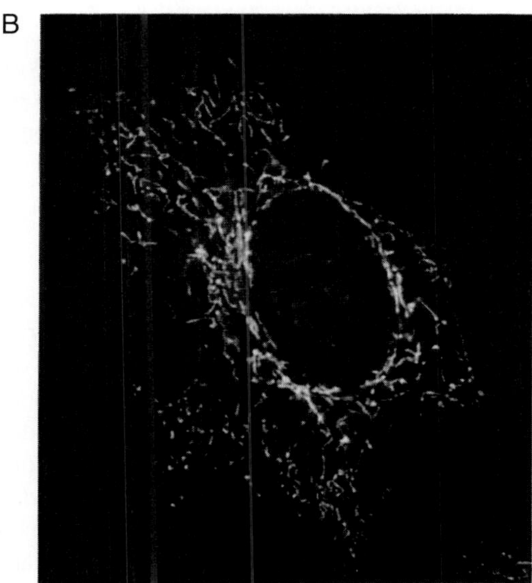

Fig. 2 (A) Schematic map of a reporter protein-glycerol phosphate dehydrogenase (GPD) targeted to the mitochondrial intermembrane space. (B) Immunolocalization of aequorin targeted to the mitochondrial intermembrane space (mimsAEQ), carried out as in Fig. 1.

II. Calcium

A. Mitochondria and Calcium Uptake: A Brief Overview

The outer and inner mitochondrial membranes (OMM and IMM, respectively) are markedly different, architecturally as well as functionally. It is noteworthy that the IMM is highly impermeable to ions in general, including Ca^{2+}. It has been demonstrated that upon physiological stimulation with agonists capable of generating inositol 1,4,5-triphosphate (InsP3) (Rizzuto *et al.,* 1993; Rutter *et al.,* 1993), a wide variety of cell types undergo major changes in mitochondrial matrix Ca^{2+} concentration ($[Ca^{2+}]_m$). In turn, the rise in $[Ca^{2+}]_m$ stimulates the activity of Ca^{2+}-dependent enzymes of the Krebs cycle and an increase in mitochondrial ATP production (Jouaville *et al.,* 1999). Calcium uptake into mitochondria, although facilitated by the strong electrochemical potential across the IMM (approximately -200 mV on the matrix side), is dependent on the activity of a ruthenium red-sensitive electrogenic calcium uniporter. This Ca^{2+} uniporter exhibits a very low affinity ($K_m \sim 5-10\ \mu M$) for the cation (Carafoli, 1987; Gunter *et al.,* 1994; Pozzan *et al.,* 1994).

How do mitochondria accumulate Ca^{2+} given their low-affinity uptake system? The answer to this apparent paradox may reside in simple morphological details: in some regions of the cell the ER and mitochondria are in close proximity (<80 nm between the two structures) (Rizzuto *et al.,* 1998b). Release of Ca^{2+} from the ER at such a microdomain is sensed by adjacent mitochondria as a significant increase in cytoplasmic Ca^{2+} concentration ($[Ca^{2+}]_c$), which enables the uptake of Ca^{2+} by mitochondria, before the signal extends to the rest of the cytosol. The localized $[Ca^{2+}]_c$ then dissipates quickly, preventing excessive mitochondrial Ca^{2+} accumulation and collapse of the mitochondrial proton-motive force due to Ca^{2+} overload.

B. Mitochondrial Calcium Measurements Using Aequorin

Aequorin, isolated from the jellyfish *Aequorea victoria,* is composed of a 21-kDa apoprotein and a hydrophobic prosthetic group, coelenterazine (molecular mass ~400 Da). The two components must be associated for the Ca^{2+}-triggered light emission to occur. The holoprotein contains three high-affinity Ca^{2+}-binding sites (homologous to the sites present in other Ca^{2+}-binding proteins, such as calmodulin). Upon binding of Ca^{2+} ions, aequorin undergoes an irreversible reaction in which one photon is emitted (Fig. 3A). The rate of this reaction depends on the $[Ca^{2+}]$ to which the photoprotein is exposed. In particular, at $[Ca^{2+}]$ between 10^{-7} and $10^{-5} M$ (the concentration normally occurring in the cytoplasm of living cells), there is a direct relationship between $[Ca^{2+}]$ and the fractional rate of consumption of the photoprotein. Figure 3B shows the Ca^{2+} response curve of aequorin at physiological conditions of pH, temperature, and ionic strength. The fractional rate of aequorin consumption expressed as the ratio between the emission of light at a defined Ca^{2+} concentration (L) and the maximal rate of the light emission at saturating $[Ca^{2+}](L_{max})$ is proportional to the second and third power of $[Ca^{2+}]$. This is the basis of the use of aequorin as a Ca^{2+} probe. Indeed, if all the light emitted by the photoprotein throughout an experiment, as well as that discharged at the end, is collected, it is possible to estimate L_{max} and then back-calculate the $[Ca^{2+}]$ to which the photoprotein was exposed at each point in time.

C. Advantages

1. Development of Targeted Probes

Prior to the isolation of the aequorin cDNA, the use of aequorin was limited to the cell types in which the photoprotein could be microinjected. Now it is possible not

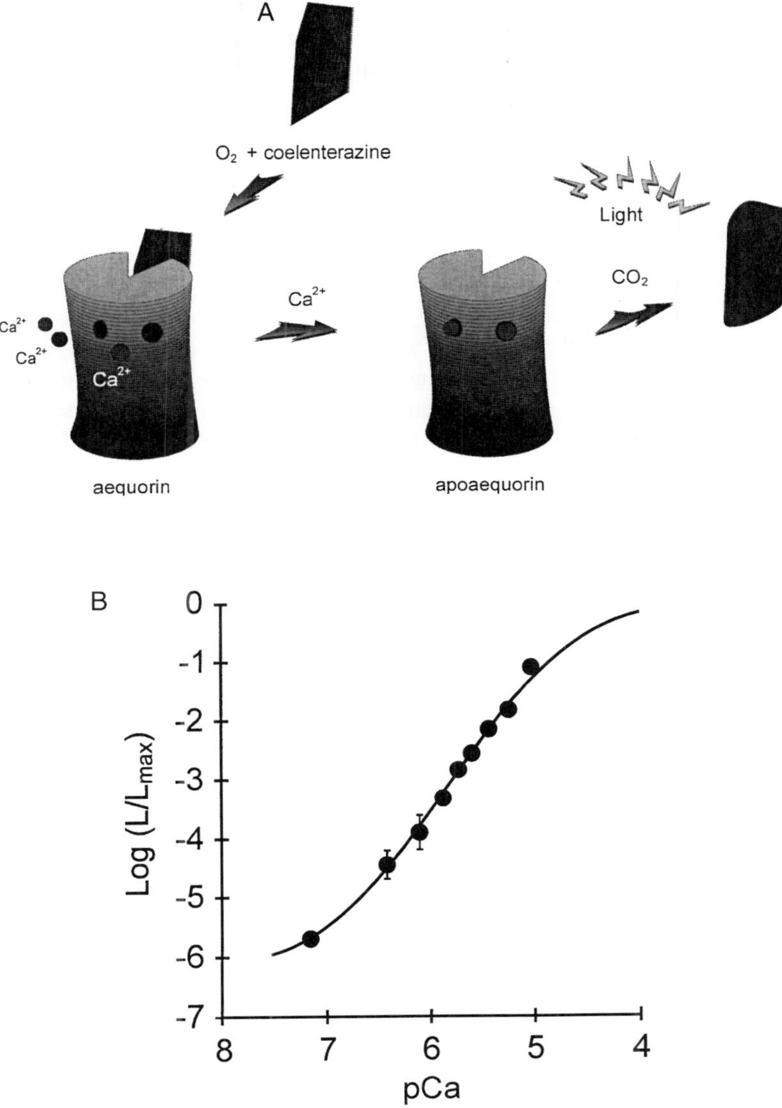

Fig. 3 (A) Schematic representation of the Ca^{2+}-dependent photon emission of aequorin. The functional probe is generated from the apoaequorin moiety (synthesized by the cell) and coelenterazine (imported from the surrounding medium). Binding of Ca^{2+} ions liberates the prosthetic group, with concomitant light emission and irreversible discharge of a molecule of aequorin. (B) Relationship between calcium concentration and light emission rate by active aequorin (L and L_{max} are, respectively, the instant and maximal rates of light emission). Note the direct relationship at calcium concentrations between 10^{-5} and 10^{-7} M.

only to express this calcium-sensing protein in the cytosol of a variety of cell types, but also to design Ca^{2+} probes specifically targeted to defined subcellular locations by adding specific targeting sequences. With this strategy, we constructed aequorin chimeras that are transported to a variety of cell locations, including the mitochondrial matrix (Rizzuto *et al.,* 1992), intermembrane space (Rizzuto *et al.,* 1998b), nucleus (Brini *et al.,* 1993), endoplasmic reticulum (Montero *et al.,* 1995), sarcoplasmic reticulum (Brini *et al.,* 1997), Golgi apparatus (Pinton *et al.,* 1998b), and subplasmamembrane region (Marsault *et al.,* 1997). This chapter focuses on those chimeras with a selective intracellular distribution to mitochondria as verified by immunocytochemistry.

2. Excellent Signal-to-Noise Ratio

Virtually all mammalian cells show a very dim luminescent background. Given the steep response of aequorin, an excellent signal-to-noise ratio is obtained, and minor variations in the amplitude of agonist-induced $[Ca^{2+}]$ changes can be detected easily.

3. Low Buffering Capacity

The binding of Ca^{2+} by aequorin could, in principle, affect intracellular Ca^{2+} homeostasis by acting as a buffer system. In practice, however, recombinant aequorin exhibits a very low Ca^{2+}-buffering capacity, as it is usually expressed at a concentration in the range of 0.1–1 μM. Even at this low concentration (which is two to three orders of magnitude lower than that of the commonly used fluorescent indicators), the excellent signal-to-noise ratio of aequorin enables measurements to be made in cell populations (Brini *et al.,* 1995).

4. Wide Dynamic Range

Aequorin is capable of measuring calcium concentrations across a wide spectrum. Figure 3b shows that accurate measurements of $[Ca^{2+}]$ can be obtained from 0.5 to 10 μM, reaching values at which most fluorescent indicators are already saturated.

5. Possibility of Coexpression with Other Proteins

A powerful approach for investigating the role and properties of the various molecular components of the Ca^{2+}-signaling apparatus is overexpression of the protein of interest, and analysis of possible effects on intracellular calcium in the molecularly modified cell.

D. Disadvantages

1. Low-Light Emission

In contrast to some fluorescent dyes that emit up to 10^4 photons without photobleaching, only 1 photon can be emitted by an aequorin molecule. Moreover, the principle of the use of aequorin for Ca^{2+} measurements is that only a small fraction of the total pool emits its photon every second. This is not a major limitation in population studies; conversely, single cell imaging requires very high expression and special apparatuses (Rutter *et al.,* 1996).

2. Expression May Be Inefficient and/or Slow

In the case of recombinant aequorin, transfection is the simplest loading procedure. A wide range of procedures have been developed, including calcium phosphate, liposomes, electroporation, and the gene gun. Nonetheless, some cell lines are resistant to transfection. Moreover, expression of recombinant aequorin may be inefficient. As a result, time for protein expression must elapse before Ca^{2+} measurements are carried out.

3. Overestimation of Average Value in a Nonhomogeneous Environment

The slope of the Ca^{2+} response curve of aequorin is steep. Therefore, if the increase of the $[Ca^{2+}]$ is not homogeneous, the average estimate will be biased toward the highest values.

E. Procedure

We obtained good aequorin expression in a wide variety of cell types (e.g., HeLa, CHO, COS, neurons, myocytes). In all cases, the cells are seeded on circular glass coverslips (13 mm in diameter) and allowed to grow until about 50% confluency. Cells are then transfected with 4 μg of mitochondrially targeted aequorin (mtAEQ) or 0.5 μg of mitochondrial intermembrane space (mims)AEQ using the calcium phosphate procedure. We noticed that high levels of mimAEQ expression had deleterious effects on some transfected cell populations. This problem was solved by reducing the quantity of DNA used for the transfection procedure. After 36 h, aequorin is reconstituted by adding the prosthetic group to the incubation medium [5 μ*M* coelenterazine for 2 h in DMEM supplemented with 1% fetal calf serum (FCS) at 37°C in 5% CO_2 atmosphere]. Approximately 2 h after reconstitution, cells are washed and transferred to the perfusion chamber of the measuring system. They are then perfused with KRB saline solution (Krebs–Ringer modified buffer: 125 m*M* NaCl, 5 m*M* KCl, 1 m*M* Na_3PO_4, 1 m*M* $MgSO_4$, 5.5 m*M* glucose, 20 m*M* HEPES, pH 7.4, 37°C).

The schematic representation of the measuring system is depicted in Fig. 4. In this system, the perfusion chamber, on top of a hollow cylinder, thermostatted by water jacket, is perfused continuously with buffer via a peristaltic pump; agonists and drugs are added to the same medium. The cell coverslip is placed a few millimeters from the surface of a low noise phototube. The photomultiplier is kept in a dark box maintained at 4°C. An amplifier discriminator is built in the photomultiplier housing. The pulses generated by the discriminator are captured by a Thorn EMI photon counting board, installed in an

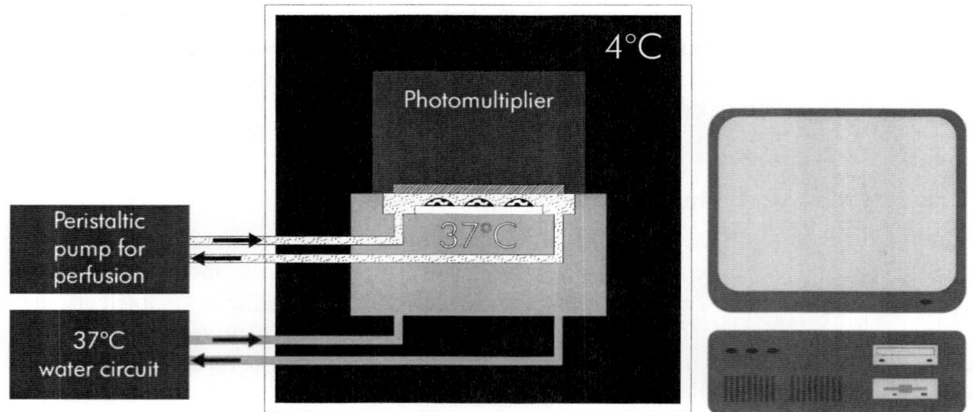

Fig. 4 Schematic representation of a custom-built luminometer. Cells loaded with a functional aequorin probe are incubated in a perfusion chamber at 37°C in close proximity to a photon-counting tube. The complete assembly is kept at 4°C, in the dark, to minimize extraneous signals. Acquisition of data and subsequent calculations to transform light emission into $[Ca^{2+}]$ are performed by a computer algorithm.

IBM-compatible computer. The board allows storage of data in the computer memory for further analysis.

An algorithm has been developed to calibrate the crude luminescent signal in terms of $[Ca^{2+}]$ that takes into account the instantaneous rate of photon emission and the total number of photons that can be emitted by aequorin in the sample (Brini *et al.,* 1995). To obtain the latter value, cells are lysed at the end of each experiment by perfusion with hypo-osmotic medium containing 10 m*M* $CaCl_2$ and a detergent (100 μM digitonin). This perfusion discharges all aequorin that was not consumed during the experiment.

F. Results

Some typical results of $[Ca^{2+}]_m$ monitoring are shown in Figs. 5 and 6. In these experiments, HeLa cells expressing the appropriate aequorin chimera were stimulated with an agonist, histamine (100 μM), which acts on a receptor coupled to the generation of inositol 1,4,5,-triphosphate and causes release of Ca^{2+} from intracellular stores.

The biphasic kinetics of $[Ca^{2+}]_c$ in HeLa cells transfected with unmodified aequorin (which is cytosolic) after stimulation with histamine is shown in Fig. 5. The release of Ca^{2+} from the stores causes an initially rapid, but temporary, increase in $[Ca^{2+}]_c$, to values of approximately 2.5 μM, followed by a sustained increase in $[Ca^{2+}]_c$ above normal basal levels, which is maintained throughout the stimulation period.

Figure 5 shows measurements obtained in HeLa cells transiently expressing mtAEQ. Upon stimulation with histamine, there is a rapid increase in $[Ca^{2+}]_m$ to approximately 10 μM, after which there is an equally rapid decrease to basal values. Given the targeting signal, it is expected that this chimeric protein is localized in the mitochondrial matrix.

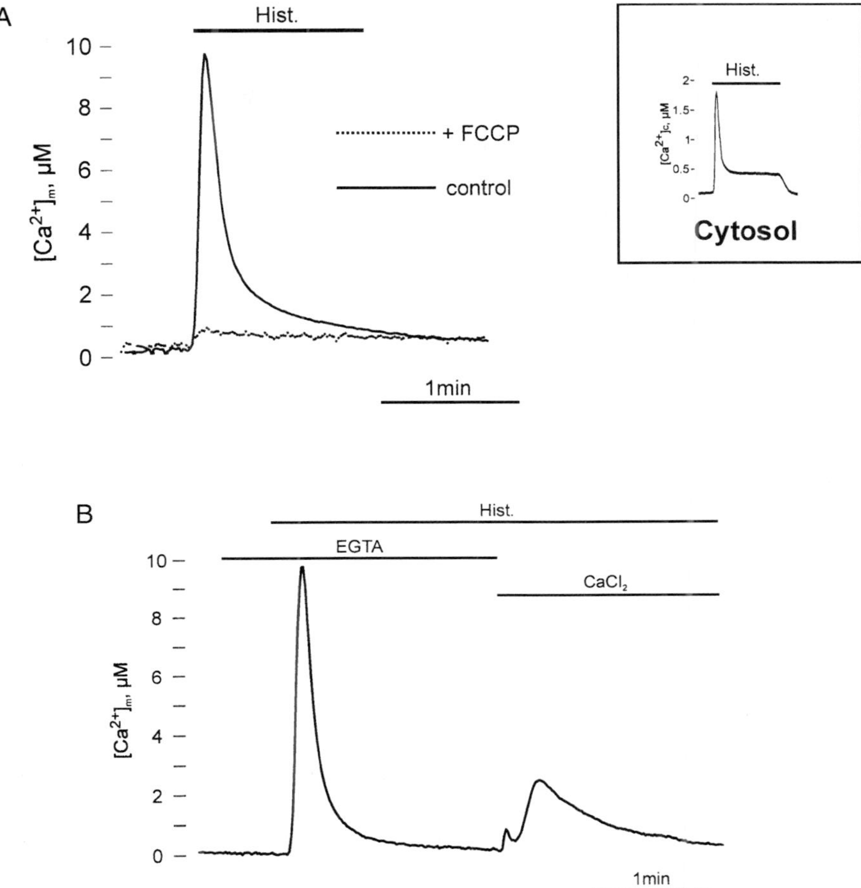

Fig. 5 (A) Effect of agonist stimulation on $[Ca^{2+}]_m$ of HeLa cells expressing aequorin targeted to the mitochondrial matrix (mtAEQ). After reconstitution of the photoprotein, coverslips with the cells were transferred to the thermostatted (37°C) chamber of the luminometer and perfused with modified Krebs–Ringer buffer (KRB; 125 m*M* NaCl, 5 m*M* KCl, 1 m*M* Na_3PO_4, 1 m*M* $MgSO_4$, 5.5 m*M* glucose, 20 m*M* HEPES, pH 7.4, 37°C). Where indicated, the cells were treated with 100 μ*M* histamine (Hist.) and/or with 5 μ*M* carbonylcyanide *p*-(trifuoromethoxy) phenylhydrazone (FCCP), added to KRB. In this and the following aequorin experiments, the traces are representative of at least five experiments. (B) The effect of release of Ca^{2+} from intracellular stores and Ca^{2+} influx on $[Ca^{2+}]_m$. Where indicated (EGTA), the cells were washed with 100 μ*M* EGTA and then treated with 100 μ*M* histamine (Hist.) added to KRB/EGTA. The medium was then switched with KRB (with 1 m*M* $CaCl_2$) and, where indicated (Hist.), the cells were challenged with 100 μ*M* histamine (added to KRB).

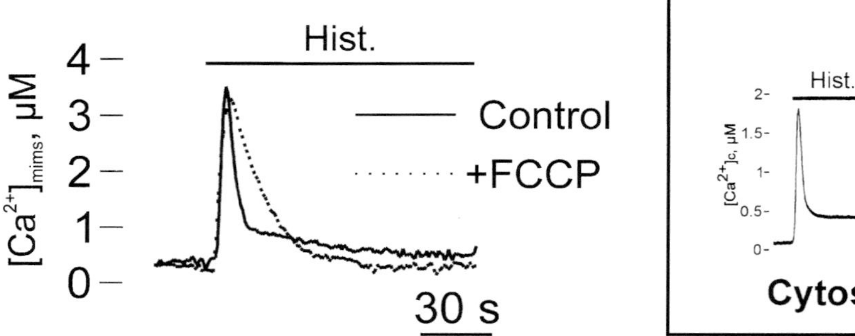

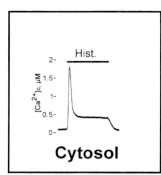

Fig. 6 Effect of agonist stimulation on $[Ca^{2+}]_{mims}$ of HeLa cells expressing aequorin targeted to the mitochondrial intermembrane space (mimsAEQ). All conditions as in Fig. 5.

Immunofluorescence data can discriminate the exact localization of mtAEQ within the organelle. To ensure that our $[Ca^{2+}]$ measurements were indeed due to changes in the mitochondrial matrix, we analyzed the effect of uncouplers on measured $[Ca^{2+}]$ levels. Because Ca^{2+} entry into the mitochondrial matrix is made possible by the electrochemical gradient across the IMM, collapse of this gradient with an uncoupler should abolish the changes in $[Ca^{2+}]$ measured by our probe. Using FCCP (5 μM), which dissipates the proton gradient across the IMM, we were unable to detect a significant response to the histamine stimulus, as shown in Fig. 5A. Further confirmation that our probe is indeed localized in the mitochondrial matrix came from experiments with ruthenium red, which is an inhibitor of the calcium uniporter of the IMM. As expected, we were unable to detect a rise in $[Ca^{2+}]_m$ upon stimulation (data not shown).

To determine the relative contribution of each of the two pathways for $[Ca^{2+}]_c$ (i.e., Ca^{2+} release from intracellular stores and Ca^{2+} influx from the extracellular medium), we incubated HeLa cells in a medium without $[Ca^{2+}]$ (with EGTA 100 μM) before histamine stimulation. Under such conditions the increase of the $[Ca^{2+}]_m$ is due only to the release of Ca^{2+} from the intracellular stores. Next, we reintroduced Ca^{2+} into the extracellular medium; in this case the rise in $[Ca^{2+}]_m$ is due to the influx through the plasma membrane channels. Data presented in Fig. 5B show that the release of Ca^{2+} from intracellular stores causes a much larger and faster increase in $[Ca^{2+}]_m$ than the influx through the plasma membrane channels. The larger effect caused by Ca^{2+} relased from the ER suggests that a close proximity between these two organelles could play a key role in the control of mitochondrial Ca^{2+} homeostasis, and thus organelle function.

To test for microdomains of close contact between ER and mitochondria, we constructed a new aequorin chimera, targeted to the IMS (mimsAEQ). Because the OMM is freely permeable to ions and small molecules, aequorin molecules present between the two mitochondrial membranes are located in a region that is in rapid equilibrium with the cytosolic portion in contact with the organelle. Such a chimera is thus sensitive to changes in Ca^{2+} in the region immediately adjacent to mitochondria.

HeLa cells transfected with mimsAEQ and stimulated with histamine show a biphasic response, as shown in Fig. 6. An initial rise in $[Ca^{2+}]_{mims}$ to ~3.5 μ*M* is followed by a rapid decrease, which gradually levels out to values above the initial $[Ca^{2+}]$. As explained previously, this type of response is due to the two mechanisms of action of the agonist: the release of Ca^{2+} from intracellular stores and the entry of Ca^{2+} from the extracellular medium.

Comparison of $[Ca^{2+}]_c$ and $[Ca^{2+}]_{mims}$ responses to histamine shows a clear difference only in the initial phase, which is due to the release of Ca^{2+} from intracellular stores. These data support the hypothesis that the opening of InsP3-sensitive channels in close proximity to mitochondria generates microdomains of high $[Ca^{2+}]$. Indeed, in such microdomains the $[Ca^{2+}]$ is much higher than the average $[Ca^{2+}]_c$. In these circumstances, the low-affinity systems present in mitochondria are capable of accumulating Ca^{2+} in the matrix of the organelle.

To determine the exact location of mimsAEQ, we performed the following experiments. We had previously determined that the collapse of the proton gradient drastically reduces the accumulation of Ca^{2+} in the matrix. If mimsAEQ were localized within the matrix, the rise in $[Ca^{2+}]$ should be abolished in the presence of uncouplers. Conversely, if mimsAEQ were localized in the space between the two mitochondrial membranes, then the presence of uncouplers should not affect the $[Ca^{2+}]$ values obtained. As shown in Fig. 6, HeLa cells transfected with mimsAEQ and treated with FCCP do not show a significant difference in $[Ca^{2+}]$ dynamics when compared to tightly coupled cells, strongly suggesting that the aequorin moiety lies in the IMS.

III. ATP

A. Measuring Free ATP Concentration Dynamically within Mitochondria of Living Cells

In most normal, aerobic cells, mitochondria are the source of the majority of cellular ATP. Thus, knowledge of intramitochondrial ATP concentration, $[ATP]_m$, is of central importance to understanding the bioenergetics of the living cell and its regulation by nutrients, hormones, and other stimuli. Of particular importance is the role of changes in intramitochondrial Ca^{2+} concentration $[Ca^{2+}]_m$ (see earlier discussion), which are likely to activate mitochondrial oxidative metabolism (Denton and McCormack, 1980) through the stimulation of mitochondrial dehydrogenases (Rutter, 1990). Conversely, decreases in ATP concentration within mitochondria may be an important and possibly early event in programmed cell death, resulting from a catastrophic collapse of the mitochondrial membrane potential (Orrenius *et al.*, 1997).

Studies with isolated mitochondria have demonstrated that the free cytosolic and mitochondrial ATP concentrations are likely to be markedly different. Thus, newly synthesized ATP is exported from the mitochondrial matrix electrogenically by ATP^{4-}: ADP^{3-} exchange (LaNoue *et al.*, 1978; Nicholls, 1982). Given a potential of as much as 180 mV across the inner mitochondrial membrane, this is predicted to set values of the ATP/ADP ratio at close to 100 in the cytosol, but nearer to 1.0 in the mitochondrial matrix.

Rapid fractionation techniques (Soboll *et al.,* 1978), applied to hepatocytes and other cell types, have lent support for the existence of this predicted difference in mitochondrial and cytosolic ATP concentration ($[ATP]_c$ and $[ATP]_m$, respectively) in living cells. Here, measurements of total cellular ATP content are made after disruption of the plasma membrane and protein denaturation in acid, using the luciferase activity of *Photinus pyralis* firefly tails (Stanley and Williams, 1969; Soboll *et al.,* 1978). However, such disruptive approaches are fraught with experimental and interpretive difficulties, linked to the inability to determine the *free* ATP concentration within the living cell. Thus, estimates of mitochondrial ATP are confounded by (1) potential contamination with other cellular compartments (e.g., secretory vesicles, lysosomes, Golgi apparatus), (2) binding to proteins and metal ions, and (3) intercellular heterogenity. Hence, measurements of total cellular ATP content provide a limited picture of the dynamic behavior of free [ATP] within mitochondria of a living cell. Further, they provide no information on the subcompartmentalization of mitochondrial ATP, i.e., the free ATP concentration in different mitochondria (or parts of the mitochondrial reticulum) (Rizzuto *et al.,* 1998b).

These problems have led us to search for means by which ATP may be measured dynamically in subcompartments and cytosolic domains of single living cells. One theoretical possibility is the use of ^{31}P-NMR (Scholz *et al.,* 1995). However, this requires large tissue samples and does not provide information on [ATP] changes within cellular organelles. Thus, the potential existence of domains of locally high or low ATP concentration could be missed entirely. Within the cytosol, such domains may exist, e.g., in the vicinity of membranes rich in ATP synthase (i.e., the mitochondrial inner membrane) or rich in ATPase activity (e.g., the plasma Na^+/K^+ ATPase, Ca^{2+} pumps, motor proteins).

At present, then, the best available biosensor for ATP (i.e., one capable of binding the nucleotide and producing a readily detectable signal) has proved to be firefly luciferase. Although purified luciferase protein has been microinjected into cells and used to measure cytosolic free ATP concentration (Bowers *et al.,* 1992), this simple approach precludes targeting of the reporter to the lumen of cellular organelles or its attachment to intracellular membranes. For this, expression of the protein from introduced cDNA is necessary. Firefly luciferase was thus cloned in the late 1980s (De Wet *et al.,* 1987), and versions of the gene, optimized for thermostability and expression in mammalian cells, were generated (Rutter *et al.,* 1998). The enzyme uses an oxidizable substrate, luciferin, which is converted to an AMP adduct before final oxidation and the release of a photon of light (see Fig. 7).

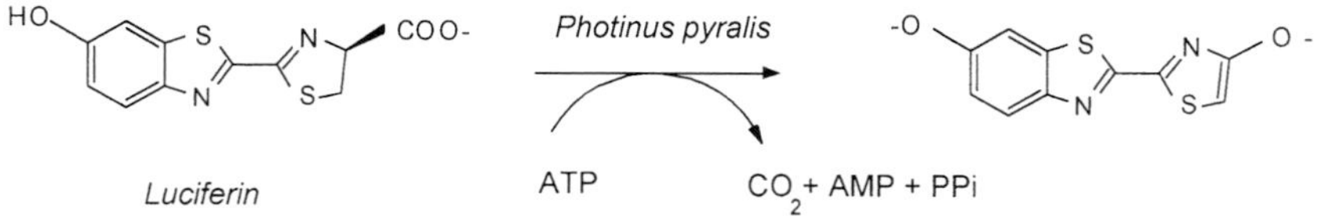

Fig. 7 The ATP-dependent luminescence reaction of luciferase.

B. Dynamic *in Vivo* Measurement of Cytosolic ATP

cDNA encoding luciferase can be introduced readily into most mammalian (and other) cell types by conventional transfection techniques, as discussed for aequorin and GFP, as well as by more sophisticated techniques, including microinjection (Rutter *et al.*, 1995) or the use of adenoviral vectors (see later). While originally intended for detection of the total luciferase amount in cell homogenates, luciferase light output can also be quantitated readily from single living cells, after the addition of the (reasonably cell-permeant) cofactor, luciferin. Under most conditions, O_2 and cofactors other than ATP are not limiting. Further, it can be calculated that the contribution of ATP consumption by luciferase represents only a tiny fraction of total cellular ATP turnover (<0.1%, even at relatively high levels of luciferase expression, e.g., 1×10^6 molecules/cell) and thus nonperturbing for normal cellular ATP homeostasis. Light output from single cells is clearly detectable even after expression from weak promoters, using highly sensitive photon-counting devices and long integration times (minutes) (White *et al.*, 1994; Rutter *et al.*, 1995, 1998; Alekseev *et al.*, 1997).

This technology has been extended to the detection of changes in free intracellular ATP concentration (Kennedy *et al.*, 1999; Jouaville *et al.*, 1999). Here, constitutively high levels of luciferase are expressed from strong viral promoters (e.g., the cytomegalovirus immediate early gene promoter, CMV-IE) so that small fluctuations in free ATP concentrations can be monitored. Relatively rapid imaging (1 data point per 1–10 s) can be performed, most conveniently using the technique of "time-resolved imaging," where the time and location (in 2D) of each single photon event is recorded (Photek, St Leonard's-on-sea, UK). At the single cell level, intensified photon counting cameras perform well (see, *e.g.*, www.photek.co.uk/, photekinc@compuserve.com or Hamamatsu at www.hpk.co.jp/products/producte.htm), although integrating charge-coupled device (CCD) cameras, and even back-illuminated CCD cameras, are usually unable to provide sufficient "detectivity" at physiologically relevant rates of data acquisition. Luciferase-expressing cells are maintained on the microscope stage, and additions are made to the medium in complete darkness. For cell populations, the detection of luciferase luminescence is readily achievable with the photon-counting tube apparatus, as described earlier for the detection of aequorin luminescence (Jouaville *et al.*, 1999; Maechler *et al.*, 1998).

Luciferase displays a K_m for ATP close to 1 mM when assayed in cell homogenates under approximate *in vivo* conditions of pH and physiological ionic strength (Kennedy *et al.*, 1999) (compared to the low micromolar range under optimal *in vitro* conditions) (DeLuca *et al.*, 1979). Confirming these values in living cells is complicated due to the distinct kinetics of the enzyme in the living cell ("glow" versus "flash" kinetics). Although the basis for this difference is not fully understood, it may reflect a lack of the accumulation of inhibitory end product (oxyluciferin) in the cell or a decreased sensitivity to this (or other inhibitors) mediated by other cellular cofactors (notably CoA). Whatever the mechanism, this makes monitoring [ATP] constantly in the living cell relatively straightforward, given sensitive photon detection equipment (see earlier discussion). Permeabilization of the cells with digitonin at varying ATP

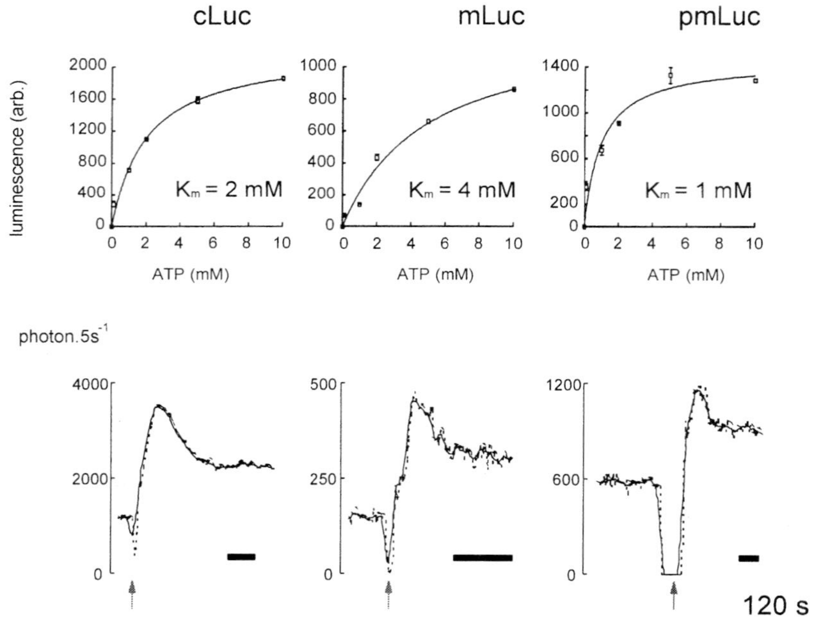

Fig. 8 Calibration of luciferases (*top*) and their use to measure free [ATP] in subcellular compartments of β-cell populations (*bottom*) (from Kennedy *et al.,* 1999). MIN6 β cells were transfected with the constructs shown (c, m, and pm, cytosolic, mitochondrial, and plasma membrane-targeted, respectively) before cell homogenization with detergent and assay of luciferase activity in a conventional luminometer. In the lower traces, luminescence was monitored in intact cells before the addition of digitonin plus a high (10 m*M*) concentration of Mg^{2+}-ATP, to achieve near-saturation of luciferase. Calculated values of [ATP] prior to cell permeabilization were close to 1 m*M* in both cytosolic compartments and the mitochondrial matrix.

concentrations allows calibration of the system and reveals ATP concentrations close to 1 m*M* (Fig. 8) for differently targeted luciferases.

Others (Maechler *et al.,* 1998) have used bacterial toxins to permeabilize cells and perform these calibrations via the addition of progressively increasing concentrations of ATP. However, we have found this a difficult procedure, likely complicated by gradual cell breakage and loss of luciferase. This may account for the discrepancy in K_m values of luciferase for ATP that have been reported (as high as 10 m*M*).

C. Engineering Luciferase for Targeting to Mitochondria

Like aequorin (see earlier discussion), luciferase cDNA can conveniently be fused 3′ to cDNA encoding the mitochondrial import sequence of cytochrome *c* oxidase subunit VIII (Fig. 9). This leads to exclusive mitochondrial localization of the probe (detected with antiluciferase antibodies).

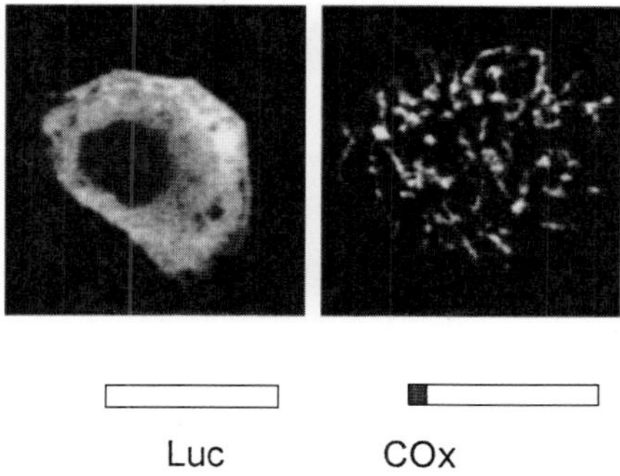

Fig. 9 Immunolocalization of expressed recombinant luciferases within MIN6 islet β cells (from Kennedy *et al.*, 1999). COx, cytochrome *c* oxidase VIII leader, for mitochondrial targeting.

Mitochondrial luciferase displays a very similar K_m value to the cytosolic (untargeted) enzyme, as expected given the removal of the presequence after mitochondrial import. Intriguingly, the free ATP within mitochondria ($[ATP]_m$) was found to be closely similar to that in the cytosol of resting β cells, at around 1.0 mM (Fig. 8). This may reflect either (1) a very low mitochondrial membrane potential in the resting cells or (2) a large difference (out < in) in free ADP concentrations, representing a dramatic "phosphorylation potential" in the cytosolic compartment ([ATP]/[ADP]/[Pi]) (LaNoue *et al.,* 1978). It should be stressed that luciferase displays relatively high selectivity for ATP over ADP so that the latter is only a weak inhibitor of activity. As a result, luciferase activity within the cell is likely to report largely [ATP] and not [ATP]/[ADP] ratio (or phosphorylation potential).

Studies with recombinant targeted luciferases have allowed us to monitor changes in free cytosolic ATP concentrations under a number of situations where this is perturbed either by changes in fuel supply to the cell or through cell stimulation with receptor agonists (Kennedy *et al.,* 1999; Jouaville *et al.,* 1999). The latter include hormones that mobilize intracellular Ca^{2+} and alter mitochondrial metabolism via Ca^{2+} accumulation in the mitochondrial matrix and stimulation of intramitochondrial dehydrogenases (Denton and McCormack, 1980; Rutter, 1990). These studies demonstrated that Ca^{2+} increases in the cytosol (and hence mitochondria) are able to activate mitochondrial ATP synthesis, elevating $[ATP]_c$. Strikingly, in the islet β cell, glucose caused much more sustained increases in $[ATP]_m$ than in $[ATP]_c$ (Fig. 10, see color plate). This observation may explain the more sustained increase in [ATP] beneath the plasma membrane, if it is created by localized mitochondria, which are strongly stimulated by locally high Ca^{2+} caused by activated Ca^{2+} influx across the plasma membrane.

Are $[ATP]_m$ changes homogeneous throughout the cell? At present, this question requires greater sensitivity than that achievable with currently available detectors and levels of luciferase expression. Indeed, it has not been possible clearly to resolve single mitochondria nor the mitochondrial reticulum by imaging. Nevertheless, with slower

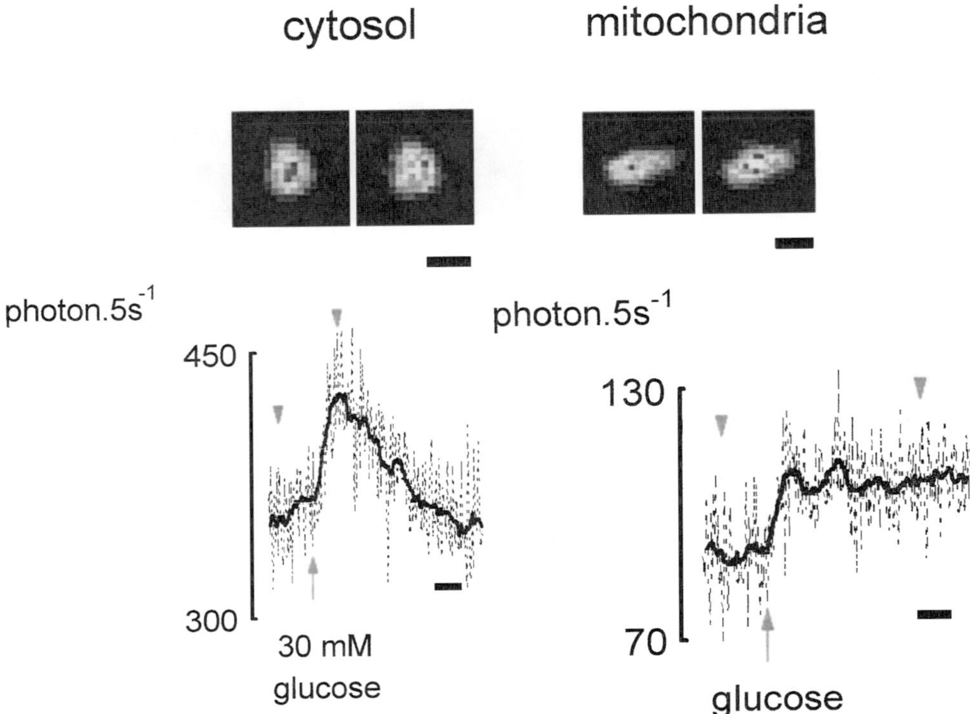

Fig. 10 Glucose-induced changes in free ATP concentration in the cytosol and mitochondrial matrix of MIN6 β cells. Images were produced at the points indicated by the gray arrowheads by integrating photon events over a 30-s interval. Traces show changes recorded from a single cell, with detection of luminescence made every 1 s (broken line) or after integration at each point for 10 s. Scale bar: 5 μm. Time bar: 120 s. From Kennedy *et al.* (1999). Note that individual mitochondria are *not* resolved readily. (See Color Plate.)

data acquisition and improved objective lenses, this may become possible. Similarly, it should be feasible to achieve resolution not only in the x, y but also the z plane. Although confocality is not feasible with a bioluminescent protein (there is no guidable laser, or point of coincidence for multiphoton approaches), digital deconvolution approaches may well become feasible if greater photon detection can be achieved (Rizzuto *et al.*, 1998a).

One potential problem in the use of luciferase is interference from other metabolites. While the concentration of many of these (e.g., phosphate, CoA) changes little or not at all under most conditions, changes in intracellular pH often occur. This may be an especially acute problem with mitochondrial luciferase, as increases in pH enhance luciferase activity, and may occur during the activation of mitochondrial dehydrogenases and the respiratory chain during increases in intracellular [Ca^{2+}]. Molecular engineering of luciferase, whose 3D structure was solved to 2.8 Å (Conti *et al.,* 1996) may offer approaches to address this problem.

D. Generating and Using Adenoviral Luciferase Vectors

A limitation of the use of expressed luciferases is that of achieving both adequate efficiency of transfection (2–10% is normal with conventional procedures) and adequate levels of expression in individual cells. Further, many cell types (especially primary cells) are difficult to transfect by conventional means (Ca-phosphate, lipoamines, electroporation, etc.). For this reason, much interest has shifted to the use of retro- and, more particularly, replication-deficient adenoviruses as a means of DNA introduction. The latter are particularly attractive, as >95% of cells of *any* mammalian species/embryological origin can typically be infected by low (entirely nontoxic) viral titers. This is also especially valuable for the study of intact organs, microorgans (e.g., islet β cells), or tissue slices (e.g., hippocampal neuron cultures), where conventional transfection is essentially impossible. For example, the power of this approach is revealed by the fact that >90% of cells within an intact pancreatic islet (a spherical microorgan of diameter approximately 200 μm) are rapidly infected to express GFP or luciferase. Furthermore, measurements of ATP using luciferase within the intact animal are in principle feasible (and luciferase has been imaged in living transgenic mice, albeit with relatively long integration times and sedated animals; Contag *et al.,* 1998). The procedure for adenoviral synthesis has been enhanced greatly with the ability to perform homologous recombination within bacteria (He *et al.,* 1998). It should now be possible to generate adenoviral versions of both cytosolic and mitochondrially targeted luciferase with which to address a range of biological questions in a range of primary cell types.

E. Future Prospects for the Use of Mitochondrially Targeted Luciferases

Targeted recombinant luciferase represents a dramatic advance in the ability to measure ATP dynamically and *in vivo* and in single cells. Although we have focused largely on the use of this tool to measure ATP concentration within the cytosol or the mitochondrial lumen, exploration of further mitochondrial subdomains may well yield surprises about mitochondrial function. For example, how does [ATP] change in the intermembrane space or at the surface of the outer mitochondrial membrane and is there a gradient between these two "compartments"? Indeed, luciferase has already been targeted to the OMM in yeast cells (Aflalo, 1990), but no measurements have yet been made in higher eukaryotes. Is the concentration of ATP the same in the "bulk" mitochondrial matrix as it is immediately beneath the mitochondrial inner membrane? Is it feasible to make multiple simultaneous recordings from luciferases (e.g., with altered luminescence spectra) and targeted to different domains or to image luciferase simultaneously with other fluorescent or luminescent reporters (such as aequorin)? These questions should be addressed in the near future given the targeting, gene delivery, and low light level imaging technologies now in existence. Other details are available at http://www.bch.bris.ac.uk/staff/rutter/index.html.

IV. pH

A. Intracellular pH

Intracellular pH (pH_i) regulates many cellular processes, including cell metabolism (Roos and Boron, 1981), gene expression (Isfort *et al.*, 1993) cell–cell coupling (Orchard and Kentish, 1990), cell adhesion (Tominga *et al.*, 1998), and cell death (Gottlied *et al.*, 1996).

The pH-sensitive fluorochromes 2′,7′-bis(carboxyethyl)-5,6-carboxyfluorescein (BCECF) and the most recently developed carboxyseminaphthorhodafluor-1 (SNARF1) and 5′,6′-carboxy seminaphthofluorescein (SNAFL) have been widely employed to monitor pH_i in both cell populations and single cells. These probes, synthesized as acetoxymethyester derivatives, are cell permeant and have the property to shift their light emission or excitation spectrum as a function of pH. Moreover, excitation wavelengths of SNARF and SNAFL enable their use with a laser confocal microscope (Zhou *et al.*, 1995). Measurements of pH_i with these dyes present some problems: during the course of experiments a significant leakage of the pH indicator occurs, which can be considerable in cells incubated at 37°C. In addition, some cells are resistant to loading because of physical barriers, such as a cell wall (bacteria, yeast, and plants) or the thickness of the tissue preparation. The most important limitation of the use of these compounds is that pH_i measurements are restricted to the cytosol and nucleus, whereas pH changes occurring in other intracellular organelles cannot be directly determined in intact cells.

This restriction is prominent for organelles such as mitochondria and the Golgi, where a proton gradient is built up by specific H^+ transport mechanisms. In mitochondria, it is universally accepted that the H^+ electrochemical potential ($\Delta\mu_{H+}$), generated by electron transport across the inner membrane coupled to H^+ ejection on the redox H^+ pumps, is used to drive ATP synthesis by ATP-synthase. In addition to ATP synthesis, $\Delta\mu_{H+}$ supports a variety of mitochondrial processes, some of which are a prerequisite for respiration and ATP synthesis, such as (1) uptake of respiratory substrates and phosphate, of ADP in exchange for ATP, and of Ca^{2+} ions; (2) the transhydrogenation reaction, and (3) the import of respiratory chain and ATP synthase subunits encoded by nuclear genes. $\Delta\mu_H^+$ comprises both the electrical ($\Delta\psi$) and the chemical component of the H^+ gradient (ΔpH). It is well established that under physiological conditions, $\Delta\psi$ represents the dominant component of $\Delta\mu_H^+$, whereas the ΔpH gradient is small. The relative contribution of $\Delta\psi$ and ΔpH across the mitochondrial membrane can be changed by the redistribution of permeant ions, such as phosphate, Ca^{2+}, or K^+, in the presence of the ionophore valinomycin. Several techniques have been developed for the determination of $\Delta\psi$ and ΔpH in isolated mitochondria (Nicholls and Fergusson, 1992). Attempts have also been carried out to measure $\Delta\psi_m$ in intact cells [for a critical evaluation of these methods, see Bernardi *et al.* (1999)]. However, measurements of matrix pH have been elusive due to the difficulty of separating the mitochondrial signal from its surrounding cytoplasm (Chacon *et al.*, 1994).

One of the most promising tools developed to overcome this difficulty is the use of recombinant pH-sensitive fluorescent proteins targeted specifically to the mitochondrial compartments.

B. pH-Sensitive Fluorescent Proteins

The GFP from *A. victoria* has been widely used as a noninvasive fluorescent reporter for gene expression (Chalfie *et al.,* 1994), protein localization (Rizzuto *et al.,* 1995), and protein trafficking (Pines, 1995). Its small size (approximately 27 kDa) and its ability to retain its intrinsic fluorescence when fused with other peptides have made GFP a very useful tool for cellular biology studies (Tsien, 1998).

Structural analysis by X-ray crystallography revealed that the cylindrical fold of GFP is made up of an 11-stranded β-barrel, threaded by an α-helix running up the axis of the cylinder. The chromophore is attached to the α-helix and is buried in the center of the cylinder. The chromophore is formed from residues 65–67, which are Ser-Tyr-Gly in the native protein, which undergo intramolecular autocatalytic cyclization and oxidation (Ormö *et al.,* 1996).

The *in vitro* spectral characteristics of both purified and recombinant wild-type GFP and of a number of other GFP mutants are influenced by pH, suggesting a pH; sensing activity for this class of proteins (Ward and Bokman, 1982; Patterson *et al.,* 1997). Therefore, the possibility has been explored that GFP mutants might be used as a pH sensor in living cells. This possibility was first investigated by Kneen *et al.* (1998), who reported that mutations S65T and F64L/S65T of wild-type GFP exhibited indistinguishable fluorescent spectra and pH titration data, with a pK_a value of 5.98. The pH-dependent absorbance spectrum of *in vitro* S65T-GFP is reported in Fig. 11A. It is apparent that the peak at 490 nm increased from pH 4 to 7.5. The finding that the titration curve for S65T-GFP was similar to that of fluorescein suggested the involvement of a single amino acid residue in its pH-sensitive mechanism (Kneen *et al.,* 1998). Crystallographic analysis of this GFP mutant at both basic (pH 8.0) and acidic (pH 4.6) pH, identify a model, likely

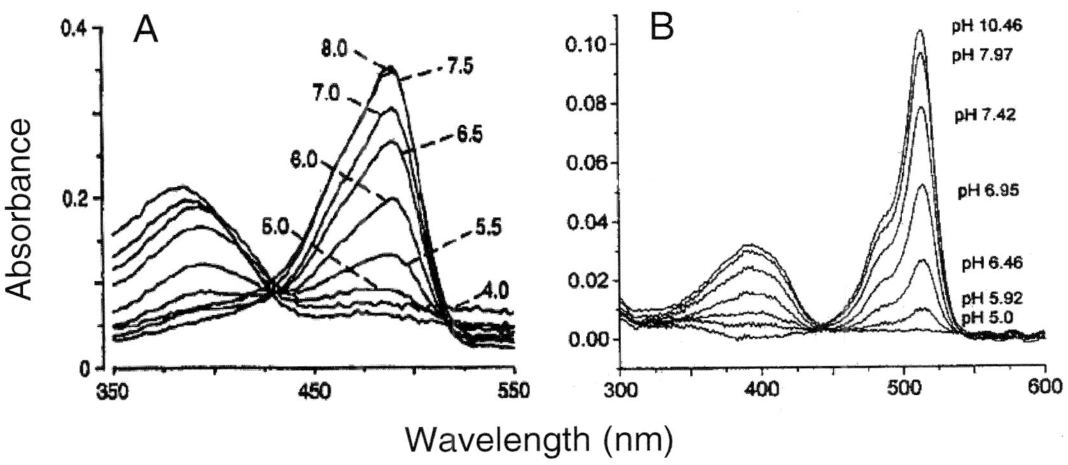

Fig. 11 pH-dependent absorbance spectra of purified S65T-GFP (A) and EYFP (B), modified from Kneen *et al.* (1998) and Llopis *et al.* (1998), respectively.

valid for other mutants as well, where the phenolic hydroxyl deriving from Tyr66 is the site of protonation (Elsliger *et al.,* 1999).

The same GFP mutant (F64L/S65T), termed GFPmut1, was expressed heterologously in the cytosol and nuclear compartments of BS-C-1 cells or rabbit proximal tubule cells. In this study, comparison of GFPmut1 and the pH-sensitive dye BCECF has been carried out, showing uniform agreement between pH_i estimates with the two methods (Robey *et al.,* 1998).

Another pH-sensitive GFP mutant, called enhanced yellow fluorescent protein (EYFP), was made by introducing the amino acid substitutions S65G/S72A/T203Y. The absorbance spectrum of EYFP *in vitro* is reported in Fig. 11B, showing that the absorbance of the peak at 514 nm increased from pH 5 to 8 (Llopis *et al.,* 1998). The apparent pK_a value of 7.1 suggests that this protein should be suitable for pH_i measurements in most subcellular compartments, particularly in the mitochondrial matrix, which is expected to be alkaline.

C. Mitochondrial-Targeted pH-Sensitive GFP

1. Available GFP Chimeras

We currently employ as pH sensors two targeted versions of the GFP mutant (S65G/S72A/T203Y), described earlier. The first is targeted to the mitochondrial matrix (mtYFP) by fusion to the targeting sequence of COX8. As Tsien (1998) described, a matrix-targeted chimera is appropriately sorted and faithfully reports the pH changes of this domain. The suitability of these GFP mutants as pH indicators in the mitochondrial matrix of living cells was evaluated in CHO (Kneen *et al.,* 1998), in HeLa cells, and in rat neonatal cardiomyocytes (Llopis *et al.,* 1998). In CHO cells expressing recombinant GFPmut1 targeted to the mitochondrial matrix, a qualitative reversible acidification of mitochondria was observed after addition of the protonophore CCCP. However, the value of mitochondrial matrix pH could not be determined accurately because of the low pK_a of this GFP mutant. Therefore, GFPmut1 does not seem to be a suitable pH indicator in this compartment, whereas it would be advantageous in more acidic organelles, such as cytoplasm and *trans*-Golgi cisternae (Kneen *et al.,* 1998). Conversely, expression of EYFP allowed quantitative determination of changes of mitochondrial matrix pH. The resting value, which was 7.98 in HeLa cells and 7.91 in rat neonatal cardiomyocytes, was collapsed rapidly to about pH 7 by protonophore addition, indicating that this GFP mutant provides a good tool for mitochondrial matrix pH measurements (Llopis *et al.,* 1998).

As yet no information is available on the intermembrane space of mitochondria. It is reasonable to assume that the pH of this compartment would be in equilibrium with the cytosol through the H^+-permeable outer membrane. However, the intermembrane space might be influenced by pH changes occurring in the matrix as a result of variations in the membrane potential ($\Delta\psi$), and microdomains exhibiting different H^+ concentrations from cytosol might exist under physiological and pathological conditions. During the process of apoptosis, for instance, the intermembrane space likely undergoes a dramatic

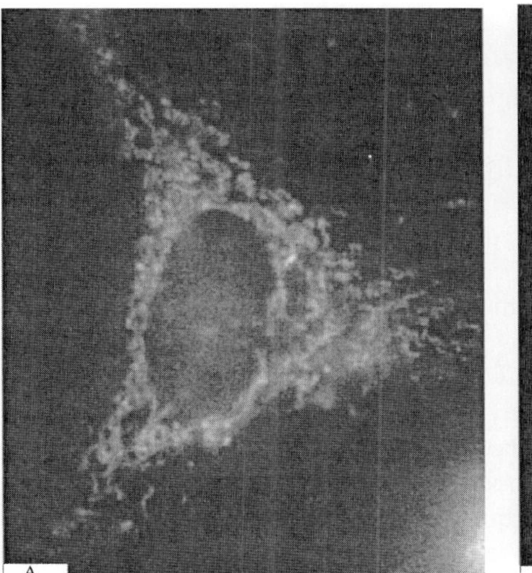

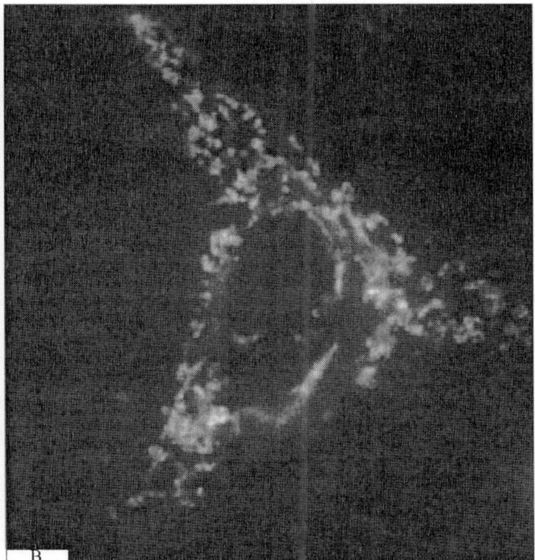

Fig. 12 HeLa cells expressing MIMS-EYFP and clamped at pH 6.5 (A) and 7.5 (B). Cells were observed 36 h after transfection using a back-illuminated CCD camera (Princeton Instruments) and Metamorph software (Universal Imaging).

change, as a few proteins normally resident in this compartment, such as cytochrome *c* and "apoptosis inducing factor," are released into the cytoplasm, where they play a crucial role in the effector phase of cell death.

EYFP targeted to the IMS has been constructed using the same cloning strategy employed for MIMS-targeted aequorin (Pinton *et al.,* 1998a). Transfection of HeLa cells with mimsEYFP yields the typical pattern of mitochondrial labeling, as shown in the image reported in Fig. 12, obtained with a digital imaging microscope (Rizzuto *et al.,* 1998a).

D. Detecting and Calibrating the Signal of Targeted GFP Mutants

1. Recombinant Expression

The procedures employed for expressing mtEYFP and mimsEYFP are the same as those described previously for the corresponding aequorin and luciferase chimeras, i.e., in most cases a simple transfection via the calcium phosphate protocol. The only difference is that the cells are seeded onto a 25-mm coverslip and transfected with 8 μg DNA/coverslip. In the case of mimsEYFP, the amount of DNA can be reduced to 2–4 μg/coverslip to prevent mitochondrial damage. It is our experience that in many cell types high levels of expression of a recombinant protein retained in the IMS (i.e., not only YFP, but also aequorin; see earlier discussion) cause a morphological rearrangement of mitochondria.

2. Visualizing GFP

Thirty-six hours after transfection the coverslip with the cells is transferred to the thermostatted stage of a digital imaging system, as described previously (Rizzuto *et al.*, 1998a). We describe our systems here briefly, but obviously the setup can be assembled with similar components from different manifacturers. Our two setups are composed of an inverted fluorescence microscope (a Nikon Eclipse 300 and a Zeiss Axiovert), an excitation filter wheel with shutter from Sutter Instruments, the dedicated YFP filter sets of Chroma (however, comparable results can be obtained with a traditional FITC set), a back-illuminated camera from Princeton Instruments, and the acquisition/analysis software Metafluor of Universal Imaging. For each time point, the fluorescence image from the selected field of view and the fluorescence intensity of the region(s) of interest can be obtained. The sensitivity of the camera results in reduction of the time of exposure to 50–100 ms in many applications. Under these conditions, prolonged experiments can be carried out with modest bleaching of the fluorophore (e.g., $<20\%$ in a 5-min experiment with image acquisition at 1- to 2-s intervals). Throughout the experiment, the variation in fluorescence intensity caused by agonists and drugs is revealed by the gray-scale representation of the fluorescence image. From the images of the same cell clamped at pH 6.5 and pH 7.5, shown in Fig. 12, it is apparent that the fluorescence signal increases with pH.

3. Converting the YFP Signal into pH Values

Coverslips with mimsEYFP-transfected cells were mounted in the thermostatted chamber (37°C) and incubated in a KCl-based saline solution containing 125 m*M* KCl, 20 m*M* NaCl, 5.5 m*M* D-glucose, 1 m*M* $CaCl_2$, 1 m*M* $MgSO_4$, 1 m*M* K_2HPO_4, 20 m*M* Na-HEPES (pH 6.5), and the ionophores nigericin and monensin (5 μM) to equalize intracellular and extracellular pH. pH was increased by additions of small amounts of NaOH while measuring the pH with a semimicrombination pH electrode. mimsEYFP fluorescence in selected areas of cells was linear with pH over the range between pH 6.5 and 8, with a linear regression coefficient between 0.9 and 0.97 in three experiments.

Quantitative determination of fluorescence in selected areas of cells as a function of pH allowed construction of the titration curve shown in Fig. 13. The estimated pH value in the IMS of HeLa cells was 7.21 ± 0.15 (six cells from three experiments), which is close to the pH_i value of 7.31 ± 0.2 (four cells from two experiments) determined with EYFP targeted to the cytosol (not shown).

V. Conclusion

Two powerful tools are now available to monitor the changes in pH in the mitochondrial intermembrane space and the matrix *in situ* in intact cells, and the relationship existing between the pH of these compartments and that of the cytosol. These probes have the great advantage of being specifically targeted to the two mitochondrial compartments and,

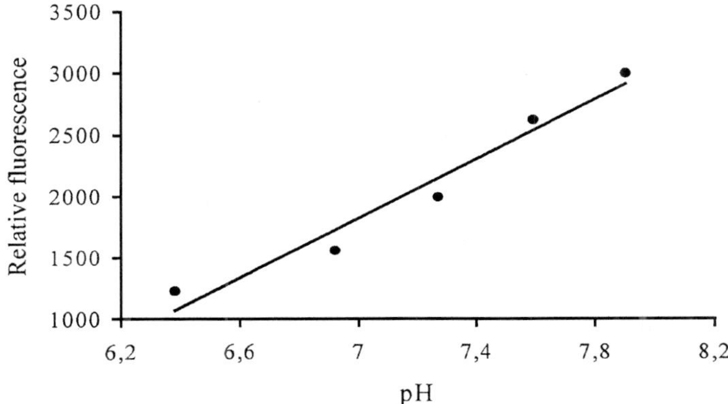

Fig. 13 *In situ* titration curve of MIMS-EYFP fluorescence as a function of pH. The curve is representative of three experiments.

through use of digital fluorescence microscopy, allow acquisition of three-dimensional images of cells in real time and with high spatial resolution. It would be possible to assess whether some regions of the mitochondrial matrix are more active than others in H^+ translocation or whether pH gradients or changes are present in the intermembrane space during cell activation by cell surface receptors or during the onset of apoptosis. The stability of these proteins and their lack of toxicity make them ideal visual indicators for a wide array of cellular applications *in vivo.*

Acknowledgments

We gratefully acknowledge grant support from Telethon, the Italian University and Health Ministries to R.R., from the Medical Research Council, U.K. (Post doctoral Fellowship to E.K.A., and the Bristol MRC Cell Imaging Facility), the Biotechnology and Biological Sciences Research Council, British Diabetic Association, and Wellcome Trust to E.K.A. and G.R. and from PRIN "Bionergetica e trasporto di membrana" to A.M.P. and M.R.

References

Aflalo, C. (1990). Targeting of cloned firefly luciferase to yeast mitochondria. *Biochemistry* **29,** 4758–4766.

Alekseev, A. E., Kennedy, M. E., Navarro, B., and Terzic, A. (1997). Burst kinetics of co-expressed Kir6.2/SUR1 clones: Comparison of recombinant with native ATP-sensitive K^+ channel behavior. *J. Membr. Biol.* **159,** 161–168.

Bernardi, P., Scorrano, L., Colonna, R., Petronilli, V., and Di Lisa, F. (1999). Mitochondrial and cell death: Mechanistic aspects and methodological issues. *Eur. J. Biochem.* **264,** 687–701.

Bowers, K. C., Allshire, A. P., and Cobbold, P. H. (1992). Bioluminescent measurement in single cardiomyocytes of sudden cytosolic ATP depletion coincident with rigor. *J. Mol. Cell Cardiol.* **24,** 213–218.

Brini, M., Marsault, R., Bastianutto, C., Alvarez, J., Pozzan, T., and Rizzuto, R. (1995). Transfected aequorin in the measurement of cytosolic Ca^{2+} concentration ($[Ca^{2+}]_c$): A critical evaluation. *J. Biol. Chem.* **270,** 9896–9903.

Brini, M., Murgia, M., Pasti, L., Picard, D., Pozzan, T., and Rizzuto, R. (1993). Nuclear Ca^{2+} concentration measured with specifically targeted recombinant aequorin. *EMBO J.* **12,** 4813–4819.

Brini, M., De Giorgi, F., Murgia, M., Marsault, R., Massimino, M. L., Cantini, M., Rizzuto, R., and Pozzan, T. (1997). Subcellular analysis of Ca^{2+} homeostasis in primary cultures of skeletal myotubes. *Mol. Cell. Biol.* **8,** 129–143.

Carafoli, E. (1987). Intracellular calcium homeostasis. *Annu. Rev. Biochem.* **56,** 395–433.

Chacon, E., Reece, J. M., Nieminen, A., Zahrebelski, G., Herman, B., and Lemasters, J. J. (1994). Distribution of electrical potential, pH, free Ca^{2+}, and volume inside cultured adult rabbit cardiac myocytes during chemical hypoxia: A multiparameter digitized confocal microscopic study. *Biophys. J.* **66,** 942–952.

Chalfie, M., Tu, Y., Euskirchen, G., Ward, W. W., and Prasher, D. C. (1994). Green fluorescent protein as a marker for gene expression. *Science* **263,** 802–805.

Contag, P. R., Olomu, I. N., Stevenson, D. K., and Contag, C. H. (1998). Bioluminescent indicators in living mammals. *Nature Med.* **4,** 245–247.

Conti, E., Franks, N. P., and Brick, P. (1996). Crystal structure of firefly luciferase throws light on a superfamily of adenylate-forming enzymes. *Structure* **4,** 287–298.

DeLuca, M., Wannlund, J., and McElroy, W. D. (1979). Factors affecting the kinetics of light emission from crude and purified firefly luciferase. *Anal. Biochem.* **95,** 194–198.

Denton, R. M., and McCormack, J. G. (1980). On the role of the calcium transport cycle in the heart and other mammalian mitochondria. *FEBS Lett.* **119,** 1–8.

De Wet, J. R., Wood, K. V., and DeLuca, M., *et al.* (1987). Firefly luciferase gene: Structure and expression in mammalian cells. *Mol. Cell. Biol.* **7,** 725–737.

Elsliger, M.-A., Wachter, R. M., Hanson, G. T., Kallio, K., and Remington, S. J. (1999). Structural and spectral response of green fluorescent protein variants to changes in pH. *Biochemistry* **38,** 5296–5301.

Gottlieb, R. A., Grauol, D. L., Zhu, J. Y., and Engler, R. L. (1996). Preconditioning rabbit cardiomyocytes: Role of pH, vacuolar proton ATPase and apoptosis. *J. Clin. Invest.* **97,** 2391–2398.

Gunter, K. K., and Gunter, T. E. (1994). Transport of calcium by mitochondria. *J. Bioenerg. Biomembr.* **26,** 471–485.

Halestrap, A. P. (1989). The regulation of the matrix volume of mammalian mitochondria in vivo and in vitro and its role in the control of mitochondrial metabolism. *Biochim. Biophys. Acta* **973,** 355–382.

He, T. C., Zhou, S., da Costa, L. T., Yu, J., Kinzler, K. W., and Vogelstein, B. (1998). A simplified system for generating recombinant adenoviruses. *Proc. Natl. Acad. Sci. USA* **95,** 2509–2514.

Isfort, R. J., Cody, D. B., Asquith, T. N., Ridder, G. M., Stuard, S. B., and LeoBoeuf, R. A. (1993). Induction of protein phosphorylation, protein synthesis: Immediate-early-gene expression and cellular proliferation by intracellular pH modulation. Implications for the role of hydrogen ions in signal transduction. *Eur. J. Biochem.* **213,** 349–357.

Jouvaille, L. S., Pinton, P., Bastianutto, C., Rutter, G. A., and Rizzuto, R. (1999). Regulation of mitochondrial ATP synthesis by calcium: Evidence for a long-term metabolic priming. *Proc. Natl. Acad. Sci. USA* **96,** 13807–13812.

Kennedy, H. J., Pouli, A. E., Jouvaille, L. S., Rizzuto, R., and Rutter, G. A. (1999). Glucose-induced ATP microdomains in single islet beta-cells. *J. Biol. Chem.* **274,** 13281–13291.

Kneen, M., Farinas, J., Li, Y., and Verkman, A. S. (1998). Green fluorescent protein as a noninvasive intracellular pH indicator. *Biophys. J.* **74,** 1591–1599.

LaNoue, K., Mizani, S. M., and Klingenberg, M. (1978). Electrical imbalance of adenine nucleotide transport across the mitochondrial membrane. *J. Biol. Chem.* **253,** 191–198.

Llopis, J., McCaffery, J. M., Miyawaki, A., Farquhar, M., and Tsien, R. Y. (1998). Measurement of cytosolic, mitochondrial, and Golgi pH in single living cells with green fluorescent proteins. *Proc. Natl. Acad. Sci. USA* **95,** 6803–6807.

Maechler, P., Wang, H., and Wollheim, C. B. (1998). Continuous monitoring of ATP levels in living insulin secreting cells expressing cytosolic firefly luciferase. *FEBS Lett.* **422,** 328–332.

Marsault, R., Murgia, M., Pozzan, T., and Rizzuto, R. (1997). Domains of high Ca^{2+} beneath the plasma membrane of living A7r5 cells. *EMBO J.* **16,** 1575–1581.

Montero, M., Brini, M., Marsault, R., Alvarez, J., Sitia, R., Pozzan, T., and Rizzuto, R. (1995). Monitoring dynamic changes in free Ca^{2+} concentration in the endoplasmic reticulum of intact cells. *EMBO J.* **14,** 5467–5475.

Nicholls, D. G. (1982). "Bioenergetics: An Introduction to the Chemiosmotic Theory." Academic Press, London

Nicholls, D. G., and Fergusson, S. F. (1992). Quantitative bioenergetics: The measurements of driving forces *In* "Bioenergetics 2," p. 39–63. Academic Press, London

Orchard, C. H., and Kentish, J. C. (1990). Effects of changes of pH on the contractile function of cardiac muscle. *Am. J. Physiol.* **258,** C967–C981.

Ormö, M., Cubitt, A. B., Kallio, Gross, L. A., Tsien, R. Y., and Remington, S. J. (1996). Crystal structure of the *Aequorea victoria* green fluorescent protein. *Science* **273,** 1392–1395.

Orrenius, S., Burgess, D. H., Hampton, M. B., and Zhivotovsky, B. (1997). Mitochondira as the focus of apoptosis research. *Cell Death Differ.* **4,** 427–428.

Patterson, G. H., Knobel, S. M., Sharif, W. D., Kain, S. R., and Piston, D. W. (1997). Use of the green fluorescent protein and its mutants in quantitative fluorescence microscopy. *Biophys. J.* **73,** 2782–2790.

Pines, J. (1995). GFP in mammalian cells. *Trends Genet.* **11,** 326–327.

Pinton, P., Brini, M., Bastianutto, C., Tuft, R. A., Pozzan, T., and Rizzuto, R. (1998a). New light on mitochondrial calcium. *BioFactors* **8,** 243–253.

Pinton, P., Pozzan, T., and Rizzuto., R. (1998b). The Golgi apparatus is an inositol 1,4,5 trisphosphate Ca^{2+} store, with distinct functional properties from the endoplasmic reticulum. *EMBO J.* **18,** 5298–5308.

Pozzan, T., Rizzuto, R., Volpe, P., and Meldolesi, J. (1994). Molecular and cellular physiology of intracellular Ca^{2+} stores. *Physiol. Rev.* **74/3,** 595–636.

Rizzuto, R., Brini, M., Murgia, M., and Pozzan, T. (1993). Microdomains of cytosolic Ca^{2+} concentration sensed by strategically located mitochondria. *Science* **262,** 744–747.

Rizzuto, R., Brini, M., Pizzo, P., Murgia, M., and Pozzan, T. (1995). Chimeric green fluorescent protein: A new tool for visualizing subcellular organelles in living cells. *Curr. Biol.* **5,** 635–642.

Rizzuto, R., Carrington, W., and Tuft, R. A. (1998a). Digital imaging microscopy of living cells. *Trends Cell Biol.* **8,** 288–292.

Rizzuto, R., Pinton, P., Carrington, W., Fay, F. S., Fogarty, K. E., Lifshitz, L. M., Tuft, R. A., and Pozzan, T. (1998b). Close contacts with the endoplasmic reticulum as determinants of mitochondrial Ca^{2+} responses. *Science* **280,** 1763–1766.

Rizzuto, R., Simpson, A. W. M., Brini, M., and Pozzan, T. (1992). Rapid changes of mitochondrial Ca^{2+} revealed by specifically targeted recombinant aequorin. *Nature* **358,** 325–328.

Robb-Gaspers, L. D., Burnett, P., Rutter, G. A., Denton, R. M., Rizzuto, R., and Thomas, A. P. (1998). Integrating cytosolic calcium signals into mitochondrial metabolic responses. *EMBO J.* **17,** 4987–5000.

Robey, R. B., Ruiz, O., Santos, A. V. P., Ma, J., Kear, F., Wang, L.-J., Li, C.-J., Bernardo, A. A., and Arruda, J. A. L. (1998). pH-dependent fluorescence of a heterologously expressed *Aequorea* green fluorescent protein mutant: In situ spectral characteristics and applicability to intracellular pH estimation. *Biochemistry* **37,** 9894–9901.

Roos, A., and Boron, W. F. (1981). Intracellular pH. *Physiol. Rev.* **61,** 296–434.

Rutter, G. A. (1990). Ca^{2+}-binding to citrate cycle enzymes. *Int. J. Biochem.* **22,** 1081–1088.

Rutter, G. A., Burnett, P., Rizzuto, R., Brini, M., Murgia, M., Pozzan, T., Tavaré, J. M., and Denton, R. M. (1996). Subcellular imaging of intramitochondrial Ca^{2+} with recombinant targeted aequorin: Significance for the regulation of pyruvate dehydrogenase activity. *Proc. Natl. Acad. Sci. USA* **93,** 5489–5494.

Rutter, G. A., Kennedy, H. J., Wood, C. D., White, M. R. H., and Tavare, J. M. (1998). Quantitative realtime imaging of gene expression in single cells using multiple luciferase reporters. *Chem. Biol.* **5,** R285–R290. [Abstract]

Rutter, G. A., Theler, J. M., Murgia, M., Wollheim, C. B., Pozzan, T., and Rizzuto, R. (1993). Stimulated Ca^{2+} influx raises mitochondrial free Ca^{2+} to supramicromolar levels in a pancreatic β-cell line. *J. Biol. Chem.* **268,** 22385–22390.

Rutter, G. A., White, M. R. H., and Tavare, J. M. (1995). Non-invasive imaging of luciferase gene expression in single living cells reveals the involvement of MAP kinase in insulin signalling. *Curr. Biol.* **5,** 890–899.

Scholz, T. D., Laughlin, M. R., Balaban, R. S., Kupriyanov, V. V., and Heineman, F. W. (1995). Effect of substrate on mitochondrial NADH, cytosolic redox state, and phosphorylated compounds in isolated hearts. *Am. J. Physiol.* **268,** H82–H91.

Soboll, S., Scholz, R., and Heldt, H. W. (1978). Subcellular metabolite concentrations: Dependence of mitochondrial and cytosolic ATP systems on the metabolic state of perfused rat liver. *Eur. J. Biochem.* **87,** 377–390.

Stanley, P. E., and Williams, S. G. (1969). Use of the liquid scintillation spectrometer for determining adenosine triphosphate by the luciferase enzyme. *Anal. Biochem.* **29,** 381–392.

Tsien, R. Y. (1998). The green fluorescent protein. *Annu. Rev. Biochem.* **67,** 509–544.

Ward, W. W., and Bokman, S. H. (1982). Reversible denaturation of *Aequorea* green fluorescent protein: Physical separation and characterization of the renatured protein. *Biochemistry* **21,** 4535–4540.

White, M. R. H., Masuko, M., Amet, L., Elliott, G., Braddock, M., Kingsman, A. J., and Kingsman, S. M. (1994). Real-time analysis of the transcriptional regulation of HIV and hCMV promoters in single mammalian cells. *J. Cell Sci.* **108,** 441–455.

Zhou, Y., Marcus, E. M., Haugland, R. P., and Opas, M. (1995). Use of a new fluorescent probe, seminaphthofluorescein-calcein, for determination of intracellular pH by simultaneous dual-emission imaging laser scanning confocal microscopy. *J. Cell Physiol.* **146,** 9–16.

CHAPTER 21

Genetic Transformation of *Saccharomyces cerevisiae* Mitochondria

Nathalie Bonnefoy* and Thomas D. Fox†

* Centre de Génétique Moléculaire
Laboratoire propre du CNRS associé à l'Université Pierre et Marie Curie
91198 Gif-sur-Yvette Cedex, France

† Department of Molecular Biology and Genetics
Cornell University
Ithaca, New York 14853

I. Introduction

A key feature of the yeast nuclear genetic system that has made it a preeminent tool for genetic and cell biological research is the fact that DNA transformed into the nuclear chromosomes of *Saccharomyces cerevisiae* is incorporated into the genome only via homologous recombination (Hinnen *et al.,* 1978). This fact allows the researcher to add, subtract, and alter genetic information in a highly controlled fashion and essentially rewrite the yeast genome at will (Rothstein, 1991).

In *S. cerevisiae,* and to date only in that species, similar manipulations based on homologous recombination have also been carried out on the mitochondrial genome. This article briefly summarizes some basic features of yeast mitochondrial genetics, describes current methods for delivery of DNA into the organelle, and outlines strategies employing homologous recombination that allow one to create directed mutations in mitochondrial genes and to insert new genes into mitochondrial DNA (mtDNA). These subjects have been reviewed previously and the reader is referred to several previous articles for more detailed discussions of the mitochondrial genetics underlying the transformation strategies discussed here (Butow *et al.,* 1996; Dujon, 1981; Fox *et al.,* 1991; Perlman *et al.,* 1979; Pon and Schatz, 1991). General methods for yeast genetics have been compiled by Guthrie and Fink (1991).

II. Important Features of *Saccharomyces cerevisiae* Mitochondrial Genetics

A. Phenotypes Associated with Mitochondrial Gene Expression

The common phenotype of mutations that affect mitochondrial genes or their expression is the inability to grow on nonfermentable carbon sources. Wild-type *S.cerevisiae* strains grow well on complete medium containing nonfermentable carbon sources, such as ethanol and glycerol [YPEG:1% yeast extract (w/v), 2% peptone (w/v), 3% ethanol (v/v) plus 3% glycerol (v/v)]. Mutants that lack a functioning oxidative phosphorylation system cannot grow on such nonfermentable medium, but grow relatively well on medium containing fermentable carbon sources such as glucose [YPD: 1 % yeast extract (w/v), 2% peptone (w/v), 2% dextrose (w/v)].

Respiratory growth of wild-type yeast can be impaired by several inhibitors of bacterial protein synthesis, and mutations conferring resistance can serve as genetic markers. Mutations in mitochondrial ribosomal RNA genes can lead to resistance to chloramphenicol (Dujon, 1980), erythromycin (Sor and Fukuhara, 1982), and paromomycin (Li *et al.,* 1982). Mutations causing resistance to the ATP synthase inhibitor oligomycin (Ooi *et al.,* 1985; Sebald *et al.,* 1979) and the cytochrome *b* inhibitor diuron (di Rago *et al.,* 1986) also provide mitochondrial genetic markers. However, it is important to note that these drug-resistant phenotypes arise spontaneously and can only be observed on nonfermentable medium in strains that respire. Thus, they are not ideal for use as selective markers in transformation experiments.

Novel mitochondrial phenotypes have been generated by placing foreign genes into mtDNA. Phenotypes based on foreign genes can serve as mitochondrial genetic markers independently of respiratory function. One such phenotype is based on the fact that nuclear genes such as *URA3* and *TRP1* cannot be expressed when inserted into mtDNA, but will escape from mitochondria to the nucleus at high frequency, leading to detectable growth phenotypes that can be scored on petri plates (Thorsness and Fox, 1990, 1993).

At least some nuclear genes can be expressed phenotypically within mitochondria if they are rewritten in the *S. cerevisiae* mitochondrial genetic code (Fox, 1987), providing novel selectable markers. Expression of the synthetic gene *ARG8*^{m} within mitochondria allows nuclear *arg8* mutants to grow without arginine (Steele *et al.*, 1996). This protein, Arg8p, is normally imported into mitochondria from the cytoplasm, but also functions when synthesized within the organelle. Thus, Arg^{+} prototrophy can become a phenotype dependent on mitochondrial gene expression. This new mitochondrial marker provides a convenient way to disrupt endogenous mitochondrial genes (Sanchirico *et al.*, 1998) as well being a useful reporter for studying mitochondrial gene expression (Bonnefoy and Fox, 2000; Steele *et al.*, 1996) and genetic instability (Sia *et al.*, 2000). In addition, because *ARG8*^{m} specifies a soluble protein, translational fusions to endogenous mitochondrial genes can create chimeric proteins useful for the study of targeting and membrane translocation of mitochondrial translation products (He and Fox, 1997, 1999). A visible reporter phenotype based on mitochondrial gene expression can also be generated by insertion into mtDNA of a synthetic gene, *GFP*^{m}, encoding the green fluorescent protein rewritten in the yeast mitochondrial code (J. S. Cohen and T. D. Fox, unpublished results).

B. Replication of mtDNA

Replication of yeast mtDNA is a complex and poorly understood process. Cells typically contain between 50 and 100 genome equivalents of mtDNA (Dujon, 1981), which are organized into a smaller number of nucleoid structures. The nucleoids, which are visible by fluorescence microscopy, are the genetic elements transmitted to daughter cells during cell division (Lockshon *et al.*, 1995; Newman *et al.*, 1996). Replication of complete, or *rho*$^{+}$, yeast mitochondrial genomes is thought to depend on a limited number of specific sites in the chromosome (de Zamaroczy *et al.*, 1984; Schmitt and Clayton, 1993) and, for unknown reasons, requires mitochondrial protein synthesis (Myers *et al.*, 1985).

The most frequent mutants in wild-type *S. cerevisiae* strains are the nonrespiring *rho*$^{-}$, or cytoplasmic petite mutants. These strains have large deletions of mtDNA that destroy the organellar gene expression machinery by deleting components of its translation system (Dujon, 1981). The DNA sequences retained in *rho*$^{-}$ mutants are typically reiterated, such that the *rho*$^{-}$ cell contains roughly the same amount of mtDNA as the wild type. Replication of *rho*$^{-}$ mtDNA differs from that of *rho*$^{+}$ mtDNA in that it does not require mitochondrial protein synthesis. Interestingly, the mtDNA sequences replicating in *rho*$^{-}$ strains can be derived from any portion of the chromosome, demonstrating that there is no clear requirement for a specific replication origin sequence in *rho*$^{-}$ mtDNAs. This is advantageous in creating mitochondrial transformants containing defined mtDNAs, as *rho*^{0} yeast strains, entirely lacking mtDNA, can be transformed with bacterial plasmid DNAs that subsequently propagate as "synthetic" *rho*$^{-}$ molecules (Fox *et al.,* 1988).

C. Recombination and Segregation of mtDNA

Unlike the highly differentiated situation in animals and plants, there is true equality of the sexes in yeast mating. Haploid cells mate by fusion, cytoplasms are mixed, and mitochondria of the haploid cells fuse to form an essentially continuous compartment (Azpiroz and Butow, 1993; Nunnari *et al.,* 1997). Homology-dependent recombination between the parental mtDNAs occurs at a high rate (Dujon, 1981) in the medial portion of the zygote (Azpiroz and Butow, 1993). Of great importance to the manipulation of mitochondrial DNA is the fact that *rho*$^{-}$ mtDNA sequences recombine readily with complete *rho*$^{+}$ genomes. Thus, if a *rho*$^{-}$ mtDNA contains wild-type genetic information in a particular region, it can recombine with a *rho*$^{+}$ mtDNA bearing a mutation in that region. The result is that the mating of the nonrespiring *rho*$^{-}$ with the nonrespiring *rho*$^{+}$ mutant yields respiring recombinants at high frequency, whose growth can be selected on nonfermentable medium.

Yeast cells containing two different kinds of mtDNA can be created by mating or by mutation. Such heteroplasmic cells rapidly give rise to homoplasmic progeny, a phenomenon known as mitotic segregation (Dujon, 1981). [However, exceptional cases of stable heteroplasmy in *S. cerevisiae* have been reported (Lewin *et al.,* 1979).] *S. cerevisiae* differs in this regard from plant and animal cells, which can maintain heteroplasmic states for extended periods of growth (Hanson and Folkerts, 1992; Wallace, 1992). In this connection it is important to note that, unlike most animal and plant cells, *S. cerevisiae* divides by a highly asymmetric budding process. New buds receive relatively few copies of mtDNA from mother cells (Zinn *et al.,* 1987), facilitating mitotic segregation. Heteroplasmic cells generated by transformation and subsequent homologous recombination also produce pure recombinant clones (Johnston *et al.,* 1988).

Taken together, these features of the *S. cerevisiae* mitochondrial genetic system allow DNA introduced from outside the cell to be propagated within the organelle as a plasmid, and the plasmid-borne mitochondrial sequences to recombine homologously with complete *rho*$^{+}$ mtDNA (Fox *et al.,* 1988).

III. Delivery of DNA to the Mitochondrial Compartment of *rho*^{0} Cells and Detection of Mitochondrial Transformants

A. Overview of Transformation Procedure

Exogenous DNA can be introduced in mitochondria of yeast cells via microprojectile bombardment (Johnston *et al.,* 1988). The standard device for microprojectile bombardment is the PDS-1000/He System, available from Bio-Rad, Inc. (http://www.bio-rad.co/templates/html/64343_products.html). This instrument uses a helium shock wave in an evacuated chamber to accelerate microscopic metal particles toward a lawn of cells on a petri plate. The shock wave is generated by rupture of a membrane at high pressure and accelerates a second membrane (the macrocarrier or flying disk), carrying the metal particles, toward the plate. Some cells on the plate are penetrated by particles and survive. DNA precipitated on the particles is thus introduced into cells and is taken up readily by the nucleus. In addition, mitochondria of a small fraction of such transformants also take up DNA. The PDS-1000 functions reproducibly for transformation of *S. cerevisiae* mitochondria.

In a typical mitochondrial transformation experiment, a large number of *rho*^{0} cells are bombarded randomly by a large number of particles (Fig. 1). In the first step, cells that

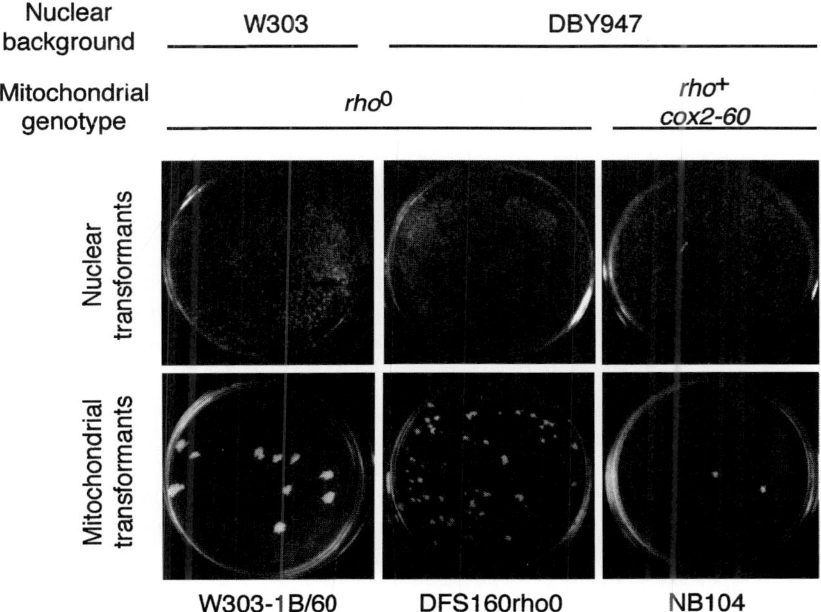

Fig. 1 Nuclear transformants and mitochondrial cotransformants obtained by bombardment of different yeast strains. The nuclear *LEU2* plasmid Yep351 (Hill *et al.*, 1986) and the *COX2* plasmid pNB69 (Bonnefoy and Fox, 2000) were precipitated together onto tungsten particles and bombarded on lawns of the *rho*^{0} strains W303-1B/60 [*MAT*α, *ade2-1, ura3-1, his3-11,15, trp1-1, leu2-3,112, can1-100* (*rho*^{0})] [a *rho*^{0} derivative of-W303-1B (Thomas and Rothstein, 1989) and DFS160 (*MAT*α *ade2-101, leu2*Δ, *ura3-52, arg8*Δ::*URA3, kar1-1* (*rho*^{0})] (Steele *et al.*, 1996), or on lawns of the *rho*$^+$ strain NB104. W303-1B/60 was derived from W303 (ATCC 200060). DFS160 was derived from DBY947 (Neff *et al.*, 1983). NB104 *rho*$^+$ mtDNA carries a 129-bp deletion, *cox2-60*, located around the *COX2* first codon (Bonnefoy and Fox, 2000) and is isonuclear to DFS160. The top plates correspond to minimal medium supplemented with sorbitol and lacking leucine. Typical plates showing about 3000 nuclear transformants for each strain are shown. Nuclear transformants were crossed by replica plating to the nonrespiring tester strain (NB160), carrying a mutation of the *COX2* initiation codon (Bonnefoy and Fox, 2000), and mitochondrial transformants (bottom plates) were detected by replica plating the mated cells onto nonfermentable medium.

have been hit and that survived are allowed to make colonies on the petri plates by selecting for a nuclear genetic marker that is included in the DNA precipitated on the particles. Mitochondrial transformants are identified among these colonies by genetic tests for the presence of new genetic information in the mitochondrial genome. This new information is typically a portion of the wild-type mtDNA sequence that can rescue a known mitochondrial marker mutation by recombination, after the transformants are mated to an appropriate *rho*$^+$ tester strain, resulting in recombinants with a detectable growth phenotype. The new wild-type sequence may be an unaltered region of the gene of interest or it may be another piece of wild-type mtDNA incorporated into a vector. Such marker rescue can work with as little as 50 bp of homologous sequence flanking the site of the mutation in the tester mtDNA. As shown in Fig. 1, transformation of *rho*$^+$ mutants can also be detected using this marker rescue strategy (or, as discussed in Section V, by directly selecting for a phenotypic change). Transformants can also be identified by scoring for expression of complete mitochondrial genes that function in *trans* (Butow *et al.,* 1996; Fox *et al.,* 1988).

B. Experimental Details for Transformation and Identification of Mitochondrial Transformants

1. Yeast Strains

Strain background is an important factor affecting the efficiency of transformation (Fig. 1). We have obtained the best results with strains in the S288c background, in particular those derived from DBY947 (Neff *et al.,* 1983). Strains derived from W303 (ATCC 200060) (Thomas and Rothstein, 1989) give lower but satisfactory efficiencies, whereas strains in the D273-10B (ATCC 24657) background are difficult to transform. Excellent hosts for mitochondrial transformation, derived from DBY947, can be obtained from the American Type Culture Collection: MCC109rh0 [*MAT*α, *ade2-101, ura3-52, kar1-1 (rho*^{0})] (Costanzo and Fox, 1993) and MCC123rho0, which is the identical strain with *MAT*a, available as ATCC 201440 and 201442, respectively.

2. Preparation of Cells

a. Grow the *rho*^{0} (or *rho*$^+$) strain to be bombarded for 2 to 3 days (stationary phase) at 30°C with agitation in complete liquid medium (YP) containing either 2% raffinose or 2% galactose. These media may be supplemented with 0.1 % glucose (to accelerate growth) and/or 100 mg/ml adenine (for Ade$^-$ auxotrophs).

b. Harvest cells and concentrate 40 to 100 times in liquid YPD medium to reach a cell density of 1 to 5 × 10^9 cells/ml.

c. Spread 0.1 ml of cells onto minimal glucose medium (0.67% yeast nitrogen base, 5% glucose, 100 mg/ml adenine, 3.3% agar) containing 1 *M* sorbitol and supplemented to provide the appropriated prototrophic selection.

3. Preparation of Microprojectiles and Precipitation of DNA

We use 0.5 μm tungsten powder obtained from Alfa-Aesar which, unfortunately, is no longer available; 0.4- and/or 0.7-μm tungsten particles available from Bio-Rad, Cat. Nos. 165-2265 and 165-2266, respectively, should be essentially equivalent. The following describes our current procedure using tungsten particles. With the exception of the sterilization step described later, the same procedure gives comparable results with gold particles (0.6 μm, Bio-Rad Cat. No. 165-2262) in our hands. A slightly different procedure for the preparation of gold particles has been described previously (Butow *et al.*, 1996) and gives equivalent results in our hands.

a. Sterilize up to 50 mg of tungsten particles by suspension in 1.5 ml of 70% ethanol in a microfuge tube and incubation at room temperature for 10 min. Wash the particles with 1.5 ml of sterile water and resuspend at 60 mg/ml in sterile 50% glycerol. Particles can be kept frozen for several months. Gold particles should be sterilized in 100% ethanol (Butow *et al.*, 1996).

b. In a microfuge tube, mix 5 μg of plasmid carrying the nuclear marker and a nuclear replication origin with 15 to 30 μg of plasmid carrying the mitochondrial DNA of interest, in a total volume of 15–20 μl. Add and mix 100 μl of tungsten particles, 4 μl of 1 *M* spermidine free base, and 100 μl of ice-cold 2.5 *M* $CaCl_2$. Incubate for 10 min on ice with occasional vortexing.

c. Spin briefly and remove the supernatant. Resuspend the particles thoroughly in 200 μl of 100% ethanol, taking extreme care to fragment aggregates of particles, using the pipette tip. Repeat least once until the particles resuspend easily.

d. Spin briefly, remove the supernatant, and add 50–60 μl of 100% ethanol. Distribute the resulting suspension evenly at the center of six macrocarriers (flying disks) placed in their holders, allowing the ethanol to evaporate (there is no need to prewash the macrocarriers or desiccate them after coating).

4. Bombardment

a. Follow the manufacturer's instructions carefully for use of the PDS-1000 apparatus. Place the rupture disk in its retaining cap and tighten using a torque wrench. Rupture disks of 1100 to 1350 psi can be used for the efficient transformation of yeast, although in our hands 1100 psi disks tend to give better results.

b. Load the macrocarrier in its holder into the assembly system. Interestingly, we have found that simply allowing the carrier disk to fly to the surface of the petri plate, by not assembling the stopping screen, yields more transformants than if the stopping screen is employed.

c. Place the open petri plate carrying the lawn of cells at 5 cm from the macrocarrier assembly. Shorter distances result in very high colony densities in the center of the plate with few colonies at the periphery, whereas longer distances decrease the transformation efficiency.

d. Evacuate the vacuum chamber to a reading of 29 to 29.5 in. Hg on the gauge of the PDS-1000. We have found that failure to draw the greatest vacuum possible reduces transformation efficiency. Cell viability is not affected significantly by a prolonged stay under these vacuum conditions.

e. Fire.

f. Remove any fragments of the macrocarrier disk with a sterile forceps. Incubate the plate at 30°C for 4 to 5 days until colonies appear (between 1000 and 10,000 per plate for S288c-related strains).

5. Identification of Mitochondrial Transformants

a. During the incubation of the bombarded plates, set up a liquid YPD culture of an appropriate *rho*$^+$ mutant (*mit*$^-$) tester strain.

b. Replica plate the transformants onto a lawn of the tester strain freshly spread on a YPD plate.

c. Incubate at 30°C for 2 days to allow mating and recombination.

d. Print to YPEG medium (or another appropriate selection medium) to detect respiring diploids. In cases where a high number of nuclear transformants are present, it may be useful to also replicate the mated cells on medium that selects for the diploids, as comparison of the resulting plates may facilitate the identification of the desired transformant on the original bombarded plate.

e. Pick colonies off the bombarded plate that correspond to the position of respiring recombinants. Streak these colonies on YPD and repeat the marker rescue with the tester strain as described earlier. Such subcloning and retesting must usually be done three times before pure stable synthetic *rho*$^-$ clones are obtained. (Cells usually lose the nuclear marker plasmid during these subcloning steps if no selection is applied for its maintenance.)

IV. Strategies for Gene Replacement in *S. cerevisiae* mtDNA

In cases where the mitochondrial gene under study encodes an active RNA molecule, such as the mitochondrial RNase P RNA, it may be possible to assay the activity of wild-type and mutant genes in the primary synthetic *rho*$^-$ transformants (Sulo *et al.*, 1995). More commonly, however, mutations affecting protein-coding genes must be placed into *rho*$^+$ mtDNA by a double recombination event.

A. Integration of Altered mtDNA Sequences by Homologous Double Crossovers

The most basic method for putting a mutant version of a mitochondrial gene, or a foreign piece of DNA flanked by mtDNA sequences, into the chromosome is to first introduce the altered sequence into a *rho*^{0} strain to create a synthetic *rho*$^-$. This donor transformant (identified as described in Section III) is then (in a second step) mated with a wild-type *rho*$^+$ recipient strain. As a result of this second mating, mitochondria from the two strains fuse, and recombination between the two mtDNAs produces recombinant *rho*$^+$ strains in which the new mtDNA sequence is integrated by double crossover events. Pure recombinant strains are generated by subsequent mitotic segregation. Because mitochondrial DNA recombination and segregation are so frequent, this simple procedure typically yields the desired integrants at frequencies between 1 and 50% of clones derived from zygotes.

If one of the strains in such a cross carries the karyogamy-defective mutation *karl-1* (Conde and Fink, 1976), which allows efficient mitochondrial fusion but reduces nuclear fusion greatly, haploid mitochondrial mutant cytoductants can be isolated after such a mating. This simple strategy has been used successfully with variations specific to each study in several laboratories (Boulanger *et al.*, 1995; Folley and Fox, 1991; Henke *et al.*, 1995; Mulero and Fox, 1994; Speno *et al.*, 1995; Szczepanek and Lazowska, 1996; Thorsness and Fox, 1993).

B. Experimental Details for Mating and Isolation of Recombinant Cytoductants

1. Mating

a. Grow cultures of the subcloned synthetic *rho*$^-$ strain and the recipient wild-type *rho*$^+$ strain overnight in liquid YPD. At least one of these two strains (usually the synthetic *rho*$^-$) must carry the *kar1-1* mutation.

b. If the synthetic *rho*$^-$ donor and the *rho*$^+$ recipient strains share nuclear markers, and therefore cannot be distinguished selectively on glucose medium, mating mixtures should contain equal numbers of cells of both strains. If nuclear auxotrophic or drug resistance markers allow selection against the synthetic *rho*$^-$ donor strain, then the mating mixture should contain a fivefold excess of donor cells.

c. We have successfully used two different mating protocols for producing cytoductants.

1. Mix 0.5 ml of each parent (alternatively 1 ml of synthetic *rho*$^-$ and 0.2 ml of wild-type *rho*$^+$) in a microfuge tube, spin, remove the supernatant, resuspend in residual liquid, and spread the mixture onto a YPD plate. Incubate at 30°C for 4 to 5 h. Check zygote formation microscopically. Scrape the mating cells from the plate and use them to inoculate fresh YPD liquid medium. Incubate at 30°C with agitation for a few hours to overnight.

2. Alternatively, mix both parents, in proportions as just described, in 10 ml of liquid YPD and shake at 30°C for 3 h. Spin the culture in a tube and incubate the pellet at 30°C for 1 h without removing the medium. Resuspend by vortexing, transfer to a fresh flask, and incubate at 30°C with agitation for at least 3 h.

2. Isolation of Cytoductants

a. Dilute the culture to obtain single colonies and plate on minimal medium, selecting for the recipient nuclear genotype and against the donor nuclear genotype, if possible.

Alternatively, plate on YPD medium. Densities of 50 to 200 colonies per plates should be obtained.

b. Replica plate the colonies thus obtained to medium that will reveal the altered phenotype of the recipient strain as a result of integration of the mutant donor sequences into its mtDNA. For example, print to YPEG to identify clones that have acquired a mutation preventing respiratory growth.

c. Mate nonrespiring candidate clones to a *rho*$^+$ tester mutant whose mitochondrial mutation is located outside of the region carried by the synthetic *rho*$^-$. The desired *rho*$^+$ recombinant cytoductants will produce respiring diploids after mating to this tester strain. (This step eliminates cytoductants that simply acquired the mtDNA of the donor strain.)

C. Streamlining the Integration of Multiple Mutations in a Short Region by Use of *rho*$^+$ Recipients Containing Defined Deletion

In situations where many nonfunctional mutations are to be placed in the same region of mtDNA, the strategy just described can be altered and made more efficient by using a *rho*$^+$ recipient that has a defined deletion in the region of interest (Costanzo and Fox, 1993; Mittelmeier and Dieckmann, 1993). It is, of course, often necessary to isolate such a recipient strain using the simple recombination method described previously. However, once the recipient is in hand, the nonrespiring recombinant cytoductants from crosses between nonrespiring synthetic *rho*$^-$ strains and the nonrespiring recipient can be identified by a positive marker rescue screen employing an appropriate tester strain (Fig. 2). Following mating between the synthetic *rho*$^-$ (carrying the experimental mutation "e" in Fig. 2) and the recipient, the cell population is plated on medium selecting for the recipient nuclear genotype. Recombinant cytoductants, unaltered recipient cells, and diploid cells will form colonies. However, only the recombinant cytoductants will be able to form respiring diploids when mated to a *rho*$^+$ tester strain carrying a marker mutation ("m" in Fig. 2) located in the deleted region (and distinct from the experimentally induced mutation).

D. Experimental Details for Identification of Nonrespiring Cytoductants by Marker Rescue

The following steps assume that one already has in hand a *rho*$^+$ recipient with a deletion mutation in the region of interest, as well as a *rho*$^-$ strain that carries wild-type information and that can recombine with the recipient deletion mutant.

1. Carry out a mating between the synthetic *rho*$^-$ donor and the *rho*$^+$ recipient containing a defined deletion to generate cytoductants as described in Section IV,B,1 (at least one of these two strains must carry the *kar1-1* mutation), and spread dilutions of the cell mixture as described in Section IV,B,2. Some of these colonies will be *rho*$^+$ recombinants that have the deleted region restored and contain the desired mutation (Fig. 2A).

2. Replica plate colonies from the mating mixture onto a YPD plate bearing a freshly spread lawn of a *rho*$^+$ tester strain that has a marker mutation within the deleted region of the recipient but distinct from the new mutation to be introduced (Fig. 2B).

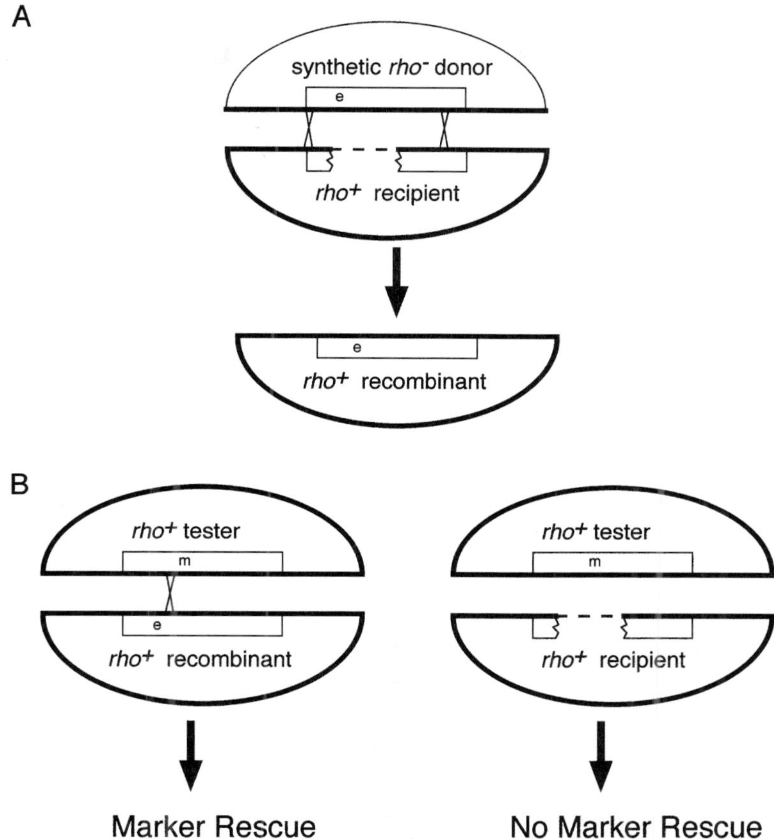

Fig. 2 Schematic diagram of recombination events that allow identification of nonrespiring recombinant cytoductants by marker rescue. Thick lines represent mtDNA sequences, and thin lines represent vector DNA. The box represents a gene under study. (A) A karyogamy defective (*kar1-1*) synthetic *rho*$^-$ donor containing an experimentally induced mutation "e" is mated to a *rho*$^+$ recipient strain with a deletion in the region of interest. The desired *rho*$^+$ recombinant cytoductants are among the cells present in the mixture after mating. (B) To distinguish the desired *rho*$^+$ recombinant cytoductants from the unaltered recipient cells and other cell types present, clones derived from the mating mixture are mated to a *rho*$^+$ tester strain bearing the marker mutation "m." The desired *rho*$^+$ recombinant cytoductants can yield respiring recombinants when mated to this tester by a crossover between "e" and "m" (and a second resolving crossover anywhere else). The ability to produce respiring recombinants identifies the desired cytoductant clones. Unaltered recipient clones, and other cells types present, cannot yield such respiring recombinants.

3. After 2 days of incubation at 30°C, print the mated cells to YPEG. Identify haploid cytoductant clones on the plates from step 1 that correspond to respiring diploids in the mating of step 2.

4. Test the candidate cytoductant clones to make sure that they are *rho*$^+$ mutants by checking to see that they yield respiring diploids when mated to a *rho*$^-$ strain carrying

wild-type information in the region deleted in the original recipient strain. This will eliminate cytoductants that simply contain the unrecombined mtDNA of the synthetic *rho*$^-$ donor.

V. Transformation of *rho*$^+$ Cells with Plasmids or Linear DNA Fragments

The first demonstration of the ability of the microprojectile bombardment to deliver DNA into yeast mitochondria depended on integration of the transforming DNA directly into a mutant *rho*$^+$ strain, converting a nonrespiring mutant into a respiring transformant (Johnston *et al.,* 1988). In addition to selection for restoration of respiratory growth, *rho*$^+$ transformants can also be obtained with DNA sequences causing expression of *ARG8*^{m} and selection for Arg$^+$ growth. In cases where a mutated DNA sequence provides a function that can be selected for phenotypically, direct transformation of *rho*$^+$ strains bearing deletions in the region of interest can be used to integrate mutations into mtDNA (Bonnefoy and Fox, 2000).

In our hands, mitochondrial transformation is 10 to 20 times more efficient during bombardment of a *rho*^{0} strain than of an isogenic *rho*$^+$ strain containing a small deletion in mtDNA (Fig. 1). This effect could be due either to physiological differences between the strains, such as the properties of the mitochondrial inner membrane, or to an advantage in establishing an incoming DNA molecule in the absence of endogenous mtDNA. Effects consistent with the latter notion have been observed in comparisons of mtDNA behavior after *rho*$^+$ × *rho*$^+$ matings as opposed to *rho*$^+$ × *rho*^{0} matings (Azpiroz and Butow, 1993).

Nevertheless, the relative inefficiency of transformation of *rho*$^+$ hosts is offset in some situations by its convenience, because *rho*$^+$ recombinants may be obtained more quickly. In addition, this strategy can be extended to transformation with linear DNA molecules obtained either from plasmid clones or by polymerase chain reaction (PCR) amplification. For example, we have found that linear DNA fragments having as little as 260 bp of homologous sequence flanking each side of a deletion mutation in a *rho*$^+$ recipient were able to yield respiring transformants at frequencies similar to those obtained with circular plasmids. The ability to use PCR-generated fragments to transform defined mtDNA deletion recipients can accelerate strain construction substantially.

Selection for transformation by DNA fragments with wild-type or near-wild-type function is straightforward by selecting first for nuclear transformants on glucose medium, as described in Section III, and then screening for the mitochondrial phenotype by replica plating (Fig. 3) (Johnston *et al.,* 1988). Interestingly, two- to fivefold fewer transformants are detected by this phenotypic selection for the integration of transforming DNA into *rho*$^+$ mtDNA than are detected by mating of the *rho*$^+$ colonies to a tester strain and scoring for respiring recombinants (as was done in the experiment of Fig. 1). Thus it appears that only a fraction of the plasmids entering a *rho*$^+$ mitochondrion during bombardment manage to integrate and become expressed phenotypically.

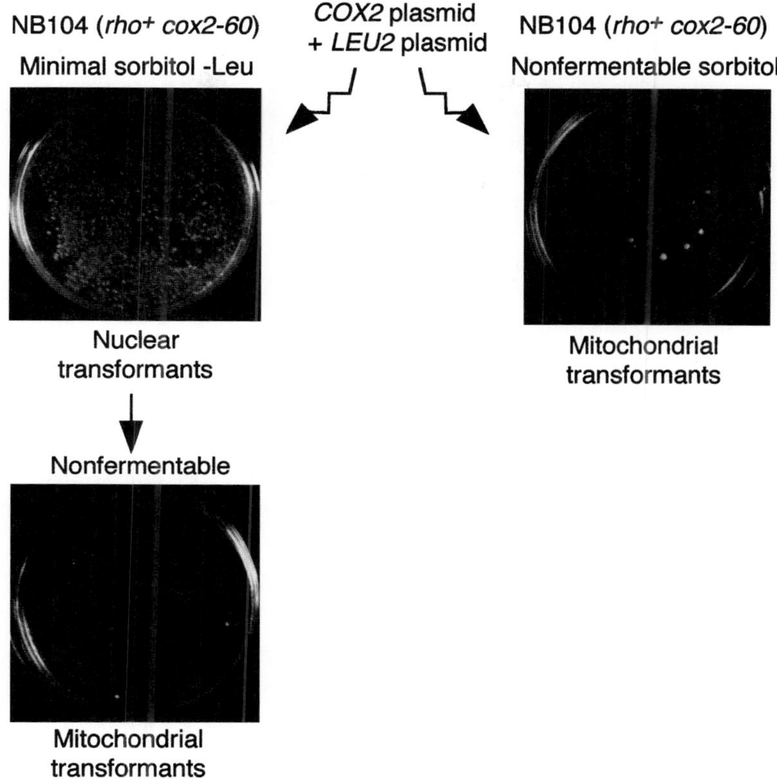

Fig. 3 Selection of *rho*$^+$ mitochondrial transformants directly after bombardment. The nuclear shuttle vector Yep351 (*LEU2*) and plasmid pNB69 (*COX2*) were bombarded together onto lawns of the *rho*$^+$ *cox2-60* strain NB104 (see legend of Fig. 1). The bombarded lawns had been spread either on minimal medium supplemented with sorbitol but lacking leucine (top right) or on nonfermentable YPEG medium supplemented with sorbitol and 0.1% glucose (top left). Leu$^+$ transformants were replica plated to nonfermentable medium (bottom plate) to select for mitochondrial transformants.

Selection of respiring transformants from nonrespiring host strains generally requires a period of outgrowth on a fermentable carbon source to allow establishment of the respiratory phenotype prior to selection for it. This phenotypic lag affects selection for respiring transformants with both nuclear (Müller and Fox, 1984) and mitochondrial genes (Johnston *et al.*, 1988). However, we have been able to successfully select respiring mitochondrial transformants in a single step by directly bombarding lawns of mutant *rho*$^+$ cells spread on YPEG plates supplemented with 0.1 % glucose (to allow a brief period of outgrowth) and 1 *M* sorbitol (Fig. 3). We have also been able to select for *rho*$^+$ transformants expressing *ARG8*^{m} by directly bombarding lawns spread on appropriate minimal glucose medium (see Section III,B,2).

The detection of transformants of a *rho*$^+$ recipient that have incorporated a DNA sequence bearing a nonfunctional mutation is difficult, but possible, using essentially the same strategy as outlined in Fig. 2. If the recipient strain has a defined deletion, and integration of the transforming DNA restores the deleted sequence, then the desired integrants can be detected by their ability to rescue a known marker mutation in the deleted region.

VI. Concluding Remarks

The methods developed for manipulation of the *S. cerevisiae* mitochondrial genome should provide a useful model for other systems. Indeed, in *Chlamydomonas*, another single-celled eukaryote in which nonreverting deletion mutations of mtDNA have been isolated, mitochondrial transformation and integration of wild-type DNA by homologous recombination have been achieved (Boynton and Gillham, 1996). Thus, as appropriate selectable markers are developed for other species, it seems likely that their mitochondrial genomes will become amenable to *in vivo* experimental analysis.

Acknowledgments

We thank David M. MacAlpine and Ronald A. Butow for advice on the use of gold particles. N.B. was a Human Frontier Science Program Organization long-term fellow (LT22/96) during the early stages of this work and is currently supported by the Association Française contre les Myopathies. T.D.F. is supported by a grant from the U.S. National Institutes of Health (GM29362).

References

Azpiroz, R., and Butow, R. A. (1993). Patterns of mitochondrial sorting in yeast zygotes. *Mol. Biol. Cell.* **4,** 21–36.

Bonnefoy, N., and Fox, T. D. (2000). *In vivo* analysis of mutated initiation codons in the mitochondrial *COX2* gene of *Saccharomyces cerevisiae* fused to the reporter gene *ARG8*^{m} reveals lack of downstream reinitiation. *Mol. Gen. Genet.* **262,** 1036–1046.

Boulanger, S. C., Belcher, S. M., Schmidt, U., Dib-Hajj, S. D., Schmidt, T., and Perlman, P. S. (1995). Studies of point mutants define three essential paired nucleotides in the domain 5 substructure of a group II intron. *Mol. Cell. Biol.* **15,** 4479–4488.

Boynton, J. E., and Gillham, N. W. (1996). Genetics and transformation of mitochondria in the green alga *Chlamydomonas. Methods Enzymol.* **264,** 279–296.

Butow, R. A., Henke, M., Moran, J. V., Belcher, S. M., and Perlman, P. S. (1996). Transformation of *Saccharomyces cerevisiae* mitochondria by the biolistic gun. *Methods Enzymol.* **264,** 265–278.

Conde, J., and Fink, G. R. (1976). A mutant of *S. cerevisiae* defective for nuclear fusion. *Proc. Natl. Acad. Sci. USA* **73,** 3651–3655.

Costanzo, M. C., and Fox, T. D. (1993). Suppression of a defect in the 5′-untranslated leader of the mitochondrial *COX3* mRNA by a mutation affecting an mRNA-specific translational activator protein. *Mol. Cell. Biol.* **13,** 4806–4813.

de Zamaroczy, M., Faugeron-Fonty, G., Baldacci, G., Goursot, R., and Bernardi, G. (1984). The *ori* sequences of the mitochondrial genome of a wild-type yeast strain: Number, location, orientation and structure. *Gene* **32,** 439–457.

di Rago, J. P., Perea, X., and Colson, A. M. (1986). DNA sequence analysis of diuron-resistant mutations in the mitochondrial cytochrome *b* gene of *Saccharomyces cerevisiae. FEBS Lett.* **208,** 208–210.

Dujon, B. (1980). Sequence of the intron and flanking exons of the mitochondrial 21S rRNA gene of yeast strains having different alleles at the *omega* and *rib-1* loci. *Cell* **20,** 185–197.

Dujon, B. (1981). Mitochondrial genetics and functions. *In* "The Molecular Biology of the Yeast *Saccharomyces*, Life Cycle and Inheritance" (J. N. Strathern, E. W. Jones, and J. R. Broach, eds.), pp. 505–635. Cold Spring Harbor Laboratory Press, Cold Spring Harbor, NY.

Folley, L. S., and Fox, T. D. (1991). Site-directed mutagenesis of a *Saccharomyces cerevisiae* mitochondrial translation initiation codon. *Genetics* **129,** 659–668.

Fox, T. D. (1987). Natural variation in the genetic code. *Annu. Rev. Genet.* **21,** 67–91.

Fox, T. D., Folley, L. S., Mulero, J. J., McMullin, T. W., Thorsness, P. E., Hedin, L. O., and Costanzo, M. C. (1991). Analysis and manipulation of yeast mitochondrial genes. *Methods Enzymol.* **194,** 149–165.

Fox, T. D., Sanford, J. C., and McMullin, T. W. (1988). Plasmids can stably transform yeast mitochondria lacking endogenous mtDNA. *Proc. Natl. Acad. Sci. USA* **85,** 7288–7292.

Guthrie, C., and Fink, G. R. (eds.) (1991). Guide to yeast genetics and molecular biology. *Methods Enzymol.* **194**.

Hanson, M. R., and Folkerts, O. (1992). Structure and function of the higher plant mitochondrial genome. *Int. Rev. Cytol.* **141,** 129–172.

He, S., and Fox, T. D. (1997). Membrane translocation of mitochondrially coded Cox2p: Distinct requirements for export of amino- and carboxy-termini, and dependence on the conserved protein Oxa1p. *Mol. Biol. Cell.* **8,** 1449–1460.

He, S., and Fox, T. D. (1999). Mutations affecting a yeast mitochondrial inner membrane protein, Pnt1p, block export of a mitochondrially synthesized fusion protein from the matrix. *Mol. Cell. Biol.* **19,** 6598–6607.

Henke, R. M., Butow, R. A., and Perlman, P. S. (1995). Maturase and endonuclease functions depend on separate conserved domains of the bifunctional protein encoded by the group I intron aI4 alpha of yeast mitochondrial DNA. *EMBO J.* **14,** 5094–5099.

Hill, J. E., Myers, A. M., Koerner, T. J., and Tzagoloff, A. (1986). Yeast/*E. coli* shuttle vectors with multiple unique restriction sites. *Yeast* **2,** 163–167.

Hinnen, A., Hicks, J. B., and Fink, G. R. (1978). Transformation of yeast. *Proc. Natl. Acad. Sci. USA* **75,** 1929–1933.

Johnston, S. A., Anziano, P. Q., Shark, K., Sanford, J. C., and Butow, R. A. (1988). Mitochondrial transformation in yeast by bombardment with microprojectiles. *Science* **240,** 1538–1541.

Lewin, A. S., Morimoto, R., and Rabinowitz, M. (1979). Stable heterogeneity of mitochondrial DNA in grande and petite strains of *S. cerevisiae. Plasmid* **2,** 474–484.

Li, M., Tzagoloff, A., Underbrink-Lyon, K., and Martin, N. C. (1982). Identification of the paromomycin-resistance mutation in the 15S rRNA gene of yeast mitochondria. *J. Biol. Chem.* **257,** 5921–5928.

Lockshon, D., Zweifel, S. G., Freeman-Cook, L. L., Lorimer, H. E., Brewer, B. J., and Fangman, W. L. (1995). A role for recombination junctions in the segregation of mitochondrial DNA in yeast. *Cell* **81,** 947–955.

Mittelmeier, T. M., and Dieckmann, C. L. (1993). In vivo analysis of sequences necessary for CBP1-dependent accumulation of cytochrome *b* transcripts in yeast mitochondria. *Mol. Cell. Biol.* **13,** 4203–4213.

Mulero, J. J., and Fox, T. D. (1994). Reduced but accurate translation from mutant AUA initiation codon in the mitochondrial *COX2* mRNA of *Saccharomyces cerevisiae. Mol. Gen. Genet.* **242,** 383–390.

Müller, P. P., and Fox, T. D. (1984). Molecular cloning and genetic mapping of the PET494 gene of *Saccharomyces cerevisiae Mol. Gen. Genet.* **195,** 275–280.

Myers, A. M., Pape, L. K., and Tzagoloff, A. (1985). Mitochondrial protein synthesis is required for maintenance of intact mitochondrial genomes in *Saccharomyces cerevisiae. EMBO J.* **4,** 2087–2092.

Neff, N. F., Thomas, J. H., Grisafi, P., and Botstein, D. (1983). Isolation of the β-tubulin gene from yeast and demonstration of its essential function in vivo. *Cell* **33,** 211–219.

Newman, S. M., Zelenaya-Troitskaya, O., Perlman, P. S., and Butow, R. A. (1996). Analysis of mitochondrial DNA nucleoids in wild-type and a mutant strain of *Saccharomyces cerevisiae* that lacks the mitochondrial HMG box protein Abf2p. *Nucleic Acids Res.* **24,** 386–393.

Nunnari, J., Marshall, W. F., Straight, A., Murray, A., Sedat, J. W., and Walter, P. (1997). Mitochondrial transmission during mating in *Saccharomyces cerevisiae* is determined by mitochondrial fusion and fission and the intramitochondrial segregation of mitochondrial DNA. *Mol. Biol. Cell.* **8,** 1233–1242.

Ooi, B. G., Novitski, C. E., and Nagley, P. (1985). DNA sequence analysis of the *oli1* gene reveals amino acid changes in mitochondrial ATPase subunit 9 from oligomycin-resistant mutants of *Saccharomyces cerevisiae Eur. J. Biochem.* **152,** 709–714.

Perlman, P. S., Birky, C. W., Jr., and Strausberg, R. L. (1979). Segregation of mitochondrial markers in yeast. *Methods Enzymol.* **56,** 139–154.

Pon, L., and Schatz, G. (1991). Biogenesis of yeast mitochondria. *In* "The Molecular and Cellular Biology of the Yeast *Saccharomyces*: Genome Dynamics, Protein Synthesis and Energetic" (J. R. Broach, J. R. Pringle, and E. W. Jones, eds.), Vol. 1, pp. 333–406. Cold Spring Harbor Laboratory Press, Cold Spring Harbor, NY.

Rothstein, R. (1991). Targeting, disruption, replacement and allele rescue: Integrative DNA transformation in yeast. *Methods Enzymol.* **194,** 281–301.

Sanchirico, M. E., Fox, T. D., and Mason, T. L. (1998). Accumulation of mitochondrially synthesized *Saccharomyces cerevisiae* Cox2p and Cox3p depends on targeting information in untranslated portions of their mRNAs. *EMBO J.* **17,** 57961–5804.

Schmitt, M. E., and Clayton, D. A. (1993). Conserved features of yeast and mammalian mitochondrial DNA replication. *Curr. Opin. Genet. Dev.* **3,** 769–774.

Sebald, W., Wachter, E., and Tzagoloff, A. (1979). Identification of amino acid substitutions in the dicyclohexylcarbodiimide-binding subunit of the mitochondrial ATPase complex from oligomycin-resistant mutants of *Saccharomyces cerevisiae Eur. J. Biochem.* **100,** 599–607.

Sia, E. A., Butler, C. A., Dominska, M., Greenwell, P., Fox, T. D., and Petes, T. D. (2000). Analysis of microsatellite mutations in the mitochondrial DNA of *Saccharomyces cerevisiae. Proc. Natl. Acad. Sci. USA* **97,** 250–255.

Sor, F., and Fukuhara, H. (1982). Identification of two erythromycin resistance mutations in the mitochondrial gene coding for the large ribosomal RNA in yeast. *Nucleic Acids Res.* **10,** 6571–6577.

Speno, H., Taheri, M. R., Sieburth, D., and Martin, C. T. (1995). Identification of essential amino acids within the proposed CuA binding site in subunit II of cytochrome c oxidase. *J. Biol. Chem.* **270,** 25363–25369.

Steele, D. F., Butler, C. A., and Fox, T. D. (1996). Expression of a recoded nuclear gene inserted into yeast mitochondrial DNA is limited by mRNA-specific translational activation. *Proc. Natl. Acad. Sci. USA* **93,** 5253–5257.

Sulo, P., Groom, K. R., Wise, C., Steffen, M., and Martin, N. (1995). Successful transformation of yeast mitochondria with RPM1: An approach for in vivo studies of mitochondrial RNase P RNA structure, function and biosynthesis. *Nucleic Acids Res.* **23,** 856–860.

Szczepanek, T., and Lazowska, J. (1996). Replacement of two non-adjacent amino acids in the *S.cerevisiae* bi2 intron-encoded RNA maturase is sufficient to gain a homing-endonuclease activity. *EMBO J.* **15,** 3758–3767.

Thomas, B. J., and Rothstein, R. (1989). Elevated recombination rates in transcriptionally active DNA. *Cell* **56,** 619–630.

Thorsness, P. E., and Fox, T. D. (1990). Escape of DNA from mitochondria to the nucleus in *Saccharomyces cerevisiae. Nature* **346,** 376–379.

Thorsness, P. E., and Fox, T. D. (1993). Nuclear mutations in *Saccharomyces cerevisiae* that affect the escape of DNA from mitochondria to the nucleus. *Genetics* **134,** 21–28.

Wallace, D. C. (1992). Diseases of the mitochondrial DNA. *Annu. Rev. Biochem.* **61,** 1175–1212.

Zinn, A. R., Pohlman, J. K., Perlman, P. S., and Butow, R. A. (1987). Kinetic and segregational analysis of mitochondrial DNA recombination in yeast. *Plasmid* **17,** 248–256.

CHAPTER 22

Transmitochondrial Technology in Animal Cells

Carlos T. Moraes,* Runu Dey,* and Antoni Barrientos†

* Department of Neurology
University of Miami School of Medicine
Miami, Florida 33136

† Department of Biological Sciences
Columbia University
New York, New York 10027

I. Introduction

Studies of vertebrate mitochondrial DNA (mtDNA) maintenance and function have relied heavily on somatic cell experimentation in culture because we are unable to manipulate mtDNA sequences in animal cells. Initial patterns of mitochondrial segregation and species-specific compatibility were performed using somatic hybrid and cybrid cells (Clayton *et al.*, 1971; De Francesco *et al.*, 1980; Giles *et al.*, 1980; Ziegler and Davidson,

METHODS IN CELL BIOLOGY, VOL. 65

1981; Hayashi *et al.,* 1983). These studies became easier to interpret with the development of cell lines devoid of mtDNA (also termed ρ^0). These cell lines can be repopulated easily with exogenous mtDNA, resulting in transmitochondrial cybrids (King and Attardi, 1989). A large number of pathogenic mtDNA mutations have been studied using this procedure, including rearrangements (Hayashi *et al.,* 1991), point mutations in tRNA (Chomyn *et al.,* 1991; King *et al.,* 1992; Hao and Moraes 1997), rRNA (Inoue *et al.,* 1996), and protein-coding genes (Bruno *et al.,* 1999; Trounce *et al.,* 1994; Jun *et al.,* 1996). This technology has also been used to study evolutionary interactions between nuclear and mitochondrial genomes by closely related species (Kenyon and Moraes, 1997). Different tissue types and techniques have been utilized for the generation of transmitochondrial cell lines in animals. Basically, a cell line devoid of mtDNA or with poisoned mitochondria function as the nuclear donor, whereas enucleated or fragmented cells function as the mtDNA donor (Fig. 1). Upon cell membrane fusion and appropriate selection, transmitochondrial cells are generated. This chapter describes current protocols used to generate transmitochondrial animal cell lines in culture.

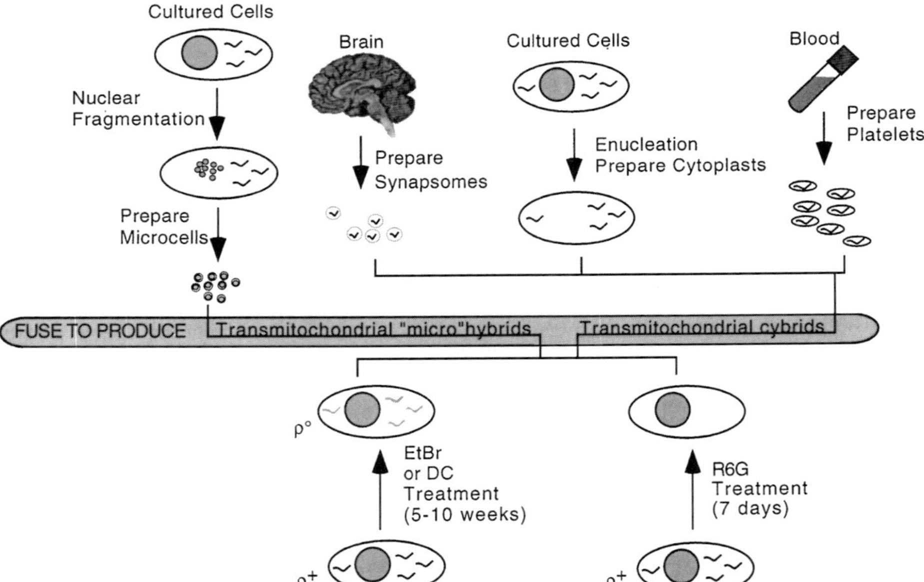

Fig. 1 Overview of methods to prepare transmitochondrial animal cell lines. Transmitochondrial cell lines are obtained by the fusion of cell lines devoid of mtDNA (ρ^0 cells, nuclear donors) and cellular fragments containing mitochondria (mtDNA donors). Different cells or tissues are suitable for preparing cytoplasmic bodies that can function as mtDNA donors. By using microcells and appropriate selection markers, one can introduce mtDNA and single chromosomes simultaneously into ρ^0 cells. Although ρ^0 cells are optimum nuclear donors, mitochondrial poisons such as rhodamine 6G (R6G) can also be used to eliminate mitochondria from cells that would function as nuclear donors. All of these procedures are described in the text.

II. Generation of ρ^0 Cells

DNA intercalating agents, such as ethidium bromide (EtBr), affect mtDNA replication in doses that are too low to affect nuclear DNA significantly (Smith *et al.,* 1971; Nass, 1972). Blocking mtDNA replication in culture during exponential growth reduces the total amount of mtDNA by half every cell doubling. To keep ethidium bromide-treated vertebrate cells growing exponentially, Morais and colleagues (1980) showed that the medium had to be supplemented with pyrimidines. The reason for this requirement is related to the dependence of dihydroorotate dehydrogenase (a mitochondrial protein) on a functional mitochondrial electron transport chain, and hence on mtDNA, for the synthesis of pyrimidines (Gregorie *et al.,* 1984).

Established cells made completely devoid of mtDNA by ethidium bromide treatment (termed ρ^0 cells) were first described by Desjardin *et al.* (1986). King and Attardi (1989) isolated human ρ^0 cells lines by treating cells with 50 ng/ml ethidium bromide and found that they were auxotrophic not only pyrimidines but also for pyruvate. The dependence on pyruvate is not understood, but it is possibly due to an increased metabolism of pyruvate to lactate in ρ^0 cells. This would reduce the availability of pyruvate for the Krebs cycle (King and Attardi, 1989). Therefore, attempts to generate ρ^0 cell lines should be performed in high glucose medium (4500 mg/ml) supplemented with fetal calf serum (FCS 5–10%), 50 μg/ml uridine, and 1 m*M* sodium pyruvate.

Although ρ^0 cells have been generated in mouse by treatment with 5 μg/ml EtBr (Tiranti *et al.,*1998), we find that most cell lines cannot grow in the presence of this concentration of ethidium bromide. Thus, one must find the maximum concentration of EtBr that can be tolerated by the specific cell line. In addition, for reasons that are not understood, certain cell lines, such as mouse-derived lines, are refractory to complete mtDNA depletion by EtBr. Hayashi and colleagues found that a different intercalating agent (ditercalinium or DC) was more efficient than EtBr in generating ρ^0 mouse as well as other cell lines (Inoue *et al.,* 1997b). They used a concentration of 56 ng/ml–1.5 μg/ml, depending on the sensitivity to the drug. We found that DC is extremely toxic to cells, and had to use relatively low doses to produce ρ^0 cells (20–50 ng/ml). Finally, the duration of treatment varies from cell to cell. EtBr-treated cells display rapid loss of mtDNA (King and Attardi, 1996; Moraes *et al.,* 1999); however, residual molecules can be retained if the treatment is shorter than 25–30 days. If the DNA intercalating agent is removed before complete depletion of mtDNA, cells repopulate with the residual genomes. We have found that smaller molecules (i.e., mtDNA with partial deletions) repopulate cells five to eight times faster than full-length genomes (Moraes *et al.,* 1999; Fig. 2).

Typically, for ρ^0 generation, an exponentially growing cell line is treated for 2 months with EtBr or DC in media supplemented with uridine (50 μg/ml) and pyruvate (110 μg/ml). EtBr is prepared as a filter-sterilized aqueous 10-mg/ml solution that is kept refrigerated and protected from light. DC availability is still limited, and information regarding its source can be obtained from laboratories working with it. After a 2-months treatment, the DNA intercalating agent is removed and the cells are allowed to grow in medium supplemented with uridine (50 μg/ml) and pyruvate (110 μg/ml) for at least 15 days before

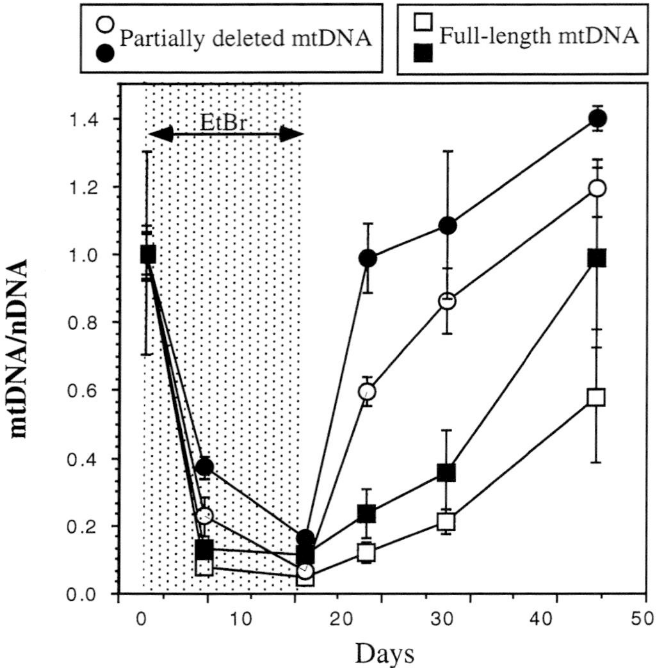

Fig. 2 Mitochondrial DNA depletion and repopulation after treatment with ethidium bromide. Cell lines essentially homoplasmic for wild-type mtDNA (squares) or partially deleted mtDNA (7.5-kb deletion, circles) were treated with ethidium bromide for 15 days (shaded time period), after which the drug was removed from the medium and the cells were allowed to grow for a total of 45 days. Cell samples were collected at different time points, had their DNA extracted, and the ratio of mtDNA to the nuclear 18S rDNA gene was determined by Southern blot, as described (Moraes *et al.*, 1999). Note that the mtDNA repopulates cells at rates that depend on their size.

tests for the presence of mtDNA are performed. It is useful to plate cells at low density and isolate single clones at this point, as some cells may still retain a few molecules of mtDNA and will repopulate their mtDNA. It is recommended to freeze some cells for regrowth before the treatment is stopped in case the mtDNA depletion was not complete. Southern blot and polymerase chain reaction (PCR) assays are useful to assess the ρ^0 status, but a more reliable assay is to test their auxotrophy to uridine. Once the ρ^0 phenotype is well established, the cells can be used as recipients of exogenous mtDNA. Because DNA intercalating agents have the potential to cause nuclear DNA mutations, ρ^0 cells should be repopulated with a "wild-type" mtDNA to be used as control. Table I describes some of the ρ^0 cells produced and used in transmitochondrial cybrid generation.

Because most ρ^0 cells are established for use in transmitochondrial cybrid generation, it is useful to choose a cell line that has a recessive nuclear marker. Thymidine kinase (TK^-) or hypoxanthine guanine phosphorybosyl transferase ($HPRT^-$) deficiency

Table I
Examples of ρ^0 Cells

Cell line	Cell type	Species	Treatment[a]	ρ^0 phenotype confirmed[b]	Reference
143B206	Osteosarcoma	Human	50 ng/ml EtBr	S/Ur/Pyr	King and Attardi (1989)
U937	Promonocytic leukemia	Human	5 ng/ml EtBr	PCR	Gamen *et al.* (1995)
C2	Myoblast	Mouse	1.5 μg/ml DC	S/PCR Ur/Pyr	Inoue *et al.* (1997a)
MIN 6	Pancreatic β cell	Mouse	56 ng/ml DC	S Ur/Pyr	Inoue *et al.* (1997a)
B82cap	Fibroblast	Mouse	1.5 μg/ml DC	S/PCR Ur/Pyr	Inoue *et al.* (1997b)
NIH3T3	Fibroblast	Mouse	1.5 μg/ml DC	S/PCR Ur/Pyr	Inoue *et al.* (1997b)
SH-SY5Y/ρ^0 64/5	Neuroblastoma	Human	5 μg/ml EtBr	Biochemically, mitochondrial protein synthesis, Pyr	Miller *et al.* (1996)
GM10611A-ρ^0	Hybrid	Hamster/human chromosome 9	5 μg/ml EtBr	PCR/COX activity	Tiranti *et al.* (1998)
L929	Connective tissue	Mouse	5 μg/ml EtBr	PCR/COX activity	Tiranti *et al.* (1998)

[a] EtBr, ethidium bromide; DC, ditercalinium.
[b] S, Southern blot; Ur, uridine dependence; Pyr, pyruvate dependence; PCR, polymerase chain reaction; COX,

are good examples of recessive nuclear markers that can be obtained by chemical mutagenesis and selection in the presence of drugs that are toxic to wild-type cells (i.e., bromodeoxyuridine in the case of TK$^-$ cells and 8-azaguanine in the case of HPRT$^-$ cells). Chemical mutagenesis and selection procedures for nuclear recessive mutations have been described elsewhere (Shapiro and Varshaver, 1975; Anderson, 1995).

III. Generation of Transmitochondrial Cybrids Using ρ^0 Cells as Nuclear Donors

As described in Section I, phenotypes linked to alterations in the mitochondrial genome can be assessed by transferring the mtDNA to a different nuclear background. The advantage of such a procedure is that major effects of nuclear genes in the phenotype can be ruled out. At present, animal mtDNA can only be introduced in a ρ^0 cell as intact mitochondria and several cell types are being developed to serve as mitochondria and mtDNA donors for these studies. Mitochondria can be injected directly into cells, but this procedure is very inefficient (King and Attardi, 1996). The use of cytoplasts (i.e., cells devoid of nuclei) has facilitated the introduction of mitochondria into ρ^0 cells by using standard membrane fusion techniques followed by growth in selective media.

It is difficult to introduce highly defective mtDNA in ρ^0 cells because the selection procedure for transmitochondrial cybrids requires some level of oxidative phosphorylation (OXPHOS) function, but we have found that very low respiratory capacity can be sufficient to confer uridine independence to some transmitochondrial cybrids (Hao and Moraes, 1996). However, when the mtDNA to be transferred is completely defective, heteroplasmic clones will have to be isolated in a first step and subsequently manipulated to achieve homoplasmy (see Section VI). Galactose sensitivity is also a useful selection system for cells with defective OXPHOS. In galactose-containing media, very little lactate and pyruvate accumulate because of the restricted flow of galactose through glucose 6-phosphate (Robinson, 1996). Galactose selection is more stringent than selection for uridine auxotrophs, and cells with partial defects in OXPHOS that can survive in uridine-free medium may not survive in galactose medium.

A. Cultured Cells as Mitochondrial (mt)DNA Donors from Adherent and Suspension Cultures

1. Adherent Cells

Cultured cells (adherent or in suspension) are excellent mtDNA donors, but the enucleation and fusion procedures are different, depending on the growth substrates. We commonly use the method described by King and Attardi (1989) to produce transmitochondrial cybrids using adherent cells as mtDNA donors. We will describe an example protocol in detail.

Example Protocol

1. In addition to culture media and cells, prepare and autoclave the following items before starting the fusion procedure: (i) a small glass bottle (e.g., 100 ml) containing 4.7 g of PEG 1500, (ii) a 30-cm forceps wrapped in aluminum foil, and (iii) two empty 250-ml centrifuge bottles with caps with sealing rings (loose in the autoclave).

2. Adherent cells grown in 35-mm dishes should be approximately 70% confluent on the day of the experiment. It is recommended to grow these small dishes inside 100-mm dishes to avoid contaminating the external surface of the 35-mm dish and to facilitate handling during microscopic examination. Because it is difficult to plate the exact number of cells and because it is better to plate the cells to be enucleated at least 24 h before enucleation (to ensure strong attachment to the surface), we recommend plating a few dishes with different numbers of cells so that the most appropriate one(s) can be used for fusion.

3. Cells attached to the 35-mm dishes are enucleated after treatment with 10 μg/ml cytochalasin B and centrifugation. Both procedures are performed simultaneously in centrifuge bottles. Remove the lid of the 35-mm dish and insert the half with cells upside down in a 250-ml centrifuge bottle (e.g., Cat. No. 21007-165, VWR Scientific) containing 30 ml of prewarmed (37°C) enucleation medium using sterile forceps. Both the bottle/cap and the forceps should be kept sterile. The lid of the 35-mm dish is kept inside the original empty 100-mm dish for later use. The enucleation medium consists of DMEM high glucose (GIBCO BRL Cat. No. 11965-092) supplemented with 5% calf serum, 10 μg/ml cytochalasin B (Sigma C-6762), and antibiotics (0.02 mg/ml, gentamicin). It is almost impossible to remove all the air trapped inside the inverted 35-mm dish, but we have found that a relatively small "air bubble" will not interfere with the enucleation

procedure. The bottles are placed in a prewarmed fixed angle rotor and centrifuged at 8000*g* for 25 min at 35°C. The rotor and the centrifuge chamber can be warmed by centrifugation without refrigeration. The temperature inside the chamber will reach approximately 35°C using this procedure, which is usually adequate for enucleation. This procedure may not be sufficient to enucleate all types of cells. A good method to assess the enucleation efficiency is to stain a "test enucleation" dish with 1 μg/ml Hoechst 33342 (Sigma) and 200 n*M* of MitoTracker (CmxRos, Molecular Probes) in phosphate-buffered saline (PBS). Even relatively low efficiency enucleations can yield transmitochondrial cybrids, but if cells without nuclei are not identified, it is unlikely that true cybrids will be formed.

4. During the time that it takes to enucleate the mtDNA donors, ρ^0 cells should be trypsinized, counted, and placed in a falcon tube, to be used after the enucleation of the mtDNA donor cells is complete. The enucleating 35-mm dish is removed from the bottle with the help of the long forceps and is placed on top of an autoclaved paper towel. All the procedures involving handling of the dish are performed inside a biological cabinet in sterile conditions. At this point the operator will have to handle the small dish. Rinse your gloved (preferably sterile) fingers thoroughly with ethanol 70%, hold the dish, and wipe its outside with sterile paper soaked with ethanol 70%. Place the clean medium-less dish back into the 100-mm empty dish and cover it with its sterile lid. Inspect the dish under the microscope to assess the enucleation. It should show the presence of cytoplasts, which resemble piece of membrane. Enucleation efficiency varies depending on the cell type, and the conditions described here may not be able to produce cytoplasts from all cell lines.

5. Immediately after a quick inspection of the enucleated cells, add 1.5×10^6 ρ^0 cells in 2.5 ml of complete medium supplemented with uridine. Normally, cells from one confluent T75 flask of ρ^0 cells should be enough for at least two fusion experiments. The dish containing cytoplasts and the ρ^0 cells are placed in the CO_2 incubator for 2.5–3 h, during which the ρ^0 cells attach to the dish and make contacts with the attached cytoplasts.

6. Approximately 4.7 g of PEG 1500 are autoclaved in a 100-ml glass bottle either the day before or just before the fusion. While the PEG is still liquid (i.e., warm but not very hot), add 4 ml of DMEM and 1 ml of dimethyl sulfoxide (DMSO). This produces a solution that is 47% PEG (v/v). If the PEG hardens before adding the medium, slowly warm up the bottle by rotating it on a flame. Once the PEG is mixed with medium and DMSO, it will remain liquid. The diluted PEG solution should be used within 24 h. The pH of the final fusion solution is slightly acidic, as observed by its yellow color. The pH can be raised by the addition of approximately 1 μl of 10 *N* NaOH. Depending on the origin of the reagents, this amount may change, but one should be careful not to add too much NaOH. If this happens, the solution will turn purple due to high pH and should be discarded.

7. After the 2.5- to 3-h incubation described in step 5, the membrane fusion is performed with the PEG solution. The mixture of cells and cytoplasts is washed three times with serum-free DMEM medium. After the last wash has been aspirated, the dish should

be placed at a 45° angle for approximately 45–60 s for the remaining medium to drain to the bottom of the dish and removed by aspiration. It is important to have all medium removed for an efficient PEG fusion. After all media have been removed, add 2.5 ml of the final PEG solution to the dish, swirl for 10 s, and leave the solution in contact with the cells for exactly 60 s. The PEG solution is aspirated and the dish is washed three times with DMEM supplemented with 10% DMSO and once with DMEM.

8. After fusion, 2.5 ml of complete medium supplemented with uridine is added to the dish and the cells are returned to the CO_2 incubator. Change the medium on the following day when a considerable amount of dead cells will be observed. If fewer than 30% of the cells remain attached, the fusion conditions may have been too severe and the concentration of PEG should be reduced (e.g., to 44%). As mentioned earlier, the sensitivity to PEG varies depending on the cell type. Fungizone (250 μg/ml amphotericin B) should not be used before, during, or immediately after the fusion, as we have observed that it increases PEG toxicity significantly.

9. Selection for transmitochondrial cybrids starts between 16 and 30 h after fusion. Selection for the presence of mtDNA is provided by removing uridine from the medium and by using dialyzed serum (regular FCS contains small levels of uridine). Selection against nonenucleated mtDNA donors, as well as hybrids, is provided by drugs that are not toxic to the specific ρ^0 cells (i.e, bromodeoxyuridine or 8-azaguanine).

10. Discrete clones (50–100 cells) should be observed 10 days after selection starts, and good size clones are usually observed by 15–20 days. We isolate clones with the aid of cloning rings (Bellco Glass Cat. No. 2090) and trypsin. Clones can be isolated and, if desired, part of it can be used immediately for PCR typing, as has been described previously for single muscle fibers (Moraes and Schon, 1996). We find that BrdU-resistant cells may contain chromosomes from the mtDNA donor. Although this is rare, it is advisable to test for the presence of additional nuclear markers (e.g., microsatellites) to assure that the BrdU-resistant clones are true cybrids.

2. Suspension Cells

Suspension cells have to be enucleated by centrifugation through a density gradient. Trounce and Wallace (1996) described two procedures that are efficient in enucleating suspension cells. A Percoll gradient is formed by mixing approximately 10^7 cells in 10 ml of medium with 10 ml of a 1 : 1 mix of complete medium and Percoll that has been preequilibrated overnight in a 37°C, 5% CO_2 incubator. The Final mixture is transferred to a sterile centrifuge tube, and 100 μl of a 2-mg/ml cytochalasin B stock (in DMSO) is added (final concentration of 20 μg/ml). The mixture is centrifuged using a fixed angle rotor at 44,000*g* for 70 min at 37°C. At the end of the run a distinct band can be identified one-third of the distance from the bottom. This band contains a mix of cytoplasts and karyoplasts that are appropriate for cybrid production if a nuclear recessive marker is used. Above this band there is a less well-defined band, which seems to be enriched in cytoplasts (Trounce *et al.,* 1996). The well defined band is removed and diluted with 10 ml of medium, centrifuged at 650*g*, and resuspended in 5 ml of medium. This preparation can be used in fusions as described earlier, with the difference that it should be performed in a test tube (see Section III,B). Nevertheless, the procedure is essentially the same. We also have used this procedure for adherent cells (after trypsinization) that did not enucleate well by the standard procedure.

B. Platelets as mtDNA Donors

Chomyn and colleagues described a simple procedure to obtain transmitochondrial cybrids using platelets as the mtDNA donor (Chomyn *et al.,* 1994; Chomyn, 1996). Platelets are good mtDNA donors because they are nucleus-free fragments of megakaryocytes. Moreover, platelets are easy to obtain and can be frozen in DMSO-containing media like cultured cells for future use in fusion experiments. The procedure has been described in detail (Chomyn, 1996). It consists of isolating platelets (in sterile conditions) after pelleting red and white blood cells by low-speed centrifugation. The platelets are washed in PBS, mixed with ρ^0 cells in a sterile tube, washed further with serum-free medium, and fused with the same 47% (v/v) PEG solution described earlier. After fusion, cellular products can be treated as described previously for cultured cells. In theory, the use of platelets as mtDNA donors makes the selection for nuclear markers unnecessary. However, it is possible to have white blood cells contaminating the platelet preparation, and analysis of the nuclear background is recommended.

Platelets from patients with neurodegenerative disorders, such as Parkinson and Azheimer disease, have been used extensively in transmitochondrial cybrid production (Swerdlow *et al.,* 1997; Gu *et al.,* 1998).

Example Protocol

1. Collect 7–20 ml of blood in heparin tubes (green top Vacutainer) and mix it with 0.1 volumes each of 0.15 *M* NaCl and 0.1 *M* trisodium citrate (pH 7.0). Centrifuge at 150*g* for 15 min.

2. Remove the top three-fourths of the platelet-rich plasma and centrifuge for 35 min at 2500*g*.

3. Resuspend the pellet in 10 ml of 0.15 *M* NaCl, 0.015 *M* Tris–HCl (pH 7.4). Approximately $1\text{–}4 \times 10^7$ platelets are obtained by this procedures. if desired, aliquots can be frozen after centrifugation and resuspension of a platelet aliquot in cell culture freezing medium (e.g., DMEM supplemented with 30% FCS and 10% DMSO).

4. Mix approximately 10^7 platelets with 10^6 ρ^0 cells and spin at 200*g* for 5 min. Aspirate all supernatant and resuspend the pellet thoroughly with 100 μl of a solution containing 4.7g PEG 1500, 1 ml of 10% DMSO and 4 ml of DMEM (serum free) prepared as described earlier. After 1 min at room temperature, dilute the sample with complete medium (i.e., DMEM with 10% FCS and 50 μg/ml uridine). If the ρ^0 cell has a nuclear marker such as TK^-, BrdU can be added to the medium to eliminate any potential nucleated blood cells. Distribute the cells into ten 100-mm dishes. If a large

amount of clones is not required, discard half of the mix and plate the other half into five 100-mm dishes.

5. Three days later replace the medium with DMEM supplemented with 10% dialyzed FCS and a drug related to the recessive nuclear marker of the ρ^0 mitochondrial recipient (e.g., BrdU). Do not add uridine or pyruvate. Isolated clones should be visible 15 days after the fusion.

C. Synaptosomes as mtDNA Donors

Hayashi and colleagues have described a procedure to use synaptosomes as mtDNA donors (Ito *et al.,* 1999). A mouse brain was washed in phosphate-buffered saline and homogenized in medium containing 0.25 *M* sucrose, 50 m*M* HEPES, pH 7.5, and 0.1 m*M* EDTA in a Teflon–glass Potter–Elvehjem homogenizer. The homogenate was centrifuged at 1000*g* for 10 min at 4°C, and the supernatant was centrifuged at 17,000*g* for 20 min at 4°C. The pellet containing synaptosomes was mixed with 5×10^6 mouse ρ^0 cells, and fusion was carried out in the presence of 50% (v/v) PEG 1500. The fusion mixture was cultivated in selective medium (without pyruvate and uridine). Fourteen to 20 days after fusion, the cybrid clones growing in the selection medium were isolated by the glass cylinder method.

IV. Generation of Transmitochondrial Cybrids Using ρ^+ Cells Treated with Rhodamine 6G as Nuclear Donors

In 1981, Ziegler and Davidson showed that pretreatment of cells with the mitochondria poison rhodamine 6G (R6G) increased the efficiency of interspecific hybrid formation, probably because it eliminated the deleterious interference of heteroplasmy between interspecific mtDNAs. Trounce and Wallace (1996) adapted the technique for the production of transmitochondrial cybrids. The procedure consists of treating the mtDNA recipient cells with 2–5 μg/ml R6G for 6–10 days. The sensitivity varies with cell type and should be standardized; the highest concentration that does not kill the cells in 6–10 days will be optimum. After the treatment, cells should die even in complete medium supplemented with uridine and pyruvate, as they lose not only OXPHOS, but all mitochondrial functions, a feature that is incompatible with life in animal cells. After R6G treatment of the mtDNA recipient, the medium is changed to R6G-free medium for 3–5 h. The R6G toxic effect is not reversible and dysfunctional mitochondria should not recover. After this "washing" period, treated cells are fused to cytoplasts of the mtDNA donor as described earlier. In theory, no nuclear selection is necessary, as R6G-treated cells that have not fused to cytoplasts should die a few days after fusion. A control fusion without cytoplasts is recommended to assure that the R6G treatment was toxic enough to kill all cells that do not receive functional mitochondria. We have had mixed results with this procedures. For xenomitochondrial cybrid production, we were frustrated by finding after several weeks that not all mtDNA from the nuclear donor had been eliminated. Nevertheless, we had positive results with "same species" cybrid production.

Example Protocol

1. Grow two T75 flasks of cells, which will function as the mtDNA recipient, with complete medium supplemented with 50 μg/uridine, 110 μg/ml pyruvate, and 3 μg/ml R6G. R6G (Sigma Cat No. R-4127) is prepared as a stock solution in sterile water and filter sterilized. R6G-treated cells do not grow as fast as untreated cells, and trypsinization should be avoided. Treat cells for 7 days, changing medium every 1–2 days. In the morning of the fusion day, change the medium to complete medium without R6G and leave the cells in this medium for 3–4 h before fusion.
2. As described earlier for regular transmitochondrial cybrid production, grow the mtDNA donor in 35-mm dishes so that they will be approximately 70% confluent at the time of fusion.
3. The mtDNA donors are enucleated by centrifugation in 250-ml centrifuge bottles in the presence of cytochalasin B (10 μg/ml), as described earlier.
4. Meanwhile, the R6G-treated cells are trypsinized and counted. Approximately 2×10^6 cells are mixed with enucleated cells and fused with the 47% PEG (v/v) solution described earlier. The remainder of the procedure is identical to what was described for fusions with ρ^0 cells.
5. After fusion, 2.5 ml of complete medium supplemented with uridine and pyruvate are added to the dish and the cells are returned to the CO_2 incubator. Change the medium on the following day, when a considerable amount of dead cells will be observed.
6. Selection for transmitochondrial cybrids starts approximately 30 h after fusion. Selection for the presence of mtDNA is not necessary because R6G-treated cells should die in a few days. However, if a functional mtDNA is being transferred, it is recommended to remove uridine and pyruvate from the medium. Selection against nonenucleated mtDNA donors, as well as hybrids, is provided by drugs that are not toxic to the nuclear donor (i.e., bromodeoxyuridine or 8-azaguanine).
7. Isolation and analysis of growing clones can be performed as described previously.

V. Generation of Transmitochondrial Hybrid Cells by Microcell-Mediated Chromosome and mtDNA Transfer

Microcell-mediated chromosome transfer has been used to transfer intact chromosomes from one mammalian cell to another (Ege and Ringertz, 1974; Fournier and Ruddle, 1977; McNeill and Brown, 1980). Using the microcell hybridization system, we have been able to transfer a limited number of chromosomes and mtDNA simultaneously to a mtDNA-less human cell (Barrientos and Moraes, 1998). The structure of mitochondria-containing microcells showed that they consisted of a single or a few micronuclei and a thin ring of cytoplasm surrounded by an intact plasma membrane. It was shown that karyoplasts produced by cytochalasin B-induced enucleation contained approximately 11% of the mitochondrial volume of whole cells (Zorn *et al.,* 1979). The cytoplasmic region of microcells contains a variable number of mitochondria, and the composition of microcell preparations is very heterogeneous, presenting small cytoplasts containing mitochondria, microcells containing mitochondria and a single micronucleus, microcells containing mitochondria and a few micronuclei, and, in rare cases, microcells containing no mitochondria. The proportion of micronuclei-containing cell fragments is approximately 30% (Barrientos and Moraes, 1998). This aspect should be taken into account when studies on the restoration of normal maintenance and expression of mtDNA in defective cells are carried out using the microcell-mediated chromosome(s) transfer system.

The proportion of hybrid clones receiving a particular chromosome to the clones selected only for the presence of mtDNA is approximately 1%. The number of clones selected for the presence of different chromosome markers was essentially the same whether selection for the presence of mtDNA was included or not, suggesting that microcells containing selected chromosomes also contained mtDNA. Using microcell-mediated chromosome transfer, hybrids are likely to receive donor chromosome randomly, but it is possible that some chromosomes are transferred preferentially. Our experience suggests that different selectable chromosomes are transferred at a similar rate. Although most microcells contain one single chromosome, the presence of a few microcells containing several micronuclei suggests that clones containing more than one donor chromosome could also be generated during fusion. In some cases, selective pressure (i.e., no uridine) is necessary for the repopulation of mtDNA introduced into cells by this system (Barrientos and Moraes, 1998).

Example Protocol

1. Microcells are isolated essentially as described (Fournier, 1981) by sequential treatment with 50 ng/ml colcemid (GIBCO-BRL) (36–48 h, depending on cell type), which arrests the cells in mitosis and results in the formation of micronuclei (Fig. 3), and 10 μg/ml cytochalasin B (Sigma Chemical) (20 min), which disrupts microfilaments. Enucleation is performed by centrifugation at in a fixed angle rotor at 11,000*g* for 60 min at 35°C. T25 flasks are placed in the rotor after adding water to cushion the flask. Many flasks can crack during centrifugation. As recommended by Fournier (1981), we used Costar flask No. 3025.

2. The microcell pellet is placed into 100 ml of DMEM and is filtered sequentially through 12-, 8-, 5-, and 5-μm Nucleopore filters (Costar, Cambridge, MA). Both the filter and filter holders (Costar) are autoclaved. The purified microcells can be used for hybridization experiments.

3. Purified microcells are centrifuged at 500*g* in a fixed angle rotor for 10 min at 4°C and microcells are resuspended in 3 ml of lectin solution (50 μg/ml phytothemagglutinin PHA-P; Sigma L9132) to promote microcell adherence to recipient cells. A ρ^0 cell line,

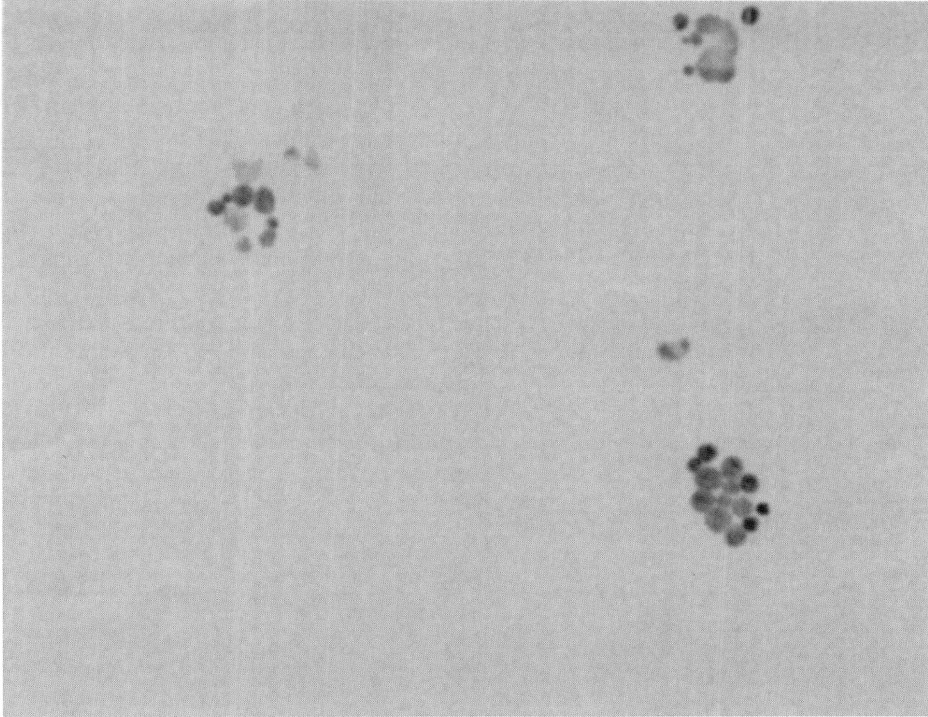

Fig. 3 Microenucleation of normal fibroblasts. Microcells containing single or a limited number of chromosomes and mtDNA can be prepared by treating cells with colcemid followed by enucleation and size selection. The first step of microenucleation of human fibroblasts is shown. Purified microcells can be fused to ρ^0 cells to generate transmitochondrial "microhybrids" if appropriate nuclear markers are available.

used as recipient cells, is grown in six 60-mm culture dishes. The recipient cells are washed in DMEM and incubated with the microcell solution (1 ml per dish) for 30 min at 37°C.

4. Cell fusion is performed by treatment with freshly prepared 46% PEG (v/v) solution described earlier for 1 min, followed by three washes in DMEM, 5% DMSO. As described for adherent cells, the hybrids are allowed to grow overnight in complete medium and are selected for the presence of appropriate markers on the next day.

VI. Manipulating Heteroplasmy

As shown in Fig. 2, mtDNA levels can be decreased temporarily by ethidium bromide treatment, providing an efficient approach to alter the heteroplasmic distribution. This procedure has been used to obtain homoplasmic-mutated clones harboring large deletions (King, 1996; Moraes *et al.,* 1999) and point mutations (Hao and Moraes, 1997). Because partially deleted mtDNA repopulates cells faster than full-length mtDNA (Moraes *et al.,* 1999), it is easier to obtain homoplasmic deletion mutant clones compared to clones bearing point mutations from heteroplasmic cells. Cells should be treated with 50 ng/ml ethidium bromide (in medium supplemented with uridine) for 15–22 days to reduce their mtDNA levels to very few molecules. The EtBr solution should be filter sterilized at high concentrations (10 mg/ml). Stocks should be protected from light and stored at 4°C. After treatment, EtBr is removed from the medium and cells are allowed to repopulate themselves with the residual mtDNA for 10–15 days. Depending on the amount and type of mtDNA remaining in the cell, homoplasmic clones can be generated after repopulation. When dealing with pathogenic mtDNA mutations it is important to maintain uridine and pyruvate supplementation during the repopulation period. Although this procedure has worked successfully for some cell lines, it can also be very frustrating, as it is a very long procedure, and at the end of 2 months one may find out that homoplasmic clones could not be obtained. Nevertheless, it is useful in cases where homoplasmic clones do not exist and studies addressing genotype-phenotype correlations are required.

Acknowledgments

This work was supported by grants from the National Institutes of Health (NIGMS and NEI), the Muscular Dystrophy Association, and the National Parkinson Foundation (through a gift from Mr. and Mrs. Wolf R. Nichols).

References

Anderson, P. (1995). Mutagenesis. *Methods Cell Biol.* **48,** 31–58.

Barrientos, A., Kenyon, L., and Moraes, C. T. (1998). Human xenomitochondrial cybrids. Cellular models of mitochondrial complex I deficiency. *J. Biol. Chem.* **273,** 14210–14217.

Barrientos, A., and Moraes, C. T. (1998). Simultaneous transfer of mitochondrial DNA and single chromosomes in somatic cells: A novel approach for the study of defects in nuclear-mitochondrial communication. *Hum. Mol. Genet.* **7,** 1801–1808.

Bruno, C., Martinuzzi, A., Tang, Y., Andreu, A. L., Pallotti, F., Bonilla, E., Shanske, S., Fu, J., Sue, C. M., Angelini, C., DiMauro, S., and Manfredi, G. (1999). A stop-codon mutation in the human mtDNA cytochrome c oxidase I gene disrupts the functional structure of complex IV. *Am. J. Hum. Genet.* **65,** 611–620.

Chomyn, A. (1996). Platelet-mediated transformation of human mitochondrial DNA-less cells. *Methods Enzymol.* **264,** 334–339.

Chomyn, A., Lai, S. T., Shakeley, R., Bresolin, N., Scarlato, G., and Attardi, G. (1994). Platelet-mediated transformation of mtDNA-less human cells: Analysis of phenotypic variability among clones from normal individuals—and complementation behavior of the tRNALys mutation causing myoclonic epilepsy and ragged red fibers. *Am. J. Hum. Genet.* **54,** 966–974.

Chomyn, A., Meola, G., Bresolin, N., Lai, S. T., Scarlato, G., and Attardi, G. (1991). In vitro genetic transfer of protein synthesis and respiration defects to mitochondrial DNA-less cells with myopathy-patient mitochondria. *Mol. Cell. Biol.* **11,** 2236–2244.

Clayton, D. A., Teplitz, R. L., Nabholz, M., Dovey, H., and Bodmer, W. (1971). Mitochondrial DNA of human-mouse cell hybrids. *Nature* **234,** 560–562.

De Francesco, L., Attardi, G., and Croce, C. M. (1980). Uniparental propagation of mitochondrial DNA in mouse-human cell hybrids. *Proc. Natl. Acad. Sci. USA* **77,** 4079–4083.

Desjardins, P., de Muys, J. M., and Morais, R. (1986). An established avian fibroblast cell line without mitochondrial DNA. *Somat. Cell Mol. Genet.* **12,** 133–139.

Ege, T., and Ringertz, N. R. (1974). Preparation of microcells by enucleation of micronucleate cells. *Exp. Cell Res.* **87,** 378–382.

Fournier, R. E. (1981). A general high-efficiency procedure for production of microcell hybrids. *Proc. Natl. Acad. Sci. USA* **78,** 6349–6353.

Fournier, R. E., and Ruddle, F. H. (1977). Microcell-mediated transfer of murine chromosomes into mouse, Chinese hamster, and human somatic cells. *Proc. Natl. Acad. Sci. USA* **74,** 319–323.

Gamen, S., Anel, A., Montoya, J., Marzo, I., Pineiro, A., and Naval, J. (1995). mtDNA-depleted U937 cells are sensitive to TNF and Fas-mediated cytotoxicity. *FEBS Lett.* **376,** 15–18.

Giles, R. E., Stroynowski, I., and Wallace, D. C. (1980). Characterization of mitochondrial DNA in chloramphenicol-resistant interspecific hybrids and a cybrid. *Somat. Cell Genet.* **6,** 543–554.

Gregoire, M., Morais, R., Quilliam, M. A., and Gravel, D. (1984). On auxotrophy for pyrimidines of respiration-deficient chick embryo cells. *Eur. J. Biochem.* **142,** 49–55.

Gu, M., Cooper, J. M., Taanman, J. W., and Schapira, A. H. (1998). Mitochondrial DNA transmission of the mitochondrial defect in Parkinson's disease. *Ann. Neurol.* **44,** 177–186.

Hao, H., and Moraes, C. T. (1996). Functional and molecular mitochondrial abnormalities associated with a C→T transition at position 3256 of the human mitochondrial genome. The effects of a pathogenic mitochondrial tRNA point mutation in organelle translation and RNA processing. *J. Biol. Chem.* **271,** 2347–2352.

Hao, H., and Moraes, C. T. (1997). A disease-associated G5703A mutation in human mitochondrial DNA causes a conformational change and a marked decrease in steady-state levels of mitochondrial tRNA(Asn). *Mol. Cell. Biol.* **17,** 6831–6837.

Hayashi, J., Ohta, S., Kikuchi, A., Takemitsu, M., Goto, Y., and Nonaka, I. (1991). Introduction of disease-related mitochondrial DNA deletions into HeLa cells lacking mitochondrial DNA results in mitochondrial dysfunction. *Proc. Natl. Acad. Sci. USA* **88,** 10614–10618.

Hayashi, J., Tagashira, Y., Yoshida, M. C., Ajiro, K., and Sekiguchi, T. (1983). Two distinct types of mitochondrial DNA segregation in mouse-rat hybrid cells. Stochastic segregation and chromosome-dependent segregation. *Exp. Cell Res.* **147,** 51–61.

Inoue, K., Ito, S., Takai, D., Soejima, A., Shisa, H., LePecq, J. B., Segal-Bendirdjian, E., Kagawa, Y., and Hayashi, J. I. (1997a). Isolation of mitochondrial DNA-less mouse cell lines and their application for trapping mouse synaptosomal mitochondrial DNA with deletion mutations. *J. Biol. Chem.* **272,** 15510–15515.

Inoue, K., Takai, D., Hosaka, H., Ito, S., Shitara, H., Isobe, K., LePecq, J. B., Segal-Bendirdjian, E., and Hayashi, J. (1997b). Isolation and characterization of mitochondrial DNA-less lines from various mammalian cell lines by application of an anticancer drug, ditercalinium. *Biochem. Biophys. Res. Commun.* **239,** 257–260.

Inoue, K., Takai, D., Soejima, A., Isobe, K., Yamasoba, T., Oka, Y., Goto, Y., and Hayashi, J. (1996). Mutant mtDNA at 1555 A to G in 12S rRNA gene and hypersusceptibility of mitochondrial translation to streptomycin can be co-transferred to rho 0 HeLa cells. *Biochem. Biophys. Res. Commun.* **223,** 496–501.

Ito, S., Ohta, S., Nishimaki, K., Kagawa, Y., Soma, R., Kuno, S. Y., Komatsuzaki, Y., Mizusawa, H., and Hayashi, J. (1999). Functional integrity of mitochondrial genomes in human platelets and autopsied brain tissues from elderly patients with Alzheimer's disease. *Proc. Natl. Acad. Sci., USA* **96,** 2099–2103.

Jun, A. S., Trounce, I. A., Brown, M. D., Shoffner, J. M., and Wallace, D. C. (1996). Use of transmitochondrial cybrids to assign a complex I defect to the mitochondrial DNA-encoded NADH dehydrogenase subunit 6 gene mutation at nucleotide pair 14459 that causes Leber hereditary optic neuropathy and dystonia. *Mol. Cell. Biol.* **16,** 771–777.

Keynon, L., and Moraes, C. T. (1997). Expanding the functional human mitochondrial DNA database by the establishment of primate xenomitochondrial cybrids. *Proc. Natl. Acad. Sci. USA* **94,** 9131–9135.

King, M. P. (1996). Use of ethidium bromide to manipulate ratio of mutated and wild-type mitochondrial DNA in cultured cells. *Methods Enzymol.* **264,** 339–344.

King, M. P., and Attardi, G. (1989). Human cells lacking mtDNA: Repopulation with exogenous mitochondrial by complementation. *Science* **246,** 500–503.

King, M. P., and Attadi, G. (1996). Mitochondria-mediated transformation of human rho(0) cells. *Methods Enzymol.* **264,** 313–334.

King, M. P., Koga, Y., Davidson, M., and Schon, E. A. (1992). Defects in mitochondrial protein synthesis and respiratory chain activity segregate with the tRNA(Leu(UUR)) mutation associated with mitochondrial myopathy, encephalopathy, lactic acidosis, and strokelike episodes. *Mol. Cell. Biol.* **12,** 480–490.

McNeill, C. A., and Brown, R. L. (1980). Genetic manipulation by means of microcell-mediated transfer of normal human chromosomes into recipient mouse cells. *Proc. Natl. Acad. Sci. USA* **77,** 5394–5398.

Miller, S. W., Trimmer, P. A., Parker, W. D., Jr., and Davis, R. E. (1996). Creation and characterization of mitochondrial DNA-depleted cell lines with "neuronal-like" properties. *J Neurochem.* **67,** 1897–1907.

Moraes, C. T., Kenyon, L., and Hao, H. (1999). Mechanisms of human mitochondrial DNA maintenance; The determining role of primary sequence and length over function. *Mol. Biol. Cell.* **10,** 3345–3356.

Moraes, C. T., and Schon, E. A. (1996). Detection and analysis of mitochondrial DNA and RNA in muscle by in situ hybridization and single-fiber PCR. *Methods Enzymol.* **264,** 522–540.

Morais, R., Gregoire, M., Jeannotte, L., and Gravel, D. (1980). Chick embryo cells rendered respiration-deficient by chloramphenicol and ethidium bromide are auxotrophic for pyrimidines. *Biochem. Biophys. Res. Commun.* **94,** 71–77.

Nass, M. M. (1972). Different effects of ethidium bromide on mitochondrial and nuclear DNA synthesis in vivo in cultured mammalian cells. *Exp. Cell Res.* **72,** 211–222.

Robinson, B. H. (1996). Use of fibroblast and lymphoblast cultures for detection of respiratory chain defects. *Methods Enzymol.* **264,** 454–464.

Shapiro, N. I., and Varshaver, N. B. (1975). Mutagenesis in cultured mammalian cells. *Methods Cell Biol.* **10,** 209–234.

Smith, C. A., Jordan, J. M., and Vinograd, J. (1971). In vivo effects of intercalating drugs on the superhelix density of mitochondrial DNA isolated from human and mouse cells in culture. *J. Mol. Biol.* **59,** 255–272.

Swerdlow, R. H., Parks, J. K., Cassarino, D. S., Maguire, D. J., Maguire, R. S., Bennett, J. P., Jr., Davis, R. E., and Parker, W. D., Jr. (1997). Cybrids in Alzheimer's disease: A cellular model of the disease? *Neurology* **49,** 918–925.

Tiranti, V., Hoertnagel, K., Carrozzo, R., Galimberti, C., Munaro, M., Granatiero, M., Zelante, L., Gasparini, P., Marzella, R., Rocchi, M., Bayona-Bafaluy, M. P., Enriquez, J. A., Uziel, G., Bertini, E., Dionisi-Vici, C., Franco, B., Meitinger, T., and Zeviani, M. (1998). Mutations of SURF-1 in Leigh disease associated with cytochrome c oxidase deficiency. *Am. J. Hum. Genet.* **63,** 1609–1621.

Trounce, I., Neill, S., and Wallace, D. C. (1994). Cytoplasmic transfer of the mtDNA nt 8993 T $\rightarrow$ G (ATP6) point mutation associated with Leigh syndrome into mtDNA-less cells demonstrates cosegregation with a decrease in state III respiration and ADP/O ratio. *Proc. Natl. Acad. Sci. USA* **91,** 8334–8338.

Trounce, I. A., Kim, Y. L., Jun, A. S., and Wallace, D. C. (1996). Assessment of mitochondrial oxidative phosphorylation in patient muscle biopsies, lymphoblasts, and transmitochondrial cell lines. *Methods Enzymol.* **264,** 484–509.

Trounce, I., and Wallace, D. C. (1996). Production of transmitochondrial mouse cell lines by cybrid rescue of rhodamine-6G pre-treated L-cells. *Somat. Cell. Mol. Genet.* **22,** 81–85.

Ziegler, M. L., and Davidson, R. L. (1981). Elimination of mitochondrial elements and improved viability in hybrid cells. *Somat. Cell Genet.* **7,** 73–88.

Zorn, G. A., Lucas, J. J., and Kates, J. R. (1979). Purification and characterization of regenerating mouse L929 karyoplasts. *Cell* **18,** 659–672.

CHAPTER 23

Diagnostic Assays for Defects in Mitochondrial DNA Replication and Transcription in Yeast and Human Cells

Bonnie L. Seidel-Rogol and Gerald S. Shadel

Department of Biochemistry
Emory University School of Medicine
Atlanta, Georgia 30322

I. Introduction

Mitochondrial (mt) transcription and mtDNA replication are dependent on nucleus-encoded factors that are synthesized in the cytoplasm and imported into the organelle. These molecules include mitochondrial RNA polymerase and requisite transcription factors, mtDNA polymerase (pol γ) and accessory replication proteins, and RNA-processing enzymes (Shadel and Clayton, 1997). Studies of numerous model systems have elucidated basic machinery required for these processes, yet the full complement of necessary factors remains to be identified and characterized. In addition, how mitochondrial gene expression and mtDNA replication are regulated in accordance with metabolic, developmental, and tissue-specific demands remains largely undetermined. The yeast model system has provided a wealth of information regarding the structure and function of mitochondrial transcription and replication proteins and has also begun to serve as a convenient model for certain human mitochondrial disease processes. The utility of using the yeast *Saccaromyces cerevisiae* to study mitochondrial genetics comes from the ability of so-called petite mutants to survive on specific carbon sources in the absence of respiratory function. Thus, these strains are ideal for the study of nuclear gene products required for mitochondrial gene expression and mtDNA maintenance.

The involvement of mitochondrial dysfunction in human disease has long been appreciated, and the number of disease states either caused or compounded by the loss of mitochondrial activities continues to rise steadily. It is now clear that mtDNA mutations cause human disease through loss of expression of mitochondrial-encoded oxidative phosphorylation (OXPHOS) components. In addition, respiration defects also result from mutations in nuclear genes that encode proteins required for function or assembly of the OXPHOS system, or factors that participate in replication or expression of the mitochondrial genome. Examples of the latter include nuclear gene mutations responsible for mtDNA depletion syndrome and disorders that result in an increased incidence of multiple mtDNA deletions in patient cells. The essential nature of this organelle and its genome underscores the need for understanding how the nucleus and mitochondria communicate in order to regulate mitochondrial gene expression.

This chapter describes several simple assays for the determination of relative mtDNA copy number (as a measure of mtDNA replication and stability) and steady-state levels of mitochondrial transcripts (as a measure of an intact mitochondrial transcription apparatus) in human and yeast cells. These diagnostic assays allow potential defects in mtDNA replication and transcription to be identified in a time- and labor-efficient manner. Once a defect of this type is postulated based on the preliminary assays described here, we encourage use of more intensive *in vitro* and *in organello* methods to address the molecular nature of the defect more precisely (see Section IV).

II. Diagnosis of mtDNA Replication Defects in Yeast and Human Cells

A. Fundamental Aspects of mtDNA Replication

The process of mtDNA replication has been well characterized in mammalian cells, where it occurs by an unusual asynchronous mechanism involving two physically separated, unidirectional origins (termed origin of heavy-strand, O_H, and origin of light-strand, O_L) (Clayton, 1982). The human mtDNA molecule is a 16.5-kb, double-stranded circle that contains a major noncoding locus called the D-loop regulatory region. It is within this region where transcription of each of the two coding strands is initiated and transcripts derived from the L-strand promoter (LSP) are processed to prime H-strand mtDNA replication at O_H. Much progress has been made toward understanding this transcription-dependent mtDNA replication mechanism, and a wealth of data in vertebrate systems support a general model for the initiation of H-strand mtDNA replication, described by Shadel and Clayton (1997). In this model, RNA transcripts are initiated at the LSP by human mtRNA polymerase in the presence of the transcription factor h-mtTFA. Next, mtRNA polymerase traverses a series of conserved sequence blocks at O_H, where the elongating LSP transcript forms a stable RNA/DNA hybrid. This hybrid serves as a substrate for RNA-processing enzymes, such as RNase MRP (Chang and Clayton, 1987; Lee and Clayton, 1998), that cleave the RNA in the hybrid to generate mature RNA primers that are utilized by DNA polγ to begin a productive mtDNA replication event. The degree of conservation of this mtDNA replication mechanism in nonvertebrates has been addressed in *S. cerevisiae,* and substantial evidence indicates that priming of wild-type mtDNA by transcription also occurs in yeast. Issues pertaining to the extent of similarity that exists between human and yeast mtDNA replication have been discussed (Shadel, 1999).

B. Methods for Detecting mtDNA and Measuring Copy Number

We routinely use two standard methods for detecting mtDNA in yeast and human cells: staining with nucleic acid-specific dyes (e.g., 4′, 6-diamidino-2-phenylindole; DAPI) and Southern hybridization with mtDNA-specific probes. In recent years, polymerase chain reaction (PCR)-based protocols have also become a standard means to detect mtDNA. This section describes a quantitative competitive PCR method that allows mtDNA copy number to be assessed in human cells. Lack of detection of mtDNA or a reduced mtDNA signal in any of these assays can be taken as preliminary evidence of a mtDNA replication defect (see Section IV).

Because relic mtDNA sequences (mitochondrial-derived pseudogenes) are often present in the nuclear genome of eukaryotes (Wallace *et al.,* 1997), it should be emphasized that Southern hybridization or PCR-based methods performed on total cellular DNA, or inadequately purified mtDNA, are prone to artifactual, false-positive signals. To ensure that the observed signals in such assays are in fact derived from mtDNA, control cell lines with isogenic nuclear backgrounds that are devoid of mtDNA (rho°) should also be tested routinely for lack of a signal. In many cases, such cell lines can be generated by extended growth in the presence of ethidium bromide (King and Attardi, 1996). If generation of rho° lines is not possible, the test mtDNA should be isolated from mitochondria that have been purified to an extent that nuclear DNA does not contaminate the preparation.

1. Detection of mtDNA by Staining with DAPI

Staining mtDNA with nucleic acid specific fluorescent dyes is a convenient means to assess quickly whether mtDNA is present in cells (Fig. 1). Our standard protocol for staining yeast mtDNA with DAPI is as follows. Yeast cells are grown to mid-exponential growth phase in synthetic dextrose medium (Bacto-yeast nitrogenous base, 6.7 g/liter; dextrose, 20 g/liter) with required amino acids supplements as described (Sherman, 1991). Cells ($\sim 1.0 \times 10^7$) are harvested by centrifugation (30 s, 16,000*g*) in a sterile 1.5-ml microcentrifuge tube, resuspended in 70% ethanol, and incubated for 10 min on ice. Cells are then washed twice with 1 ml of deionized water, resuspended in 100 μl of 50 ng/ml DAPI (Sigma Chemical; diluted from a 1 mg/ml stock that can be stored at $-20°$C) in phosphate-buffered saline (PBS), and incubated for an additional 10 min.

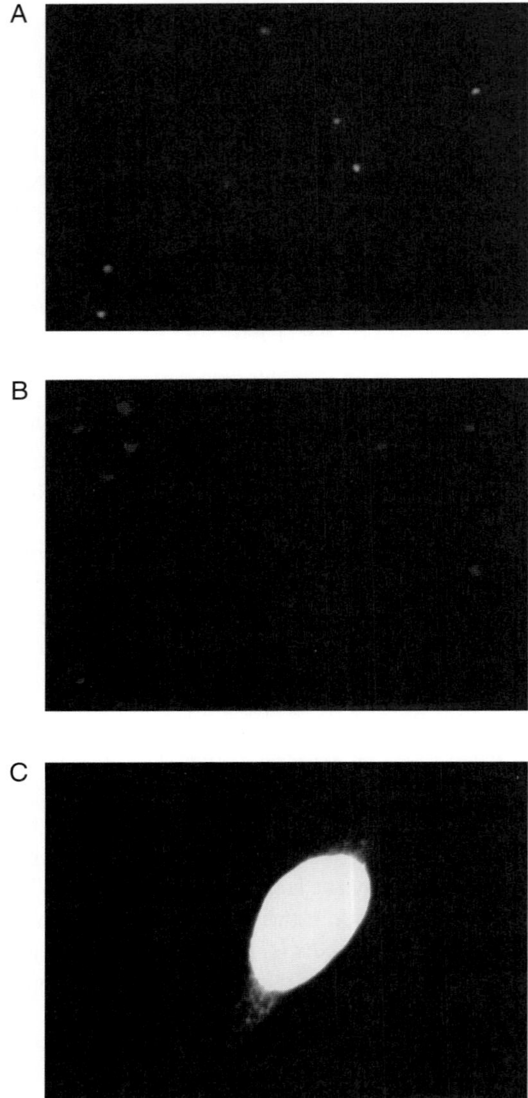

Fig. 1 Visualization of mtDNA in yeast and human cells using DAPI. The yeast strain GS125 (Wang and Shadel, 1999) was depleted of mtDNA after extensive growth in the presence of ethidium bromide to create the strain GS125 rho°. DAPI staining of GS125 rho° reveals only nuclear DNA staining (A), whereas the identically treated isogenic GS125 strain exhibits characteristic punctate cytoplasmic mtDNA staining in addition to the nuclear DNA signal (B). (C) A HeLa Tet-On cell stained with DAPI according to the procedure described in Spector *et al.* (1998). In order to visualize the punctate mtDNA signal throughout the cytoplasm in these cells, longer photographic exposures were required that "bleached out" the nuclear DNA signal (bright oval shape in the center of the cell).

DAPI fluorescence is observed using a fluorescence microscope equipped with a UV filter set; we use an Olympus BX60 epifluorescence microscope. Both a nuclear DNA signal and a mtDNA signal are observed in wild-type cells (Fig. 1B). Alternatively, it is also possible to stain mtDNA in growing cell cultures by adding DAPI (1 μg/ml) to the growth medium. In our hands, the nuclear DNA signal is markedly reduced and variable under these conditions. In addition, DAPI can also be used to stain mtDNA in fixed human cells. We routinely use the protocol described in Spector *et al.* (1998) for this purpose (Fig. 1C).

2. Detection of mtDNA by Southern Hybridization

Because specific DNA sequences are detected by Southern hybridization, this technique can be used not only to detect and compare the relative amount of mtDNA in two samples, but to address whether the genome is intact. For example, the presence of mtDNA, as indicated by fluorescence microscopy, does not assess whether these genomes are full-length or mutated. This can be problematic, particularly in yeast, where mutant strains (*rho*$^-$ petite strains) arise spontaneously that propagate extensively deleted, repetitive mtDNA molecules. The following protocol is used for analyzing human mtDNA by Southern hybridization; yeast mtDNA can be analyzed in a similar manner using yeast mtDNA probes.

a. Southern Hybridization of mtDNA from Human Cell Lines

The following protocol was developed for HeLa cells, but should be generally applicable to any human cell line harboring an intact mtDNA genome. A HeLa Tet-On cell line was obtained from Clontech and cells were grown in monolayer in Dulbecco's Minimal Essential Medium (DMEM) (Mediatech Cellgro) containing 4.5 g/liter glucose, 10% fetal bovine serum, 2 mM L-glutamine, and 100 μg/ml G418 with subculturing every 3–4 days. Cells were grown in 75-cm^2 culture flasks to ~80% confluency (~1×10^7 cells) and after trypsinization were harvested by centrifugation in a clinical centrifuge. Total cellular DNA was isolated using an RNA/DNA purification protocol exactly as described by the manufacturer (RNA/DNA Midi Kit, Qiagen Inc.). Five micrograms of total cellular DNA was digested to completion with *Nco*I, separated by electrophoresis on a 0.8% agarose gel, and transferred to a nylon membrane by capillary flow as described (Sambrook *et al.*, 1989). The DNA was then cross-linked to the membrane with UV light using a Stratalinker apparatus (Stratagene, Inc.). Nucleic acid hybridization was accomplished using Rapid-Hyb buffer exactly as described by the manufacturer (Amersham, Inc.) in a rolling-bottle hybridization oven (Techne Hybridiser, HB-1D). The mtDNA-specific probe was generated by PCR using total cellular DNA as a template (forward PCR primer: 5′-AGCAAACCACAGTTTCATGCCCA-3′; reverse primer: 5′-TAATGTTAGTAAGGGTGGGGAAG-3′), which results in a 5456-bp product corresponding to nucleotides (nt) 8188–13,644 (according to Anderson *et al.*, 1981) of the mtDNA sequence. The PCR product was gel purified and radiolabeled with Ready-to-Go labeling beads according to the protocol provided by the manufacturer (Amersham, Inc.). A control nuclear DNA probe was generated in the same fashion, except that a

381-bp PCR product derived from the nuclear 28S rRNA gene locus was used as a template (forward PCR primer: 5′-GCCTAGCAGCCGACTTAGAACTGG-3′; reverse PCR primer: 5′-GGCCTTTCATTATTCTACACCTC-3′). After Southern hybridization, the mtDNA probe hybridized to a 7555-bp *Nco*I restriction fragment of human mtDNA and the 28S probe hybridized to a 2588-bp *Nco* I restriction fragment of nuclear DNA, corresponding to repeats from the ribosomal DNA locus. The mtDNA signal and the nuclear DNA signal were then quantitated by densitometry of exposed X-ray film (using a Bio-Rad Fluor-S MultiImager) or directly using a Phosphoimager system (FujiX BAS1000). The nuclear DNA signals were used to normalize the mtDNA signals for loading were differences, allowing the relative amounts of mtDNA in each sample to be compared.

3. Quantition of mtDNA by Competitive Polymerase Chain Reaction

A useful assay to determine more precisely the mtDNA copy number in cells is competitive PCR (e.g., see Zhang *et al.* 1994). This method involves the generation of a control template that consists of an internally deleted version of the gene sequence that is to be analyzed in the experiment. This control template, which is amplified using the same primers as the test DNA, is added to the experimental DNA samples in known amounts, where it competes with the test template for primer binding and amplification. A series of competitive PCR reactions is generated consisting of several reactions with the same amount of test DNA, but differing in the amount of competitor template that is present. This series is amplified under identical PCR conditions, and the products are analyzed by agarose gel electrophoresis. Quantitation of the experimental and control products allows the calculation of an equivalence point (where the test and control signal are equal) that can be used to determine the concentration of test DNA in the sample (Freeman *et al.,* 1999).

We have applied this technique to assay mtDNA copy number in human cells. A 494-bp region of the mitochondrial ND2 gene can be amplified by PCR using the following primers: forward 5′-GGCCCAACCCGTCATCTAC- 3′ and reverse 5′-CCACCTCAAC-TGCCTGCTATG-3′. To generate the competitive template for this assay, this ND2 fragment was first ligated into a pGEM-T vector (Promega, Inc.) and its nucleotide sequence was confirmed. We then generated an internal deletion by digesting the resulting plasmid with *Bst*EII, which cleaves a unique internal site in the ND2 gene fragment, and treating the linearized DNA with Bal31 as described (Sambrooke, 1989). After ligation of the deleted products, a control plasmid with an internal deletion of 204 bp of the ND2 gene (pND2Δ290) was isolated. Using ND2 gene-specific primers, this plasmid produces a 290-bp PCR product that, after electrophoresis, is easily distinguished from the 494-bp fragment generated from the wild-type ND2 gene (Fig. 2A).

Increasing amounts (0.026–2.6 ng) of pND2Δ290 are added to samples containing a constant amount (200 ng) of total HeLa Tet-On cellular DNA, and PCR amplification is performed (see legend of Fig. 2 for PCR conditions). The PCR products are precipitated in ethanol, separated on a 1% agarose gel, and visualized by staining with ethidium bromide (Fig. 2A). In our experiments, quantitation of the fluorescent products is usually accomplished using a Bio-Rad Fluor-S MultiImager. These data are plotted and analyzed as described by Freeman *et al.* (1999) to determine the amount of mtDNA present in the

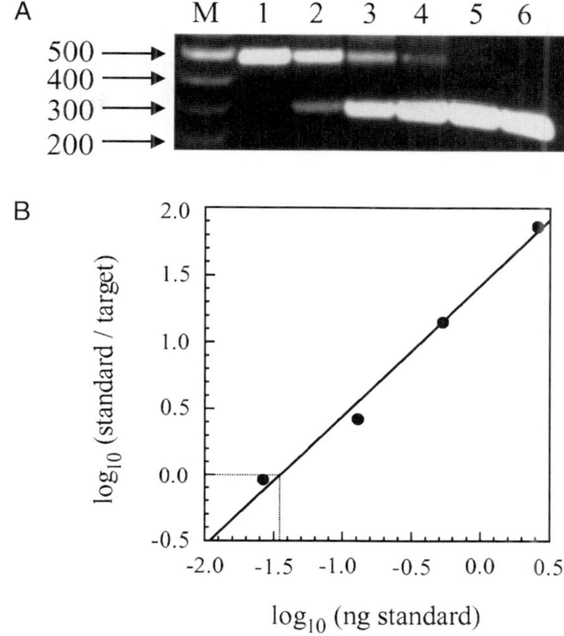

Fig. 2 Competitive PCR analysis of mtDNA from HeLa Tet-On cells. (A) Agarose gel electrophoretic analysis of competitive PCR products. Lanes 1–5 indicate products resulting from PCR reactions that contained 200 ng of total cellular DNA and differing amounts of external competitor DNA (pND2Δ290) as follows: lane 1, 0 ng; lane 2, 0.26 ng; lane 3, 0.39 ng; lane 4, 0.52 ng; and lane 5, 2.6 ng. Lane 6 contains 2.6 ng pND2Δ290 only. DNA molecular weight standards run with the PCR reactions are indicated on the left. PCR was performed in a final volume of 100 μl containing 3.5 m*M* $MgCl_2$, 10 m*M* of each dNTP, 100 ng of each primer, and 2.5 units of *Taq* DNA polymerase (Bethesda Reseach Laboratories, Inc.). *Taq* DNA polymerase was added after a 5-min denaturation at 95°C, and amplification was for 25 cycles (1 min at 95°C, 1 min at 50°C, and 3 min at 72°C) followed by one 10-min elongation period at 72°C. (B) Quantitative analysis of competitive PCR products. The amount of each PCR product was determined by fluorescence using a Bio-Rad Fluor-S MultiImager, and these data are plotted as described by Freeman *et.al.* (1999), which should yield a straight line. The equivalence point (indicated as dashed lines on the graph) is determined by calculating the value of X, where $Y = 0$ using the equation for this line. This value of X can be used to estimate the amount of target mtDNA in the sample. In this example, an estimate of 35 pg of mtDNA was obtained from a value of $X = -1.46$ (at $Y = 0$).

sample (Fig. 2B). Knowledge of the total number of cells used for the DNA preparation allows a ready estimation of the average mtDNA copy number/per cell.

III. Analyzing Mitochondrial Transcripts *in vivo*

A. Mitochondrial Transcription Initiation

Compared to other characterized transcription systems, including bacterial systems, the mitochondrial transcription apparatus is relatively simple in terms of protein requirements for initiation. This situation has allowed faithful *in vitro* transcription systems to be developed for several species, including yeast, *Xenopus,* mouse, and human, which have facilitated the study of protein–protein and protein–nucleic acid interactions required for transcription in the organelle. While interesting species-specific differences exist with regard to transcription (Shadel and Clayton, 1993), several basic principles are conserved from yeast to humans that are summarized here.

Perhaps the best understood mitochondrial transcription system is that from the yeast *S. cerevisiae.* Core yeast mitochondrial RNA polymerase (sc-mtRNA polymerase) is a 150-kDa protein encoded by the *RPO41* gene and is homologous to the single-subunit RNA polymerases of the bacteriophages T7, T3, and SP6 (Masters *et al.,* 1987). In addition, it has a unique amino-terminal extension of ~400 amino acids that harbors a mitochondrial matrix localization signal (aa 1-29) and comprises at least one additional functional domain required for mtDNA replication or stability (Wang and Shadel, 1999). Unlike bacteriophage RNA polymerases, which are themselves sufficient for transcription initiation, sc-mtRNA, polymerase requires a transcription factor, sc-mtTFB, to bind specifically to a simple nonanucleotide promoter ATATAAGTA(+1) and initiate transcription. Analysis of the nucleotide sequence of the *MTF1* gene, which encodes sc-mtTFB, revealed regions with apparent similarity to bacterial sigma factors (Jang and Jaehning, 1991). Indeed, certain properties of sc-mtTFB, observed during transcription *in vitro* (Mangus *et al.,* 1994), and aspects of its interaction with the core sc-mtRNA polymerase (Cliften *et al.,* 1997) suggest that sc-mtTFB functions in a manner analogous to this class of proteins. However, mutational analysis (Shadel and Clayton, 1995) and sequence comparisons of the three known yeast mtTFB homologs (Carrodeguas *et al.,* 1996) point to some potentially important differences between sigma factors and mtTFB in both structure and function. In addition to sc-mtTFB, yeast mitochondria also possess an abundant mtDNA-binding protein (sc-mtTFA) that contains two high mobility group (HMG) box DNA-binding domains (Diffley and Stillman, 1991). This protein, which is encoded by the *ABF2* gene, is part of the mtDNA nucleoid (Newman *et al.,* 1996), has been implicated in stabilizing recombination intermediates (MacAlpine *et al.,* 1998), and can stimulate or inhibit transcription *in vitro* depending on its concentration (Parisi *et al.,* 1993). Mutations that disrupt the *RPO41* or *MTF1* genes, as well as the *ABF2* gene (under certain growth conditions), result in instability and ultimately loss of mtDNA, consistent with a role for each of these factors in mtDNA replication and/or stability as well as transcription.

As in yeast, core human mtRNA (h-mtRNA) polymerase is encoded by a nuclear gene (Tiranti *et al.,* 1997), is homologous to bacteriophage RNA polymerases, and requires at least one transcription factor, human mitochondrial transcription factor A (h-mtTFA). The h-mtTFA protein was identified in mitochondrial extracts through its ability to promote efficient specific transcription initiation by human mtRNA polymerase *in vitro* by binding to a specific site at human mtDNA promoters (Fisher and Clayton, 1988). Like its yeast homolog (sc-mtTFA), h-mtTFA contains two HMG-box DNA-binding domains, but in addition contains a basic C-terminal tail that imparts specific DNA-binding and transcriptional activation capabilities to the protein (Dairaghi *et al.,* 1995). The importance of mtTFA in mammals has been underscored through disruption of its gene in mice, which leads to early embryonic lethality, presumably through lack of mtDNA gene expression and replication (Larsson *et al.,* 1998). However, deciphering the precise role of mtTFA in transcription and mtDNA replication, as well as the molecular basis for the differential transcription factor requirements of yeast and humans (Shadel and Clayton, 1993), awaits establishment of mammalian *in vitro* transcription and replication systems that utilize recombinant mtRNA polymerase.

B. Standard Assays for Mitochondrial Transcription

In general, the methodologies for assaying mitochondrial transcript levels are similar, if not identical, to those utilized to analyze nucleus-derived transcripts. For example, Northern hybridization, nuclease protection (e.g., RNase protection), and reverse-transcriptase polymerase chain reaction (RT-PCR) assays have all been used successfully to monitor steady-state levels of specific mitochondrial transcripts in cultured human and yeast cells. Because these methods are standard in modern molecular biology laboratories and are well documented in other methods volumes (e.g., Sambrook, 1989), a detailed treatment of these common techniques will not be presented here. However, there are aspects of mitochondrial genome organization and expression that must be considered when these types of assays are performed on mitochondrial transcripts. For example, it should be appreciated that four different types of mature RNA molecules are synthesized in yeast and human mitochondria: mRNA, tRNA, rRNA, and origin-associated RNAs. These RNA molecules display a range of half-lives *in vivo.* In addition, with the exception of origin RNAs in yeast, many of these RNA species are synthesized as polycistronic RNA transcripts that must be processed into mature molecules. Thus, steady-state levels of these transcripts can be affected dramatically by alterations in RNA-processing activity as well as by transcription. Finally, the inherent and induced instability of the mitochondrial genome, especially in yeast, presents additional complications for transcript analysis *in vivo.* Although we outline assays in human and yeast cells in this chapter, the basic methodology should be applicable to any mitochondrial system where genome sequence information is available.

1. Assaying Mitochondrial Transcripts in Human Cells

Because human mtDNA has two major transcription units, one for each strand, and the H-strand transcript is subject to a site-specific termination event (Kruse *et al.,* 1989), we suggest employment of a set of at least four gene-specific probes to estimate whether mitochondrial transcription is normal in a human cell. The minimal set should consist of the following: one probe for the L-strand (e.g., ND6); two probes for the H-strand, one before (e.g., 12S or 16S rRNA) and one after (e.g., ND1 or COX I) the transcription termination site; and a control probe, corresponding to a nucleus-encoded RNA that does not change under the experimental conditions being tested (commonly, rRNA or glyceraldehyde 6-phosphate dehydrogenase, GAPDH). Because mtDNA transcripts are relatively abundant, whole cell RNA preparations can be used routinely, obviating the need to prepare RNA from purified mitochondria. However, when possible, control experiments should be carried out on cells lacking mtDNA (rho°), or with highly purified mitochondria, to ensure that the observed signals are of mitochondrial origin. To compare samples under different conditions, the mitochondrial and nuclear transcript signals are quantitated, allowing a ratio of mitochondrial to nuclear signal to be calculated, with the condition that the control signal does not change during the course of the experiment. In this manner, the relative steady-state amount of L-strand and H-strand transcripts can be determined and compared between samples. Comparison of the two H-strand signals may also provide information regarding the relative amount of transcription termination that is occurring at the termination site. For transcript analysis from cultured human cells, we generally use Qiagen RNA/DNA kits for RNA isolation. In our hands, these RNA preparations are suitable for analysis by Northern hybridization, RNase protection, and RT-PCR.

2. Assaying Mitochondrial Transcripts in Yeast

The strategy for analyzing transcript levels in yeast is virtually identical to that described earlier for human cells. However, the following considerations take into account unique features of the yeast mitochondrial genetic system, including the presence of ~20 independent transcription units and several transcripts that can harbor introns.

A conserved nonanucleotide promoter is associated with all the known yeast mitochondrial transcription units, suggesting that transcripts may respond similarly to changes in transcription initiation capacity. However, variability in nucleotides at +1 and +2 relative to the start site of transcription is known to affect promoter efficiency, making it possible that differential promoter regulation is operative under certain circumstances. Thus, estimation of whether transcription is occurring normally in yeast requires employment of a set of gene-specific probes that assay not only the different types of mature RNA molecules, but also different promoter contexts. A minimal set should include at least one of each the following: a mRNA (e.g., *COB1, COX1, COX2,* or *COX3*); a rRNA (e.g., 14S or 21S); a tRNA (e.g., $tRNA^{Glu}$ or $tRNA^{fMet}$); and a control probe, corresponding to a nucleus-encoded RNA that does not change under the experimental conditions being tested (e.g., 18S rRNA). To be more thorough, one could also include probes for the *RPM1* gene, which encodes the RNA component of mitochondrial RNase P, and *VAR1,* which encodes a mitochondrial ribosomal protein. When designing these probes, care should be taken to avoid intron sequences often contained in the *COX1, COB1,* and the 21S rRNA genes. Examples of how this type of "global" transcript analysis has been used to assess changes in mitochondrial transcript levels that occur in yeast in response to glucose repression have been reported (Ulery, 1994; Mueller and Getz, 1986).

3. Assaying Mitochondrial Transcription in Yeast Under Condition of Wild-type Genome Instability

Yeast mtDNA is inherently unstable and mutations in many nuclear genes can enhance this instability (Shadel, 1999). For example, disruption of genes encoding the yeast mitochondrial transcription apparatus (i.e., *RPO41* and *MTF1*) leads to mtDNA instability and eventual loss of the genome. This makes assaying mitochondrial transcript levels under such conditions problematic because the mtDNA is either not wild type or is absent altogether. We have utilized a class of mtDNA mutants called hypersuppressive (HS) *rho*$^-$ petites to circumvent this problem of template loss in order to assay transcription initiation capability *in vivo*. In particular, we have used the HS3324 *rho*$^-$ mitochondrial genome, which harbors a 966-bp, head-to-tail repeat of *ori5* (a putative yeast mitochondrial origin), that is stable in *RPO41* null backgrounds (Lorimer *et al.*, 1995). That is, the mtDNA present in this strain is able to replicate by a mechanism that does not require intact mtRNA polymerase activity. However, the *ori5* repeat does contain a nonanucleotide promoter that allows production of a specific RNA transcript to be assayed *in vivo*. The utility of such strains stems from the ability to introduce the HS3324 genome (or any HS genome) into strains of interest by a process known as cytoduction (Lancashire and Mattoon, 1979). In this manner, strains that cannot maintain wild-type mtDNA due to a nuclear DNA mutation can be assayed for the presence of an intact transcription apparatus. For example, we have used an *rpo41* null HS3324 strain to determine that mutated alleles of *RPO41* encode proteins that are capable of transcription *in vivo* (Fig. 3), but nonetheless are defective for wild-type mtDNA maintenance (Wang and Shadel, 1999). In principle, this method can be used to analyze mutants in other mitochondrial transcription components (e.g., sc-mtTFB) to determine their effects on transcription *in vivo* in the presence of a stable template. As is the case for mtRNA polymerase, additional functions for such factors may be elucidated in this manner.

IV. Additional Considerations

The assays described in this chapter provide an initial assessment of whether mtDNA replication or mitochondrial transcription is occurring normally in human or yeast cells. While these assays are helpful for initial diagnostic purposes, it is noted that a reduction in the steady-state level of mtDNA or mtRNA in a cell, which is the end point of all the assays described in this chapter, may be due to perturbations in the cell other than those that affect mtDNA replication or transcription per se. For example, a reduction in the steady-state level of mtDNA could, in principle, be caused by increased mtDNA instability. The situation in yeast is complicated further by the fact that mutations in a large number of genes, including those that abolish mitochondrial translation, result in mtDNA instability through mechanisms that remain unclear (Shadel, 1999). Thus, if a reduction in mtDNA copy number is detected in one of the assays described earlier, additional methods should be employed to determine if a DNA replication defect is the source of the problem. Examples of more direct assays for mtDNA replication include labeling of mtDNA with 5-bromo-2-deoxy uridine (Berk and Clayton, 1973; Davis and Clayton, 1996; Meeusen *et al.*, 1999) and ligation-mediated PCR to detect nascent mtDNA strands (Kang *et al.*, 1997).

Similar complications influence the interpretation of the transcription assays described herein. For example, a decrease in the steady-state level of a specific mtRNA species

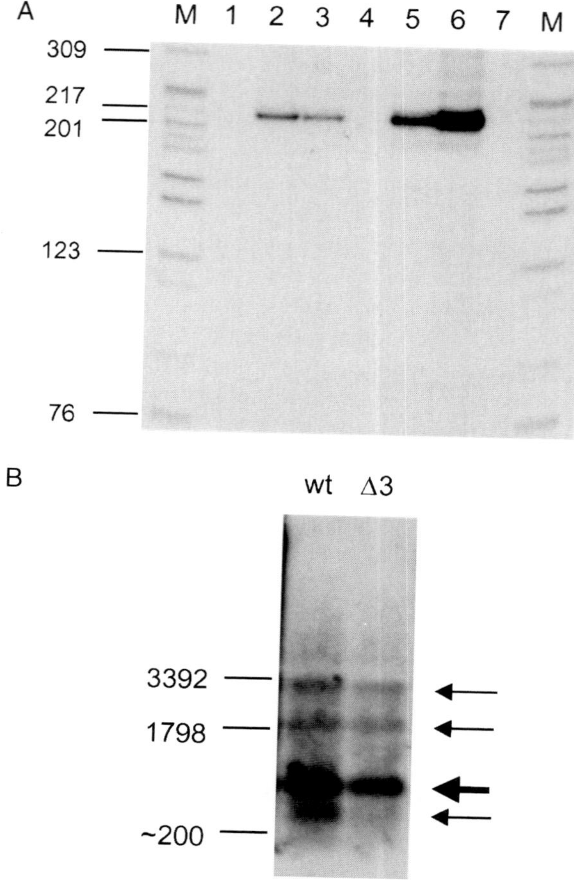

Fig. 3 Analysis of mitochondrial *ori5* transcripts in a hypersupptressive *rho*$^-$ yeast strain. The haploid yeast strain BS127 HS3324 ΔRPO41 contains an *ori5* repeat *rho*$^-$mitochondrial genome and lacks mtRNA polymerase due to a chromosomal disruption of the *RPO41* locus (Lorimer *et al.,* 1995). This strain was transformed with plasmids that harbor either a wild-type *RPO41* gene (pGS348) or the temperature-sensitive allele *rpo41Δ3* (pGS348Δ3). After growth at 30 and 37°C for ~20 generations, total cellular RNA was isolated from these strains as described (Schmitt *et al.,* 1990), as well as from the parental strain that lacks mtRNA polymerase activity. (A) Results of an S1 nuclease protection assay using an *ori5*-specific probe as described (Wang and Shadel, 1999). RNA analyzed from strains grown at 30°C (lanes 1–3) and from strains grown at 37°C (lanes 4–7) are shown. The parental strain (devoid of mtRNA polymerase) did not yield any *ori5* transcript signal (lanes 1 and 4), whereas strains harboring either the wild-type *RPO41*-containing plasmid (lanes 2 and 5) or an *rpo41Δ3*-containing plasmid (lanes 3 and 6) each produced *ori5* transcripts that initiated at the predicted transcription start site. Lane 7 is a control lane that is the same as lane 6, except the sample was treated with RNase A prior to analysis, indicating that the observed signals were derived from RNA. M, *Hpa*II-digested pBR322 DNA markers (length in nucleotides is indicated on the left). (B) Northern analysis of *ori5* transcripts in *RPO41* (wt) and *rpo41Δ3* (Δ3) plasmid-containing strains grown at 37°C performed as described (Wang and Shadel, 1999). The major *ori5* RNA transcript is indicated by the bold arrow. Two additional minor transcripts of ~2 and ~3 kb, corresponding to two and three *ori5* repeat lengths, are indicated by thin arrows, as is a subrepeat length transcript that was observed only in the wild-type strain. Indicated on the left is the position where known RNA species migrated.

may be caused by a decrease in RNA stability or improper RNA processing rather than a reduction in transcription capacity. Again, additional assays have been developed that can be employed to address some of these issues more directly. For example, *in organello* methods for analyzing mitochondrial transcripts have been established for yeast (Poyton *et al.,* 1996) and vertebrate cells (Enriquez et al., 1996; Gaines, 1996). By examining many transcripts simultaneously in this manner, such analyses could, in theory, differentiate between RNA processing and RNA transcription defects; the presence of incompletely processed precursor species indicates the former possibility. As already noted, faithful *in vitro* transcription assays have been developed for both human and yeast systems (Mangus and Jaehning, 1996; Shadel and Clayton, 1996). These systems can be employed to assay mitochondrial transcription initiation, elongation, or termination activity (Fernandez-Silva *et al.,* 1996).

Acknowledgment

This work was supported by Grant HL-59655 from the National Institutes of Health awarded to G.S.S.

References

Anderson, S., Bankier, A. T., Barrell, B. G., de Bruijn, M. H., Coulson, A. R., Drouin, J., Eperon, I. C., Nierlich, D. P., Roe, B. A., Sanger, F., Schreier, P. H., Smith, A. J., Staden, R., and Young, I. G. (1981). Sequence and organization of the human mitochondrial genome. *Nature* **290,** 457–465.

Berk, A. J., and Clayton, D. A. (1973). A genetically distinct thymidine kinase in mammalian mitochondria: Exclusive labeling of mitochondrial deoxyribonucleic acid. *J. Biol. Chem.* **248,** 2722–2729.

Carrodeguas, J. A., Yun, S., Shadel, G. S., Clayton, D. A., and Bogenhagen, D. F. (1996). Functional conservation of yeast mtTFB despite extensive sequence divergence. *Gene Expr.* **6,** 219–230.

Chang, D. D., and Clayton, D. A. (1987). A novel endoribonuclease cleaves at a priming site of mouse mitochondrial DNA replication. *EMBO J.* **6,** 409–417.

Clayton, D. A. (1982). Replication of animal mitochondrial DNA. *Cell* **28,** 693–705.

Cliften, P. F., Park, J. Y., Davis, B. P., Jang, S. H., and Jaehning, J. A. (1997). Identification of three regions essential for interaction between a sigma-like factor and core RNA polymerase. *Genes Dev.* **11,** 2897–2909.

Dairaghi, D. J., Shadel, G. S., and Clayton, D. A. (1995). Addition of a 29 residue carboxyl-terminal tail converts a simple HMG box-containing protein into a transcriptional activator. *J. Mol. Biol.* **249,** 11–28.

Davis, A. F., and Clayton, D. A. (1996). In situ localization of mitochondrial DNA replication in intact mammalian cells. *J. Cell Biol.* **135,** 883–893.

Diffley, J. F. X., and Stillman, B. (1991). A close relative of the nuclear, chromosomal high-mobility group protein HMG1 in yeast mitochondria. *Proc. Natl. Acad. Sci. USA* **88,** 7864–7868.

Enriquez, J. A., Perez-Martos, A., Lopez-Perez, M. J., and Montoya, J. (1996). In organello RNA synthesis system from mammalian liver and brain. *Methods Enzymol.* **264,** 50–57.

Fernandez-Silva, P., Micol, V., and Attardi, G. (1996). Mitochondrial DNA transcription initiation and termination using mitochondrial lysates from cultured human cells. *Methods Enzymol.* **264,** 129–139.

Fisher, R. P., and Clayton, D. A. (1988). Purification and characterization of human mitochondrial transcription factor 1. *Mol. Cell. Biol.* **8,** 3496–3509.

Freeman, W. M., Walker, S. J., and Vrana, K. E. (1999). Quantitative RT-PCR: Pitfalls and potential. *Biotechniques* **26,** 124–125.

Gaines, G. L., III (1996). In organello RNA synthesis system from HeLa cells. *Methods Enzymol.* **264,** 43–49.

Jang, S. H., and Jaehning, J. A. (1991). The yeast mitochondrial RNA polymerase specificity factor, MTF1, is similar to bacterial sigma factors. *J. Biol. Chem.* **266,** 22671–22677.

Kang, D., Miyako, K., Kai, Y., Irie, T., and Takeshige, K. (1997). *In vivo* determination of replication origins of human mtiochondrial DNA by ligation-mediated PCR. *J. Biol. Chem.* **272,** 15275–15279.

King, M. P., and Attardi, G. (1996). Isolation of human cell lines lacking mitochondrial DNA. *Methods Enzymol.* **264,** 304–313.

Kruse, B., Narasimhan, N., and Attardi, G. (1989). Termination of transcription in human mitochondria: Identification and purification of a DNA binding protein factor that promotes termination. *Cell* **58,** 391–397.

Lancashire, W. E., and Mattoon, J. R. (1979). Cytoduction: A tool for mitochondrial genetic studies in yeast. Utilization of the nuclear-fusion mutation kar 1-1 for transfer of drug r and mit genomes in *Saccharomyces cerevisiae. Mol. Gen. Genet.* **170,** 333–344.

Larsson, N. G., Wang, J., Wilhelmsson, H., Oldfors, A., Rustin, P., Lewandoski, M., Barsh, G. S., and Clayton, D. A. (1998). Mitochondrial transcription factor A is necessary for mtDNA maintenance and embryogenesis in mice. *Nature Genet.* **18,** 231–236.

Lee, D. Y., and Clayton, D. A. (1998). Initiation of mitochondrial DNA replication by transcription and R-loop processing. *J. Biol. Chem.* **273,** 30614–30621.

Lorimer, H. E., Brewer, B. J., and Fangman, W. L. (1995). A test of the transcription model for biased inheritance of yeast mitochondrial DNA. *Mol. Cell. Biol.* **15,** 4803–4809.

MacAlpine, D. M., Perlman, P. S., and Butow, R. A. (1998). The high mobility group protein Abf2p influences the level of yeast mitochondrial DNA recombination intermediates *in vivo. Proc. Natl. Acad. Sci. USA* **95,** 6739–6743.

Mangus, D. A., Jang, S. H., and Jaehning, J. A. (1994). Release of the yeast mitochondrial RNA polymerase specificity factor from transcription complexes. *J. Biol. Chem.* **269,** 26568–26574.

Mangus, D. A., and Jaehning, J. A. (1996). Transcription *in vitro* with *Saccharomyces cerevisiae* mitochondrial RNA-polymerase. *Methods Enzymol.* **264,** 57–66.

Masters, B. S., Stohl, L. L., and Clayton, D. A. (1987). Yeast mitochondrial RNA polymerase is homologous to those encoded by bacteriophages T3 and T7. *Cell* **51,** 89–99.

Meeusen, S., Tieu, Q., Wong, E., Weiss, E., Schieltz, D., Yates, J. R., and Nunnari, J. (1999). Mgm101p is a novel component of the mitochondrial nucleoid that binds DNA and is required for the repair of oxidatively damaged mitochondrial DNA. *J. Cell Biol.* **145,** 291–304.

Mueller, D. M., and Getz, G. S. (1986). Steady state analysis of mitochondrial RNA after growth of yeast *Saccharomyces cerevisiae* under catabolite repression and derepression. *J. Biol. Chem.* **261,** 11816–11822.

Newman, S. M., Zelenaya-Troitskaya, O., Perlman, P. S., and Butow, R. A. (1996). Analysis of mitochondrial DNA nucleoids in wild-type and a mutant strain of *Saccharomyces cerevisiae* that lacks the mitochondrial HMG box protein Abf2p. *Nucleic Acids Res.* **24,** 386–393.

Parisi, M. A., Xu, B., and Clayton, D. A. (1993). A human mitochondrial transcriptional activator can functionally replace a yeast mitochondrial HMG-box protein both *in vivo* and *in vitro. Mol. Cell. Biol.* **13,** 1951–1961.

Poyton, R. O., Bellus, G., McKee, E. E., Sevarino, K. A., and Goehring, B. (1996). In organello mitochondrial protein and RNA synthesis systems from *Saccharomyces cerevisiae. Methods Enzymol.* **264,** 36–42.

Sambrook, J., Fritsch, E. F., and Maniatis, T. (1989). *In* "Molecular Cloning: A Laboratory Manual," 2nd Ed. Cold Spring Harbor Press, Cold Spring Harbor, NY.

Schmitt, M. E., Brown, T. A., and Trumpower, B. L. (1990). A rapid and simple method for preparation of RNA from *Saccharomyces cerevisiae. Nucleic Acids Res.* **18,** 3091–3092.

Shadel, G. S. (1999). Yeast as a model for human mtDNA replication. *Am. J. Hum. Genet.* **65,** 1230–1237.

Shadel, G. S., and Clayton, D. A. (1993). Mitochondrial transcription initiation: Variation and conservation. *J. Biol. Chem.* **268,** 16083–16086.

Shadel, G. S., and Clayton, D. A. (1995). A *Saccharomyces cerevisiae* mitochondrial transcription factor, sc-mtTFB, shares features with sigma factors but is functionally distinct. *Mol. Cell. Biol.* **15,** 2101–2108.

Shadel, G. S., and Clayton, D. A. (1996). Mapping promoters in displacement-loop region of vertebrate mitochondrial DNA. *Methods Enzymol.* **264,** 139–148.

Shadel, G. S., and Clayton, D. A. (1997). Mitochondrial DNA maintenance in vertebrates. *Annu. Rev. Biochem.* **66,** 409–435.

Sherman, F. (1991). Getting started with yeast. *Methods Enzymol.* **194,** 3–21.

Spector, D. L., Goldman, R. D., and Leinwand, L. A. (1998). *In* "Cells, a Laboratory Manual," Vol. 3, p. 101.10. Cold Spring Harbor Press, Cold Spring Harbor, NY.

Tiranti, V., Savoia, A., Forti, F., D'Apolito, M. F., Centra, M., Rocchi, M., and Zeviani, M. (1997). Identification of the gene encoding the human mitochondrial RNA polymerase (h-mtRPOL) by cyberscreening of the expressed sequence tags database. *Hum. Mol. Genet.* **6,** 615–625.

Ulery, T. L., Jang, S. H., and Jaehning, J. A. (1994). Glucose repression of yeast mitochondrial transcription: Kinetics of derepression and role of nuclear genes. *Mol. Cell. Biol.* **14,** 1160–1170.

Wallace, D. C., Stugard, C., Murdock, D., Schurr., T., and Brown, M. D. (1997). Ancient mtDNA sequences in the human nuclear genome: A potential source of errors in identifying pathogenic mutations. *Proc. Natl. Acad. Sci. USA* **94,** 14900–14005.

Wang, Y., and Shadel, G. S. (1999). Stability of the mitochondrial genome requires an amino-terminal domain of yeast mitochondrial RNA polymerase. *Proc. Natl. Acad. Sci. USA* **96,** 8046–8051.

Zhang, H., Cooney, D. A., Sreenath, A., Zhan, Q., Agbaria, R., Stowe, E. E., Fornace, A. J., Jr., and Johns, D. G. (1994). Quantitation of mitochondrial DNA in human lymphoblasts by a competitive polymerase chain reaction method: Application to the study of inhibitors of mitochondrial DNA content. *Mol. Pharmacol.* **46,** 1063–1069.

CHAPTER 24

Analysis of Mitochondrial Translation Products *in Vivo* and *in Organello* in Yeast

Benedikt Westermann, Johannes M. Herrmann, and Walter Neupert

Institut fur Physiologische Chemie der Ludwig-Maximilians-Universitat Munchen
80336 Munchen, Germany

I. Introduction

Mitochondria are semiautonomous organelles possessing their own genome and their own transcription and translation machinery (for review, see Gray *et al.,* 1999). The mitochondrial DNA (mtDNA) encodes a limited number of proteins and RNAs essential for the formation of respiratory-competent mitochondria. Among the mitochondrially encoded proteins are components of respiratory complexes I (NADH-ubiquinone oxidoreductase), II (succinate-ubiquinone oxidoreductase), III (ubiquinol-cytochrome *c* oxidoreductase), IV (cytochrome *c* oxidase), and V (ATP synthase) and components of the mitochondrial ribosomes. Furthermore, ribosomal RNAs and, in some organisms,

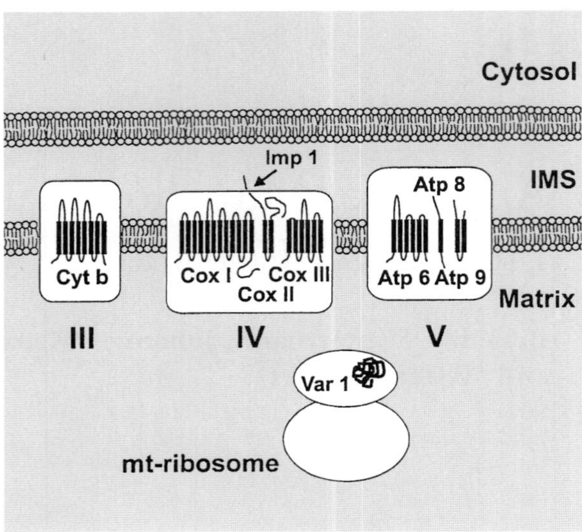

Fig. 1 Location and topology of the proteins synthesized in yeast mitochondria. III, respiratory chain complex III (ubiquinol-cytochrome *c* oxidoreductase); IV, respiratory chain complex IV (cytochrome *c* oxidase); V, respiratory chain complex V (ATP synthase); mt-ribosome, mitochondrial ribosome. The N-terminal end of the transmembrane proteins is drawn at the left-hand side. See text for details.

some tRNAs are encoded by mtDNA. The size and gene content of mitochondrial genomes are highly variable among different species. Mitochondrial genome sizes range from <6-kbp in *Plasmodium falciparum* to >366-kbp pairs in *Arabidopsis thaliana*. Despite the capacity of mitochondria to encode and synthesize proteins, the vast majority of mitochondrial proteins is encoded in the nucleus, synthesized in the cytosol, and imported in a co- or posttranslational manner (for review, see Attardi and Schatz, 1988; Costanzo and Fox, 1990; Neupert, 1997; Poyton and McEwen, 1996).

The mitochondrial genome of the yeast *Saccharomyces cerevisiae* is roughly 80 kbp in size and encodes eight major proteins: Cytochrome *b* is a subunit of complex III; CoxI, CoxII, and CoxIII are subunits of complex IV; Atp6, Atp8, and Atp9 are subunits of the F_0 part of the ATP synthase (complex V); and Var1, the only soluble mitochondrially encoded protein, is a component of the small subunit of the mitochondrial ribosome. The location and topology of the proteins synthesized in yeast mitochondria are depicted in Fig. 1. Upon export to the intermembrane space, the amino terminus of CoxII is processed by the inner membrane peptidase, Imp1.

This chapter describes procedures for the *in vivo* and *in organello* labeling of mitochondrial translation products in the yeast *S. cerevisiae*. Applications for the analysis of mitochondrial translation for the investigation of various aspects of organelle biogenesis, such as protein folding and assembly, protein degradation, and protein sorting, are discussed.

II. Labeling of Mitochondrial Translation Products *in Vivo*

The labeling of mitochondrial translation products *in vivo* (Douglas and Butow, 1976) is a simple and straightforward method to analyze the pattern of mitochondrial protein synthesis in living cells. In brief, yeast cells are harvested from a logarithmically growing culture, cytosolic translation is stopped by the addition of cycloheximide, cells are incubated in the presence of [^{35}S]methionine, labeling of mitochondrial translation is stopped by the addition of chloramphenicol and an excess of cold methionine, and a total cell extract is prepared and analyzed by sodium dodecyl sulfate–polyacrylamide gel electrophoresis (SDS–PAGE) and autoradiography.

Yeast cells from a fresh plate are inoculated in 20 ml of minimal medium containing 0.67% (w/v) yeast nitrogen base, 2% (w/v) galactose, and—if necessary—30 mg/liter of the required amino acids, adenine, or uracil (Sherman *et al.,* 1986). The culture is incubated under agitation (120 rpm) at the appropriate temperature overnight and grown to an optical density (OD_{578}) of 0.5 to 2.0. After harvesting the cells by centrifugation for 5 min at 1000*g* at room temperature (Beckman JA20 rotor), the OD_{578} is adjusted to 3. From this cell suspension, 250 μl is transferred to a microfuge tube, and 5 μl of a freshly prepared cycloheximide stock solution (7.5 mg/ml in water) is added. After a 1-min incubation, 8 μl of an amino acid stock solution (2 mg/ml of each amino acid except methionine) and 2 μl [^{35}S]methionine (10 mCi/ml) are added. The samples are incubated under agitation at the desired temperature (standard temperature is 30°C). Labeling is stopped by the addition of 0.5 mg/ml chloramphenicol (stock solution 65 mg/ml in ethanol) and 4 m*M* unlabeled methionine. Incubation is continued for another 10 min. Cells are pelleted in a microfuge and resuspended in 500 μl water. After addition of 75 μl lysis solution [1.85 *M* NaOH; 7.5% (v/v) 2-mercaptoethanol], the sample is vortexed and incubated for 10 min at room temperature. For precipitation of proteins, 600 μl 50% (w/v) trichloroacetic acid is added, the sample is incubated on ice for 30 min, and is then centrifuged at 30,000*g* for 30 min at 4°C. The supernatant is removed by aspiration, and the pellet is washed with ice-cold acetone and dried at 37°C. The sample is resuspended in 50 μl sample buffer [2% (w/v) SDS, 10% (v/v) glycerol, 2% (v/v) 2-mercaptoethanol; 0.02% (w/v) bromphenol blue, 60 m*M* Tris–HCl, pH 6.8] by 30 min agitation at room temperature. It is important not to boil the samples! Ten to 20 μl is used per lane for SDS–PAGE (see Section V).

III. Labeling of Mitochondrial Translation Products *in Organello*

The labeling of mitochondrial translation products in isolated organelles (Herrmann *et al.,* 1994a; McKee and Poyton, 1984) is the method of choice when mitochondrially translated proteins are analyzed further after the labeling reaction. In brief, isolated mitochondria are incubated in the presence of [^{35}S]methionine and an energy-regenerating system, labeling is stopped by the addition of an excess of cold methionine, organelles are collected by centrifugation, washed, and either directly analyzed by SDS–PAGE and autoradiography, or manipulated further.

Mitochondria are isolated by differential centrifugation (see Chapter 2) and frozen in aliquots. To prepare fresh 1.5× translation buffer, add 375 μl 2.4 *M* sorbitol, 225 μl 1 *M* KCl, 22.5 μl 1 *M* KP_i buffer, pH 7.2, 30 μl 1 *M* Tris–HCl, pH 7.2, 19 μl 1 *M* Mg_2SO_4, 45 μl 100 mg/ml fatty acid-free bovine serum albumin, 30 μl 200 m*M* ATP, 15 μl 50 m*M* GTP, 1.7 mg α-ketoglutarate, 3.5 mg/ml phosphoenolpyruvate, and 9.1 μl amino acid stock solution (consisting of 2 mg/ml of all proteinogenic amino acids except methionine) and adjust the volume with distilled water to 1 ml. Mix 20 μl 1.5× translation buffer with 1.5 μl pyruvate kinase (0.5 mg/ml), 5.5 μl distilled water, and 2 μl mitochondria (10 mg protein/ml, in 250 m*M* sucrose, 1 m*M* EDTA, 10 m*M* MOPS-KOH, pH 7.2) and incubate for 2 min at 30°C. Then, 1 μl [^{35}S]methionine (10 mCi/ml) is added, and mitochondrial translation is allowed to proceed for 20 min at 30°C. Labeling is stopped by the addition of 10 μl 0.2 *M* cold methionine and incubation is continued for 5 min at 30°C. Mitochondria are collected by centrifugation for 5 min at 9000*g* at 2°C and washed with 200 μl washing buffer (0.6 *M* sorbitol, 1 m*M* EDTA, 5 m*M* methionine). Mitochondria are either resuspended in an adequate buffer for further analysis (see Section IV) or solubilized in 20 μl sample buffer (see Section II) and subjected to SDS–PAGE (see Section V).

IV. Use of Mitochondrial Translation for the Study of Various Aspects of Mitochondrial Biogenesis

After synthesis, mitochondrially translated proteins have to acquire their native folded conformation and assemble into larger protein complexes consisting mainly of nuclear-encoded proteins. The components of respiratory complexes have to be integrated into the mitochondrial inner membrane, and the amino-terminal end of CoxII has to be exported to the intermembrane space, where it becomes processed (see Fig. 1). Misfolded and nonassembled proteins are degraded. This section reviews some work from our laboratory in order to illustrate how mitochondrial translation might be used to study these processes.

A. Protein Folding

Var1 is the only soluble protein synthesized in yeast mitochondria. The newly synthesized polypeptide chain is bound by mitochondrial chaperone proteins that keep it in a soluble state until the protein assembles into the ribosomal 38S subunit. The soluble state of Var1 can be observed by a simple aggregation assay. In brief, mitochondria are lysed with detergent after *in organello* translation, aggregated proteins are pelleted by centrifugation, soluble proteins are precipitated from the supernatant, and both fractions are analyzed by SDS–PAGE and autoradiography. Using this protocol, mitochondrial Hsp70 (mt-Hsp70, also termed Ssc1p) and mitochondrial DnaJ, Mdj1p, have been identified to play a role in maintaining newly synthesized Var1 in a soluble conformation (Herrmann *et al.,* 1994b; Westermann *et al.,* 1996).

After *in organello* translation (see Section III), mitochondria are resuspended in 100 μl lysis buffer [150 m*M* NaCl, 5 m*M* EDTA, 10 m*M* Tris-HCl, pH 7.4, 0.1% (v/v) Triton X-100] and lysed for 15 min on ice. Aggregated material is pelleted by centrifugation for 10 min at 17,500*g* at 2°C. Soluble proteins are precipitated from the supernatant by the addition of 10 μl 100% (w/v) trichloroacetic acid, incubation on ice for 30 min, and centrifugation at 30,000*g* for 30 min at 2°C. The pellet is washed with ice-cold acetone and dried at 37°C. Both samples are resuspended in 15 μl sample buffer (see Section II) and analyzed by SDS–PAGE and autoradiography (see Section V). As a control, an aliquot of total *in organello* translation should be loaded. Note that Var1 is especially sensitive to aggregation at elevated temperatures (e.g., 37°C).

B. Protein Assembly

Var1, Atp6, and Atp9 have been used to study the assembly of proteins into larger complexes (Arlt *et al.,* 1996; Herrmann *et al.,* 1994b). Here, it is important to treat the yeast culture with 2 mg/ml chloramphenicol (stock solution 65 mg/ml in ethanol) for 2 h before the cells are harvested for isolation of mitochondria. This treatment depletes mitochondrial translation products by inhibition of organellar protein synthesis, whereas cytosolic protein translation remains unaffected. Therefore, precomplexes lacking the mitochondrially encoded subunits accumulate and are available for assembly with labeled proteins after *in organello* protein synthesis.

To analyze the assembly of Var1 into the ribosomal 38S subunit, mitochondria are lysed with detergent after *in organello* translation, ribosomes are resolved on a sucrose gradient, proteins are precipitated and analyzed by SDS–PAGE, and ribosomal RNA is analyzed by recording the absorption at 260 nm. A detailed protocol is described in Herrmann *et al.* (1994b).

After *in organello* translation employing mitochondria isolated from a chloramphenicol-treated culture, two bands of higher molecular weight become apparent. A 48-kDa band represents an oligomer of Atp9, and a 54-kDa band represents the 48-kDa oligomer with an associated Atp6 subunit. Both oligomers are resistant to 2% SDS, however, are disrupted by trichloroacetic acid treatment, and cannot be found in mitochondria isolated from cycloheximide-treated cultures (see later). The formation of these fragments has been used to investigate the role of mt-Hsp70 in the assembly of oligomeric membrane protein complexes (Herrmann *et al.,* 1994b).

C. Protein Degradation

Mitochondrially synthesized proteins that are not assembled into their complexes are subject to degradation by mitochondrial proteases. These nonassembled translation products can be accumulated by treating the yeast culture with 150 mg/l cycloheximide for 2 h before harvesting the cells for preparation of mitochondria. This treatment inhibits cytosolic protein synthesis and therefore depletes the nuclear-encoded assembly partners of mitochondrial translation products.

For the assay of protein degradation, it is important that puromycin (50 μg/ml) is used to stop mitochondrial translation after the labeling reaction. Chloramphenicol must not be used because this antibiotic would result in the inhibition of the proteolytic breakdown (Langer *et al.,* 1995). It should be noted that incompletely synthesized polypeptide chains are released from the ribosomes by the puromycin treatment, whereas they remain bound after chloramphenicol treatment.

The degradation of completely synthesized mitochondrial translation products can be monitored in a pulse-chase reaction. *In organello* translation is carried out essentially as described in Section III and stopped with puromycin. Proteolysis is allowed to occur at 37°C, and aliquots are withdrawn at various time points. Mitochondria are reisolated by centrifugation, washed, and analyzed by SDS–PAGE and autoradiography. As degradation proceeds, the bands of labeled proteins will disappear during further incubation at 37°C.

For the generation of incompletely synthesized polypeptide chains, *in organello* translation is carried out in the presence of 2.5 μg/ml puromycin. Translation is stopped by the addition of 20 m*M* unlabeled methionine and 50 μg/ml puromycin. After extensive washes, mitochondria are resuspended in translation buffer, and degradation is allowed to proceed at 37°C. Aliquots are withdrawn at various time points, supplemented with 12.5% trichloroacetic acid (TCA), and incubated on ice for 30 min. The TCA-soluble fraction (containing the [^{35}S]methionine that becomes released from polypeptide chains) is separated from the TCA-insoluble fraction (containing intact or partially hydrolyzed proteins) by centrifugation for 10 min at 25,000*g* at 2°C. The radioactivity in both fractions is measured by scintillation counting. Detailed protocols for the degradation of completely and incompletely synthesized mitochondrial translation products associated with the mitochondrial inner membrane have been reported by Langer *et al.* (1995).

Employing the assays just described, it was shown that mt-Hsp70 is required for the degradation of nonassembled Atp6 (Herrmann *et al.,* 1994b). Furthermore, the YTA10-12 complex, an AAA protease in the inner membrane of mitochondria, was shown to be required for the assembly of the ATP synthase and to mediate the ATP-dependent degradation of nonassembled polypeptides in the inner membrane (Arlt *et al.,* 1996; Pajic *et al.,* 1994). Moreover, it was demonstrated that mitochondrial proteases are required for the activity of intron-encoded mRNA maturases. Protease mutants exhibit a defect in the expression of intron-containing genes in mitochondria, as becomes evident from the observation of an altered pattern of mitochondrial translation products (Arlt *et al.,* 1998; van Dyck *et al.,* 1998).

D. Protein Sorting

After synthesis on mitochondrial ribosomes, a precursor form of CoxII (pCoxII) is exported from the matrix, and its N-terminal end is processed by the Imp1 protease in the intermembrane space (see Fig. 1). A defect in export or processing becomes apparent by the accumulation of pCoxII, which has a slightly lower mobility in SDS–PAGE. In order to discriminate between defects in export and processing, mitochondria can be converted to mitoplasts by selectively opening the outer membrane by hypotonic swelling and subsequent treatment with protease (see Chapter 2). It was shown that *in organello*-translated pCoxII is still protease resistant in mitoplasts obtained from a mutant defective in the inner membrane protein Oxa1 (Hell *et al.,* 1997). These experiments contributed to the identification of Oxa1 as a component required for the export of proteins from the mitochondrial matrix.

V. Analysis of Mitochondrial Translation Products by Sodium Dodecyl Sulfate–Polyacrylamide Gel Electrophoresis

In vivo- or *in organello*-translated proteins are resolved by SDS–PAGE and visualized by autoradiography. Although this is a standard technique, some critical points should be considered. First of all, mitochondrial translation products must not be boiled in sample buffer. Boiling of the very hydrophobic proteins leads to an irreversible formation of protein aggregates that do not enter the gel. Second, the running behavior of the mitochondrially synthesized proteins depends very much on the gel system used. We recommend the use of the following system to obtain a good resolution of all translation products. Other gel systems might result in a different migration of mitochondrial translation products relative to each other. A representative example of an autoradiography of *in organello*-translated proteins is shown in Fig. 2. Note that there is a considerable amount of variability of the relative intensities of the protein bands depending on strains and experimental conditions.

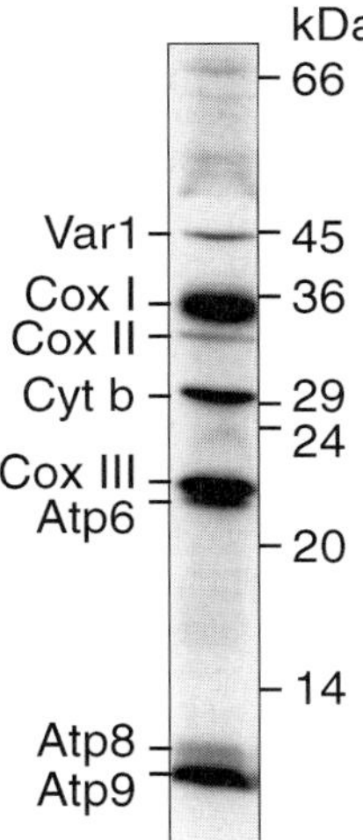

Fig. 2 Pattern of mitochondrial translation products synthesized *in organello*. Isolated mitochondria of wild-type strain W303 were preincubated for 10 min at 25°C in translation buffer and labeled for 30 min at 25°C as described. After a 5-min chase, organelles were reisolated by centrifugation, washed, and solubilized in sample buffer. Labeled proteins were separated by SDS–PAGE. After staining and destaining the dried gel was exposed to an X-ray film overnight. Var1, a protein of the small ribosomal subunit; CoxI–III, subunits I–III of the cytochrome *c* oxidase; Cyt *b*, cytochrome *b* of the ubiquinol-cytochrome *c* oxidoreductase; Atp6, 8, and 9, subunits 6, 8, and 9 of the ATP synthase. The mobility of the protein standards used is indicated.

Mitochondrially synthesized proteins are identified by their mobility in SDS–PAGE. The largest mitochondrial translation product is Var1. This protein runs as a sharp band with an apparent molecular mass of about 45 kDa. It should be noted that there is a strain-dependent variation in size of up to 7% and that the size of the protein can even vary within the same strain due to recombinational events (Butow *et al.,* 1985). CoxI runs as a fuzzy band close to the 36-kDa marker. The intensity of CoxI labeling varies very much depending on yeast strains and experimental conditions. For CoxII, usually only the mature-sized species is observed. Some processing- or export-deficient strains, however, might show the slightly larger precursor, pCoxII (see, e.g., Hell *et al.,* 1997). Both CoxII and pCoxII run as relatively sharp bands below CoxI. Cytochrome *b* (Cyt b) usually gives a strong signal and runs above the 29-kDa marker. CoxIII and Atp6 run between the 24- and 20-kDa markers with the upper of these two bands being CoxIII. Atp8 and Atp9 run very close to the front. Atp9 shows a slightly higher mobility in the gel; however, these two proteins are sometimes difficult to resolve. Occasionally, two higher molecular mass bands might be observed: a 48-kDa complex consisting of an Atp9 oligomer and a 54-kDa complex consisting of an Atp9/Atp6 oligomer. These SDS-resistant complexes are especially abundant in mitochondria with surplus nuclear-encoded precomplexes. They can be disassembled by precipitation with trichloroacetic acid prior to gel electrophoresis (see also Section IV,B.). Furthermore, additional bands might be observed in strains with splicing abnormalities.

We use an SDS gel with a separating gel (90 × 150 × 1 mm) consisting of 16% (w/v) acrylamide, 0.1% (w/v) bisacrylamide, 385 m*M* Tris–HCl, pH 8.8, 0.1% (w/v) sodium dodecyl sulfate, 0.05% (w/v) ammonium persulfate, 0.035% (v/v) *N,N,N′,N′*,-tetramethylethylendiamine (TEMED) and a stacking gel (10 × 150 × 1 mm) consisting of 5% (w/v) acrylamide, 0.033% (w/v) bisacrylamide, 60 m*M* Tris–HCl, pH 6.8, 0.1% (w/v) sodium dodecyl sulfate, 0.05% (w/v) ammonium persulfate, and 0.1% (v/v) TEMED. The running buffer contains 10 g/liter sodium dodecyl sulfate, 144 g/liter glycine; and 30 g/liter Tris base. Electrophoresis is performed in a vertical chamber at constant 25 mA.

After electrophoresis, the gel is either stained with 0.1% (w/v) Coomassie brilliant blue in 50% (v/v) methanol; 10% (v/v) acetic acid, destained with 50% (v/v) methanol; 10% (v/v) acetic acid and dried, or proteins are transferred onto nitrocellulose. Radioactively labeled mitochondrial translation products are detected by autoradiography.

VI. Discussion

The biogenesis of a functional protein requires the transcription of its gene, maturation of the mRNA, translation to polypeptide, prevention of aggregation of the newly synthesized polypeptide chain emerging from the ribosome, targeting to its proper location, and folding to its native conformation. In many cases, the newly synthesized protein has to be integrated into a membrane, matured by a processing peptidase, and/or assembled into a higher molecular protein complex. Finally, misfolded and nonassembled polypeptides have to be degraded by proteases. The proteins synthesized in yeast mitochondria are an excellent model system to study these processes because they can be labeled and identified easily without the need of antibody production or enzymatic assays.

Even though only eight major proteins are encoded by the mitochondrial genome in yeast, they provide a complete set of model proteins for the investigation of all of the just-mentioned aspects of protein biogenesis. The yeast mitochondrial genome harbors introncontaining genes, and defects of mRNA maturation were observed to result in an altered expression of Cyt b and CoxI (Arlt *et al.,* 1998; van Dyck *et al.,* 1998). Var1 is a soluble protein sensitive to aggregation after its synthesis and can be used as a substrate for various processes involving the action of molecular chaperones in the mitochondrial matrix (Herrmann *et al.,* 1994b; Westermann *et al.,* 1996). pCoxII has a cleavable signal sequence for export across the inner membrane. Because it is synthesized in the matrix (and does not have to be imported from the cytosol), it is an ideal candidate to study protein export out of the mitochondrial matrix. The number of transmembrane segments of the mitochondrially encoded respiratory chain components ranges from 1 (Atp8) to 12 (CoxI). Thus, these proteins provide a wide range of different topologies for the study of membrane integration processes. All mitochondrially synthesized proteins have to assemble into higher molecular complexes. To investigate this process, mitochondria have the unique advantage that precomplexes consisting of nuclear-encoded proteins can be accumulated by the chloramphenicol treatment of yeast cultures prior to the isolation of mitochondria (Herrmann *et al.,* 1994b). Similarly, nonassembled mitochondrial translation products can be accumulated by the cycloheximide treatment of yeast cultures in order to study protein degradation processes.

Several components have been identified to play a role in the biogenesis of mitochondrially synthesized proteins. Most notably, it was shown by cross-linking and coimmunoprecipitation approaches (see Chapter 12 for a discussion of these techniques) that the chaperones mt-Hsp70 and Mdj1p interact physically with newly synthesized proteins in mitochondria (Herrmann *et al.,* 1994b; Westermann *et al.,* 1996). Furthermore, it was demonstrated that the inner membrane protein Oxa1 is in direct contact with mitochondrially synthesized proteins exported across the inner membrane (Hell *et al.,* 1998). In conclusion, the analysis of mitochondrial translation *in vivo* and *in organello* has turned out to be a valuable experimental system for the study of many processes not only important for the biogenesis of mitochondria, but also for the biogenesis of proteins in general. Some of the involved components have been identified recently and many more are likely to be discovered.

References

Arlt, H., Steglich, G., Perryman, R., Guiard, B., Neupert, W., and Langer, T. (1998). The formation of respiratory chain complexes in mitochondria is under the proteolytic control of the m-AAA protease. *EMBO J.* **17,** 4837–4847.

Arlt, H., Tauer, R., Feldmann, H., Neupert, W., and Langer, T. (1996). The YTA10-12 complex, an AAA protease with chaperone-like activity in the inner membrane of mitochondria. *Cell* **85,** 875–885.

Attardi, G., and Schatz, G. (1988). Biogenesis of mitochondria. *Annu. Rev. Cell. Biol.* **4,** 289–333.

Butow, R. A., Perlman, P. S., and Grossman, L. I. (1985). The unusual Var1 gene of yeast mitochondrial DNA. *Science* **228,** 1496–1501.

Costanzo, M. C., and Fox, T. D. (1990). Control of mitochondrial gene expression in *Saccharomyces cerevisiae. Annu. Rev. Genet.* **24,** 91–113.

Douglas, M. G., and Butow, R. A. (1976). Variant forms of mitochondrial translation products in yeast: Evidence for location of determinants on mitochondrial DNA. *Proc. Natl. Acad. Sci. USA* **73,** 1083–1086.

Gray, M. W., Burger, G., and Lang, B. F. (1999). Mitochondrial evolution. *Science* **283,** 1476–1481.

Hell, K., Herrmann, J., Pratje, E., Neupert, W., and Stuart, R. A. (1997). Oxa1p mediates the export of the N- and C-termini of pCoxII from the mitochondrial matrix to the intermembrane space. *FEBS Lett.* **418,** 367–370.

Hell, K., Herrmann, J. M., Pratje, E., Neupert, W., and Stuart, R. A. (1998). Oxa1p, an essential component of the N-tail export machinery in mitochondria. *Proc. Natl. Acad. Sci. USA* **95,** 2250–2255.

Herrmann, J. M., Fölsch, H., Neupert, W., and Stuart, R. A. (1994a). Isolation of yeast mitochondria and study of mitochondrial protein translation. *In* "Cell Biology: A Laboratory Handbook" (J. E. Celis, ed.), pp. 538–544. Academic Press, San Diego.

Herrmann, J. M., Stuart, R. A., Craig, E. A., and Neupert, W. (1994b). Mitochondrial heat shock protein 70, a molecular chaperone for proteins encoded by mitochondrial DNA. *J. Cell Biol.* **127,** 893–902.

Langer, T., Pajic, A., Wagner, I., and Neupert, W. (1995). Proteolytic breakdown of membrane-associated polypeptides in mitochondria of *Saccharomyces cerevisiae. Methods Enzymol.* **260,** 495–503.

McKee, E. E., and Poyton, R. O. (1984). Mitochondrial gene expression in *Saccharomyces cerevisiae*. I. optimal conditions for protein synthesis in isolated mitochondria. *J. Biol. Chem.* **259,** 9320–9331.

Neupert, W. (1997). Protein import into mitochondria. *Annu. Rev. Biochem.* **66,** 863–917.

Pajic, A., Tauer, R., Feldmann, H., Neupert, W., and Langer, T. (1994). Yta10p is required for the ATP-dependent degradation of polypeptides in the inner membrane of mitochondria. *FEBS Lett.* **353,** 201–206.

Poyton, R. O., and McEwen, J. E. (1996). Crosstalk between nuclear and mitochondrial genomes. *Annu. Rev. Biochem.* **65,** 563–607.

Sherman, F., Fink, G. R., and Hicks, J. (1986). *Methods in Yeast Genetics: A Laboratory Course.* Cold Spring Harbor Laboratory Press, Cold Sping Harbor, NY.

van Dyck, L., Neupert, W., and Langer, T. (1998). The ATP-dependent PIM1 protease is required for the expression of intron-containing genes in mitochondria. *Genes Dev.* **12,** 1515–1524.

Westermann, B., Gaume, B., Herrmann, J. M., Neupert, W., and Schwarz, E. (1996). Role of the mitochondrial DnaJ homologue Mdj1p as a chaperone for mitochondrially synthesized and imported proteins. *Mol. Cell. Biol.* **16,** 7063–7071.

CHAPTER 25

Numerical Methods for Handling Uncertainty in Microarray Data: An Example Analyzing Perturbed Mitochondrial Function in Yeast

Charles B. Epstein, Walker Hale IV, and Ronald A. Butow

Department of Molecular Biology
University of Texas Southwestern Medical Center
Dallas, Texas 75390

I. Introduction

Microarray technology is in an explosive growth phase, and several recent reviews have described the essentials of the method (Cheung *et al.,* 1999; Duggan *et al.,* 1999; Eisen and Brown, 1999). This chapter discusses methodological issues related to the processing and evaluation of numerical results from microarray experiments, which have not been thoroughly treated elsewhere in the literature. In addition, we illustrate the

METHODS IN CELL BIOLOGY, VOL. 65

application of microarray technology to the study of how perturbations of mitochondrial function affect gene expression in yeast.

Our experience suggests that the "wet" steps in spotted DNA microarray methodology [including slide washing and coating, polymerase chain reaction (PCR) product manipulation and printing, and mRNA labeling and hybridization] are well described in recent publications (DeRisi *et al.,* 1997; Eisen and Brown, 1999; see also the Pat Brown laboratory microarray guide on the internet at http://cmgm.stanford.edu/pbrown/protocols/index.html). These procedures give excellent results and can be followed essentially without modification. A unique exception to this rule concerns the interval between coating slides and printing. We obtain optimal results when we use slides as soon as possible after coating with poly-L-lysine. After as little as a month, the coating becomes less effective and signal intensities fall markedly. Our current procedure is to coat slides immediately prior to printing and to steam treat and UV irradiate them immediately after printing. In this regard, we differ somewhat from the recommendation in Eisen and Brown (1999). Succinic anhydride treatment may be done shortly before the arrays are used and should be done using materials that are as fresh as possible. We have no current estimate of the shelf life of arrays treated in this fashion, but have noted no deterioration over a period of 3 months.

II. Quantitative Methods in Microarray Research

A. Ratiometric Method

Microarray experiments are performed as follows: two different mRNA preparations are independently labeled using reverse transcriptase, to make cDNAs incorporating spectrally distinguishable fluorochrome-dNTP conjugates. The cDNAs are mixed and competitively hybridized to a solid-phase array of spotted PCR products. After hybridization has proceeded 4–16 h, the array is washed and scanned with a laser-powered scanner, and the mean pixel intensities are computed for each gene in each of the two channels.

The raw values obtained in each channel for each gene span a wide range, from no signal to saturation of scanner. The values obtained are critically dependent on differences in gene-specific mRNA abundance prior to the reverse transcriptase reaction; this is the foundation of the microarray method. However, the mean pixel intensities for each spot are not dependent solely on mRNA abundance. Additional sources of variability include the effect of the concentrations of spotted PCR products, variability in gene length, variability in spot diameter, and potentially sequence-specific effects on the efficiency of incorporation of fluorochrome-conjugated dNTPs by reverse transcriptase. In effect, the mapping from mRNA abundance to signal intensity is influenced by poorly controlled factors, which differ from gene to gene in a microarray experiment.

To address these uncertainties, microarray analysis proceeds by computing the *ratio* of intensities between the two colors for each gene. The ratios are more informative than the raw intensities in each color, because in the ratio, many of the unknown sources of variability cancel out. Although the ratiometric method has been quite successful, it does have some limitations. Foremost among these is the constraint it imposes on the potential to compare results among separate hybridizations, particularly if independent RNA preparations are used. An additional shortcoming of the ratiometric method is that it tends to discard information related to the relative quality (or uncertainty) of measurements: ratios obtained from relatively low-intensity pixels are inherently subject to more methodological noise than equivalent ratios from high-intensity pixels. Nevertheless, the method is already a proven source of new insights into diverse aspects of cell physiology and pathology, and the versatility of the microarray method will undoubtedly improve as additional refinements are made.

B. Image Acquisition

After microarrays are hybridized and washed, they are scanned using a suitable laser-powered scanner. A number of excellent commercial scanners are now available supplied with confocal imaging optics and photomultiplier tube (PMT)-based detection systems. We have obtained good results with two such scanners: a Scan Array 3000 (General Scanning) and a GenePix 4000A (Axon Instruments). The GenePix is a more recent design with excellent speed, freedom from artifactual detection of the platter holding the slide, and clear images with low variance in background pixel intensity.

Three criteria must be considered to obtain optimal two channel microarray scans: (1) avoidance of signal saturation, (2) matching the two scans to one another, and (3) avoidance of image acquisition at 100% PMT sensitivity. Scans are usually recorded as 16-bit images, indicating that each pixel can assume a value no higher than 2^{16}, or 65,536. It is therefore essential to operate the scanner at a sensitivity that minimizes spot saturation so as to obtain meaningful quantitative information for high-intensity spots.

Microarrays are hybridized with two spectrally distinct fluorochromes. The scan of each fluorochrome has a characteristic signal-to-noise (S/N) ratio, which is controlled both by how well a particular RNA was labeled and by the background noise of the particular hybridization. When the S/N ratios are similar between the two fluors on a single microarray, it is then possible to adjust the sensitivity of the two scans so that the histograms of pixel intensity for the two fluors are essentially equivalent. Well-matched scans are desirable because they reduce or eliminate the need to normalize data (see later) in one channel during a subsequent stage of analysis. The operating software of the GenePix scanner conveniently displays the histogram of pixel intensity, facilitating the matching of the two scans. Note that the procedure of matching scans is based on realized pixel intensities, and not on nominal scanner PMT settings, reflecting intrinsic differences in the brightness of the two fluorochromes, differences in the sensitivity of the scanner in each of the two channels, and differences in the efficiency of cDNA synthesis.

The optimal scan is nearly always obtained by reducing PMT sensitivity below 100% because PMTs introduce additional noise when used at full sensitivity. In contrast, the lasers in commercial scanners may be used at full power without degradation of fluorochrome; indeed, the lasers in the GenePix are preset to always operate at full power.

C. Image Processing and Data Analysis

1. Computing Ratios

Once array images have been obtained, they are converted to numerical values using image analysis software. There are a number of dedicated packages for this purpose, including at least one program available free to the academic community (see http://rana.stanford.edu/software/). All of these programs enable one to specify efficiently the location in the image of the array of DNA spots, and then proceed to define "signal" and "background" regions for each spot. Because background intensity may vary across an array, it is customary to derive background intensities for each spot independently using interspot pixels near the corresponding spot. Typically, there is also a "dead zone" of unused pixels between signal and background regions. This simplifies computation in the presence of variable spot diameters or irregular spot placements. Mean, median, and/or total pixel intensities are computed in signal and background regions for each spot.

The goal in array analysis is to compute the ratio (i.e., fluor1/fluor2) of local background subtracted (LBS) integrated signals (ΣS_{LBS}) for each gene. Normally, there are different numbers of pixels in signal and background regions. Suppose that in each channel there are N pixels in the signal region and M pixels in the background region for a given spot. Then the background-subtracted integrated signal (ΣS_{LBS}) is the integrated pixel intensity of the signal region (ΣS) minus the integrated pixel intensity of the background region (ΣB), scaled by the relative area of the two regions. In algebraic notation,

$$\Sigma S_{LBS} = \Sigma S - (N/M)\, \Sigma B \tag{1}$$

This is equally true for both channels. Therefore,

$$\frac{\Sigma S_{LBS}^{fluor1}}{\Sigma S_{LBS}^{fluor2}} = \frac{\Sigma S^{fluor1} - (N/M)\Sigma B^{fluor1}}{\Sigma S^{fluor2} - (N/M)\Sigma B^{fluor2}} \tag{2}$$

gives the desired ratio. Curiously, N and M need not be explicitly known and need not be the same from spot to spot on the array. For any given spot, it is possible to multiply the numerator and denominator of Eq. (2) by $1/N$. Denoting mean signal as S_{avg} and mean background as B_{avg}, this yields

$$\frac{\Sigma S_{LBS}^{fluor1}}{\Sigma S_{LBS}^{fluor2}} = \frac{S_{avg}^{fluor1} - B_{avg}^{fluor1}}{S_{avg}^{fluor2} - B_{avg}^{fluor2}} \tag{3}$$

giving the desired ratio entirely as a function of mean values of signal and background regions in each color. This is true despite the fact that the mean signal is computed from all of the pixels inside the designated signal region, including some background pixels falling between the DNA spot and the signal region outer boundary.

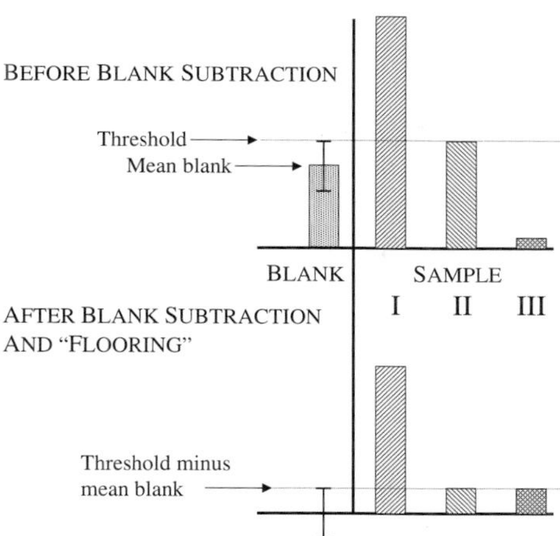

Fig. 1 Flooring method. A graphic representation of our routine approach to microarray data analysis. Numerous negative controls ("blank" spots containing exogenous DNA or SSC only) are printed on our array. Mean and standard deviation are computed for the blanks. The mean blank is typically a small, positive number. Values at or above mean blank plus 2 SD are considered to be above "threshold" and are subjected to blank subtraction. Values at or below threshold are replaced by a "floor" value, equivalent to the difference between threshold and mean blank. For values at threshold, the two methods of data treatment are equivalent.

2. Low Value Rejection or Substitution

Nearly all arrays have the property that many spots have very low intensity. It is reasonable to impose the requirement that a gene be distinguishable from "off" in at least one of the two channels, prior to including data from that gene in the results of the experiment; otherwise, these very dim spots, potentially representing unexpressed genes, will contaminate data with their highly irreproducible, and possibly large (but erroneous), expression ratios.

Suppose a gene is "off" in one channel and bright in the other channel. How then does one compute the expression ratio for that gene? The number in the "off" condition is somewhat arbitrary and is highly influenced by variability in background pixels for the array in question. To address this problem, we have adopted a method called "flooring", i.e., the substitution of a "floor" value for very low values (Fig. 1). The objective of this method is to avoid the spurious computation of arbitrarily large expression ratios when considering genes that have undergone a transition from "undetectable" to "detectable." Our method also effectively excludes from consideration ratios based on genes that are weak in both channels.

In our typical yeast arrays, we print a moderate number of non-yeast DNA sequences (which do not cross-hybridize with sequences in the yeast genome) and a large number of blank spots containing just 3× SSC. We compute mean and standard deviation for these negative control features and use these statistics to estimate the minimum measurable number that is distinguishable from a negative control. Typically, we use mean blank (after local background subtraction) plus two standard deviations as the minimum detectable value, which we call "threshold." The mean blank is always a small, positive number.

If a signal ($S_{avg}-B_{avg}$) exceeds threshold, we subtract mean blank from that signal. Mean blank is defined as the mean over all blanks of the term ($S_{avg}^{BLANK} - B_{avg}^{BLANK}$), which is first computed independently for each blank. This is done in both channels, prior to computation of expression ratios. If a signal is equal to threshold, we also subtract mean blank. This results in the conversion of a threshold value to the difference between threshold and mean blank, which from the definition of threshold, depends on the variability in the blanks. This difference is equivalent to the floor value. Finally, if a value is below threshold, rather than subtracting anything from it, we substitute the floor value. For any gene that is below threshold in both channels, we logically substitute an expression ratio of 1 (log ratio = 0) and do not compute its expression ratio as floor (fluor1)/floor (fluor2) because that ratio would merely reflect the differing variability in the blanks in the two channels. As noted, the floor value in this method depends on the variability in the blanks; when the variability in the blanks is highly discrepant between the two channels, this method must be used with due caution, as it tends to generate large, potentially spurious expression ratios for genes that are weak in both channels, if they fall above threshold in the channel with low variability in the blanks but fall below threshold in the channel with high variability in the blanks.

3. Normalization of Microarray Data

Microarray data sets consist of ratios of background-subtracted intensities derived from two images. The numerical values of these ratios depend on the scanner sensitivities used in each of the two channels scanned, the relative amounts of cDNA used during hybridization, and the ratios of gene expression. (Only the last term has any biological significance.) Thus even if the cDNA for the two channels is added in equal amounts, all results will depend on the (somewhat arbitrary and unknown) scanner sensitivity in the two channels. When log ratios are computed for a control and experimental sample, one typically finds non-zero values for the global sum as well as the modal value of the log ratios. Both of these improbable results have been achieved despite the use of equivalent amounts of mRNA for the two fluorescent-labeling reactions. Given the critical dependence of ratios on scanner sensitivity, and the desirability of computing ratios that are as comparable as possible between experiments, it is a common practice to "normalize" the two channels algebraically after extracting quantitative values. The resulting normalized values are then taken as estimators of the actual gene expression ratios.

The objective of normalization is to remove the artifactual effect of scanner sensitivity on the computed values of expression ratios. Normalization can be done to satisfy various criteria, but the two most common criteria are either that (1) the global sum of log ratios is zero or (2) the sum of log ratios for a predefined subset of genes (either "housekeeping" genes or exogenously spiked controls) is zero. Whichever objective is to be satisfied, the value of the normalization coefficient is computed as the antilog of the mean log ratio of the raw values, where the mean is taken to be (1) the global mean or (2) the mean of a limited subset of the genes, respectively. The normalization is done by multiplying the denominator of each ratio by the normalization coefficient. After normalization, log ratios will sum to zero over whatever set of genes was used to compute the normalization.

D. Variability in Microarray Data

The strength of the microarray method is that it provides information on a genome wide scale. However, microarray data are subject to a significant amount of variability. The amount of variability differs widely from spot to spot, depending on the strengths of the signals forming the expression ratio. This shortcoming is not a fatal flaw; it merely constrains somewhat the appropriate range of applications of these data.

It is probably most appropriate to use microarray data in cluster analysis (Eisen *et al.,* 1998; Khan *et al.,* 1998; Alon *et al.,* 1999; Golub *et al.,* 1999; Tamayo *et al.,* 1999; Tavazoie *et al.,* 1999), in which one poses the questions (1) which genes resemble which genes and (2) which experiments resemble which experiments? Cluster analysis is an elegant response to the presence of noise, because, by combining the results of numerous experiments, the errors associated with any particular measurement are overwhelmed by the preponderance of data. Array technology is also useful for answering questions such as "what genes responded the most to a particular perturbation" and "did this perturbation affect a particular class of genes?" It is probably least appropriate to use microarray data to find out what happened to a single gene in a single experiment, particularly if that gene was represented by weak signals or did not experience a large change in expression between the two conditions being compared.

Variability in microarray data may be found on at least three levels. First, if a given gene is printed more than once on a microarray, the expression ratios estimated from replicate spots will show similar, but discrepant values in a single array hybridization. Second, if a given pair of RNA preparations is fluorescently labeled twice and hybridized to two distinct microarrays, the two experiments will usually show substantial correlation (correlation coefficients near 0.8 are common), but there will also be discrepancies in the expression ratios deduced for nearly all genes. This discrepancy stems from systematic errors, such as slide-to-slide differences in mean signal intensities and mean background intensities. Finally, if a given RNA is prepared twice, from two independent cultures and compared to itself, many genes will manifest a ~2-fold change, and we have occasionally seen that up to 3% of the genes may show a 3-fold change. This introduces the real biological variability that may exist between any pair of cultures (even the identical strain prepared twice in the identical fashion) and underscores the importance of striving for consistency in the culture conditions (media, optical density at harvest, temperature, aeration, etc.) of all strains to be compared. When designing experiments with microarrays, it will be most useful to recall each of these different kinds of variability.

We wished to estimate the variability associated with replicate measurements of expression ratios in single yeast array hybridizations. We developed a method based on the naturally occurring internal redundancies of the yeast genome. There are 51 pairs of genes that are present as exact duplicates in the yeast genome. In addition, the official set of yeast open reading frames (ORFs) (Cherry *et al.,* 1998) contains 307 gene pairs overlapping by at least 200 nucleotides. These are typically bona fide (transcribed) genes that overlap a fortuitous ORF on the opposite strand. The fortuitous ORF may be nested entirely within the real gene, or the region of overlap may be confined to the mutual 3′ or 5′ ends. Typical yeast arrays include both members of numerous overlapping ORF pairs, and given that arrays are printed with double-stranded PCR products, both of the DNA spots can hybridize the fluorescent cDNA copied from the transcripts of the transcribed member of the ORF pair. A list of pairs of overlapping and identical ORFs in the yeast genome is available at http://hamon.swmed.edu/butow_array/orf_overlap.html.

Overlapping ORFs are not an ideal tool for estimating the reproducibility of array measurements because they do not sample the reverse transcripts with probes of equal length. Additionally, overlapping ORFs whose overlap is confined to the 5′ end of each gene suffer the shortcoming that reverse transcripts primed from the poly(A) sequence at the 3′ end of the transcribed gene may not extend as far as the region of overlap. However, overlapping ORFs afford the advantage that they are present on most yeast arrays, including microarrays fabricated using oligonucleotides from Research Genetics, Inc., as well as Affymetrix, Inc. Gene Chips. As such, they can be used to compute an index of reproducibility of internal, replicate measurements, which can be applied to comparisons between experiments done within a laboratory, as well as to comparisons between the array efforts of different laboratories using different methods and different array platforms. It should be appreciated, however, that the analysis of overlapping ORFs may be expected to overestimate the error that would be derived from an analysis based entirely on true replicates.

We examined the log expression ratios of all the duplicated and overlapping ORFs in several microarray hybridizations and plotted results for each gene against its paired partner (Fig. 2). To numerically compare different slides, we computed the antilog of the root mean square (RMS) discrepancy between paired measurements. This statistic conveys an estimate of the average discrepancy of expression ratios, expressed as a fold difference between two ratio measurements, and usually equals approximately 1.5 (Fig. 2, legend). This indicates that, on average, if a gene has an expression ratio of 1 in one measurement, it will have an expression ratio of 1.5 (i.e., induction) or 1/1.5 (i.e., repression) in the other measurement. This error is similar between microarrays (DeRisi *et al.,* 1997; Spellman *et al.,* 1998; C. B. Epstein and R. A. Butow, unpublished results) and Affymetrix Gene Chips (Cho *et al.,* 1998).

III. Application of Microarrays to the Study of Cellular Responses to Perturbed Mitochondrial Function

A. Introduction

Microarray technology has proved to be a useful tool for studying how cells adapt their metabolism to changing environmental conditions. For example, one of the earliest analyses evaluated the suite of genes induced by growth in galactose (Lashkari *et al.,*

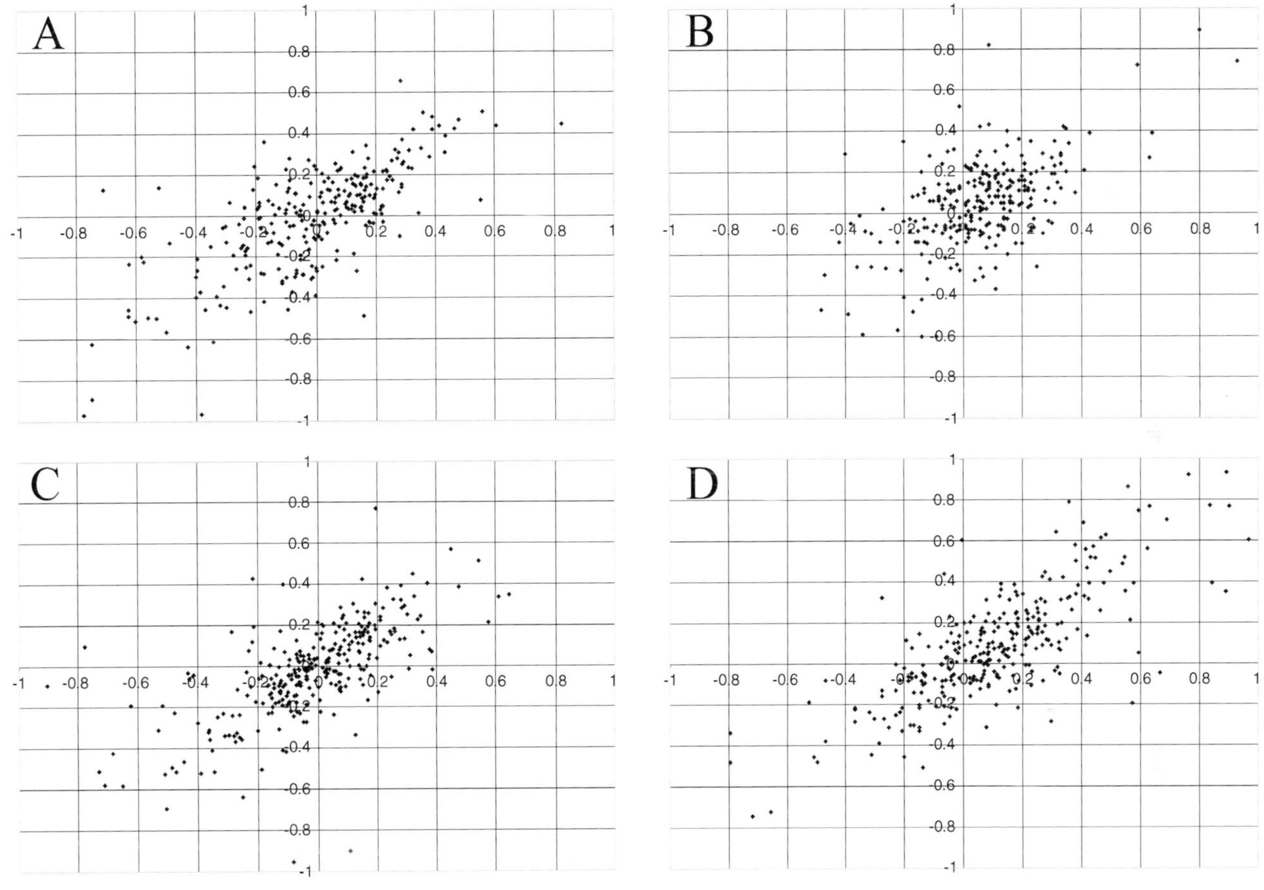

Fig. 2 Analysis of internally redundant genes on yeast arrays. (A) Treatment of yeast with chloramphenicol, $t = 9$ h, sample size (N) = 258, antilog root mean square (RMS) deviation = 1.55; (B) α factor synchronous cell cycle, $t = 84$ min, $N = 272$, antilog RMS deviation = 1.56; (Spellman *et al.*, 1998); (C) *cdc28* synchronous cell cycle, $t = 60$ min, $N = 312$, antilog RMS deviation = 1.53 (Cho *et al.*, 1998); (D) diauxic shift, seventh time point, $N = 287$, antilog RMS deviation = 1.53 (DeRisi *et al.*, 1997). Although all of the yeast genomic arrays are approximately the same size, the number of points on the graphs varies due to the exclusion of points that had a log ratio of zero in one or both of the two genes under comparison.

1997). A subsequent study characterized the reconfiguration of gene expression during the diauxic shift, in which yeast switch from fermentative to oxidative metabolism (DeRisi *et al.*, 1997). These studies have shown that the changes in metabolic pathways are reflected to a significant extent by changes in patterns of gene expression.

Our laboratory has investigated how cells adapt their metabolism in response to changes in mitochondrial function. These studies have led to the identification of a novel interorganelle signaling pathway from mitochondria to the nucleus, which we have called retrograde regulation (Liao and Butow, 1993; Jia *et al.*, 1997). For example, using conventional Northern blot analysis, we showed that as mitochondrial function becomes compromised in cells, expression of genes encoding enzymes catalyzing the first three steps of the tricarboxylic acid cycle switch from *HAP* to *RTG* gene control (Liu and Butow, 1999). To provide a genome-wide view of cellular responses to mitochondrial dysfunction, we have initiated studies using microarray analysis to analyze gene expression in cells in which mitochondrial function has been perturbed by mutation, or the use of mitochondrial inhibitors. The following sections discuss the general methodology of this approach and give a specific example of the effects of treating cells with chloramphenicol, a specific inhibitor of mitochondrial protein synthesis.

B. Cell Culture Conditions

In any study employing inhibitors, it is important to establish in advance a suitable concentration at which the compound may be used. This serves two purposes. First, it confirms that the strain being used has not suffered a cryptic mutation that renders it insensitive to the effects of the inhibitor. Second, it enables one to establish the minimal concentration necessary to completely inhibit the process under study. Because all yeast genes are being interrogated, use of excessive concentrations of inhibitors may lead to the detection of pleiotropic effects, complicating the analysis. In the case of inhibitors of respiratory ATP synthesis, we determine the minimal inhibitory concentration by titration of growth on glycerol medium, as glycerol is an obligatory respiratory carbon source. In experiments with inhibitors or with strains bearing mutations that affect mitochondrial function, we typically culture cells in medium containing 3% raffinose as a carbon source. Raffinose is a fermentable but nonrepressing carbon source allowing cells lacking respiratory function to grow, without imposing catabolite repression. As a further precaution, whenever we compare a wild-type strain to a mutant or inhibited strain, we preculture the wild-type strain in glycerol medium to select against any respiratory deficient cells (petites) that may have accumulated.

C. Treatment of Cells with Chloramphenicol

Chloramphenicol is a specific inhibitor of mitochondrial protein synthesis. Thus, its effect on mitochondrial function is elaborated over the course of several cell doublings during which the pool of mitochondrially synthesized proteins is diluted out in the culture. Long-term treatment (15 to 20 generations) of cultures with chloramphenicol results in the formation of cytoplasmic petites. We exposed cells to 3 mg/ml of chloramphenicol for no longer than 9 h so as to avoid the production of petites and then prepared RNA from treated and control cells and analyzed the RNA on a microarray printed using a robotic spotter designed and fabricated locally. The chloramphenicol treatment spanned approximately four cell doublings, potentially causing a 16-fold reduction in the levels of mitochondrial translation products. Table I presents all genes that had an absolute log expression ratio of at least 0.6 after several hours of exposure to chloramphenicol. This corresponds to a 4-fold induction or repression of gene expression. The numbers are derived from the average of the 5- and 9-h time points, which showed excellent correlation overall. The full data set is available on the internet (http://hamon.swmed.edu/~butow_array/array_data.html).

1. Genes Upregulated by Chloramphenicol

The single gene induced most strongly and consistently by chloramphenicol treatment is formate dehydrogenase (YOR388C, YPL276W, and YPL275W). YOR388C is characterized as a formate dehydrogenase based on structural considerations. Formate dehydrogenase normally carries out the oxidation of formate to carbon dioxide and reduces NAD+ to NADH in the process.

Chloramphenicol-exposed cells may be suffering the effects of defects in mitochondrial electron transport and may respond with diverse genes that contribute to the establishment of a normal balance between oxidized and reduced NAD cofactors. Note that the three ORFs mentioned earlier consist of one bona fide gene (YOR388C) and two probable pseudogenes. YPL275W and YPL276W are adjacent ORFs whose sequence is nearly identical to YOR388C. Collectively, they span the length of YOR388C, although there is a frame shift between them, consistent with a pattern of gene duplication followed by conversion of one copy into a pair of adjacent pseudogenes. These three ORFs provide an interesting example of how the internal redundancy of the yeast genome affords the opportunity to evaluate the reproducibility of replicate samples.

Another major response to chloramphenicol includes the induction of components of the stress response. *HSP26, HSP12,* and *YRO2* (a protein with homology to *HSP30*) are all in the set of genes that are strongly induced. Genes involved in energy metabolism are also strongly induced, including *GPM2* (phosphoglycerate mutase), possibly reflecting an increased reliance on glycolysis as oxidative function gradually decreases in these cultures.

2. Genes Downregulated by Chloramphenicol

Genes downregulated by chloramphenicol include a number of genes involved in carbohydrate metabolism. *HXT2,* one of several high-affinity hexose transporters in yeast (Kruckeberg and Bisson, 1990), is strongly repressed. Curiously, two of the genes most strongly repressed are adjacent ORFs, although they are separated by 520 nucleotides and do not share sequence homology (YDL037C and YDL038C). The former has sequence similarity to glucan 1,4-α-glucosidase, an extracellular enzyme involved in starch

Table I
Log_{10}-Transformed Expression Ratios from the Comparison of Chloromphenicol-Treated to -Untreated Yeast[a]

ORF	Gene	log ratio	Description
Genes upregulated in chloramphenicol			
YPL276W		1.484	Protein with similarity to formate dehydrogenases
YBR072W	*HSP26*	1.396	Heat shock protein 26
YPL275W		1.316	Protein with similarity to formate dehydrogenases
YOR388C	*FDH1*	1.257	Similar to formate dehydrogenases
YIL057C		1.185	Protein of unknown function
YLR142W	*PUT1*	1.051	Proline oxidase
YFL014W	*HSP12*	0.987	12-kDa heat shock protein
YER150W	*SPI1*	0.963	Similar to Sed1; highly expressed in stationary phase
YGL121C		0.881	Protein of unknown function
YLR327C		0.851	Protein with strong similarity to Stf2p
YER067W		0.763	Protein of unknown function
YPL201C		0.713	Protein of unknown function
YDR533C		0.667	Protein of unknown function
YBR128C	*APG14*	0.657	Involved in autophagy
YBR054W	*YR02*	0.657	Homolog to HSP30 heat shock Protein YR01 (*S. cerevisiae*)7
YKL026C		0.651	Protein with similarity to Hyr1p, Ybr244p, and glutathione peroxidases
YGR243W		0.647	Protein of unknown function
YDR536W	*STL1*	0.639	Sugar transporter-like protein
YLR460C		0.638	Protein with strong similarity to Ycr102p
YOR391C		0.636	Protein of unknown function
YPL185W		0.635	Protein of unknown function
YDL021W	*GPM2*	0.633	Phosphoglycerate mutase, involved in glycolysis
YMR322C		0.628	Protein of unknown function
YBR147W		0.623	Protein of unknown function; has seven potential transmembrane segments
YER066C-A		0.610	Protein of unknown function
Genes downregulated in chloramphenicol			
YNR053C		−0.600	Protein with similarity to human breast tunor-associated autoantigen
YEL040W	*UTR2*	−0.603	Protein of unknown function
YNL234W		−0.616	Hemoprotein with similarity to mammalian globins
YLR063W		−0.623	Protein of unknown function
YNL049C	*SFB2*	−0.625	Putative zinc finger protein
YBL012C		−0.626	Protein of unknown function
YDR432W	*NPL3*	−0.626	Nuclear shuttling protein with an RNA recognition motif
YOR321W	*PMT3*	−0.635	Dolichyl phosphate-D-mannose:Protein *O*-D-mannosyltransferase
YGR160W		−0.645	Protein of unknown function
YOR107W		−0.665	Protein with weak similarity to humaan G_0/G_1 switch regulatory protein 8
YMR120C	*ADE17*	−0.667	5-Aminoimidazole-4-carboxamide ribonucleotide (AICAR) transformylase VIMP cyclohydrolase
YBR105C	*VID24*	−0.671	Peripheral vesicle membrane protein
YGR079W		−0.675	Protein of unknown function
YNL112W	*DBP2*	−0.693	ATP-dependent RNA helicase of DEAD box family
YOR302W		−0.702	Carbamoylphosphate synthase of the arginine biosynthetic pathway
YDR433W		−0.725	Protein of unknown function; questionable ORF
YGR159C	*NSR1*	−0.765	Nuclear localization sequence-binding protein
YNL065W		−0.825	Member of the multidrug-resistance 12-spanner family of the major facilitator superfamily (MFS-MDR)
YOR047C	*STD1*	−0.870	Homologous to MTH1; interacts with the SNF1 protein kinase and TBP in two-hybrid and in *in vitro*-binding studies
YBR238C		−0.943	Protein of unknown function
YOR338W		−1.120	Protein of unknown function
YDL038C		−1.143	Protein of unknown function
YMR011W	*HXT2*	−1.467	High-affinity hexose transporter-2
YDL037C		−1.532	Protein with similarity to glucan 1,4-α-glucosidase

[a]Cells were exposed to 3 mg/ml of chloramphenicol for 5 and 9 h. Hybridizations were done at both time points, and expression ratios from the two arrays were averaged, as they showed excellent overall correlation. All genes having an absolute $\log_{10}$ expression ratio of at least 0.6 are shown, corresponding to a ~4-fold change. Functional descriptions were supplied by Proteome, Inc. (Hodges *et al.*, 1999).

degradation. Finally, YBR238C, like YDL038C, shows extremely strong, consistent downregulation in cells exposed to chloramphenicol. Nevertheless, both of these genes encode proteins "of unknown function" (Hodges *et al.,* 1999). Array analyses can suggest interesting genes for further study, and multidimensional cluster analyses can provide clues to gene function based on correlated response with genes of known function. Nevertheless, conventional genetic, biochemical, and molecular biological experiments will be required to define the role of many of the genes detected using a microarray.

References

Alon, U., Barkai, N., Notterman, D. A., Gish, K., Ybarra, S., Mack, D., and Levine, A. J. (1999). Broad patterns of gene expression revealed by clustering analysis of tumor and normal colon tissues probed by oligonucleotide arrays. *Proc. Natl. Acad. Sci. USA* **96,** 6745–6750.

Cherry, J. M., Adler, C., Ball, C., Chervitz, S. A., Dwight, S. S., Hester, E. T., Jia, Y., Juvik, G., Roe, T., Schroeder, M., Weng, S., and Botstein, D. (1998). SGD: *Saccharomyces* genome database. *Nucleic Acids Res.* **26,** 73–79.

Cheung, V. G., Morley, M., Aguilar, F., Massimi, A., Kucherlapati, R., and Childs, G. (1999). Making and reading microarrays. *Nature Genet.* **21,** 15–19.

Cho, R. J., Campbell, M. J., Winzeler, E. A., Steinmetz, L., Conway, A., Wodicka, L., Wolfsberg, T. G., Gabrielian, A. E., Landsman, D., Lockhart, D. J., and Davis, R. W. (1998). A genome-wide transcriptional analysis of the mitotic cell cycle. *Mol. Cell* **2,** 65–73.

DeRisi, J. L., Iyer, V. R., and Brown, P. O. (1997). Exploring the metabolic and genetic control of gene expression on a genomic scale. *Science* **278,** 680–686.

Duggan, D. J., Bittner, M., Chen, Y., Meltzer, P., and Trent, J. M. (1999). Expression profiling using cDNA microarrays. *Nature Genet.* **21,** 10–14.

Eisen, M. B., and Brown, P. O. (1999). DNA arrays for analysis of gene expression. *Methods Enzymol.* **303,** 179–205.

Eisen, M. B., Spellman, P. T., Brown, P. O., and Botstein, D. (1998). Cluster analysis and display of genome-wide expression patterns. *Proc. Natl. Acad. Sci. USA* **95,** 14863–14868.

Golub, T. R., Slonim, D. K., Tamayo, P., Huard, C., Gaasenbeek, M., Mesirov, J. P., Coller, H., Loh, M. L., Downing, J. R., Caligiuri, M. A., Bloomfield, C. D., and Lander, E. S. (1999). Molecular classification of cancer: class discovery and class prediction by gene expression monitoring. *Science* **286,** 531–537.

Hodges, P. C. E., McKee, A. H., Davis, B. P., Payne, W. E., and Garrels, J. I. (1999). The yeast proteome database (YPD): A model for the organization and presentation of genome-wide functional data. *Nucleic Acids Res.* **27,** 69–73.

Jia, Y., Rothermel, B., Thornton, J., and Butow, R. A. (1997). A basic helix-loop-helix-leucine zipper transcription complex in yeast functions in a signaling pathway from mitochondria to the nucleus. *Mol. Cell. Biol.* **17,** 1110–1117.

Khan, J., Simon, R., Bittner, M., Chen, Y., Leighton, S. B., Pohida, T., Smith, P. D., Jiang, Y., Gooden, G. C., Trent, J. M., and Meltzer, P. S. (1998). Gene expression profiling of alveolar rhabdomyosarcoma with cDNA microarrays. *Cancer Res.* **58,** 5009–5013.

Kruckeberg, A. L., and Bisson, L. F. (1990). The HXT2 gene of *Saccharomyces cerevisiae* is required for high-affinity glucose transport. *Mol. Cell. Biol.* **10,** 5903–5913.

Lashkari, D. A., DeRisi, J. L., McCusker, J. H., Namath, A. F., Gentile, C., Hwang, S. Y., Brown, P. O., and Davis, R. W. (1997). Yeast microarrays for genome wide parallel genetic and gene expression analysis. *Proc. Natl. Acad. Sci. USA* **94,** 13057–13062.

Liao, X., and Butow, R. A. (1993). RTG1 and RTG2: Two yeast genes required for a novel path of communication from mitochondria to the nucleus. *Cell* **72,** 61–71.

Liu, Z., and Butow, R. A. (1999). A transcriptional switch in the expression of yeast tricarboxylic acid cycle genes in response to a reduction or loss of respiratory function. *Mol. Cell. Biol.* **19,** 6720–6728.

Spellman, P. T., Sherlock, G., Zhang, M. Q., Iyer, V. R., Anders, K., Eisen, M. B., Brown, P. O., Botstein, D., and Futcher, B. (1998). Comprehensive identification of cell cycle-regulated genes of the yeast *Saccharomyces cerevisiae* by microarray hybridization. *Mol. Biol. Cell.* **9,** 3273–3297.

Tamayo, P., Slonim, D., Mesirov, J., Zhu, Q., Kitareewan, S., Dmitrovsky, E., Lander, E. S., and Golub, T. R. (1999). Interpreting patterns of gene expression with self-organizing maps: Methods and application to hematopoietic differentiation. *Proc. Natl. Acad. Sci. USA* **96,** 2907–2912.

Tavazoie, S., Hughes, J. D., Campbell, M. J., Cho, R. J., and Church, G. M. (1999). Systematic determination of genetic network architecture. *Nature Genet.* **22,** 281–285.

APPENDIX 1

Basic Properties of Mitochondria

Kammy Fehrenbacher
Department of Anatomy and Cell Biology
Columbia University

Organism	Mitochondria per cell	% of total cell volume	Genomes per organelle	Genomes per cell	Size (µm)	Volume (μm^3)
Yeast *(Saccharomyces cerevisiae)*	1 (A*)	12 (C**)	2–50 (D*)	44; 88 (C*)	0.16 (C*)	0.14 (C*)
	10; 22 (C*)	12; 4 (B**)	4 (C)	50; 100 (E*)		
	<10; 30–50 (B*)					
Yeast *(Neurospora crassa)*	50–100 (F)	11.9 (G*)	7 µg/mg mito prote	2.45 mg/g dry cell wt	–	0.131 (H)
Xenopus oocyte	480–390,000 (I*)	–	5–10 (D)	1.92×10^8 (K)	–	0.30 (J*)
	$16 \pm 2 \times 10^6$ (J*)	–	12 (J*/**)	–	–	–
Mouse oocyte	14–29 (L*)	5.29–7.16 (L*/**)	–	–	–	vg 0.21–0.26 (L*^)
Mouse L cell	250 ± 70 (M)	–	2–6 (M,N)	1,800 (O)	.36 diameter (P*^)	–
Mouse fertilized embryo	–	–	–	119,000 (HH)	–	–
Rat liver cell	1,400–2,200 (Q*)	28 (S)	5–10 (D)	7,000–22,000 (U)	–	3.83–0.64 (Q*)
	2,601 (R*)	18.4 (T*)	–	13,005–26,010 (V)	–	–
Pig oocyte	0.1/μm^3 (W*/**)	6 (W**)	–	–	0.62 (W**^)	–
Human HeLa cell	383–882 (X*)	9.4–11 (X*)	–	8,800 (O)	–	0.27–0.31 (X*)
Human liver cell	2,200 (Y)	17.6 (Y)	–	–	–	0.84 (Y)
Human osteosarcoma cell 143B	–	–	–	9,100 ± 1,600 (Z)	–	7.0 pl (Z)
Human fibrosarcoma cell HT1080	–	–	–	9,700 ± 1,800 (Z)	–	6.1 pl (Z)
Human melanoma cell	–	–	–	10,300 ± 2,100 (Z)	–	4.1 pl (Z)
VA_2 = B/CAP23						
Human HeLa S3 cell	–	–	–	7,900 ± 1,700 (Z)	–	4.9 pl (Z)
Human cultured A2780 cells	avg 107 (AA)	–	1–15, avg 4.6 (AA)	500 (AA)	–	–
Human fibroblast cell lines (6)	–	–	–	2,400 ± 6,000 (BB)	–	–
Human transformed cell lines (4)	–	–	–	1,600 ± 7,200 (BB)	–	–
Human platelets	4 (CC)	–	1 (CC)	4 (CC)	–	–
Human oocytes	–	–	–	30,000 ± 80,000 (DD)	–	–
Human sperm	–	–	–	100 (EE)	–	–
Bovine oocytes	–	–	–	100,000 (FF)	–	–
Bovine sperm	–	–	–	75 (GG)	–	–
Chicken embryo fibroblast (CEF)	–	–	–	604 ± 134 (II)	–	–
Chicken liver tumor DU24	–	3.8 (JJ)	–	360 ± 80 (JJ)	–	–
Quail cell line LSCC–H32	–	4.5 (JJ)	–	276 ± 88 (JJ)	–	–

References:

(A) Hoffmann and Avers (1973). *Science* **181,** 749–751.
(B) Stevens (1977). *Cell. Biol.* **28,** 37–56.
(C) Grimes *et al.* (1974), *J. Cell. Biol.* **61,** 565–574.
(D) Alberts *et al.* (1994). *Molecular Biology of the Cell, 3rd Ed.*, pp. 553 and 706, Garland, New York.
(E) Williamson (1970). *Cold Spr. Harb. Sym. Quant. Biol.* **24,** 247–276.
(F) Prokisch (1999). *Institut Physiol. Chem. Univ. Munich.* personal communication.
(G) Hawley and Wagner (1967). *J. Cell. Biol.* **35,** 489–499.
(H) Luck and Reich (1962). *Proc. Natl. Acad. Sci. USA* **52,** 931–938.
(I) Marinos and Billett (1981). *J. Embryol. Exp. Morphol.* **62,** 395–409.
(J) Marinos (1985). *Cell. Diff.* **16,** 139–143.
(K) Based on 1.6×10^7 organelles/cell and 12 mtDNAs/organelle. (Ref. J)
(L) Nogawa *et al.* (1988). *J. Morph.* **195,** 225–234.
(M) Nass (1969). *J. Mol. Biol.* **42,** 521–528.
(N) Nass (1966). *Proc. Natl. Acad. Sci. USA* **56,** 1215–1232.
(O) Bogenhagen and Clayton (1974). *J. Biol. Chem.* **249,** 7991–7995.
(P) King *et al.* (1972). *J. Cell Biol.* **53,** 127–142.
(Q) David (1979). *Exp. Pathol.* **17,** 359–373.
(R) Pieri *et al.* (1975). *Exp. Gerontol.* **10,** 291–304.
(S) Blouin *et al.* (1977). *J. Cell Biol.* **72,** 441–455.
(T) Loud (1962). *J. Cell. Biol.* **15,** 481–487.
(U) Based on a range of 5–10 (ref. D) to 1400–2200 (ref. Q) mtDNAs/organelle.
(V) Based on a range of 5–10 (ref. D) to 2601 (ref. R) organelles/cell.
(W) Cran (1985). *J. Reprod. Fertil.* **74,** 237–245.
(X) Posakony *et al.* (1977). *J. Cell Biol.* **74,** 468–491.
(Y) Rohr *et al.* (1976). *In "Progress in Liver Diseases"* (Propper and Schaffner, eds.), Vol. 5, pp. 24–34. Grune and Stratton, New York.
(Z) King and Attardi (1989). *Science* **246,** 500–503.
(AA) Satoh and Kuroiwa (1991). *Exp. Cell. Res.* **196,** 137–140.
(BB) Shmookler Reis and Goldstein (1983). *J. Biol. Chem.* **258,** 9078–9085.
(CC) Shuster *et al.* (1988). *Biochem. Biophys. Res. Commun.* **155,** 1360–1365.
(DD) Chen *et al.* (1995). *Am. J. Hum. Genet.* **57,** 239–247.
(EE) Birky (1983). *Science* **222,** 468–475.
(FF) Michaels and Hauswirth (1982). *Dev. Biol.* **94,** 246–251.
(GG) Bahr and Engler (1970). *Exp. Cell. Res.* **60,** 338–340.
(HH) Piko and Taylor (1987). *Dev Biol.* **123,** 364–374.
(II) Desjardins *et al.* (1985). *Mol. Cell. Biol.* **5,** 1163–1169.
(JJ) Morials *et al.* (1988). *In Vitro Cell. Dev. Biol.* **24,** 649– 658.

Notes:

^ Stereologic and morphometric analysis allows for a three-dimensional calculation from two-dimensional measurements. Areas and diameters are precursors to these kinds of calculations, but cannot give the same information without necessary coefficients and parameters.

A* Values are from analysis of a diploid iso-N strain grown in liquid nutrient containing glucose or ethanol as energy sources.

B* Values are the number of mitochondria in an exponentially growing culture in glycerol or glucose and in a stationary culture, respectively.

B** Values represent the total volume of mitochondria per cell volume in derepressed respiring cells and in glucose-repressed cells, respectively.

C* Values are based on haploid and diploid cells, respectively, growing exponentially on lactate medium (without catabolite repression).

C** Values are constant regardless of ploidy.

E* Values are in haploid and diploid cells, respectively.

G* Value is the average of time points taken over 30 h of growth and is the percentage mitochondrial volume/cytoplasmic volume, assuming sphericity.

I* Calculation based on average number of mitochondria profiles/100-mm^2 section of EM and calculated size of mitochondria in previtellogenic oocytes. Mitochondrial number was only calculated in oocytes with diameters between 50 and 250 mm. Range correlates with increase in oocyte size during development.

J* Calculation based on mature oocyte and a standard ellipsoid mitochondrion diameter of 0.2–0.4 mm.

J** Value based on an average number of 2×10^8 mtDNAs in a mature oocyte, and 25×10^4 ng mtDNA/mitochondrion in a mature oocyte.

L* Value is mean number of mitochondria per section, based on two-dimensional analysis of single EM sections (i.e., not three-dimensionally extrapolated).

L** Value is the ratio of the area of mitochondria to the area of cytoplasm in a two-dimensional EM section.

P* Value based on 2-dimensional analysis of single EM sections (i.e., not three-dimensionally extrapolated).

Q* Values are a range of mean calculations from postnatal rats, 2 to 6 months of age, respectively.

R* Value is number of mitochondria per cell after 27 months of age.

T* Value is percentage total cytoplasmic area occupied by the mitochondria of an average normal rat liver cell.

W* Value is mean number of mitochondria/μm^3 of oocyte volume, based on a 110-μm oocyte diameter and a 0.62-μm mean mitochondrial diameter.

W** Values based on mature oocytes 40 h after stimulation with hCG, after which fertilization can be achieved.

X* Values are a range over entire cell cycle (beginning with time 0 after mitosis) of HeLa F-315 cells. Values are based only on two-dimensional analysis.

APPENDIX 2

Linearized Maps of Circular Mitochondrial Genomes from Representative Organisms

Compiled by Eric A. Schon
Columbia University

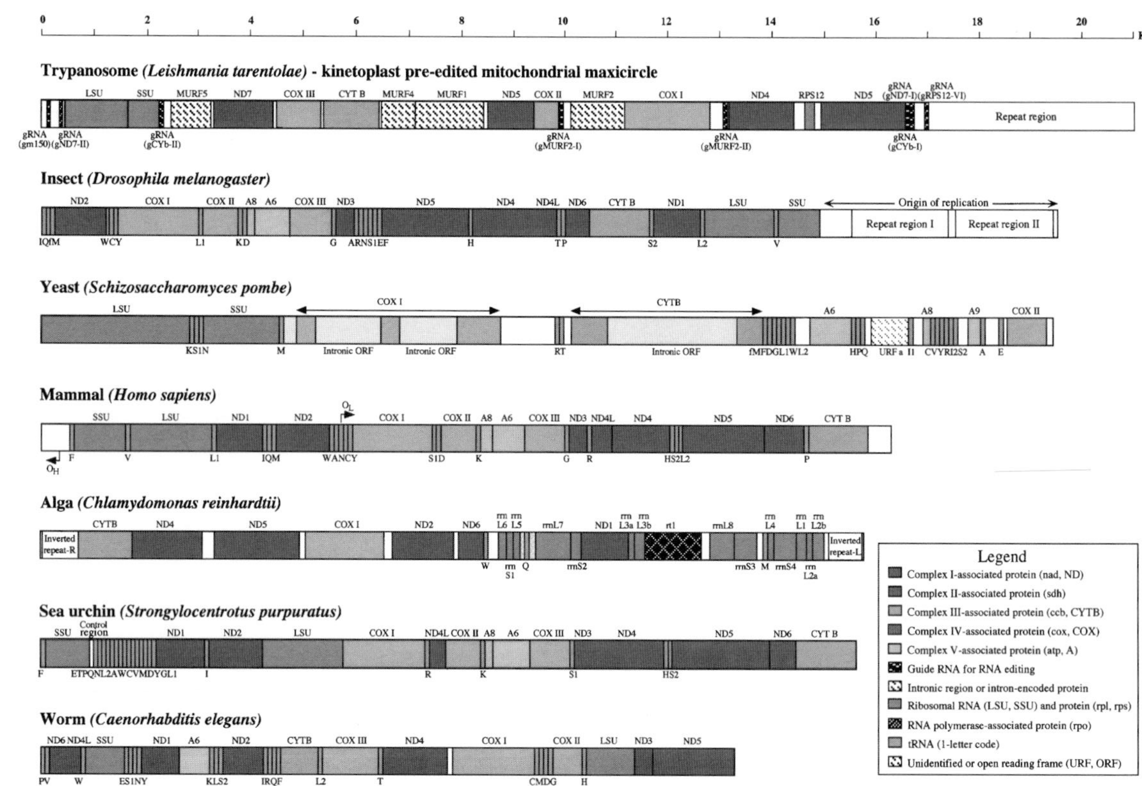

Kb
Trypanosome (Leishmania tarentolae) - kinetoplast pre-edited mitochondrial maxicircle
LSU SSU MURF5 ND7 COX III CYT B MURF4 MURF1 ND5 COX II MURF2 COX I ND4 RPS12 ND5
gRNA (gND7-I) gRNA (gRPS12-VI)
Repeat region
gRNA (gm150) gRNA (gND7-II) gRNA (gCYb-II) gRNA (gMURF2-I) gRNA (gMURF2-II) gRNA (gCYb-I)
Insect (Drosophila melanogaster)
ND2 COX I COX II A8 A6 COX III ND3 ND5 ND4 ND4L ND6 CYT B ND1 LSU SSU
Origin of replication
Repeat region I
Repeat region II
IQfM WCY L1 KD G ARNS1EF H TP S2 L2 V
Yeast (Schizosaccharomyces pombe)
LSU SSU COX I CYTB A6 A8 A9 COX II
KS1N M Intronic ORF Intronic ORF RT Intronic ORF fMFDGL1WL2 HPQ URF a I1 CVYRI2S2 A E
Mammal (Homo sapiens)
SSU LSU ND1 ND2 COX I COX II A8 A6 COX III ND3 ND4L ND4 ND5 ND6 CYT B
F V L1 IQM WANCY S1D K G R HS2L2 P
Alga (Chlamydomonas reinhardtii)
Inverted repeat-R
CYTB ND4 ND5 COX I ND2 ND6 rrn L6 rrn L5 rrnL7 ND1 rrn L3a rrn L3b rtl rrnL8 rrn L4 rrn L1 rrn L2b
Inverted repeat-L
W rrn S1 Q rrnS2 rrnS3 M rrnS4 rrn L2a
Sea urchin (Strongylocentrotus purpuratus)
SSU Control region ND1 ND2 LSU COX I ND4L COX II A8 A6 COX III ND3 ND4 ND5 ND6 CYT B
F ETPQNL2AWCVMDYGL1 I R K S1 HS2
Worm (Caenorhabditis elegans)
ND6 ND4L SSU ND1 A6 ND2 CYTB COX III ND4 COX I COX II LSU ND3 ND5
PV W ES1NY KLS2 IRQF L2 T CMDG H
Legend
Complex I-associated protein (nad, ND)
Complex II-associated protein (sdh)
Complex III-associated protein (ccb, CYTB)
Complex IV-associated protein (cox, COX)
Complex V-associated protein (atp, A)
Guide RNA for RNA editing
Intronic region or intron-encoded protein
Ribosomal RNA (LSU, SSU) and protein (rpl, rps)
RNA polymerase-associated protein (rpo)
tRNA (1-letter code)
Unidentified or open reading frame (URF, ORF)

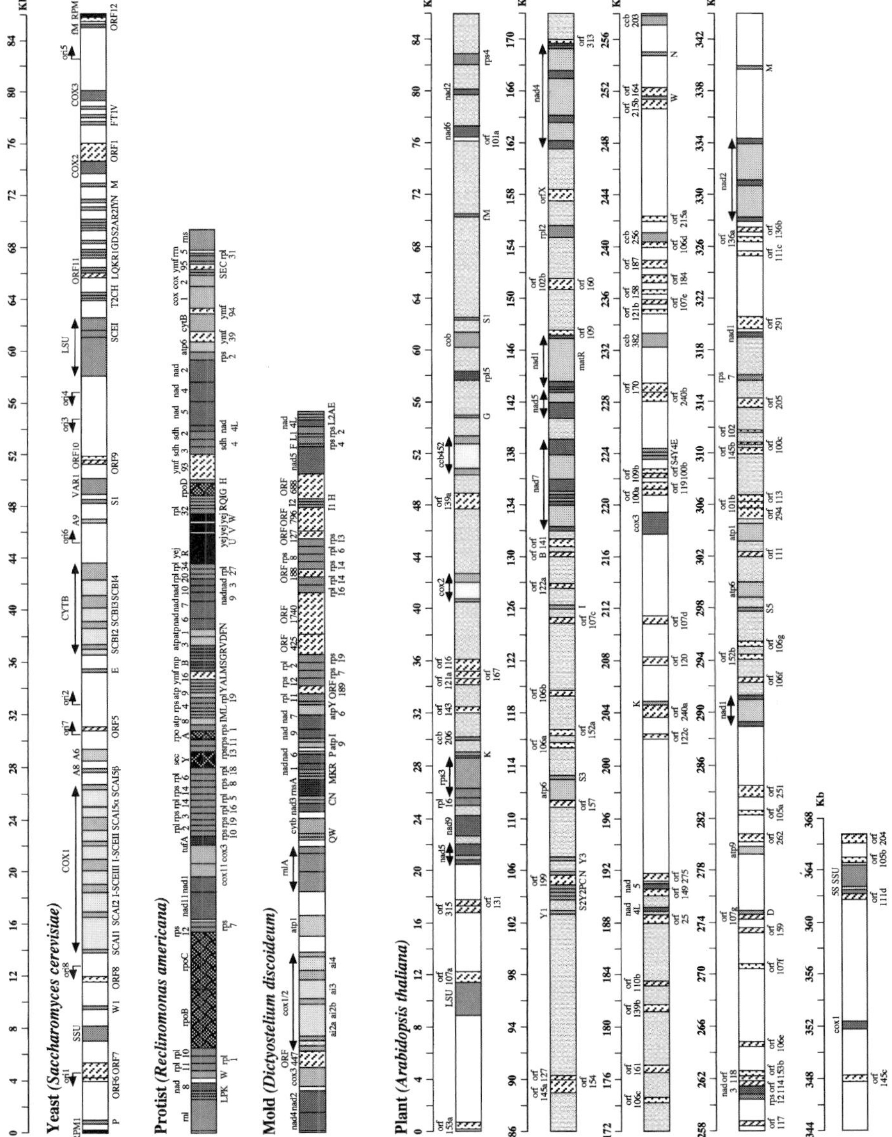

Yeast (Saccharomyces cerevisiae)
Protist (Reclinomonas americana)
Mold (Dictyostelium discoideum)
Plant (Arabidopsis thaliana)
Kb

APPENDIX 3

Mitochondrial Genetic Codes in Various Organisms

Compiled by Eric A. Schon
Columbia University

The standard genetic code											
UUU	F	Phe	UCU	S	Ser	UAU	Y	Tyr	UGU	C	Cys
UUC	F	Phe	UCC	S	Ser	UAC	Y	Tyr	UGC	C	Cys
UUA	L	Leu	UCA	S	Ser	UAA	*	Ter	UGA	*	Ter
UUG	L	Leu	UCG	S	Ser	UAG	*	Ter	UGG	W	Trp
CUU	L	Leu	CCU	P	Pro	CAU	H	His	CGU	R	Arg
CUC	L	Leu	CCC	P	Pro	CAC	H	His	CGC	R	Arg
CUA	L	Leu	CCA	P	Pro	CAA	Q	Gln	CGA	R	Arg
CUG	L	Leu	CCG	P	Pro	CAG	Q	Gln	CGG	R	Arg
AUU	I	Ile	ACU	T	Thr	AAU	N	Asn	AGU	S	Ser
AUC	I	Ile	ACC	T	Thr	AAC	N	Asn	AGC	S	Ser
AUA	I	Ile	ACA	T	Thr	AAA	K	Lys	AGA	R	Arg
AUG	M	Met	ACG	T	Thr	AAG	K	Lys	AGG	R	Arg
GUU	V	Val	GCU	A	Ala	GAU	D	Asp	GGU	G	Gly
GUC	V	Val	GCC	A	Ala	GAC	D	Asp	GGC	G	Gly
GUA	V	Val	GCA	A	Ala	GAA	E	Glu	GGA	G	Gly
GUG	V	Val	GCG	A	Ala	GAG	E	Glu	GGG	G	Gly

		Mitochondrial genetic codes: Deviations from the standard code												
	Standard	(a)	(b)	(c)	(d)	(e)	(f)	(g)	(h)	(i)	(j)	(k)	(l)	(m)
UAG	Stop								Ala	Leu	Leu			
UGA	Stop	Trp	Trp	Trp	Trp	Trp	Trp		Trp			Cys		Trp
UCA	Ser										Stop			
UUA	Leu												Stop	
CUU	Leu		Thr											
CUC	Leu		Thr											
CUA	Leu		Thr											
CUG	Leu		Thr											
AUA	Ile	Met	Met		Met		Met							Met
AAA	Lys					Asn		Asn	Asn					Asn
AGA	Arg	Stop			Ser	Ser	Gly		Ser					Ser
AGG	Arg	Stop			Ser	Ser	Gly		Ser					Ser

Note: The standard code is present in the mitochondria of land plants and green algae.

(a) Includes all known vertebrates.

(b) Includes Fungi *(Candida, Hansenula, Kluveromyces, Saccharomyces, Schizosaccharomyces).*

(c) Includes Fungi *(Ascobolus, Aspergillus, Candida, Neurospora, Podospora);* Coelenterata; Mycoplasma; Protozoa *(Leishmania, Paramecia, Tetrahymena, Trypanosoma);* Spiroplasma; also a red alga *(Chondrus).*

(d) Includes Inverebrata, such as Crustaceae *(Artemia);* Insecta *(Apis, Drosophila, Locusta);* Mollusca; Nematoda *(Ascaris, Caenorhabditis);* Acoelomorpha and Catenulida (related to flatworms).

(e) Includes Echinodermata, such as Asterozoa (starfish) and Echinozoa (sea urchins).

(f) Includes Ascidiae (sea squirts).

(g) Includes Rhabditophora (flatworms); see Telford et al. (2000) Proc Natl Acad Sci USA 97:11359-11364.

(h) Includes Chlorophycean algae *(Hydrodictyon, Pediastrum, Tetraedron);* see Hayashi-Ishimaru et al. (1996) Curr Genet 30:29-33.

(i) Includes Chlorophycean algae *(Coelastrum, Scenedesmus).*

(j) Includes a green alga *(Scenedesmus obliquus).*

(k) Includes Ciliata *(Euplotes).*

(l) Includes a protist *(Thraustochytrium aureum).*

(m) Includes Trematoda.

APPENDIX 4

Gene Products Present in Mitochondria of Yeast and Animal Cells[a]

Compiled by Eric A. Schon
Columbia University

Gene product	Function/comment	LOC.	E.C. #	Yeast Symbol	Yeast GenBank	Animal Symbol	Animal GenBank
AMINO ACID AND NITROGEN METABOLISM							
Acetohydroxy acid isomeroreductase	Ile/Val syn; mtDNA recomb	MAT?	1.1.1.86	ILV5	X04969.1	-	-
Acetolactate synthase	Ile/Val syn	MAT?	4.1.3.18	ILV2	X02549.1	B0334.3a (w)	Z66519.1 (w)
Acetolactate synthase, regulatory subunit	Ile/Val syn	MAT?	4.1.3.18	ILV6	X59720.1	-	-
Acetylglutamate kinase/acetylglutamyl phosphate reductase	Arg/ornithine syn	MAT?	2.7.2.8	ARG5,6	X57017.1	-	-
Acetylglutamate synthase	Arg/ornithine syn	MAT?	2.3.1.1	ARG2	Z49346.1	-	-
Acetylornithine acetyltransferase bifunctional protein	Arg/ornithine syn	MAT?	2.3.1.35	ARG7	U90438.1	-	-
Acetylornithine aminotransferase	Arg/ornithine syn	MAT?	2.6.1.11	ARG8	X84036.1	-	-
Acyl-CoA dehydrogenase-8 (isobutyryl-CoA dehydrogenase)	Val catabolism	MAT?	-	-	-	[ACAD8]	AF126245.1
Agmatinase	Agmatine deg	MAT	3.5.3.11	-	-	None	AW237392.1 (e)
Alanine aminotransferase	Ala metab	MAT?	2.6.1.2	YLR089c	Z73261.1	AAT1	D10355.1 (i)
Aralkyl acyl-CoA:amino acid N-acyltransferase	Gly metab	MAT?	2.3.1.13	-	-	None	AJ223301.1 (b)
Arginase II	Arg deg/NOS regulation	MAT?	3.5.3.1	-	-	ARG2	D86724.1
Arginine decarboxylase	Arg deg	NC	4.1.1.19	-	-	[ADC] (r)	S79271.1 (r, i)
Arylacetyl acyl-CoA amino acid N-acyltransferase	Gly metab	MAT?	2.3.1.13	-	-	None	AF022073.1 (b)
Aspartate aminotransferase	Asp/Glu transamination	MAT	2.6.1.1	AAT1	X68052.1	GOT2	M22632.1
Beta-alanine pyruvate aminotransferase	Ala metab	MAT?	2.6.1.44	-	-	AGT2 (r)	AB002584.1 (r)
Betaine aldehyde dehydrogenase (aldehyde dehydrogenase E3)	Gly syn	MAT	1.2.1.3	-	-	ALDH9	U34252.1
Branched chain alpha-ketoacid dehydrogenase kinase	Leu/Ile/Val deg	MAT	2.7.1.115	YIL042c	Z46861.1	BCKDK	AF026548.1
Branched chain amino acid aminotransferase	Leu/Ile/Val metab	MAT?	2.6.1.42	BAT1	X78961.1	BCAT2	U68418.1
Carbamoyl phosphate synthetase 1	Urea cycle	MAT?	6.3.4.16	-	-	CPS1	Y15793.1
Choline dehydrogenase	Gly syn	MIM	1.1.99.1	-	-	CHDH	AJ272267.1 (i)
2-Dehydro-3-deoxyphosphoheptonate aldolase (DAHPS)	Aromatic aa syn, 1st step	NC	4.1.2.15	ARO3	X13514.1	-	-
Delta 1-pyrroline-5-carboxylate synthetase	Pro syn (first 2 steps in human)	MIM	1.2.1.41	PRO2	U43565.1	PYCS	X94453.1
Dihydroxy acid dehydratase (DAD)	Ile/Val syn	MAT?	4.2.1.9	ILV3	Z49516.1	-	-
Dimethylglycine dehydrogenase (Me2GlyDH)	Gly metab; binds folate	MAT	1.5.99.2	-	-	DMGDH	AF111858.1
Gamma-aminobutyrate aminotransferase	Glu metab	MAT	2.6.1.19	-	-	ABAT	L32961.1
Glutamate dehydrogenase 1	Glu metab	MAT	1.4.1.3	-	-	GLUD1	M20867.1
Glutamate dehydrogenase 2 (NAD[P]+)	Glu metab	MAT	1.4.1.3	-	-	GLUD2	U08997.1
Glutaminase, kidney isoform (also alt spliced isoforms)	Glu metab	MIM	3.5.1.2	-	-	KIAA0838	AB020645.1
Glutaminase, liver isoform (also alt spliced isoform)	Glu metab	MIM	3.5.1.2	-	-	None	AF223944.1
Glutamine synthetase (glutamate ammonia ligase)	Glu/Gln metab	MAT?	6.3.1.2	-	-	GLNS (c)	S45408.1 (c)
Glutaryl-CoA dehydrogenase (GCD)	Lys/Trp metab	MAT	1.3.99.7	-	-	GCDH	U69141.1
Glycine C-acetyltransferase	Gly metab	MAT?	2.3.1.29	-	-	GCAT	AF077740.1
Glycine amidinotransferase (GATM)	Creatine syn	MIM	2.1.4.1	-	-	AGAT	S68805.1
Glycine decarboxylase H protein (aminomethyltransferase)	Gly deg	MAT	1.4.4.2	GCV3	U12980.1	GCSH	M69175.1
Glycine decarboxylase L protein (dihydrolipoyl dehydrogenase)	Gly deg	MAT	1.8.1.4	LPD1	M20880.1	DLD	J03490.1
Glycine decarboxylase P protein (dehydrogenase)	Gly deg	MAT	1.4.4.2	GCV2	U20641.1	GCSP	D90239.1
Glycine decarboxylase T protein (aminomethyltransferase)	Gly deg	MAT	2.1.2.10	GCV1	L41522.1	GCST	D13811.1
Homoaconitase	Lys syn	MAT	4.2.1.36	LYS4	X93502.1	-	-
Homoisocitrate dehydrogenase	Lys syn	MAT	1.1.1.155	LYS12	Z46728.1	None	BE250293.1 (e)
3-Hydroxymethyl-3-methylglutaryl-CoA lyase	Leu deg	MAT	4.1.3.4	-	-	HMGCL	L07033.1
2-Isopropylmalate synthase	Leu syn	MAT	4.1.3.12	LEU4	M12893.1	-	-
2-Isopropylmalate synthase homolog, potentially mitochondrial	Function unknown	NC	-	YOR108w	Z75016.1	-	-
Isovaleryl-CoA dehydrogenase	Leu deg	MAT	1.3.99.10	-	-	IVD	M34192.1
Kynurenine/alpha-aminoadipate aminotransferase	Glu metab	MAT?	2.6.1.7/39	-	-	None	AF097994.1
Kynurenine aminotransferase/glutamine transaminase K	Glu metab	MAT?	2.6.1.7	YJL060w	Z49335.1	None (r)	Z49696.1 (r)
Kynurenine 3-monooxygenase (kynurenine 3-hydroxylase)	Trp deg	MOM	1.14.13.9	-	-	None	Y13153.1
Methylenetetrahydrofolate bifunctional enzyme (NAD)	Met metab	MAT	1.5.1.15	-	-	MTHFD2	X16396.1
Methylenetetrahydrofolate reductase	Met metab	MAT?	1.5.1.20	MET12	U36624.1	-	-
Methylenetetrahydrofolate reductase	Met metab	MAT?	1.5.1.20	MET13	Z72647.1	-	-
Methylenetetrahydrofolate trifunctional enzyme (NADP)	Met metab	MAT	1.5.1.5	MIS1	J03724.1	MTHFD1	J04031.1
Methylmalonate-semialdehyde dehydrogenase	Val/pyrimidine metab	MAT?	1.2.1.27	-	-	MMSDH (b)	L08643.1 (b)
Monoamine oxidase A	Amine neurotransmitter deg	MOM	1.4.3.4	-	-	MAOA	M68840.1
Monoamine oxidase B	Amine neurotransmitter deg	MOM	1.4.3.4	-	-	MAOB	M69177.1

Ornithine aminotransferase	Arg/ornithine syn	MAT	2.6.1.13			OAT	M12267.1
Ornithine carbamoyltransferase	Arg/ornithine syn	MAT?	2.1.3.3	ARG3	M11946.1	OTC	K02100.1
Ornithine decarboxylase	Arg deg	NC	4.1.1.17	-	-	ODC1	M31061.1
Proline oxidase	Pro/Glu metab	MAT	1.5.3.-	PUT1	M18107.1	PRODH	U82381.1
Propionyl-CoA carboxylase, alpha chain	Met deg	MAT	6.4.1.3	-	-	PCCA	X14608.1
Propionyl-CoA carboxylase, beta chain	Met deg	MAT	6.4.1.3	-	-	PCCB	X73424.1
1-Pyrroline-5-carboxylate dehydrogenase	Pro/Glu metab	MAT	1.5.1.12	PUT2	M10029.1	ALDH4	U24266.1
Sarcosine dehydrogenase	Gly metab	MAT	1.5.99.1	-	-	[SARDH]	AJ223317.1
Serine hydroxymethyltransferase	Ser/gly metab	MAT	2.1.2.1	SHM1	Z36131.1	SHMT2	U23143.1
Serine-pyruvate aminotransferase	Ala/Ser metab	MAT?	2.6.1.51	-	-	AGT1 (r)	X06357.1 (r)
Serine/threonine deaminase, possibly mitochondrial	Ser/Thr metab	NC	4.2.1.13	CHA1	M85194.1	SDH2 (r)	J03863.1 (r)
Threonine deaminase	Ile syn	MAT?	4.2.1.16	ILV1	M36383.1	-	-
APOPTOSIS							
Apoptosis inducing factor (AIF)	DNA laddering	IMS	-	-	-	PDCD8	AF100928.1
B-cell lymphoma 2 protein (Bcl-2) (2 splice variants)	Anti-apoptotic protein	MIM	-	-	-	BCL2	M13994.1
Bcl2/adeno E1B 19-kD-interacting protein 3 (Bnip3)	Pro-apoptotic protein	MOM?	-	-	-	BNIP3	U15174.1
Bcl2/adeno E1B 19-kD-interacting protein 3-like (Bnip3L)	Alters membrane permeability	MOM?	-	-	-	BNIP3L	AF079221.1
Bcl2-antagonist of cell death (Bad)	Binds Bcl-x/Bcl-2	MOM?	-	-	-	BAD	AF031523.1
Bcl2-antagonist/killer 1 (Bak)	Pro-apoptotic protein	MOM?	-	-	-	BAK1	U16811.1
Bcl2-antagonist/killer 2 (Bak2)	Pro-apoptotic protein	MOM?	-	-	-	BAK2	U16812.1
Bcl2-associated X protein, alpha isoform (Bax alpha)	Pro-apoptotic protein	MOM	-	-	-	BAX	L22473.1
Bcl2 family-1, *Drosophila* ortholog (Drob1)	Pro-apoptotic protein	ND	-	-	-	DROB1 (f)	AB032430.1 (f)
Bcl2-interacting killer, apoptosis-inducing (Bik)	Pro-apoptotic protein	MOM?	-	-	-	BIK	U34584.1
Bcl2-interacting mediator of cell death (Bim) (3 isoforms)	Pro-apoptotic protein	MOM?	-	-	-	[BIM]	AF032458.1
Bcl2-like 1 (Bcl-x) (at least 3 isoforms, including Bcl-xL, Bcl-xS)	Anti-apoptotic protein	MOM?	-	-	-	BCL2L1	Z23115.1
Bcl2-related ovarian killer, *Drosophila* homolog (DBok)	Pro-apoptotic protein	ND	-	-	-	[DBOK] (f)	AF228044 (f)
Bcl2-related protein A1 (hematopoetic-specific)	Anti-apoptotic protein	MOM?	-	-	-	BCL2A1	U27467.1
Bcl2-associated athanogene 1 (Bag-1), short form	Bcl-2 binding protein	ND	-	-	-	BAG1	Z35491.1
Bcl-w apoptosis regulator	Anti-apoptotic protein	ND	-	-	-	BCLW	U59747.1
BH3 interacting domain death agonist, truncated (tBID)	Triggers cyt c release	MOM	-	-	-	BID	AF042083.1
Bifunctional apoptosis regulator (BAR)	Anti-apoptotic protein	ND	-	-	-	BAR	AF173003.1
Caspase 2 precursor	Pro-apoptotic protein	ND	3.4.22.-	-	-	Casp2 (m)	Y13085.1 (m)
Caspase 3 precursor (2 isoforms)	Pro-apoptotic protein	IMS	3.4.22.-	-	-	CPP32	U13737.1
Caspase 8 precursor	Pro-apoptotic protein	MOM?	3.4.22.-	-	-	CASP8	AF102146.1
Caspase 9 precursor (2 isoforms)	Pro-apoptotic protein	IMS	3.4.22.-	-	-	CASP9	AB019205.1
Cell death effector (CED-4) (cytosolic in humans)	Pro-apoptotic protein	ND	-	-	-	ced-4 (w)	X69016.1 (w)
Cell death activator EGL-1	Pro-apoptotic protein	ND	-	-	-	egl-1 (w)	AF057309.1 (w)
Cell death inducing DFF45-like effector B (CIDE-B)	Pro-apoptotic protein	ND	-	-	-	[CIDEB]	AF218586.1
c-Jun NH2-terminal kinase 1	Inactivates Bcl-xL	NC	2.7.1.-	-	-	JNK1	L26318.1
c-Jun NH2-terminal kinase 2	Inactivates Bcl-xL	NC	2.7.1.-	-	-	JNK2	L31951.1
Cytomegalovirus-encoded vMIA fr unspliced exon I UL37 mRNA	Inhibitor of apoptosis	MOM	-	-	-	HCMVUL37	X17403.1
Death associated protein 3	Pro-apoptotic protein	MAT	-	-	-	DAP3	X83544.1
Direct IAP binding protein with low PI (Smac/DIABLO)	Pro-apoptotic protein	ND	-	-	-	DIABLO	AF203914.1 (m)
Extracellular signal-related kinase 1 (Erk1)	Phosphorylates Bcl-2	MOM	2.7.1.-	-	-	ERK1	X60188.1
Extracellular signal-related kinase 2 (Erk2)	Phosphorylates Bcl-2	MOM	2.7.1.-	-	-	ERK2	M84489.1
Gelsolin (actin regulatory protein)	Anti-apoptotic protein	NC	-	-	-	GSN	X04412.1
Hepatitis B virus transactivating protein X	Binds VDAC3	MOM?	-	-	-	HBX	X69798.1
Hid protein	Interacts with Bcl-xL	ND	-	-	-	HID (f)	U31226.1 (f)
HIV transactivating regulatory protein Tat	Induces apoptosis	ND	-	-	-	TAT	M11840.1
HIV viral protein R (at least 2 isoforms)	Induces apoptosis	MOM?	-	-	-	VPR	U81843.1
l(2)tid protein, long isoform (hTid-1L); homolog of yeast Mdj1	Increases apoptosis	MAT	-	-	-	TID1	AF061749.1
l(2)tid protein, short isoform (hTid-1S); homolog of yeast Mdj1	Suppresses apoptosis	MAT	-	-	-	TID1	AF061749.1
Myeloid cell leukemia sequence 1 (Mcl-1)	Anti-apoptotic protein	MIM?	-	-	-	MCL1	AF118124.1
Noxa ("damage"), mediator of p53-dependent apoptosis	Pro-apoptotic protein	ND	-	-	-	[NOXA]	Q13794
p53-regulated Apoptosis-Inducing Protein 1 (3 isoforms)	Proapoptotic protein	ND	-	-	-	[p53AIP1]	AB045830.1
Porin (PorB) from *Neisseria meningitidis* (Nm)	Anti-apoptotic protein	NC	-	-	-	PorB (Nm)	X65530.1 (Nm)

Appendix 4—*continued*

Gene product	Function/comment	LOC.	E.C. #	Yeast		Animal	
				Symbol	GenBank	Symbol	GenBank
Protein kinase Calpha	Phosphorylates Bcl-2	ND	2.7.1.-	-	-	PRKCA	X52479.1
Protein kinase Cdelta	Induces apoptosis	MOM	2.7.1.-	-	-	PRKCD	D10495.1
Protein phosphatase 2A, catalytic subunit	Dephosphorylates Bcl-2	ND	3.1.3.16	-	-	PPP2CA	X12646.1
RIP-like protein kinase RIP3	Pro-apoptotic protein	ND	-	-	-	RIP3	AF156884.1
Septin-like protein (ARTS) (2 isoforms)	Translocates to nucleus	IMS?	-	-	-	PNUTL2	AF073312.1
SM-20 (similar to *C. elegans* egl-9)	Proapoptotic protein	ND	-	-	-	[SM20] (r)	U06713.1 (r)
Thyroid hormone [orphan] nuclear receptor TR3	Translocates to mitos	MOM?	-	-	-	[HMR]	L13740.1
Tumor suppressor protein p53	Proapoptotic protein	ND	-	-	-	TP53	M14695.1
Vacuolating cytotoxin VacA of *Helicobacter pylori* (Hp)	Pro-apoptotic protein	ND	-	-	-	VACA (Hp)	U29401.1 (Hp)
WW domain-containing oxidoreductase (WOX1)	Proapoptotic protein	ND	-	-	-	WOX1	AF187015.1
CARBOHYDRATE METABOLISM							
Acetate non-utilizing protein (Acn9)	Carbon assimilation	IMS	-	YDR511w	U33057.1	None	AW404075.1 (e)
Acetyl-CoA C-acetyltransferase 1	Ketone body metab	MAT?	2.3.1.9	-	-	ACAT1	D90228.1
Acetyl-CoA synthetase	TCA cycle-associated	MAT?	6.2.1.1	ACS1	X66425.1	[AceCS2] (m)	AB046742.1 (m)
Aconitate hydratase (aconitase)	TCA cycle	MAT	4.2.1.3	ACO1	M33131.1	ACO2	U80040.1
Aconitase homolog, potentially mitochondrial	Function unknown	NC	4.2.1.3	YJL200c	Z49475.1	-	-
Alcohol dehydrogenase III	Ethanol syn	MAT	1.1.1.1	ADH3	K03292.1	-	-
Alcohol dehydrogenase IV, potentially mitochondrial	Ethanol syn	NC	1.1.1.1	ADH4	Z72778.1	-	-
Aldehyde dehydrogenase	Ethanol metab	MAT	1.2.1.3	ALD5	U56605.1	ALDH5	M63967.1
Aldehyde dehydrogenase (NAD+)	Ethanol metab	MAT	1.2.1.3	ALD4	Z75282.1	ALDH2	Y00109.1
Aldehyde dehydrogenase (NAD+)	Ethanol metab	MAT	1.2.1.3	ALD1	M57887.1	-	-
Aldehyde dehydrogenase 2 (NAD+)	Ethanol metab	MAT	1.2.1.3	ALDH2	Z17314.1	-	-
Aldose reductase, potentially mitochondrial (found in mito 2-D gels)	Ethanol metab	NC	1.1.1.21	-	-	ALDR1	X15414.1
Alpha-ketoglutarate dehydrogenase complex E1 component	TCA cycle-related	MAT	1.2.4.2	KGD1	M26390.1	OGDH	D10523.1
Alpha-ketoglutarate dehydrogenase complex E2 component	Keto acid decarboxylation	MAT	2.3.1.61	KGD2	M34531.1	DLST	D16373.1
Alpha-ketoglutarate dehydrogenase complex E3 component	Dihydrolipoamide dehydrog.	MAT	1.8.1.4	LPD1	M20880.1	DLD	J03490.1
Arabinono-1,4-lactone oxidase	Ascorbic acid syn	ND	1.1.3.37	ALO1	U40390.1	-	-
Biotin apo-protein ligase	Biotinylates carboxylases	ND	6.3.4.-	-	-	HLCS	D23672.1
Biotin synthase, putatively mitochondrial	Biotin syn	NC	2.8.1.6	BIO2	Z73071.1	-	-
Carbonic anhydrase VA (also called V)	C-compound metab	MAT?	4.2.1.1	-	-	CA5A	L19297.1
Carbonic anhydrase VB	C-compound metab	MAT?	4.2.1.1	-	-	CA5B	AB021660.1
Carbonyl reductase (NADPH)	Carbonyl cpd metab	MAT	1.1.1.184	-	-	Cbr2 (m)	D26123.1 (m)
Citrate synthase	TCA cycle	MAT	4.1.3.7	CIT1	Z23259.1	CS	AF047042.1
Citrate synthase 2 (peroxisomal), potentially mito (has cryptic MTS)	TCA cycle?	MAT?	4.1.3.7	CIT2	M14686.1	-	-
Citrate synthase 3	TCA cycle	MAT	4.1.3.7	CIT3	X88846.1	-	-
Cytochrome b2 (lactate cytochrome c oxidoreductase)	Lactate oxdn	IMS	1.1.2.3	CYB2	X03215.1	None	AW025423.1 (e,i)
Dihydrolipoamide S-acetyltransferase (PDHC-E2)	TCA cycle-related	MAT	2.3.1.12	LAT1	J04096.1	DLAT	J03866.1
Dihydrolipoamide branched chain transacylase E2 component	TCA cycle-related	MAT	2.3.1.-	-	-	DBT	X66785.1
D-Lactate dehydrogenase, sperm-specific isozyme C4	Lactate/pyruvate metab	ND	1.1.1.27	-	-	Ldhc (m)	M17587.1 (m)
D-Lactate dehydrogenase (cytochrome)	Lactate/pyruvate metab	MIM	1.1.2.4	DLD1	X66052.1	CG3835 (f)	AE003423.1 (f)
D-lactate dehydrogenase (actin-interacting protein)	Function unclear	MAT	1.1.2.4	AIP2	Z74226.1	-	-
DRAP deaminase (Rib2) homolog, potentially mitochondrial	Riboflavin syn?	NC	4.2.1.70	YGR021w	Z72806.1	None	AW968801.1 (e)
DRAP deaminase (Rib2) homolog, potentially mitochondrial	Riboflavin syn?	NC	-	YDL036c	Z74084.1	K07E8.7 (w)	AF016678.1 (w)
DRAP deaminase (Rib2) homolog, potentially mitochondrial	Riboflavin syn?	NC	-	YGR169c	Z72954.1	-	-
Folic acid syn trifunctional protein, potentially mitochondrial	Dihydrofolate syn	NC	4.1.2.25	FOL1	Z71532.1	-	-
Folylpolyglutamate synthase	Folate metab	MAT?	6.3.2.17	-	-	FPGS	M98045.1
Fumarate hydratase (fumarase)	TCA cycle	MAT	4.2.1.2	FUM1	J02802.1	FH	U59309.1
Fumarate reductase	TCA cycle-related	MAT	1.3.99.1	OSM1	Z49551.1	-	-
Galactoside alpha(2-3) sialyltransferase	Glycosylation	MOM	2.4.99.4	-	-	SIAT4	L29555.1
Galactoside alpha(2-6) sialyltransferase	Glycosylation	MOM	2.4.99.1	-	-	SIAT1	X17247.1
Glucokinase (hexokinase 4) (3 isoforms in human)	Glycolysis	MOM	2.7.1.1	-	-	GCK	M90299.1
Glycerol kinase (3 isoforms in human)	Glycerol metab	ND	2.7.1.30	GUT1	X69049.1	GK	L13943.1
Glycerol-3-phosphate dehydrogenase (FAD)	Glycerol metab	MIM	1.1.99.5	GUT2	X71660.1	GPD2	U12424.1
Glycerol-3-phosphate dehydrogenase (NAD)	Glycerol metab	MIM	1.1.1.8	GPD2	Z74801.1	-	-

Hexokinase I	Glycolysis	MOM?	2.7.1.1	HXK1	M14410.1	HK1	M75126.1
Hexokinase II	Glycolysis	MOM?	2.7.1.1	HXK2	M11181.1	HK2	Z46376.1
Hexokinase III	Glycolysis	MOM	2.7.1.1	-	-	HK3	U51333.1
3-Hydroxybutyrate dehydrogenase	Ketone body metab	MAT	1.1.1.30	-	-	BDH (r)	M89902.1 (r)
3-Hydroxyisobutyrate dehydrogenase (HIBADH)	Br chain oxo-acid metab	MAT	1.1.1.31	-	-	None	J04628.1 (r,i)
Isocitrate dehydrogenase (NAD+) subunit 1	TCA cycle	MAT	1.1.1.41	IDH1	M95203.1	IDH3G	Z68907.1
Isocitrate dehydrogenase (NAD+) subunit 2	TCA cycle	MAT	1.1.1.41	IDH2	M74131.1	IDH3A	U07681.1
Isocitrate dehydrogenase 3 (NAD+) subunit beta, isoform B	TCA cycle	MAT	1.1.1.41	-	-	IDH3B	U49283.1
Isocitrate dehydrogenase (NADP+)	TCA cycle	MAT	1.1.1.42	IDP1	M57229.1	IDH2	X69433.1
Isocitrate dehydrogenase (NADP+)	TCA cycle	MAT?	1.1.1.42	-	-	None	U52144.1
Lipoate protein ligase, potentially mitochondrial	Lipoate biosynthesis	NC	-	YLR239c	U19027.1	CG9804 (f)	AE003606.2 (f)
Lipoate protein ligase A homolog, potentially mitochondrial	Lipoate syn/transp?	NC	-	YJL046w	Z49321.1	CG8446 (f)	AE003807.2 (f)
Lipoic acid synthase	Lipoate syn	MAT?	-	LIP5	Z75104.1	LAS	AJ224162.1 (i)
Malate dehydrogenase	Malate-aspartate shuttle	MAT	1.1.1.37	MDH1	J02841.1	MDH2	AF047470.1
Malic enzyme, NAD(+)-dependent	TCA cycle-related	MAT	1.1.1.40	-	-	ME2	M55905.1
Malic enzyme, NADP(+)-dependent	TCA cycle	MAT	1.1.1.38	MAE1	Z28029.1	ME3	X79440.1
2-Methylisocitrate lyase	Propionyl-CoA/Thr metab	MAT	4.1.3.1	ICL2	Z48951.1	-	-
Methylmalonyl Co-A mutase	TCA cycle-related	MAT	5.4.99.2	-	-	MUT	M65131.1
3-Methyl-2-oxobutanoate dehydrogenase (lipoamide) E1-alpha	Br chain keto acid metab	MAT	1.2.4.4	-	-	BCKDHA	Z14093.1
3-Methyl-2-oxobutanoate dehydrogenase (lipoamide) E1-beta	Br chain keto acid metab	MAT	1.2.4.4	-	-	BCKDHB	M55575.1
Phenacrylic acid decarboxylase, potentially mitochondrial	Cinnamic acid resistance	NC	4.1.1.-	PAD1	L09263.1	-	-
Phosphoenolpyruvate carboxykinase 2	Gluconeogenesis	MAT?	4.1.1.32	-	-	PCK2	X92720.1
Pyruvate carboxylase	TCA cycle-related	MAT	6.4.1.1	-	-	PC	S72370.1
Pyruvate dehydrogenase (lipoamide) E1-alpha subunit	TCA cycle-related	MAT	1.2.4.1	PDA1	M29582.1	PDHA1	M24848.1
Pyruvate dehydrogenase (lipoamide) E1-alpha subunit 2	TCA cycle-related	MAT	1.2.4.1	-	-	PDHA2	M86808.1
Pyruvate dehydrogenase (lipoamide) E1-beta subunit	TCA cycle-related	MAT	1.2.4.1	PDB1	M98476.1	PDHB	J03576.1
Pyruvate dehydrogenase complex protein X	Binds E3 to E2 core	MAT	1.2.4.1	PDX1	Z72978.1	PDX1	AF001437.1
Pyruvate dehydrogenase kinase isozyme 1	Regulation of PDHC	MAT	2.7.1.99	-	-	PDK1	L42450.1
Pyruvate dehydrogenase kinase isozyme 2	Regulation of PDHC	MAT	2.7.1.99	-	-	PDK2	L42451.1
Pyruvate dehydrogenase kinase isozyme 3	Regulation of PDHC	MAT	2.7.1.99	-	-	PDK3	L42452.1
Pyruvate dehydrogenase kinase isozyme 4	Regulation of PDHC	MAT	2.7.1.99	-	-	PDK4	U54617.1
Pyruvate dehydrogenase kinase homolog, potentially mitochondrial	Function unknown	NC	-	YGL059w	Z72581.1	-	-
Pyruvate dehydrogenase phosphatase, cat subunit, isoform 1	Regulation of PDHC	MAT	3.1.3.43	-	-	[PDP1] (r)	AF062740.1 (r)
Pyruvate dehydrogenase phosphatase, cat subunit, isoform 2	Regulation of PDHC	MAT	3.1.3.43	-	-	[PDP2] (r)	AF062741.1 (r)
Pyruvate dehydrogenase phosphatase, regulatory subunit	Regulation of PDHC	MAT	3.1.3.43	-	-	[PDPR] (b)	AF026954.1 (b)
Riboflavin kinase (flavokinase)	Flavoenzyme syn	MIM	2.7.1.26	FMN1	Z49701.1	-	-
Succinic semialdehyde dehydrogenase, NAD-dependent	GABA deg	MAT?	1.2.1.24	-	-	ALDH5A1	Y11192.1
Succinyl CoA: 3-oxoacid CoA transferase	Ketone body metab	MAT	2.8.3.5	-	-	SCOT	U62961.1
Succinyl-CoA synthetase, subunit alpha	TCA cycle	MAT	6.2.1.4	LSC1	Z75050.1	SUCLA1	AF104921.1
Succinyl-CoA synthetase, subunit beta	TCA cycle	MAT	6.2.1.4	LSC2	Z73029.1	SUCLA2	AB035863.1
Succinyl CoA synthetase, subunit beta	TCA cycle	MAT	6.2.1.5	-	-	Suclg2 (m)	AF058956.1 (m,i)
Thiamine biosynthetic protein	Thiamine syn; mtDNA repair	MAT?	-	THI4	Z72929.1	-	-
CARRIERS AND TRANSPORTERS							
ABC-mitochondrial erythroid transporter (ABC-me)	Function unknown	MIM	-	-	-	None	AF266284 (m)
ABC protein, similar to MDR transporter, putatively mitochondrial	Function unknown	NC	-	MDL1	L16958.1	ABCB2	L21204.1
ABC protein, sub-family B (MDR/TAP), member 8	Function unknown	ND	-	-	-	ABCB8	AF047690.1
ABC protein, sub-family C (CFTR/MRP), member 6, short form	Function unknown	NC	-	-	-	ABCC6	X95715.1
ADP/ATP carrier (adenine nucleotide translocator) isoform 1	ADP/ATP import/export	MIM	-	AAC1	M12514.1	ANT1	J02966.1
ADP/ATP carrier (adenine nucleotide translocator) isoform 2	ADP/ATP import/export	MIM	-	PET9	J04021.1	ANT2	M57424.1
ADP/ATP carrier (adenine nucleotide translocator) isoform 3	ADP/ATP import/export	MIM	-	AAC3	M34076.1	ANT3	J03592.1 (i)
ADP/ATP carrier homolog, putative (MCF protein)	Function unknown	MIM?	-	-	-	None	AC003083.1 (i)
CAB1, similar to StAR	Cholesterol transp	ND	-	-	-	MLN64	X80198.1
Calsequestrin, skeletal muscle isoform (calmitine)	Calcium binding protein	MIM	-	-	-	CASQ1	S73775.1
Carnitine/acylcarnitine translocase	Ac-carnitine transp to mitos	MIM	-	CRC1	Z75008.1	CACT	Y10319.1
Carnitine O-acetyltransferase	Fatty acid transp	MIM	2.3.1.7	CAT2	Z14021.1	CRAT	X78706.1
Carnitine O-acetyltransferase	Fatty acid transp	MIM	2.3.1.7	YAT1	X74553.1	-	-

Appendix 4—*continued*

Gene product	Function/comment	LOC.	E.C. #	Yeast		Animal	
				Symbol	GenBank	Symbol	GenBank
Carnitine O-acetyltransferase, similar to Yat1p	Fatty acid transp	ND	-	YER024w	U18778.1	-	-
Carnitine palmitoyltransferase I, liver type	Fatty acid transp	MOM	2.3.1.21	-	-	CPT1A	L39211.1
Carnitine palmitoyltransferase I, muscle type	Fatty acid transp	MOM	2.3.1.21	-	-	CPT1B	U62733.1
Carnitine palmitoyltransferase II (carnitine O-acetyltransferase 2)	Fatty acid transp	MIM	2.3.1.21	-	-	CPT2	U09648.1
Cationic amino acid transporter-2A, potentially mitochondrial	Cationic aa transp	NC	-	-	-	ATRC2	U76368.1
Chloride ion channel 27 (found in mito 2-D gels)	Chloride transp?	NC	-	-	-	CLIC1	U93205.1
Chloride intracellular channel p64H1	Chloride transp	ND	-	-	-	CLIC4	AF097330.1
Citrin	Citrulline transp	MIM	-	-	-	SLC25A13	AF118838.1
Cobalt transporter	Co transp	MIM	-	COT1	M88252.1	-	-
Diazepam binding inhibitor (acyl-CoA binding protein [ACBP])	Acyl-CoA ester transp	MOM	-	-	-	DBI	M14200.1
Dicarboxylate carrier	Dicarboxylic acid transp	MIM	-	DIC1	U79459.1	DIC	AJ131613.1
FAD carrier protein (MCF protein)	FAD (flavin) transp	MIM	-	FLX1	L41168.1	None	AI431745 (e*)
FAD carrier protein homolog, putative, similar to Flx1p	Function unknown	MIM	-	YIL006w	Z38113.1	-	-
Folate transporter, similar to yeast Flx1p	Folate transp	MIM	-	-	-	[HMFT]	AF283645.1
Frataxin	Iron export	MAT?	-	YFH1	Z74168.1	FRDA	U43747.1
Glucose transporter, type 3 (brain isoform), potentially mitochondrial	Glucose transp	ND	-	-	-	GLUT3 (r)	D13962.1 (r)
Iron accumulation protein	Fe transp?	MIM	-	MMT1	Z49808.1	-	-
Iron accumulation protein	Fe transp?	MIM	-	MMT2	Z73580.1	-	-
Iron transporter (ABC)	FeS protein export	MIM	-	ATM1	X82612.1	[MTABC3]	AF076775.1
Iron transporter homolog similar to Atm1p (ABC)	Function unknown	ND	-	-	-	[M-ABC1]	AF047690.1
Iron transporter homolog similar to Atm1p (ABC)	Function unknown	ND	-	-	-	ABC7	AF038950.1
Major facilitator family member, potentially mitochondrial	Function unknown	NC	-	YBR293w	Z36162.1	-	-
Manganese transporter	Mn transp	ND	-	SMF1	U15929.1		
Manganese transporter, probable	Mn transp?	MIM	-	SMF2	U00062.1	-	-
MCF protein	Function unknown	MIM	-	RIM2	Z36061.1	-	-
MCF protein	Function unknown	MIM	-	YDL119c	Z74167.1	-	-
MCF protein	Function unknown	MIM	-	YEL006w	U18530.1	-	-
MCF protein	Function unknown	MIM	-	YFR045w	D50617.1	-	-
MCF protein	Function unknown	MIM	-	YGR257c	Z73042.1	None	AC003083.1
MCF protein	Function unknown	MIM	-	YHR002w	U10555.1	None	AC004143.1
MCF protein	Function unknown	MIM	-	YMC1	X67122.1	COLT (f)	Y12495.1 (f)
MCF protein	Function unknown	MIM	-	YMC2	Z35973.1	-	-
MCF protein	Function unknown	MIM	-	YMR166c	Z49705.1	-	-
MCF protein	Function unknown	MIM	-	YNL083w	Z71359.1	None	AW023500 (e*)
MCF protein (also in peroxisomes)	Function unknown	ND	-	YPR128c	U40829.1	None	AA918972 (e*)
MCF protein CGI-69	Function unknown	ND	-	-	-	None	AF151827.1
MCF protein (Graves disease carrier protein)	Function unknown	MIM	-	-	-	SLC25A16	M31659.1
MCF protein, similar to Rim2p	Function unknown	MIM	-	YHM2	Z48756.1	-	-
MCF protein, similar to SLC25A16	Function unknown	MIM	-	YGR096w	Z72881.1	None	AA397670.1 (e*)
MCF 25, member 12 (Aralar1)	Calcium binding carrier	MIM?	-	YPR021c	Z71255.1	SLC25A12	Y14494.1
MCF 25, member 17	Function unknown	MIM	-	-	-	SLC25A17	Y12860.1
Mitochondrial carrier homolog	Function unknown	MIM	-	YPR011c	Z71255.1	MTCH (f)	AF176011.1 (f)
Mitochondrial carrier homolog 1, isoform a	Function unknown	ND	-	-	-	Mtch1a (m)	AF176007.2 (m)
Mitochondrial carrier homolog 1, isoform b	Function unknown	ND	-	-	-	Mtch1b (m)	AF192559.2 (m,i)
Mitochondrial carrier homolog 2	Function unknown	ND	-	-	-	[MTCH2]	AF176008.1
Monocarboxylate transporter (MCT1)	Pyruvate transp	ND	-	ESBP6	X84903.1 (a)	SLC16A1	L31801.1
Ornithine transporter	Arginine syn	MIM	-	ORT1	X87414.1	ORNT1	AF112968.1
Oxaloacetate transporter	Oxaloacetate transp	MIM	-	OAC1	AJ238698.1	-	-
Oxodicarboxylate carrier, isoform 1	Oxoglutarate/oxoadipate transp	MIM	-	ODC1	U43703.1	ODC	R29313 (e,i)
Oxodicarboxylate carrier, isoform 2	Oxoglutarate/oxoadipate transp	MIM	-	ODC2	Z75130.1	-	-
2-Oxoglutarate/malate carrier (MCF protein)	Oxoglutarate/malate exchange	MIM	-	-	-	SLC25A11	AF070548.1
Peripheral benzodiazepine receptor	Cholesterol transp	MOM?	-	-	-	BZRP	U12421.1
Peripheral benzodiazepine receptor interacting protein	Function unknown	ND	-	-	-	[PRAX1]	AF039571.1
Phosphate carrier (p32) (MCF protein)	Phosphate carrier (p32)	MIM	-	YHM1	Z74246.1	-	-
Phosphate transporter (MCF protein) (import receptor p32)	Phosphate carrier (PIC)	MIM	-	MIR1	X57478.1	None	BE741585.1 (e,i)

Phosphate transporter (MCF protein), putative	Phosphate carrier (PIC)?	MIM?	-	YER053c	U18796.1	PHC	X77337.1
Potassium channel (IRKA; Kir4.1), putatively mitochondrial	ATP-sensitive K+ channel	ND	-	-	-	KCNJ10	U73192.1
Succinate/fumarate transporter	Succinate-fumarate transp	MIM	-	SFC1	Z25485.1	None	AI568562.1 (e,i)
Sodium/calcium exchanger, potentially mitochondrial	Ca++/H+ antiporter	ND	-	-	-	SLC8A1	M91368.1 (a)
Sodium/hydrogen exchanger	Na+/H+ antiporter	MIM?	-	NHX1	U33007.1	SLC9A6	AF030409.1
Sodium/hydrogen exchanger, potentially mitochondrial	Na+/H+ antiporter	ND	-	-	-	SLC9A1	S68616.1
Sodium/potassium/calcium exchanger, potentially mitochondrial	Na+/Ca++, K+ antiporter	ND	-	-	-	SLC24A1	AF026132.1
Steroidogenic acute regulatory protein (StAR)	Sterol transp	ND	-	-	-	STAR	U17280.1
Sterol carrier protein 2	Lysosome -> mito sterol transfer	ND	2.3.1.16	-	-	SCP2	M75883.1
Thiamine transporter, putatively mitochondrial	Thiamine transp	ND	-	YOR192c	Z75100.1	SLC19A2	AF135488.1
Tricarboxylate carrier family transporter	Function unknown	MIM	-	YOR271c	Z75179.1	-	-
Tricarboxylate transporter	Citrate transp	MIM	-	CTP1	U17503.1	SLC25A1	U25147.1
Uncoupling protein 1 (brown fat)	Proton carrier	MIM	-	-	-	UCP1	U28480.1
Uncoupling protein 2	Proton carrier	MIM	-	-	-	UCP2	U76367.1
Uncoupling protein 3	Proton carrier	MOM	-	-	-	UCP3	U84763.1
Uncoupling protein 4	Proton carrier	MOM	-	-	-	UCP4	AF110532.1
Uncoupling protein 5 (3 isoforms) (also called BMCP1)	Proton carrier	MIM	-	-	-	UCP5	AF078544.1
Vesicular inhibitory amino acid transporter, potentially mitochondrial	GABA transp	NC		-	-	Viaat (m)	AJ001598.1 (m)
Voltage-dependent anion channel 1 (Porin)	Anion import	MOM	-	POR1	M34907.1	VDAC1	L06132.1
Voltage-dependent anion channel 2 (Porin)	Anion import	MOM	-	POR2	Z38125.1	VDAC2	L08666.1
Voltage-dependent anion channel 3	Anion import	MIM	-	-	-	VDAC3	AF038962.1
Voltage-dependent anion channel 4	Anion import	MIM	-	-	-	VDAC4	S75651.1 (i)
Wilson disease protein (mito localization is controversial)	Copper transp	ND	3.6.1.36	-	-	ATP7B	U11700.1
LIPID METABOLISM							
Cholesterol, steroid, and xenobiotic metabolism							
Acetoacetyl-CoA thiolase (cytosolic in yeast [Erg10])	Mevalonate/sterol syn, 1st step	MAT?	2.3.1.9	-	-	ACAT1	D90228.1
Adrenodoxin (cholesterol side-chain-cleavage system)	Steroid and FeS cluster syn	MAT	-	YAH1	Z73608.1	FDX1	J03548.1
Adrenodoxin/ferredoxin reductase	Pregnenolone syn	MAT	1.18.1.2	ARH1	U38689.1	FDXR	J03826.1
3-Beta hydroxy-5-ene steroid dehydrogenase type I	Steroid syn	MIM?	1.1.1.145	-	-	HSD3B1	M27137.1
3-Beta hydroxy-5-ene steroid dehydrogenase type II	Steroid syn	MIM?	1.1.1.145	-	-	HSD3B2	M67466.1
3-Beta hydroxy-5-ene steroid dehydrogenase type III	Steroid syn	MIM?	1.1.1.145	-	-	Hsd3b3 (m)	M77015.1 (m)
3-Beta hydroxy-5-ene steroid dehydrogenase type IV	Steroid syn	MIM?	1.1.1.145	-	-	Hsd3b4 (m)	L16919.1 (m)
3-Beta hydroxy-5-ene steroid dehydrogenase type V	Steroid syn	MIM?	1.1.1.145	-	-	Hsd3b5 (m)	L41519.1 (m)
3-Beta hydroxy-5-ene steroid dehydrogenase type VI	Steroid syn	MIM?	1.1.1.145	-	-	Hsd3b6 (m)	AF031170.1 (m)
Cytochrome P450 11 beta, polypeptide 1 (11B1)	Steroid 11-beta-hydroxylase	ND	1.14.15.4	-	-	CYP11B1	X55764.1
Cytochrome P450 11 beta, polypeptide 2 (11B2)	Steroid 11-beta-hydroxylase	ND	1.14.15.4	-	-	CYP11B2	X54741.1
Cytochrome P450 1A1 (derived from cytP4501A1), 2 isoforms	Xenobiotic metab	ND	1.14.14.1	-	-	CYP1A1 (r)	K02246.1 (r)
Cytochrome P450 2B1, phosphorylated @ Ser-128	Xenobiotic metab	ND	1.14.14.1	-	-	CYP2B1 (r)	J00719.1 (r)
Cytochrome P450 2B2	Xenobiotic metab	ND	1.14.14.1	-	-	CYB2B2 (r)	J00728.1 (r)
Cytochrome P450 2E1	Xenobiotic/ketone body metab	ND	1.14.14.1	-	-	CYP2E1	J02625.1
Cytochrome P450 3A1	Xenobiotic metab	ND	1.14.14.1	-	-	CYP3A1 (r)	M10161.1 (r)
Cytochrome P450 3A2	Xenobiotic metab	ND	1.14.14.1	-	-	CYP3A2 (r)	M13646.1 (r)
Cytochrome P450 25-hydroxy vitamin D-1 alpha hydroxylase	Vitamin D metab	ND	1.14.-.-	-	-	CYP27B1	AB005038.1
Cytochrome P450-SCC side chain cleavage enzyme	Cholesterol monooxygenase	ND	1.14.15.6	-	-	CYP11A1	M14565.1
Cytochrome P450 sterol 27-hydroxylase	Vitamin D metab	ND	1.14.14.-	-	-	CYP27A1	M62401.1
Cytochrome P450 vitamin D(3) 24-hydroxylase (CC24)	Vitamin D metab	ND	1.14.-.-	-	-	CYP24	L13286.1
Cytochrome P450 vitamin D(3) hydroxylase-associated protein	Amidase	MIM	-	-	-	[VDHAP]	U00694.1
Diacylglycerol O-acyltransferase homolog (ACAT-related protein)	Cholesterol metab	ND	2.3.1.26	-	-	DGAT	AF059202.1
S-Adenosyl-methionine:delta-24-C-sterol methyltransferase	Ergosterol syn	ND	2.1.1.41	ERG6	X74249.1	-	-
Fatty acid metabolism							
ACP synthase (phosphopantetheine protein transferase)	Pantetheinylation of ACP	MAT?	2.7.8.7	PPT2	Y16253.1	-	-
Acetyl-CoA carboxylase 2	Long chain fatty acid syn	MOM	6.4.1.2	ACC1	M92156.1	ACC2	U89344.1
Acyl-carrier protein (ACP) (also in mammalian complex I)	Type II fatty acid synthase	MAT	1.6.5.3	ACP1	Z28192.1	NDUFAB1	AF087660.1
Acyl-CoA dehydrogenase, long chain-specific (LCAD)	Beta-oxdn of fatty acids	MAT	1.3.99.13	-	-	ACADL	M74096.1
Acyl-CoA dehydrogenase, medium chain-specific (MCAD)	Beta-oxdn of fatty acids	MAT	1.3.99.3	-	-	ACADM	M91432.1
Acyl-CoA dehydrogenase, short/branched chain-specific	Beta-oxdn of fatty acids	MAT	1.3.99.-	-	-	ACADSB	U12778.1

Appendix 4—*continued*

Gene product	Function/comment	LOC.	E.C. #	Yeast		Animal	
				Symbol	GenBank	Symbol	GenBank
Acyl-CoA dehydrogenase, short chain-specific (SCAD)	Beta-oxdn of fatty acids	MAT	1.3.99.2	-	-	ACADS	M26393.1
Acyl-CoA dehydrogenase, very long chain-specific (VLCAD)	Beta-oxdn of fatty acids	MIM	1.3.99.-	-	-	ACADVL	D43682.1
Acyl-CoA thioesterase (MT-ACT48)	Activation of fatty acids	ND	3.1.2.2	-	-	[Act48.2] (m)	AJ238894.1 (m)
Acyl-CoA thioesterase, very-long chain-specific	Beta-oxdn of fatty acids	MAT	3.1.2.2	-	-	[MTE-1] (r)	Y09333.1 (r)
Alpha-methylacyl-Coa racemase	Beta-oxidation of fatty acids	ND	5.1.99.4	-	-	RM	AF158378.1
Beta-ketoacyl-ACP synthase	Type II fatty acid synthase	MAT?	2.3.1.41	CEM1	X73488.1	F10G8.9 (w)	Z80216.1 (w)
Ceramidase, non-lysosomal	Sphingolipid metab	ND	3.5.1.23	-	-	[CDase]	AA913512.1 (e)
Cytochrome b5	Fatty acid/sterol syn	MOM	-	CYB5	L22494.1	CYB5	M22865.1
Cytochrome b5 reductase (diaphorase)	Desaturation of fatty acids	MOM	1.6.2.2	CBR1	Z28365.1	DIA1	Y09501.1
Delta3,5-delta2,4-dienoyl-CoA isomerase (HPXEL)	Beta-oxdn of fatty acids	MAT?	5.3.3.-	-	-	ECH1	U16660.1
2,4-Dienoyl-CoA reductase	Beta-oxdn of unsat fatty acids	MAT?	1.3.1.34			DECR1	L26050.1
Dodecenoyl-CoA delta isomerase	Beta-oxdn of unsat fatty acids	MAT	5.3.3.8	-	-	DCI	Z25820.1
Electron transfer flavoprotein, subunit alpha (ETF-alpha)	Electron transfer from FAD	MAT	-	YPR004c	Z71255.1	ETFA	J04058.1
Electron transfer flavoprotein, subunit beta	Electron transfer from FAD	MAT	-	ETF-beta	Z72992.1	ETFB	X71129.1
Enoyl CoA hydratase, short chain 1	Beta-oxdn of fatty acids	MAT	4.2.1.17	YDR036c	Z74332.1	ECHS1	D13900.1
Farnesyl diphosphate synthase homolog (KIAA1293)	Isoprenoid/mevalonate syn	ND	2.5.1.10	-	-	FDPS	D14697.1
Fatty acid-CoA ligase, long-chain 1	Beta-oxdn of fatty acids	MOM	6.2.1.3	-	-	FACL1	L09229.1
Fatty acid-CoA ligase, long-chain 2	Beta-oxdn of fatty acids	MOM	6.2.1.3	-	-	FACL2	D10040.1
3-Hydroxyacyl-CoA dehydrogenase, short chain (SCHAD)	Beta-oxdn of fatty acids	MAT	1.1.1.35	-	-	HADHSC	AF095703.1
3-Hydroxyacyl-CoA dehydrogenase, type II (SCHAD)	Beta-oxdn of fatty acids	MAT	1.1.1.35	-	-	HADH2	AF037438.1
Hydroxymethylglutaryl-CoA synthase (HMG-CoA synthase)	Mevalonate syn	ND	4.1.3.5	ERG13	X96617.1	HMGCS2	X83618.1
3-Ketoacyl-CoA thiolase (acetyl-CoA acyltransferase 2)	Beta-oxdn of fatty acids	ND	2.3.1.16	-	-	ACAA2	D16294.1
Malonyl CoA:ACP transferase ([ACP] S-malonyltransferase)	Type II fatty acid synthase	MAT?	2.3.1.39	MCT1	Z75129.1	None	AI536031.1 (e,i)
Malonyl-CoA decarboxylase	Regulates fatty acid metab	MAT?	4.1.1.9	-	-	MLYCD	AF090834.1
NADH-cytochrome b5 reductase, IMS form, 32 kDa	Electron transfer to cyt c	IMS	1.6.2.2	MCR1	X81474.1	-	-
NADH-cytochrome b5 reductase, MOM form, 34 kDa	Reduces cytochrome b5	MOM	1.6.2.2	MCR1	X81474.1	-	-
3-Oxoacyl-[ACP] reductase	Type II fatty acid synthase	MAT?	1.-.-.-	OAR1	Z28055.1	HEP27	U31875.1
Trifunctional protein, alpha subunit	Beta-oxdn of fatty acids	MAT?	1.1.1.35	-	-	HADHA	D16480.1
Trifunctional protein, beta subunit	Beta-oxdn of fatty acids	MAT?	2.3.1.16	-	-	HADHB	D16481.1
Phospholipid metabolism							
Cardiolipin synthase	Cardiolipin syn	MIM	2.7.8.-	CRD1	Z74190.1	F23H11.9 (w)	AF003389.1 (w)
CDP-alcohol phosphatidyltransferase, probably mitochondrial	Cardiolipin syn	NC	-	-	-	DJ967N21.6	AL035461.11
CDP-diacylglycerol synthase	Phospholipid syn	MIM/MOM	2.7.7.41	CDS1	Z35898.1	CDS2	Y16521.1
CDP-diacylglycerol synthase (highly expressed in retina)	Phospholipid syn	MIM/MOM	2.7.7.41	-	-	CDS1	U60808.1
Diacylglycerol cholinephosphotransferase	Phospholipid syn	MOM?	2.7.8.2	CPT1	J05203.1	-	-
Glycerol-3-phosphate acyltransferase	Phospholipid syn	ND	2.3.1.15	-	-	Gpam (m)	M77003.1 (m)
Glycerolphosphate acyltransferase	Phospholipid syn	MOM	2.3.1.51	SLC1	L13282.1	AGPAT1	Y09565.1
Glycerolphosphate acyltransferase beta, potentially mitochondrial	Phospholipid syn	ND	2.3.1.51	-	-	AGPAT2	AF000237.1
Phosphatidylglycerolphosphate (PG-P) synthase	Cardiolipin syn	MOM	2.7.8.8	PGS1	AJ012047.1	PGS1 (h)	AB016930.1 (h)
Phosphatidylinositol synthase	Phospholipid syn	MOM	2.7.8.11	PIS1	J02697.1	PIS1	AF014807.1
Phosphatidylserine decarboxylase 1	Phospholipid syn	MIM	4.1.1.65	PSD1	L20973.1	PSSC (h)	M62722.1 (h,i)
Phosphatidylserine synthase 1	Phospholipid syn	MOM	2.7.8.8	CHO1	X05944.1	PSSA	D14694.1
Phosphatidylserine synthase 2	Phospholipid syn	MOM?	2.7.8.-	-	-	PSSB (h)	AB004109.1 (h)
Phospholipase A2, group IB	Phospholipid metab	MIM	3.1.1.4	-	-	PLA2G1B	M21054.1
Phospholipase A2, group IIA	Phospholipid metab	MIM	3.1.1.4	-	-	PLA2G2A	M22430.1
Phospholipase A2, group V	Phospholipid metab	MIM	3.1.1.4	-	-	PLA2G5	U03090.1
NUCLEIC ACID METABOLISM							
Nucleotide and phosphate metabolism							
Adenylate kinase 1 (ATP:AMP phosphotransferase)	Adenine nucleotide metab	IMS	2.7.4.3	ADK1	Y00413.1	ADK2	U84371.1
Adenylate kinase 2 (ATP:AMP phosphotransferase)	Adenine nucleotide metab	MIM/MAT	2.7.4.3	ADK2	X65126.1	ZK673.2 (w)	Z48585.1 (w)
Adenylate kinase 3 (GTP:AMP phosphotransferase)	Adenine nucleotide metab	MAT	2.7.4.10	-	-	AK3	X60673.1
Adenylate kinase 3 (GTP:AMP phosphotransferase) homolog	Function unknown	ND	2.7.4.10	-	-	None	AK001553.1
Creatine kinase, mitochondrial 1 (ubiquitous)	High-energy phosphate transfer	MIM	2.7.3.2	-	-	CKMT1	J04469.1
Creatine kinase, mitochondrial 2 (sarcomeric)	High-energy phosphate transfer	MIM	2.7.3.2	-	-	CKMT2	J05401.1

Deoxyguanosine kinase	Purine syn	MAT	2.7.1.113	-	-	DGUOK	U41668.1
5'(3')-Deoxyribonucleotidase	Thymine/uracil metab	ND	3.1.3.5	-	-	[DNT2]	AF210652.1
Dihydroorotate dehydrogenase	Pyrimidine syn	MIM	1.3.3.1	-	-	DHODH (r)	X80778.1 (r)
Inorganic pyrophosphatase	Phosphate metab	MIM?	3.6.1.1	PPA2	M81880.1	None	BE882442.1 (e,i)
NAD(P) transhydrogenase	Redox of NAD/NADP	MIM	1.6.1.1	-	-	NNT	U40490.1
Nucleoside diphosphate kinase (nm23-H4)	Non-ATP NTP syn	IMS	2.7.4.6	-	-	NME4	Y07604.1
Nucleoside diphosphate kinase (nm23-H6)	Non-ATP NTP syn	IMS?	2.7.4.6	-	-	NM23H6	AF051941.1
Thymidine kinase	dTTP syn	MAT?	2.7.1.21	-	-	TK2	U77088.1
DNA replication							
Cruciform cutting endonuclease	Holliday junction resolvase	MIM	3.1.-.-	CCE1	M65275.1	-	-
D-loop binding protein	Function unknown	MAT?	-	-	-	MTDBP (u)	AJ011076.1 (u)
DNA exonuclease, 3'->5'	DNA editing	MAT	-	REX2	Z73231.1	[SFNalpha]	AL110239.1
DNA helicase	DNA unwinding	MIM	-	HMI1	Z74837	-	-
DNA helicase	DNA unwinding	MAT	-	PIF1	X05342.1	None	AI652391 (e*)
DNA ligase	Ligates free DNA ends	MAT	6.5.1.1	CDC9	X03246.1	LIG3	X84740.1
DNA polymerase gamma, accessory subunit	Increases processivity	MAT?	2.7.7.7	-	-	POLG2	AF177201.1
DNA polymerase gamma, catalytic subunit	Replication of mtDNA	MAT?	2.7.7.7	MIP1	J05117.1	POLG	U60325.1
High mobility group protein	mtDNA-binding protein	MAT?	-	ABF2	M73753.1	CG4217 (f)	AE003731.1 (f)
Single-stranded DNA-binding protein	Coats ss-mtDNA	MAT	-	RIM1	S43128.1	SSBP	M94556.1
Topisomerase II	Decatenates circles	MAT	5.99.1.3	-	-	topA (s)	D82024.1 (s)
DNA plasticity, recombination, and repair							
Apurinic/apyrimidinic endonuclease, potentially mitochondrial	Excises AP sites	NC	-	-	-	XTH2	AJ011311.1
DNA endonuclease (mtDNA-encoded)	Helicase	MAT?	-	SCAI5-alpha	AJ011856.1	-	-
DNA endonuclease I-SceI (21S rRNA) (mtDNA-encoded)	Transposase; intron homing	MAT	-	SECY	M11280.1	-	-
DNA endonuclease I-SceII (aI4) (mtDNA-encoded)	COX1 ai4 intron homing	MAT?	-	SCAI4	V00694.1	-	-
DNA endonuclease I-SceIII (aI3) (mtDNA-encoded)	COX1 ai3 intron homing	MAT?	-	SCAI3	V00694.1	-	-
DNA endonuclease III	Removes thymine glycols (BER)	MAT	-	NTG1	L05146.1	NTH1	AC005600.1
DNA endonuclease G	DNA endo/exonuclease	MAT?	3.1.30.-	NUC1	X06670.1	ENDOG	X79444.1
DNA endonuclease G-like protein 1, probably mitochondrial	DNA endo/exonuclease	NC	3.1.30.-	-	-	ENDOGL1	AB020523.1
DNA repair protein for interstrand crosslinks, potentially mito	Incises crosslinks	NC	-	PSO2	X64004.1	-	-
dUTP pyrophosphatase	Prevents U incorp. in DNA (BER)	MAT?	3.6.1.23	DUT1	X74263.1	DUT	AF018432.1
KU80 autoantigen, C-terminal truncated isoform, 68 kDa	DNA end-binding activity	ND	-	-	-	XRCC5	M30938.1
Maturase/reverse transcriptase aI1 (mtDNA-encoded)	COX1 ai1 intron homing	MAT?	-	SCAI1	L36897.1	-	-
Maturase/reverse transcriptase aI2 (mtDNA-encoded)	COX1 ai2 intron homing	MAT?	-	SCAI2	V00694.1	-	-
Mitochondrial genome maintenance protein	Homologous recombination	MAT?	-	MHR1	AB016430.1	-	-
Mitochondrial genome maintenance protein	mtDNA maintenance	ND	-	MMF1	Z38060.1	PSP	X95384.1
Mitochondrial genome maintenance protein	Function unknown	NC	-	PET18	X59720.1	-	-
Mitochondrial genome maintenance protein (in mito nucleoid)	Repairs oxidative damage	MAT?	-	MGM101	X68482.1	-	-
Mitochondrial inner membrane nuclease (ExoI)	DNA damage-inducible gene	MIM	-	DIN7	X90707.1	F45G2.3 (w)	Z93382.1 (w)
MutM homolog (8 mito isoforms)	8-oxo-G DNA glycosylase (BER)	MIM	-	-	-	OGG1	AB019528.1
MutS homolog	DNA mismatch repair	MAT?	-	MSH1	M84169.1	-	-
MutT homolog, p26 isoform	8-oxo-dGTPase (BER)	MIM	3.1.6.-	-	-	MTH1	AB025240.1
MutY homolog (3 mito isoforms)	A and 2-OH-A glycosylase	MAT	-	-	-	MYH	AB032920.1
Photolyase-like protein (cryptochrome 2)	Cuts cyclobutane dimers	MAT?	4.1.99.3	PHR1	X03183.1	Cry2 (m)	AF156987.1 (m)
Protein associated w. mito. biogenesis/cell cycle reg	mtDNA maintenance?	MAT?	-	ERV2	Z68111.1	-	-
Uracil-DNA glycosylase	Removes U from DNA (BER)	MAT?	3.2.2.-	UNG1	J04470.1	UNG1	X15653.1 (g*)
UV-damaged DNA endonuclease *(S. pombe)* (Sp)	UV-dependent excision repair	ND	-	UVDE (Sp)	D78571.1 (Sp)	-	-
RNA transcription, processing, and maturation							
AU-specific RNA-binding enoyl-CoA hydratase	mt-RNA deg	ND	-	-	-	AUH	X79888.1
COX RNA-associated protein	Maturase for COX1	MAT?	-	MSS1	X69481.1	None	R13025 (e*)
COX1 mRNA splicing protein	Splices COX1 intron aI5-beta	MAT?	-	MSS18	X07650.1	-	-
Cyt b pre-mRNA processing protein	Cyt b mRNA processing	MAT?	-	CBT1	Z28208.1	-	-
Cyt b pre-mRNA processing protein 1	Cyt b mRNA processing	MAT?	-	CBP1	K02647.1	-	-
Cyt b pre-mRNA processing protein 2	Splices COB aI5 intron	MAT?	-	CBP2	K00138.1	-	-
Maturation protein of pre-rRNA	RNA processing	MIM	-	PRP12	S92205	-	-
Mitochondrial protein aI5-beta (mtDNA-encoded)	Function unknown	MAT?	-	SCAI5-beta	AJ011856.1	-	-
mRNA maturase bI2 for Cyt b (mtDNA-encoded)	Splices COB bI2 intron	MAT	-	SCBI2	AJ011856.1	[HFBR2]	AL037712.1 (e)

Appendix 4—*continued*

Gene product	Function/comment	LOC.	E.C. #	Yeast		Animal	
				Symbol	GenBank	Symbol	GenBank
mRNA maturase bI3 for Cyt b (mtDNA-encoded)	Splices COB bI3 intron	MAT	-	SCBI3	AJ011856.1	[HUTE1]	AL047790.1 (e)
mRNA maturase bI4 for Cyt b (mtDNA-encoded)	Splices COB bI4 intron	MAT	-	SCBI4	AJ011856.1	None	AV711052.1 (e)
Nuclear cleavage/polyadenylation factor I, component	pre-mRNA 3' end processing	ND	-	RNA14	M73461.1	-	-
Poly(A) binding protein 2, potentially mitochondrial	Polyadenylates mt-RNAs?	NC	-	-	-	PABP2	AF026029.1
Ribonuclease H (type II), probably mitochondrial	RNA deg	NC	3.1.26.4	-	-	None	AF039652.1
Ribonuclease II	RNA degradosome	MAT?	-	MSU1	U15461.1	-	-
RNA helicase (ATP-dependent)	RNA degradosome	MAT?	-	SUV3	M91167.1	SUV3	AF042169.1
RNA helicase (DEAD box)	Group II intron splicing in yeast	MAT	-	MSS116	Z48784.1	None (r)	U25746.1 (r)
RNA helicase (DEAD box), potentially mitochondrial	Function unnown	NC	-	YGL064c	Z72586.1	-	-
RNA polymerase II	Transcription	MAT?	2.7.7.6	RPO41	M17539.1	POLRMT	U75370.1
RNAse MRP processing protein, 15.8 kDa, potentially mitochondrial	Not clear	NC	-	POP7	Z36036.1	-	-
RNAse MRP processing protein, 22.6 kDa, potentially mitochondrial	Not clear	NC	-	POP3	X95844.1	-	-
RNAse MRP, protein component	Endoribonuclease	MAT	-	SNM1	Z37982.1	-	-
RNAse MRP, RNA component	Endoribonuclease	MAT	-	NME1	Z14231.1	RMRP	X51867.1
RNAse P, protein component	5' processing of mt-tRNAs	MAT?	3.1.26.5	RPM2	L06209.1	-	-
RNAse P, RNA component (mtDNA-encoded 9S RNA in yeast)	5' processing of mt-tRNAs	MAT?	-	RPM1	U46121.1	[H1 RNA]	X15624.1
RNA splicing protein	Splices COB intron bI3	MAT?	-	MRS1	X05509.1	-	-
RNA splicing protein	Splices Group II introns	MAT?	-	MRS2	M82916.1	-	-
RNA splicing protein (MCF protein)	Splicing of mt-RNAs	MIM	-	MRS3	Z49408.1	None	AI133696.1 (e,i)
RNA splicing protein (MCF protein)	Splicing of mt-RNAs	MIM	-	MRS4	Z28277.1	None	AW341177.1 (e)
RNA splicing protein, putative, similar to Mrs3p and Mrs4p	Function unknown	MIM	-	PET8	U02536.1	None	BE787836.1 (e)
rRNA (guanosine-2'-O)-methyltransferase	Ribose methyltransferase	MAT?	2.1.1.-	PET56	L19947.1	Y45F3A.9 (w)	AL032621.1 (w)
Transcription factor 1	Mito transcription initiation	MAT?	-	MTF1	X13513.1	TCF6L1	M62810.1
Transcription factor 2	Mito transcript stability	MAT	-	MTF2	X14719.1	-	-
Transcription termination factor	Terminates at 16S rRNA	MAT	-	-	-	MTERF	Y09615.1
ORGANELLAR MORPHOLOGY AND INHERITANCE							
Actin	Rescues mdm20 mutations	MOM?	-	ACT1	V01288.1	ACTB	M10277.1
Annexin I	Function unknown	ND	-	-	-	ANX1	X05908.1
Annexin V (found in mito 2-D gels)	Function unknown	NC	-	-	-	ANX5	X12454.1
Annexin VI	Function unknown	ND	-	-	-	ANX6	Y00097.1
Dynamin-related GTPase	Inner membrane remodeling	IMS	-	MGM1	X62834.1	[OPA1]	AB011139.1
Dynamin-related protein (called DLP1/ DRP1/ dymple in mammals)	GTPase for mito fission	ND	-	DNM1	L40588.1	[DLP1]	AF061795.1
Fuzzy onions protein	GTPase for mito fusion	MOM	-	FZO1	Z36048.1	FZO1 (f)	U95821.1 (f)
Kinesin heavy chain	Microtubule motor protein	MOM	-	-	-	KIF5B	X65873.1
Kinesin light chain, isoform B	Microtubule motor protein	MOM	-	-	-	KLC (r)	M75148.1
Kinesin-like protein KIF1B	Microtubule motor protein	MOM	-	-	-	Kif1b (m)	AF090190.1 (m)
Latent transforming growth factor-beta 1 (TGF-beta 1)	Growth/differentiation factor	ND	-	-	-	TGFB1 (r)	X52498.1 (r)
Latent TGF-beta binding protein 1 (LTBP-1)	Growth/differentiation factor	ND	-	-	-	Ltbp1 (m)	AF022889.1 (m)
Mitochondrial capsule selenoprotein	Sperm mito structure	MOM	-	-	-	MCSP	X89960.1
Mitochondrial division protein	Binds Dnm1	MOM	-	MDV1	Z49387.1	LIS1	L13385.1
Mitochondrial fission-associated protein	Binds Dnm1	MOM	-	FIS1	Z38060.1	-	-
Myosin VI (unconventional myosin)	Molecular motor protein	NC	-	-	-	MYO6	U90236.2
Outer membrane protein OMP25	Mito clustering	MOM	-	-	-	OMP25 (r)	AF107295.1 (r)
Plectin ("cytolinker" protein)	Binds intermediate filaments	MOM	-	-	-	None	Z54367.1
Protein associated w. mito. inheritance	Function unknown	MOM	-	MDM12	U62252.1		
Protein associated w. mito. inheritance	Function unknown	NC	-	MDM20	U54799.1	-	-
Protein associated w. mito. morphology and inheritance	Couples mitos to actin	MOM	-	MDM10	X80874.1	-	-
Protein associated w. mito. morphology and inheritance	Ubiquitin protein ligase	NC	6.3.2.-	RSP5	U18916.2	[WWP2]	U96114.1
Protein associated w. mito. shape and structure	Couples mitos to actin	MOM	-	MMM1	L32793.1	-	-
Protein associated w. sperm meiotic spindle assembly/cytokinesis	Function unknown	ND	-	-	-	DES1 (f)	X94180.1 (f)
Spermatogenesis associated factor (SPAF), potentially mito	AAA family protease	NC	-	-	-	Spaf (m)	AF049099.1 (m)
Sulfhydryl oxidase, FAD-linked	Loss of MIM and mito structure	MAT	-	ERV1	X60722.1	-	-
Sulfhydryl oxidase homolog, similar to Erv1p	mtDNA maintenance?	MAT?	-	ERV2	Z68111.1	-	-
Translation initiation factor eIF3, 135-kDa subunit	Mito clustering	NC	-	CLU1	AF004911.1	F55H2.6 (w)	Z27080.1 (w)

Tropomyosin 1	Rescues mdm20 mutations	NC	-	TPM1	M25501.1	-	-
Vesicle-associated membrane protein-1B	Fuses transport vesicles	MOM	-	-	-	[VAMP1B]	AF060538.1
PROTEIN SORTING							
Outer membrane components							
Cytochrome c import factor	Importation of yeast Cyc1/Cyc7	ND	-	CYC2	L28428.1	-	-
Metaxin 1	Translocase, binds metaxin 2	MOM	-	MSP1	X68055.1	MTX1	U46920.1
Metaxin 2	Translocase, binds metaxin 1	MOM	-	-	-	MTX2	AF053551.1
Mitochondrial import stimulation factor (MSF) L subunit	Targets precursor to mitos	MOM	-	-	-	YWHAE	U28936.1
Mitochondrial import stimulation factor (MSF) S1 subunit	Targets precursor to mitos	MOM	-	-	-	YWHAZ	U28964.1
Outer membrane import receptor, putative	Not clear	MOM	-	-	-	[hTOM]	AF026031.1
Outer membrane import receptor subunit, 5 kDa	Subunit of GIP	MOM	-	TOM5	U40829.1	-	-
Outer membrane import receptor subunit, 6 kDa	Assembly of GIP complex	MOM	-	TOM6	Z22815.1	-	-
Outer membrane import receptor subunit, 7 kDa	Dissociation of GIP complex	MOM	-	TOM7	Z71346.1	-	-
Outer membrane import receptor subunit, 20 kDa	Translocase receptor	MOM	-	TOM20	X75319.1	[TOM20]	D13641.1
Outer membrane import receptor subunit, 22 kDa	Translocase receptor	MOM	-	TOM22	X82405.1	[TOM22]	AB040119.1
Outer membrane import receptor subunit, 37 kDa	Binds Tom70	MOM	-	TOM37	Z49703.1	TOM34	U58970.1
Outer membrane import receptor subunit, 40 kDa	Translocase channel	MOM	-	TOM40	X56885.1	TOM40	AF043250.1
Outer membrane import receptor subunit, 70 kDa	Translocase receptor	MOM	-	TOM70	X05585.1	-	-
Outer membrane import receptor subunit, 72 kDa	Function unknown	MOM	-	TOM72	U00059.1	None	AA843594 (e*)
Intermembrane space components							
Inner membrane import receptor subunit, 8 kDa	Binds Tim9/Tim10	IMS	-	TIM8	Z49636.1	TIMM8A	U66035.1
Inner membrane import receptor subunit, 8 kDa	Mediates import across IMS	IMS	-	-	-	TIMM8B	AF152350.1
Inner membrane import receptor subunit, 13 kDa	Binds Tim9/Tim10	IMS	-	TIM13	Z72966.1	TIMM13A	AF152351.1
Inner membrane import receptor subunit, 13 kDa	Mediates import across IMS	IMS	-	-	-	TIMM13B	AF144700.1
Inner membrane import receptor subunit, putative	Mediates import across IMS	IMS?	-	-	-	TIM9B	AF150105.1
Subunit of the TIM22 complex	Binds Tim10	IMS	-	TIM9	AF093244.1	TIMM9	AF152353.1
Subunit of the TIM22 complex (also called Mrs11p)	Binds Tim9	IMS	-	TIM10	Z80875.1	TIMM10	AF152354.1
Subunit of the TIM22 complex (also called Mrs5p)	Binds Tim22/Tim54	IMS/MIM	-	TIM12	M90689.1	-	-
Inner membrane components							
Heat shock protein Mdj2	Protein folding chaperone	MIM	-	MDJ2	Z71604.1	-	-
Inner membrane import translocase subunit, 16.5 kDa	Forms channel with Tim23	MIM	-	TIM17	X77796.1	TIM17A	AF106622.1
Inner membrane import translocase subunit, 16.5 kDa	Forms channel with Tim23	MIM	-	-	-	TIM17B	AF034790.1
Inner membrane import translocase subunit, 22 kDa	In Tim54-Tim22 complex	MIM	-	TIM18	Z75205.1	-	-
Inner membrane import translocase subunit, 23 kDa	Connects MOM and MIM	MOM+MIM	-	TIM23	X74161.1	TIM23	AF030162.1
Inner membrane import receptor subunit, 44 kDa	Binds Ssc1/Mge1/HSP70	MIM	-	TIM44	X67276.1	TIM44	AF041254.1
Inner membrane peptidase (IMP), subunit 1	Cuts signal seq in IMS	MIM	3.4.99.-	IMP1	S55518.1	None	AA977197 (e*)
Inner membrane peptidase (IMP), subunit 2	Cuts signal seq in IMS	MIM	3.4.99.-	IMP2	Z49213.1	None	T67154 (e*)
Inner membrane peptidase (IMP), putative subunit 3	Req for Imp1 function	MIM	-	SOM1	X90459.1	-	-
Inner membrane protease	Processing/export from MAT	MIM	-	YIM1	Z47071.1	-	-
Inner membrane translocase	In translocation channel	MIM	-	TIM22	Z74265.1	TIM22	AF155330.1
Inner membrane translocase	In translocation channel	MIM	-	TIM54	Z49329.1	-	-
N-terminal tail export protein	N-tail export protein to IMS	MIM	-	OXA1	X77558.1	OXA1L	AJ001981.1
Pentamidine-resistance protein	Protein export	MIM	-	PNT1	U15217.1	-	-
Matrix components							
Cyclophilin (peptidyl-prolyl cis-trans isomerase)	Cyclosporin-sensitive	MAT	5.2.1.8	CPR3	M84758.1	CYP3	M80254.1
Heat shock protein 10	Associates w Hsp60	MAT	-	HSP10	X75754.1	HSPE1	U07550.1
Heat shock protein 60	Associates w Hsp10	MAT	-	HSP60	M33301.1	HSPD1	M34664.1
Heat shock protein 70 family member	Import motor w Tim44	MAT	-	SSC1	M27229.1	HSPA9B	L15189.1
Heat shock protein 70 family member	Import protein	MAT	-	ECM10	U18530.1	[C37H5.8] (w)	U88315.1 (w)
Heat shock protein Mdj1p	Binds Ssc1	MAT	-	MDJ1	Z28336.1	-	-
Mitochondrial processing peptidase (MPP), alpha subunit	Cleaves MTS	MAT	3.4.24.64	MAS2	X14105.1	MPPA	D50913.1
Mitochondrial processing peptidase (MPP), beta subunit	Cleaves MTS	MAT	3.4.24.64	MAS1	X07649.1	MPPB	AF054182.1
Mitochondrial intermediate peptidase (MIP)	Cleaves MTS	MAT	3.4.24.59	OCT1	U10243.1	MIPEP	U80034.1
Nucleotide exchange/release factor	Binds Ssc1, Ecm10	MAT	-	MGE1	Z75140.1	DROE1 (f)	U34903.1 (f)
Protein associated w. Intramitochondrial sorting	Function unknown	ND	-	YDR185c	Z46727.1	-	-
Protein associated w. Intramitochondrial sorting, similar to YDR185c	Function unknown	ND	-	YLR168c	X70279.1	-	-

Appendix 4—*continued*

Gene product	Function/comment	LOC.	E.C. #	Yeast Symbol	Yeast GenBank	Animal Symbol	Animal GenBank
Protein associated w. Intramitochondrial sorting, similar to YDR185c	Function unknown	ND	-	YLR193c	U14913.1	-	-
PROTEIN TRANSLATION AND STABILITY							
Ribosomal components - large subunit							
Ribosomal protein (no E. coli homolog found)	Component of large subunit	MAT	-	-	-	MRP-L3	Not found
Ribosomal protein (no E. coli homolog found)	Component of large subunit	MAT	-	-	-	MRP-L5	Not found
Ribosomal protein (no E. coli homolog found)	Component of large subunit	MAT	-	-	-	MRP-L22	AA307896.1 (e)
Ribosomal protein (no E. coli homolog found)	Component of large subunit	MAT	-	-	-	MRP-L31	W25509.1 (e)
Ribosomal protein L2 (no E. coli homolog found)	Component of large subunit	MAT	-	-	-	MRP-L2	Not found
Ribosomal protein L10, putatively mitochondrial	Component of large subunit	ND		RPL10	Z73247.1	RPL10	M73791.1
Ribosomal protein L12 (homolog of E. coli L7/L12)	Component of large subunit	MAT	-	YGL068w	Z72591.1	MRP-L12	X79865.1
Ribosomal protein L18, putative (homolog of E. coli L24)	Component of large subunit	MAT	-	-	-	MRP-L18	AL038493 (e)
Ribosomal protein L19, putative (no E. coli homolog found)	Component of large subunit	MAT	-	IMG1	X59720.1	-	-
Ribosomal protein L36 (homolog of E. coli L36/B)	Component of large subunit	MAT	-	YPL183w-a	Z73539.1	-	-
Ribosomal protein L25 (homolog of E. coli L22)	Component of large subunit	MAT	-	-	-	MRP-L25	H87659.1 (e)
Ribosomal protein MRP49 (no E. coli homolog found)	Component of large subunit	MAT	-	MRP49	Z28167.1	-	-
Ribosomal protein MRP-L15/RLX1 (homolog of E. coli L19)	Component of large subunit	MAT	-	-	-	MRP-L15	D14660.1
Ribosomal protein, putative (homolog of E. coli L1)	Component of large subunit	MAT	-	YDR116c	Z48758.1	-	-
Ribosomal protein, putative (homolog of E. coli L34)	Component of large subunit	MAT	-	YDR115w	Z48758.1	-	-
Ribosomal protein RML2 (homolog of E. coli L2)	Component of large subunit	MAT	-	RML2	U18779.1	MRP-L14	AF132956.1
Ribosomal protein YmL2 (homolog of E. coli L27)	Component of large subunit	MAT	-	MRP7	Z71281.1	None	W69555 (e*)
Ribosomal protein YmL3 (no E. coli homolog found)	Component of large subunit	MAT	-	MRPL3	Z49211.1	-	-
Ribosomal protein YmL4 (homolog of E. coli L29)	Component of large subunit	MAT	-	MRPL4	Z30582.1	-	-
Ribosomal protein YmL5/YmL7 (homolog of E. coli L5)	Component of large subunit	MAT	-	MRPL7	Z49701.1	-	-
Ribosomal protein YmL6 (homolog of E. coli L4)	Component of large subunit	MAT	-	YML025c	Z46659.1	-	-
Ribosomal protein YmL8 (homolog of E. coli L17)	Component of large subunit	MAT	-	MRPL8	X53841.1	MRP-L26	AA312160.1 (e)
Ribosomal protein YmL9 (homolog of E. coli L3)	Component of large subunit	MAT	-	MRPL9	X65014.1	MRL3	X06323.1
Ribosomal protein YmL10/YmL18 (homolog of E. coli L15)	Component of large subunit	MAT	-	MRPL10	Z71560.1	MRP-L7	W00599.1 (e)
Ribosomal protein YmL11 (homolog of E. coli L10)	Component of large subunit	MAT	-	MRPL11	Z74250.1	MRP-L8	R56369.1 (e)
Ribosomal protein YmL13 (no E. coli homolog found)	Component of large subunit	MAT	-	MRPL13	X73673.1	-	-
Ribosomal protein YmL14/YmL24 (homolog of E. coli L28)	Component of large subunit	MAT	-	MRPL24	Z47815.1	-	-
Ribosomal protein YmL15 (no E. coli homolog found)	Component of large subunit	MAT	-	MRPL15	U20618.1	-	-
Ribosomal protein YmL16 (homolog of E. coli L6)	Component of large subunit	MAT	-	MRPL6	X69480.1	-	-
Ribosomal protein YmL17/YmL30 (no E. coli homolog found)	Component of large subunit	MAT	-	MRPL17	Z71528.1	None	AI571752.1 (e*)
Ribosomal protein YmL19 (homolog of E. coli L11)	Component of large subunit	MAT	-	MRPL19	Z71461.1	None	BE395676.1 (e,i)
Ribosomal protein YmL20 (no E. coli homolog found)	Component of large subunit	MAT	-	MRPL20	X53840.1	-	-
Ribosomal protein YmL23 (homolog of E. coli L13)	Component of large subunit	MAT	-	MRPL23	Z75058.1	None	AA430750 (e*)
Ribosomal protein YmL25 (no E. coli homolog found)	Component of large subunit	MAT	-	MRPL25	X56106.1	-	-
Ribosomal protein YmL27 (no E. coli homolog found)	Component of large subunit	MAT	-	MRPL27	Z36151.1	MRP-L27	AI056473.1 (e)
Ribosomal protein YmL28 (no E. coli homolog found)	Component of large subunit	MAT	-	MRPL28	M88597.1	-	-
Ribosomal protein YmL31 (no E. coli homolog found)	Component of large subunit	MAT	-	MRPL31	X15099.1	-	-
Ribosomal protein YmL32 (no E. coli homolog found)	Component of large subunit	MAT	-	MRPL32	S48552.1	-	-
Ribosomal protein YmL33 (homolog of E. coli L30)	Component of large subunit	MAT	-	MRPL33	D90217.1	MRP-L28	AA005080.1 (e)
Ribosomal protein YmL34/Yml38 (homolog of E. coli L14)	Component of large subunit	MAT	-	MRPL38	Z28169.1	MRP-L32	C02170.1 (e)
Ribosomal protein YmL35 (no E. coli homolog found)	Component of large subunit	MAT	-	MRPL35	U32517.1	-	-
Ribosomal protein YmL36 (no E. coli homolog found)	Component of large subunit	MAT	-	MRPL36	Z35991.1	-	-
Ribosomal protein YmL37 (no E. coli homolog found)	Component of large subunit	MAT	-	MRPL37	Z36137.1	-	-
Ribosomal protein YmL39 (no E. coli homolog found)	Component of large subunit	MAT	-	MRPL39	Z49810.1	-	-
Ribosomal protein YmL40 (no E. coli homolog found)	Component of large subunit	MAT	-	MRPL40	Z73529.1	-	-
Ribosomal protein YmL41 (homolog of E. coli L23)	Component of large subunit	MAT	-	MRP20	M81696.1	L23MRP	U26596.1
Ribosomal protein YmL44 (no E. coli homolog found)	Component of large subunit	MAT	-	MRPL44	X17552.1	-	-
Ribosomal protein YmL47 (homolog of E. coli L16)	Component of large subunit	MAT	-	MRPL16	Z35799.1	-	-
Ribosomal protein YmL49 (no E. coli homolog found)	Component of large subunit	MAT	-	MRPL49	Z49371.1	-	-
Ribosomal RNA, large subunit (mtDNA-encoded)	Component of large subunit	MAT	-	rRNA21S	AJ011856.1	MTRNR2	J01415.1
Ribosomal components - small subunit							

Ribosomal protein (no E. coli homolog found)	Component of small subunit	MAT	-	MRP8	Z28142.1	-	-
Ribosomal protein (no E. coli homolog found)	Component of small subunit	MAT	-	MRP10	Z74093.1	-	-
Ribosomal protein (no E. coli homolog found)	Component of small subunit	MAT	-	MRP21	Z35851.1	-	-
Ribosomal protein (no E. coli homolog found)	Component of small subunit	MAT	-	MRP51	U43503.1	-	-
Ribosomal protein (no E. coli homolog found)	Component of small subunit	MAT	-	-	-	MRP-S13	AA477685.1 (e)
Ribosomal protein (no E. coli homolog found)	Component of small subunit	MAT	-	-	-	MRP-S28	AF070663.1
Ribosomal protein (homolog of E. coli RPS3) (mtDNA-encoded)	Component of small subunit	MAT	-	VAR1	V00705.1	MRP-S4	Not found
Ribosomal protein MRP1 (no E. coli homolog found)	Component of small subunit	MAT	-	MRP1	M15160.1	MRP-S5	Not found
Ribosomal protein MRP13 (no E. coli homolog found)	Component of small subunit	MAT	-	MRP13	Z72869.1	-	-
Ribosomal protein MRP-S9 (homolog of E. coli S9)	Component of small subunit	MAT	-	MRPS9	Z36015.1	-	-
Ribosomal protein MRPS28 (homolog of E. coli S15)	Component of small subunit	MAT	-	MRPS28	X55977.1	-	-
Ribosomal protein NAM9 (homolog of E. coli S4)	Component of small subunit	MAT	-	NAM9	M60730.1	-	-
Ribosomal protein PET123 (no E. coli homolog found)	Component of small subunit	MAT	-	PET123	X52362.1	-	-
Ribosomal protein S2 (homolog of E. coli S2)	Component of small subunit	MAT	-	MRP4	M82841.1		
Ribosomal protein S5 (homolog of E. coli S5)	Component of small subunit	MAT	-	MRPS5	Z36120.1	E02A10.1 (w)	Z78063.1 (w)
Ribosomal protein S7 (homolog of E. coli S7)	Component of small subunit	MAT	-	RSM7	Z49613.1	MRP-S7	AF077042.1
Ribosomal protein S10, putative (homolog of E. coli S10)	Component of small subunit	MAT	-	RSM10	Z54075.1	-	-
Ribosomal protein S10 (no E. coli homolog found)	Component of small subunit	MAT?	-	-	-	[MRP-S10]	AW960852.1 (e)
Ribosomal protein S12 (homolog of E. coli S12)	Component of small subunit	MAT	-	YNR036c	Z71651.1	RPMS12	Y11681.1
Ribosomal protein S12 (no E. coli homolog found)	Component of small subunit	MAT	-	-	-	MRP-S12	U80813.1 (e)
Ribosomal protein S13 (homolog of E. coli S13)	Component of small subunit	MAT	-	YNL081c	Z71357.1	-	-
Ribosomal protein S14 (homolog of E. coli S14)	Component of small subunit	MAT	-	MRP2	M15161.1	[MRP-S14]	AL049705.1
Ribosomal protein S17-like, putatively mitochondrial	Component of small subunit?	NC	-	YMR188c	Z49808.1		
Ribosomal protein S18 (no E. coli homolog found)	Component of small subunit	MAT	-	RSM18	U18796.1	-	-
Ribosomal protein S19 (homolog of E. coli S19)	Component of small subunit	MAT	-	RSM19	Z71652.1	-	-
Ribosomal protein S22 (no E. coli homolog found)	Component of small subunit	MAT?	-	-	-	[MRP-S22]	AA773362.1 (e)
Ribosomal protein S23 (no E. coli homolog) (CGI-138)	Component of small subunit	MAT?	-	-	-	[MRP-S23]	AF151896.1
Ribosomal protein S24 (homolog of E. coli S16)	Component of small subunit	MAT	-	YPL013c	U33335.1	[[CGI-132]	AF151890.1
Ribosomal protein S24 (no E. coli homolog found)	Component of small subunit	MAT?	-	-	-	[MRP-S24]	AF161453.1 (i)
Ribosomal protein S25 (no E. coli homolog found)	Component of small subunit	MAT	-	-	-	[MRP-S25]	BE279536.1 (e)
Ribosomal protein S50 (no E. coli homolog found)	Component of small subunit	MAT	-	RSM50	Z28155.1	-	-
Ribosomal protein S51 (no E. coli homolog found)	Component of small subunit	MAT?	-	RSM51	D10263.1	-	-
Ribosomal protein S52 (no E. coli homolog found)	Component of small subunit	MAT	-	RSM52	Z46727.1	-	-
Ribosomal protein S53 (no E. coli homolog found)	Component of small subunit	MAT?	-	RSM53	Z46728.1	-	-
Ribosomal protein S54 (no E. coli homolog found)	Component of small subunit	MAT?	-	RSM54	Z49601.1	-	-
Ribosomal protein S55 (no E. coli homolog found)	Component of small subunit	MAT	-	RSM55	Z73000.1	-	-
Ribosomal protein YMR-31 (no E. coli homolog found)	Component of small subunit	MAT	-	YMR31	X17540.1	-	-
Ribosomal protein YmS2 (no E. coli homolog found)	Component of small subunit	MAT	-	MRPS2	U10556.1	None	AA234494 (e*)
Ribosomal protein YmS16 (no E. coli homolog found)	Component of small subunit	MAT	-	MRP17	X58362.1	-	-
Ribosomal protein YmS18 (homolog of E. coli S11)	Component of small subunit	MAT	-	YNL306w	Z71582.1	W04D2.5 (w)	Z75552.1 (w)
Ribosomal RNA, small subunit (mtDNA-encoded)	Component of small subunit	MAT	-	rRNA15S	AJ011856.1	MTRNR1	J01415.1
Ribosomal RNA, 5S	Function unknown	MAT?	-	-	-	[5SRNA]	X71799.1
Aminoacyl-tRNA synthetases							
Alanyl-tRNA synthetase	Charging of tRNA-Ala	MAT?	6.1.1.7	-	-	ALAS (f)	AF188716.1 (f)
Arginyl-tRNA synthetase	Charging of tRNA-Arg	MAT	6.1.1.19	MSR1	L39019.1	None	AI697717.1 (e*)
Aspariginyl-tRNA synthetase	Charging of tRNA-Asn	MAT	6.1.1.22	YCR024c	X59720.1	-	-
Aspartyl-tRNA synthetase	Charging of tRNA-Asp	MAT	6.1.1.12	MSD1	M26020.1	-	-
Glutamyl-tRNA synthetase	Charging of tRNA-Glu	MAT	6.1.1.17	MSE1	L39015.1	T07A9.2 (w)	AF036706.1 (w)
Glycyl-tRNA synthetase	Charging of tRNA-Gly	MAT?	6.1.1.14	GRS1	Z35990.1	None	D30658.1
Histidyl-tRNA synthetase	Charging of tRNA-His	MAT	6.1.1.21	HTS1	M14048.1	HARSR	U18937.1
Isoleucyl-tRNA synthetase	Charging of tRNA-Ile	MAT	6.1.1.5	ISM1	L38957.1	Z81038.1 (w)	C25A1.7A (w)
Leucyl tRNA synthetase	Charging of tRNA-Leu	MAT	6.1.1.4	NAM2	J03495.1	KIAA0028	D21851.1
Lysyl-tRNA synthetase	Charging of tRNA-Lys	MAT	6.1.1.6	MSK1	X57360.1	None	AF285758.1
Methionyl-tRNA formyltransferase	Formylation of tRNA-Met	MAT	2.1.2.9	FMT1	Z35774.1	[MTFMT] (b)	AB004316.1 (b,i)
Methionyl-tRNA synthetase	Charging of tRNA-Met	MAT	6.1.1.10	MSM1	X14629.1	None	AA280602.1 (e,i)
Phenylalanyl-tRNA synthetase alpha chain	Charging of tRNA-Phe	MAT	6.1.1.20	MSF1	J02691.1	FARS1	AF097441.1
Seryl-tRNA synthetase, potentially mitochondrial	Charging of tRNA-Ser	NC	6.1.1.11	DIA4	U10400.1	SERRSMT	AB029948.1

Appendix 4—*continued*

Gene product	Function/comment	LOC.	E.C. #	Yeast		Animal	
				Symbol	GenBank	Symbol	GenBank
Threonyl-tRNA synthetase	Charging of tRNA-Thr	MAT	6.1.1.3	MST1	M12087.1	C47D12.6 (w)	Z69902.1 (w)
Tryptophanyl-tRNA synthetase	Charging of tRNA-Trp	MAT	6.1.1.2	MSW1	M12081.1	WARS2	AJ242739.1
Tyrosyl-tRNA synthetase	Charging of tRNA-Tyr	MAT	6.1.1.1	MSY1	L42333.1	K08F11.4 (w)	U70855.1 (w)
Valyl-tRNA synthetase	Charging of tRNA-Val	MAT	6.1.1.9	VAS1	J02719.1	None	BE314033.1 (e,i)
Transfer RNAs							
ATP(CTP):tRNA nucleotidyltransferase	Adds CCA to 3' end of tRNAs	MAT	2.7.7.25	CCA1	M59870.1	-	-
Dimethylguanine tRNA methyltransferase	Methylates G34 in tRNAs	MIM	2.1.1.32	TRM1	M17193.1	None	AC005546.1 (e*)
Glutamyl-tRNA(Gln) amidotransferase, subunit A	Charging Gln-tRNA(Gln)	MIM?	6.3.5.-	YMR293c	X80836.1	-	-
Glutamyl-tRNA(Gln) amidotransferase, subunit B	Charging Gln-tRNA(Gln)	MIM	6.3.5.-	PET112	L22072.1	PET112	AF026851.1
Pseudouridine synthase	Pseudouridine-38,39 in tRNAs	ND	4.2.1.70	DEG1	D44600.1	E02H1.3 (w)	Z47075.1 (w)
Pseudouridine synthase	Pseudouridine-55 in tRNAs	ND	4.2.1.70	PUS4	Z71568.1	None	AW020963 (e*)
Pseudouridine synthase (PUS5)	Pseudouridine-2819 in 21S rRNA	ND	-	YLR165c	U51921.1	-	-
tRNA isopentenyltransferase	Isopentenylation of tRNAs	MAT	2.5.1.8	MOD5	M15991.1	None	BE315223.1 (e,i)
tRNA-Ala (mtDNA-encoded)	Transfer RNA - Ala	MAT	-	tA(TGC)Q	AJ011856.1	MTTA	J01415.1
tRNA-Arg1 (mtDNA-encoded) (codon AGR in yeast)	Transfer RNA - Arg (TCT)	MAT	-	tR(TCT)Q1	AJ011856.1	MTTR	J01415.1
tRNA-Arg2 (mtDNA-encoded) (codon CGN in yeast)	Transfer RNA - Arg (ACG)	MAT	-	tR(ACG)Q2	AJ011856.1	-	-
tRNA-Asn (mtDNA-encoded)	Transfer RNA - Asn	MAT	-	tN(GTT)Q	AJ011856.1	MTTN	J01415.1
tRNA-Asp (mtDNA-encoded)	Transfer RNA - Asp	MAT	-	tD(GTC)Q	AJ011856.1	MTTD	J01415.1
tRNA-Cys (mtDNA-encoded)	Transfer RNA - Cys	MAT	-	tC(GCA)Q	AJ011856.1	MTTC	J01415.1
tRNA-fMet (mtDNA-encoded)	Transfer RNA - fMet	MAT	-	tM(CAT)Q2	AJ011856.1	-	-
tRNA-Gln (mtDNA-encoded)	Transfer RNA - Gln	MAT	-	tQ(TTG)Q	AJ011856.1	MTTQ	J01415.1
tRNA-Glu (mtDNA-encoded)	Transfer RNA - Glu	MAT	-	tE(TTC)Q	AJ011856.1	MTTE	J01415.1
tRNA-Gly (mtDNA-encoded)	Transfer RNA - Gly	MAT	-	tG(TCC)Q	AJ011856.1	MTTG	J01415.1
tRNA-His (mtDNA-encoded)	Transfer RNA - His	MAT	-	tH(GTG)Q	AJ011856.1	MTTH	J01415.1
tRNA-Ile (mtDNA-encoded)	Transfer RNA - Ile	MAT	-	tI(GAT)Q	AJ011856.1	MTTI	J01415.1
tRNA-Leu (mtDNA-encoded) (codon UUR in human)	Transfer RNA - Leu	MAT		tL(TAA)Q	AJ011856.1	MTTL1	J01415.1
tRNA-Leu2 (mtDNA-encoded) (codon CUN in human)	Transfer RNA - Leu (CUN)	MAT	-	-	-	MTTL2	J01415.1
tRNA-Lys (mtDNA-encoded)	Transfer RNA - Lys	MAT	-	tK(TTT)Q	AJ011856.1	MTTK	J01415.1
tRNA-Lys (CUU)	Transfer RNA - Lys	MAT	-	tRNA-K1	K00286.1	-	-
tRNA-Met (mtDNA-encoded)	Transfer RNA - Met	MAT	-	tM(CAT)Q1	AJ011856.1	MTTM	J01415.1
tRNA-Phe (mtDNA-encoded)	Transfer RNA - Phe	MAT	-	tF(GAA)Q	AJ011856.1	MTTF	J01415.1
tRNA-Pro (mtDNA-encoded)	Transfer RNA - Pro	MAT	-	tP(TGG)Q	AJ011856.1	MTTP	J01415.1
tRNA-Ser1 (mtDNA-encoded) (codon UCN in human)	Transfer RNA - Ser (TGA)	MAT	-	tS(TGA)Q2	AJ011856.1	MTTS1	J01415.1
tRNA-Ser2 (mtDNA-encoded) (codon AGY in human)	Transfer RNA - Ser (GCT)	MAT	-	tS(GCT)Q1	AJ011856.1	MTTS2	J01415.1
tRNA-Thr1 (mtDNA-encoded)	Transfer RNA - Thr (XXX)	MAT	-	tT(xxx)Q2	AJ011856.1	MTTT	J01415.1
tRNA-Thr2 (mtDNA-encoded) (codon ACN in yeast)	Transfer RNA - Thr (TGT)	MAT	-	tT(TGT)Q1	AJ011856.1	-	-
tRNA-Trp (mtDNA-encoded)	Transfer RNA - Trp	MAT	-	tW(TCA)Q	AJ011856.1	MTTW	J01415.1
tRNA-Tyr (mtDNA-encoded)	Transfer RNA - Tyr	MAT	-	tY(GTA)Q	AJ011856.1	MTTY	J01415.1
tRNA-Val (mtDNA-encoded)	Transfer RNA - Val	MAT	-	tV(TAC)Q	AJ011856.1	MTTV	J01415.1
Translation factors							
ATPase 9 RNA-associated protein, 59 kDa	Req. for ATP9 translation	ND	-	AEP1	M80615.1	-	-
ATPase 9 RNA-associated protein, 67.5 kDa	Accum of ATP9 mRNA	ND	-	AEP2	M59860.1	-	-
COX1 RNA-associated protein	Stability/transl of COX1 mRNA	MIM	-	PET309	L06072.1	-	-
GTPase involved in expression of COX1	Proofreading in translation	MAT?	-	MSS1	X69481.1	F39B2.7 (w)	Z92834.1 (w)
Polypeptide chain release factor 1	Terminates at UAA/UAG	MAT	-	MRF1	X60381.1	[MRF1]	AF072934.1
Protein involved in expression of COX2	Exports Cox2 to MIM	MAT?	-	MSS2	X81477.1	-	-
Protein required for protein synthesis	Function unknown	MAT?	-	PET130	U37712.1	-	-
Regulator of protein synthesis	Interacts w. transl apparatus	ND	-	MBR1	M63309.1	-	-
Regulator of protein synthesis	Interacts with Mbr1	ND	-	ISF1	X72671.1	-	-
Ribosome recycling factor, putative	Function unclear	MAT	-	FIL1	AB016033.1	-	-
Translational accuracy factor	Proofreading in translation	MAT?	-	MTO1	Z72758.1	F52H3.2 (w)	Z66512.1 (w)
Translational activator of COB mRNA	Cyt b mRNA processing	MAT	-	CBP6	M10154.1	-	-
Translational activator of COB mRNA	23-kDa protein	MIM	-	CBS1	X15650.1	-	-
Translational activator of COB mRNA	45-kDa protein	MIM	-	CBS2	X13523.1	-	-
Translational activator of COX1 mRNA	Translation of COX1 mRNA	MIM	-	MSS51	J01487.1	-	-

Translational activator of COX2 mRNA	Translation of COX2 mRNA	MIM	-	PET111	M17143.1	-	-
Translational activator of COX3 mRNA	Translation of COX3 mRNA	MIM	-	PET54	X13427.1	-	-
Translational activator of COX3 mRNA	Translation of COX3 mRNA	MIM	-	PET122	X07558.1	-	-
Translational activator of COX3 mRNA	Translation of COX3 mRNA	MIM	-	PET127	X87331.1	-	-
Translational activator of COX3 mRNA	Translation of COX3 mRNA	MIM	-	PET494	K03520.1	-	-
Translational elongation factor G	Translocation from A to P site	MAT	-	MEF1	X58378.1	None	L14684.1 (r)
Translational elongation factor G2	Translocation from A to P site	MAT	-	MEF2	L25088.1	None	BE891390.1 (e)
Translational elongation factor Ts	GDP release from EF-Tu	MAT	-	-	-	TSFM	AF110399.1
Translational elongation factor Tu	Delivers tRNAs to A site	MAT	-	TUF1	Z75095.1	TUFM	L38995.1
Translational elongation factor, potentially mitochondrial	LepA family GTP binding protein	NC	-	GUF1	U22360.1	ZK1236.1 (w)	L13200.1 (w)
Translational initiation factor 2	Forms initiation complex	MAT	-	IFM1	X58379.1	MTIF2	L34600.1
Translational release factor 1	Releases polypeptides	MAT?	-	-	-	MTRF1	AF072934.1
Protein degradation and stability							
Clp protease, ATP-binding subunit	Chaperone/protease subunit	MAT	-	MCX1	Z36096.1	CLPX	AJ006267.1
Clp protease, proteolytic subunit	General cleavage protease	MAT?	3.4.21.92	-	-	CLPP	Z50853.1
LON protease	ATP-dependent protease	MAT	3.4.21.-	PIM1	X74544.1	[LON]	U02389.1
Metalloprotease of the AAA family	MIM protein deg, with Yta12	MIM	3.4.24.-	AFG3	X76643.1	AFG3L1	AW248384.1 (e,i)
Metalloprotease of the AAA family	MIM protein deg, with Afg3	MIM	3.4.24.-	YTA12	U09358.1	AFG3L2	Y18314.1
Metalloprotease of the AAA family (paraplegin)	Unassembled MIM protein deg	MIM	3.4.24.-	YME1	L14616.1	SPG7	Y16610.1
Metalloprotease of the AAA family (paraplegin-like)	MIM protein deg?	ND	-	-	-	YME1L1	AJ132637.1
O-sialoglycoprotein endopeptidase, potentially mitochondrial	Hydrolyzes O-sialoglycoproteins	NC	3.4.24.57	QRI7	Z74152.1	C01G10.10 (w)	Z81030.1 (w)
Prohibitin complex, subunit 1	Inhibits m-AAA protease	MAT?	-	PHB1	U16737.1	PHB	S85655.1 (g*)
Prohibitin complex, subunit 2	Inhibits m-AAA protease	MAT?	-	PHB2	Z73016.1	BAP37 (m)	X78683.1 (m)
Proteasome (26S) regulatory subunit, possibly mitochondrial	Protein deg	NC	-	MPR1	X79561.1	POH1	U86782.1
Protein disulfide isomerase (found in mito 2-D gels)	Rearranges S-S bonds	NC	5.3.4.1	-	-	ERP60	D83485.1
Protein disulfide isomerase-related protein P5 (in mito 2-D gels)	Rearranges S-S bonds	NC	5.3.4.1	-	-	None	D49489.1
Protein disulfide isomerase/prolyl hydroxylase (in mito 2-D gels)	Rearranges S-S bonds	NC	5.3.4.1	-	-	PDI	M22806.1
Sulfite oxidase	Degrades S-containing aa's	IMS	1.8.3.1	-	-	SUOX	L31573.1
Thimet oligopeptidase (yeast saccharolysin; rat neurolysin)	Metalloendopeptidase	IMS	3.4.24.37	PRD1	X76504.1	None (r)	X87157.1 (r)
X-prolyl aminopeptidase P homolog	Releases Pro-linked aa's	ND	3.4.11.9	YLL029w	Z73134.1	XPNPEPL	X95762.1
RESPIRATORY CHAIN/OXIDATIVE PHOSPHORYLATION							
Complex I (NADH dehydrogenase, ubiquinone [NDU])							
NADH dehydrogenase, external, rotenone-insensitive	Oxidizes cyto NADH	MIM	-	NDE1	Z47071.1	-	-
NADH dehydrogenase, internal, rotenone-insensitive	Oxidizes cyto NADH in MAT	MIM	1.6.5.3	NDI1	X61590.1	-	-
NADH dehydrogenase, rotenone-insensitive	Oxidizes cyto NADH in IMS	MIM	-	NDE2	Z74133.1	-	-
NDU subunit 1 (mtDNA-encoded)	Structural subunit	MIM	1.6.5.3	-	-	MTND1	J01415.1
NDU subunit 2 (mtDNA-encoded)	Structural subunit	MIM	1.6.5.3	-	-	MTND2	J01415.1
NDU subunit 3 (mtDNA-encoded)	Structural subunit	MIM	1.6.5.3	-	-	MTND3	J01415.1
NDU subunit 4 (mtDNA-encoded)	Structural subunit	MIM	1.6.5.3	-	-	MTND4	J01415.1
NDU subunit 4L (mtDNA-encoded)	Structural subunit	MIM	1.6.5.3	-	-	MTND4L	J01415.1
NDU subunit 5 (mtDNA-encoded)	Structural subunit	MIM	1.6.5.3	-	-	MTND5	J01415.1
NDU subunit 6 (mtDNA-encoded)	Structural subunit	MIM	1.6.5.3	-	-	MTND6	J01415.1
NDU subunit NDUFA1, 7.5 kDa	Structural subunit	MIM	1.6.5.3	-	-	NDUFA1	U54993.1
NDU subunit NDUFA2, 8 kDa	Structural subunit	MIM	1.6.5.3	-	-	NDUFA2	AF047185.1
NDU subunit NDUFA3, 9 kDa	Structural subunit	MIM	1.6.5.3	-	-	NDUFA3	AF044955.1
NDU subunit NDUFA4, 9 kDa	Structural subunit	MIM	1.6.5.3	-	-	NDUFA4	U94586.1
NDU subunit NDUFA5, 13 kDa	Structural subunit	MIM	1.6.5.3	-	-	NDUFA5	U53468.1
NDU subunit NDUFA6, 14 kDa	Structural subunit	MIM	1.6.5.3	-	-	NDUFA6	AF047182.1
NDU subunit NDUFA7, 14.5 kDa	Structural subunit	MIM	1.6.5.3	-	-	NDUFA7	AF054178.1
NDU subunit NDUFA8, 19 kDa	Structural subunit	MIM	1.6.5.3	-	-	NDUFA8 (b)	X59697.1 (b)
NDU subunit NDUFA9, 39 kDa	Structural subunit	MAT	1.6.5.3	-	-	NDUFA9	L04490.1
NDU subunit NDUFA10,42 kDa	Structural subunit	MAT	1.6.5.3	-	-	NDUFA10	AF087661.1
NDU subunit NDUFAB1, 8 kDa (acyl carrier protein [ACP])	Structural subunit	MIM	1.6.5.3	-	-	NDUFAB1	AF087660.1
NDU subunit NDUFB1, 7 kDa	Structural subunit	MIM	1.6.5.3	-	-	NDUFB1	AF054181.1
NDU subunit NDUFB2, 8 kDa	Structural subunit	MIM	1.6.5.3	-	-	NDUFB2	AF050639.1
NDU subunit NDUFB3, 12 kDa	Structural subunit	MIM	1.6.5.3	-	-	NDUFB3	AF035839.1

Appendix 4—*continued*

Gene product	Function/comment	LOC.	E.C. #	Yeast		Animal	
				Symbol	GenBank	Symbol	GenBank
NDU subunit NDUFB4, 15 kDa	Structural subunit	MIM	1.6.5.3	-	-	NDUFB4	AF044957.1
NDU subunit NDUFB5, 16 kDa	Structural subunit	MIM	1.6.5.3	-	-	NDUFB5	AF047181.1
NDU subunit NDUFB6, 17 kDa	Structural subunit	MIM	1.6.5.3	-	-	NDUFB6	AF035840.1
NDU subunit NDUFB7 (B18)	Structural subunit	MIM	1.6.5.3	-	-	NDUFB7	M33374.1
NDU subunit NDUFB8, 19 kDa	Structural subunit	MIM	1.6.5.3	-	-	NDUFB8	AF044958.1
NDU subunit NDUFB9, 22 kDa	Structural subunit	MIM	1.6.5.3	-	-	NDUFB9	AF044956.1
NDU subunit NDUFB10, 22 kDa	Structural subunit	MIM	1.6.5.3	-	-	NDUFB10	AF044954.1
NDU subunit NDUFC1, 6 kDa	Structural subunit	MIM	1.6.5.3	-	-	NDUFC1	AF047184.1
NDU subunit NDUFC2, 14.5 kDa	Structural subunit	MIM	1.6.5.3	-	-	NDUFC2	AF087659.1
NDU subunit NDUFS1, 75 kDa	Structural subunit	MIM	1.6.5.3	-	-	NDUFS1	X61100.1
NDU subunit NDUFS2, 49 kDa	Structural subunit	MIM	1.6.5.3	-	-	NDUFS2	AF013160.1
NDU subunit NDUFS3, 30 kDa	Structural subunit	MIM	1.6.5.3	-	-	NDUFS3	AF067139.1
NDU subunit NDUFS4, 18 kDa	Structural subunit	MIM	1.6.5.3	-	-	NDUFS4	AF020351.1
NDU subunit NDUFS5, 15 kDa	Structural subunit	MIM	1.6.5.3	-	-	NDUFS5	AF047434.1
NDU subunit NDUFS6, 13 kDa	Structural subunit	MIM	1.6.5.3	-	-	NDUFS6	AF044959.1
NDU subunit NDUFS7, 20 kDa	Structural subunit	MIM	1.6.5.3	-	-	NDUFS7 (b)	X65020.1 (b)
NDU subunit NDUFS8, 23 kDa	Structural subunit	MIM	1.6.5.3	-	-	NDUFS8	U65579.1
NDU subunit NDUFV1, 51 kDa	Structural subunit	MIM	1.6.5.3	-	-	NDUFV1	AF053070.1
NDU subunit NDUFV2, 24 kDa	Structural subunit	MIM	1.6.5.3	-	-	NDUFV2	M22538.1
NDU subunit NDUFV3, 9 kDa	Structural subunit	MIM	1.6.5.3	-	-	NDUFV3	X99726.1
Complex II (succinate dehydrogenase [SDH])							
SDH assembly protein (similar to Hsp60p)	SDH-specific chaperone	MIM	-	TCM62	Z35913.1	-	-
SDH cytochrome b large subunit (CII-3)	Anchor w Sdh4; binds Ub	MIM	1.3.5.1	SDH3	X73884.1	SDHC	D49737.1
SDH cytochrome b small subunit (CII-4)	Anchor w Sdh3; reduces Ub	MIM	1.3.5.1	SDH4	L26333.1	SDHD	AB006202.1
SDH flavoprotein (Fp) subunit	Catalytic subunit w Sdh2	MIM	1.3.5.1	SDH1	M86746.1	SDHA	D30648.1
SDH flavoprotein (Fp) subunit homolog (SDH1b)	Function unknown	MIM	1.3.5.1	YJL045w	Z49320.1	C34B2.7 (w)	AF043693.1 (w)
SDH Iron-sulfur protein (Ip) subunit	Catalytic subunit w Sdh1	MIM	1.3.5.1	SDH2	Z73146.1	SDHB	U17248.1
Complex III (ubiquinone cytochrome c reductase [UCR])							
UCR assembly protein	Assembles Rieske ISP	MIM	-	BCS1	S47190.1	BCS1	AF026849.1
UCR assembly protein, 20 kDa	Function unknown	MIM	-	CBP4	U10700.1	-	-
UCR assembly protein, 34.5 kDa	Function unknown	MIM	-	CBP3	J04830.1	-	-
UCR assembly protein, 51 kDa	Function unknown	MIM	-	ABC1	X59027.1	CG7616 (f)	AE003545.1 (f)
UCR assembly-like protein similar to Abc1, potentially mitochondrial	Function unknown	NC	-	YLR253w	U20865.1	CG3608 (f)	AE003464.1 (f)
UCR subunit 1, 47 kDa in human	Core protein 1	MIM	1.10.2.2	COR1	J02636.1	UQCRC1	L16842.1
UCR subunit 2, 45 kDa in human	Core protein 2	MIM	1.10.2.2	QCR2	X05120.1	UQCRC2	J04973.1
UCR subunit 3, 35 kDa in human (mtDNA-encoded)	Cytochrome b	MIM	1.10.2.2	CYTB	AJ011856.1	MTCYB	J01415.1
UCR subunit 4, 28 kDa in human	Cytochrome c1	MIM	1.10.2.2	CYT1	X00791.1	CYC1	J04444.1
UCR subunit 5, 22 kDa in human (196-aa mature polypeptide)	Rieske iron-sulfur protein	MIM	1.10.2.2	RIP1	M23316.1	UQCRFS1	L32977.1
UCR subunit 6, 13.4 kDa in human (also called subunit 8)	Cyt b-associated protein	MIM	1.10.2.2	QCR7	X00256.1	UQCRB	M22348.1
UCR subunit 7, 9.5 kDa in human	Core-associated protein	MIM	1.10.2.2	QCR8	X05550.1	None	D50369.1
UCR subunit 8, 9.2 kDa in human (also called subunit 6)	Hinge protein	MIM	1.10.2.2	QCR6	X00551.1	UQCRH	M36647.1
UCR subunit 9, 8.0 kDa in human (78-aa MTS from subunit 5)	Reiske ISP targeting peptide	MIM	1.10.2.2	-	-	UQCRFS1	L32977.1
UCR subunit 10, 7.2 kDa in human	Cyt c1-associated protein	MIM	1.10.2.2	QCR9	M59797.1	[UCRX]	AL036801.3 (e)
UCR subunit 11, 6.4 kDa in human	Reiske ISP-associated protein	MIM	1.10.2.2	QCR10	U07275.1	[UQCR]	D55636.1
Complex IV (cytochrome c oxidase [COX])							
COX assembly protein	Heme A farnesyltransferase	MIM	2.5.1.29	COX10	M55566.1	COX10	U09466.1
COX assembly protein	Copper transport	MIM	-	COX11	X55731.1	COX11	AF044321.1
COX assembly protein	Function unknown	MIM	-	COX14	U15040.1	-	-
COX assembly protein	Heme O metab	MIM	-	COX15	L38643.1	COX15	AF044323.1
COX assembly protein	Copper transport	IMS	-	COX17	L75948.1	COX17	L77701.1
COX assembly protein	Targets Cox2 to MIM	MIM?	-	COX18	U59742.1	-	-
COX assembly protein	COX II maturation/assembly	MIM	-	COX20	Z48612.1	-	-
COX assembly protein	Function unknown	MIM?	-	PET117	L06066.1	None	AL050321.8 (e)
COX assembly protein	Function unknown	ND	-	PET100	U91943.1	-	-
COX assembly protein	Function unknown	MIM?	-	PET191	L06067.1	-	-

COX assembly protein	Copper transport	MIM	-	SCO1	X17441.1	SCO1	AF026852.1
COX assembly protein	Copper transport	MIM	-	SCO2	Z35893.1	SCO2	AF177385.1
COX assembly protein	Function unknown	MIM	-	SHY1	Z72897.1	SURF1	Z35093.1
COX subunit I (mtDNA-encoded)	Structural subunit	MIM	1.9.3.1	COX1	AJ011856.1	MTCO1	J01415.1
COX subunit II (mtDNA-encoded)	Structural subunit	MIM	1.9.3.1	COX2	AJ011856.1	MTCO2	J01415.1
COX subunit III (mtDNA-encoded)	Structural subunit	MIM	1.9.3.1	COX3	AJ011856.1	MTCO3	J01415.1
COX subunit IV in yeast (subunit Vb in human)	Structural subunit	MIM	1.9.3.1	COX4	X01418.1	COX5B	M19961.1
COX subunit Va in yeast (subunit IV in human)	Structural subunit	MIM	1.9.3.1	COX5A	X02561.1	COX4	M21575.1
COX subunit Vb in yeast (subunit IV in human)	Structural subunit	MIM	1.9.3.1	COX5B	M17799.1	COX4	M21575.1
COX subunit VI in yeast (subunit Va in human)	Structural subunit	MIM	1.9.3.1	COX6	M10138.1	COX5A	M22760.1
COX subunit VIa in yeast (subunit VIa-L in human)	Structural subunit	MIM	1.9.3.1	COX13	X72970.1	COX6A1	X15341.1
COX subunit VIa-H in human	Structural subunit	MIM	1.9.3.1	-	-	COX6A2	U66875.1
COX subunit VIb in yeast (subunit VIb in human)	Structural subunit	MIM	1.9.3.1	COX12	M98332.1	COX6B	X13923.1
COX subunit VII in yeast (subunit VIIa-H in human)	Structural subunit	MIM	1.9.3.1	COX7	X51506.1	COX7A1	M83186.1
COX subunit VIIa in yeast (subunit VIc in human)	Structural subunit	MIM	1.9.3.1	COX9	J02633.1	COX6C	X13238.1
COX subunit VIIa-L in human	Structural subunit	MIM	1.9.3.1	-	-	COX7A2	X15822.1
COX subunit VIIa-related protein in human	Function unknown	NC	-	-	-	COX7RP	AB007618.1
COX subunit VIIb in human	Structural subunit	MIM	1.9.3.1	-	-	COX7B	Z14244.1
COX subunit VIII in yeast (subunit VIIc in human)	Structural subunit	MIM	1.9.3.1	COX8	J02634.1	COX7C	X16560.1
COX subunit VIIIa-L in mouse	Structural subunit	MIM	1.9.3.1	-	-	Cox8a (m)	U37721.1 (m)
COX subunit VIIIa-H in mouse	Structural subunit	MIM	1.9.3.1	-	-	Cox8b (m)	U15541.1 (m)
Respiratory chain assembly protein	Bypass complex IV/V assembly	MIM	-	MBA1	Z36054.1	-	-
Complex V (ATP synthase [ATPase])							
ATPase assembly protein	Function unknown	MIM	-	ATP10	J05463.1	-	-
ATPase assembly protein	Interacts w. subunit beta	MAT	-	ATP11	M87006.1	-	-
ATPase assembly protein	Interacts w. subunit alpha	MAT	-	ATP12	M61773.1	-	-
ATPase coupling factor B	Structural subunit of F0	MIM	3.6.1.34	-	-	None	U79253.1
ATPase expression protein	Assoc w A6/A8 expression	ND	-	NCA2	AAB68052.1	-	-
ATPase expression protein	Assoc w A6/A8 expression	ND	-	NCA3	L20786.1	-	-
ATPase inhibitor protein, 10 kDa	ATPase inhibitor	MIM	-	INH1	D00443.1	[ATPI]	AB029042.1
ATPase proteolipid 68MP homolog, 6.8 kDa	Function unknown	ND	-	-	-	C14ORF2	AF054175.1
ATPase stabilizing factor, 9 kDa	Function unknown	MIM	-	STF1	D00347.1	-	-
ATPase stabilizing factor, 15 kDa	Function unknown	MIM	-	STF2	D00444.1	-	-
ATPase stabilizing factor, 18 kDa	Stabilizes/folds Atp12	MAT	-	FMC1	Z38125.1	-	-
ATPase stabilizing factor Stf2p homolog, potentially mitochondrial	Function unknown	NC	-	YLR327c	U20618.1	-	-
ATPase subunit alpha (heart isoform)	Structural subunit of F1	MIM	3.6.1.34	ATP1	J02603.1	ATP5A1	D14710.1
ATPase subunit alpha (liver isoform)	Structural subunit of F1	MIM	3.6.1.34	-	-	ATP5A2 (b)	M19680.1 (b,i)
ATPase subunit beta	Structural subunit of F1	MIM	3.6.1.34	ATP2	M12082.1	ATP5B	M27132.1
ATPase subunit gamma	Structural subunit of F1	MIM	3.6.1.34	ATP3	U08318.1	ATP5C1	D16561.1
ATPase subunit delta	Structural subunit of F1	MIM	3.6.1.34	ATP16	Z21857.1	ATP5D	X63422.1
ATPase subunit epsilon	Structural subunit of F1	MIM	3.6.1.34	ATP15	X64767.1	ATP5E	AF077045.1
ATPase subunit 4 (human subunit b)	Structural subunit of F0	MIM	3.6.1.34	ATP4	X06732.1	ATP5F1	X60221.1
ATPase subunit 5 (OSCP)	Structural subunit of F0	MIM	3.6.1.34	ATP5	X12356.1	ATP5O	X83218.1
ATPase subunit 6 (subunit a in E. coli) (mtDNA-encoded)	Structural subunit of F0	MIM	3.6.1.34	ATP6	AJ011856.1	MTATP6	J01415.1
ATPase subunit 7 (human subunit d)	Structural subunit of F0	MIM	3.6.1.34	ATP7	Z28016.1	ATP5JD	AF087135.1
ATPase subunit 8 (mtDNA-encoded)	Structural subunit of F0	MIM	3.6.1.34	ATP8	AJ011856.1	MTATP8	J01415.1
ATPase subunit 9 (mtDNA-encoded in yeast)	Structural subunit of F0	MIM	3.6.1.34	ATP9	AJ011856.1	-	-
ATPase subunit 9 (human subunit c), P1 isoform	Structural subunit of F0	MIM	3.6.1.34	-	-	ATP5G1	D13118.1
ATPase subunit 9 (human subunit c), P2 isoform	Structural subunit of F0	MIM	3.6.1.34	-	-	ATP5G2	D13119.1
ATPase subunit 9 (human subunit c), P3 isoform	Structural subunit of F0	MIM	3.6.1.34	-	-	ATP5G3	U09813.1
ATPase subunit e	Structural subunit of F0	MIM	3.6.1.34	TIM11	U32517.1	ATP5I	D50371.1
ATPase subunit f	Structural subunit of F0	MIM	3.6.1.34	ATP17	U72652.1	ATP5J2	AF088918.1
ATPase subunit g	Structural subunit of F0	MIM	3.6.1.34	ATP20	Z49919.1	ATP5JG	AF092124.1
ATPase subunit h (human subunit F6 [coupling factor 6])	Structural subunit of F0	MIM	3.6.1.34	ATP14	U51673.1	ATP5J	M37104.1
ATPase subunit I (also called subunit j)	Structural subunit of F0	MIM	3.6.1.34	ATP18	AF073791.1	-	-
ATPase subunit k (ORF YOL077w-a)	Dimerizes w subunits e/g	MIM	3.6.1.34	ATP19	Z74820.1	-	-

Appendix 4—*continued*

Gene product	Function/comment	LOC.	E.C. #	Yeast		Animal	
				Symbol	GenBank	Symbol	GenBank
Heme/cytochrome/FeS group metabolism							
5-Aminolevulinate synthase	Heme syn	MAT	2.3.1.37	HEM1	M26329.1	ALAS1	X56351.1
5-Aminolevulinate synthase 2, erythroid-specific	Heme syn	MAT	2.3.1.37	-	-	ALAS2	X56352.1
Chaperone, homolog of E. coli Hsc20	FeS cluster syn/assembly	ND	-	JAC1	Z72540.1	None	A1796137 (e*)
Coproporphyrinogen III oxidase	Heme syn	IMS	1.3.3.3	-	-	CPO	Z28409.1
Cysteine desulfurase (homolog of E. coli NifS/IscS)	Provides S to FeS clusters	MAT	-	NFS1	M98808.1	NIFS	AF097025.1
Cytochrome c heme lyase (CCHL)	Cytochrome c syn	MIM	4.4.1.17	CYC3	X04776.1	HCCS	U36787.1
Cytochrome c, isoform 1	Electron transfer	IMS	-	CYC1	V01298.1	CYC	M22877.1
Cytochrome c, isoform 2	Electron transfer	IMS	-	CYC7	V01299.1	ZC116.2 (w)	Z74046.1 (w)
Cytochrome c1 heme lyase (CC1HL)	Cytochrome c1 syn	MIM	4.4.1.-	CYT2	X67017.1	None	AA699571.1 (e,i)
Ferrochelatase	Heme syn	MIM	4.99.1.1	HEM15	J05395.1	FECH	D00726.1
Heat shock protein 70 family member	FeS cluster syn/assembly	MAT?	-	SSQ1	U19103.1	None	BE273464.1 (e,i)
Iron metabolism protein (homolog of E. coli IscU)	FeS cluster syn/assembly	MAT	-	ISU1	U43703.1	[ISU1]	AV705042.1 (e)
Iron metabolism protein (homolog of E. coli NifU)	FeS cluster syn/assembly	MAT	-	ISU2	Z75133.1	ISCU2	U47101.1
Iron metabolism protein (homolog of E. coli NifU)	FeS cluster syn/assembly	MAT	-	NFU1	Z28040.1	R10H10.1 (w)	Z70686.1 (w)
Iron-sulfur assembly protein	FeS cluster syn/assembly	MAT	-	ISA1	Z73132.1	Y39B6.ee (w)	AL132896.1 (w)
Iron-sulfur assembly protein	FeS cluster syn/assembly	MAT	-	ISA2	Z71255.1	Y54G11A.9 (w)	AL034488.1 (w)
Protoporphyrinogen oxidase	Heme syn	MIM	1.3.3.4	HEM14	Z71381.1	PPOX	D38537.1
Protoporphyrinogen oxidase homolog, potentially mitochondrial	Function unknown	NC	-	YNL063w	Z71339.1	CG9531 (f)	AE003613.1 (f)
Ubiquinone metabolism							
Electron transfer flavoprotein dehydrogenase	Reduction of ubiquinone	MIM	1.5.5.1	YOR356w	Z75264.1	ETFDH	S69232.1
Hexaprenyl dihydroxybenzoate methyltransferase	Ubiquinone syn	MIM	2.1.1.114	COQ3	M73270.1	Y57G11C.11 (w)	Z99281.1 (w)
Hexaprenyl pyrophosphate synthetase	Ubiquinone syn	MIM	2.5.1.-	COQ1	J05547.1	[TPT]	AF118395.1
p-Hydroxybenzoate:polyprenyltransferase	Ubiquinone syn	MIM	2.5.1.-	COQ2	M81698.1	CG9613 (f)	AE003678.1 (f)
Protein involved in synthesis of coenzyme Q	Function unknown	ND	-	COQ4	AF005742.1	CG3877 (f)	AE003524.2 (f)
Ubiquinone C-methyltransferase	Ubiquinone syn	MIM	2.1.1.-	COQ5	Z97052.1	CG2453 (f)	AE003491.2 (f)
Ubiquinone monooxygenase	Ubiquinone syn	ND	1.14.13.-	COQ6	AF003698.1	CG7277 (f)	AE003610.2 (f)
Ubiquinone monooxygenase or hydroxylase	Ubiquinone syn	MIM	-	COQ7	X82930.1	COQ7	AF032900.1
SIGNAL TRANSDUCTION							
A-kinase anchoring protein 1	Binds protein kinase	MOM	-	-	-	AKAP1	U34074.1
A-RAF protein kinase	MAP kinase signalling pathway	IMS, MAT	2.7.1.-	-	-	ARAF1	X04790.1
cAMP-dependent protein kinase, regulatory subunit RII alpha	Membrane assoc in sperm	MOM	-	-	-	PRKAR2A	X14968.1
Cellular retinoic acid-binding protein, type I	Retinoid metab?	ND	-	-	-	CRABP1	S74445.1
Embryonic differentiation protein DIF-1 (MCF protein)	Embryonic differentiation	ND	-	-	-	DIF-1 (w)	Z48240.1 (w)
Glucocorticoid hormone receptor	Function unknown	ND	-	-	-	GRL	M10901.1
Growth factor receptor-bound protein 10	Binds RAF1	MOM	-	-	-	GRB1	AF000017.1
HS1-associated protein	Binds tyrosine kinase	MOM?	-	-	-	HAX1	U68566.1
Neurofibromin	Stimulates RAS GTPase	ND	-	-	-	NF1	M82814.1
Nitric oxide synthase, neuronal isoform (nNOS)	Syn of NO; neurotransmission	MOM	1.14.13.39	-	-	NOS1	U17327.1
Phosphatidylinositol 4-kinase 230	Intracellular signalling	MOM	2.7.1.67	-	-	PI4K230	AF012872.1
Raf-1 oncogene, cytosolic serine/threonine kinase	Targeted to mitos by Bcl-2	MOM	-	-	-	RAF1	X03484.1
Retinoblastoma-like protein 2, E1a-associated (p130)	May regulate cell cycle	ND	-	-	-	RBL2	X76061.1
Serine/threonine phosphatase 2c, potentially mitochondrial	Delays mito inheritance?	ND	3.1.3.16	PTC5	Z74998.1	-	-
Serine/threonine protein kinase, potentially mitochondrial	Signal transduction?	NC	2.7.1.-	YGR052w	Z72837.1	-	-
Serine/threonine protein kinase, potentially mitochondrial	Signal transduction?	NC	2.7.1.-	KIN82	X59720.1	RSK3	X85106.1
Serine/threonine protein kinase PCTAIRE-2 (brain-specific) (CDK)	Signal transduction?	ND	2.7.1.-	-	-	PCTK2 (r)	AB005540.1 (r)
Sorcin (22-kDa calcium-binding protein), possibly mitochondrial	Calcium signaling	MOM?				SRI	L12387.1
Transforming protein RhoA (found in mito 2-D gels)	Function unknown	NC	-	-	-	RHOA	L25080.1
Triiodothyronine (T3) receptor c-ErbAalpha1-related protein p43	Binds mtDNA	MAT	-	-	-	THRA1	X55005.1
Tudor protein	Signal transduction?	ND	-	-	-	TUD (f)	X62420.1 (f)
Tudor repeat associator with PCTAIRE 2 (Trap)	Signal transduction?	ND	-	-	-	TRAP (r)	AB030644.1 (r)
Tumor necrosis factor type 1 receptor associated protein TRAP-1	HSP-90 homolog	MAT	-	-	-	TRAP1 (f)	AF115775.2 (f)

STRESS RESPONSE							
Cytochrome c peroxidase	Hydrogen peroxide deg	IMS	1.11.1.5	CCP1	X62422.1	-	-
Glutathione peroxidase 1	Oxidizes glutathione	MAT?	1.11.1.9	-	-	GPX1	X13709.1
Glutathione peroxidase 4	Oxidizes glutathione	MAT?	1.11.1.9	-	-	GPX4	X71973.1
Glutathione reductase	Reduces glutathione	ND	1.6.4.2	GLR1	L35342.1	GSR	X15722.1
Glutathione S-transferase	Glutathione conjugation	MOM	-	-	-	MGST1	J03746.1
Glutathione S-transferase, subunit 13	Glutathione conjugation	MAT?	2.5.1.18	-	-	None	AF070657.1
Glyoxalase II	Glutathione reduction	MAT	3.1.2.6	GLO4	Z74948.1	CG4365 (f)	AE003592.1 (f)
Heat shock protein 78 (similar to ClpB/HSP104)	Mito thermotolerance	MAT	-	HSP78	L16533.1	Skd3 (m)	U09874.1 (m)
Histatin 5 (HSN-5) antifungal salivary peptide	Binds *Candida albicans* mitos	ND	-	-	-	HTN3	M26665.1
Mercaptopyruvate sulfurtransferase	Cyanide detoxification	ND	2.8.1.2	-	-	TST2	X59434.1
NADH/NADPH thyroid oxidase p138-tox, potentially mitochondrial	Generates hydrogen peroxide	NC	-	-	-	None	AF181972.1
Superoxide dismutase 2 (MnSOD)	Scavenges free radicals	MAT	1.15.1.1	SOD2	X02156.1	SOD2	M36693.1
Thiol peroxidase of the 1-Cys family (mTPx)	Thiol-specific antioxidant?	ND	-	YBL064c	Z35825.1	R07E5.2 (w)	Z32683.1 (w)
Thioredoxin	Reduces disulfides	MAT?	-	TRX3	X59720.1	TXN2	U78678.1
Thioredoxin-dependent peroxide reductase	Antioxidant protein	MAT?	-			AOP1	D49396.1
Thioredoxin-dependent peroxide reductase (found in mito 2-D gels)	Antioxidant protein	NC	1.11.1.7	-	-	[AOP2]	D14662.1
Thioredoxin peroxidase type II	Mn2+ homeostasis	ND	-	AHP1	Z73281.1	AOEB166	AF110731.1
Thioredoxin peroxidase AOE372 (found in mito 2-D gels)	Antioxidant protein	NC	-	-	-	None	U25182.1
Thioredoxin reductase 2	Reduces disulfides	MAT?	1.6.4.5	TRR2	U00059.1	[TRXR2]	AF044212.1
Thiosulfate sulfurtransferase (rhodanese)	Cyanide detoxification	MAT	2.8.1.1	YOR251c	Z75159.1	TST	D87292.1
MISCELLANEOUS							
CDC6-related protein, potentially mitochondrial	Function unknown	NC	-	-	-	HSCDC6	U77949.1
C21orf2, 21.3-kDa protein	Function unknown	ND	-	-	-	C21ORF2	Y11392.1
Cytokeratin type II	Intermediate filament protein	MIM	-	-	-	K6HF	Y17282.1
DJ-1 protein (found in mito 2-D gels)	Oncogene	NC	-	-	-	[DJ1]	D61380.1
Electron transport protein homolog	Function unknown	NC	-	YTP1	U22109.1	-	-
ES-1 protein homolog, KNP-Ia (long form) and KNP-Ib (short form)	Function unknown	ND	-	-	-	HES1	D86061.1
Four-and-a-half LIM-only protein 2 (in skeletal muscle; heart)	Cytoskeletal remodeling	ND	-	-	-	FHL2	U29332.1
Four-and-a-half LIM-only protein 3 (in skeletal muscle)	Cytoskeletal remodeling	ND	-	-	-	FHL3	U60116.2
Galectin, 16-kDa isoform (CG-16)	Binds beta-galactoside	ND	-	-	-	None (c)	M57240.1 (c)
Galectin-3 internal gene, ORF2 (nt 421-786 in M36682.1)	Alt splicing of LGALS3	ND	-	-	-	[GALIG]	M36682.1
Globin, beta chain (found in mito 2-D gels)	Function unknown	NC	-	-	-	HBB	L26463.1
Globin, gamma-A chain (found in mito 2-D gels)	Function unknown	NC	-	-	-	HBG1	M91036.1
Heat shock protein 22	Function unknown	MAT	-	-	-	HSP22 (f)	X03888.1 (f)
HTLV-1 ORF X-II (protein p13II), 87 aa	Binds huntingtin	ND	-	-	-	None	L08433.1
Huntingtin-associated protein 1 (HAP1)	Function unknown	ND	-	-	-	HAP1	AJ224877.1
Hypothetical protein, potentially mitochondrial	Function unknown	NC	-	YPL103c	U43281.1	-	-
Hypothetical protein, potentially mitochondrial	Function unknown	NC	-	YNR020c	Z71635.1	-	-
Hypothetical protein, potentially mitochondrial	Function unknown	NC	-	YJR122w	Z49622.1	CG8043 (f)	AE003680.2 9f)
Hypothetical protein, potentially mitochondrial	Function unknown	NC	-	YHR059w	U00061.1	-	-
Hypothetical protein, potentially mitochondrial	GTP-binding protein	NC	-	YLF2	Z29089.1	[PTD004]	AF078859.1
Hypothetical protein, potentially mitochondrial	Function unknown	NC	-	YGL226w	Z72748.1	-	-
Hypothetical protein, potentially mitochondrial	Function unknown	NC	-	YGL107c	Z72629.1	-	-
Hypothetical protein, potentially mitochondrial	Function unknown	NC	-	AMI3	Z74802.1	-	-
Hypothetical protein, potentially mitochondrial	Function unknown	NC	-	YDR196c	Z48784.1	-	-
Hypothetical protein, potentially mitochondrial	Function unknown	NC	-	YGR265w	Y07777.1	-	-
Hypothetical protein, potentially mitochondrial	Function unknown	NC	-	YMR158w	Z49705.1	-	-
Hypothetical protein, potentially mitochondrial	Function unknown	NC	-	YNL184c	Z71461.1	-	-
Hypothetical protein, potentially mitochondrial	Function unknown	NC	-	YOL008w	Z74750.1	-	-
Hypothetical protein, potentially mitochondrial	Function unknown	NC	-	YOL071w	Z74813.1	Y57A10A.29 (w)	AL117195.1 (w)
Hypothetical protein, probably mitochondrial	Function unknown	ND	-	YNL213c	Z71489.1	-	-
Hypothetical protein, probably mitochondrial	Function unknown	ND	-	-	-	KIAA0141	D50931.1
Leucine-zipper protein, potentially mitochondrial	Function unknown	NC	-	-	-	FEZ1	AF123659.1
Mitochondrial acidic matrix protein (human P32 [gC1q-R])	Ca binding?	MAT	-	MAM33	Z38060.1	[C1QBP]	L04636.1
Mitofilin, heart motor protein (HMP)	Function unknown	ND	-	-	-	IMMT	D21094.1

Appendix 4—*continued*

Gene product	Function/comment	LOC.	E.C. #	Yeast		Animal	
				Symbol	GenBank	Symbol	GenBank
MTCP1 oncogene, alt spliced 8-kDa protein	Function unknown	ND	-	-	-	MTCP1	Z24459.1
Myocilin (trabecular inducible glucocorticoid response protein)	Mutated in glaucoma	ND	-	-	-	MYOC	U85257.1
Neighbor of COX4 (NOC4)	Function unknown	ND	-	-	-	COX4AL	AF005888.1
ORF Q0010 (mtDNA-encoded)	Function unknown	ND	-	[ORF6]	AJ011856.1	-	-
ORF Q0017 (mtDNA-encoded)	Function unknown	ND	-	[ORF7]	AJ011856.1	-	-
ORF Q0032 (mtDNA-encoded)	Function unknown	ND	-	[ORF8]	AJ011856.1	-	-
ORF Q0092 (mtDNA-encoded)	Function unknown	ND	-	[ORF5]	AJ011856.1	-	-
ORF Q0142 (mtDNA-encoded)	Function unknown	ND	-	[ORF9]	AJ011856.1	-	-
ORF Q0143 (mtDNA-encoded)	Function unknown	ND	-	[ORF10]	AJ011856.1	-	-
ORF Q0144 (mtDNA-encoded)	Function unknown	ND	-	[Q0144]	AJ011856.1	-	-
ORF Q0167 (mtDNA-encoded)	Function unknown	ND	-	RF2	AJ011856.1	-	-
ORF Q0182 (mtDNA-encoded)	Function unknown	ND	-	[ORF11]	AJ011856.1	-	-
ORF Q0255 (mtDNA-encoded)	Function unknown	ND	-	[ORF1]	AJ011856.1	-	-
ORF Q0297 (mtDNA-encoded)	Function unknown	ND	-	[ORF12]	AJ011856.1	-	-
Pilin-like transcription factor, potentially mitochondrial	Function unknown	NC	-	YCL033c	X59720.1	[PILB]	AF122004.1
Protein involved in mitochondrial metabolism, 73 kDa	Function unknown	ND	-	SLS1	Z48452.1	-	-
Protein involved in respiration, putatively mitochondrial	Function unknown	NC	-	YKR016w	Z28241.1	-	-
Protein of the outer mitochondrial membrane, 45 kDa	Function unknown	MOM	-	OM45	M31796.1	-	-
Protein of unknown function, potentially mitochondrial	Protease?	NC	-	YDR430c	U33007.1	NRD1	U64898.1
Protein of unknown function	Function unknown	ND	-	YAL011w	L05146.1	-	-
Protein required for respiratory growth, potentially mitochondrial	Function unknown	NC	-	YKL137w	Z28137.1	-	-
Von Hippel-Lindau tumor suppressor, 18-kDa isoform	Function unknown	ND	-			VHL	AF010238.1
TOTAL NUMBER OF GENES				**584**		**734**	

[a] Names of gene products usually follow the yeast nomenclature. For those animal gene products that have alternatively-spliced isoforms, only one representative isoform is usually indicated. Gene symbols within brackets are unofficial. All yeast loci are from *Saccharomyces cerevisiae,* except for one from *Schizosaccharomyces pombe.* All animal loci are from human, unless otherwise indicated: b, bovine; c, chicken; f, fly *(D. melanogaster);* h, Chinese hamster; m, mouse; p, pig; r, rat; s, slime mold *(D. discoideum);* u, sea urchin; w, worm *(C. elegans).* Other abbreviations: a, alternative reading frame or alternative AUG start; ABC, ATP-binding cassette; BER, base excision repair; deg, degradation; e, unannotated human expressed sequence tag (EST); e*, unannotated human EST found by Rötig et al. (2000) *Mol. Genet. Metab.* 69:223-232; g*, annotated gene found by Rötig et al. (2000) *Mol. Genet. Metab.* 69:223-232; GIP, general import pore; i, incomplete partial sequence; IMS, intermembrane space; ISP, iron-sulfur (FeS) protein; LOC, location in mitochondria; MAT, matrix; metab, metabolism; MIM, mitochondrial inner membrane; MOM, mitochondrial outer membrane; MTS, mitochondrial targeting signal; NC, mitochondrial localization not checked; ND, localized to mitochondria, but specific compartment not determined; NOS, nitric oxide synthase; syn, synthesis; transp, transport; ?, presumed mitochondrial localization within mitochondria, but not verified experimentally.

APPENDIX 5

Direct and Indirect Inhibitors of Mitochondrial ATP Synthesis

Nanette Orme-Johnson
Tufts University School of Medicine

A few important notes about these tables. They are in no sense complete with respect to the compounds listed or the references cited for their use. I have tried to include the more commonly used inhibitors and others that are of either historical interest or that have been found recently and may become more widely used. The frequency of use of a compound may be assessed by the number of references indicated, i.e., * (fewer than 10 references), ** (10 to 100 references), or *** (greater than 100 references). Additionally, because only recent references are cited if the latest reference for a compound is, e.g., 1975, it may be concluded that this compound is probably not used frequently now.

Table 5a
Inhibitors of Electron Transport Chain

The family of complex I inhibitors is large and structurally diverse. A number of excellent review articles are available. The review by Degli Esposti (1) lists complex I inhibitors with references to their use in beef heart mitochondria and/or submitochondrial particles, citing the concentrations used in that system and the efficacy of the inhibitors compared to that of rotenone. The adjacent review by Miyoshi (2) discusses the relation of the structure of complex I inhibitors to their efficacy. The review by Alali *et al.* (3) and the papers by Landolt *et al.* (4), Zafra-Polo *et al.* (5), and Miyoshi *et al.* (6) discuss in detail the annonaceous acetogenins. Reference (7) by Miyoshi *et al.* and references therein discusses pyridinium-type inhibitors. Certain quinone antagonists may inhibit both complex I and complex III (8), but at very different concentrations; the inhibition that occurs at lower inhibitor concentration is cited here. A number of inhibitors of the respiratory chain are potential or actual antibiotics, antiparasitics, chemotherapeutic agents, and/or insecticides. The review by Lümmen (9) discusses complex I inhibitors as insecticides and acaricides.

Inhibitor	Site[a]	References[b,c]	Notes[d]
Amiodarone*	Complex I	**M:** Hamster lung and liver (21); mouse liver (22); rat liver (23)	(a) Also inhibits complex V (24) and CPT-1 in rat heart mitochondria (25)
Amytal** (Amobarbitol)	Complex I	**T/C:** Cardiomyocytes (26); ROC-1 cells (27); saponin-permeabilized muscle fibers (28). **M:** *C. parapsilosis* (29); digitonin-treated homogenates of mouse brain (30); rat liver (31); potato tuber (32). **SMP:** Rat heart (31)	
Annonaceous acetogenins** Annonacin, Annonin, Asimicin, Bullatacin, Gigantetrocins, Molvazarin, Otivarin, Rolliniastatins, Squamocin, Trilobactin	Complex I	**T/C:** Insect Sf9 cells (Bullatacin and analogs) (33); solid tumor cell lines, see reference (3) for a review. **M:** beef heart and *N. crassa* (annonin VI) (34); European corn borer (35): rat liver and midgut of *Manduca sexta* larvae (Bullatacin and analogs) (33), rat liver (20 compounds, including Annonacin, Asimicin, Gigantetrocins, Bullatacin, Molvazarin, Trilobactin) (4). **SMP:** beef heart (Molvazarin, Otivarin, Rolliniastatin-1 and -2, Squamocin) (36); (Cherimolin-1, Rolliniastatin-2) (37); (11 compounds, including Bullatacin, Gigantetrocin, Squamocins) (6); (6 compounds, including Guanacone and derivatives, Motrillin, Rolliniastatin-2) (38); (Bullatacin) (39); (6 compounds, including Annonacin, Corossolin, Rolliniastatin-1) (40); (Crossolin, Rolliniastatin-1 and -2) (41); blowfly flight muscle (42). **IC:** beef heart (Bullatacin and analogs) (33); (Molvazarin, Otivarin, Rolliniastatin-1 and -2, Squamocin) (36)	(a) Very large class of natural products (tetrahydrofuran acetogenins) isolated from Annonaceae plants. Those compounds tested inhibit complex I. A much larger number of these compounds have been found to be toxic to transformed cells, presumably, at least in part, due to this inhibition (b) Structure–function studies are reviewed in (3-6)
Antimycin A***	Complex III	**T/C:** *G. graminis* mycelia (also inhibits AOX) (43); mouse fibrosarcoma cells (44); mouse C2C12 myocytes (45); *S. cerevisiae* (46); transgenic tobacco cells (47); *P. Yoelii* erythrocytes (48); permeabilized tachyzoites of *T. gondii* (49); *T. brucei brucei* (50). **M:** *G. graminis* (43); *L. mexicana* promastigote (51); potato tuber (52); lugworm (53); rat brain (54); rat liver (55); soy bean seedling roots (56); tobacco (47); *T. brucei brucei* (50). **SMP:** beef heart (57, 58); rat liver (59). **IC:** beef heart and *N. crassa* (60); beef heart (crystal structure of inhibitor complex) (61)	(a) Center N inhibitor (62)

(*Continues*)

Table 5a
(Continued)

Inhibitor	Site[a]	References[b,c]	Notes[d]
Atovaquone* (Mepron)	Complex III	**T/C:** *P. yoelii* in intact erythrocytes (48); *P. carinii* (63); tachyzoites (permeabilized) of *Toxoplasma gondii* (49). **M:** *P. falciparum* and *P. yoelii* (64); rat liver (49); rat and human liver (65)	(a) Inhibits mammalian as well as parasite dihydroorotate dehydrogenase, but the mammalian enzyme is much less sensitive than the parasite enzyme
Aureothin*	Complex I	**M:** Beef heart and *N. crassa* (34); rat liver (66). **IC:** *N. crassa* (34)	(a) Formerly known as mycolutien (67)
Azide***	Complex IV	**T/C:** *C. albicans* (68): CRI–G1 insulinoma cells (69); crayfish neurons (70); human kidney cells (71); mouse C2C12 myocytes (45): rat hepatocytes (72); *S. cerevisiae* (46); SY5Y cells (73). **M:** digitonin–treated homogenates of the mouse brain (30); rat liver (74); *L. mexicana* promastigote (51). **SMP:** beef heart (75)	(a) Also inhibits complex V
β-carboline derivatives**	Complex I	**T/C:** *T. cruzi* epimastigotes (76). **M:** beef heart (β-carbolines and β-carboliniums) (77); rat liver (4 compounds) (78), (7 compounds) (79) and (TaClo) (78)	(a) 5H-pyrido[4,3-β]indoles (b) Inhibits the synthesis of 1 Me TIQ (80)
β-methoxyacrylate derivatives** Melithiazol, myxothiazol, MOA, oudemansin, pterulone, stilbene, strobilurin	Complex III	**T/C:** cardiomyocytes (myxothiazol) (26, 81); *P. yoelii* in intact erythrocytes (myxothiazol) (48); rat hepatocytes (myxothiazol) (82). **M:** *L. mexicana* promastigote (myxothiazol) (51); soybean seedling roots (myxothiazol) (56). **SMP:** beef heart (83); (MOA stilbene or myxothiazol) (57, 84), (melithiazols) (85); rat liver (myxothiazol) (86). **IC:** beef Heart (myxothiazol) (60, 87–89), (crystal structure for inhibitor complex with MOA stilbene or myxothiazol) (61); *N. crassa* (myxothiazol) (60); *S. cerevisiae* mutants (MOA stilbene or myxothiazol) (90)	(a) Also called MOAs (b) Bind at Q(o) site (88) (c) Reference (91) discusses an MOA photoaffinity label (d) Melithiazol and myxothiazol are products of myxobacteria (e) Strobilurin is also called mucidin (92)
Capsaicin*	Complex I	**M:** beef heart and potato tuber (93); rat liver (94). **SMP:** beef heart (95). **IC:** Beef heart (also several analogs) (96)	(a) Isolated from *Capsicum* (b) Capsaicin site overlaps the rotenone site, but not the piericidin A site (95)
Cyanide***	Complex IV	**T/C:** *G. graminis* mycelia (43); *P. yoelii* in intact erythrocytes (48); *S. cerevisiae* (46); *T. gondii* Tachyzoites (permeabilized) (49); *T. brucei brucei* (50). **M:** *G. graminis* (43); *L. mexicana* promastigote (51); lugworm (53); rat brain (54); soybean seedling roots (56); potato tuber (52); soybeans (20). **SMP:** Rat liver (59)	
Diphenylene-Iodonium**	Complex I (AOX)	**T/C:** cultured chick cardiomyocytes (26); cultured fibroblasts from normals and from LHON patients (97); tobacco cells (17) (AOX). **M:** beef heart (98); potato tuber and cuckoo pint spadix (99). **SMP:** potato tuber and pea leaves (100). **IC:** beef heart (101)	
Ethoxyformic anhydride*	Complex III	**M:** pigeon heart (102). **SMP:** beef heart (57, 103); rat liver (59)	(a) Modifies Histidines in Rieske Fe–S protein

(Continues)

Table 5a
(Continued)

Inhibitor	Site[a]	References[b,c]	Notes[d]
Fenazaquin*	Complex I	**T/C:** insect Sf9 cells (104); MCF-7 human breast cancer cells (105). **SMP:** beef heart (95); rat liver (104). **IC:** beef heart (104)	(a) Reference (106) discusses quinazoline photoaffinity labels
Fenpyroximate*	Complex I	**M:** beef heart and *N. crassa* (34); rat liver and spider mites (107). **SMP:** beef heart (95)	
Flavonoids** Flavone	Complex I	**T/C:** complex I-deficient Chinese hamster CCL 16-B2 cells transfected with NDI1 of *S. cerevisiae* (flavone) (108). **M:** beef heart (structure function studies) (109, 110); *L. mexicana* promastigote (flavone) (51); rat brain (flavone) (111); rat liver (8 compounds) (112). **SMP:** beef heart (11 compounds) (113). **IC:** *S. cerevisiae* (flavone) (114)	(a) See also rotenoids. (b) Some are free radical scavengers and antioxidants (c) Some inhibit complex V (e.g., quercetin)
Funiculosin**	Complex III	**M:** rat liver (115). **SMP:** beef heart (58). **IC:** *N. crassa* and beef heart (60)	(a) Fish mitochondria are resistant (116) (b) Binds Qi site (58); center N inhib (62)
HQNO**	Complex III	**T/C:** *B. bovis* (117); mouse LA9 cells (118); U937 cells (119). **M:** *L. mexicana* promastigote (51); rat liver (120). **SMP:** beef heart (121)	(a) Binds at Qi site (58, 119)
Isoquinolines or Isoquinoliniums**	Complex I	**T/C:** PC12 and SK-N-MC dopaminergic cells (isoquinoline and TIQ) (122); endothelial cells (salsolinol) (123). **M:** rat brain (TIQ) (124); mouse brain (N-Me-TlQ and N-Me-IQ+) (125); rat liver (15 isoquinoline derivatives structurally related to MPTP) (126); mouse liver (MIQ+) (127). **SMP:** beef heart (9 N-methyl quinolinium analogues) (128); rat forebrain (11 isoquinolines, 2 dihydroisoquinolines, salsolinol and 8 other 1,2,3,4-tetrahydroisoquinolines) (129)	(a) Inhibits α-ketoglutartate but not glutamate dehydrogenase (six isoquinoline derivatives) (130) (b) Some prevent the synthesis of 1MeTlQ (80). (c) Inhibits monoamine oxidase from human brain synaptosomal mitochondria (131)
Malonate***	Complex II	**T/C:** PC12 cells (11); rat marrow stroma (132). **M:** beef heart, potato tuber and rat liver (uncoupled) (133); *L. mexicana* promastigote (51); rat brain (134: Zini, 1999 #2414); rat heart (135); rat liver (136); *T. brucei brucei* (50). **SMP:** beef heart (137); rat liver (59)	(a) Protects against covalent modification by ethoxyformic anhydride and rose bengal (138)
MPP+***	Complex I	**T/C:** human SH-SY5Y neuroblastoma cells (139); rat cerebellar granule cells (140). **M:** beef heart and rat liver (141); rat brain and liver (142); rat brain and NB69 cells (143); synaptosomal cells from different rat brain areas (144). **SMP:** beef heart (145, 146), (MPP^+, six *N*-methyl pyridinium analogs and nine *N*-methyl quinolinium analogs) (128) and (MPP^+, five 4′-MPP[s] analogs and five 4′-phenylpyridine analogues) (147); beef heart (MPP^+, also used structural analogs of MPP^+) (148); fragments from rat forbrain (MPP^+) (129); rat liver inverted membrane preparations (MPP^+) (149). **IC:** beef heart (MPP^+, five 4′-MPP^+ analogs, and five 4′-phenylpyridine analogs) (147)	(a) Neurotoxic metabolite, produced by the action of monoamine oxidase B on MPTP (150, 151) (b) Inhibits α-ketoglutarate dehydrogenase (130, 152, 153) (c) Inhibits the synthesis of 1MeTIQ (80) (d) Inhibit synthesis of mitochondrial DNA (154), not by inhibiting DNA polymerase but by destabilizing the D loop (155)

(Continues)

Table 5a
(Continued)

Inhibitor	Site[a]	References[b,c]	Notes[d]
Myxobacterial products*** Aurachins, melithiazol, myxalamid, myxothiazol, phenalamid, phenoxan, stigmatellin, thiagazole	Complex III	**M:** cuckoo pint spadices (aurachin C and analogs) (also inhibits AOX) (19); beef heart and *N. crassa* (aurachins A and B, myxalamid, phenalamid, phenoxan, thiagazole) (34). **SMP:** beef heart (myxothiazol) (58) and (stigmatellin) (57, 121, 156); rat liver (myxothiazol) (86). **IC:** beef heart (myxothiazol) (60, 87–89); *N. crassa* (myxothiazol) (60) and (aurachins) (34); beef Heart (crystal structure for inhibitor complex with stigmatellin) (61)	(a) Isolated from myxobacterium (b) Aurachin binds at Qi site (19) (c) Myxothiazol, and stigmatellin bind at Qo site (58) (d) Melithiazol and myxothiazol are β-methoxyacrylates and are listed as such
3-Nitropropionic acid**	Complex II	**T/C:** murine embryonal carcinoma cells (10); PC12 cells (11); rat cerebellar granule cells and astrocytes (12); rat hippocampal slices (13) and neurons (14); rat isolated, beating atria (15). **M:** PC6 cells and TNF-treated neurons (16); rat heart (15)	
Piericidin**	Complex I	**M:** beef heart and *N. crassa* (piericidin A) (34); rat liver (157); *Y. lipolytica* (158). **SMP:** beef heart (6, 36, 95, 146)]. **IC:** *N. crassa* (34)	(a) Piericidin A site overlaps the rotenone site but not the capsaicin site (95) (b) Ref (159) studied structural analogs
n-Propylgallate**	AOX	**T/C:** harpin-treated tobacco cells (17); several types of marine phytoplankton and *C. reinhardtii* (18). **M:** cuckoo pint spadices (also *n*-propyl gallate analogs) (19); soybean cotyledon (19, 20).	
Pyridaben*	Complex I	**T/C:** MCF-7 human breast cancer cells (105). **M:** rat liver and tobacco hook worm midgut (104). **SMP:** beef heart (39, 113). **IC:** beef Heart (105)	(a) (Trifluoromethyl) diazirinyl[^{3}H] pyridaben ([^{3}H]TDP) is a photoaffinity label (39) (b) Inhibits TPA-induced ornithine decarboxylase in cultured cells (105)
Rhein**	Complex I	**T/C:** rat hepatocytes (160). **M:** rat liver (161, 162). **SMP:** beef heart (163)	
Rotenoids*** Includes Deguelin Rotenone	Complex I	**T/C:** astrocytes (rotenone) (164); cultured mouse epidermal cells (deguelin) (165); *T. brucei brucei* (rotenone) (50). **M:** beef heart (rotenone) (36); beef heart and *N. crassa* (rotenone) (34); *L. mexicana* promastigote (rotenone) (51); lugworm (rotenone) (53); rat brain (rotenone) (54); *T. brucei brucei* (rotenone) (50). **SMP:** beef heart (rotenone) (146); beef heart and potato tuber (rotenone and 16 analogs) (166); beef heart (rotenone and 9 analogs, deguelin and 18 analogs) (113); Rat heart (rotenone) (31). **IC:** *N. crassa* (rotenone) (34)	(a) Isoflavonoids from isolated from *Leguminosae* plants (b) Structural features of rotenoids needed for inhibition of complex I are similar to those for blocking phorbol ester-induced ornithine decarboxylase (105, 113, 167) (c) The rotenone site overlaps both the piericidin A and the capsaicin site, the latter two sites do not overlap (95)

(Continues)

Table 5a
(Continued)

Inhibitor	Site[a]	References[b,c]	Notes[d]
Salicylhydroxamic acid**	AOX	**T/C:** *G. graminis* mycelia (43); tea leaf discs (52); several types of marine phytoplankton and *C. reinhardtii* (18); *T. brucei brucei* Beattie, 1996 #1369]. **M:** lugworm (53); potato tuber (52)	(a) Also called SHAM
Tannins*	Complex I	**M:** rat liver (168). **IC:** beef heart (169)	
TTFA**	Complex II	**T/C:** germinating seeds of wheat and maize (170); human cervical carcinoma C33A cells (171); mouse fibrosarcoma (44); rat renal proximal tubular cells (172). **M:** bone marrow-derived cell line, 32D (173); rat brain rain nonsynaptic tissue (174); *L. mexicana* promastigote (51); rat brain (175). **SMP:** beef heart (176); *S. ratti* (177)	(a) Thenoyltrifluoroacetone
UHDBT**	Complex III	**T/C:** trypomastigote forms of *T. brucei brucei* (178). **M:** Jerusalem artichoke tuber and cuckoo pint spandices (also inhibits AOX) (179); rat liver (180). **IC:** beef heart (181); chicken heart (88); *S. cerevisiae* (182); beef heart (crystal structure of inhibitor complex) (61)	(a) Synthetic analog of ubiquinone (b) Binds at Q(o) site (87, 88)

Table 5b
Inhibitors of F_0F_1-ATPase (Complex V) or the Adenine Nucleotide Translocator

Inhibitor	Site[a]	References[b,c]	Notes
Atractyloside***	ANT	**T/C:** neuronally differentiated PC12 cells (153); permeabilized guinea pig heart muscle (183); rat hepatocytes (82); sperm of the telost fishes (184). **M:** *L. mexicana* promastigotes (51); rat brain (142); rat heart (185); rat liver (186); rat liver and skeletal muscle (187); *S. cerevisiae* (188). **SMP:** beef heart (189). **IC:** beef heart (190); rat heart mitochondria (142); *S. cerevisiae* (191)	(a) The difference in binding of atractylate and carboxy-atractylate to rat liver mitochondria is discussed in reference (192)
Aurovertin**	Complex V	**T/C:** AS-30D rat hepatoma cells (193); ρ^0 HeLa S3 cells (194). **M:** beef heart (195); *L. mexicana* promastigotes (51); pea leaves (196). **SMP:** beef heart (197). **IC:** beef heart in reconstituted liposomes (197); human F1-ATPase (194); *T. cruzi* (198)	
Azide	Complex V	**T/C:** o human tumor cells (199); ρ^0 HeLa S3 cells (194). **SMP:** beef heart (75); **IC:** human F1-ATPase (194)	(a) Also inhibits complex IV
Bongkrekic***	ANT	**T/C:** cerebrocortical neurons (200); ρ^0 HeLa S3 cells (194); HepG2 cells and rat primary hepatocytes (201); Jurkat cells (202); murine pre-B-cell line BAF3 (203); ρ^0 HeLa S3 cells (194). **M:** rat heart (204); rat liver (205); rat liver and skeletal muscle (187); *S. cerevisiae* (188). **SMP:** beef heart (189, 206). **IC:** beef heart (190, 207); *S. cerevisiae* (191)	(a) Produced by *Burkholderia cocovenenans*

(Continues)

Table 5b
(Continued)

Inhibitor	Site[a]	References[b,c]	Notes
Carboxy-atractyloside	ANT	**M:** beef heart (208); rat heart from normal and cold-exposed rats (209); rat liver (210); rat liver and skeletal muscle (187); *S. cerevisiae* (188). **SMP:** beef heart (206). **IC:** beef heart (190, 207); *S. cerevisiae* (191)	
DCCD***	Complex V	**M:** rat liver (211); rat liver (from rats infected with *F. hepatica*) (212); *S. cerevisiae* (213). **SMP:** beef heart (214), inner membrane (215); *H. diminuta* (216). **IC:** beef heart (217); spinach leaves (218)	(a) Modifies COOH groups of Glu/Asp (b) Also inhibits complex III (219) and complex IV (220, 221) (c) Binds Asp-160 of yeast cyt *b* (222) (d) Inhibits mito VDAC channel (223)
Efrapeptin	Complex V	**T/C:** *A. castellanii* (224); CHO cells (225). **M:** beef heart (226, 227). **SMP:** beef heart (228); rat testis (229). **IC:** beef heart (230); *T. cruzi* (198)	(a) Block exocytic but not endocytic trafficking of proteins (231) (b) Binds to a unique site in the central cavity of F_1-ATPase
Leucinostatins**	Complex V	**T/C:** CHO cells (225). **M:** rat liver (211)	(a) Uncouples at higher amounts (211)
Oligomycin***	Complex V	**T/C:** *A. castellanii* (224); AS–30D rat hepatoma cells (193); CHO cells (225); neuronally differentiated PC12 cells (153); rat hepatocytes (82); *S. cerevisiae* (46). **M:** rat brain (54); *L. mexicana* promastigotes (51); rat heart (209); rat liver (210). **SMP:** beef heart (156, 214); rat heart (232). **IC:** human F_1-ATPase (194); spinach leaves (218)	
Venturicidin**	Complex V	**M:** pea leaves (196); potato tuber (233); rat liver (211). **SMP:** beef heart (234, 235). **IC:** beef heart in lipid bilayer (236); beef heart F_0 (237)	

Table 5c
Uncouplers

These compounds decrease the yield of ATP by decreasing one or both of the two terms that contribute to the protomotive force ($\Delta\mu_H$). Thus, they decrease the membrane potential ($\Delta\Psi$), the pH gradient (ΔpH), or both.

Inhibitor	References[b,c]	Notes
A23187**	**T/C:** rabbit thoracic aorta rings (238); rat pituitary (GH3, GC, and GH3B6), adrenal (PC12), and mast (RBL-1) cell lines (239); rat hepatocytes (240). **M:** beef heart (241); rat liver (242, 243). **SMP:** *L. braziliensis* promastigotes (244)	(a) Principally a divalent cation ionophore; preference $Ca^{2+} > Mg^{2+}$ (b) Also called calcimycin
CCCP***	**T/C:** bovine sperm (245); filarial nematode (246); human cervical carcinoma C33A cells (171); Jurkat and CEM cells (247); mouse C2C12 myocytes (45); rat anterior pituitary GH(3) cells (248); rat cerebellar neurons (249); spinach leaf cells (250). **M:** *H. diminuta* (216); mouse C2C12 myocytes (45); rat brain (251, 252); rat heart and liver (253); spinach leaves (250). **SMP:** beef heart (246, 254). **IC:** beef heart complex III (255)	(a) Protonophore

(Continues)

Table 5c
(Continued)

Inhibitor	References[b,c]	Notes
Dinitrophenol (DNP)***	**T/C:** Jurkat and CEM cells (247); telost fish sperm (184); *S. cerevisiae* (256). **M:** beef liver and pea stem (257); rat heart, liver and skeletal muscle (253); rat liver (258); rat liver and skeletal muscle (187). **SMP:** rat heart (253)	(a) Protonophore (b) Alkylated (2-alkyl-4, 6-dinitrophenol) inhibits complex II (259, 260)
FCCP***	**T/C:** *T. brucei brucei* (261); hamster brown fat cells (262); rat hepatocytes (82); rat renal proximal tubule cells (263). **M:** beef liver and pea stem (257); hamster brown fat (264); *Leishmania mexicana* (51); rat heart and liver (253); rat liver (265); rat liver and skeletal muscle (187); rodent brown adipose (266). **SMP:** beef heart (83); *H. diminuta* (216); *L. braziliensis* promastigotes (244)	(a) Protonophore
Gramicidin**	**T/C:** bloodstream forms of *T. brucei brucei* (267); rat cerebellar neurons (249); rat hepatocytes (82); *S. cerevisiae* (268). **M:** rat heart and liver (253); rat liver and skeletal muscle (187). **SMP:** beef heart (269)	(a) Pore forming (b) Polypeptide antibiotic (c) Ref. (270) studied pore forming by gramicidin analogs
Monensin**	**T/C:** rat brain synaptosomes (271); rat hepatocytes (272); rat liver and lung slices (272); rat renal proximal tubule cells (263). **M:** hamster brown fat (264); rat liver and skeletal muscle (187); rat liver (273)	(a) Causes disintegration of the ER-Golgi system in spinach leaves (250) (b) Electroneutral monovalent antiporter, especially Na^+/H^+ (c) Complex polyether antibiotic structurally related to nigericin
Nigericin**	**T/C:** bloodstream forms of *T. brucei brucei* (267); rat renal proximal tubule cells (263). **M:** hamster brown fat (264); rat liver (273)	(a) Electroneutral monovalent antiporter, especially K^+/H^+ (b) Complex polyether antibiotic, structurally related to monensin
SF6847**	**M:** Cells overexpressing human Bcl-2 or from livers of Bcl-2 transgenic mice (274); rat heart and liver (253); normal and cold-exposed rat heart (209). **SMP:** beef heart (275)	(a) Protonophore
Valinomycin***	**T/C:** bloodstream forms of *T. brucei brucei* (267); mouse C2C12 myocytes (45); murine pre-B BAF3 cells (203); rat cerebellar neurons (249); rat renal proximal tubule cells (263); *S. cerevisiae* (268). **M:** hamster brown fat (264); rat heart (276); rat liver (273). **SMP:** beef heart (83). **IC:** beef heart complex III in asolectin liposomes (277) and in potassium-loaded phospholipid vesicles (219)	(a) Monovalent uniporter for K^+

Compounds

Common name or abbreviation	Chemical name
1MeTIQ	1-Methyl-1,2,3,4-tetrahydroisoquinoline
A23187	6*S*-[6α92*S**, 3*S**), 8β(*R**), 9β, 11α]-5-(methylamino)-2-([3,9,11-trimethyl-8-[1-methyl-2-oxo-2-(1*H*-pyrrol-2-yl)ethyl]-1,7-dioxaspiro[5,5]undec-2-yl]methyl)-4-benzoxazolecarboxylic acid
Amiodarone or AMD	(2-Butyl-3-benzofuranyl)[4-[2-(diethylamino)ethoxy]-3,5-diiodophenyl]methanone

(Continues)

Table 5c
(*Continued*)

Compounds

Common name or abbreviation	Chemical name
Amobarbital or Amytal	5-Ethyl-5-(3-methylbutyl)-2,4,6-(1*H*,3*H*,5*H*)-pyrimidinetrione
Antimycin	3-Methylbutanoic acid 3-([3-(formylamino)-2-hydroxybenzoyl]amino)-8-hexyl-2,6-dimethyl-4, 9-dioxo-1,5-dioxonan-7-yl ester
Arylazido-β-alanine ADP-ribose	*N*-(4-Azido-2-nitrophenyl) β alanine
Atovaquone	2-[*trans*-4- (4′-Chlorophenyl)cyclohexyl]-3-hydroxy-1,4-naphthoquinone
Aureothin	2-Methoxy-3,5-dimethyl-6-(tetrahydro-4-[2-methyl-3-(4-nitrophenyl)-2-propenylidene]-2-furanyl)-4H-pyran-4-one
β-Carboline alkaloids	9*H*-Pyrido-[3,4-b]-indole derivatives
CCCP	Carbonyl cyanide *m*-chlorophenylhydrazone
DCCD	Dicyclohexylcarbodiimide
DNP	Dinitrophenol
FCCP	Carbonyl cyanide *p*-trifluoromethoxyphenylhydrazone
Fenazaquin	2-Decyl-4-quinazolinylamine
Fenpyroximate	4(*cis*-4-*tert*-Butylcyclohexylamino)5-chloro-6-ethylpyrimidine
Flavone	2-Phenyl-4*H*-1-benzopyran-4-one
HQNO	*n*-Heptyl-4-hydroxyquinoline-*N*-oxide
MIQ^+	1-Methyl-isoquinolone
MOA stilbene	3-Methoxy-2(2-styrylphenyl)propenic acid-methylester
MPTP	1-Methyl-4-phenyl-1,2,3,6-tetrahydropyridine
MPP^+	1-Methyl-4-phenylpyridinium
$N\text{-}Me\text{-}IQ^+$	*N*-Methyl-isoquinolinium ion
N-Me-TIQ	*N*-Methyl-1,2,3,4-tetrahydroisoquinoline
Piercidin	2,3-Dimethoxy-4-hydroxy-5-methyl-6-polyprenyl-pyridine
Rhein	4,5-Dihydroxyanthraquinone-2-carboxylic acid
Salsolinol	1-Methyl-6,7-dihydroxy-1,2,3,4-tetrahydroisoquinoline
TaClo	1-Trichloromethyl-1,2,3,4-tetrahydro-β-carboline
THIQ	1,2,3,4-Tetrahydroisoquinoline; also called TIQ
TIQ	1,2,3,4-Tetrahydroisoquinoline; also called THIQ
TTFA	4,4,4-Trifluoro-1-(2-thienyl)-butane-1,3-dione
TTFB	4,5,6,7-Tetrachloro-2-trifluoromethylbenzimidazole
UHDBT	5-*n*-Undecyl-6-hydroxy-4,7-dioxobenzothiazole

Organisms

Abbreviation	Full name	Abbreviation	Full name	Abbreviation	Full name
A. castellanii	*Acanthamoeba castellanii*	*H. diminuta*	*Hymenolepis diminuta*	*S. cerevisiae*	*Saccharomyces cerevisiae*
B. bovis	*Babesia bovis*	*L. braziliensis*	*Leishmania braziliensis*	*S. ratti*	*Strongyloides ratti*
C. albicans	*Candida albicans*	*L. mexicana*	*Leishmania mexicana*	*T. gondii*	*Toxoplasma gondii*
C. parapsilosis	*Candida parapsilosis*	*N. crassa*	*Neurospora crassa*	*T. brucei brucei*	*Trypanosoma brucei brucei*
C. reinhardtii	*Chlamydomonas reinhardtii*	*P. falciparum*	*Plasmodium falciparum*		
F. hepatica	*Fasciola hepatica*	*P. yoelii*	*Plasmodium yoelii*	*T. cruzi*	*Trypanosoma cruzi*
G. graminis	*Gaeumannomyces graminis*	*P. carinii*	*Pneumocystis carinii*	*Y. lipolytica*	*Yarrowia lipolytica*

[a]Complex I, NADH:ubiquinone reductase; complex II, succinate:ubiquinone reductase; complex III, ubiquinol:cytochrome *c* reductase; complex IV, cytochrome *c* oxidase; complex V, F_0F_1-ATPase; AOX, alternative oxidase; ANT, adenine nucleotide translocator.

[b]Abbreviations used: **T/C,** tissue of cells; M, mitochondria; **SMP,** submitochondrial particles; **IC,** isolated complex. The source is indicated in each.

[d]Center N [equivalently, the Qi, Q(i), or Qi site] is the ubiquinone reduction site in complex III. Center P [equivalently, the Qo, Q(o), or Qo site] is the ubiquinol oxidation site in complex III. Center N inhibitors include *n*-heptyl-4-hydroxyquinoline-*N*-oxide (HQNO), antimycins, and aurachins. Center P inhibitors include 5-*n*-undecyl-6-hydroxy-4,7-dioxobenzothiazole (UHDBT), β-methoxyacrylates, stigmatellin, and myxothiazol.

References listed in table

1. Degli Esposti M (1998) *Biochim. Biophys. Acta* **1364,** 222–235.
2. Miyoshi H (1998) *Biochim. Biophys. Acta* **1364,** 236–244.
3. Alali FQ *et al.* (1999) *J Nat. Prod.* **62,** 504–540.
4. Landolt JL *et al.* (1995) *Chem. Biol. Ineract.* **98,** 1–13.
5. Zafra-Polo MC *et al.* (1996) *Phytochemistry* **42,** 253–271.
6. Miyoshi H *et al.* (1998) *Biochim. Biophys. Acta* **1365,** 443–452.
7. Miyoshi H *et al.* (1998) *J. Biol. Chem.* **273,** 17368–17374.
8. Degli Esposti M *et al.* (1993) *Biochem. Biophys. Res. Commun.* **190,** 1090–1096.
9. Lummen P (1998) *Biochem. Biophys. Acta* **1364,** 287–96.
10. Pass MA *et al.* (1994) *Nat Toxins* **2,** 386–394.
11. Keller JN *et al.* (1998) *J. Neurosci.* **18,** 4439–4450.
12. Olsen C *et al.* (1999) *Brain Res.* **850,** 144–149.
13. Riepe MW *et al.* (1996) *Exp. Neurol.* **138,** 15–21.
14. Brorson JR *et al.* (1999) *J. Neurosci.* **19,** 147–158.
15. Lopez PS *et al.* (1998) *Toxicol. Appl. Pharmacol.* **148,** 1–6.
16. Bruce-Keller AJ *et al.* (1999) *J. Neuroimmunol.* **93,** 53–71.
17. Xoe Z *et al.* (2000) *Mol. Plant. Microbe Interact,* **13,** 183–190.
18. Eriksen NT *et al.* (1999) *Aqu. Microbial Ecol.* **17,** 145–152.
19. Hoefnagel MH *et al.* (1995) *Eur. J. Biochem.* **233,** 531–537.
20. Hoefnagel MH *et al.* (1995) *Arch. Biochem. Biophys.* **318,** 394–400.
21. Card JW *et al.* (1998) *Toxicol. Lett.* **98,** 41–50.
22. Fromenty B *et al.* (1990) *J. Pharmacol. Exp. Ther.* **255,** 1377–1384.
23. Ribeiro SM *et al.* (1997) *Cell. Biochem. Funct.* **15,** 145–152.
24. Chen CL *et al.* (1994) *Res Commun. Mol. Pathol. Pharmacol.* **85,** 193–208.
25. Kennedy JA *et al.* (1996) *Biochem. Pharmacol.* **52,** 273–280.
26. Becker LB *et al.* (1999) *Am. J. Physiol.* **277,** H2240–H2246.
27. Jurkowitz MS *et al.* (1998) *J. Neurochem.* **71,** 535–48.
28. Wiedemann FR *et al.* (1998) *J. Neurol. Sci.* **156,** 65–72.
29. Guerin M *et al.* (1989) *Biochimie* **71,** 887–902.
30. Kunz WS *et al.* (1999) *J. Neurochem.* **72,** 1580–1585.
31. Wyatt KM *et al.* (1995) *Biochem Pharmacol.* **50,** 1599–1606.
32. Railson JK *et al.* (1973) *J. Bioenerg.* **4,** 409–422.
33. Ahammadsahib KI *et al.* (1993) *Life Sci.* **53,** 1113–1120.
34. Friedrich T *et al.* (1994) *Eur. J. Biochem.* **219,** 691–698.
35. Lewis MA *et al.* (1993) *Pestic Biochem. Physiol.* **45,** 15–23.
36. Degli Esposti M *et al.* (1994) *Biochem. J.* **301,** 161–167
37. Estornell E *et al.* (1997) *Biochem. Biophys. Res. Commun.* **240,** 234–238.
38. Gallardo T *et al.* (1998) *J. Nat. Prod.* **61,** 1001–1005.
39. Schuler F *et al.* (1999) *Proc. Natl. Acad. Sci. USA* **96,** 4149–4153.
40. Tormo JR *et al.* (1999) *Chem. Biol. Interact.* **122,** 171–183.
41. Tormo JR *et al.* (1999) *Arch. Biochem. Biophys.* **369,** 119–126.
42. Londershausen M *et al.* (1991) *Pestic. Sci.* **33,** 427–438.
43. Joseph-Hone T *et al.* (1998) *J. Biol. Chem.* **273,** 11127–11133.
44. Schulze-Osthoff K *et al.* (1992) *J. Biol. Chem.* **267,** 5317–5323.
45. Biswas G *et al.* (1999) *EMBO J.* **18,** 522–533.
46. Grant CM *et al.* (1997) *FEBS Lett.* **410,** 219–222.
47. Maxwell DP *et al.* (1999) *Proc. Natl. Acad. Sci. USA* **96,** 8271–8276.
48. Srivastava IK *et al.* (1997) *J. Biol. Chem.* **272,** 3961–3966.
49. Vercesi AE *et al.* (1998) *J. Biol. Chem.* **273,** 31040-31047.
50. Beattie DS *et al.* (1996) *Eur. J. Biochem.* **241,** 888–894.
51. Bermudez R *et al.* (1997) *Mol. Biochem. Parasitol.* **90,** 43–54.
52. Kumar S *et al.* (1999) *Anal. Biochem.* **268,** 89–93.
53. Volkel S *et al.* (1996) *Eur. J. Biochem.* **235,** 231–237.
54. Zini R *et al.* (1999) *Drugs Exp. Clin. Res.* **25,** 87–97.
55. Ohkohchi N *et al.* (1999) *Transplantation* **67,** 1173–1177.
56. Miller AH *et al.* (1993) *FEBS Lett.* **329,** 259–262.
57. Matsuno-Yagi A *et al.* (1999) *J. Biol. Chem.* **274,** 9283–9288.
58. Rich PR *et al.* (1990) *Biochim. Biophys. Acta* **1018,** 29–40.
59. Moreno-Sanchez R *et al.* (1999) *Eur. J. Biochem.* **264,** 427–433.
60. Bechmann G *et al.* (1992) *Eur. J. Biochem.* **208,** 315–325.
61. Yu CA *et al.* (1998) *Biochim. Biophys. Acta* **1365,** 151–158.
62. Brasseur G *et al.* (1994) *FEBS Lett.* **354,** 23–29.
63. Cushion MT *et al.* (2000) *Antimicrob. Agents Chemother.* **44,** 713–719.
64. Fry M *et al.* (1992) *Biochem. Pharmacol.* **43,** 1545–1553.
65. Knecht W *et al.* (2000) *FEBS Lett.* **467,** 27–30.
66. Magae J *et al.* (1993) *Biosci. Biotechnol. Biochem.* **57,** 1628–1631.
67. Schwartz JL *et al.* (1976) *J. Antibiot.* (*Tokyo*) **29,** 236–241.
68. Maesaki S *et al.* (1998) *J. Antimicrob. Chemother.* **42,** 747–753.
69. Harvey J *et al.* (1999) *Br. J. Pharmacol.* **126,** 51–60.
70. Nguyen PV *et al.* (1997) *J. Neurophysiol.* **78,** 281–294.

71. Zager Ra *et al.* (1997) *Kidney Int.* **51,** 728–738.
72. Nakagawa Y *et al.* (1996) *Toxicology* **114,** 135–145.
73. Cassarino DS *et al.* (1998) *Biochem. Biophys. Res. Commun.* **248,** 168–173.
74. Starkov AA *et al.* (1997) *Biochim. Biophys. Acta* **1318,** 173–183.
75. Galkin MA *et al.* (1999) *FEBS Lett.* **448,** 123–126.
76. Rivas P *et al.* (1999) *Comp. Biochem. Physiol. C Pharm. Tox. Endocrinol.* **122,** 27–31.
77. Krueger MJ *et al.* (1993) *Biochem. J.* **291,** 673–676.
78. Janetzky B *et al.* (1995) *J. Neural Transm. Suppl.* **46,** 265–273.
79. Albores R *et al.* (1990) *Proc. Natl. Acad. Sci. USA* **87,** 9368–9372.
80. Yamakawa T *et al.* (1999) *Neurosci. Lett.* **259,** 157–160.
81. Yao Z *et al.* (1999) *Am. J. Physiol.* **277,** H2504-H2509.
82. Chandel NS *et al.* (1997) *J. Biol. Chem.* **272,** 18808–18816.
83. Galkin As *et al.* (1999) *FEBS Lett.* **451,** 157–161.
84. Junemann S *et al.* (1998) *J. Biol. Chem.* **273,** 21603–21607.
85. Sasse F *et al.* (1999) *J. Antibiot (Tokyo)* **52,** 721–729.
86. Kozlov AV *et al.* (1999) *FEBS Lett.* **454,** 127–130.
87. Kim H *et al.* (1998) *Proc. Natl. Acad. Sci. USA* **95,** 8026–8033.
88. Crofts AR *et al.* (1999) *Biochemistry* **38,** 15807–15826.
89. Crofts AR *et al.* (1999) *Proc. Natl. Acad. Sci. USA* **96,** 10021–10026.
90. Geier BM *et al.* (1992) *Eur. J. Biochem.* **208,** 375–380.
91. Mansfield RW *et al.* (1990) *Biochim. Biophys. Acta* **1015,** 109–115.
92. Von Jagow G *et al.* (1986) *Biochemistry* **25,** 775–778.
93. Satoh T *et al.* (1996) *Biochim Biophys Acta* **1273,** 21–30.
94. Chudapongse P *et al.* (1981) *Biochem. Pharmarcol.* **30,** 735–740.
95. Okun JG *et al.* (1999) *J. Biol. Chem.* **274,** 2625–2630.
96. Shimomura Y *et al.* (1989) *Arch. Biochem. Biophys.* **270,** 573–577.
97. Cock HR *et al.* (1999) *J. Neurol. Sci.* **165,** 10–17.
98. Ragan CI *et al.* (1977) *Biochem. J.* **163,** 605–615.
99. Roberts TH *et al.* (1995) *FEBS Lett.* **373,** 307–309.
100. Bykova NV *et al.* (1999) *Biochem. Biophys. Res. Commun.* **265,** 106–111.
101. Majander A *et al.* (1994) *J. Biol. Chem.* **269,** 21037–21042.
102. Herrero A *et al.* (1997) *Mech. Aging Dev.* **98,** 95–111.
103. Anderson WM *et al.* (1995) *Biochim. Biophys. Acta* **1230,** 186–193.
104. Hollingworth RM *et al.* (1994) *Biochem. Soc. Trans.* **22,** 230–233.
105. Rowlands JC *et al.* (1998) *Pharmacol. Toxicol.* **83,** 214–219.
106. Latli B *et al.* (1996) *Chem. Res. Toxicol.* **9,** 445–450.
107. Obata T *et al.* (1992) *Pestic. Sci.* **34,** 133–138.
108. Seo BB *et al.* (1998) *Proc. Natl. Acad. Sci. USA* **95,** 9167–9171.
109. Hodnick WF *et al.* (1988) *Biochem. Pharmacol.* **37,** 2607–2611.
110. Hodnick WF *et al.* (1994) *Biochem. Pharmacol.* **47,** 573–580.
111. Ratty AK *et al.* (1988) *Biochem. Med. Metab. Biol.* **39,** 69–79.
112. Santos AC *et al.* (1998) *Free Radic. Biol. Med.* **24,** 1455–1461.
113. Fang N *et al.* (1998) *Proc. Natl. Acad. Sci. USA* **95,** 3380–3384.
114. de Vries S *et al.* (1988) *Eur. J. Biochem.* **176,** 377–384.
115. Pietrobon D *et al.* (1981) *Eur. J. Biochem.* **117,** 389–394.
116. Degli Esposti M *et al.* (1992) *Arch. Biochem. Biophys.* **295,** 198–204.
117. Gozar MM *et al.* (1992) *Int. J. Parasitol.* **22,** 165–171.
118. Howell N *et al.* (1988) *J. Mol. Biol.* **203,** 607–618.
119. Brambilla L *et al.* (1998) *FEBS Lett.* **431,** 245–249.
120. Halestrap AP (1987) *Biochim. Biophys. Acta* **927,** 280–290.
121. Matsuno-Yagi A *et al.* (1996) *J. Biol. Chem.* **271,** 6164–6171.
122. Seaton TA *et al.* (1997) *Brain Res.* **777,** 110–118.
123. Melzig MF *et al.* (1993) *Neurochem. Res.* **18,** 689–693.
124. Suzuki K *et al.* (1989) *Biochem. Biophys. Res. Commun.* **162,** 1541–1545.
125. Suzuki K *et al.* (1992) *J. Neurol. Sci.* **109,** 219–223.
126. McNaught KS *et al.* (1996) *Biochem. Pharmacol.* **51,** 1503–1511.
127. Aiuchi T *et al.* (1996) *Neurochem. Int.* **28,** 319–323.
128. Miyoshi H *et al.* (1997) *J. Biol. Chem.* **272,** 16176–16183.
129. McNaught KS *et al.* (1995) *Biochem. Pharmacol.* **50,** 1903–1911.
130. McNaught KS *et al.* (1995) *Neuroreport* **6,** 1105–1108.
131. Minami M *et al.* (1993) *J. Neural. Transm Gen. Sect.* **92,** 125–135.
132. Klein BY *et al.* (1996) *J. Cell Biochem.* **60,** 139–147.
133. Brand MD *et al.* (1994) *Eur. J. Biochem.* **226,** 819–829.
134. Vogel R *et al.* (1999) *Neurosci. Lett.* **275,** 97–100.
135. Korshunov SS *et al.* (1997) *FEBS Lett.* **416,** 15–18.
136. Chien LF *et al.* (1996) *Biochem. J.* **320,** 837–845.
137. Catia Sorgato M *et al.* (1985) *FEBS Lett.* **181,** 323–327.
138. Hederstedt L *et al.* (1986) *Arch. Biochem. Biophys.* **247,** 346–354.
139. Fall CP *et al.* (1999) *J. Neurosci. Res.* **55,** 620–628.
140. Camins A *et al.* (1997) *J. Neural Transm.* **104,** 569–577.
141. Murphy MP *et al.* (1995) *Biochem. J.* **306,** 359–365.

142. Cassarino DS *et al.* (1999) *Biochim. Biophys. Acta* **1453,** 49–62.
143. Pardo B *et al.* (1995) *J. Neurochem.* **64,** 576–582.
144. Bougria M *et al.* (1995) *Eur. J. Pharmacol.* **291,** 407–415.
145. Pecci L *et al.* (1994) *Biochem. Biophys. Res. Commun.* **205,** 264–268.
146. Singer TP *et al.* (1994) *Biochim. Biophys. Acta* **1187,** 198–202.
147. Gluck MR *et al.* (1994) *J. Biol. Chem.* **269,** 3167–3174.
148. Hasegawa E *et al.* (1997) *Arch. Biochem. Biophys.* **337,** 69–74.
149. Sablin SO *et al.* (1996) *J. Biochem. Toxicol.* **11,** 33–43.
150. Castagnoli N. Jr. *et al.* (1985) *Life Sci.* **36,** 225–230.
151. Heikkila RE *et al.* (1985) *Neurosci. Lett.* **62,** 389–394.
152. Mizuno Y *et al.* (1988) *J. Neurol. Sci.* **86,** 97–110.
153. Chalmers-Redman *et al.* (1999) *Biochem. Biophys. Res. Commun.* **257,** 440–447.
154. Miyako K *et al.* (1999) *Eur. J. Biochem.* **259,** 412–418.
155. Umeda S *et al.* (2000) *Eur. J. Biochem.* **267,** 200–206.
156. Matsuno-Yagi A *et al.* (1997) *J. Biol. Chem.* **272,** 16928–16933.
157. Bhuvaneswaran C *et al.* (1972) *Experientia* **28,** 777–778.
158. Kerscher SJ *et al.* (1999) *J. Cell Sci.* **112,** 2347–2354.
159. Chung KH *et al.* (1989) *Zeit Naturfor Sec C J Biosci.* **44,** 609–616.
160. Bironaite D *et al.* (1997) *Chem. Biol. Interact.* **103,** 35–50.
161. Miccadei S *et al.* (1993) *Anticancer Res.* **13,** 1507–1510.
162. Floridi A *et al.* (1994) *Biochem. Pharmacol.* **47,** 1781–1788.
163. Glinn MA *et al.* (1997) *Biochim. Biophys. Acta* **1318,** 246–254.
164. Di Monte DA *et al.* (1999) *Toxicol. Appl. Pharmacol.* **158,** 296–302.
165. Gerhauser C *et al.* (1997) *Cancer Res.* **57,** 3429–3435.
166. Ueno H *et al.* (1996) *Biochim. Biophys. Acta* **1276,** 195–202.
167. Udeani GO *et al.* (1997) *Cancer Res.* **57,** 3424–3428.
168. Konishi K *et al.* (1993) *Biol. Pharm. Bull.* **16,** 716–718.
169. Konishi K *et al.* (1999) *boil. Pharm. Bull.* **22,** 240–243.
170. Igamberdiev AU *et al.* (1995) *FEBS Lett.* **367,** 287–290.
171. Suzuki S *et al.* (1999) *Oncogene* **18,** 6380–6387.
172. van de Water B *et al.* (1995) *Mol. Pharmacol.* **48,** 928–937.
173. Berridge MV *et al.* (1993) *Arch. Biochem. Biophys.* **303,** 474–482.
174. Barja G *et al.* (1998) *J. Bioenerg. Biomembr.* **30,** 235–243.
175. Suno M *et al.* (1989) *Arch. Gerontol. Geriatr.* **8,** 291–297.
176. Yu L *et al.* (1987) *J. Biol. Chem.* **262,** 1137–1143.
177. Armson A *et al.* (1995) *Int. J. Parasitol.* **25,** 257–260.
178. Turrens JF *et al.* (1986) *Mol. Biochem. Parasitol.* **19,** 259–264.
179. Cook ND *et al.* (1985) *Arch. Biochem. Biophys.* **240,** 9–14.
180. Trumpower BL *et al.* (1980) *J. Bioenerg. Biomembr.* **12,** 151–164.
181. Zhang L *et al.* (1999) *FEBS Lett.* **460,** 349–352.
182. Ljungdahl PO *et al.* 1989) *J. Biol. Chem.* **264,** 3723–3731.
183. Kohnke D *et al.* (1997) *Mol. Cell. Biochem.* **174,** 101–113.
184. Lahnsteiner F *et al.* (1999) *J. Exp. Zool.* **284,** 454–465.
185. Mildaziene V *et al.* (1995) *Arch. Biochem. Biophys.* **324,** 130–134.
186. Elimadi A *et al.* (1997) *Br. J. Pharmacol.* **121,** 1295–1300.
187. Andreyev A *et al.* (1989) *Eur. J. Biochem.* **182,** 585–592.
188. Roucou X *et al.* (1997) *Biochim. Biophys. Acta* **1324,** 120–132.
189. Ziegler M *et al.* (1993) *J. Biol. Chem.* **268,** 25320–25328.
190. Brandolin G *et al.* (1993) *J. Bioenerg. Biomembr.* **25,** 459–472.
191. Fiore C *et al.* (2000) *Protein Expr. Purif.* **19,** 57–65.
192. Klingenberg M *et al.* (1975) *Eur. J. Biochem.* **52,** 351–363.
193. Nakashima RA *et al.* (1984) *Cancer Res.* **44,** 5702–5706.
194. Buchet K *et al.* (1998) *J. Biol. Chem.* **273,** 22983–22989.
195. Vazquez-Laslop N *et al.* (1990) *J. Biol. Chem.* **265,** 19002–19006.
196. Valerio M *et al.* (1994) *Eur. J. Biochem.* **221,** 1071–1078.
197. Persson B *et al.* (1987) *Biochim. Biophys. Acta* **894,** 239–257.
198. Cataldi de Flombaum MA *et al.* (1981) *Mol. Biochem. Parasitol.* **3,** 143–155.
199. Appleby RD *et al.* (1999) *Eur. J. Biochem.* **262,** 108–116.
200. Budd SL *et al.* (2000) Proc. Natl. Acad. Sci. USA **97,** 6161–6166.
201. Pastorino JG *et al.* (2000) *Hepatology* **31,** 1141–1152.
202. Stridh H *et al.* (1999) *Chem. Res. Toxicol.* **12,** 874–882.
203. Furlong IJ *et al.* (1998) *Cell Death Differ.* **5,** 214–221.
204. Griffiths EJ *et al.* (1993) *Biochem. J.* **290,** 489–495.
205. Hermesh O *et al.* (2000) *Biochim. Biophys. Acta* **1457,** 166–174.
206. Majima E *et al.* (1995) *J. Biol. Chem.* **270,** 29548–29554.
207. Brustovetsky N *et al.* (2000) *J. Neurosci.* **20,** 103–113.
208. Huber T *et al.* (1999) *Biochemistry* **38,** 762–769.
209. Simonyan RA *et al.* (1998) *FEBS Lett.* **436,** 81–84.
210. Valenti D *et al.* (1999) *FEBS Lett.* **444,** 291–295.

211. Shima A *et al.* (1990) *Cell Struct. Funct.* **15,** 53–58.
212. Lenton LM *et al.* (1994) *Biochim. Biophys. Acta* **1186,** 237–242.
213. Guerin B *et al.* (1994) *J. Biol. Chem.* **269,** 25406–25410.
214. Gaballo A *et al.* (1998) *Biochemistry* **37,** 17519–17526.
215. Hekman C *et al.* (1991) *J. Biol. Chem.* **266,** 13564–13571.
216. Mercer NA *et al.* (1999) *Exp. Parasitol.* **91,** 52–58.
217. Walker JE *et al.* (1991) *Biochemistry* **30,** 5369–5378.
218. Hamasur B *et al.* (1992) *Eur. J. Biochem.* **205,** 409–416.
219. Miki T *et al.* (1994) *J. Biol. Chem.* **269,** 1827–1833.
220. Wikstrom M *et al.* (1985) *J. Inorg Biochem.* **23,** 327–334.
221. Prochaska LJ *et al.* (1986) *Biochemistry* **25,** 781–787.
222. Beattie DS (1993) *J. Bioenerg. Biomembr.* **25,** 233–244.
223. Shoshan-Barmatz V *et al.* (1996) *FEBS Lett.* **386,** 205–210.
224. Edwards SW *et al.* (1982) *Biochem. J.* **202,** 452–458.
225. Simmons WA *et al.* (1983) *Somatic Cell Genet.* **9,** 549–566.
226. Lardy H *et al.* (1975) *Fed. Proc.* **34,** 1707–1710.
227. Bossard MJ *et al.* (1981) *J. Biol. Chem.* **256,** 1518–1521.
228. Cross RL *et al.* (1978) *J. Biol. Chem.* **253,** 4865–4873.
229. Vazquez-Memije ME *et al.* (1984) *Arch. Biochem. Biophys.* **232,** 441–449.
230. Abrahams JP *et al.* (1996) *Proc. Natl. Acad. Sci. USA* **93,** 9420–9424.
231. Muroi M *et al.* (1996) *Biochem. Biophys. Res. Commun.* **227,** 800–809.
232. Panov AV *et al.* (1995) *Arch Biochem. Biophys.* **316,** 815–820.
233. Valerio M *et al.* (1993) *FEBS Lett.* **336,** 83–86.
234. Matsuno-Yagi A *et al.* (1993) *J. Biol. Chem.* **268,** 6168–6173.
235. Matsuno-Yagi A *et al.* (1993) *J. Biol. Chem.* **268,** 1539–1545.
236. Miedema H *et al.* (1994) *Biochem. Biophys. Res. Commun.* **203,** 1005–1012.
237. Collinson IR *et al.* (1994) *Biochemistry* **33,** 7971–7978.
238. Cappelli-Bigazzi M *et al.* (1997) *J. Mol. Cell. Cardiol.* **29,** 871–879.
239. Pizzo P *et al.* (1997) *J. Cell Biol.* **136,** 355–366.
240. Marsh DC *et al.* (1993) *Hepatology* **17,** 91–98.
241. Jung DW *et al.* (1996) *Arch. Biochem. Biophys.* **332,** 19–29.
242. Panov A *et al.* (1996) *Biochemistry* **35,** 12849–12856.
243. Meinicke AR *et al.* (1998) *Arch. Biochem. Biophys.* **349,** 275–280.
244. Benaim G *et al.* (1990) *Mol. Biochem. Parasitol.* **39,** 61–68.
245. Breitbart H *et al.* (1990) *Biochim. Biophys. Acta.* **1026,** 57–63.
246. Hayes DJ *et al.* (1999) *Mol. Biochem. Parasitol.* **38,** 159–168.
247. Linsinger G *et al.* (1999) *Mol. Cell. Biol.* **19,** 3299–3311.
248. Wu SN *et al.* (1999) *J. Pharmacol. Exp. Ther.* **290,** 998–1005.
249. Sureda FX *et al.* (1997) *Cytometry* **28,** 74–80.
250. Wanke M *et al.* (2000) *Biochim. Biophys. Acta* **1463,** 188–194.
251. Curti C *et al.* (1999) *Mol. Cell. Biochem.* **199,** 103–109.
252. Vogel R *et al.* (1999) *Neurosci. Lett.* **275,** 97–100.
253. Starkov AA *et al.* (1997) *Biochim. Biophys. Acta* **1318,** 159–172.
254. Tran TV *et al.* (1991) *Biochim. Biophys. Acta* **1059,** 265–274.
255. Cocco T *et al.* (1992) *Eur. J. Biochem.* **209,** 475–481.
256. Oyedotun KS *et al.* (1999) *J. Biol. Chem.* **274,** 23956–23962.
257. Vianello A *et al.* (1995) *FEBS Lett.* **365,** 7–9.
258. Colleoni M *et al.* (1996) *Pharmacol. Toxicol.* **78,** 69–76.
259. Tan AK *et al.* (1993) *J. Biol. Chem.* **268,** 19328–19333.
260. Yankovskaya V *et al.* (1996) *J. Biol. Chem.* **271,** 21020–21024.
261. Xiong ZH *et al.* (1997) *J. Biol. Chem.* **272,** 31022–31028.
262. Mohell N *et al.* (1987) *Am. J. Physiol.* **253,** C301–C308.
263. Rodeheaver DP *et al.* (1993) *J. Pharmacol. Exp. Ther.* **265,** 1355–1360.
264. Jezek P *et al.* (1998) *Am. J. Physiol.* **275,** C496–C504.
265. Catisti R *et al.* (1999) *FEBS Lett.* **464,** 97–101.
266. Rial E *et al.* (1999) *EMBO J.* **18,** 5827–5833.
267. Ruben L *et al.* (1991) *J. Biol. Chem.* **266,** 24351–24358.
268. Haass-Mannle H *et al.* (1997) *J. Photochem. Photobiol. B* **41,** 90–102.
269. Kotlyar AB *et al.* (1998) *Biochim. Biophys. Acta.* **1365,** 53–59.
270. Rottenberg H *et al.* (1989) *Biochemistry* **28,** 4355–4360.
271. Erecinska M *et al.* (1996) *Brain Res.* **726,** 153–159.
272. Kawanishi T *et al.* (1991) *J. Biol. Chem.* **266,** 20062–20069.
273. Grijalba MT *et al.* (1998) *Free Radic. Res.* **28,** 301–318.
274. Shimizu S *et al.* (1998) *Proc. Natl. Acad. Sci. USA* **95,** 1455–1459.
275. Matsuno-Yagi A *et al.* (1989) *Biochemistry* **28,** 4367–4374.
276. Holmuhamedov EL *et al.* (1999) *J. Physiol.* (*Lond*) **519** Pt 2, 347–360.
277. Tolkatchev D *et al.* (1996) *J. Biol. Chem.* **271,** 12356–12363.

INDEX

D

J

K

L

M

R

S

VOLUMES IN SERIES

Founding Series Editor
DAVID M. PRESCOTT

Volume 1 (1964)
Methods in Cell Physiology
Edited by David M. Prescott

Volume 2 (1966)
Methods in Cell Physiology
Edited by David M. Prescott

Volume 3 (1968)
Methods in Cell Physiology
Edited by David M. Prescott

Volume 4 (1970)
Methods in Cell Physiology
Edited by David M. Prescott

Volume 5 (1972)
Methods in Cell Physiology
Edited by David M. Prescott

Volume 6 (1973)
Methods in Cell Physiology
Edited by David M. Prescott

Volume 7 (1973)
Methods in Cell Biology
Edited by David M. Prescott

Volume 8 (1974)
Methods in Cell Biology
Edited by David M. Prescott

Volume 9 (1975)
Methods in Cell Biology
Edited by David M. Prescott

Volume 10 (1975)
Methods in Cell Biology
Edited by David M. Prescott

Volume 11 (1975)
Yeast Cells
Edited by David M. Prescott

Volume 12 (1975)
Yeast Cells
Edited by David M. Prescott

Volume 13 (1976)
Methods in Cell Biology
Edited by David M. Prescott

Volume 14 (1976)
Methods in Cell Biology
Edited by David M. Prescott

Volume 15 (1977)
Methods in Cell Biology
Edited by David M. Prescott

Volume 16 (1977)
Chromatin and Chromosomal Protein Research I
Edited by Gary Stein, Janet Stein, and Lewis J. Kleinsmith

Volume 17 (1978)
Chromatin and Chromosomal Protein Research II
Edited by Gary Stein, Janet Stein, and Lewis J. Kleinsmith

Volume 18 (1978)
Chromatin and Chromosomal Protein Research III
Edited by Gary Stein, Janet Stein, and Lewis J. Kleinsmith

Volume 19 (1978)
Chromatin and Chromosomal Protein Research IV
Edited by Gary Stein, Janet Stein, and Lewis J. Kleinsmith

Volume 20 (1978)
Methods in Cell Biology
Edited by David M. Prescott

Advisory Board Chairman
KEITH R. PORTER

Volume 21A (1980)
Normal Human Tissue and Cell Culture, Part A: Respiratory, Cardiovascular, and Integumentary Systems
Edited by Curtis C. Harris, Benjamin F. Trump, and Gary D. Stoner

Volume 21B (1980)
Normal Human Tissue and Cell Culture, Part B: Endocrine, Urogenital, and Gastrointestinal Systems
Edited by Curtis C. Harris, Benjamin F. Trump, and Gray D. Stoner

Volume 22 (1981)
Three-Dimensional Ultrastructure in Biology
Edited by James N. Turner

Volume 23 (1981)
Basic Mechanisms of Cellular Secretion
Edited by Arthur R. Hand and Constance Oliver

Volume 24 (1982)
The Cytoskeleton, Part A: Cytoskeletal Proteins, Isolation and Characterization
Edited by Leslie Wilson

Volume 25 (1982)
The Cytoskeleton, Part B: Biological Systems and *in Vitro* Models
Edited by Leslie Wilson

Volume 26 (1982)
Prenatal Diagnosis: Cell Biological Approaches
Edited by Samuel A. Latt and Gretchen J. Darlington

Series Editor
LESLIE WILSON

Volume 27 (1986)
Echinoderm Gametes and Embryos
Edited by Thomas E. Schroeder

Volume 28 (1987)
Dictyostelium discoideum: Molecular Approaches to Cell Biology
Edited by James A. Spudich

Volume 29 (1989)
Fluorescence Microscopy of Living Cells in Culture, Part A: Fluorescent Analogs, Labeling Cells, and Basic Microscopy
Edited by Yu-Li Wang and D. Lansing Taylor

Volume 30 (1989)
Fluorescence Microscopy of Living Cells in Culture, Part B: Quantitative Fluorescence Microscopy—Imaging and Spectroscopy
Edited by D. Lansing Taylor and Yu-Li Wang

Volume 31 (1989)
Vesicular Transport, Part A
Edited by Alan M. Tartakoff

Volume 32 (1989)
Vesicular Transport, Part B
Edited by Alan M. Tartakoff

Volume 33 (1990)
Flow Cytometry
Edited by Zbigniew Darzynkiewicz and Harry A. Crissman

Volume 34 (1991)
Vectorial Transport of Proteins into and across Membranes
Edited by Alan M. Tartakoff

Selected from Volumes 31, 32, and 34 (1991)
Laboratory Methods for Vesicular and Vectorial Transport
Edited by Alan M. Tartakoff

Volume 35 (1991)
Functional Organization of the Nucleus: A Laboratory Guide
Edited by Barbara A. Hamkalo and Sarah C. R. Elgin

Volume 36 (1991)
Xenopus laevis: Practical Uses in Cell and Molecular Biology
Edited by Brian K. Kay and H. Benjamin Peng

Series Editors
LESLIE WILSON AND PAUL MATSUDAIRA

Volume 37 (1993)
Antibodies in Cell Biology
Edited by David J. Asai

Volume 38 (1993)
Cell Biological Applications of Confocal Microscopy
Edited by Brian Matsumoto

Volume 39 (1993)
Motility Assays for Motor Proteins
Edited by Jonathan M. Scholey

Volume 40 (1994)
A Practical Guide to the Study of Calcium in Living Cells
Edited by Richard Nuccitelli

Volume 41 (1994)
Flow Cytometry, Second Edition, Part A
Edited by Zbigniew Darzynkiewicz, J. Paul Robinson, and Harry A. Crissman

Volume 42 (1994)
Flow Cytometry, Second Edition, Part B
Edited by Zbigniew Darzynkiewicz, J. Paul Robinson, and Harry A. Crissman

Volume 43 (1994)
Protein Expression in Animal Cells
Edited by Michael G. Roth

Volume 44 (1994)
Drosophila melanogaster: Practical Uses in Cell and Molecular Biology
Edited by Lawrence S. B. Goldstein and Eric A. Fyrberg

Volume 45 (1994)
Microbes as Tools for Cell Biology
Edited by David G. Russell

Volume 46 (1995) (in preparation)
Cell Death
Edited by Lawrence M. Schwartz and Barbara A. Osborne

Volume 47 (1995)
Cilia and Flagella
Edited by William Dentler and George Witman

Volume 48 (1995)
Caenorhabditis elegans: Modern Biological Analysis of an Organism
Edited by Henry F. Epstein and Diane C. Shakes

Volume 49 (1995)
Methods in Plant Cell Biology, Part A
Edited by David W. Galbraith, Hans J. Bohnert, and Don P. Bourque

Volume 50 (1995)
Methods in Plant Cell Biology, Part B
Edited by David W. Galbraith, Don P. Bourque, and Hans J. Bohnert

Volume 51 (1996)
Methods in Avian Embryology
Edited by Marianne Bronner-Fraser

Volume 52 (1997)
Methods in Muscle Biology
Edited by Charles P. Emerson, Jr. and H. Lee Sweeney

Volume 53 (1997)
Nuclear Structure and Function
Edited by Miguel Berrios

Volume 54 (1997)
Cumulative Index

Volume 55 (1997)
Laser Tweezers in Cell Biology
Edited by Michael P. Sheez

Volume 56 (1998)
Video Microscopy
Edited by Greenfield Sluder and David E. Wolf

Volume 57 (1998)
Animal Cell Culture Methods
Edited by Jennie P. Mather and David Barnes

Volume 58 (1998)
Green Fluorescent Protein
Edited by Kevin F. Sullivan and Steve A. Kay

Volume 59 (1998)
The Zebrafish: Biology
Edited by H. William Detrich III, Monte Westerfield, and Leonard I. Zon

Volume 60 (1998)
The Zebrafish: Genetics and Genomics
Edited by H. William Detrich III, Monte Westerfield, and Leonard I. Zon

Volume 61 (1998)
Mitosis and Meiosis
Edited by Conly L. Rieder

Volume 62 (1999)
Tetrahymena Thermophila
Edited by David J. Asai and James D. Forney

Volume 63 (2000)
Cytometry, Third Edition, Part A
Edited by Zbigniew Darzynkiewicz, J. Paul Robinson, and Harry Crissman

Volume 64 (2000)
Cytometry, Third Edition, Part B
Edited by Zbigniew Darzynkiewicz, J. Paul Robinson, and Harry Crissman

ISBN 0-12-544169-X

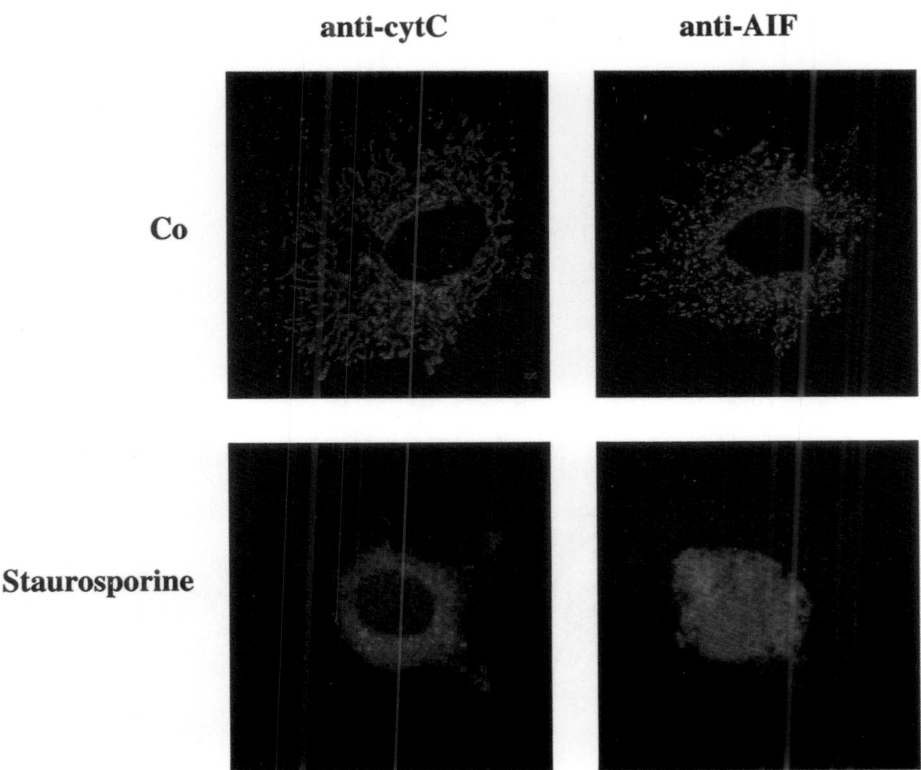

Fig. 8.2 Analysis of AIF and cytochrome *c* release from mitochondria. Mouse embryonic fibroblasts (MEF) cells, in the absence (Co) and presence of staurosporin (8 h, 2 μM), were stained for AIF (red fluorescence) or cytochrome *c* (green fluorescence), followed by confocal analysis. In these experimental conditions, staurosporin-treated cells have released both AIF and cytochrome *c* from the punctate mitochondrial localization. Note that only AIF translocates to the nucleus, where it induces fragmentation of the DNA into high molecular weight fragments (not shown). Immunostaining protocol: A rabbit antiserum was generated against a mixture of three peptides derived from the mAIF amino acid (aa) sequence (aa 151–170, 166–185, 181–200, coupled to keyhole limpet hemocyanine, generated by Syntem, Nîmes, France). This antiserum (ELISA titer ~10,000) was used (diluted 1/1000) on paraformaldehyde (4%, w/v) and picric acid-fixed (0.19% v/v) cells (cultured on 100-μm coverslips; 18 mm diameter; Superior, Germany), and revealed with a goat antirabbit IgG conjugated to phycoerythrine (Southern Biotechnology, Birmingham, AL). Cells were counterstained for the detection of cytochrome *c* (mAb 6H2.B4 from Pharmingen), revealed by a goat antimouse IgG fluorescein isothiocyanate conjugate (Southern Biotechnology).

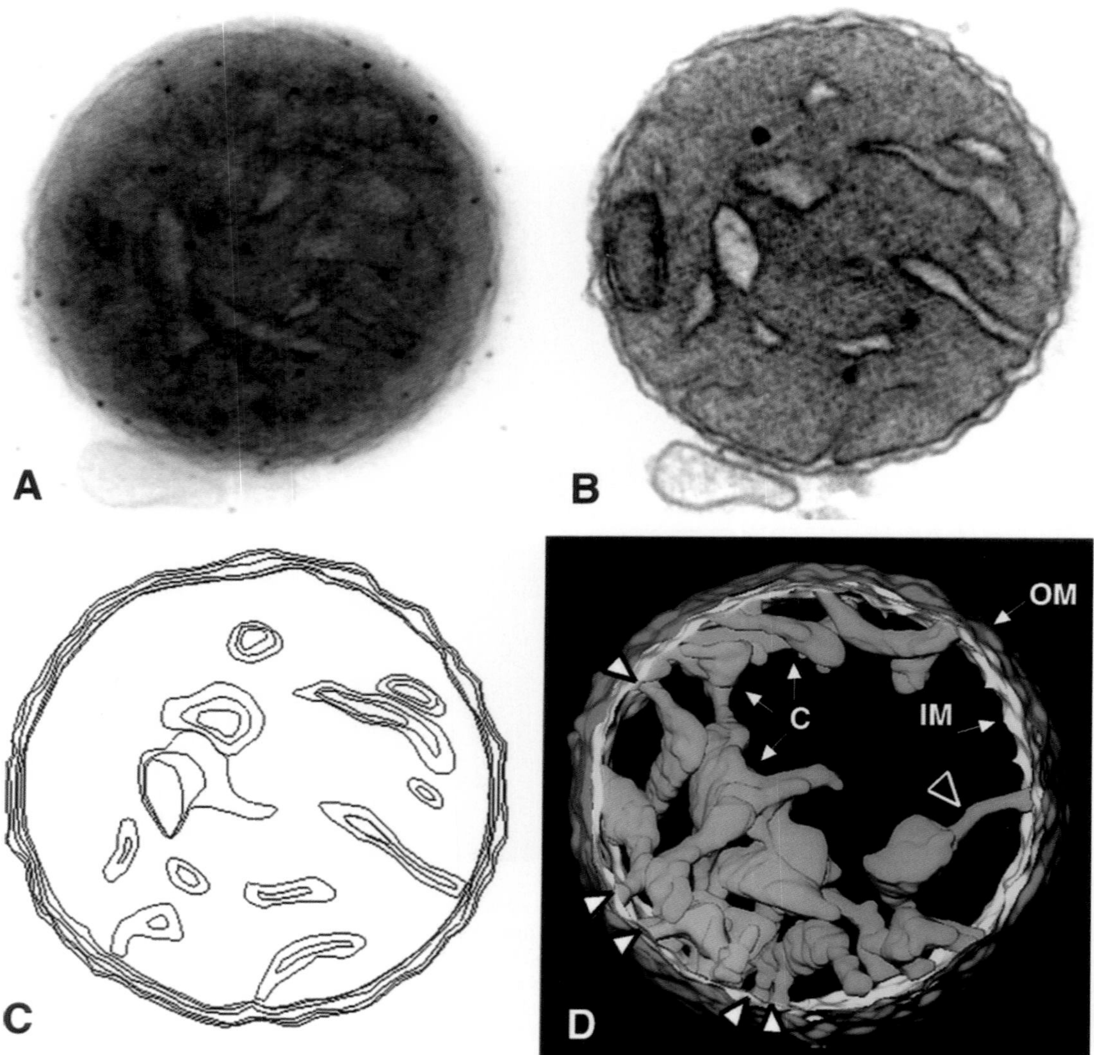

Fig. 14.1 Tomographic reconstruction of a conventionally fixed and plastic-embedded rat liver mitochondrion. (A) Projection image recorded at 1000 kV of a 0.5-μm-thick section of an isolated, partially "condensed" rat liver mitochondrion 1.5 μm in diameter. (B) Five-nanometer slice through a tomogram reconstructed by weighted back-projection of a double-tilt series of images from the mitochondrion in A by the procedure of Penczek *et al.* (1995). (C) Contours drawn of membrane profiles in three slices from the 3D reconstruction, 15 nm apart. (D) Surface-rendered model of the mitochondrion generated from a full set of membrane contours showing the outer membrane (OM), inner boundary membrane (IM), and selected cristae (C). Arrowheads point to narrow (ca. 30 nm) tubular regions of the cristae that connect the internal compartments to each other and to the inner boundary membrane. Reproduced with permission from Mannella (2000).

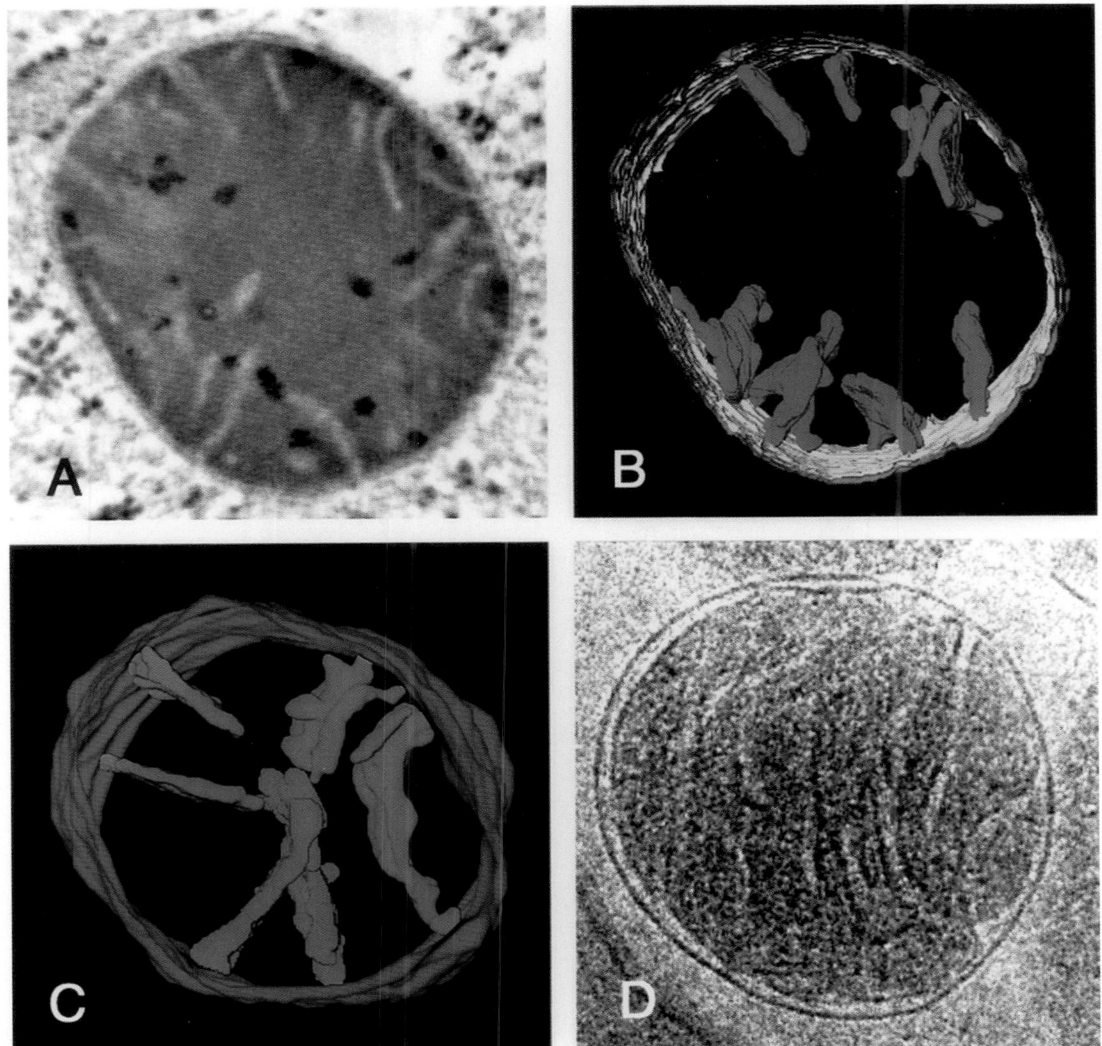

Fig. 14.2 Application of cryotechniques to mitochondrial imaging. (A) A 5-nm slice through a tomogram of an "orthodox" mitochondrion (700 nm across, 220 nm thick) in a specimen of high-pressure frozen, cryosubstituted rat liver. (B) Surface-rendered model of the mitochondrion in A showing outer and inner boundary membranes and a few of the cristae. A and B reproduced with permission from Mannella *et al.* (1998). (C) Three-dimensional reconstruction of a mitochondrion (900 nm across, 250 nm thick) in a section of rat liver prepared by the procedure of Tokuyasu (1989). Reproduced with permission from Mannella *et al.* (1997). (D) Cryoelectron microscopic image of an unstained, frozen-hydrated intact mitochondrion (500 nm diameter) isolated from *Neurospora crassa.*

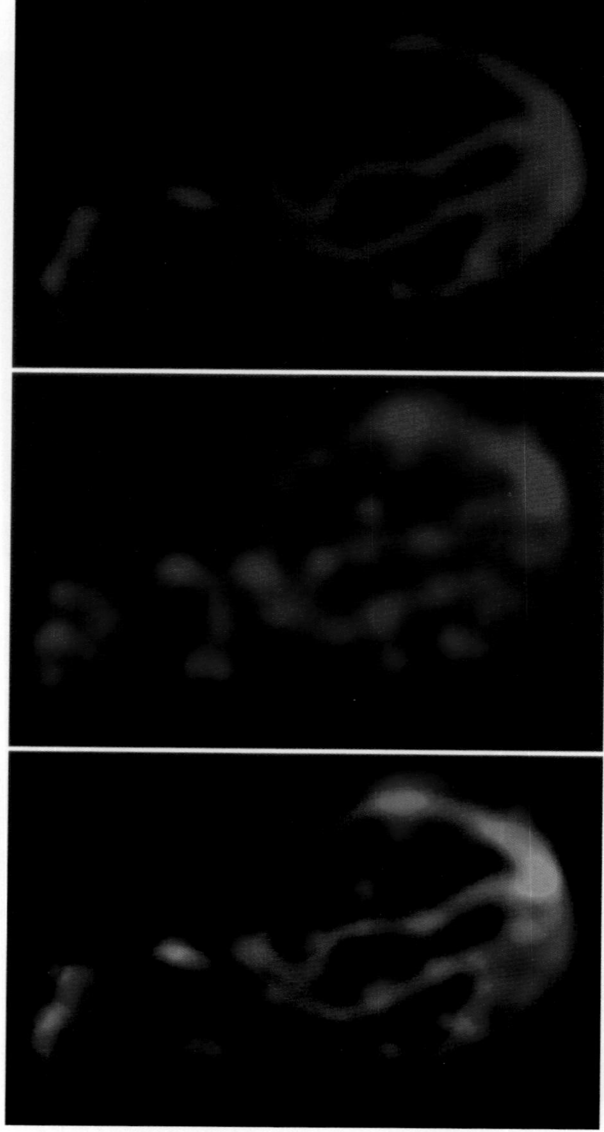

Fig. 15.4 Indirect immunofluorescence of a *S. cerevisiae* cell fixed with paraformaldehyde showing staining of a cytoplasmic protein, Arp2p, that is targeted to mitochondria. The green image shows fluorescence staining obtained with a CS1-GFP marker, which targets GFP to the mitochondrial matrix via a 52 amino acid citrate synthase leader sequence. The red image shows staining obtained using a primary rabbit antibody raised against a unique peptide within the Arp2 protein sequence, and a secondary rhodamine-conjugated goat–antirabbit antibody. The combined image shows Arp2p punctate structures coalign with mitochondrial tubules.

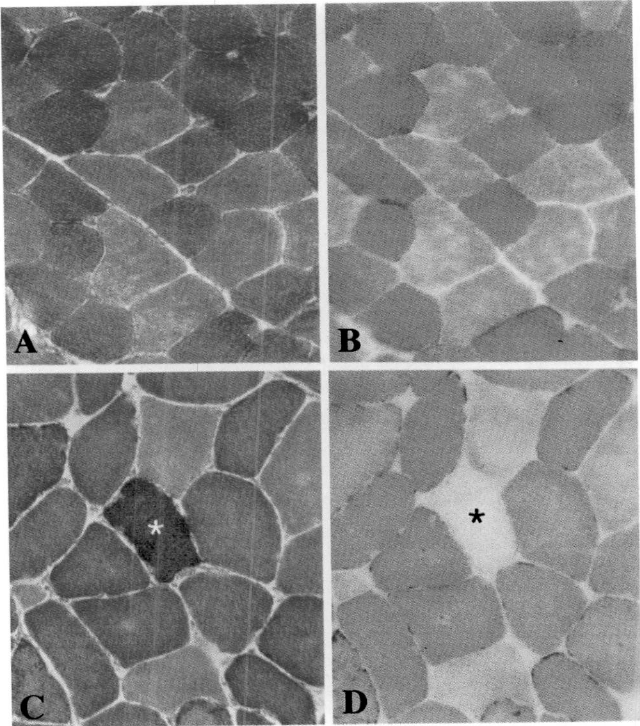

Fig. 18.1 Histochemical stains for SDH (A) and COX (B) activities on serial muscle sections from a normal and a KSS patient. Material from the unaffected individual shows a checkerboard pattern with both enzymes (A and B). In one section, KSS samples show one RRF (asterisk) by SDH stain (C). The same fiber on the serial section (asterisk) shows reduced COX activity (D).

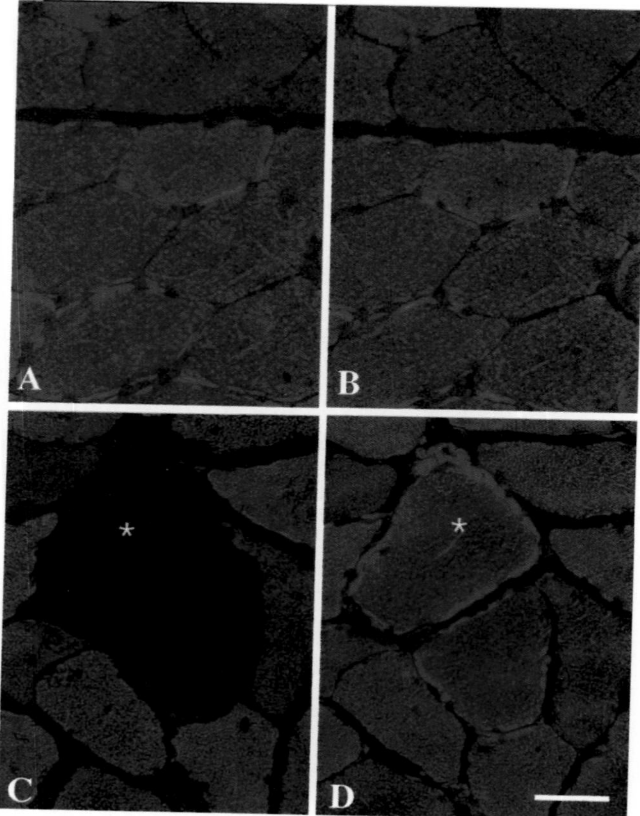

Fig. 18.2 Immunolocalization of COX II (A) and COX IV (B) on serial muscle sections from a normal and a KSS patient. The normal muscle shows an almost identical mitochondrial network for COX II and for COX IV (A and B). KSS samples show one RRF (asterisk) that lacks COX II immunostain in one section (C), and the same fiber on the serial section (asterisk) shows enhanced stain for COX IV (D). Bar: 50 μm.

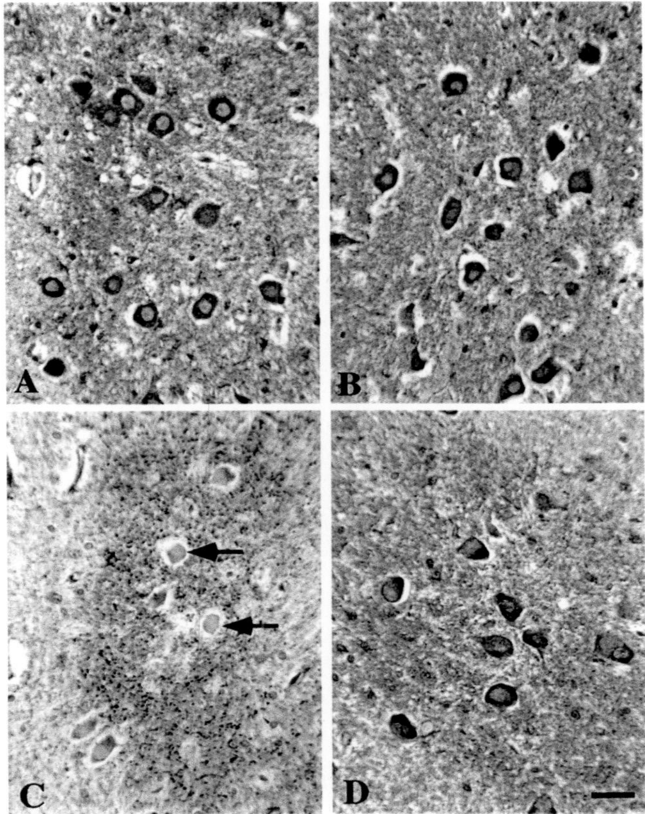

Fig. 18.3 Immunostaining of sections of the olivary nucleus from a control (A and B) and from a MERRF patient (C and D) for the localization of the mtDNA-encoded COX II subunit of complex IV (A and C) and the nDNA-encoded FeS subunit of complex III (B and D). The patient shows a marked decrease of immunostain in neurons for COX II (arrows), but normal stain for FeS. Bar: 50 μm.

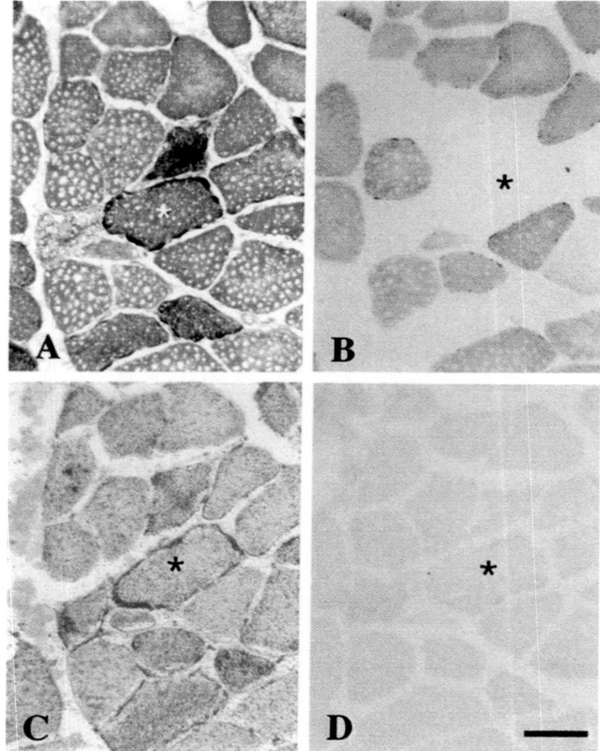

Fig. 18.4 Cellular localization of mtDNA by ISH. Serial muscle sections from a patient with MERRF were stained for SDH (A) and COX (B) activity, and for mtDNA localized by ISH (C). The ISH signal is seen as red material. Note the strong mtDNA signal in a RRF (asterisk) characterized by increased SDH activity and lack of COX activity. A control section subjected to ISH without the denaturing step (D) shows no hybridization signal. Bar: 50 μm.

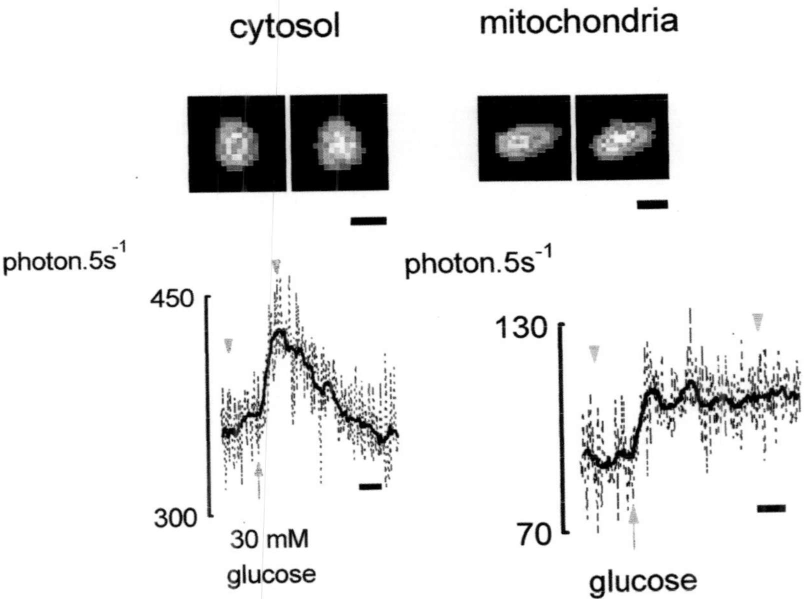

Fig. 20.10 Glucose-induced changes in free ATP concentration in the cytosol and mitochondrial matrix of MIN6 β cells. Images were produced at the points indicated by the gray arrowheads by integrating photon events over a 30-s interval. Traces show changes recorded from a single cell, with detection of luminescence made every 1 s (broken line) or after integration at each point for 10 s. Scale bar: 5μm Time bar: 120 s. From Kennedy *et al.* (1999). Note that individual mitochondria are *not* resolved readily.

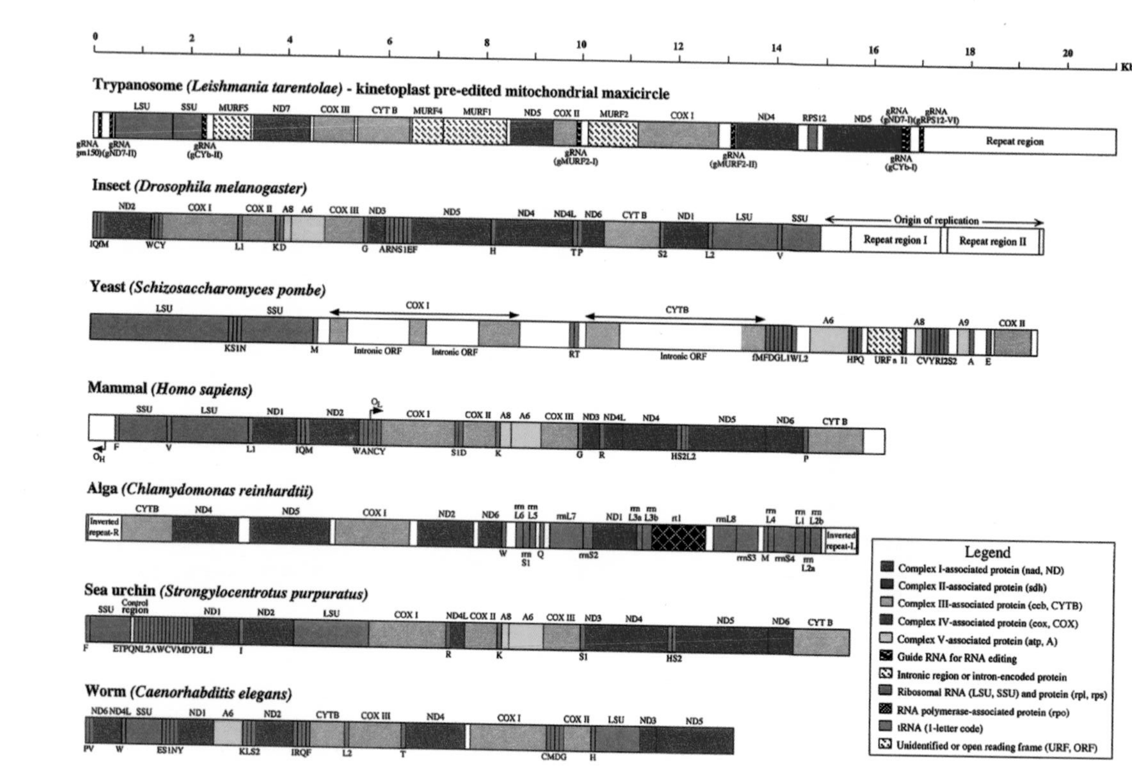

Kb
Trypanosome (Leishmania tarentolae) - kinetoplast pre-edited mitochondrial maxicircle
Repeat region
Insect (Drosophila melanogaster)
Origin of replication
Repeat region I
Repeat region II
Yeast (Schizosaccharomyces pombe)
Intronic ORF
Mammal (Homo sapiens)
Alga (Chlamydomonas reinhardtii)
Inverted repeat-R
Inverted repeat-L
Sea urchin (Strongylocentrotus purpuratus)
Control region
Worm (Caenorhabditis elegans)
Legend
Complex I-associated protein (nad, ND)
Complex II-associated protein (sdh)
Complex III-associated protein (cob, CYTB)
Complex IV-associated protein (cox, COX)
Complex V-associated protein (atp, A)
Guide RNA for RNA editing
Intronic region or intron-encoded protein
Ribosomal RNA (LSU, SSU) and protein (rpl, rps)
RNA polymerase-associated protein (rpo)
tRNA (1-letter code)
Unidentified or open reading frame (URF, ORF)

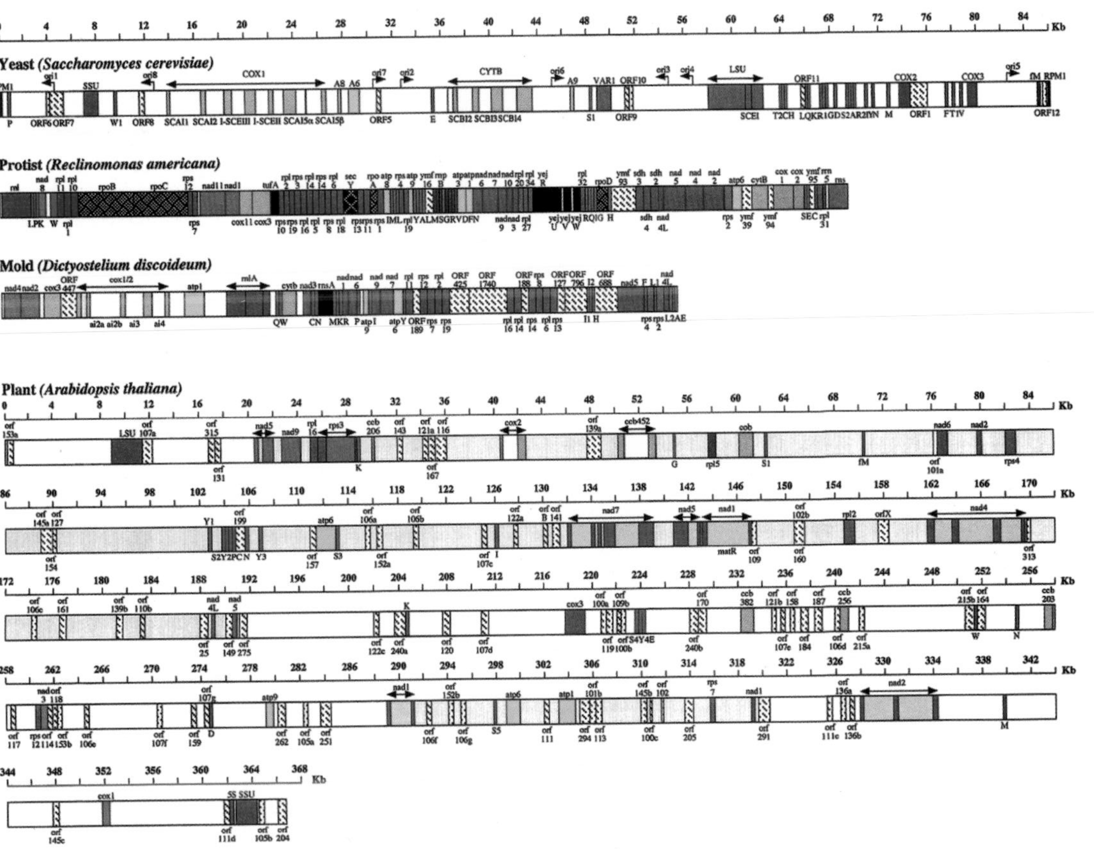

Appendix 2, Fig. 1 Linearized maps of circular mitochondrial genomes from representative organisms.